Glacimarine Environments: Processes and Sediments

Geological Society Special Publications
Series Editor J. BROOKS

GEOLOGICAL SOCIETY SPECIAL PUBLICATION NO 53

Glacimarine Environments: Processes and Sediments

EDITED BY

JULIAN A. DOWDESWELL

Scott Polar Resarch Institute
University of Cambridge
UK

JAMES D. SCOURSE

School of Ocean Sciences
University College of North Wales
Bangor
UK

1990

Published by
The Geological Society
London

THE GEOLOGICAL SOCIETY

The Geological Society of London was founded in 1807 for the purposes of 'investigating the mineral structures of the earth'. It received its Royal Charter in 1825. The Society promotes all aspects of geological science by means of meetings, special lectures and courses, discussions, specialist groups, publications and library services.

It is expected that candidates for Fellowship will be graduates in geology or another earth science, or have equivalent qualifications or experience. All Fellows are entitled to receive for their subscription one of the Society's three journals: *The Quarterly Journal of Engineering Geology*, the *Journal of the Geological Society* or *Marine and Petroleum Geology*. On payment of an additional sum on the annual subscription, members may obtain copies of another journal.

Membership of the specialist groups is open to all Fellows without additional charge. Enquiries concerning Fellowship of the Society and membership of the specialist groups should be directed to the Executive Secretary, The Geological Society, Burlington House, Piccadilly, London W1V 0JU.

Published by the Geological Society from:
The Geological Society Publishing House
Unit 7
Brassmill Enterprise Centre
Brassmill Lane
Bath
Avon BA1 3JN
UK
(*Orders:* Tel. 0225 445046)

Distributor
USA
AAPG Bookstore
PO Box 979
Tulsa
Oklahoma 74101–0979
USA
(*Orders*: Tel: (918)584–2555)

First published 1990

British Library Cataloguing in Publication Data
Glacimarine environments.
1. Glacial. Marine sediments
I. Dowdeswell, J.A. (Julian A) *1957*– II. Scourse, J.D.
III. Series
551.46083

ISBN 0–903317–54–0

Contents

Preface

Convincing interpretations of many ancient glacimarine sequences depend on studies of analogous processes active in modern glacimarine environments. Studies of both ancient glacimarine sequences and modern glacimarine environments have increased in recent years. The time therefore seemed ripe to hold a meeting to bring workers on the contemporary and ancient record together, and so the meeting which generated this volume was conceived. The two-day meeting, organized on behalf of the Marine Studies Group of the Geological Society of London, in conjunction with the International Glaciological Society, and promoted under the aegis of IGCP Project 260, was held in March 1989 at Burlington House in Piccadilly, the London home of the Geological Society. Many of the papers presented at the meeting are contained within this volume.

It was a pleasure to welcome contributions from North America, and from many of the countries of continental Europe, in addition to those from Britain. A blend of papers on modern and ancient glacimarine environments was achieved, and we found organizing the meeting a stimulating and satisfying experience. We had initially intended dividing the volume into two more-or-less equal halves devoted to 'modern' and 'ancient', but we were pleasantly surprised by the number of contributions which transcended this boundary, and so felt that to impose such a division would have been artificial. Following an introductory paper on approaches to the study of glacimarine environments by the editors, Boulton's review paper is concerned with general principles in both modern and ancient environments. This is followed by a series of papers concerned largely with processes in modern glacimarine environments (Powell, Cowan & Powell, Domack, Gilbert, Dowdeswell & Murray, Carlson *et al.*, Aitken, and Syvitski *et al.*). The paper by Syvitski *et al.* is followed by a review by Fairchild & Spiro, as both are concerned with geochemical aspects of glacimarine sediments. A number of studies of ancient sequences are then organized by age and latitude; there is a paper from Canada (Woodworth-Lynas & Guigné), three concerning the Nordic and Barents Seas (Bischof *et al.*, Solheim *et al.* and Vorren *et al.*), one from Greenland (Huddart & Peacock), two from Sweden (Stevens and Lord) and two from the continental shelf to the west of the British Isles (Scourse *et al.* and Stoker). The two final papers are concerned with earlier Cenozoic (Eyles & Lagoe) and pre-Cenozoic (Moncrieff & Hambrey) glacimarine environments.

We would like to thank the Secretary, staff and housekeepers of the Geological Society for help in the organization of the meeting and the publication of this volume. The successful organization of the meeting also depended in large part on the very helpful advice of Dr Colin Summerhayes, Chairman of the Marine Studies Group at the time of the conception of the meeting, and on financial support from British Petroleum and the Geological Society. Dr Chris Evans of the British Geological Survey kindly organized the core workshop and poster display. Dr Mike Hambrey, Dr Judith Allen, Bill Austin, Marianne Cromack, Gordon Hamilton, Tavi Murray, David Sexton and Ruth Watkins helped organizationally in various critical ways. Evelyn Dowdeswell compiled the index.

Finally, we thank our contributors, who not only kept up a very high standard of presentation at the meeting, but also stuck (fairly closely) to deadlines imposed by us during the course of the production of this volume. This enabled a relatively short interval between the meeting and final publication. This also reflects on our reviewers, drawn from a number of countries, who deserve considerable praise for their efforts in providing more than 50 sets of comments on the contributions quickly and fairly.

JULIAN A. DOWDESWELL
JAMES D. SCOURSE

On the description and modelling of glacimarine sediments and sedimentation

JULIAN A. DOWDESWELL[1] & JAMES D. SCOURSE[2]

[1] *Scott Polar Research Institute, University of Cambridge, Cambridge CB2 1ER, UK*

[2] *School of Ocean Sciences, University College of North Wales, Bangor, Gwynedd LL59 5EY, UK*

Abstract: Models of glacimarine environments fall into two categories: (a) qualitative summaries of the depositional record; and (b) quantitative models of the physical processes of glacimarine sedimentation and their rates of operation. Qualitative approaches include descriptive vertical lithofacies profiles and models derived from observations of two- or three-dimensional sedimentary or seismic facies changes. Qualitative models are based on observations in modern glacimarine environments where the processes and settings are relatively well constrained, or on sedimentary or seismic investigations of Cenozoic and pre-Cenozoic successions. Some represent hypothetical sedimentary facies associations based on intuitive reasoning in the absence of observations (e.g. beneath ice shelves). Quantitative process and rate models can be based on theory or empirical data. The nature and rate of sedimentation beneath ice shelves has been modelled quantitatively. Scaled-down laboratory models, for example of iceberg melting, have also been used experimentally. Process models linking different parts of the glacimarine system are an important future research area.

Glacimarine sediments occur relatively widely in the Cenozoic and more ancient geological record because their preservation potential is greater than that of glacigenic sediments on land (e.g. Hambrey & Harland 1981; Anderson 1983). We define glacimarine environments broadly, as all those areas where sediment is deposited in the sea after release from glacier ice (including grounded tidewater ice fronts, floating glacier tongues, ice shelves and icebergs) or sea ice (Powell 1984). This relatively wide definition of the glacimarine environment implies: (a) that glacimarine sediments are variable in composition from, for example, coarse-grained ice contact fans to isolated iceberg-rafted dropstones in a muddy matrix of non-glacial origin; and (b) that a wide variety of glaciological, oceanographic, sedimentary and biogenic processes are associated with the sources, transport, deposition and reworking of glacimarine sediments (Fig. 1). This paper aims to provide a background to detailed studies of glacimarine environments, by outlining and discussing a number of the approaches used to describe and model glacimarine sediments and the processes of glacimarine sedimentation.

Approaches to modelling

The notion of modelling has been used in various contexts within the scientific literature on glacimarine environments. The terms *model* and *modelling* have not been reserved for numerical approaches to the physical processes occurring in glacimarine environments and their rates of operation (e.g. Orheim & Elverhøi 1981; Powell 1981; Anderson *et al.* 1983; Molnia 1983; Fairchild & Hambrey 1984). They have been applied, in common with investigations of other sedimentary environments, more broadly to include non-quantitative, descriptive summaries and interpretations of the glacimarine sedimentary record (Fig. 2). Other workers have referred to such two- and three-dimensional representations of glacimarine sedimentary sequences as schematic diagrams rather than term them models (e.g. Syvitski *et al.* 1987). The usage of the term is much less important than the requirement, in both the mathematical and more qualitative cases, to understand the nature of the original observational dataset, the simplifications and the assumptions that are made as part of any summarizing or modelling exercise.

We can, therefore, divide models of the glacimarine environment into two broad groups. First, there are summaries or models which describe and interpret glacimarine sediments and processes qualitatively. Many of these models are based on observations of two- or three-dimensional sedimentary or seismic facies changes (e.g. Eyles *et al.* 1985; Vorren *et al.* 1990). Two-dimensional models often link facies

From Dowdeswell, J. A. & Scourse, J. D. (eds), 1990, *Glacimarine Environments: Processes and Sediments*. Geological Society Special Publication No 53, pp 1–13.

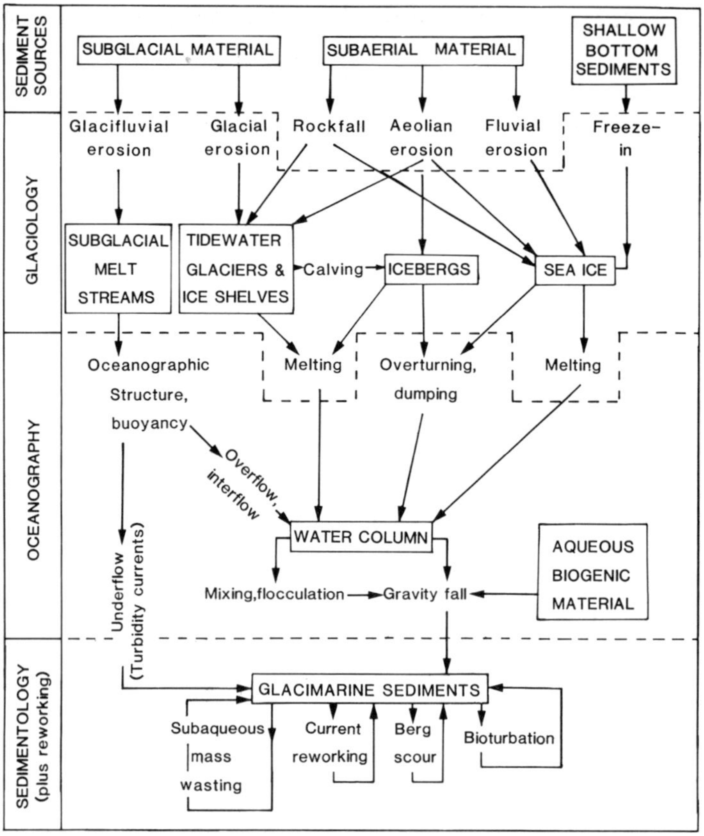

Fig. 1. The complex system making up the glacimarine sedimentary environment (from Dowdeswell 1987). Labelled boxes represent sediment stores and other labels are processes. The diagram is not comprehensive but shows the main sediment stores, flows and processes.

changes to distance from the sediment source, which is usually glacier ice. Three-dimensional models also include lateral variations in sedimentation and may contain descriptions of sea floor morphology (e.g. Powell 1981). These models often have an interpretative component relating to the environment of deposition. Other qualitative or conceptual models are essentially hypotheses, setting out intuitive arguments relating to processes assumed to be operating and their sedimentary outcome (e.g. Powell 1984). This latter approach provides useful hypotheses for future testing and discussion.

The second major category involves quantitative models of the physical processes of glacimarine sedimentation and the rates at which they operate (e.g. Drewry & Cooper 1981; Dowdeswell & Murray 1990). These quantitative models can be based on theoretical or empirical considerations, and are normally built around a set of equations set down in computer code. The term *numerical model* is reserved for equations whose solutions are derived numerically rather than algebraically, usually using finite difference or finite element methods (e.g. Syvitski 1989b). Hardware laboratory experiments form a further type of model. The output from quantitative models can, in turn, be applied to the development of descriptive models of glacimarine sedimentary architecture (e.g. Boulton 1990).

Qualitative and quantitative modelling approaches to the understanding of glacimarine environments are now discussed. Examples are given of each type of model, and the applications and limitations of these approaches are considered.

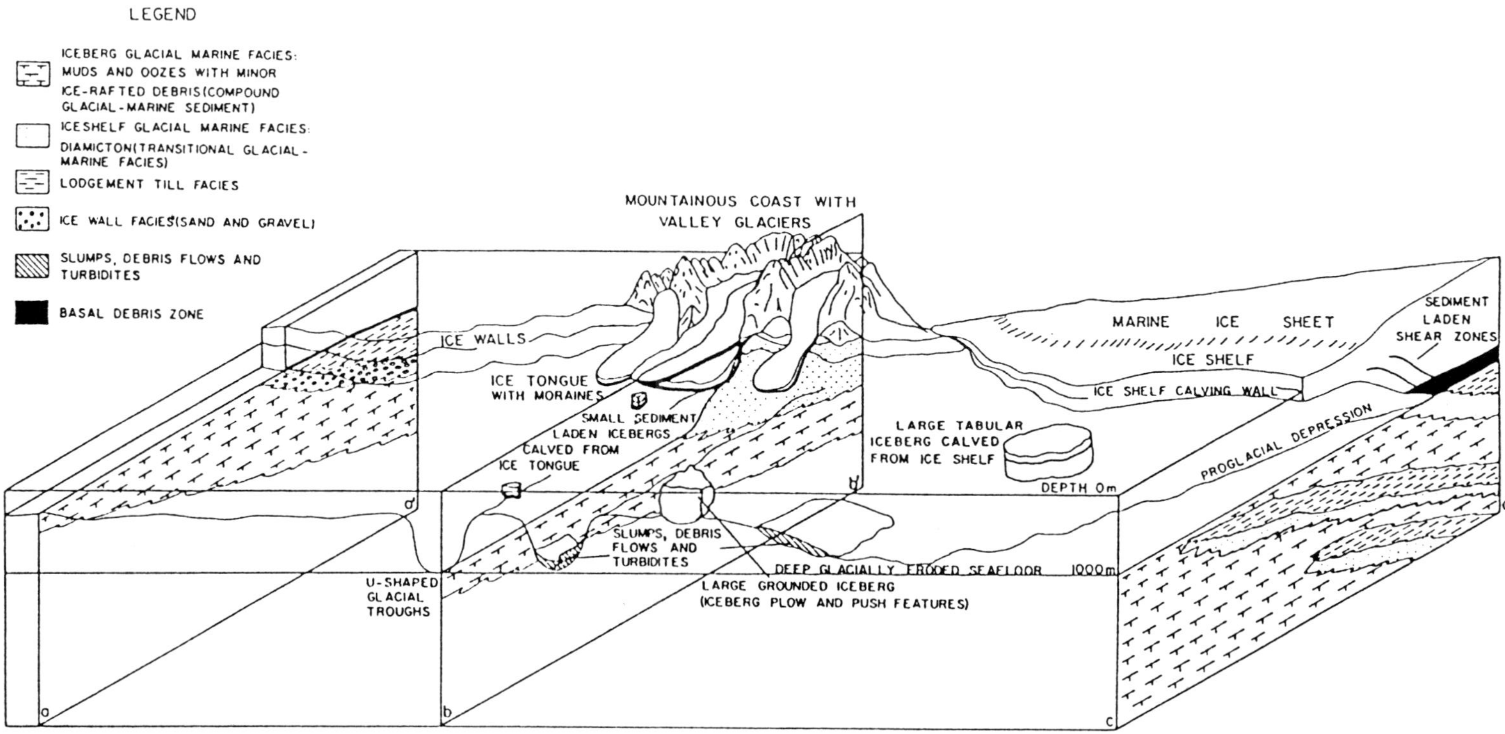

Fig. 2. A model of glacimarine environments and facies in a polar glacial setting (from Anderson *et al.* 1983).

Descriptions and qualitative models of glacimarine sediments

Vertical lithofacies profiles

Descriptive vertical lithofacies profiles have been applied widely to glacimarine depositional sequences preserved onshore, their use having evolved from similar approaches in fluvial and glacial sedimentology (Miall 1977, 1978; Eyles *et al.* 1983). Such studies include both Cenozoic and pre-Cenozoic sequences interpreted to be of glacimarine origin (e.g. Visser, 1983a; Eyles & Eyles 1984; Eyles *et al.* 1985; McCabe *et al.* 1986). The approach involves field observations of grain size, bedding, sedimentary structures, and contacts between units, and documentation via standardized lithofacies codes (Eyles *et al.* 1983). In glaciated basins a four part code is commonly used, the basic subdivisions representing facies of diamict, gravel, sand and fines (Fig. 3). Assemblages of lithofacies are then used to construct vertical lithofacies profiles (Fig. 3) and, together with the description of lateral lithofacies variability and geometry, have been used to interpret depositional environments (Walker 1984; Reading 1986).

A strength of the descriptive method is that it allows 'field properties to be depicted independent of genetic terminology and provides a firm basis for subsequent environmental interpretation and analytical work' (Eyles *et al.* 1983). The development of this inherently generic technique in glaciated terrain was stimulated in part by difficulties imposed by the loose usage of the genetic term *till* in basic field descriptions. This impeded detailed palaeoenvironmental interpretations and caused confusion about the precise nature of sediments being tested in the laboratory. Studies of glacimarine sediments using a lithofacies approach usually remain at the level of primary description. Examples where the system is used as a skeleton for secondary laboratory analysis are less common.

Two- and three-dimensional facies models

The majority of two-dimensional facies models portray proximal-distal transects from glacier source towards open ocean. Within this general

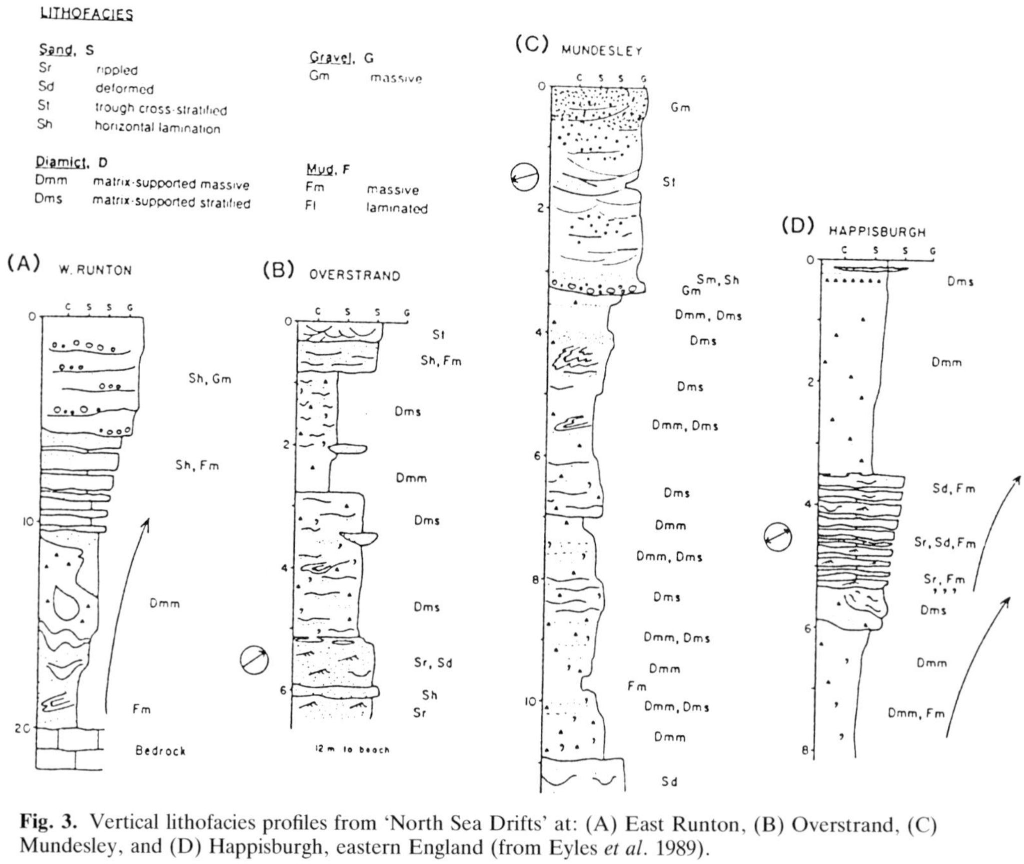

Fig. 3. Vertical lithofacies profiles from 'North Sea Drifts' at: (A) East Runton, (B) Overstrand, (C) Mundesley, and (D) Happisburgh, eastern England (from Eyles *et al.* 1989).

pattern, two types exist. The first group is based on observations of processes and facies from modern glacimarine environments, but which may include some assumptions about sediment characteristics (e.g. Elverhøi *et al.* 1983; Powell 1984; Dunbar *et al.* 1989; Fig. 4). The second category involves processes inferred from observed lithofacies assemblages, seismic stratigraphies and morphological features, from both onshore and offshore palaeo-sequences (e.g. Eyles & McCabe 1989; Larter & Barker 1989). The relationship between empirical data and its interpretation is, therefore, different in the two approaches. The first group is based largely on data concerning processes and/or direct observations of their effects, whereas the second is based on inferential evidence concerning lithofacies and seismic sections.

Whether observational or inferential, a difficulty with any two-dimensional model of glacimarine facies changes based on distance from glacier source is that observed patterns of sedimentation vary three-dimensionally. The problem is at its most severe close to tidewater glacier termini, side-entry glaciers and other sediment sources, where steep gradients in grain size and facies linked to, for example, changing meltwater velocity occur both orthogonal and parallel to the ice front (e.g. Powell 1990). Particularly where tidewater ice cliffs are of considerable length, the nature of the subglacial hydrological system is of considerable importance to the resulting lateral facies variations, through the frequency of occurrence and water discharge of basal conduits. More distally, the depositional environment becomes more homogeneous over larger areas. Other examples include the preferential build-up of sediments on the seaward-looking right margin of some northern hemisphere fjords related to Coriolis effects and the lateral deflection of turbid surface meltwater plumes in eddies (e.g. Syvitski 1989a). Despite these complications, two-dimensional glacimarine facies models provide simplified and useful summaries of the nature of sedimentation.

Three-dimensional models are able to incorporate these lateral facies changes, and are better able to depict significant landform-sediment associations (e.g. Powell 1981; Vorren *et al.* 1983; Powell & Molnia 1989). Three-dimensional models of biofacies, as well as lithofacies, have also been presented recently (e.g. Dale *et al.* 1989; Aitken 1990). As with two-dimensional facies models, some three-dimensional models are based mainly on observations of modern glacimarine processes and their contemporary environmental context, whereas others are more inferential and based on either lithofacies sequences exposed onshore (e.g. Visser 1983b; Eyles *et al.* 1989) or seismic mapping of offshore units (e.g. Vorren *et al.* 1989).

Discussion

When using models of glacimarine sediments as analogues for the interpretation of older glacigenic sequences, it is important to distinguish between models based on studies of modern and ancient environments. Models based on observations of the contemporary or near-contemporary glacimarine environment have the advantage of establishing the glaciological, oceanographic and topographic processes and settings for sedimentation. Interpretations of modern sediments are, therefore, relatively well founded and constrained. However, a problem with the modern record is that some glacimarine settings, relatively common in the past, are scarcely represented at present; for example, the large marine-based ice sheets of northern hemisphere shelf areas, and the presence of grounded glacier ice to the edge of continental shelves. The glacimarine environments we see today provide appropriate analogues for conditions during former interglacials. In addition, a number of modern glacimarine settings have been shifted poleward relative to full glacial conditions, rather than being absent altogether. A second difficulty with the contemporary record is that access to the sea floor in ice-infested waters or beneath floating ice shelves is difficult at best. Descriptions of contemporary glacimarine sediments are often based on relatively short cores collected at a relatively low density, rather than on sections which are often more extensive laterally and vertically. Fine-grained sediments are often sampled preferentially by coring, since gravels are difficult to penetrate. Where a seismic stratigraphy is available to link a set of scattered cores, those cores do not always extend beneath the uppermost reflector.

In contrast to modern glacimarine settings, a strength of investigations of ancient, uplifted glacimarine sections is that, while the environmental context of the record is often difficult to reconstruct in detail, large areas of exposed sediment are sometimes available (e.g. Eyles & Lagoe 1990; Moncrieff & Hambrey 1990). Long stratigraphies in the pre-Pleistocene record are important to, for example, studies of multiple climatic, sea level and tectonic events. There is, therefore, considerable scope for comparative work incorporating information from both modern and ancient successions in order to better specify the nature and interpretation of glacimarine sediments.

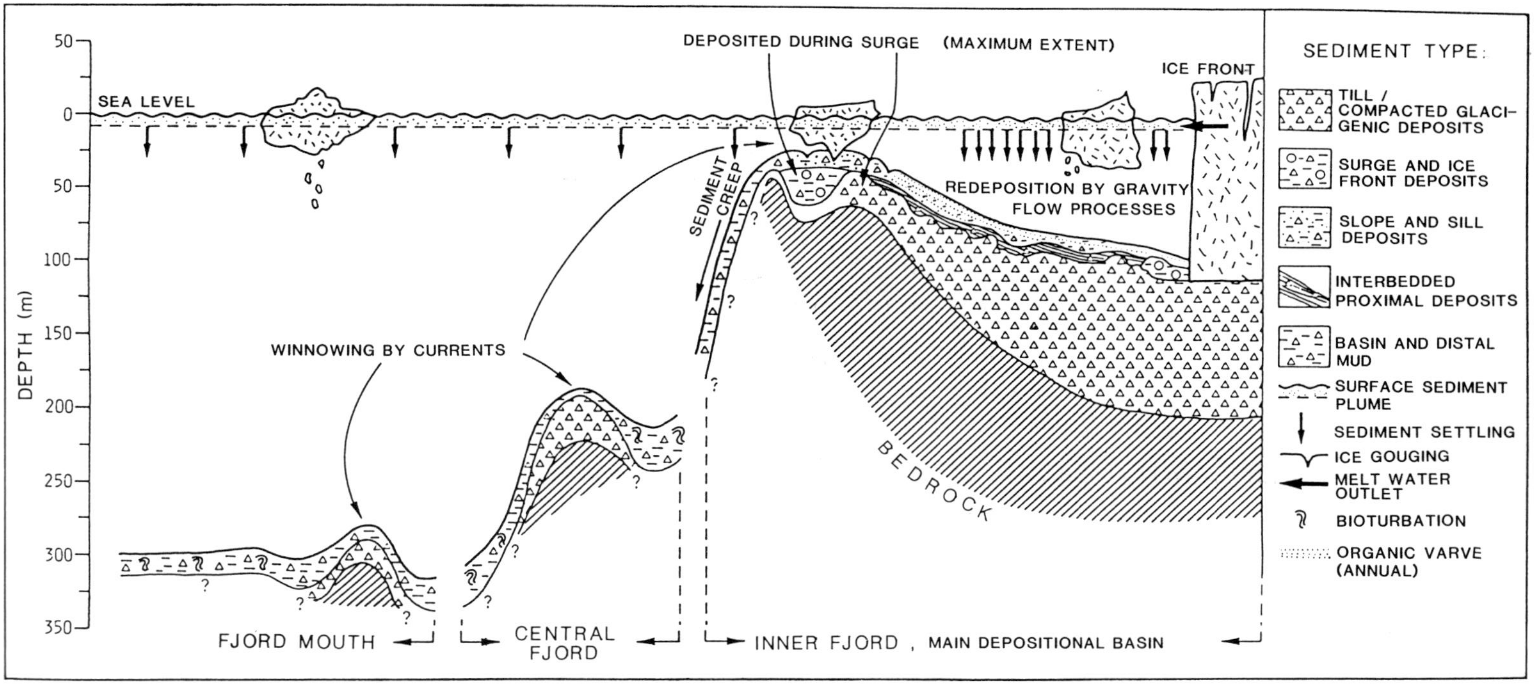

Fig. 4. Two-dimensional schematic model showing processes and sediments in a fjord fed by a tidewater glacier. The model is based on data from Kongsfjorden, Spitsbergen (from Elverhøi *et al.* 1983).

A number of qualitative models have also been put forward based on Pleistocene and more ancient sediments assumed to be of glacimarine origin. Often these are based on compilations of vertical lithofacies profiles of exposed material (e.g. Eyles *et al.* 1989). This is a useful exercise in summarizing the description and interpretation of a given sequence. Ideally, these models should use observational data based on modern environments in order to interpret the depositional processes taking place.

A particularly positive aspect of some lithofacies models has been the explicit attempt to place glacimarine and related sequences within a basin-wide context. Such large-scale investigations of major glacimarine depositional basins have resulted in radical reinterpretations of Pleistocene sedimentary sequences; for example, those around the Irish Sea and Norfolk coast of Britain (Eyles & McCabe 1989; Eyles *et al.* 1989). However, more detailed studies of individual sites are also required to validate these broader scale models. In the British example, a number of detailed studies have already been undertaken (e.g. McCabe 1986; McCabe *et al.* 1986) and areas of controversy remain (e.g. Eyles & Eyles 1984; Thomas & Dackombe 1985).

A problem may arise when lithofacies and seismic-based models derived from studies of Pleistocene sediments are applied as analogues in the environmental interpretation of another older sequence. Thus, a model based on the interpretation of a late Pleistocene sequence may later be used to interpret other non-contemporary sediments. There is little wrong with this approach, so long as the nature and limitations of the original work remain clear. An illustration is provided by the use of Gibbard's (1980) model of 'stratified till' deposition by basal melting beneath floating glacier ice (Fig. 5). The model is based on a thorough study of the sedimentology of the late Wisconsinan Catfish Creek Till at the north shore of Lake Erie. It has been used in a number of subsequent papers as a justification for assigning a similar origin to other Pleistocene and earlier stratified glacigenic deposits. The origin of the Catfish Creek Till is, however, an issue of longstanding controversy (Evenson *et al.* 1977; Dreimanis 1982) and, while Gibbard's interpretation may be correct, any subsequent invocation of the model needs to reflect this debate. However, Gibbard's paper has been used on several occasions simply in parenthesis to validate similar environmental interpretations of non-contemporary deposits.

The application of environmental interpretations of marine seismic records is also an area of possible controversy. For example, sediment wedges, termed *till tongues* by King & Fader (1986), have been observed on seismic records along the northeastern North American shelf. King & Fader interpret the material underlying and overlying the wedge as glacimarine sediment deposited through the water column beneath a floating ice shelf, and the wedge itself as subglacial till deposited when grounding takes place. Seismic records from the southern Barents Sea show similar sediment wedges,

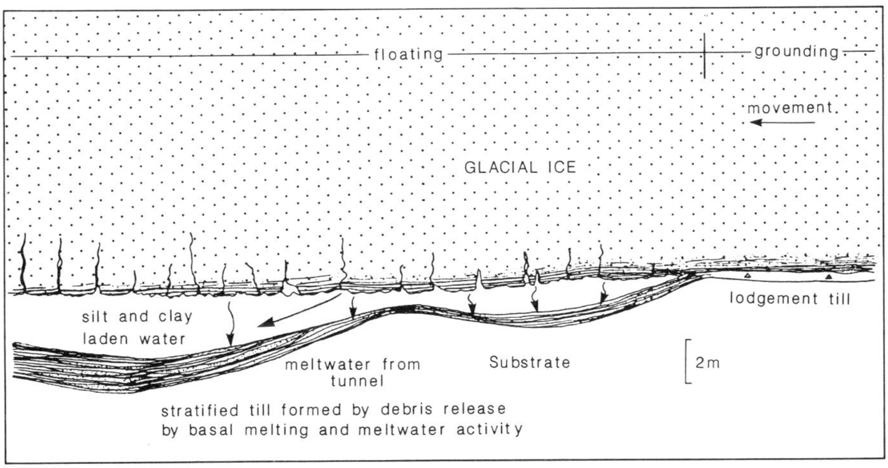

Fig. 5. Model of 'stratified till' deposition by basal melting beneath a floating ice sheet, inferred from studies of the late Wisconsinan Catfish Creek Till, Ontario (from Gibbard 1980).

which have been interpreted by Vorren *et al.* (1989) as grounding-line fans not associated with ice shelves. The interpretation of these features is therefore difficult, and they do not represent a straightforward diagnostic tool for use in subsequent environmental interpretations.

A similar difficulty is met with qualitative models which are essentially hypotheses setting out sedimentary processes and pattern thought to be taking place. They have sometimes been used by subsequent workers to interpret late Cenozoic and pre-late Cenozoic sedimentary sequences and treated in this context as if they were based on observational evidence rather than intuitive reasoning. This can lead to a continuing circularity of argument in the literature, where interpretations of ancient sequences are nominally validated by reference to models of glacimarine facies in modern environments which are, in fact, hypothetical rather than observational. This need not be an unreasonable approach in certain cases, especially where direct evidence is difficult to obtain (e.g. from beneath large ice shelves).

An example of a hypothetical sedimentary facies association which has been applied to the interpretation of older glacimarine sequences on a number of occasions is that of Drewry & Cooper (1981) for the case of an ice shelf or outlet glacier undergoing irregular changes in grounding line position (Fig. 6). The diagram represents their intuitive ideas on the nature of sedimentation through time given the oceanographic and glaciological conditions they consider to be operating. It is not based on any direct oceanographic data, sediment core or seismic reflection evidence but, nonetheless, represents a very useful hypothesis based on constrained speculation. It has, however, been used in a significant number of subsequent papers as if it were an established, well documented summary based on observational data, in order to validate interpretations of other sedimentary sequences.

Two- and three-dimensional glacimarine sedimentary facies models have been presented for a number of modern environments (e.g. Powell 1981, 1984; Elverhøi *et al.* 1983; Molnia 1983; Anderson *et al.* 1983). A major focus for future models should include hitherto poorly known environmental settings of key significance to palaeoglaciological, palaeoceanographic and palaeogeographical reconstruction. For example, fast-flowing outlet glaciers and ice streams today drain vast areas of the Greenland and Antarctic Ice Sheets, and are fundamental to ice sheet morphology and dynamics. Glaciological reconstructions of palaeo-ice sheets require that the former locations of these key

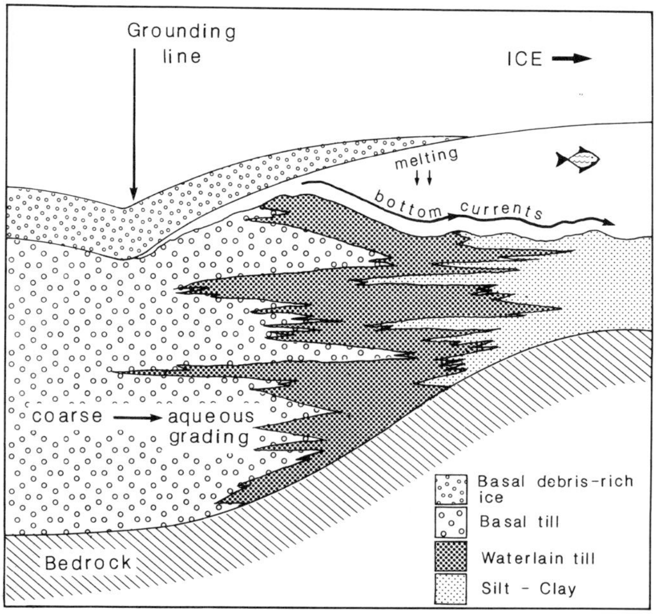

Fig. 6. Intuitive model of sedimentary facies produced by a floating ice shelf or outlet glacier exhibiting irregular oscillations of the grounding line (from Drewry & Cooper 1981).

elements of ice sheet discharge are known. Investigating and modelling sedimentary architecture at the margins of contemporary fast-flowing outlet glaciers is, therefore, of considerable environmental significance. Similarly, the nature of deposition beneath modern ice shelves and around their grounding lines is another important area for future field study and modelling. Access is clearly a major problem in this case, but several relatively small Antarctic ice shelves (e.g. Wordie Ice Shelf; Swithinbank 1988) have undergone rapid retreat in the last century, exposing the sedimentary record beneath them for possible investigation. Sedimentary facies models based on modern observations in these, and several other, environments are either unavailable or poorly specified at present, but are of considerable importance in the reconstruction of former ice sheets terminating in marine settings.

Quantitative models of glacimarine sedimentation

Physical process models

Physical process models attempt to describe the nature of processes taking place in glacimarine settings. They are usually set out as equations specifying aspects of the physics of the process involved. Glaciological, oceanographic and sedimentary processes operating in glacimarine environments have all been treated in this way. Oceanographic examples include the nature of fjord two-layer circulation and controls on the thickness of the surface plume of turbid meltwater (e.g. Long 1975; McClimans 1979; Syvitski 1989a). Syvitski *et al.* (1987) also give a series of equations on sedimentary processes, for example those relating to subaqueous slope stability in fjords, including calculations relating to geotechnical properties, stresses and failure criteria.

Models of such individual processes can also be linked together to simulate sedimentation within basins. Syvitski *et al.* (1988) outline a theoretical numerical model of the growth of prograding deltas. The processes of particle scavenging, bed load dumping at river mouths, turbidity current deposition and mass transfer by creep and slides are each specified mathematically, and linked in a numerical model of delta growth through time.

The glaciological model of Drewry & Cooper (1981) concerning melting at the base of ice shelves provides an example of an indirect solution to the problem of process modelling, because direct observations of sub-ice shelf oceanography are difficult to obtain. This model incorporates theory on the basal melting of ice shelves, and an empirical equation on debris content with distance above the ice shelf base, to yield predictions of sedimentation rates from ice shelves. The final model represents, therefore, a mixed theoretical and empirical approach to the problem of ice shelf sedimentation. Thus, basal melting/freezing (M) is calculated, based on the assumption of steady state, from the relationship (Crary *et al.* 1962):

$$M = -v\ (dH/dx) + A + He_z \qquad (1)$$

where v is horizontal ice shelf velocity, dH/dx is ice thickness gradient along an ice flow line, A is the rate of ice shelf surface accumulation, and e_z is vertical strain rate. The amount of basal melting or freeze-on for any small horizontal increment along a flow line (dx) is:

$$dy = (M\ dx)/v \qquad (2)$$

where M is derived from equation (1) and v is horizontal ice velocity. Drewry & Cooper combine computed melt rates along ice shelf flow lines with an empirical Heaviside function representing vertical variations in debris concentration to give a model of sedimentation rates with distance from an ice shelf grounding line.

This model, in common with other physical process models applicable in glacimarine settings, uses theory to account for the mechanisms leading to sedimentation. A number of these models, glaciological, oceanographic and sedimentary, also yield estimates of the rate at which the mechanisms they describe operate. The next section considers the models of the rate of operation of such processes, but emphasizes empirical rather than theoretical approaches.

Empirical models of rates of operation

Where the type of mechanism operating in a given setting is uncertain, or the processes involved are poorly understood, empirical equations are often used to specify the rate at which a given process operates. In this approach, the details of process need not be specified and, indeed, are often not well understood. Examples include the rate of calving of icebergs from tidewater glaciers (Brown *et al.* 1982), the reduction in water velocity in a fjord with distance from the mouth of a glacier-fed river (Relling & Nordseth 1979), and the decay of sedimentation rates with distance from tidewater glaciers (Görlich 1986; Görlich *et al.* 1987).

Several theories have been proposed to account for the calving of icebergs from grounded and floating ice fronts (e.g. Reeh 1968; Holdsworth 1978; Kristensen 1983). However, the data required to make calculations based on these theories, and their lack of applicability in many situations, means that empirical approaches have also been used. An empirical approach usually requires only limited knowledge of the physical mechanisms leading to an event. In this example, a simple observational relationship is developed between the rate of iceberg calving and parameters suspected to influence it. Brown *et al.* (1982) compared observed calving rates from 14 Alaskan tidewater glaciers with a number of parameters relating to the form and flow of their tidewater margins. No floating termini were included in the analysis. The best predictor of calving rate (v_c in m/yr) was found to be the depth of water (h_w) in which the glacier termini were grounded, through the equation:

$$v_c = c\, h_w \qquad (3)$$

where the coefficient c has a value of 27.1/yr. This relationship is presumably related to increasing buoyancy at the glacier tidewater termini as water deepens, but the simple empirical equation given expresses only the rate of operation of the calving process rather than the details of the mechanism or mechanisms involved.

When applying such empirical models, the nature of the observational datasets on which they are based need due consideration. Thus, the relationship between calving rate and water depth outlined above is not applicable to floating glacier tongues and ice shelves, where increases in water depth are of little relevance once the ice is fully buoyant. In this case, even though the equation itself says nothing about the calving mechanism, the likely nature of the processes involved has to be considered when evaluating the potential applicability of the model. Further, the relationship, and especially the value of the coefficient c, may not be applicable to tidewater glaciers outside southeast Alaska and even to some Alaskan tidewater glaciers (Clapperton *et al.* 1989).

Hardware laboratory experiments

Only a limited number of laboratory experiments have been undertaken concerning glacimarine processes and sediments. This is a little unexpected, given the logistical problems associated with direct observations within glacimarine settings. Laboratory models have attendant problems relating to scaling and the simplifications usually made of the more complex natural environment. Despite this, there appears to be considerable scope for developing new laboratory experiments to understand specific problems concerning the glacimarine environment.

An example of an area where hardware models have yielded results of importance to our understanding of glacimarine processes and their rates of operation is that of experimental investigations of iceberg melting. A knowledge of the relationship between iceberg melting and ocean temperature is important in estimating the rate of sedimentation from icebergs and the maximum limits to which iceberg rafted debris can be transported, yet direct field measurements of berg melting are very difficult to obtain. Several workers have placed ice blocks of the order of 1 m^3 in size and of appropriate depth–width–length ratio in water baths of varying temperature and salinity (e.g. Huppert & Josberger 1980; Russell-Head 1980). Measurements of ice melt rates and the effects of meltwater release on stratification were made. The empirical data on melt rate changes with temperature, for example, can then be applied to problems concerning the release of included debris from icebergs as they drift through ocean waters of varying temperature (e.g. Dowdeswell & Murray 1990).

There is also potential for the application of laboratory experimental approaches to the understanding and interpretation of sedimentological problems in glacimarine settings. There is a long tradition of successful laboratory work within sedimentology in general (e.g. Allen 1985). The behaviour of clouds of sediment falling through water has been modelled experimentally, allowing calculation of the area of sea floor over which such material will spread (Krishnappan 1975). These experimental results are applicable to debris dumped when ice rafts overturn or fragment (Gilbert 1990). Future experiments could include the laboratory study of dropstone structures and penetration depths in order to reconstruct the geotechnical properties of the sea floor at the time of impact. There is clearly scope for many other sedimentological experiments of relevance to the understanding and interpretation of glacimarine sediments and processes, providing that scaling difficulties can be overcome.

Discussion

The use of sets of equations to describe glacimarine processes and their rates of operation under, for example, changing climatic conditions is clearly a useful approach to the understanding and environmental interpretation of the glacimarine record. However, such models are often constrained by a combination of the known complexity of the system and a lack of observed boundary conditions. For example, we know from qualitative observations of glacier-influenced fjords that the surface sediment plume is variable in extent and turbidity through space and time. The implication is that a number of parameters are interacting to produce this spatial and temporal variability. There are two sets of problems here. The first is to specify the ways in which forcing factors such as winds, tides, currents and a Coriolis term affect plume spreading. The second is to define realistic sets of parameter values for each of these factors.

In general, a combination of sound theoretical or empirical models and a knowledge of field conditions is required for the results of quantitative modelling to be most useful. In the absence of constraining field data, model results can still be of considerable utility as indicators of the sensitivity of a system to changes in external parameters. A problem remaining in the use of theoretical, empirical and hardware models is the extrapolation of results to cases where basic assumptions are broken, or the environmental setting is inappropriate.

Some attempts to link together models of different parts of the glacimarine environment have been made. Boulton (1990), for example, has combined models of glacier extent and sea level change with a simple model of glacimarine facies change controlled by distance from glacier source and sedimentation rate. Ice sheet growth and decay, and the patterns of relative sea level change this produces, are modelled mathematically, and the patterns of sedimentation over space and through time are inferred from changes in distance from the ice mass and water depth based on the sedimentary facies models.

However, many theoretical and empirical models of processes and their rates of operation in glacimarine environments represent the action of a single or small group of processes. It is, in fact, the interaction of a relatively large number of processes, each of which is itself variable over space and through time, which goes to make up the glacimarine sedimentary record. The development of comprehensive mathematical models of the glacimarine system requires the linkage together of compatible models of individual processes (cf. Syvitski *et al.* 1988).

Summary

A number of approaches to the description and modelling of glacimarine sediments and sedimentation have been outlined. It is the nature of the simplifications and assumptions on which any model is based that must be considered carefully in any subsequent application. In particular, the use of sedimentary and seismic facies models derived from Pleistocene sequences as analogues in the interpretation of other ancient sediments must be treated with care. Wherever possible, interpretations should be pinned to observations from modern glacimarine environments, where the processes operating and their environmental setting is more closely constrained. This is not always possible, however, given the difficulties of making observations in ice-infested waters and beneath ice shelves and the lack of modern analogues for some full glacial environmental settings. Uplifted and well exposed sections of pre-Pleistocene glacimarine sediments can provide proxy evidence where modern analogues are lacking: for example, the case of grounded ice masses at the continental shelf break. Sedimentary facies models derived from the modern and ancient glacimarine record are, therefore, complementary.

Quantitative models have also been developed to investigate the nature and rates of operation of glacimarine processes. They have been particularly useful in environments where the gathering of direct observational data in modern settings is difficult. The same is true of laboratory experimental approaches, which have so far been used relatively little in studies of glacimarine sedimentation. The linking together of compatible process and rate models of different parts of the glacimarine system is an area of potential future growth in the study of the glacimarine environment.

We thank D.J. Drewry, A. Elverhøi, C.H. Eyles, N. Eyles, P.L. Gibbard, R. Gilbert, M.J. Hambrey, R.D. Powell, M.J. Sharp, A. Solheim and J.P.M. Syvitski for commenting on this paper. R. Ficken redrew several figures.

References

AITKEN, A.E. 1990. Fossilization potential of Arctic fjord and continental shelf benthic macrofaunas. *In*: DOWDESWELL, J.A. & SCOURSE, J.D. (eds), *Glacimarine Environments: Processes and Sediments*. Geological Society, London, Special Publication, **53**, 155–176.

ALLEN, J.R.L. 1985. *Principles of Physical Sedimentation*. Allen & Unwin, London.

ANDERSON, J.B. 1983. Ancient glacial-marine deposits: their spatial and temporal distribution. *In*: MOLNIA, B.F. (ed.), *Glacial-Marine Sedimentation*. Plenum, New York, 3–92.

——, BRAKE, C., DOMACK, E., MYERS, N. & WRIGHT, R. 1983. Development of a polar glacial-marine sedimentation model from Antarctic Quaternary deposits and glaciological information. *In*: MOLNIA, B.F. (ed.), *Glacial-Marine Sedimentation*. Plenum, New York, 233–264.

BOULTON, G.S. 1990. Sedimentary and sea level changes during glacial cycles and their control on glacimarine facies architecture. *In*: DOWDESWELL, J.A. & SCOURSE, J.D. (eds), *Glacimarine Environments: Processes and Sediments*. Geological Society, London, Special Publication, **53**, 15–52.

BROWN, C.S., MEIER, M.F. & POST, A.S. 1982. Calving speed of Alaska tidewater glaciers, with application to Columbia Glacier. *U.S. Geological Survey Professional Paper*, **1258**-C.

CLAPPERTON, C.M., SUGDEN, D.E. & PELTO, M. 1989. Relationship of land terminating and fjord glaciers to Holocene climatic change, South Georgia, Antarctica. *In*: OERLEMANS, J. (ed.), *Glacier Fluctuations and Climatic Change*. Kluwer, Dordrecht, 57–75.

CRARY, A.P., ROBINSON, E.S., BENNETT, H.F. & BOYD, W.W., Jr. 1962. Glaciological studies of the Ross Ice Shelf, Antarctica, 1957–60. *International Geophysical Year Glaciological Report*, 6.

DALE, J.E., AITKEN, A.E., GILBERT, R. & RISK, M.J. 1989. Macrofauna of Canadian Arctic fjords. *Marine Geology*, **85**, 331–358.

DOWDESWELL, J.A. 1987. Processes of glacimarine sedimentation. *Progress in Physical Geography*, **11**, 52–90.

—— & MURRAY, T. 1990. Modelling rates of sedimentation from icebergs. *In*: DOWDESWELL, J.A. & SCOURSE, J.D. (eds), *Glacimarine Environments: Processes and Sediments*. Geological Society, London, Special Publication, **53**, 121–137.

DREIMANIS, A. 1982. Two origins of the stratified Catfish Creek Till at Plum Point, Ontario, Canada. *Boreas*, **11**, 173–180.

DREWRY, D.J. & COOPER, A.P.R. 1981. Processes and models of Antarctic glaciomarine sedimentation. *Annals of Glaciology*, **2**, 117–122.

DUNBAR, R.B., LEVENTER, A.R. & STOCKTON, W.L. 1989. Biogenic sedimentation in McMurdo Sound, Antarctica. *Marine Geology*, **85**, 155–179.

ELVERHØI, A., LØNNE, Ø. & SELAND, R. 1983. Glaciomarine sedimentation in a modern fjord environment, Spitsbergen. *Polar Research*, **1** (new series), 127–149.

EVENSON, E.B., DREIMANIS, A. & NEWSOME, J.W. 1977. Subaquatic flow tills: a new interpretation for the genesis of some laminated till deposits. *Boreas*, **6**, 115–133.

EYLES, C.H. & EYLES, N. 1984. Glaciomarine sediments of the Isle of Man as a key to late Pleistocene stratigraphic investigations in the Irish Sea Basin. *Geology*, **12**, 359–364.

—— & LAGOE, M.B. 1990. Sedimentation patterns and facies geometries on a temperate glacially-influenced continental shelf: the Yakataga Formation, Middleton Island, Alaska. *In*: DOWDESWELL, J.A. & SCOURSE, J.D. (eds), *Glacimarine Environments: Processes and Sediments*. Geological Society, London, Special Publication, **53**, 363–386.

——, EYLES, N. & MIALL, A.D. 1985. Models of glaciomarine sedimentation and their application to the interpretation of ancient glacial sequences. *Palaeogeography Palaeoclimatology and Palaeoecology*, **51**, 15–84.

EYLES, N. & MCCABE, A.M. 1989. The Late Devensian (<22,000 BP) Irish Sea Basin: the sedimentary record of a collapsed ice sheet margin. *Quaternary Science Reviews*, **8**, 307–351.

——, EYLES, C.H. & MCCABE, A.M. 1989. Sedimentation in an ice-contact subaqueous setting: the mid-Pleistocene 'North Sea Drifts' of Norfolk, U.K. *Quaternary Science Reviews*, **8**, 54–74.

—— & MIALL, A.D. 1983. Lithofacies types and vertical profile models; an alternative approach to the description and environmental interpretation of glacial diamict and diamictite sequences. *Sedimentology*, **30**, 393–410.

FAIRCHILD, I.J. & HAMBREY, M.J. 1984. The Vendian succession of northeastern Spitsbergen: petrogenesis of a dolomite-tillite association. *Precambrian Research*, **26**, 111–167.

GIBBARD, P. 1980. The origin of stratified Catfish Creek Till by basal melting. *Boreas*, **9**, 71–85.

GILBERT, R. 1990. Rafting in glacimarine environments. *In*: DOWDESWELL, J.A. & SCOURSE, J.D. (eds), *Glacimarine Environments: Processes and Sediments*. Geological Society, London, Special Publication, **53**, 105–120.

GÖRLICH, K. 1986. Glacimarine sedimentation of muds in Hornsund Fjord, Spitsbergen. *Annales Societatis Geologorum Poloniae*, **56**, 433–477.

——, WESLAWSKI, J.M. & ZAJACZKOWSKI, M. 1987. Suspension settling effect on macrobenthos biomass distribution in the Hornsund fjord, Spitsbergen. *Polar Research*, **5** (new series), 175–192.

HAMBREY, M.J. & HARLAND, W.B. (eds), 1981. *Earth's pre-Pleistocene Glacial Record*. Cambridge University Press, Cambridge.

HOLDSWORTH, G. 1978. Some mechanisms for the calving of icebergs. *In*: HUSSEINY, A.A. (ed.), *Iceberg Utilization, Proceedings of the First International Conference, Ames, Iowa*, 160–175.

HUPPERT, H.E. & JOSBERGER, E.G. 1980. The melting of ice in cold stratified water. *Journal of Physical Oceanography*, **10**, 953–960.

KING, L.H. & FADER, G.B.J. 1986. Wisconsinan glaciation of the Atlantic continental shelf of southeast Canada. *Geological Survey of Canada, Bulletin*, **363**.

KRISHNAPPAN, B.G. 1975. *Dispersal of Granular Material Dumped in Deep Water*. Inland Waters Directorate, Canada Centre for Inland Waters, Scientific Series, 55.

KRISTENSEN, M. 1983. Iceberg calving and deterioration in Antarctica. *Progress in Physical Geography*, **7**, 313–328.

LARTER, R.D. & BARKER, P.F. 1989. Seismic stratigraphy of the Antarctic Peninsula Pacific margin. a record of Pliocene-Pleistocene ice volume and paleoclimate. *Geology*, **17**, 731–734.

LONG, R.R. 1975. Circulations and density distribution in a deep, strongly stratified, two-layer estuary. *Journal of Fluid Mechanics*, **71**, 529–540.

MCCABE, A.M. 1986. Glaciomarine facies deposited by retreating tidewater glaciers – an example from the Late Pleistocene of Northern Ireland. *Journal of Sedimentary Petrology*, **56**, 880–894.

——, HAYNES, J.R. & MACMILLAN, N. 1986. Late Pleistocene tidewater glaciers and glaciomarine sequences from north County Mayo, Republic of Ireland. *Journal of Quaternary Science*, **1**, 73–84.

MCCLIMANS, T.A. 1979. On the energetics of river plume entrainment. *Fluid Dynamics*, **13**, 67–82.

MIALL, A.D. 1977. A review of the braided river depositional environment. *Earth Science Reviews*, **13**, 1–62.

—— 1978. Lithofacies types and vertical profile models in braided rivers: a summary. *In*: MIALL, A.D. (ed.). *Fluvial Sedimentology*. Canadian Society of Petroleum Geologists Memoir, **5**, 597–604.

MOLNIA, B.F. 1983. Subarctic glacial-marine sedimentation: a model. *In*: MOLNIA, B.F. (ed.), *Glacial-Marine Sedimentation*, Plenum, New York, 95–144.

MONCRIEFF, A.C.M. & HAMBREY, M.J. 1990. Marginal-marine glacial sedimentation in the late Precambrian succession of East Greenland. *In*: DOWDESWELL, J.A. & SCOURSE, J.D. (eds), *Glacimarine Environments: Processes and Sediments*. Geological Society, London, Special Publication, **53**, 387–410.

ORHEIM, O. & ELVERHØI, A. 1981. Model for submarine glacial deposition. *Annals of Glaciology*, **2**, 123–128.

POWELL, R.D. 1981. A model for sedimentation by tidewater glaciers. *Annals of Glaciology*, **2**, 129–134.

—— 1984. Glacimarine processes and inductive lithofacies modelling of ice shelf and tidewater glaciers sediments based on Quaternary examples. *Marine Geology*, **57**, 1–52.

—— 1990. Glacimarine processes at grounding-line fans and their growth to ice-contact deltas. *In*: DOWDESWELL, J.A. & SCOURSE, J.D. (eds), *Glacimarine Environments: Processes and Sediments*. Geological Society, London, Special Publication, **53**, 53–74.

—— & MOLNIA, B.F. 1989. Glacimarine sedimentary processes, facies and morphology of the south-southest Alaska shelf and fjords. *Marine Geology*, **85**, 359–390.

READING, H.G. 1986. *Sedimentary Environments and Facies*. Blackwell, Oxford (Second Edition).

REEH, N. 1968. On the calving of ice from floating glaciers and ice shelves. *Journal of Glaciology*, **7**, 215–232.

RELLING, O. & NORDSETH, K. 1979. Sedimentation of a river suspension into a fjord basin, Gaupnefjord in western Norway. *Norsk Geografisk Tidsskrift*, **33**, 187–203.

RUSSELL-HEAD, D.S. 1980. The melting of free-drifting icebergs. *Annals of Glaciology*, **1**, 119–122.

SWITHINBANK, C.W.M. 1988. Antarctica. *In*: WILLIAMS, R.S., Jr. & FERRIGNO, J.G. (eds), *Satellite Image Atlas of Glaciers of the World. U.S. Geological Survey Professional Paper*, 1386-B.

SYVITSKI, J.P.M. 1989a. On the deposition of sediment within glacier-influenced fjords: oceanographic controls. *Marine Geology*, **85**, 301–329.

—— 1989b. The process-response model in quantitative dynamic stratigraphy. *In*: CROSS, T.A. (ed.), *Quantitative Dynamic Stratigraphy*. Prentice Hall, New York, 309–334.

——, BURRELL, D.C. & SKEI, J.M. 1987. *Fjords: Processes and Products*. Springer, New York.

——, SMITH, J.N., CALABRESE, E.A. & BOUDREAU, B.P. 1988. Basin sedimentation and the growth of prograding deltas. *Journal of Geophysical Research*, **93**, 6895–6908.

THOMAS, G.S.P. & DACKOMBE, R.V. 1985. Comment on 'Glaciomarine sediments of the Isle of Man as a key to late Pleistocene stratigraphic investigations in the Irish Sea Basin'. *Geology*, **13**, 445–446.

VISSER, J.N.J. 1983a. Submarine debris flow deposits from the Upper Carboniferous Dwyka Tillite Formation in the Kalahari Basin, South Africa. *Sedimentology*, **30**, 511–523.

—— 1983b. Glacial-marine sedimentation in the Late Paleozoic Karoo Basin, Southern Africa. *In*: MOLNIA, B.F. (ed.), *Glacial-Marine Sedimentation*. Plenum, New York, 667–701.

VORREN, T.O., HALD, M., EDVARDSEN, M. & LIND-HANSEN, O-W. 1983. Glacigenic sediments and sedimentary environments of continental shelves: general principles with a case study from the Norwegian shelf. *In*: EHLERS, J. (ed.), *Glacial Deposits of Northwest Europe*. Balkema, Rotterdam, 61–73.

——, LEBESBYE, E., ANDREASSON, K. & LARSEN, K.B. 1989. Glacigenic sediments on a passive continental margin as exemplified by the Barents Sea. *Marine Geology*, **85**, 251–272.

——, LEBESBYE, E. & LARSEN, K.B. 1990. Geometry and genesis of the glacigenic sediments in the southern Barents Sea. *In*: DOWDESWELL, J.A. & SCOURSE, J.D. (eds), *Glacimarine Environments: Processes and Sediments*. Geological Society, London, Special Publication, **53**, 269–288.

WALKER, R.D. (ed), 1984. *Facies Models*. Geoscience Canada, Reprint Series, 1 (Second Edition).

Sedimentary and sea level changes during glacial cycles and their control on glacimarine facies architecture

G.S. BOULTON

Department of Geology and Geophysics, Grant Institute, Kings Buildings, University of Edinburgh, West Mains Road, Edinburgh EH9 3JW, UK

Abstract: In part 1 a simple sedimentation model is presented in which subglacial ice-contact, proglacial ice-contact, inner and outer proximal, and distal glacimarine zones are distinguished. In these, facies associations are controlled by: (a) distance from the glacier and the diffusion of glacially-derived water masses and their suspended sedimentary (including berg-transported) debris into nearby oceanic waters; (b) the form and depth of the sea bed, and its position in relation to the continental margin (which help determine water mass character and tidal, wave-driven and geostrophic currents and the stability of sea-bottom sediments).

In part 2 a model of eustatic/isostatic sea level change in the vicinity of a generic ice sheet through a whole glacial cycle is developed. The sedimentation model from part 1 is linked to the sea level model to predict patterns of marine sedimentation near to an ice sheet through a glacial cycle, in inner shelf, outer shelf and continental slope environments.

Comparisons are made between theoretical models and observed patterns of glacimarine facies architecture produced during the last glaciation of Britain, continental Europe and Spitsbergen. The phase relationship between local ice sheet advance and retreat and that of the Laurentide ice sheet, the principal determinant of global eustatic change, is a major determinant of large scale sedimentation patterns.

Cycles of ice sheet expansion and decay produce characteristic three-dimensional patterns of glacimarine facies and relationships to glaciterrestrial and non-glacial sediments. Repetitive styles of facies architecture reflect the intimate relationships between many of the principal controls on sedimentation rate and sediment character during glacial cycles. They are a response to interactions between four coupled phenomena: the geometry of the crust; the spatial and temporal pattern of expansion and decay of the ice sheet (and its phase relation to the global glaci-eustatic cycle); isostatic response of the crust; and patterns of ocean circulation.

This article explores the consequences of these interactions for facies architecture. It does so in two parts. The first discusses the instantaneous spatial organization of facies in relation to the position of the glacier, the form of the depositional surface and ocean–water circulation. The second discusses the influence of ice sheet expansion and decay and the associated relative sea level changes in limiting and displacing facies zones and determining their overall geometry.

The large scale three-dimensional architecture of sediment sequences is a reflection of the organization of sedimentary environments in space and time. Each major environment is characterized by sets of interacting processes which produce repetitive associations of lithofacies. Individual lithofacies are not described in this paper, but rather the sets of processes typical of each environment and the genetic sediment types which they produce.

Part 1: Glacimarine facies distributions in space: (a) the ice-contact zone

Ice contact sediments and landforms in glacimarine environments comprise: (a) those which form subglacially and which are progressively exposed on the sea floor during glacier retreat (though not formed in a marine environment, they are discussed here as they form the substratum for many glacimarine sequences); (b) those which form at the glacier front.

Subglacial sediment and landforms

There is no fundamental reason why subglacial sedimentary processes should be different in kind beneath glaciers terminating in water and those terminating on land. The buoyant forces which influence the terminal zones of the former can merely be regarded as contributing to the

From DOWDESWELL, J. A. & SCOURSE, J. D. (eds), 1990, *Glacimarine Environments: Processes and Sediments*. Geological Society Special Publication No 53, pp 15–52.

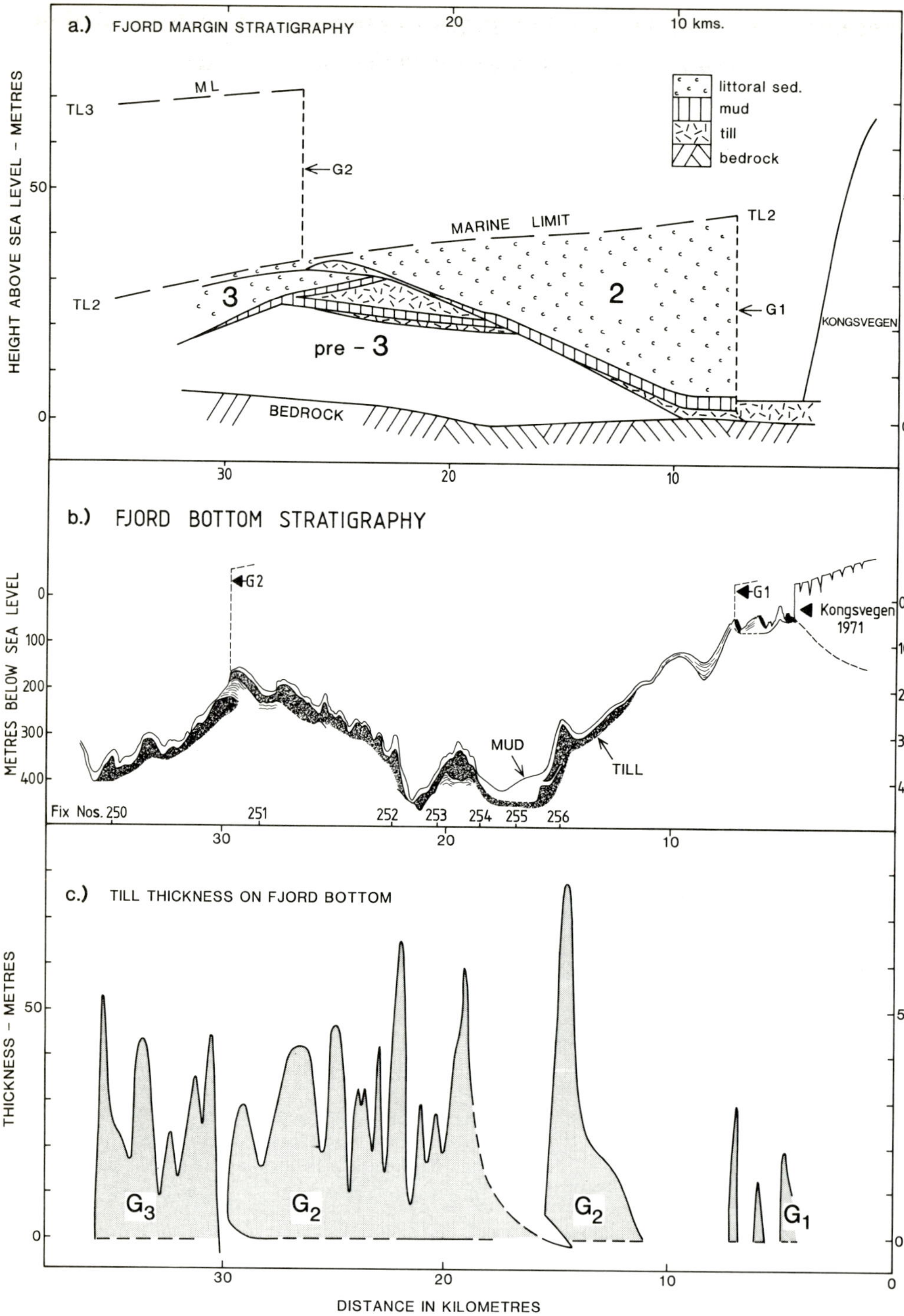

Fig. 1. Comparison of the thickness and character of the mid-fjord and fjord margin sedimentary sequences in Kongsfjorden, Spitsbergen (see Fig. 11). (a) The sequence exposed above sea level on the northern shore of Kongsfjorden. Numbers 1–3 represent glacioisostatic facies cycles as shown in Fig. 26. G1, G2 show glacier maximum positions in cycles 1 and 2. TL_2 and TL_3 show transgression limits for cycles 2 and 3. Modern sea level is the transgression limit for cycle 1. (b) Fjord bottom stratigraphy based on interpretations of sparker survey (1971). (c) Thickness of fjord bottom till produced by glacial events G1, G2 and G3 along the same line as b.

diminution of subglacial effective pressures towards the terminus in much the same way as occurs beneath the wedge-shaped termini of glaciers ending on land. We therefore expect flutes (see Solheim *et al.* 1990) drumlins and eskers to form beneath marine-based glaciers as they do on land.

There is a frequent contrast between the thickness of the tills produced by marine based glaciers compared with those produced by terrestrial glaciers. For instance, in Spitsbergen during the last century, the glacier Sefströmbreen advanced rapidly over an arm of the sea and onto an island, Cora Island, where it deposited up to 10–20 m of deformation till consisting of remoulded marine mud (De Geer 1910; Lamplugh 1911; Boulton & Van der Meer 1989). Similarly Kongsvegen, also in Spitsbergen, advanced over the eastern Løven islands (Elverhøi *et al.* 1983) and plastered on to it over 5 m of remoulded marine mud, which seismic records just offshore show to be up to 25 m in thickness (Fig. 1). Entirely terrestrially-derived tills, deposited on land from both these episodes, do not exceed 1–2 m in thickness. I suggest that a glacier flowing over mud-dominated glacimarine sediments, through which drainage will be poor, will induce high pore-water pressures in them and through deformation, remould substantial thicknesses of marine mud into deformation till. Further glacially-transported debris may be added through release of englacial debris as a consequence of melting of basal ice. On the land margin, however, coarse littoral deposits resulting from the previous glacio-isostatic cycle will tend to be more resistant to glacier-induced forces and also facilitate subglacial drainage and pore water pressure dissipation so that sediment remoulding will not be such an effective mechanism of till generation (Fig. 1- Boulton & Jones, 1979; Boulton, 1987).

Where glaciers flow into marine environments, the buoyant force of the water mass, and soft, easily-deformable subglacial sediments tend to reduce basal frictional resistance, which is conducive to a fast flow typical of ice streams, which tend to draw down ice domes in the hinterland. In many cases, such areas are prone to surging behaviour (cf. Elverhøi *et al.* 1983) and develop large longitudinal tensile stresses which cause extensive crevassing. Soft subglacial sediment is intruded into crevasses, and if the glacier tongue stagnates, as is generally the case at the end of a surge, the intruded till ridges survive after ice melting and form reticulate patterns on the sea floor. Figure 2 shows the pattern of crevasse-intruded ridges which survived after the Sefströmbreen surge of 1896. This process was first identified by de Geer (1910) and has been amplified by Lamplugh (1911), Gripp (1929), Solheim & Pfirman (1985) and others.

Proglacial sediments and land forms

Subglacial and englacial streams draining a glacier are subject to a much more dramatic check on their velocities when they enter the sea than when they drain from a terrestrial ice margin. Although suspended sediment concentrations may make some meltwater dense enough to form underflows which retain bed contact, most rises to the sea surface to form turbid brackish-water plumes (Fig. 3). Thus, coarse, fluvially-transported sediment, tends to be much more restricted to the immediate ice margin in glaciers flowing into the sea compared

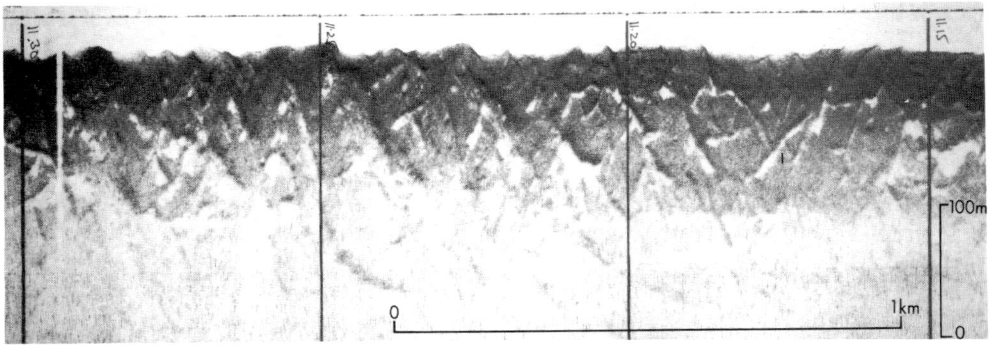

Fig. 2. Transit sonar image showing glacially-overridden glacimarine sediments which have been pressed-up into crevasses which formed during the surge of Sefströmbreen, Spitsbergen. The sediments (up to 15 m thick) were sheared along beneath the glacier during the surge.

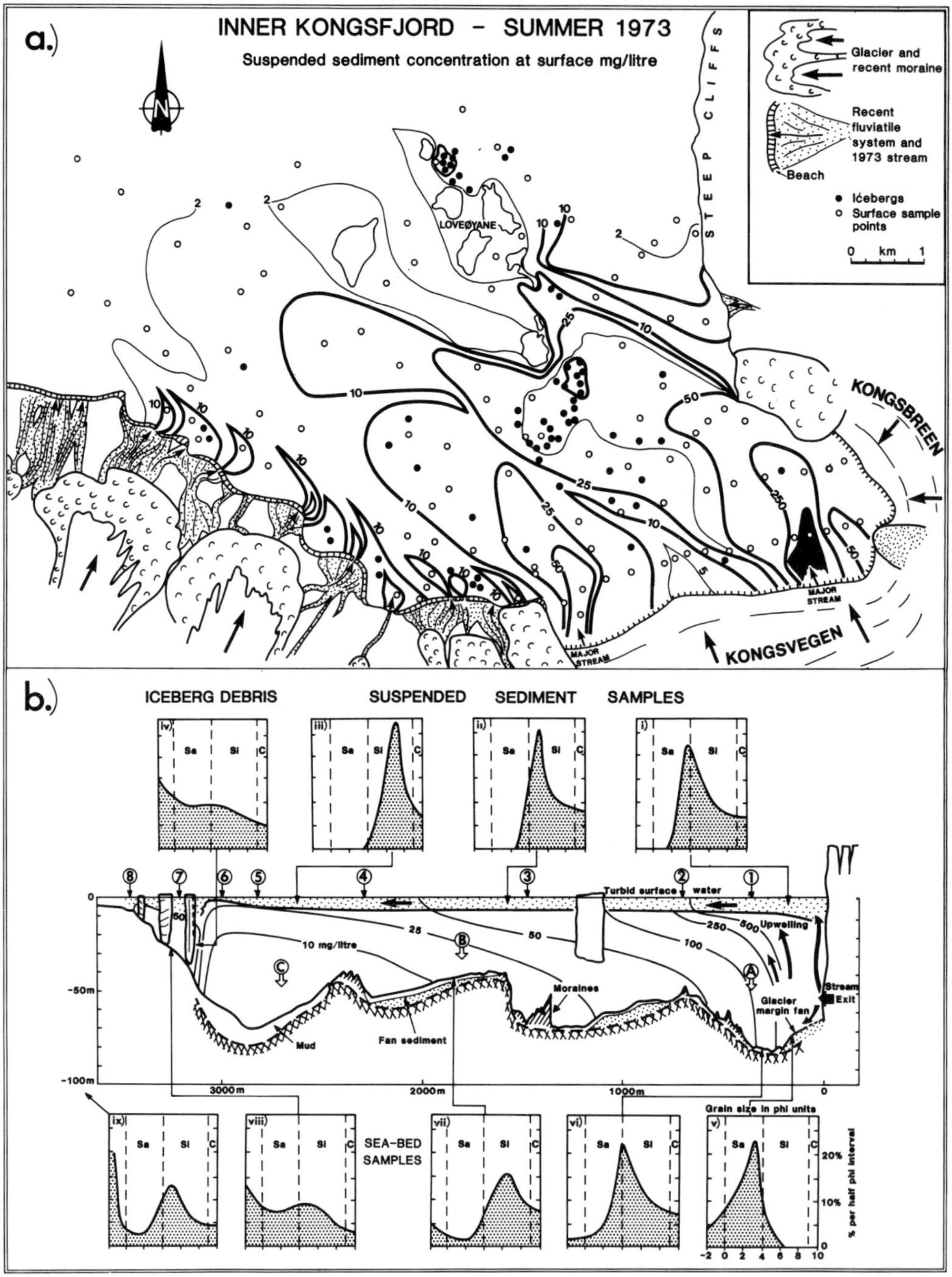

Fig. 3. (a) Suspended sediment concentrations in surface waters and the location of large icebergs in the proglacial area of Kongsvegen, Spitsbergen (Fig. 1). Measurements were taken from four boats during a two-hour period spanning high tide at a time of negligible surface wind (July 23, 1973), using a light attenuation metre calibrated by water samples. Inter-station turbidity contours were interpolated from photographs taken from nearby mountain peaks. Suspended sediment is largely derived from two dominant meltwater rivers which emerge from ice-front tunnels. (b) Section in the same area normal to the glacier front, showing fjord bottom geology, suspended sediment concentrations and grain-size distribution of sediment on the bed, in the water column and collected in sediment traps. A, B and C show the location of sediment traps (point of arrow). Fjord bottom geology (see Boulton 1986) was determined by sparker profiling and core and grab sampling. Suspended sediment concentrations were measured from profiles using a light attenuation metre at the same time as the measurements in (a). Sediment traps were emplaced for 10 days.

with their terrestrial counterparts. The coarse-grained submarine fans which they form thus tend to hug the ice margin (Powell 1981). The location of these fans is determined by the temporal pattern of ice front fluctuation and the location and stability of meltwater stream outlets (Boulton 1986). If the ice front undergoes a major readvance, or a winter readvance during general retreat, push-moraines tend to form. The origin and structure of these features has been described by Birnie (1977) for glacier margins on land, and by Boulton (1986) for marine margins.

Figure 4 shows a schematic diagram which illustrates the assemblage of ice-contact forms produced by a glacier draining into a fjord.

Fan bodies which accumulate at the margin of Kongsvegen are up to 15 m in thickness, and 300 m across (Fig. 3b). The point of entry into the fjord of the streams which feed them is extremely unstable, partially due to falling seracs which must block tunnel mouths. As a consequence fan sediment masses produced by a single unstable stream may extend for several hundred metres across the glacier front. If several streams come into the fjord, we should expect a series of transversely-oriented fan ridges from each stream, but with similar longitudinal spacings, as discussed by Boulton (1986).

If the grounding line does not coincide with the ice front, as in the case of an ice shelf, we may expect subglacial deltas to develop from sediment transported at the glacier/bed interface. This sediment may be water-transported or ice-transported. The bulk of water transport is likely to occur as channelled flows which discharge sediment from specific points at the grounding line to produce fans or deltas. Disseminated flow is likely to discharge a much smaller volume of water or sediment, either by sheet-flow at an ice bedrock interface (Weertman 1972) or as an intergranular flow through a subglacial sediment (Boulton & Hindmarsh 1987).

Ice-transported detritus may be englacial or in a subglacial deforming layer. A glacier moving over a bedrock surface with only an englacial load is unlikely to transport enough material to supply a significant delta, but a deforming subglacial bed may (Alley *et al.*

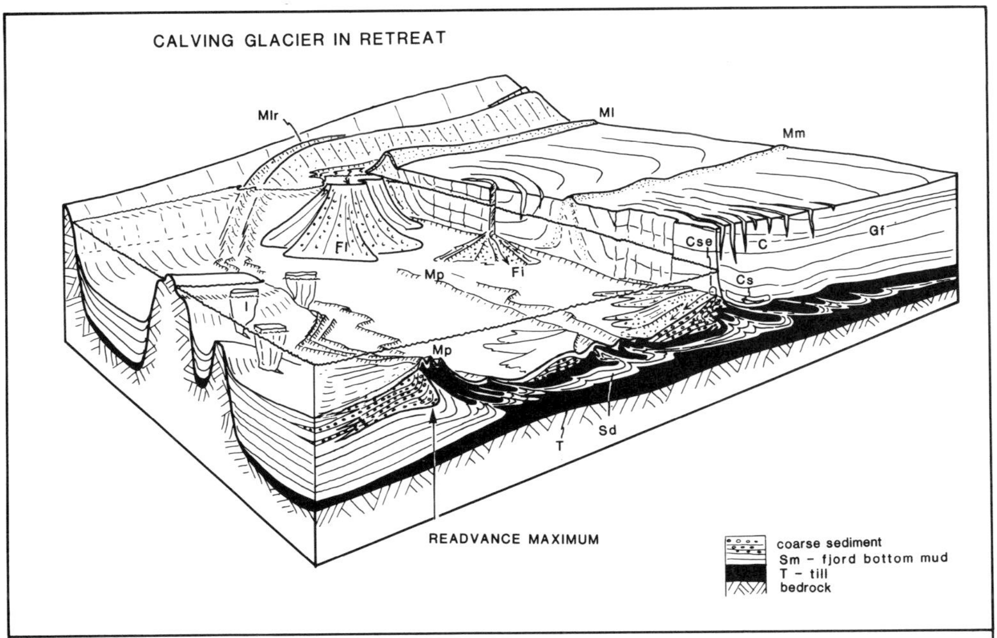

C - crevasses Mm - medial moraine Ml - lateral moraine Gf - glacier foliation Fi - ice-contact fan
Mp - push moraine Cs - subglacial stream tunnel Cse - subglacial stream exit Fl - lateral margin fan delta
I - iceberg Mlr - lateral moraines of readvance maximum Sd - deformed pre-advance sediment

Fig. 4. Schematic diagram showing the principal sea-bed features in the proglacial area of a calving glacier, and the underlying sedimentary sequences.

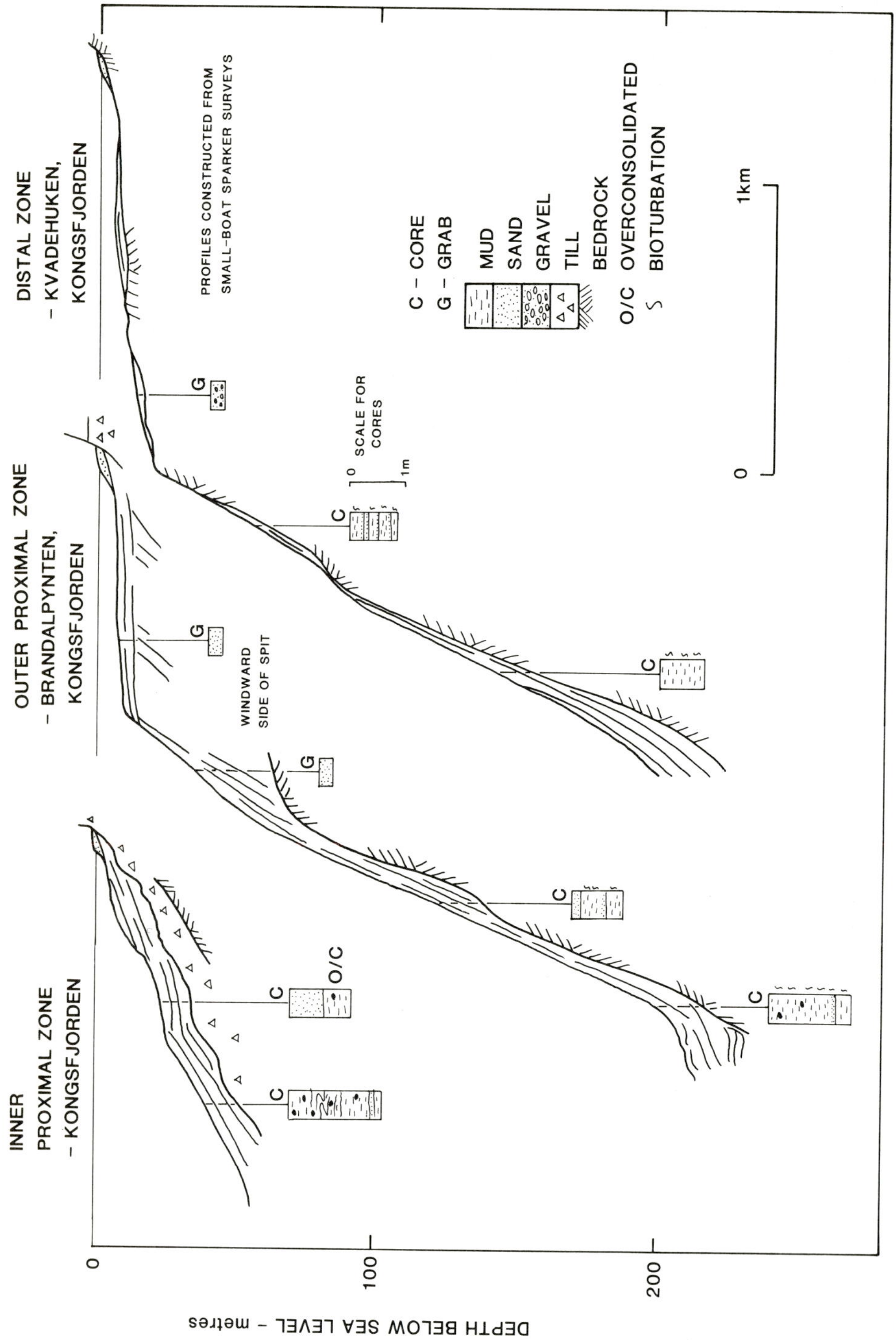

Fig. 5. Shore-face facies in a variety of proximal and distal locations.

1987), producing a two-dimensional deltaic feature all along the grounding line, provided that the bed slopes steeply enough (Fig. 6).

Sorting processes during flow and avalanching of this deformation till down the delta foresets will tend to produce stratification. Both two-dimensional *till-fed* and three-dimensional *stream-fed* deltas will respond similarly to changing sea level. If sea level is static, prograding foresets will build out, to which, in till-fed deltas, the deforming till mass will provide topsets (Fig. 6Ai). A sea level rise will cause retreat of the grounding line (Fig. 6Aii), but subsequent stabilization will cause development of a second prograding foreset/till topset couplet (Fig. 6Aiii). A sea level fall (Fig. 6Aiia) will tend to enhance development of a foreset/topset couplet at a lower level, caused by lowering of the base of the ice shelf. Where a glacier extends to the continental shelf edge, both till-fed and stream-fed deltas will tend to form at the shelf break in response to eustatic and isostatic changes in sea level (Fig. 6B).

King & Fader (1986) and King *et al.* (1987) have discussed possible subglacial processes at the shelf break through their seismostratigraphic studies of the Scotian and Mid-Norwegian shelves. They identify *till tongues* which interdigitate with glacimarine units at the shelf-break in response to changes in the position of the grounding line, in much the way shown in Fig. 6A. However, they believe that the melt-out process would be the principal source of sediment, either as till proximal to the grounding-line, or as a sediment deposited by settling through the water column distal to the grounding line. I would stress the potentially greater volumetric importance of shear transport and remoulding of subglacial sediments (Boulton & Jones 1979) as a source of till, and, in such a regime, the importance of slumping of the deformation till which is continuously discharged at

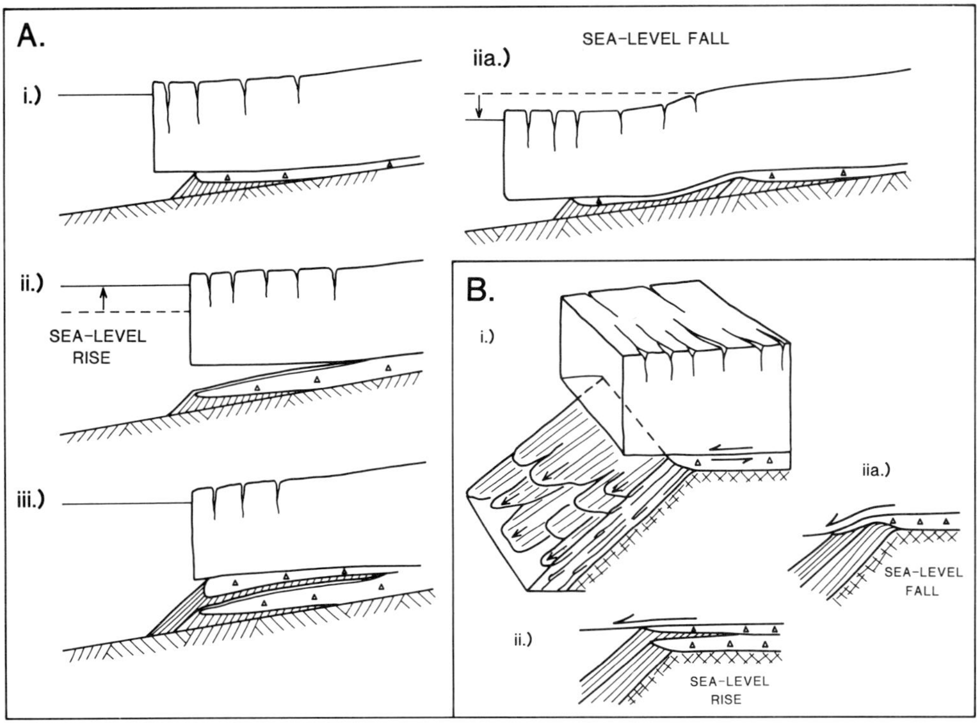

Fig. 6. Suggested sedimentary response of sub-ice shelf sediments to sea-level change. Note that deposits from suspension are omitted. (a) Sedimentation changes beneath an ice shelf on a continental shelf. i-ii-iii show the response to sea level rise, and i-iia show the response to sea level fall. (b) Sedimentation beneath an ice shelf whose grounding line occurs at the continental shelf edge. i-ii show the response to sea level rise and i-iia the response to sea-level fall. Till (assumed to form largely from shear-remoulded marine mud) is shown by triangles. Prograded foresets are shown which reflect the advance of the grounding line. Note how sea level rise leads to formation of two couplets.

the grounding line (cf. Alley *et al.* 1987). The deformation process does not require erosion of immediately subjacent materials. They may be derived from far up-glacier with deposition predominant in the terminal zone of the glacier (cf. Fig. 12).

(b) The proximal zone: inner shelf or inner fjord

The glacier proximal zone, rarely in excess of 10–20 km in width is dominated by an estuarine style of water circulation during the summer season. Large volumes of densely sediment-charged meltwater are discharged from glacier termini. This may reach the sea from meltwater rivers where the glacier terminates on land, or directly from tunnels in or beneath the ice where the glacier terminates in the sea. They give rise to surface plumes of brackish, turbid, glacier meltwaters containing large suspended sediment concentrations (Fig. 3a), and induce a thermohaline counter current at depth.

Very high rates of sedimentation from suspension have been reported in such zones. Elverhøi *et al.* (1983) report 50–100 mm a^{-1} at a distance of 10 km from Kongsvegen in Spitsbergen. Powell & Cowan (1986) report 150–3000 mg cm^2 day^{-1} within 1 km of the front of McBride Glacier in Alaska. Figure 7 shows the sedimentation rate*(S)* falls off logarithmically with distraps moored 5 m above the sea floor beyond the glacier Kongsvegen in Spitsbergen, plotted together with data from other sources. Sedimentation rate(*S*) falls off logarithmically with distance from the glacier so that:

$$S(\text{mm a}^{-1}) = \log^{-1} [100-1.33 \log(d-1)]$$

where d is the distance from the glacier in km. The proximal environments from which measurements were taken are in fjords, where bottom currents are relatively weak and sedimentation from suspension dominates, apart from in the glacier-margin zone of fan deposition (Fig. 4). However, even on open, stormy coastlines such as that of southern Iceland, which is fed by major meltwater rivers, and even though glaciers do not reach the sea, a mud-dominated proximal zone extends for some 10–20 km beyond the river mouths (Boulton *et al.* 1988).

It is possible to distinguish an Inner Proximal Zone where sedimentation rates are sufficiently high to inhibit benthic life, and bioturbation is rare, and an Outer Proximal Zone, where benthic life and bioturbation are common (Fig. 7b) (Farrow *et al.* 1983).

An intensive study of suspended sediment concentration, sedimentation rate and sediment granulometry near the front of the tidewater glacier Kongsvegen was undertaken in late July 1973. Measurements of surface water turbidity and turbidity/depth profiles were taken using light-attenuation metres (calibrated against samples) from four vessels over a two-hour period spanning a single high tide at a time of slight surface wind. Inter-station contours of surface turbidity (Fig. 3a) were interpolated from photographs taken from nearby mountain peaks. It shows sediment plumes emanating from two major ice front streams believed to debouch into the fjord near to the base of the ice cliff.

Sub-surface processes were inferred from the turbidity/depth profiles, three sediment traps anchored between 17 and 35 m above the sea bed over a 10-day period, grab and core samples, and sparker profiles (Fig. 3b). Powerful turbulence and a raised water surface near to the plume apex suggested strong stream upwelling (see Syvitski 1989). In this zone (i), sediment in the plume had a mode at the fine sand/silt boundary. Further from the glacier (ii), the mode became finer-grained with no further sand in the plume by 2 km from the glacier (iii). On the sea-bed, the distal part of the ice-front fan (v) was sand-dominated. Muds just beyond the fan margin (vi) showed a mode in the sand fraction, but this rapidly gave way to a silt mode (vii) further from the glacier.

The high rate of sedimentation from suspension in the inner proximal zone tends to overwhelm the iceberg-dropped component (Fig. 3b, vi and vii; Fig. 7b; Fig. 8a) whereas in the outer proximal zone the berg-dropped component becomes important (Fig. 3b, ix; Fig. 7b; Fig. 8b). Icebergs tend to ground and become stranded in shallow areas and it is here that clast-rich diamictons tend to be produced from a sediment-rain dominated by iceberg-derived debris (Figs 7b & 8c). High turbidity often characterizes this zone (Fig. 3b) and any differences between the grain-size distribution of debris collected from bergs and sediment on the sea floor (Fig. 3b, iv, viii) reflects such processes as current winnowing and extraneous sediment inputs. High rates of sedimentation near to the glacier front tend to produce underconsolidated muds (Fig. 7b). These tend to be unstable and may generate mass flows and secondary turbidites (Fig. 8d). Evidence for such processes is frequently seen in sediment cores in the inner proximal zone, though it is rare in the outer proximal zone, except where very high bed slopes occur.

The influence of seabed slopes can be seen

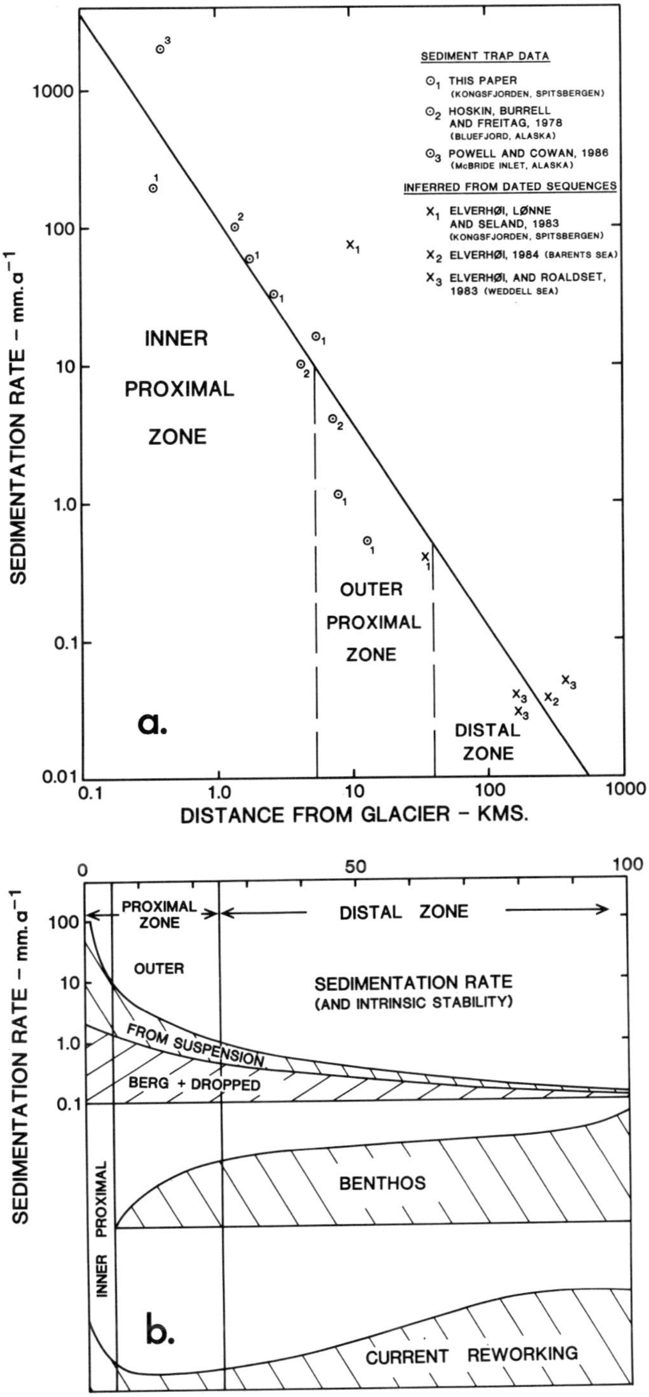

Fig. 7. (a) Relationships between sedimentation rate and distance from the glacier from a number of glacial environments. (b) Schematic diagram showing changing sediment properties with distance from the glacier.

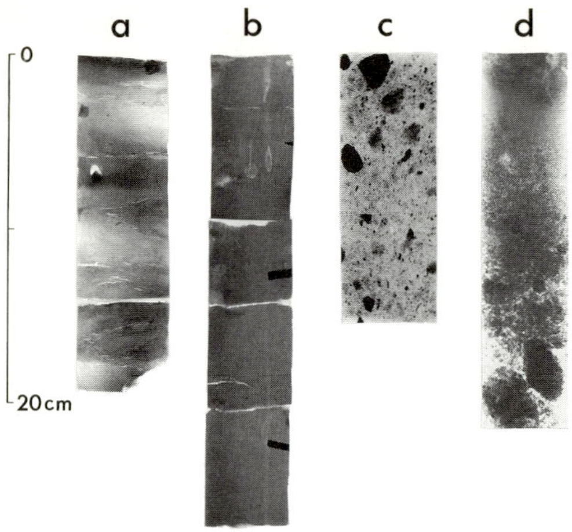

Fig. 8. X-rays from sediment cores in representative glacimarine environments. (a) Inner-proximal zone, Kongsfjord, Spitsbergen (site at 900 m, Fig. 3b). (b) Outer-proximal zone, Kongsfjord. A less silt-rich sediment showing bioturbation (150 m water depth). (c) Clast-rich diamicton from a shoal on which debris-rich icebergs ground, Kongsfjord (site at 3300 m, Fig. 3b). (d) Turbidite sequence generated by slumping from the flanks of Cambridge Fiord, Baffin Island (core CA4.2; see Fig. 9).

very clearly in the contrasts between fjord-bottom sedimentation in the outer proximal zone in many Spitsbergen and Baffin Island Fjords (Fig. 9). In the former, submarine fjord-margin slopes tend to be in the range of 5°–20°, and as a consequence there is very little evidence of gravity flows down the fjord margins. Sediments are draped over the underlying topography, and bioturbation is strong. In the latter, slopes are frequently in excess of 35–40°, and as a consequence, there is ample evidence of debris flows and secondary turbidite action. Sediments preferentially infill the fjord bottom and bioturbation is suppressed.

It is particularly important to understand sedimentary processes in the sublittoral zone as these are often important components of glacio-isostatically uplifted sequences seen on land. Figure 5 shows a series of sampled nearshore profiles from the inner proximal zone to the distal zone from the relatively sheltered environment of Kongsfjorden, Spitsbergen (including a profile from just beyond the fjord mouth). In the sheltered fjords there tend to be shallow (−5m) sub-beach platforms associated with normal wave activity and tidal fluctuations, below which, away from active sediment inputs, either bedrock is exposed (or lies at shallow depth beneath a thin sediment cover) or, in the case of inner Kongsfjorden, recently deposited till is currently being eroded. There is then a tendency for platform development at −20 to −30m which I interpret as wave base for storms, for wave height rarely exceeds 3–4 m in these sheltered environments (Syvitski *et al.* 1983). Moign (1965) reports evidence of strong sub-littoral erosion even in inner Kongsfjorden, Spitsbergen, down to a depth of 30 m below sea level, and on the open coast of Southern Iceland strong sub-littoral erosion has produced a 2 km wide bench at −20 m above the major prograding coastal sediment wedge built out during the Holocene by powerful meltwater rivers (Boulton *et al.* 1988). Thick mud units tend only to develop beneath this depth. Boulton *et al.* (1982) also concluded from the study of uplifted glacimarine sequences in Spitsbergen that −30 m was the upper level of mud accumulation.

At points of sediment input to the fjord basin, such as fan-deltas and spits, a coarse-grained sandy apron may descend to considerable depth (Fig. 5b) (e.g. Syvitski 1987; Boulton *et al.* 1985).

If we presume that the primary deposition rate of glacially-derived sediment from a fjord head glacier is a function of distance from the glacier source (Figs 3 & 7), the nature of shoreface sedimentation will depend upon the location of side inputs (Fig. 5) and the balance between primary sedimentation rate and rates of operation of different erosional agents. The latter are as follows.

Bottom currents. Strong shallow water currents can be associated with glacier-derived turbid surface plumes. The easternmost plume in Fig. 3a attains surface velocities of 6m s^{-1}, and is probably largely responsible for erosion of a prominent 200 m wide shelf at −23m in recently deposited till which was deposited on the fjord side during a neoglacial advance of Kongsvegen (fig. 5a).

Tidal currents may be particularly strong along the shoreface where fjords are constricted, and may reach to considerable depth. The profile in fig. 5b occurs at such a constriction. Sandy current winnowed units lying on shallow bedrock extend down to −70m.

Wave action. This is largely a function of exposure. In sheltered fjords, storm-wave

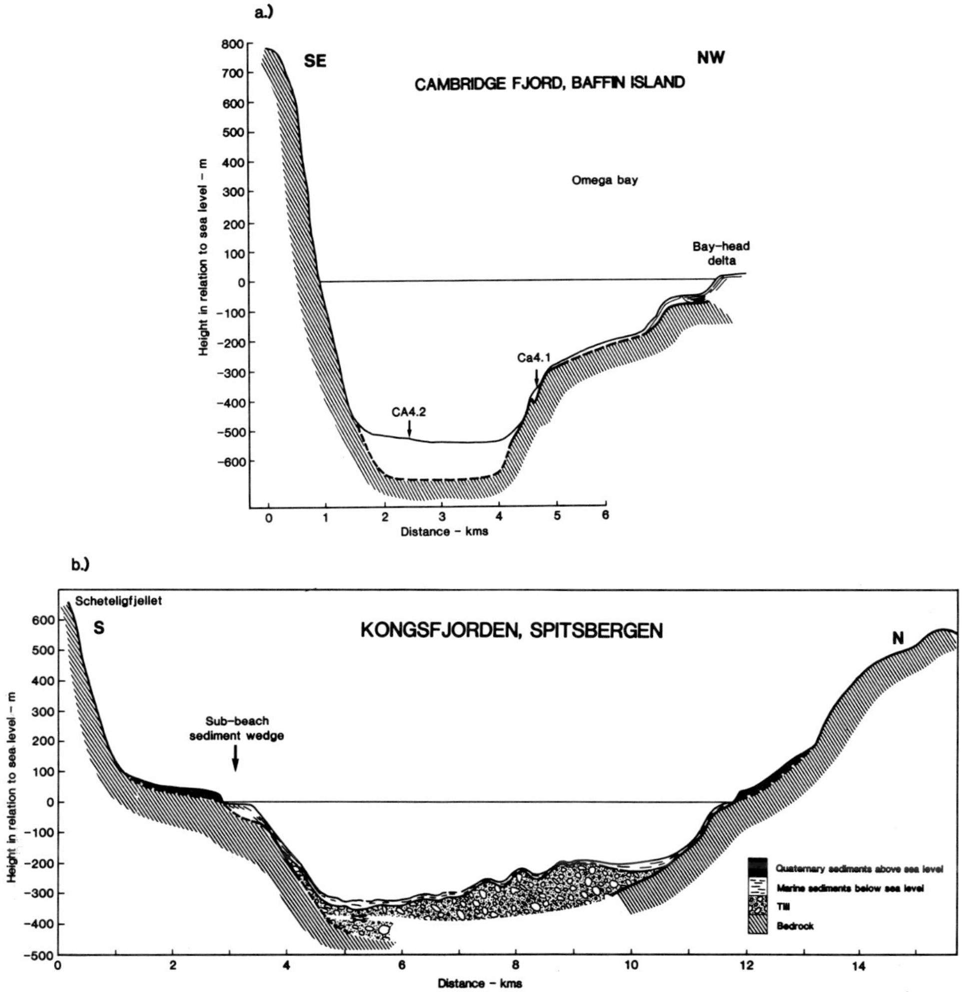

Fig. 9. Contrasts between (a) an "infill sequence" in Cambridge Fjord Baffin Island, where slumping from the steep margins (up to 40°) produces a flat-bottomed, slump and turbidite (Fig. 7d) infill sequence, and (b) a "draped-sequence" (Fig. 7b) in Kongsfjorden, Spitsbergen, where the fjord margin slopes are between 5° and 20°.

scoured platforms rarely extend to more than −20m.

Slumping and creep. The exceptionally high sedimentation rate in the inner proximal zone ensures that pore fluid pressure dissipation does not keep pace with sedimentation. Pore fluid contents 36% greater than in normally consolidated samples were found in muds at the base of 2.4 m long cores in flat bed areas of inner Kongsfjorden. As a consequence, sediment flow occurs on any significant slope in this zone, and sidescan sonographs show ubiquitous sediment creep features on the fjord sidewalls. However, although there is a strong tendency for an infill sequence to develop, high rates of sedimentation sustain a thin mud drape on the fjord walls (Fig. 5)

In the outer proximal zone, sedimentation rates are lower, sediment is more stable and draped sequences become the norm on gently sloping shore-faces. We thus expect a sequence resulting from glacier retreat to show a transition from infill to draped sequence as the inner proximal gives way to the outer proximal zone. Fig. 5b may show an example of this.

Iceberg disruption. Major sediment disruption can occur along tracks of iceberg dispersal (Barnes & Rearic 1985). Moreover small bergs aground in shallow water, rising and falling under storm wave action, may have a significant effect in eroding and reorganizing shallow shore platforms.

Benthonic life is also strongly controlled by shoreface processes (e.g. Dale *et al.* 1989). It is generally inhibited in the inner proximal zone by high rates of sedimentation, although shallow wave and current-scoured surfaces can be well colonised. In the outer proximal zone, there is a richer benthonic fauna below the level of wave and current scour.

(c) The distal zone: outer shelf or outer fjord

The distal sedimentary regime contrasts strongly with that in the proximal zone. Suspended sediment concentrations in the water column may be up to three orders of magnitude less than in the inner proximal zone (Fig. 7). Moreover, upwelling of deep water along the continental margin (Anderson *et al.* 1984; Jacobs, 1989) and/or strong long-shelf geostrophic currents (Boulton *et al.* 1988), coupled with the greater energy of long-waves associated with the outer shelf, generally ensure a more energetic sea bed environment than in the proximal zone. As a consequence, the distal zone tends to be one of sediment reworking down to considerable depth, erosion of earlier sediments and the formation of sandy and gravelly lag horizons.

Some deeper basins on outer shelves show evidence of mud sedimentation during the Holocence, although this rarely exceeds a net sedimentation rate of 10 cm/1000 years. On very broad shelves, such as the Barents Shelf, sediment reworking dominates near the shelf edge, where the oceanic influence is strongest, but in the inner part of the shelf, in water depths greater than 100 m, there has been net sedimentation of mud during the Holocene at rates of between 2 and 8 cm/1000 years (Elverhoi 1984). Enhanced nutrient supply in zones of upwelling sustains high rates of biological productivity and rich benthonic and planktonic floras and faunas, which contribute to important bioclastic horizons (Elverhoi 1984). A large proportion of the primary clastic sediment input to many outer shelf areas is of ice-rafted detritus (Orheim & Elverhoi 1981), although the input will vary strongly as iceberg tracks are strongly dependent on bathymetry and current and wind circulation.

The effects of some of these processes are shown in Fig. 10 of the Ross Sea, in Antarctica, and the western Spitsbergen continental shelf. There is strong upwelling along both margins producing strong currents and reworking at the shelf edge and a major bioclastic component in shelf-edge sediments. Away from the shelf edge, the depth of the Antarctic shelf is sufficient to inhibit reworking except on banks in less than 300 m of water (Anderson *et al.* 1984). In Spitsbergen strong long-shelf currents produce re-working on the mid-shelf, with strong sedimentation from turbid meltwater overflows in the innershelf (fjord) environment. It is possible that a similar effect may be produced in the Ross Sea from sub-ice shelf meltwater (Jacobs 1989). Sub-ice shelf melting in the Ross Sea ensures that icebergs from the ice shelf are debris-free, and thus that little berg-dropped material is deposited close to the ice shelf (Anderson *et al.* 1984). Berg-dropped material on the mid and outer shelf is derived from icebergs carried into the region from the west. In Spitsbergen, debris-transporting bergs are derived directly from the nearby ice margins as well as being transported from afar by the long-shelf current.

Most northern hemisphere glaciated shelves show a relatively simple geometry, of shallow planar banks of 100–200 m depth, crossed by glacially-eroded troughs which descend 300–400 m below sea level (Fig. 11).

On the Iceland shelf (Boulton *et al.* 1988) till is absent in the troughs and thick on their flanks and on the adjacent banks, apart from on the inner shelf where a moraine has been generated during a Late Weichselian glacier standstill or readvance. In Kongsfjorden and Isfjorden, on the western Spitsbergen Shelf, tills are absent in the troughs in the inner shelf, but do occur on the outer shelf where they thicken towards a shelf edge (Fig. 13a). In both cases however, till appears to occur across the whole shelf on the flanking banks, apart from in shallow-water zones where it has been removed by sub-littoral erosion.

A simple numerical model is used to help explain these patterns. When a glacier flows across a continental shelf showing linear, flow-parallel troughs and flanking banks, such as that of West Spitsbergen (Fig. 11) the trough depth will be inadequate to entirely channel a glacier extending near to the shelf edge, except in its terminal zone. As ice will be approximately 200 m thicker in the troughs than flanking banks, a greater driving stress will occur along the trough, leading to enhanced flow and a tendency

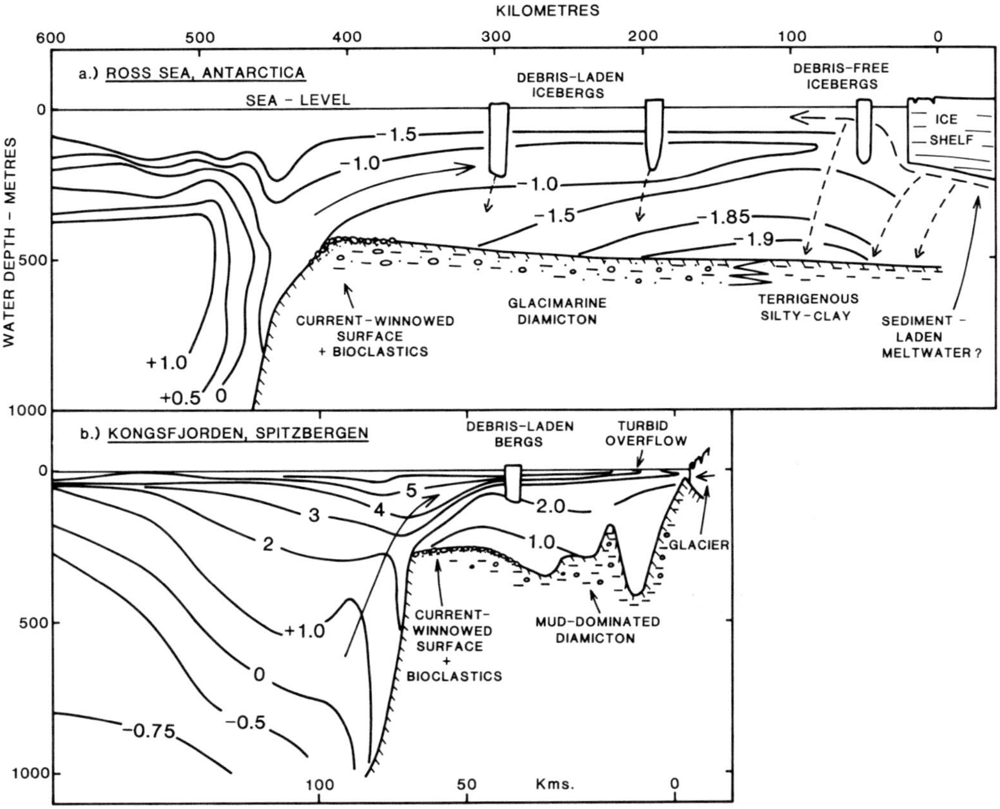

Fig. 10. Ocean thermal structure in summer in glacimarine environments from the glacier front across the continental shelf and into the deep ocean, showing the character of bottom sediments (thickness of the ornament does not represent sediment thickness). (a) Ross Sea, Antarctica (re-drawn from Anderson *et al.*, 1984). (b) Kongsfjorden and the west Spitsbergen continental shelf (NERC cruise *G.A. Reay* 1986).

for an ice stream to develop which draws down ice from the flanking banks.

At the basal boundary, we assume

$$U_b = \frac{\tau_b}{P_e^{\lambda}} \qquad (1)$$

where U_b and τ_b are basal velocity and shear stress in the x-direction, and λ is a friction parameter (Morland *et al.* 1984).

For the ice stream

$$\tau_b = \rho g(H_1 + H_2) \sin \alpha. \qquad (2)$$

For the banks adjacent to the ice stream

$$\tau_b = \rho g H_2 \sin \alpha \qquad (3)$$

where H_1 is the depth of the trough below the bank's surface, H_2 is the ice thickness over the bank and α is the slope of the glacier surface. Longitudinal stresses and shear at the margins of the ice stream are ignored.

Discharge (Q_s) of the ice stream will be

$$Q_s = \frac{W\rho g(H_1 + H_2)^2 \sin \alpha}{P_e^{\lambda}} \qquad (4)$$

where W is the width of the trough. This will be made up of two components, that due to accumulation (a) and ablation (a negative) along the ice stream.

$$Q_a = \int_0^x W_a \, dx. \qquad (5)$$

and that due to flow into the ice stream from flanking sluggish ice on the banks

$$Q_f = \int_0^x q_f \, dx. \qquad (6)$$

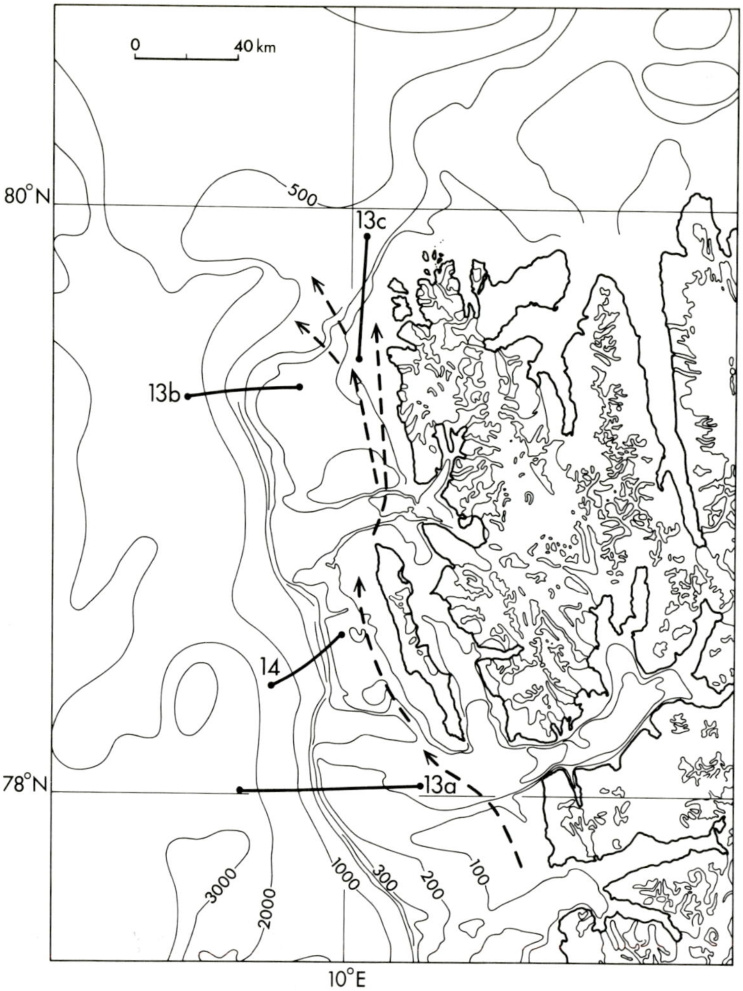

Fig. 11. Map of the west Spitsbergen continental margin showing Kongsfjorden (79°N) and Isfjorden (78°N), the glacially eroded troughs which emanate from them, and the intervening shallow banks. Large 'trough mouth fans' occur at the shelf edge. The predominant shelf currents flow north, starving the western shelf edge of sediment but feeding the shelf edge at 79.5°N to produce a zone of active progradation. Lines of sections shown in other figures are marked.

At any point, Q_s will be

$$Q_s = \int_0^x q_f \, dx + \int_0^x W_a \, dx. \qquad (7)$$

The slope across the margin of the ice stream (α_y) at any point will be

$$\sin^{-1}\alpha_y = \frac{(Q_s - Q_i)P_e^\lambda}{\rho g H_2^2}. \qquad (8)$$

Assuming the values of λ as calculated by Morland *et al.* (1984) and arbitrary and constant values of H_1, W and p_e, it is possible to reconstruct the confluent pattern of flow towards the ice stream and the distribution of basal velocity.

A simple expression is used to compute the resultant patterns of erosion and deposition (Boulton *et al.* 1984) which are determined by changes in the basal sediment transport rate, assumed to be proportional to U_b. Thus, the erosion rate ($\acute{E}$) is given by

$$\acute{E} = K_1 \frac{dU_b}{dx}. \qquad (9)$$

When this is negative, deposition occurs. I shall however assume that sediment transport occurs as a result of subglacial sediment deformation (Boulton & Hindmarsh 1987) and thus that

$$\acute{E} = K_2 \frac{d\tau_b}{dx}. \qquad (10)$$

To permit this, we expect very low values of effective pressure (p_e) because of the difficulties of drainage from the subglacial mass.

Figure 12a shows the flow paths and erosion deposition pattern for a glacier flowing over a surface containing a longitudinal trough where there is no significant calving. In a steady state, we would expect the zone of deposition near to the glacier margin to show thicker till units within the trough than on the flanking banks.

If the glacier flows into the sea, but the calving rate is small, inhibited perhaps by fixed sea ice, and if additional buoyancy in the terminal zone is enough to reduce P_e and thereby the basal shear stress, it will enhance the tendency towards subglacial deposition in the terminal zone (Fig. 12b).

If however the calving rate is very high, it may only be balanced by velocity in the high velocity zone where dU_b/dx and $d\tau_b/dx$ are positive, thus ensuring that erosion will occur as far as the glacier terminus (Fig. 12c). A further consequence will be to dump all the debris in transport immediately beyond the glacier terminus. Again we expect a strong contrast on the flanking banks where the glacier may terminate on land, with a marginal depositional zone, or in shallower water than in the trough which, because of a slower calving rate, will result in less erosion, or deposition on the banks.

If the ice stream begins to float on the shelf before the terminus, we would certainly expect a terminal depositional zone as p_e will go to zero at the grounding line (Fig. 12d). If it only floats at the shelf break, where p_e may decrease extremely rapidly, a terminal erosion zone may give way immediately to a proglacial dump (Fig. 12e).

Thus, in most circumstances we would expect a strong contrast between till thickness on the banks and in the trough.

(d) The shelf break and continental slope

No modern continental shelf break in a glacimarine environment lies in the proximal or ice-contact zone. To understand proximal and ice-contact shelf-break and continental slope processes we must refer to those localities where glaciers reached the shelf edge during glacial periods.

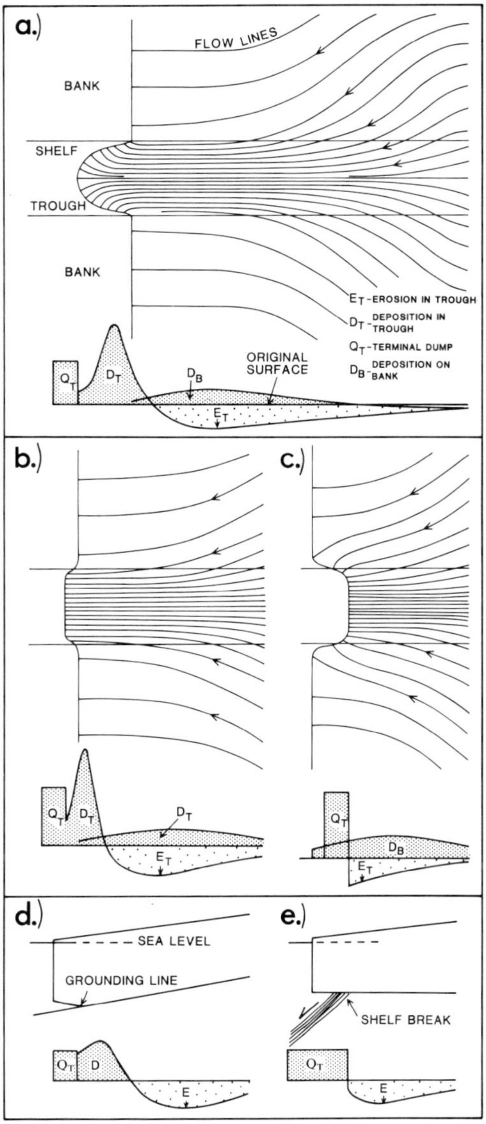

Fig. 12. Theoretical patterns of erosion and till deposition along shelf troughs and on the flanking banks. (a) Ice stream terminating on the shelf without significant calving. (b) Ice stream terminating on the shelf with some calving. (c) With very strong calving. (d) Shelf ice stream with floating terminus. (e) Ice stream floating at the shelf edge.

A major feature of the distal margins of glacially eroded troughs crossing many high-latitude shelves are '*trough-mouth fans*' where the degree of progradation of the shelf has been much greater than in the adjacent banks (Fig. 11). The sequence in such a trough mouth fan, at the mouth Isfjordrenna, in west Spitsbergen, is shown in Fig. 13a. Coring reveals the uppermost unit to be a proximal glacimarine mud, capped by a gravelly lag. Where the mud has been completely removed by erosion, we find an underlying dense, massive and unfossiliferous diamicton, which is presumed to be till. It has been suggested in previous sections that thick glacimarine muds only develop in the proximal zone, which would imply that the mud unit reflects the advance of a glacier onto the outer continental shelf. If this is so, then it seems most likely that it forms part of a couplet with the underlying till, the latter reflecting glacier advance to the shelf edge, and the former the early stages of retreat. Below, beyond and above the till, strong reflectors indicate major progradation. This may reflect discharge of sediment from the tidewater or floating margin of a fast ice stream occupying the trough.

There is a contrast on the western Spitsbergen continental slope between the style of failure in front of trough-mouth fans and the intervening banks. In the latter, where till thicknesses are

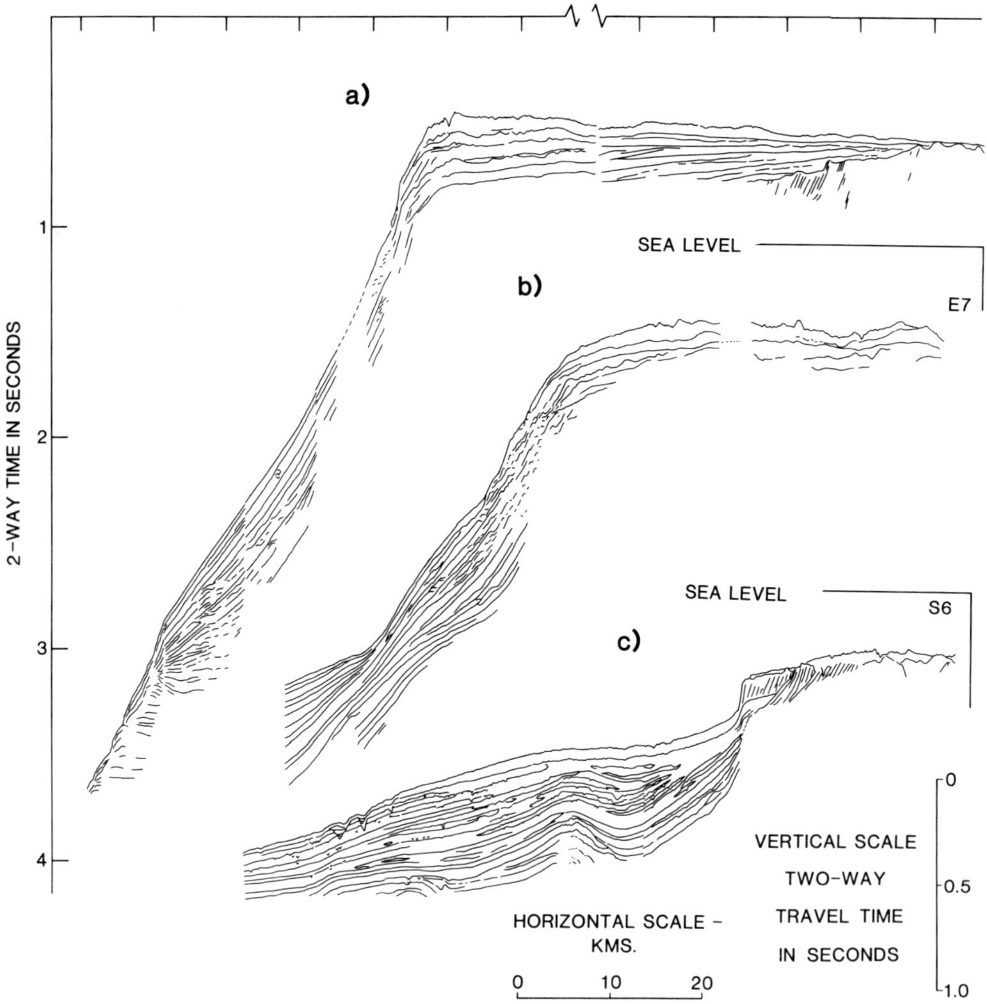

Fig. 13. Interpretation of air-gun records across the West Spitsbergen shelf edge. See Fig. 11 for location. (a) Section across the fan at the mouth of Isfjordrenna showing glacial couplets. (b) Thickening of sediment on the lower continental slope north of Kongsfjordrenna. A currently inactive margin where slumping of the upper slope feeds the lower slope. (c) Depositionally active north-western margin of the Spitsbergen shelf showing progradation on the upper slope and shelf break.

relatively large, deep-seated failure planes occur along which large sediment masses are displaced (Fig. 14). In the former, where till-tongues pass into acoustically layered sediments at the shelf edge no such large structures can be discerned, and it is presumed that mass-transfer occurs through numerous small scale failures and slumps.

It is clear from Fig 13a–b that current rates of sedimentation on the upper continental slope beyond trough-mouth fans are very low and that little sediment is discharged directly over the shelf-edge of western Spitsbergen. Geostrophic and tidal currents on the west coast of Spitsbergen produce a generally northerly drift of sea-floor sediment and surface water. This dominant long-shelf sediment drift is discharged into deep water across the northwestern shelf edge (Fig. 11), where, in contrast to the western shelf, evidence of active progradation is clearly seen on seismic profiles (Fig. 13c). This contrast between and 'active' northwestern shelf edge and continental slope and an 'inactive' western shelf edge and slope is reflected in the seismic stratigraphy. Whereas in the former the most recent seismic sedimentary units are involved in prograding, overstepping relationships on the upper slope and shelf (Fig. 13c), in the latter they show an on-lapping relationship lower on the slope (Fig. 13b). Indeed, in many profiles, this on-lapping of the lower slope can be seen to be the product of major slides and slumps on a continental slope of 3° to 6° (Fig. 14), and in some, successive major slumped units can be identified which are up to 140 m in thickness and 3–5 km in longitudinal extent.

It is suggested that advance of a glacier to the shelf edge disrupts the long-shelf sediment dispersal pathway, moves the proximal zone of surface-plume dominated sedimentation to the shelf edge, produces high rates of sediment discharge directly over the shelf edge and active shelf progradation (Fig. 13a). During periods when glaciers have retreated from the shelf, oceanic water penetrates onto the shelf and produces long-shelf sediment dispersion. High rates of primary sediment discharge over the shelf-edge cease, the glacially-oversteepened upper slope begins to fail and slumped masses accumulate low on the continental slope.

Part 2: The effect of glacial cycles and associated sea level changes on facies architecture

It is argued below that the facies architecture produced by a glacial cycle is the consequence of a temporal shift of the spatial distribution of facies described in part 1, in response to changing glacier position and changing glacially-controlled sea level. A model of glacially-controlled sea level is developed below, which is used to predict the distribution of principal facies elements in time and space through a glacial cycle. These theoretical distributions are then compared with evidence of glacially controlled sea level change and facies organization from Pleistocene and pre-Pleistocene sequences.

A model of sea-level change near to an ice sheet during a glacial cycle

The growth and decay of Pleistocene ice sheets has involved an exchange of mass of about 4×10^7 km^3 between the oceans (2.2×10^8 km^2 area) and ice sheets on northern Europe and North America (14×10^6 km^2 area – Antarctic and Greenland ice sheets underwent relatively little areal or volumetric change, at least during the last glacial cycle). This results in fluctuations of ocean level of about 120 m and changes in ice sheet thickness in mid-latitudes of up to 4 km. The cyclical concentration and dissipation of mass from mid-latitude ice sheets affects sea level in three principal ways: (i) in global eustatic changes as a direct result of ice volume change; (ii) in local isostatic crustal flexure in response to cyclical loading and unloading of the crust by ice and water; (iii) local changes in the sea surface level in response to the gravitational attraction of changing masses.

These components of sea level will have an effect on *relative sea level* (R_s), which is the local level of the sea in relation to a fixed point on the solid earth surface. Using an up-positive, down-negative convention:

$$R_s = E_s + I_s + G_s$$

where E_s, I_s and G_s are the net changes in eustatic sea level, in isostatic displacement of the crust and gravitational change in sea level surface respectively in relation to an assumed interglacial equilibrium condition. These variables are treated independently for simplicity's sake.

For the purpose of illustration a two-dimensional ice sheet model is used (Boulton *et al.* 1984) in which expansion and decay of an ice sheet is approximated by a sequence of steady states which are determined by snow-line elevations. This is used in preference to a time-dependent ice sheet model (Hindmarsh *et al.* 1987) merely because it is easier to match a

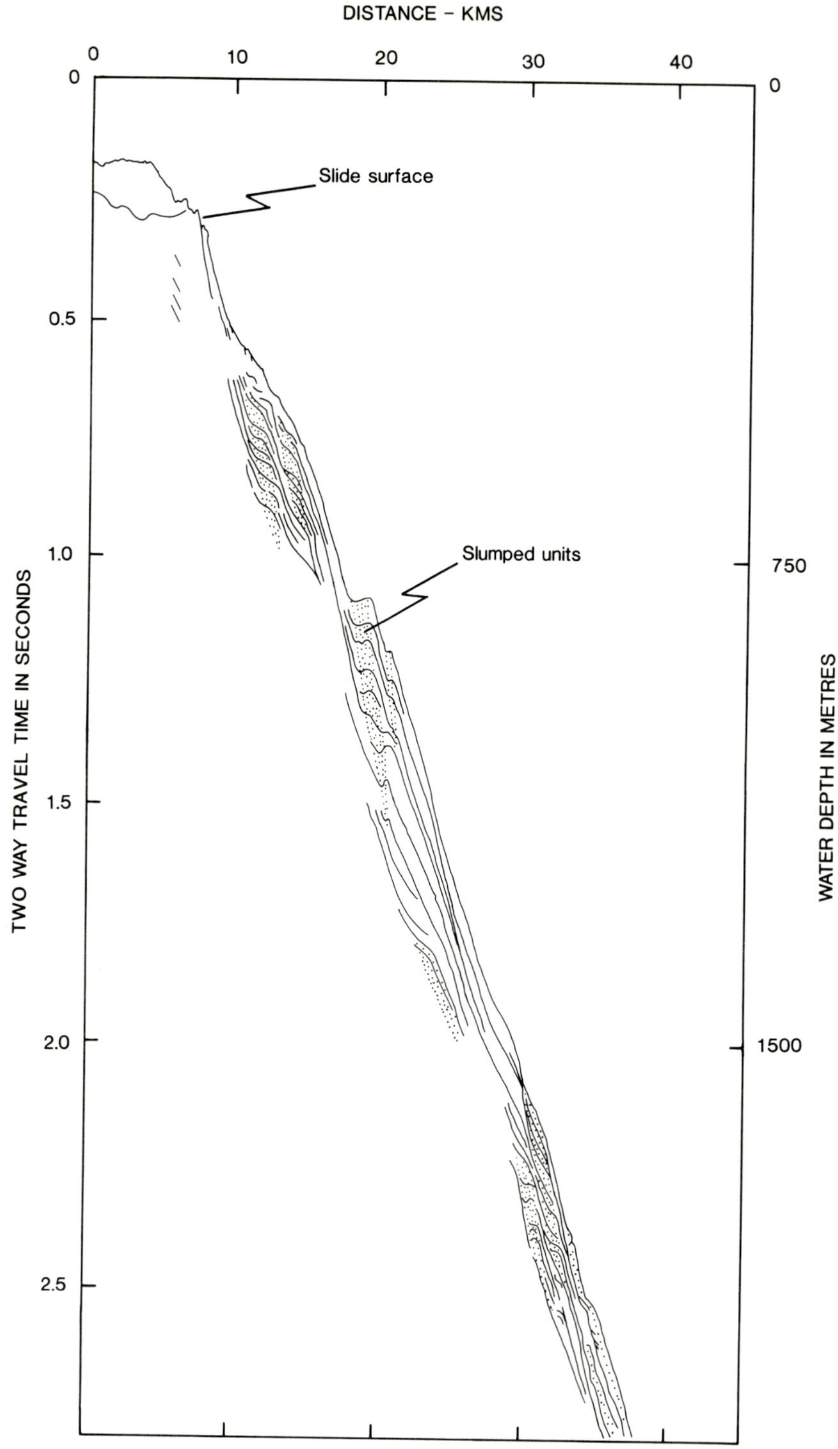

Fig. 14. Interpretation of an air-gun record showing major slumps on the upper continental slope of west-Spitsbergen. Slumps comprise two major elements, a smooth slide surface below which lies a longitudinally-compressed mass which has moved along the slide plane and accumulated at its foot. They show a typical sigmoidal reflector patterns. See Fig. 11 for locations.

prescribed pattern of glacier variation in time and space.

In computing isostatic deflection, we assume the response of the solid earth to ice sheet loading to be governed by a rigid, elastic lithosphere and an underlying viscous asthenosphere. The stiffness of the lithosphere will ensure that gravitational equilibrium, if and when achieved, will be a regional equilibrium, rather than a local one beneath every loaded point. As a consequence, downward deflection will occur both beneath and slightly beyond the loaded area, and an upward bulge will occur beyond that. The pattern of depression and uplift will be determined by lithosphere elasticity, whilst the rate of response will be determined by flow in the asthenosphere. Following Turcotte & Schubert (1982), the deflection [$W_{(x)}$ – positive downwards] beneath a point load (V_i) will be:

$$W_{(x)} = \frac{1}{8D} V_1 \alpha^3 e^{-x} (\cos x + \sin x) \quad (11)$$

where D is the flexural rigidity of the lithosphere, $x = \frac{|x|}{\alpha}$ and $\alpha = \left(\frac{4D}{\rho mg}\right)^{1/4}$ and ρm is the density of the asthenosphere. Flexure is computed by assuming the ice sheet load to consist of a series of point loads.

The rate of change of elevation of the surface of the lithosphere is calculated following Oerlemans & van der Veen (1984), and is given by

$$\frac{\delta h_b}{\delta t} = D_a \frac{\delta^2}{\delta x^2} (h_{b0} - h_b + W) \quad (12)$$

where h_{bo} is the initial elevation, D_a is the diffusivity of the asthenosphere and is given by $D_a = gH^3/3v$ where H^3 is the thickness of the asthenosphere and ν its viscosity. H and ν are not known, but Walcott (1973) has estimated values of D_a from North America and Scandinavia. Values of D_a = 1.67 m^2 s^{-1} and $D = 5 \times 10^{23}$ Nm were used.

Figure 15 shows progressive deflection of the crust in response to prescribed glacier growth and decay using the above scheme. Deflection does not achieve equilibrium with ice load during glacier expansion, and as a result, maximum crustal depression occurs during deglaciation. A proglacial bulge occurs which is progressively displaced as a wave beyond the advancing front

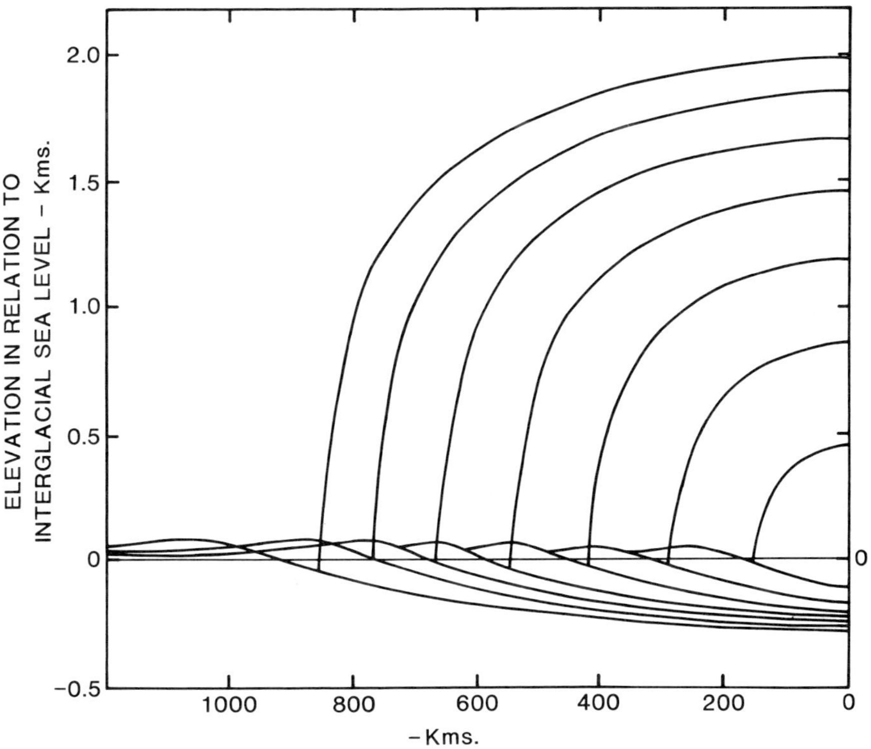

Fig. 15. Progressive crustal deflection produced by glacier growth.

of the ice sheet.

The gravitational attraction of the ice sheet mass for the water surface was calculated using the analysis of Peltier *et al.* (1978).

The pattern of relative sea level change near to an ice sheet in response to the ice sheet's expansion and decay was computed for an ice sheet cycle in phase with the global eustatic sea level cycle (Fig. 16a), and for cycles which both lead (Fig. 16c) and lag (Fig. 16b) the global eustatic cycle by 4.25 ka. The first reflects the pattern expected around the ice sheet which exhibits the largest volumetric change and thus dominates the global eustatic cycle (e.g. the Wisconsinan Laurentide ice sheet) or a smaller one in phase with it (e.g. the European Weichselian ice sheet during decay); the second is a smaller ice sheet which leads the dominant ice sheet's fluctuations (e.g. the Devensian British ice sheet during its decay phase); and the third one which lags the dominant ice sheet (e.g. the Weichselian West Spitsbergen ice sheet in its decay phase).

The relative sea level changes which would occur at a series of points at different distances from the point of ice sheet nucleation for each of these cases are shown in Figure 17. They all share similar features.

(i) In the area far beyond the maximum extent of the glacier, a symmetrical cycle of relative sea level change is presumed to have a similar phase and amplitude to the global eustatic cycle (Fig. 17–1). Nearer to the ice sheet, in the zone of the extra-marginal bulge (Fig. 17–2), the amplitude is significantly greater than that of the global eustatic cycle, whilst there is a sea level phase lead where local ice sheet growth and decay leads the global eustatic cycle.

(ii) In areas overriden by the glacier relative sea level change is asymetric, with the maximum sea level regression preceding glacier advance over any site, and the maximum transgression following retreat (Fig. 17, 3–5).

(iii) In Fig. 16 the rates of advance and decay of the northern margin of the ice sheet are prescribed to be less than those of the southern margin. As a consequence, the ice sheet residence time at any point a given distance inside the northern margin will be greater than at a similar distance from the southern margin. The isostatic effect will therefore be greater in the north, and deglaciation will tend to coincide with phases of much higher sea level (Fig. 17–5) than at the southern margin, where phases of relatively low sea level will be succeeded by phases of transgression (Fig. 17–3).

(iv) Because of the wide range of rates of glacio-isostatic uplift during deglaciation between the maximum extent of ice and the centre of glaciation, there will, at some point, be a sustained equivalence between rates of eustatic sea level rise and crustal uplift, leading to a static relative sea level (Figs 17–3b, 17–5a-c).

There are several elements of dissimilarity.

(i) The extent of the zone in which isostasy dominates relative sea level is greatest where the local glacial cycle lags the global eustatic cycle and least when the local cycle leads the global eustatic cycle;

(ii) Where the local ice sheet lags or is in phase with the global eustatic cycle, there is a relatively simple contrast between the eustatically-dominated zone of deglacial sea-level transgression (Fig. 17–1, 2) and the isostatically dominated zone of deglacial sea-level regression (Figs 17–4, 17–5ab). However where local deglaciation leads global eustasy the early decay of the isostatic component before significant eustatic rise, produces a pattern of sea level regression followed by transgression in a broad zone near to the ice sheet's maximum extent (Figs 17–3, 17–5c).

Figure 18 shows the isobases produced as a consequence of the glacial cycles shown in Fig. 16a. These represent the current disposition of planes which lay at sea level at the times given on the diagram. Their form reflects the net result of the isostatic warping of the crust and the eustatic sea level change which has occurred since those times. They are natural bounding surfaces for marine sediment packages. They show how beach sediments of a given age which formed during the glacial cycle will tend to lie above modern sea level in the isostatically-dominated zones, and how their lateral equivalents will lie well below sea level in the eustatically dominated zone. Important characteristics are as follows.

(i) The marine limit in the isostatic zone is diachronous and dips at a smaller angle than isobases. Only instantaneous ice sheet disappearance will give an isochronous marine limit.

(ii) The location of the extra-glacial isostatic 'bulge' is indicated by a depression of isobases in the eustatic zone.

(iii) They show evidence of a regression followed by transgression during glacier build up in the area within and just beyond that occupied by the glacier at its maximum extent.

(iv) They show that in the case of a glacier whose decay leads the global eustatic decay, regression followed by transgression during deglaciation occurs in a zone near to the maximum extent of the glacier.

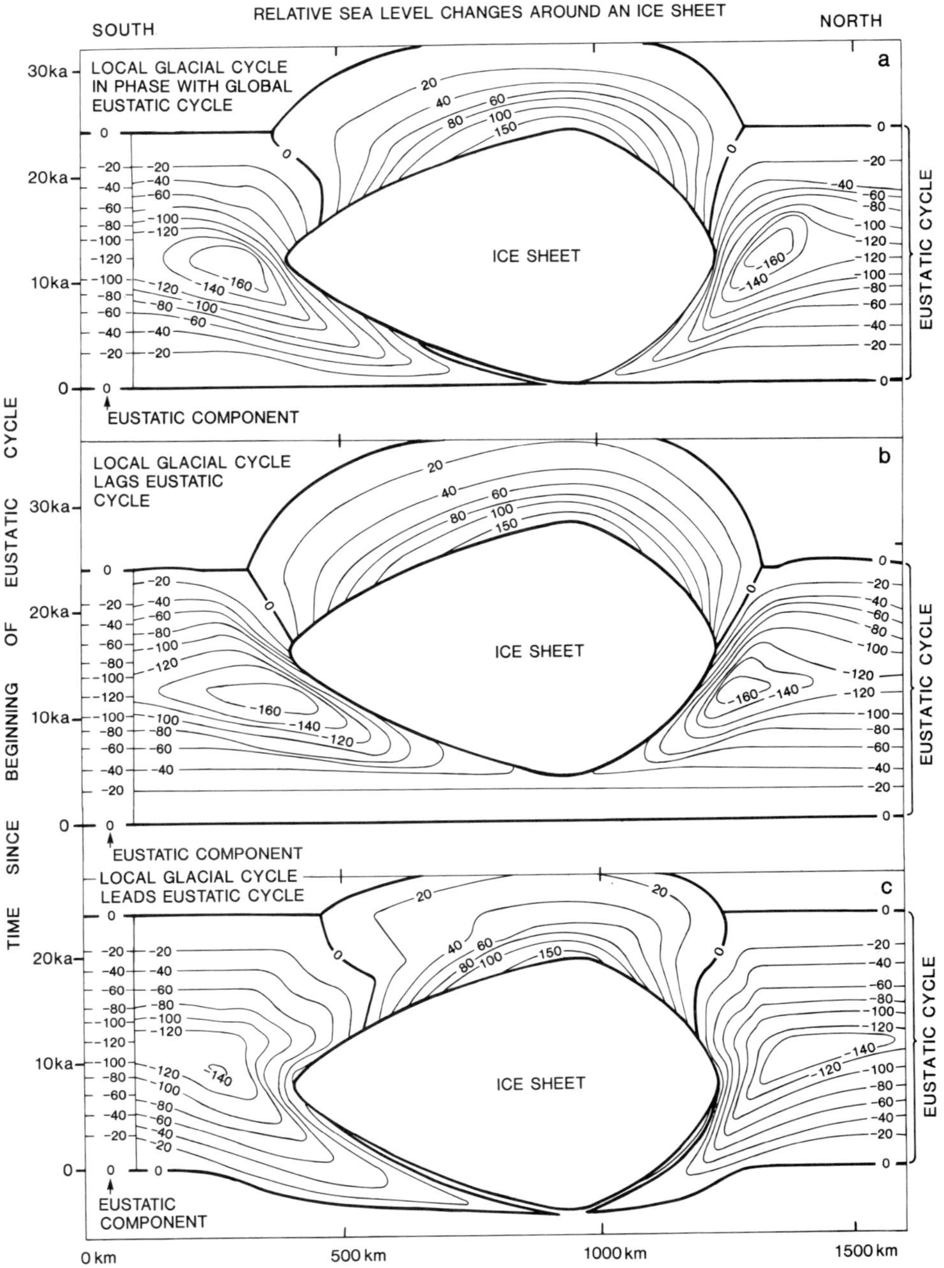

Fig. 16. Patterns of relative sea level change in space and time produced by expansion and decay of an ice sheet resulting from interaction of eustatic, isostatic and gravitational components of sea level change. (a) Local ice sheet cycle in phase with dominant global eustatic cycle. (b) Local cycle lags global cycle. (c) Local cycle leads global cycle. The component of sea level change due to the global eustatic cycle is shown on the left of the diagrams. Elapsed time is shown from the beginning of the global eustatic cycle.

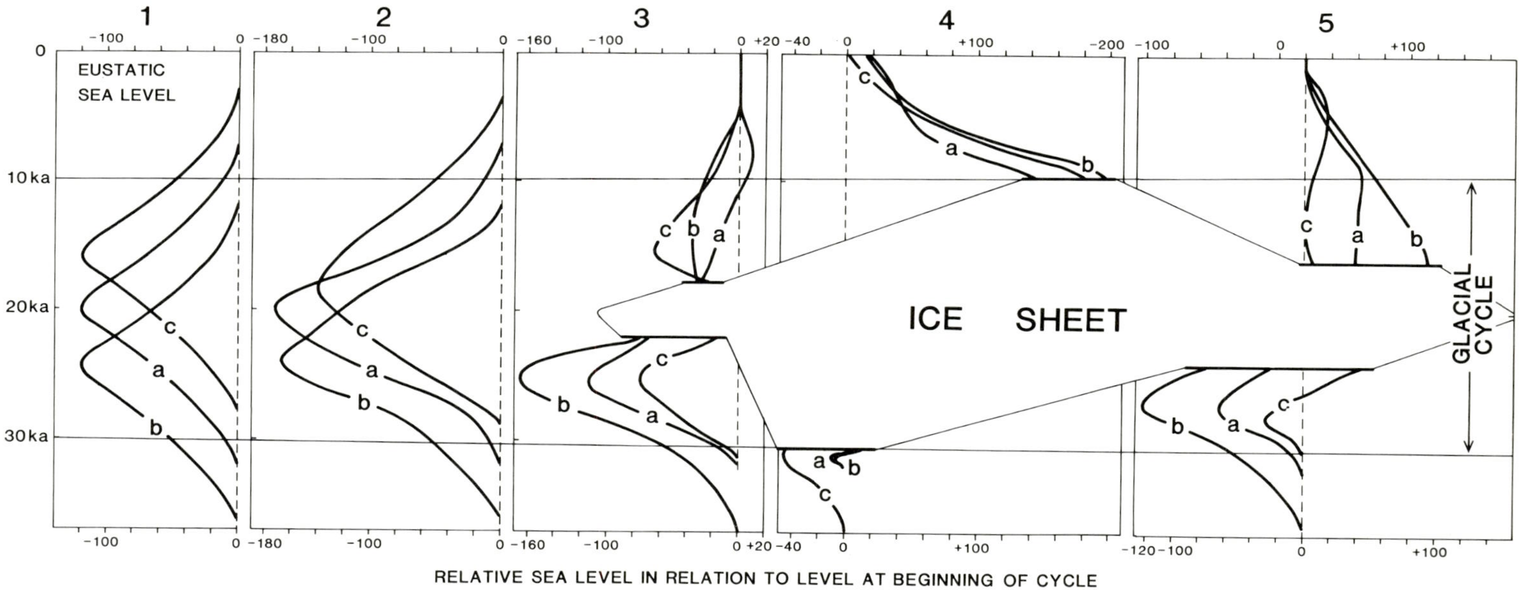

Fig. 17. Relative sea level change through a glacial cycle at localities 1–5 which occur at distances of 1 = 0 km; 2 = 275 km; 3 = 440 km; 4 = 800 km and 5 = 1190 km in Fig. 16. Curves marked *a*. are those where the local glacial cycle and global eustatic cycle are in phase, those marked *b*. where local cycle lags the global cycle, and those marked *c*. where the local cycle leads the global cycle. The glacial cycle occurs between 30 and 10 ka. Leads and lags can be seen from its relation to the eustatic cycles shown at locality 1.

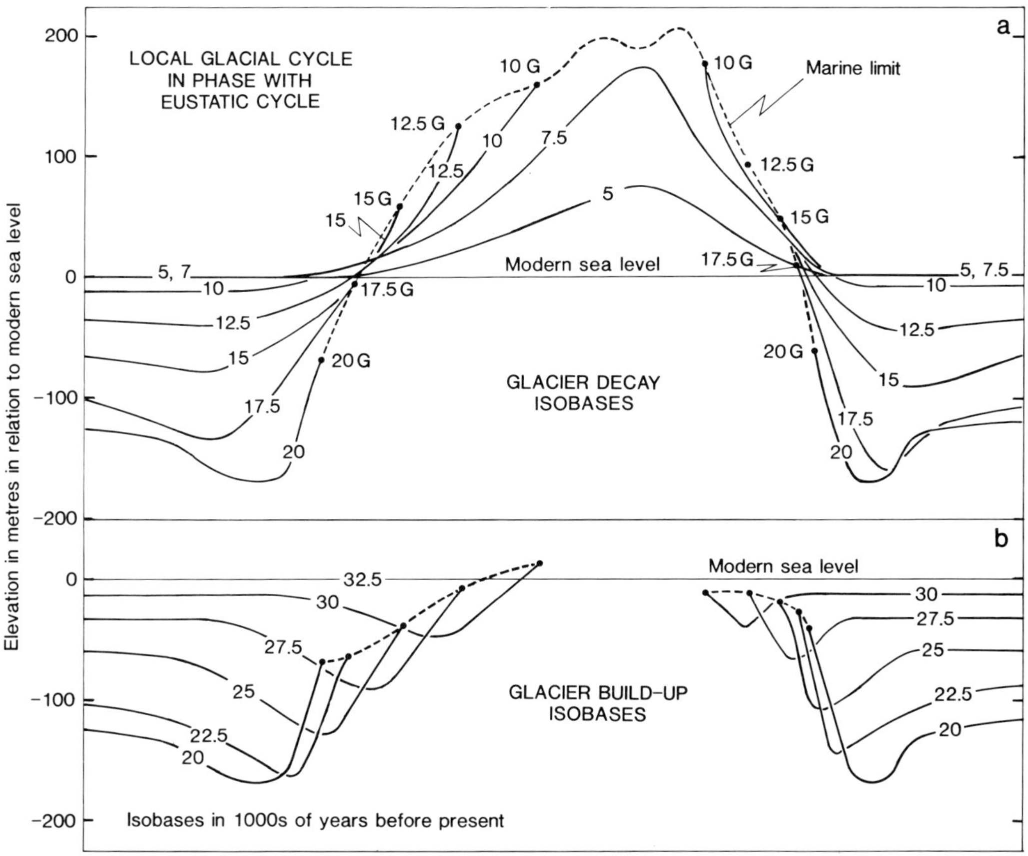

Fig. 18. Two dimensional form of isobases dating both from the period of ice sheet expansion and the phase of ice sheet decay from the model in Fig. 16 where the local glacial cycle is in phase with the global cycle. Isobases show the current form of surfaces which lay at sea level at the stated time.

Glacimarine facies architecture on the land margin and shelf: theory

The simple spatial model of glacimarine facies developed in part 1 relates the character and rate of accumulation of sediment at any site to its distance from the glacier, the slope of the substratum and its relation to sea level. If this is combined with a model of glacier extent and sea level change, such as that in Fig. 16, the distribution of sedimentary environments in space and time can be reconstructed. An example is shown in Fig. 19 for a local glacial cycle in phase with the global eustatic cycle.

Figure 20 illustrates the sedimentary sequences which would develop in response to this temporal and spatial shift of environments. Figure 20a shows the vertical sequences which would develop at given distances from the centre of glaciation at all elevations, and Fig. 20b the horizontal sequences at specific elevations. It combines the sedimentation rate model from Fig. 6 and a model of the shoreline processes in Figs 5 and 22 with the sea level/environment models in Figs 17 & 19. Uplifted marine sequences in the isostatically-dominated zone are shown by 4–10 in Fig. 20a and largely coincide with the glaciated area, apart from sequence 10 at the northern margin of the ice sheet, where longer glacier occupancy has produced stronger crustal deflection reflected in a transgression beyond the limit of glaciation. The flanking eustatically-dominated zones suffered low glacial sea levels. The transgression limit (= marine limit) and regression limit are as shown in Fig. 16. Several other attributes are worthy of note.

(i) Glacimarine muds only survive above sea level in the areas of greatest uplift and well below the marine limit.

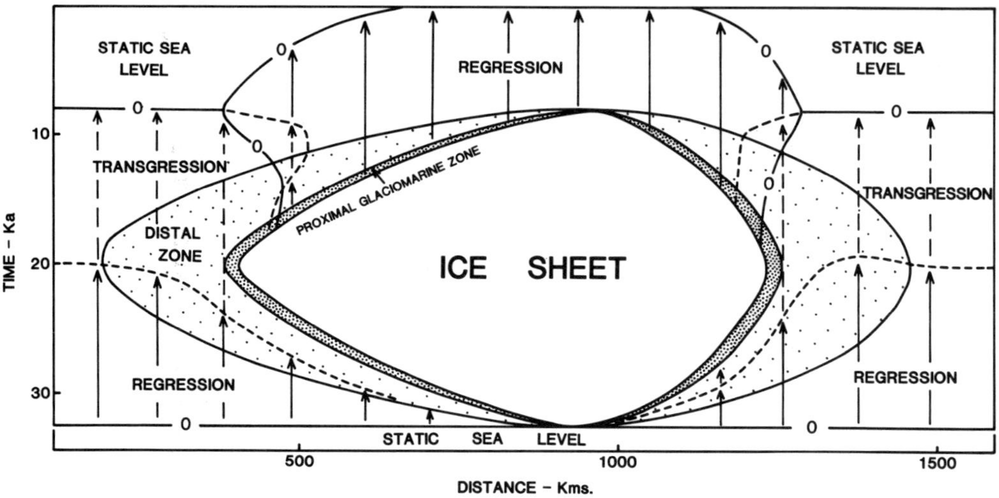

Fig. 19. Distribution of environments in space and time during a glacial cycle in which the local ice sheet is in phase with the global eustatic cycle (c.f. Fig. 16a).

(ii) No normal marine muds survive above sea level.

(iii) The regression limit pre-dates the glacial maximum and is diachronous in the isostatic zone although it is synchronous with the glacial maximum in the eustatic zone (Fig. 18). Early glacial marine sediments thus survive at a high elevation in the isostatic zone, where the regression limit is high and where it is not removed by glacier erosion (Fig. 20−9), and only at low elevations in the eustatic zone, where erosion beneath the low regression limit removes them (Figs 20−1, 2, 10, 12).

(iv) It has been argued (Boulton 1982) that the peripheral zone of an ice sheet is a zone of deposition and the internal zone one of erosion. Early glacial (and pre-glacial) sequences near to the maximum extent of the ice sheet will tend to be preserved and capped by till (Fig. 20−9) whilst those nearer the interior of the ice sheet at its maximum (Figs 20−6, 8) will tend to be removed by subglacial erosion.

(v) Major erosional notches along the shoreline are presumed to reflect long periods of stable sea level. They occur in the eustatic zone where eustatic change ceases long before the isostatic component has decayed, and in the isostatic zone where rates of isostatic uplift and eustatic sea level rise are similar. This often occurs near the marine limit (Fig. 17−5a), and may also lead to major sediment aggradations at sea level due to the high rates of sediment supply in the glacimarine environment (Fig. 20−9).

Figure 20b shows expected sequences at specific elevations above and below sea level. It utilizes an erosional/depositional theory (Boulton 1982) which predicts net erosion in the inner zone occupied by an ice sheet, net deposition in an outer zone, and deposition without erosion in an outermost zone, which is the only area where immediately pre-glacial sediments will be preserved. In deep water (−200 m) we expect a complete late-glacial and post-glacial record to be preserved. In shallow water (−60 m), we expect evidence of early glacial sea-level regression to be preserved in the outer glacial zone, but strong marine erosion of the glacial, late-glacial and post-glacial sequence. Above at increasing heights above interglacial sea level (+100 m, +150 m) we expect the lateral extent of uplifted glacimarine sediments to become progressively more restricted.

Figure 22 shows the altitudinal distribution of the major sediment facies groups resulting from a glacial cycle. The upper limit of beach and littoral facies is marked by the marine limit. Its lower limit is presumed to lie some 10 m or so below the regression limit. It is assumed that in unsheltered environments, glacimarine mud will not accumulate in less than 30 m of water (Boulton *et al.* 1982) and that its deposition will cease as glaciers withdraw from the catchment. In the isostatic zone the topographically highest glacimarine mud units will be the oldest muds and lie about 30 m below the marine limit. In the eustatic zone of deglacial marine transgres-

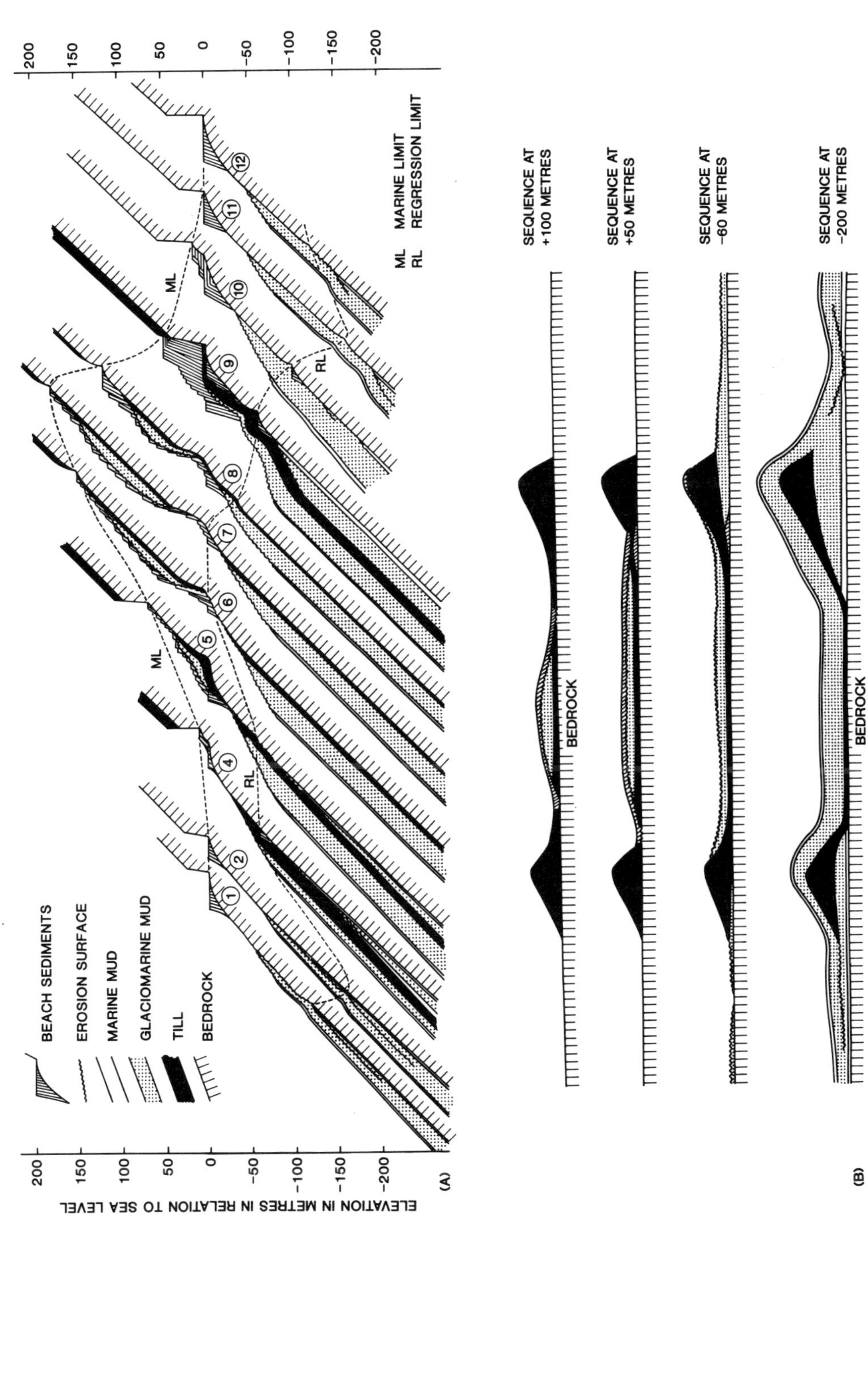

Fig. 20. Facies distribution produced by a complete glacial cycle in which the local cycle lags the global cycle (c.f. Fig. 17). (a) Onshore/offshore sections at specific distances to the north and south of the centre of glaciation. (b) Longitudinal sections at specific elevations above and below sea level.

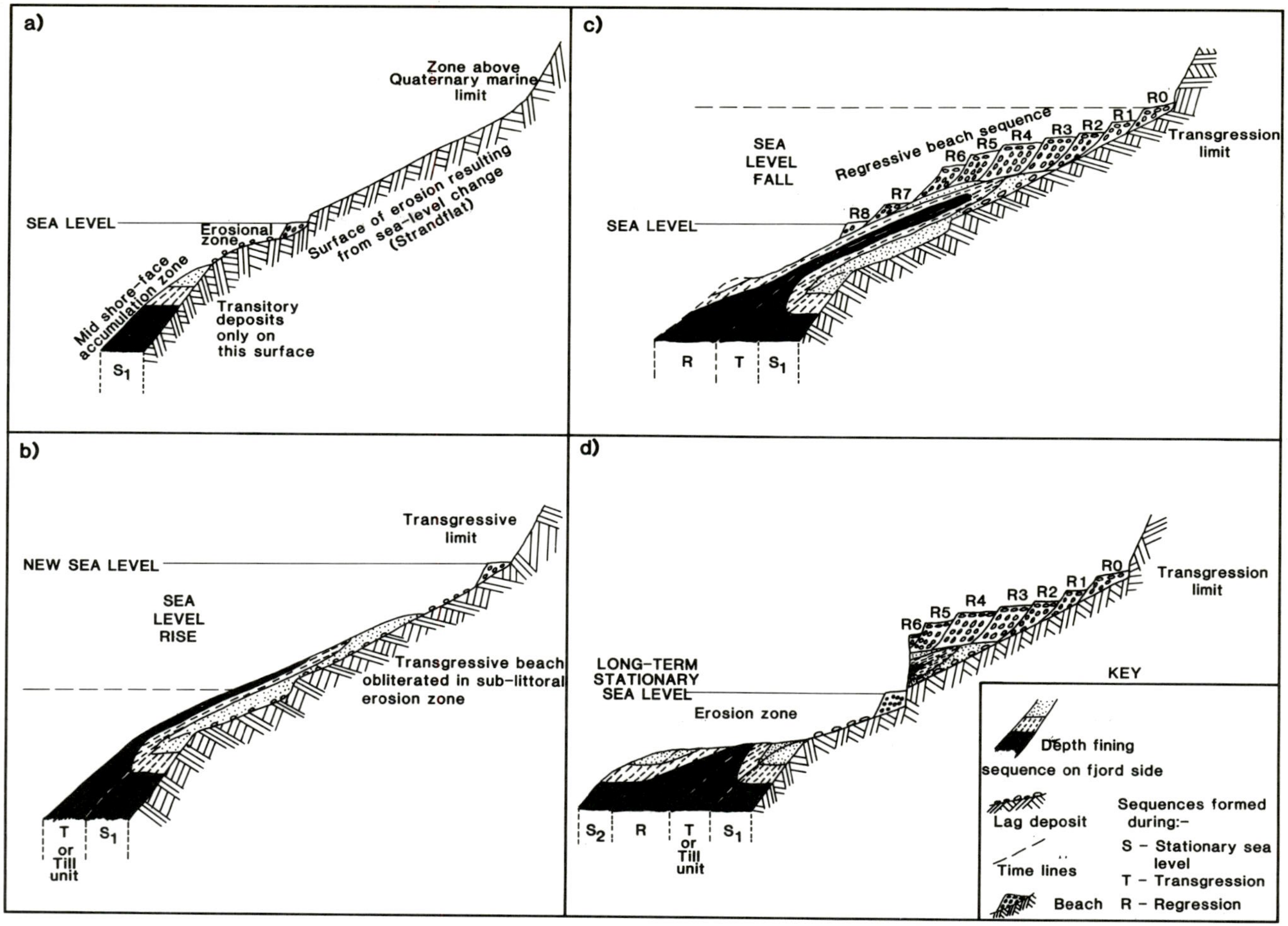

Fig. 21. Sedimentary sequence in the nearshore zone resulting from a glacial-isostatically dominated sea level cycle in a site just beyond a glacial maximum in an area where the local cycle lags the global cycle. (a) Facies distribution during interglacial. (b) Consequence of sea level rise resulting from dominant glacio-isostatic crustal depression. The transgressive unit (T) could represent a till if the site had been overridden by a glacier. (c) Late-glacial sea level regression. (d) Erosional notch produced during static interglacial sea level.

sion the topographically highest glacimarine muds will be the youngest muds, whose deposition will cease as glaciers withdraw from the catchment area. The youngest glacimarine muds at any elevation can be determined for Fig. 22 by superimposing this on the isobase plot in Fig. 18.

In the interglacial North Sea muds only accumulate in water depths greater than about 70 m. Similar upper limits for mud deposition occur on other temperate continental shelves. In the isostatic zone, the maximum elevation of interglacial muds will depend upon the net uplift which has occurred since glacier withdrawal from the catchment. In the eustatic zone, the highest interglacial muds will be youngest and lie at about 70 m below modern sea level.

The location of sea level is particularly important, for a small change in water depth in the nearshore zone can lead to rapid changes of sedimentary environment from one of mud deposition to one of strong erosion and beach sediment accumulation. The zone above and below interglacial sea level, which is swept successively by these regimes as sea levels rise and fall during glacial cycles, is one of considerable sedimentary complexity. The character of the sequences produced by a simple cycle of sea level rise and fall and consequent displacement of shore-face facies zonation in an isostatically-dominated zone is shown in Fig. 21a-d. The transgressive beach tends to be obliterated by the succeeding erosional phase, except at the transgression limit, whilst other components of the transgressive sequence tend to be removed in the erosional zone during sea level fall. Thus, only at lower elevations, where there has been a longer period of sedimentation and shorter erosional phases, are muddy sediments preserved beneath beach sediments. The sequences preserved above interglacial sea levels, which in the Quaternary are the most directly accessible, tend to reflect very short timespans during late-glacial sea level regression. Complete sequences will tend to lie in deep water.

Glacimarine facies architecture on the land margin and shelf: examples

The dominant control on global eustatic sea change during the last glacial cycle has been the fluctuation of mass of the Laurentide ice sheet. Examples of local relative sea level history and glacimarine facies architecture have been chosen from the areas of the last European ice sheet whose decay has been roughly in phase with the Laurentide, from the west Spitsbergen ice sheet margin which lagged the Laurentide and the British ice sheet which led it.

a. Patterns of relative sea level change. Figure 23 shows patterns of relative sea level change and Fig. 24 the form of isobases along radial transects across the areas of Europe, Spitsbergen and Britain occupied by ice sheets during the Late Weichselian.

Deglaciation of western Spitsbergen clearly lagged the decay of the Laurentide ice sheet (Boulton *et al.* 1982; Mangerud *et al.* 1987). As a consequence, it produced the most extensive zone of isostatically-dominated late-glacial and post-glacial sea level change as a proportion of ice sheet radius of the three ice sheets used here as examples. This effect is smaller in the case of the European ice sheet and least in the case of the British ice sheet, as we would expect from the phasing of deglaciation in relation to that of the Laurentide ice sheet (cf. Fig. 16). Major

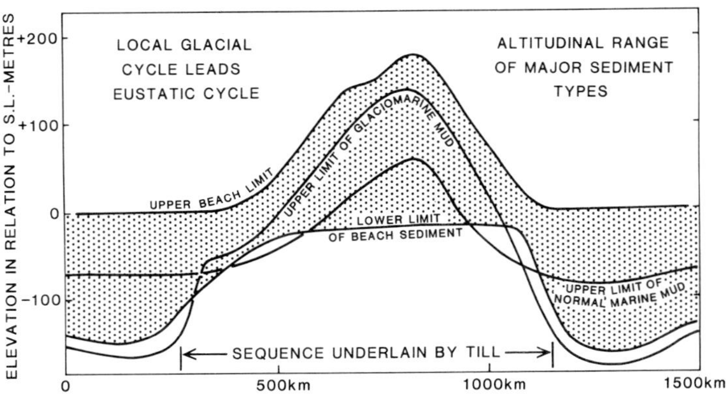

Fig. 22. Altitudinal range of major marine sediments types produced during a complete glacial cycle.

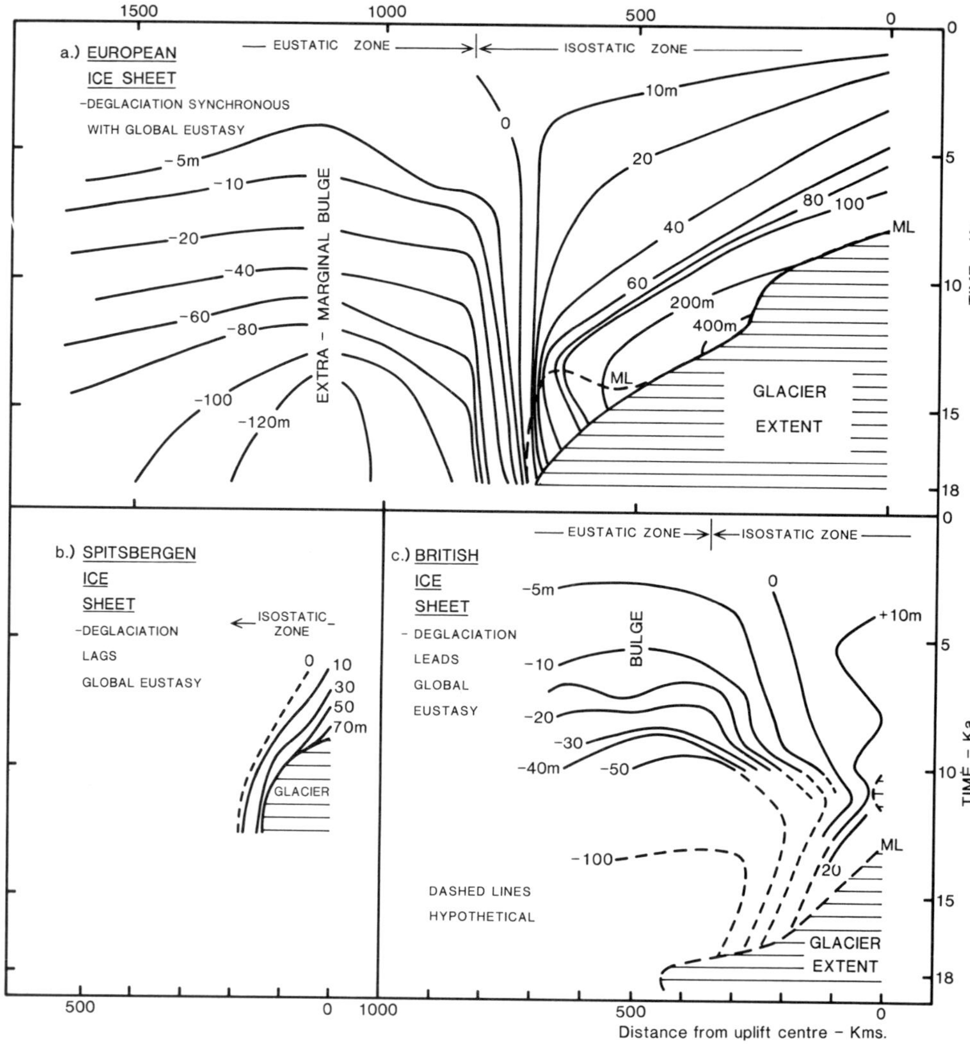

Fig. 23. Relative sea level change in time and space inferred from direct geological evidence. (a) Along a radial transect across the Weichselian area of the European ice sheet, through the western Baltic and eastern North Sea. (b) Along a radial transect across the area occupied by the Late Weichselian West Spitsbergen ice sheet (see Fig. 25 for location). (c) Along east coast of Britain across the area occupied by the Late Weichselian ice sheet.

global deglacial sea level rise begins at about 14 ka (Mix & Ruddiman 1985), whilst major European deglaciation begins at about 15 and 13 ka (Boulton *et al.* 1985; Lundqvist 1986), major deglaciation of parts of the British ice sheet had already occurred by 15 ka (Coope & Brophy 1972) and major West Spitsbergen deglaciation took place after 11–10 ka (Boulton *et al.* 1982; Mangerud *et al.* 1987).

In the case of Britain there is also evidence of a major late glacial sea level regression followed by transgression which the model (Fig. 16) predicts should occur where local deglaciation leads the global eustatic rise, due to an early decay of the glacio-isostatic component of relative sea level before a major eustatic sea level rise. In the Irish Sea basin, McCabe & Eyles (1988) and McCabe *et al.* (1986) have argued for sea levels up to 100 m above sea level at the time of deglaciation, whereas Holocene sea levels on the margin of the basin show a major transgression from −20m at 9 ka to modern sea

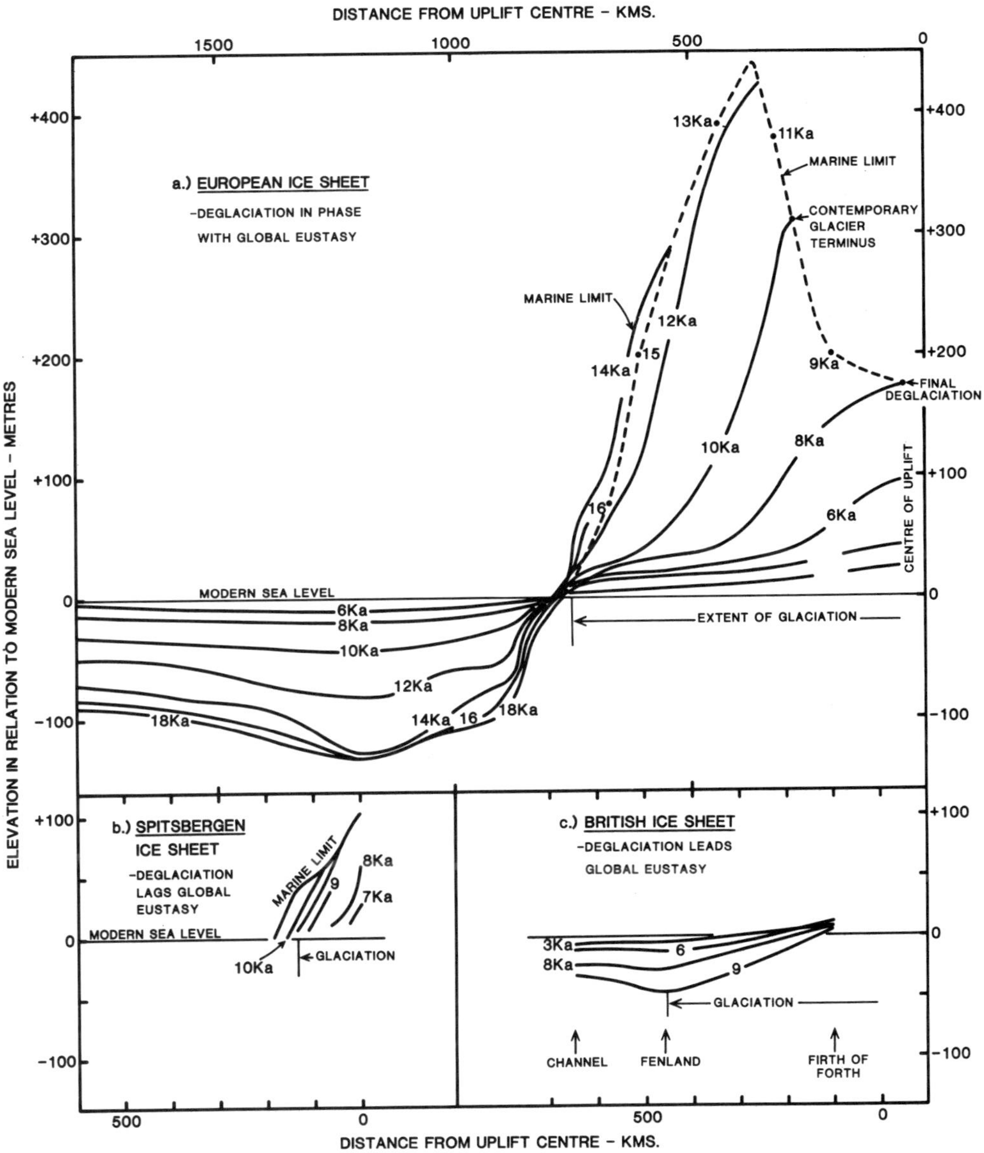

Fig. 24. Isobases along the same transect lines shown in Fig. 23.

level by 2 ka (Tooley 1978). A similar pattern can be inferred on the eastern coast of Britain (see next section) although the late glacial marine limit was significantly lower than at similar latitudes on the west coast.

The patterns of isobases for Europe and Britain (Fig. 24) also show a trough in the proximal part of the eustatic zone which probably reflects collapse of an extra-marginal bulge.

b. Patterns of facies distribution. Glacimarine facies architecture in an area where local deglaciation lagged the dominant eustatic cycle is well demonstrated in Spitsbergen. Figure 25 shows a map of the elevation of the marine limit over the Spitsbergen archipelago. It appears to reflect isostatic uplift in response to decay of two major Late Weichselian ice domes, one over the eastern part of the main islands

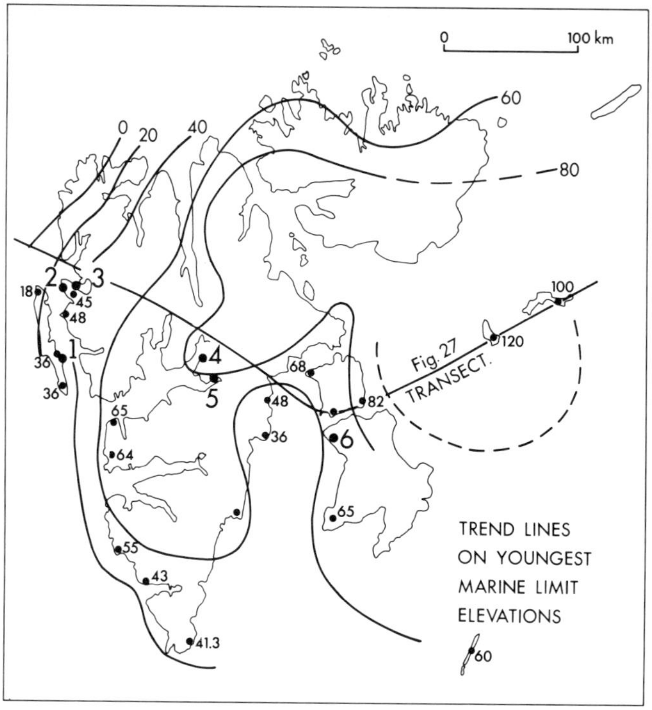

Fig. 25. Map of Spitsbergen showing the location of sites shown in Fig. 28 and the transect in Fig. 27.

of Vestspitsbergen and Nordaustlandet and another to the east of the archipelago. Fig. 26 shows a transect across these domes onto which isobases have been projected along lines parallel to the marine limit from sites near to the line of transect. If we assume that the highest relative sea level at any site is contemporary with deglaciation at that site, the intersection of isobases with the marine limit gives the age of deglaciation (c.f. Fig. 18). Thus, on the western side of the Spitsbergen ice dome, deglaciation of the west coast commenced prior to 10 ka and the ice dome in the vicinity of Kong Karls Land disappeared at about 10 ka. The marine limit is diachronous within the area covered by the ice sheet but isochronous in the area on the west of Spitsbergen not covered by the ice sheet.

A series of sections near to the transect show coarsening-upward sequences reflecting lowering of relative sea level. The highest elevations at which glacimarine muds have been found preserved generally lie between 20 and 30 m below the marine limit, which is similar to the depth below which muddy sediments tend to accumulate in exposed locations in modern fjords. Figure 27 shows sedimentary logs at several sites across the transect shown in Fig. 26. For the central area, where the marine limit is high above modern sea level, there is an upward sequence of till, followed by muddy glacimarine sediment of a variety of facies, overlain uncomfortably by coarse littoral sediments deposited during isostatically-dominated sea level fall (sites 3–6, fig. 27). At the limit of late Weichselian glaciation on the west coast of Spitsbergen, isostatic depression of the crust was still sufficiently strong to produce a high marine limit and permit mud to accumulate at some sea level sites. The lowest mud units are presumably contemporary with or may even pre-date the maximum of glaciation. Further west, the Late Weichselian marine limit is so low to have precluded deep-water mud accumulation above modern sea level, and coarse littoral deposits of the Late Weichselian/Holocene sea level regression lie directly above similar deposits of a regression from an earlier high sea level associated with a major glacial phase (sites 1–2, fig. 28). Further west still, the marine limit is presumed to fall below modern sea-level.

Sedimentary sequences produced by complete glacial cycles in sheltered fjords, well below the depth likely to be influenced by sea

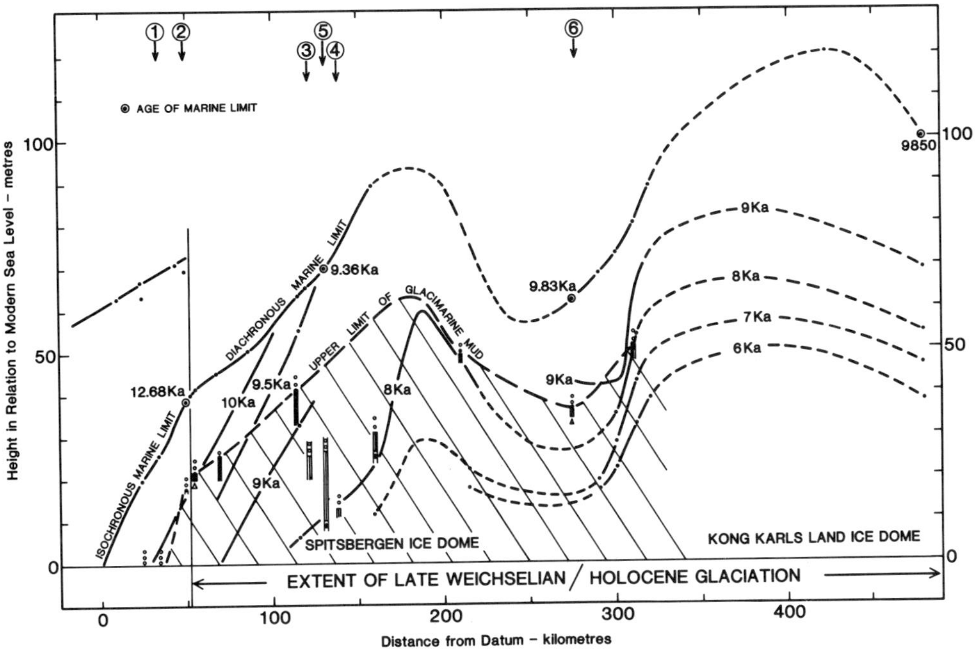

Fig. 26. Section across Spitsbergen showing the elevation of the Late Weichselian marine limit and isobases (in the west). The upper limit of glacimarine mud is shown and the location of important sections (triangles show till, vertical lines show glacimarine mud, and circles show beach sediment). Beyond the extent of Late Weichselian glaciation, an earlier, probably Early Weichselian marine limit survives, relating to uplift after glacier expansion to the shelf edge. The age of the marine limit, which is presumed to approximately date deglaciation, is based at 50 km on Mangerud *et al.*, (1987), at 140 km and 280 on Boulton *et al.*, (1982) and at 490 km on Salvigsen (1981). 1–6 show the locations of sections illustrated in Fig. 27.

level change (e.g. Fig. 20b: −200 m), have not, to my knowledge, been cored through in areas of modern glaciation. We do not expect them to undergo the same strong deep water scouring as sequences on open shelves, and if we assume that sea level change has a minor influence on sedimentation, only glacier proximity and sea bottom form will influence sedimentary sequences.

It was argued in part 1 that extensive glacimarine mud sedimentation only occurs on the middle and outer shelf when glaciers expand onto the shelf, bringing it within the zone of heavy fall-out from turbid plumes and displacing energetic ocean waters. When this occurs, we expect a full glacial cycle to show an upward sequence of till (partly made up of remoulded glacimarine sediment produced during the advance) and mud-dominated glacimarine sediment with evidence of current scour and reworking of the surface produced after deglaciation (Fig. 30). The failure of the Late Weichselian Spitsbergen ice sheet to expand over Spitsbergen's western shelf explains why there does not appear to be an extensive Late Weichselian glacimarine mud blanket on the outer shelf.

Glacimarine muds below the depth of modern wave induced scour on the West Spitsbergen Shelf, the Barents Shelf, the southern Iceland Shelf, in the North Sea, and many other high latitude shelves (Elverhøi *et al.*, 1983; Boulton *et al.*, 1988) tend to have armoured surfaces with predominantly reworked sands and muds in pockets on their surfaces. This may partly reflect low early Holocene sea levels, but the persistence of scour to the present day suggests that the extension of energetic oceanic waters onto interglacial shelves is a major cause. It is very likely that many glacial and glacimarine diamicton surfaces become armoured due to scour in late-glacial or early interglacial times, and thus that the major part of the glacial and glacimarine sediments removed from shelves in interglacials are removed at an early stage.

A quite different facies architecture to that in Spitsbergen is found in Southern Britain where deglaciation led global eustasy. On the eastern

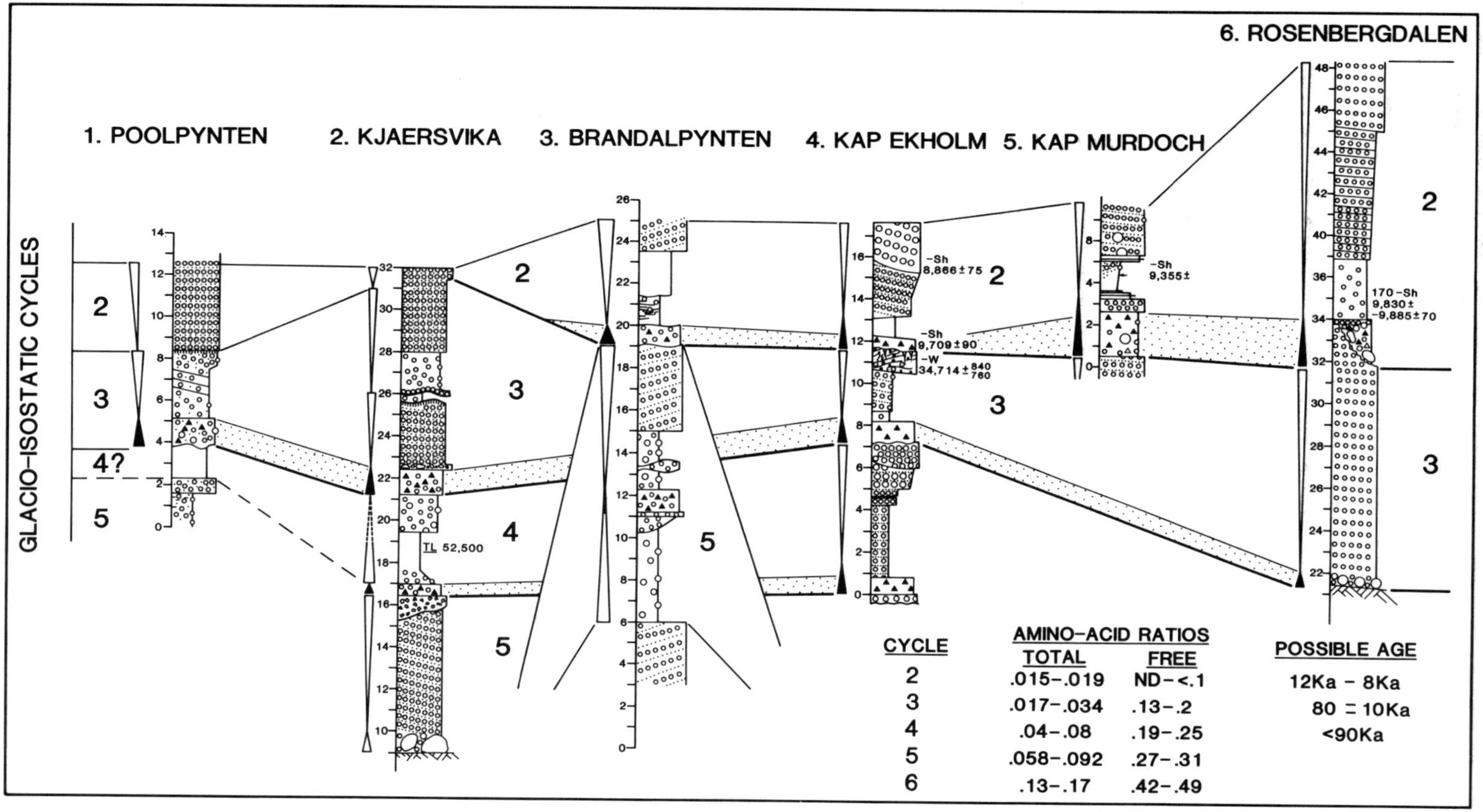

Fig. 27. Sedimentary logs through sequences of superimposed glacio-isostatic facies cycles. Individual cycles are numbered 2 to 5. Cycle 2 is of Late-Weichselian age and cycle 3 probably of Early Weichselian age. The locations of the sections 1–6 are shown in Figs 25 & 26.

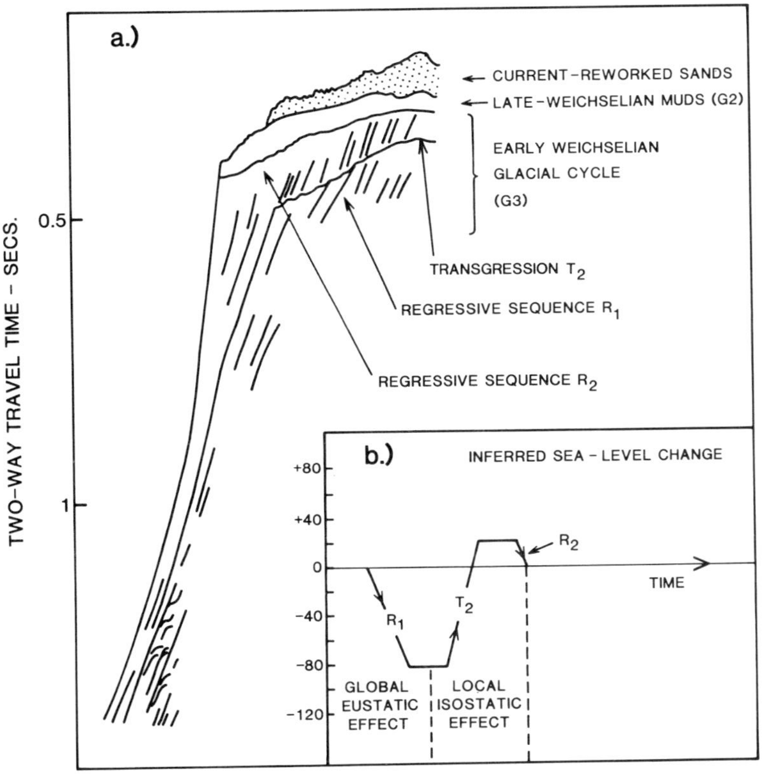

Fig. 28. (a) Detail of interpretation an of air-gun record across the trough mouth fan at the western extremity of Kongsfjordrenna (Fig. 11). (b) Interpretation of the structure in (a) in terms of sea level change through a whole glacial cycle.

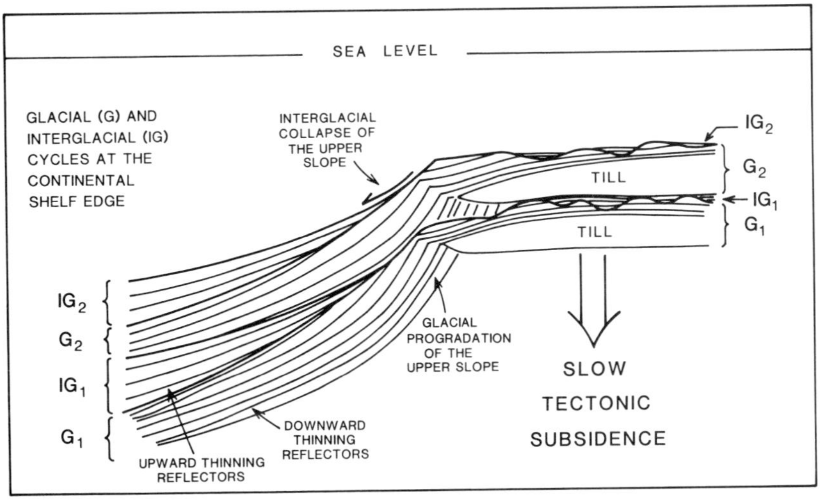

Fig. 29. Idealized structure of a shelf edge and upper slope in which glacier expansion to the shelf edge produces upper slope progradation, and phases of glacier retreat permit collapse of the upper slope and build-up of sediments lower on the slope. Tectonic subsidence permits a vertical sequence to build up which would otherwise be removed by energetic shelf processes.

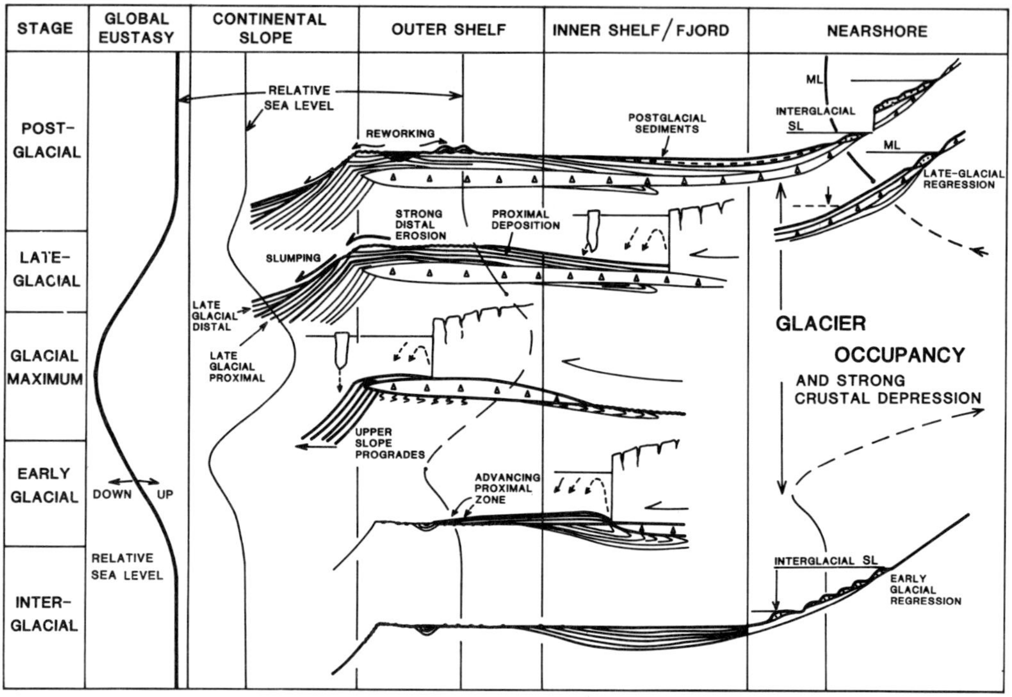

Fig. 30. A model of glacimarine architecture in space and time showing the building blocks of glacimarine facies architecture in different continental margin locations which accumulate through a whole glacial cycle. The relative sea level changes appropriate to each zone are shown.

shores of the Irish Sea and on the North Sea coast of England the last 10 000 years have seen a sea level rise from about −40 to −50 m bsl. On the northeast coast of Ireland (McCabe *et al.* 1986; McCabe & Eyles 1988) a marine limit up to +100 m asl lies above uplifted glacimarine sequences which in turn overlie tills, demonstrating an isostatically dominated high sea level immediately after deglaciation, followed by sea level regression. These patterns fit well with an ice sheet whose deglaciation leads global eustasy. The initial sea level fall from a high marine limit represents the early decay of the isostatic component and the later sea level rise represents the global eustatic rise. On the northeast coast of England the late-glacial marine limit is probably near or below sea level, reflecting the southeast axis of the Late Devensian British Ice Sheet which loaded the northern Irish Sea more than eastern England. Nonetheless, it shows a pattern of sea level change similar to that of the northern Irish Sea and reflecting early deglaciation. Flandrian sea levels appear to show a continuous rise (Tooley 1978), from −40 m at about 10 ka B.P. suggesting a low glacial sea level and a eustatically dominated postglacial transgression. However, just off the coast of northeast England, the Late Devensian glacimarine St Abb's beds occur in water depths of 40 m. implying that relative sea level during deglaciation was at least as high as −10 m. Thus, late-glacial sea level must have been isostatically-dominated, regressing from about −10 m at deglaciation to at least −40 m at 10 ka, followed by a eustatically dominated transgression up to modern sea level by about 4 ka.

c. Multiple glaciation. Figure 27 shows a series of sections across the Spitsbergen archipelago which contain evidence of pre-Late Weichselian glacial events. Some are represented by complete glacio-isostatic facies cycles, but some by only parts of cycles, either because of subglacial erosion (e.g. Brandalpynten, Rosenbergdalen, beneath cycle-2 till) or littoral and sub-littoral erosion (e.g. Kjaersvika, top of cycle 5), or because the magnitude and period of ice sheet loading was inadequate to produce strong crustal depression and sufficient marine submergence to ensure deposition of a glacimarine mud component (e.g. Poolpynten, base of cycle 2). Sites 1 and 2 in Fig. 27 lie beyond the limit of

Late Weichselian glaciation. However, they underwent a limited amount of glacio-isostatic submergence, with marine limits of 13 m and 32 m respectively. Water depth at the transgression maximum at these sites was not, however, great enough to ensure preservation of a mud sequence above sea level. The sites would be represented by a position at about 1100 km in Fig. 22.

A magnificent example of preservation of glacio-isostatic facies sequences is demonstrated by Eiriksson (1981) from the Late Cenozoic of Tjørnes, Iceland. In the volcanic area of northern Iceland, sites which emerge above sea level are likely to be covered by lava in a geologically-short time after emergence. The area has been subject to frequent glaciation during the Late Cenozoic, most of which have been large enough to produce a major isostatic effect and typical glacio-isostatic facies sequences. Shortly after emergence, the coarse littoral sediments are overlain by lavas, which are resistant to glacier erosion in the succeeding glacial phase, thus helping to preserve underlying cycles. The sequence is of remarkable thickness because of local tectonic subsidence. The mean net rate of sedimentation has clearly been equivalent to the subsidence rate as isostatically-dominated cycles recur throughout a sequence of at least 600 m. This is true however in many areas of tectonic subsidence which are glacier fed. The rate of sediment discharge is always more than adequate to keep pace with subsidence, but the sediment accumulation rate is restrained by erosional processes in the zone of strong bottom currents near to sea level, with the consequence that the net deposition rate is tectonically controlled.

Effects of multiple glaciation are also seen in marine-limit relationships. In areas of relative tectonic stability, the highest marine-limits beyond the extent of recent glaciation will be coeval with the extent of the largest and longest phase of glacier expansion. Figs 26 and 27 show such relationships in western Spitsbergen where, beyond the limits of Late Weichselian glaciation, a marine limit associated with a more extensive Mid-Early Weichselian glacial episode (cycle 3) lies between 30 and 60 m above the Late Weichselian marine limit.

Facies architecture on the continental shelf-edge and slope

The architecture of shelf-edge and upper continental slope sequences reflects the form of the continental margin, the current regime, sea level and sediment supply. Along the Spitsbergen continental margin, and glacimarine margins such as those of east Greenland and southern Iceland, there is a considerable contrast between the mouths of glaciated troughs and the shelf edge adjacent to intervening shallow banks. Within the trough mouth fans of western Spitsbergen topset units tend to comprise successive couplets of till and glacimarine mud as a result of glacier expansion to or near to the shelf edge. At least three such couplets have developed at the mouth of Isfjordrenna (Fig. 13a). Evidence from the west coast of Spitsbergen (Salvigsen 1977; Boulton *et al.* 1982; Miller 1982) suggests that the last phase of glacier expansion beyond the west coast cannot have been younger than early Weichselian (75–100 ka). If, as some have suggested (Andrews 1974; Mørner 1977; Boulton *et al.* 1982), high latitude glacial maxima develop during the early parts of glacial periods, it is possible that each couplet represents similar early phases of separate Late Pleistocene 100 ka glacial cycles, and that their stacking one above the other is facilitated by passive thermal subsidence of the Spitsbergen continental margin, which carries older units below an erosional base level prior to accumulation of succeeding units. Extrapolated subsidence rates derived from the evidence of Myhre & Eldholm (1988) of about 20 m/100 ka are compatible with this suggestion.

The detailed seismic stratigraphy of these shelf edge couplets may reveal the pattern of sea level change through a whole glacial cycle in a way which has not been reported elsewhere. Figure 28a shows a seismic section across the mouth of the Kongsfjordrenna fan. If we initially assume that at some fixed depth below sea level there is a transition from erosion to deposition, then the offlapping reflectors of R_1 would reflect a relative sea level fall of about 80 m (Fig. 28b). The subsequent upward shift of the point of progradation would reflect relative sea level rise of about 100 m, partly coeval with emplacement of the till unit by a glacier which advanced to the shelf edge. The subsequent offlapping reflectors of R_2 would reflect a relative sea fall to an approximately modern level. This sequence may represent the complete glacial cycle of sea level change, similar to that shown in Fig. 17–5b. The R_1 sequence would reflect slow eustatic sea level fall, associated with shelf progradation, initially because of stronger erosion in shallower shelf waters and subsequently because of the approach of the glacier. The lagged isostatic response to glacier loading then causes a rapid relative sea level rise. The progradation in the R_2 sequence may reflect a nearby source of glacial sediment, whilst the offlap would represent relative sea level fall as a result of deglaciation and crustal rebound. The height of the transgression prior to final sea level fall would define the marine limit. The early Weichselian glacier extension beyond the modern coastline (cycle 3 in Fig. 27), which may have been the glacial phase in which the uppermost glacial couplet at the shelf edge was deposited, is associated with a marine limit of about 70 m at the coastline (Fig. 26), extrapolating to a marine limit of 50–60 m at the shelf edge. This compares with a marine limit inferred from seismic stratigraphy of 20 m. However, we would expect the amplitude of relative sea level change associated with ice sheet expansion to be overestimated from evidence of shelf-edge sediments, as increased sediment discharges will tend to reduce the critical erosion/deposition depth, and more energetic circulation to increase this during phases of reduced glacier extent.

There is a strong contrast between sediment architecture in the trough-mouth sediment wedges and on the flanking shallow banks. Seismic textures in the sediment forming the bank topsets are typical of those which we associate with till. Major 'till' masses are separated by relatively well defined reflecting surfaces, rather than the thick inter-till units found in the trough-mouth wedges. It is suggested that this reflects the relative shallowness of these banks which, during period of glacier retreat, are subject to strong bottom-current and/or iceberg scour which either remove glacimarine mud units produced during glacial maxima or homogenize them in iceberg turbates (Barnes & Rearic 1985). Similar bank-trough contrasts are found at the shelf edge of southeast Iceland (Boulton *et al.* 1988). The most recent sediment package on the upper and middle continental slope of West Spitsbergen shows internal reflectors and bounding surfaces which diverge with depth (Fig. 13b), demonstrating that recent sedimentation rates progressively increase from the shelf break to the middle slope (basement ridges on the lower slope complicate the pattern of sedimentation in deeper waters; Fig. 13a). Major slumped masses within this unit and other signs of instability on the upper slope (Fig. 14), suggest that the downward increase in net sedimentation rate predominantly reflects slumping of pre-existing upper slope sediments. This unit overlies the lateral equivalent on the upper continental slope of the uppermost glacial couplet, which is developed as a wedge of sediment at the shelf break, and which reflects substantial building-out of the shelf edge. It becomes thinner in deeper water. On the basis of these relationships, I suggest that, during the last extension of glaciers to the shelf edge, large contemporary sediment discharges over the shelf edge led to building out of the upper continental slope. On glacier withdrawal and the re-establishment of a S–N coastal current dispersing sediment to the north, the shelf edge was relatively starved of a primary sediment supply and underwent a net mass loss by slumping, to produce thick sediment accumulations on the middle and lower slope. I suggest that this may be a repetitive pattern of evolution of the edges of many glaciated shelves (Fig. 29) although even the Spitsbergen shelf shows complications such as the deflection of sediment dispersal pathways towards the northwestern shelf edge during the present interglaciation, and consequent modern progradation of the shelf edge there.

Summary

Figure 30 summarizes the way in which the individual facies associations, which are the building blocks of the sedimentary architecture, are emplaced in time and space during a complete glacial cycle.

Some of the fieldwork reported here was done in association with J. Jarvis, N. Chroston, C.T. Baldwin, J.D. Peacock and G.H. Miller; their contributions are gratefully acknowledged. Results are also derived from the SAFE programme, and from cruises of the *G A Reay* (1986) and *Tydeman* (1988). Collaboration in these with J.P.M. Syvitski, J. Jarvis, J.J.M. van der Meer, G. Postma, R. Schüttenhelm and C. Mesdag is gratefully acknowledged. Much of the work was generously supported by NERC and the Royal Society of London. An anonymous referee's helpful comments are also acknowledged.

References

ALLEY, R.B., BLANKENSHIP, D.D., BENTLEY, C.R. & ROONEY, S.T. 1987. Till beneath ice stream B.3. Till deformation: evidence and implications. *Journal of Geophysical Research*, **92**, 8903–8912.

ANDERSON, J.B., BRAKE, C.F. & MYERS, N.C. 1984. Sedimentation on the Ross Sea Continental Shelf, Antarctica. *Marine Geology*, **57**, 295–333.

ANDREWS, J.T. 1974. Cainozoic glaciations and crustal movements of the Arctic. *In*: IVES, J.D. & BARRY, R.G. (eds.) *Arctic and Alpine Environments*. Methuen, London, 277–317.

BARNES, P.W. & REARIC, D.M. 1985. Rates of sediment disruption by sea ice as determined from characteristics of dated ice gouges created since 1975 on the inner shelf of the Beaufort Sea, Alaska. *United States Geological Survey Open File Report 85–473*.

BIRNIE, R.V. 1977. A snow-bank push mechanism for the formation of some 'annual' moraine ridges. *Journal of Glaciology*, **18**, 77–85.

BOULTON, G.S. 1982. Subglacial processes and the development of glacial bedforms. *In*: DAVIDSON-ARNOTT, R., NICKLING, W., & FAHEY, D.D. (eds.) *Glacial, Glacio-fluvial and Glacio-lacustrine Systems*. Geo Books, Norwich.

—— 1986. Push moraines and glacier-contact fans in marine and terrestrial environments. *Sedimentology*, **33**, 667–698.

—— 1987. A theory of drumlin formation by subglacial deformation. *In*: ROSE, J. & MENZIE, J. (eds) *Drumlins*, Balkema, Rotterdam, 25–81.

——, BALDWIN, C.T., PEACOCK, J.D., MCCABE, A.M., MILLER, G., JARVIS, J., HORSFIELD, B., WORSLEY, P., EYLES, N., CHROSTON, P.N., DAY, T.E., GIBBARD, P., HARE, P.E. & VAN BRANN, V. 1982. A glacioisostatic facies model and amino acid stratigraphy for late Quaternary events in Spitsbergen and the Arctic. *Nature*, **298**, 437–441.

—— & HINDMARSH, R.C.A 1987. Sediment deformation beneath glaciers: rheology and geological consequences. *Journal of Geophysical Research*, **92**, 9059–9082.

—— & JONES, A.S. 1979. Stability of temperate ice sheets resting on beds of deformable sediment. *Journal of Glaciology*, **24**, 29–43.

——, SMITH, G.D., JONES, A.S. & NEWSOME, J. 1985. Glacial geology and glaciology of the last mid-latitude ice sheets. *Journal of the Geological Society of London*, **142**, 447–474.

——, SMITH, G.D. & MORLAND, L.W. 1984. The reconstruction of former ice sheets and their mass balance characteristics using a non-linearly viscous flow model. *Journal of Glaciology*, **30**, 140–152.

——, THORS, K. & JARVIS, J. 1988. Dispersal of glacially-derived sediment over part of the continental shelf of south Iceland. *Marine Geology*, **83**, 193–223.

—— & VAN DER MEER, J.J.M. (eds) 1989. *Report on an expedition to Spitsbergen in 1984 to study glaciotectonic phenomena*. University of Amsterdam.

——, VAN DER MEER, J.J.M. & CAMERAAT, E. 1985. Side-entry systems in fiords: inputs to the fiords from their margins. *In*: SYVITSKI, J.P.M. & BLAKENEY, C. (eds) *Sedimentology of Arctic Fiords Experiment*, vol. 2 Canadian Data Report of Hydrography and Oceans Sciences, **28**, 121–126.

COOPE, G.R. & BROPHY, J.A. 1972. Late-glacial environmental changes indicated by a coleopteran succession from North Wales. *Boreas*, **1**, 97–142.

DALE, J.E., AITKEN, A.E., GILBERT, R. & RISK, M.J. 1989. Macrofauna of Canadian Arctic Fjords. *Marine Geology*, **85**, 331–358.

DE GEER, G. 1910. Guide de l'excursion au Spitzberg. Excursion Al. *XIe Congres Geologigue Internationale*, Stockholm.

EIRIKSSON, J. 1981. Lithostratigraphy of the upper Tjörnes sequence, North Iceland: the Breidavik Group. *Acta Naturalia Islandica*, **29**, 1–37.

ELVERHØI, A. 1984. Glaciogenic and associated marine sediments in the Weddell Sea, fjords of Spitsbergen and the Barents Sea: a review. *Marine Geology*, **57**, 53–88.

——, LØNNE, Ö. & SELAND, R. 1983. Glacimarine sedimentation in a modern fjord environment, Spitsbergen. *Polar Research*, **1**, 127–149.

—— & ROALDSET, E. 1983. Glacimarine sediments and suspended particulate matter, Weddell Sea, Antarctica. *Polar Research*, **1**, 1–21.

FARROW, G.E., SYVITSKI, J.P.M. & TUNNICLIFFE, V. 1983. Suspended particulate loading on the macrobenthos in a highly turbid fjord; Knight Inlet, British Columbia. *In*: GORDON, D.E. (ed.) *Turbid Water Symposium. Canadian Journal of Fisheries and Aquatic Sciences*, 132–152.

GRIPP, K. 1929. Glaciologische und Geologische Ergebnisse der Hamburgischen Spitzbergen-Expedition 1927. *Abhandlung-Naturwissenschafflicher Verein Zu Hamburg*, **XXII**, 147–249.

HINDMARSH, R.C.A., MORLAND, L.W., BOULTON, G.S. & HUTTER, K. 1987. The unsteady flow of ice sheets: a parabolic problem with two moving boundaries. *Geophysical and Astrophysical Fluid Dynamics*, **93**, 183–223.

HOPPE, G. 1948. Isrecessionen från Norrbottens Kustland i belysning av de glaciala formelementen. *Geographica*, **20**.

HOSKIN, C.M., BURRELL, D.C. and FREITAG, G.R. 1978. Suspended sediment dynamics in Blue Fjord, Western Prince William Sound, Alaska. *Estuarine and Coastal Marine Science*, **7**, 1–16.

JACOBS, S.S. 1989. Marine controls on modern sedimentation on the Antarctic continental shelf. *Marine Geology*, **85**, 121–153.

KING, L.H. & FADER, G. 1986. Wisconsinan glaciation of the continental shelf – southeast Atlantic Canada. *Geological Survey of Canada, Bulletin*, **363**.

——, ROKOENGEN, K & GUNLEIKSRUD, T. 1987. Quaternary seismostratigraphy of the Mid-Norwegian Shelf, 65°–67°31N – A till tongue stratigraphy. *Institutt for Kontinentals okkelndersökelser publication*, **114**.

LAMPLUGH, G.W. 1911. On the shelly moraine of the Sefström glacier and other Spitsbergen phenomena illustrative of British glacial conditions. *Proceedings of the Yorkshire Geological Society*, **17**, 216–241.

LUNDQUIST, J. 1986. Stratigraphy of the central area of the Scandinavian glaciation. *Quaternary Science Reviews*, **5**, 269–292.

MANGERUD, J., BOLSTAD, M., ELGERSMA, A., HELLIKSEN, D., LANDVIK, J.Yl, LYCKE, K., LØNNE, I., SALVIGSEN, O., SANDAHL, T. & SEJRUP, H.P. 1987. The Late Weichselian glacial maximum in Western Svalbard. *Polar Research*, **5**, 275–278.

MCCABE, A.M. & EYLES, N. 1988. Sedimentology of an ice-contact glacimarine delta, Carey Valley, Northern Ireland. *Sedimentary Geology*, **59**, 1–14.

——, HAYNES, J.R. & MACMILLAN, N.F. 1986. Late-

Pleistocene tidewater glaciers and glacimarine sequences from north County Mayo, Republic of Ireland. *Journal of Quaternary Science*, **1**, 73–84.

Miller, G.H. 1982. Quaternary depositional episodes, western Spitsbergen, Norway: aminostratigraphy and glacial history. *Arctic and Alpine Research*, **24**, 321–340.

Mix, A.C. & Ruddiman, W.F. 1985. Structure and timing of the last deglaciation: oxygen-isotope evidence. *Quaternary Science Reviews*, **4**, 59–108.

Moign, A. 1965. Contribution à l'étude littorale et sous-marine de la Baie du Roi, Spitsberg. *Cahiers oceanographiques*, **8**, 453–563.

Morland, L.W., Smith, G.D. & Boulton, G.S. 1984. Basal sliding relations deduced from ice sheet data. *Journal of Glaciology*, **30**, 131–139.

Myhre, A.M. & Eldholm, O. 1988. The western Svalbard margin (74°–80°N). *Marine and Petroleum Geology*, **5**, 134–156.

Mørner, N.A. 1977. Southward displacement of the distribution of glaciation during the three maxima of the last ice age. *Journal of Glaciology*, **18**, 305–308.

Oerlemans, J. & van der Veen, C.J. 1984. *Ice Sheets and Climate*. Reidel, Dordrecht.

Orheim, O. & Elverhøi, A. 1981. Model for submarine glacial deposition. *Annals of Glaciology*, **2**, 123–128.

Peltier, W.R., Farrell, W.E. & Clark, J.A. 1978. Glacial isostasy and relative sea level: a global finite element model. *Tectonophysics*, **50**, 81–110.

Powell, R.D. 1981. A model for sedimentation by tidewater glaciers. *Annals of Glaciology*, **2**, 129–134.

—— & Cowan, E.A. 1986. Depositional processes at McBride Inlet and Riggs glacier. *In*: Anderson, P.J., Goldthwait, R.P. & McKenzie, G.D. (eds) Observed Processes of Glacial Deposition in Glacier Bay, Alaska. *Ohio State University, Institute of Polar Studies, Miscellaneous Publications*, **256**, 140–156.

Salvigsen, O. 1977. Radiocarbon datings and the extension of the Weichselian ice-sheet in Svalbard. *Norsk Polarinstitutt Aarbok*, **1976**, 209–242.

Solheim, A. & Pfirman, S.L. 1985. Sea floor morphology outside a grounded, surging glacier: Bråsvellbreen, Svalbard. *Marine Geology*, **65**, 127–143.

Solheim, A., Russwurn, L., Elverhøi, A. & Neyland Berg, M. 1990. Glacial geomorphic features in the northern Barents Sea: direct evidence for grounded ice and implications for the pattern of deglaciation and late glacial sedimentation. *In*: Dowdeswell, J.A. & Scourse, A.D. (eds), *Glacimarine Environments: Processes and Sediments*. Geological Society, London, Special Publication, **53**, 253–268.

Syvitski, J.P.M. 1987. Proximal prodelta investigations at two arctic deltas: Itirbilung and Cambridge Fiords, Baffin Island. *Geological Survey of Canada Open File*, **1589**, ch. 6.

—— 1989. On the deposition of sediment within glacier-influenced fjords: oceanographic controls. *Marine Geology*, **85**, 301–329.

——, Burrell, D.C. & Skei, J.M. 1983. *Fjords: Processes and products*. Springer, New York.

Tooley, M.J. 1978. *Sea level fluctuations in North-West England during the Flandrian Stage*, Clarendon, Oxford.

Turcotte, D.L. & Schubert, G. 1982. *Geodynamics. Applications of continuum physics to geological problems*. Wiley, New York.

Vorren, T.O., Lebesbye, K., Andreassen, K. & Larssen, K-B. 1989. *Marine Geology*, **85**, 251–272.

Walcott, R.I. 1973. Structure of the earth from glacio-isostatic rebound. *Annual Review of Earth and Planetary Sciences*, **1**, 15–37.

Weertman, J. 1972. General theory of water flow at the base of a glacier or ice sheet. *Reviews of Geophysics and Space Physics*, **10**, 287–333.

Glacimarine processes at grounding-line fans and their growth to ice-contact deltas

ROSS D. POWELL

Department of Geology, Northern Illinois University, DeKalb, Illinois 60115, USA

Abstract: Grounding-line fans originate from subglacial and basal stream tunnels at grounding lines of glaciers terminating in a marine environment. During melt seasons discharge forms a turbulent jet beyond the efflux. Over the initial zone of flow establishment in the jet, it may remain in contact with the sea floor for up to 13.4*D* (*D*-tunnel diameter) along a runout distance, if velocity is constant. Plug flow in this zone deposits marine outwash sediment commonly having chaotic texture and graded and welded contacts. Texture may vary rapidly because of discharge pulses, and the jet may move laterally to produce sheet or cut-and-fill geometries. Imbricate gravel is deposited near the efflux, then farther out are sheet or weakly channelized finer-grained gravels and sands. In some instances a migrating barchanoid bar forms at the detachment zone and forms large-scale trough cross-beds. Beyond the detachment zone the turbulent jet becomes vertical and sediment cascades from it in a 'veil'. This sediment is added to by flows from continuous failures of bed load sediment in the detachment zone. Resulting deposits are interstratified sediment gravity flows and thick, coarse end-members of cyclopsams. The jet changes to a plume when inertial forces become less than buoyancy forces. Turbulent vorticies within the jet can re-entrain particles into a buoyant plume. The plume remains vertical, although continuously spreading, until reaching neutral buoyancy. Fan depocentres have been observed to accumulate at over 10^6 m^3 a^{-1} in temperate glacial areas. Eventually, at quasi-stable grounding lines, fans may aggrade to sea level to form ice-contact deltas. At incipient deltas when the delta plain is intertidal, coarse sediment is redistributed to the prodelta with each tidal cycle. Intense prodelta rhythmites mark the transition of a fan into a delta.

Discharge of water from a tunnel directly within or at the base of a standing body of water is a rare situation in natural systems. One natural occurrence is meltwater discharge from a glacier. Freshwater input is from full pipe flow rather than from an open-channel flow and the efflux is below, rather than directly at water level. These differences, especially when complicated by water and sediment discharge fluctuations from basal or subglacial streams, create a depositional environment that is different from common deltaic environments.

Deposits from these discharges have been termed subaqueous outwash fans (Rust & Romanelli 1975) as a subset of glacier contact fans (cf. Boulton 1986), and will be termed here grounding-line fans (Powell 1988). Grounding-line fans have a sediment source from a glacial tunnel that discharges meltwater where a glacier ends on a lake or sea floor. The fans can be made up of a variety of lithofacies including subaquatic outwash, which is used here to define sediment deposited from a jet of subaquatic meltwater while it is in contact with the lake or sea floor.

Marine outwash includes a range of particle sizes from gravel to clay deposited by a horizontal jet in contact with the sea floor. The rest of the fan system includes sediment gravity flow deposits from the marine outwash and deposits settled from the jet and plume of turbid meltwater after it detaches from the bed.

This paper presents descriptions of how grounding-line fans form from full tunnel glacial discharges into the marine environment. The fan systems are described using fluid dynamics with examples taken from modern and Carboniferous sedimentary sequences. Many variations can occur based on glacial regime, position of the efflux, terminus stability, terminus form (tidewater terminus or ice shelf), and rate of terminus retreat. If a terminus remains quasi-stable for some time, and high sediment accumulation rates at the grounding line are greater than terminus retreat rates, then the depocentre can build to sea level. The fan then transforms to an ice-contact delta.

From DOWDESWELL, J. A. & SCOURSE, J. D. (eds), 1990, *Glacimarine Environments: Processes and Sediments*. Geological Society Special Publication No 53, pp 53–73.

Discharge dynamics

Tunnel flow

Sources of jets forming grounding-line fans are englacial and/or subglacial stream tunnels. Because they vent subaquatically they must always be full of water, at least to the height of the piezometric surface formed by sea level. Hydrodynamically, sediment transport in the tunnels may be modelled using full pipe-flow conditions.

For flow in pipes to be turbulent, the Reynolds number, $R_e = \rho D U_t/\mu > 4000$. This condition is met in most cases for active meltwater tunnels. Here ρ is fluid density, D is tunnel diameter, U_t is mean flow velocity in tunnels or pipes, and μ is dynamic viscosity. Neglecting head losses from changes in tunnel diameter and bends, the velocity of a full-tunnel discharge can be approximated by combining the Colebrook-White equation with the Darcy–Weisbach formula (cf. Chadwick & Morfett 1986), to give:

$$U_t = -2(2gDS_o)^{0.5}\log\{(k_s/37D) + (2.51\nu)/[D(2gDS_o)^{0.5}]\} \quad (1)$$

where S_o is slope of hydraulic gradient, k_s is bed surface roughness, g is gravitational acceleration, and ν is kinematic viscosity.

Engineers have empirically modelled the flow of solids in pipes. Most sediment transport models of pipe flow have dealt with a uniform size distribution rather than the more poorly sorted sediment found in glacial tunnel systems. Consequently, flows in a glacial tunnel would be a combination of the modelled heterogeneous flows where all sediment is in transport in decreasing concentrations upward, and flow with a sediment bed (cf. Fig. 15.18, Graf 1971). The important parameters derived from the models for describing tunnel flow are particle size, pipe diameter, flow velocity, and sediment concentration (Task Committee 1970; Wiedenroth & Kirchner 1972).

The dimensionless velocity most commonly determined is the lowest mean flow that ensures deposit-free transport (U_c). For a given particle size, velocity (U_c) must increase with increasing pipe diameter and sediment concentration (Task Committee 1970). For incipient movement of particles ($0.5<d<30$ mm), U_c for pipes is greater by a factor of 2.5 to 3.0 than for open-channel flows (Novak & Nalluri 1972). Newitt *et al.* (1955) found that the total transport capacity was greater when two sands of different uniform sizes are mixed together compared with their separate capacities. They noted maximum transport of gravel (to 3.2 mm) when it was mixed with 15% fine sand (Condolius & Chapus 1963).

Many studies have noted that U_c is about 2 to 3 m s^{-1} for particle sizes at least 2 to 4 mm in diameter. Following the Task Committee's (1970) recommendations, the Durand (1953) relationship is best for heterogeneous transport, that is:

$$U_c = f\{d[2gD(s-1)]^{0.5}\} \quad (2)$$

where d is particle diameter and s is particle specific gravity. The function is a Froude number that varies slightly with concentration, and the relationship is best shown graphically (Task Committee 1970). The function neglects density of a turbid flow and viscosity variations.

For flows with bed motion, the Task Committee (1970) recommended using a relationship established by Graf & Acaroglu (1968) as a guide, but also noted that the dimensionless plot is not well constrained. The relationship is:

$$CU_cR_h[(s-1)gd^3]^{-0.5} = 10.39[d(s-1)/SR_h]^{-2.52} \quad (3)$$

where C is sediment concentration and R_h is the full pipe hydraulic radius (the ratio of cross-sectional area to wetted perimeter; not the hydraulic radius of pipe to bed depth of Wilson (1969)).

Two additional processes are important to consider because they involve blockage of pipes and fluctuating sediment discharge. First, for gravel beds, bedforms develop rapidly and are isolated at low flows. A flat-bed forms at higher velocities. The gravel bed can stop moving or 'freeze' and may build up to block the pipe eventually, only to disintegrate as pressure increases at the top of the bed (Durand 1951). Secondly, in inclined pipes sediment commonly settles at a bend, and sediment deposited as a bed can slide back down a sloping pipe to block the pipe at the bend eventually. Sediment moving in bedforms, tunnel blocking by bed 'freezing', and tunnel blocking in inclined sections are three factors that can make sediment discharge from a glacial tunnel episodic. Superimposed on these effects are diurnal, annual, precipitation event and catastrophic event fluctuations in water discharge.

Tunnels in marine-terminating glaciers may become blocked by sediment as bed load is deposited during the falling phase of discharge. However, submarine tunnels that are not blocked by sediment must still remain totally full of water as discharge falls; by comparison, terrestrial tunnels are only partially full of water at this stage. Even in winter some meltwater discharges at Alaskan tidewater termini (Cowan 1988; also cf. Hooke *et al.* 1985), indicating tunnels are full of fresh water. Glacial ice may deform to fill terrestrial tunnels during winter and during the following melt season new tunnels are formed (e.g. Hooke *et al.* 1985). However, tunnels in marine-terminating glaciers that are filled with water probably have sufficient flow to melt ice at a greater rate than it deforms, and the tunnels are maintained (cf. Röthlisberger & Lang 1987; Hooke 1989).

By considering total heads of the meltwater and sea water systems, sea water can quite easily dam fresh water in a tunnel if the piezometric surface in the glacier does not drop substantially as discharge wanes. Sea water may enter a tunnel if the fresh water/sea water interface is sloped or if the glacial piezometric surface drops to nearly sea level. All of the above conditions are favourable for maintaining the opening of a tunnel even if the grounding line advances.

In addition to the glaciological factors of basal or subglacial and englacial stream sites, the tunnel maintenance by melting could well cause reoccupation of the tunnel at the start of the next melt season. Discharges have been commonly observed to issue at the same location from one melt season to the next at modern tidewater termini (Powell & Molnia 1989). This process explains how some Pleistocene eskers feeding large grounding-line fans could form in tunnels that had survived several melt seasons to allow the fan depocentre to receive large volumes of sediment (Smith 1982; Shreve 1985).

Jet and plume flow

Once at the tunnel mouth, the discharge is best modelled as either a jet or plume (Fig. 1). In a jet, momentum flux dominates and in a plume buoyancy flux dominates. Glacial discharges almost always issue as jets. The efflux may issue directly as a forced plume during low discharge at the start and end of each melt season when sediment discharge is also low (Cowan 1988).

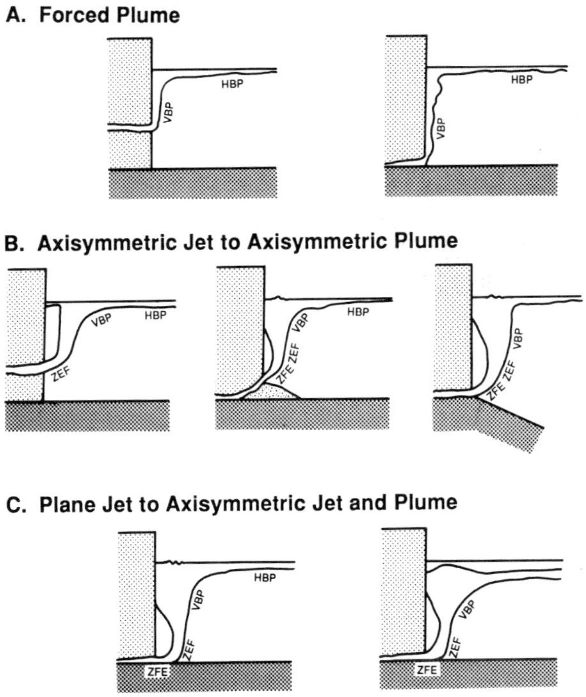

Fig. 1. Diagrammatic sketches (not to scale) of glacial meltwater discharges entering the sea as a forced plume (A) from englacial or basal/subglacial positions. A jet may initially be axisymmetric (B) if it enters from an englacial position or from a basal/subglacial position if the tunnel or the sea floor is inclined near the efflux. Most commonly, subglacial discharges form a plane jet (C) that rises to form an overflow or less commonly, an interflow. ZFE: zone of jet flow establishment, ZEF: zone of established jet flow, VBP: vertical buoyant plume, HBP: horizontal buoyant plume.

Even if the discharge starts as a jet, it transforms to a plume with increasing distance from the tunnel mouth.

Dimensionless relationships derived for evaluating whether a discharge forms a plume or a jet consider the relative importance of momentum flux versus buoyancy flux (Morton 1959). The Richardson number

$$R = QB^{0.5}M^{-1.25} \tag{4}$$

near the tunnel mouth determines whether the efflux is a turbulent buoyant jet or a plume (List 1982), where Q is discharge or volume flux, B is buoyancy flux equivalent to $g'Q$, and M is momentum flux, or $\rho_o Q U_{mo}$; $g' = g(\rho_a - \rho_o)/\rho_o$, ρ_a is the ambient fluid density, ρ_o the initial discharge density, and U_{mo} the initial mean centreline velocity. The limiting condition for jet behaviour rather than plume behaviour of axisymmetric discharges is $R \leq 0.56$ and of plane discharges is $R \leq 0.735$ (List 1982).

Discharge at the tunnel mouth is turbulent and is best modelled as a turbulent jet. When the efflux is englacial the initial jet is axisymmetric with mixing in three dimensions. However, because most glacial debris is in basal transport, major sedimentary deposits of eskers and grounding-line fans are probably produced by basal or subglacial discharges. Basal or subglacial discharges may be axisymmetric if the tunnel is inclined upwards away from the sea floor, or if the grounding line rests at the edge of a hanging valley or edge of a sea floor cliff. However, presumably the initial flow commonly behaves more like a plane jet because it discharges close to and interacts with the sea floor. Flow behaviour in axisymmetric and plane jets is similar; modelled plane jets originate from a slot and agree satisfactorily with theory (Albertson *et al.* 1950), but in glacial discharges flow is planar due to interaction with the sea floor, rather than the shape of tunnel mouth.

Models for turbulent jets show that their position, velocity, and mixing or entrainment rates depend on jet velocity, reduced gravitational effect from buoyancy, and size of the tunnel mouth (e.g. List 1982; Turner 1986). The most important effect on jet behaviour beyond the tunnel mouth is the rate of entrainment of ambient sea water. The entrainment process occurs by ring vortices which grow in an orderly manner downstream. Near the efflux a high speed jet has a laminar shear layer that is unstable and thickens very rapidly into a mixing layer between the jet and ambient fluid (Turner 1986). Vortices are the key elements in the layer because they wrap ambient fluid around themselves; fusion processes then mix the two fluids (List 1982). Beyond the vortex zone more random and chaotic turbulent flow occurs.

A jet is divided into three zones based on mixing characteristics. The initial zone has been termed the zone of flow establishment (ZFE) where the centreline velocity approximately equals the outlet velocity and the centre of the jet behaves as a plug flow (Albertson *et al.* 1950). Beyond and surrounding that zone, which extends horizontally 6.2D (all xD distances measured along the centreline of the jet) away from a circular outlet (5.2D from a slot), is a transition zone where the core velocity changes from a constant value to a standard rate of deceleration (between 4D to 8D for plane jets; Bates 1950). The 4D value of Bates is less than the 5.2D value of Albertson *et al.* because Bates considered bottom friction. Last is the zone of established flow (ZEF), where turbulent mixing occurs throughout the jet (Albertson *et al.* 1950) which, at least beyond 40D, can be modelled as a turbulent jet (List 1982). The ZEF can ideally extend to beyond 200D for axisymmetric jets and beyond 2000D for plane jets (Albertson *et al.* 1950). If bottom friction and suspended sediment are considered these ideal distances differ, as is discussed below.

At the end of the ZEF, residual axial velocity decays rapidly to zero because of large-scale turbulence. Gravity forces become larger than inertial forces (reflected in larger R values) and the jet becomes a buoyant plume. The ZEF and initial buoyant plume in the glacimarine case is near-vertical or at a high angle of inclination (Syvitski 1989). Once the plume attains a level of neutral buoyancy it spreads horizontally from the vertical buoyant plume (VBP) to a horizontal buoyant plume (HBP).

For a glacimarine discharge, the ZFE is approximately horizontal and, when basal or subglacial, is in traction with the sea floor. Ideally, the horizontal distance for a jet depends on a dimensionless parameter that considers the inertial and gravitational (here buoyancy) forces. The parameter can be represented either as a Richardson number as in equation 4 (that is, not the local parameter based on gradients of density and velocity) or, following Abraham (1965), the densimetric Froude number:

$$F = U_m^{\,2}(g'D)^{-1} \tag{5}$$

where U_m is mean centreline velocity.
As $|F| \rightarrow 0$ the jet tends to be directed vertically upward, and as $|F| \rightarrow \infty$ the jet axis tends to straight, either horizontal or along a line following the attitude of the tunnel. A continuous underflow forms if $F<0$, that is, when $\rho_a < \rho_o$. Such conditions are normal in glacilacustrine

settings, but potentially less common in glacimarine settings, having been inferred (e.g. Powell 1981, 1983; Mackiewicz *et al.* 1984), but not confirmed.

Other variables that are important geologically are the rate of velocity decay and rate of spreading of the jet. A general relationship for centreline velocity decay is (Albertson *et al.* 1950):

$$(U_m/U_{mo})\ (x/D) = 6.2 \qquad (6)$$

for axisymmetric jets, and

$$(U_m/U_{mo})\ (x/D)^{0.5} = 2.28 \qquad (7)$$

for plane jets, where D is, respectively, diameter and width of tunnel mouth and x is centreline distance from the tunnel mouth.

Other estimates for centreline velocity are summarized by List (1982), and he presented the dimensionless relationship:

$$U_m = 6.9\ M^{0.5}\ x^{-1} \qquad (8)$$

for axisymmetric jets.

However, once a jet has suspended particles, its velocity decay is faster and is suggested to follow the dimensionless relation in axisymmetric flow of Carstens & Rao (1971):

$$(U_o/U_{mo})\ (x/D)^{1.47} = 45.66 \qquad (9)$$

For this criterion the length of the ZFE would be 13.4D, for $U_o = U_{mo}$ as assumed by Albertson *et al.* (1950) for the other estimates above. Furthermore, when the plane jet experiences bottom friction the jet spreads and decelerates more rapidly (Wright 1978), such that the forces of friction and inertia are important, and:

$$U_m/U_{mo} = \exp - \{[x/D]\ [8/(f_b+f_i)] \times [\rho_o/(\rho_o-\rho_a)]\} \qquad (10)$$

where f_b and f_i are resistance coefficients for the bottom and fluid interface, respectively.

The above velocities are all for centreline flow; the distribution of mean velocity across the jet follows a Gaussian decay away from the centreline (Albertson *et al.* 1950). The rate of lateral velocity decay is also a direct function of centreline velocity (List 1982). Diffusion of sediment in the jet behaves in a similar fashion to transverse water velocity gradients but depends on the root mean square of turbulent velocities and particle settling velocity (e.g. Singamsetti 1966).

The horizontal ZFE portion of the jet has the highest sediment concentrations. In this zone, the coarsest bed load is deposited and finer gravels and sands are transported along the runout distance in contact with the sea floor in hyperconcentrated flows. Then the jet detaches to form a vertical turbulent jet (VTJ) in the ZEF (Syvitski 1989). Along the runout distance sheet deposits probably form if flow is regular, but if discharge pulses and/or the jet wavers laterally, repeated scour-and-fill structures may be produced (Rust & Romanelli 1975). A barchanoid bar may form at the detachment zone, from which sediment gravity flows are initiated and move down the mid- to lower fan (see below). Furthermore, a ‘veil’ of sediment rains from the VTJ to form sheets or weakly channelized deposits of the coarsest end-member of cyclopsams (Mackiewicz *et al.* 1984), or the ‘veil’ sediment is redeposited by sediment gravity flows. With high suspended sediment loads, fall-back of sediment at the detachment zone may increase sediment concentrations sufficiently to form an underflow.

The jet edge can build sediment into levee forms and produce internal structures that reflect the decrease of energy away from the jet (as in the Lamplugh Glacier example presented below).

Different fan geometries and facies can be produced with different sediment and water discharges (Fig. 2). The exact form in any one situation will be a combination of these varieties because of the high variability of glacial discharges.

The final aspect to consider about the jet is its terminal height, which is controlled by initial frictional forces, inertia, buoyancy, rate of ambient fluid entrainment by ring vorticies, rate of particle settling and re-entrainment, and density gradient in the ambient fluid. The critical height (z_c) at which the buoyancy forces dominate, that is, where the jet is in transition to a plume, is (Morton 1959):

$$z_c = M^{0.75}/B^{0.5}. \qquad (11)$$

Many results indicate that buoyant jets degenerate quite rapidly to plumes over several metres (Fischer *et al.* 1979).

If the ambient fluid through which the jet rises is density stratified, then that also determines the terminal height, and equation (11) can be modified to expressions that consider stratification (List 1982), such that for an axisymmetric jet:

$$z_{ja} = 3.8\ (M/N^2)^{0.5} \qquad (12)$$

and for a plane jet

$$z_{jp} = 3.6\ (M/N^2)^{0.33} \qquad (13)$$

where N^2 represents linear density stratification. Furthermore, the momentum of the jet may carry dense fluid beyond a level of neutral stab-

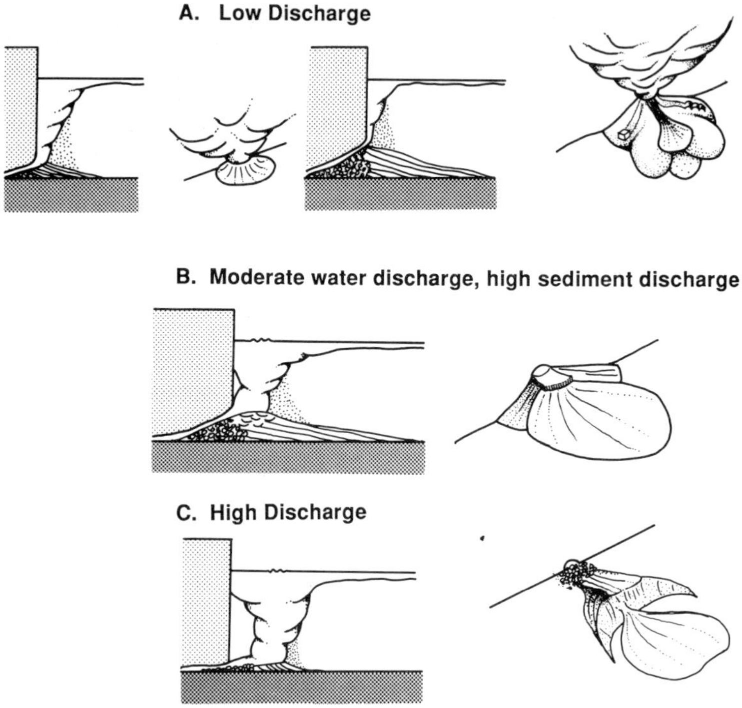

Fig. 2. Alternative fan types with different discharges. (A) At low discharges the fan grows progressively with coarse traction deposits dumped at the efflux and most deposits from settling or slumping/sliding and sediment gravity flows. (B) At higher sediment discharges the ZFE remains in traction to produce rapid sedimentation deposits in sheets or crude scour-and-fill structures. Settling and sediment gravity flow deposits make up more distal deposits. (C) At high discharge the ZFE can include a migrating barchanoid bar at the detachment zone.

ility so that fall-back occurs. If the jet has sufficient momentum, it will intersect the water surface and cause 'boiling' above the upwelling.

These terminal height estimates do not consider jets with particulate loads. Particles are continually settling from the VBJ, but a small portion is re-entrained at the edge of the flow because of vortex mixing (cf. Fox 1970). This process will be discussed in more detail for plumes, where the rate of entrainment is a factor of 1.5 faster than jets (Fischer *et al.* 1979). Consequently, plumes experience fluid dilution at a faster rate, but may entrain more sediment at their edges.

As momentum of the jet decreases, gravity forces start to dominate and the jet transforms to a VBP. Turbulence in the transition is not well understood, but turbulence intensities are higher in plumes than jets because of two processes (List 1982). First, buoyancy of plumes causes potential energy to be transferred to turbulent kinetic energy and, second, plumes have a high mean shear due to a high mean momentum flux.

Dimensional analysis of steady plumes is summarized by List (1982), who suggests centreline velocity is:

$$U_{\mathrm{m}} = k_{\mathrm{a}}(B/x)^{0.33} \tag{14}$$

for axisymmetric plumes, and

$$U_{\mathrm{m}} = k_{\mathrm{p}}\, B^{0.33} \tag{15}$$

for plane plumes, where k_{a} and k_{p} are constants of proportionality with approximate values of 4.7 and 1.65, respectively. The terminal height of an axisymmetric plume is (List 1982):

$$z_{\mathrm{pa}} = 3.8\ (B/N^3)^{0.25} \tag{16}$$

and for a plane plume is

$$z_{\mathrm{pp}} = 2.5\ (B/N^3)^{0.33}. \tag{17}$$

Plume ascent slows progressively with increasing particle concentration; however, the velocity of particulate-rich plumes decays less rapidly than pure fluid plumes because of greater inertia imparted with particles. As particles are released during flow, the buoyancy flux tends to become more like a pure fluid plume (Carey *et al.* 1988). With particles, the initial buoyancy flux of the plume is:

$$B_{po} = Qg(\rho_a - \beta)/\rho_a \quad (18)$$

where β is the bulk density of the plume, and the subsequent buoyancy flux of the plume is:

$$B_p = Q(1-C)g(\rho_a - \rho_p)/\rho_a \quad (19)$$

which will increase relatively as C, the volumetric concentration of particles, decreases. The ρ_p term is plume fluid density.

Particle concentrations in plumes have Gaussian distributions, becoming more dilute towards the edge. However, edge concentrations may be higher because of re-entrainment by vortex mixing. Except for those lost by diffusion, particles that have settling velocities much lower than the plume velocity are continually re-entrained in the plume. Coarser particles experience rapid fall-out and their lateral flow component is weak. With this settling, a 'sedimentation veil' forms around the plume in which there are sediment-rich and sediment-poor stringers and particle settling is considerably more rapid than Stokes' law predicts (Carey *et al.* 1988). Entrainment of the sediment in the veil can cause a substantial increase in mass flux with height, principally at the plume edge.

Dilute gravity flows down the plume edge have been documented under conditions similar to a natural glacimarine system: 10 gl^{-1} particle concentrations in fresh water plumes rising through salt (ρ_a = 1.021 g cm^{-3}) water (Carey *et al.* 1988). The gravity flows are partially entrained in the plume but after descent they also flow across the floor. When particle concentrations are higher, for plume density about equal to or above ambient fluid, the plume may bend over or collapse to form a density flow. In the glacimarine environment this may perhaps occur because plume density can be increased by particle settling with minor caballing effects, then become re-entrained by vortex mixing. The site where this is most likely to occur is at the end of the runout distance of the plane jet in contact with the sea floor. There, the jet detaches to form the VBJ and all larger particles that were in intermittent suspension will settle rapidly (cf. Syvitski 1989).

At lower stream discharges, where the effluent is dominated by buoyancy forces and quickly becomes a plume, the jet rises up the glacier face. As it ascends it melts the face to form a series of vertical runnels (Fig. 3) within a calving bay which has formed because these areas are also sites of rapid calving (Sikonia & Post 1979). If calving is slow, deeper vertical channels are melted into the face and are expressed at sea level by caves (Fig. 3; Gilbert 1983; Powell 1983). Melting may alter the attitude of the plume so that it is inclined and planar with some frictional drag affecting the upper boundary of the flow (cf. Ellison & Turner 1959). Now entrainment is not constant as in vertical plumes, but varies as $g'h \cos\Theta/U_p^2$, a Richardson number, where Θ is the angle of inclination to the horizontal, h is thickness of flow, and U_p is mean plume velocity. If the discharge has more inertia, the VBJ may be forced away from the glacier face and commonly can be super-elevated to 'boil' at the surface (Fig. 3).

The height to which the VBP rises depends on its density relative to ambient fluid density. Once at a position of neutral buoyancy, the buoyant plume flows laterally as a gravity current or HBP. It either forms an interflow within the water column or more commonly an overflow at the surface. A turbid plume can be considered as two gravity currents; one with particles follows the B_{po} relationship (Equation 18), and the other using fluid buoyancy is a typical barotropic flow away from the discharge. The velocity of the plume acting as a gravity current (U_g) is:

$$U_g = k_g(g'h)^{0.5} \quad (20)$$

where h is thickness of the flow and k_g is a constant of about unity (Carey *et al.* 1988).

Instabilities and mixing of the surface flow may be driven by internal turbulence (Turner 1973; p. 117) or interfacial turbulence. Turbulence generated at the sea surface is caused principally by wind-shear. Mixing at the lower boundary is a function of density differences and degree of shear. Instabilities may be generated at the interface following the Kelvin–Helmholtz relation (Turner 1973; p. 94) so that internal waves form and large 'billows' may extend into the flow. The interfacial waves are unstable and may grow exponentially if the initial instability follows:

$$(U_a - U_p)^2 > U_w g({\rho_a}^2 - {\rho_p}^2)/\omega\rho_a\rho_p \quad (21)$$

where ω is angular frequency and U_w is phase velocity. For small density differences (when buoyancy and inertia are the dominant forces) this relationship can take the form of a densimetric Froude number:

$$(U_a - U_p)^2/g'\lambda > 1/\pi = 0.32 \quad (22)$$

where λ is interfacial wavelength ($=2\pi\ U_c/\omega$).

For thicker mixed interfaces, instability occurs when the densimetric Froude number is greater than about 4.0 and the most unstable condition is where λ = 7.5h (Turner 1973;

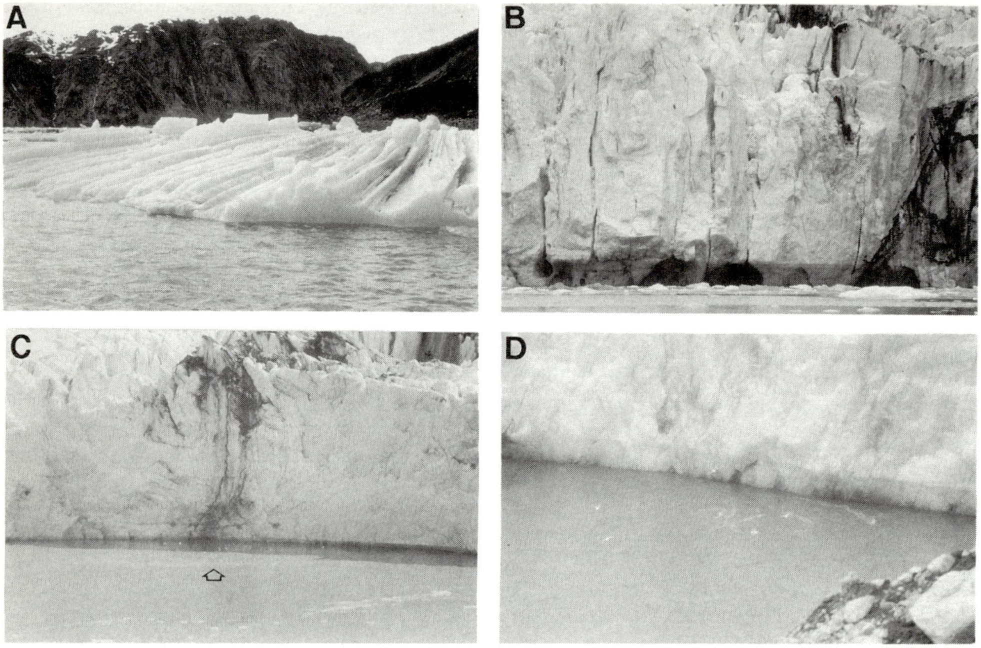

Fig. 3. (A) Features from upwelling at a tidewater cliff are vertical melt runnels, here shown (40 to 80 cm deep) on a face preserved on a recently calved bergy bit; (B) a melted cave at the tide limit (range shown about 5 m) from a subglacial discharge having risen 100 m from the sea floor; (C) a super-elevated jet upwelling (shown by arrow) at the sea surface away from the tidewater cliff. White specks at upwelling are seagulls and the cliff is about 85 m high. (D) A close-up of a similar upwelling with breaking waves and seagulls.

p. 99). Internal turbulence within the flow may be generated by residual momentum, or in a way similar to that just described, if there is mass particle settling.

The form of the transition from a VBP (or VBJ) to a HBP depends on residual momentum. Sufficient momentum may carry dense fluid beyond the level of neutral stability, so that fall-back occurs near the initiation of the plume. Some suspended particles may not be carried to the maximum plume height because of their high settling velocities, and other particles settle through the hypopycnal overflow. The net result of these effects is particle transport low in the plume or surface layer. Occasional 'billows' may cause some upwelling of particles from mixing at the lower interface (Fig. 4). With distance from the glacier, surface wind shear can enhance mixing sufficiently to give a more uniform distribution of sediment.

The base of the overflowing density current may be sharp or gradual depending on size of vortices and amount of mixing, which can be enhanced by wind effects, and degree of entrainment across the bottom of the flow (Turner 1986). If there is an interfacial shear layer where there is a sharp step in velocity, then mixing is initially diminished (Turner 1986).

Fluctuations in plume velocity and rate of sedimentation from buoyant plumes have been discussed recently by Syvitski *et al.* (1988), and particulate processes and deposits from plumes are discussed by Cowan & Powell (1990).

Field examples of submarine tunnel discharges

Modern Alaska

An example of flow directly at the tunnel efflux was recently described by Powell & Molnia (1989) for a deposit observed to form below and within the intertidal zone of McBride Inlet, Glacier Bay. Imbricate boulders (a–b planes dipping upstream) up to 2 m across were deposited within the ZFE whose runout distance extended at least 50 m from the basal/subglacial stream mouth. The boulders are edge-rounded to rounded, poorly sorted, and are spread laterally for up to 25 m. They are not channelized and appear to have been deposited in sheet

Fig. 4. Billows of sediment rising to the surface of an overflow, induced by turbulence lower in the flow. Tidewater cliff is about 300 m away.

form, probably from a plane jet which was fed from a subglacial tunnel inclined at 9° up-flow. Beyond the gravels is muddy, very coarse sand and granule gravel, in crude beds 0.5 to 1 m thick with welded contacts. These beds are interstratified with 0.2 to 0.5 m thick beds of structureless, moderately sorted, fine to medium sand; structureless muddy sand that can have rare laminae of very fine sand; and thin to medium beds of open-framework pebble gravel with rounded clasts. These are interpreted as having been deposited at and beyond the detachment zone of the jet with an occasional incursion of bed load sediment (better sorted gravels) during pulses of increased discharge.

A second modern example is from Lamplugh Glacier, Glacier Bay, where the surface of a grounding-line fan was intertidal and, at low tide, structures formed during higher tidal stages were exposed at the surface (Powell 1983). Structureless, poorly sorted gravel, interpreted as having been deposited from tractional flow of a jet core, was overlain by 0.4 m of cross-stratified sand. Bedforms on the sand surface indicated a decrease in traction at the jet edges progressively away from the efflux (Fig. 5). These would have formed at high tidal stage and thus probably reflect the increasing effects of buoyancy over inertia in the jet. They represent a case of a plane jet in traction phase. From observation at high tide, this was not a fully submarine discharge and at low tide it was observed to be a partially full pipe-flow. The horizontal jet apparently was not fully channelized and the jet edge transporting the

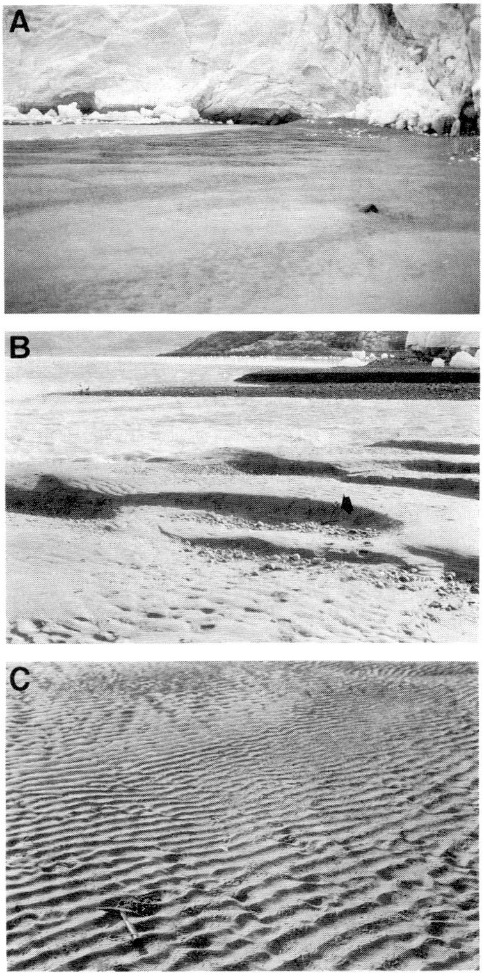

Fig. 5. Intertidal surface at Lamplugh Glacier, Glacier Bay, Alaska, showing bedforms produced by the edge of a jet from the subglacial stream discharge. Looking towards the tunnel mouth, a plane bed is in foreground, ripples in middle ground and dunes with ripples in background (A). Close-ups of the dune (B) and ripple (C) fields are described in text.

finer sediment produced the bedform sequence. The jet edge remained in traction over a runout distance of tens of metres and then detached from the sea floor.

Within 20 m of the tunnel efflux 3−*D* dunes were present with asymmetrical ripples of fine- to medium-grained sand on their stoss side, coarse-grained sand and granules on the lee face and gravel in sandy mud in their troughs. For the next 15 m the sand had sinuous asymmetrical ripples. Beyond the ripples, and extending out to the sharp brink-point of the fan surface, the sand bed was planar. The ripple bed shows remnant 3−D current ripple forms, indicating flow in the direction of jet discharge, that were modified to combined-flow ripples as the tide fell.

Tasmanian Carboniferous

During the latest Carboniferous, Tasmania was located at 75 to 80°S palaeomagnetic latitude, and was experiencing glaciation (Crowell & Frakes 1972; Banks 1981). Glacigenic formations of the lower Parmeener Supergroup crop out as a shore platform near Wynyard on the north coast of Tasmania (Fig. 6). The rocks of interest are at the base of the Parmeener

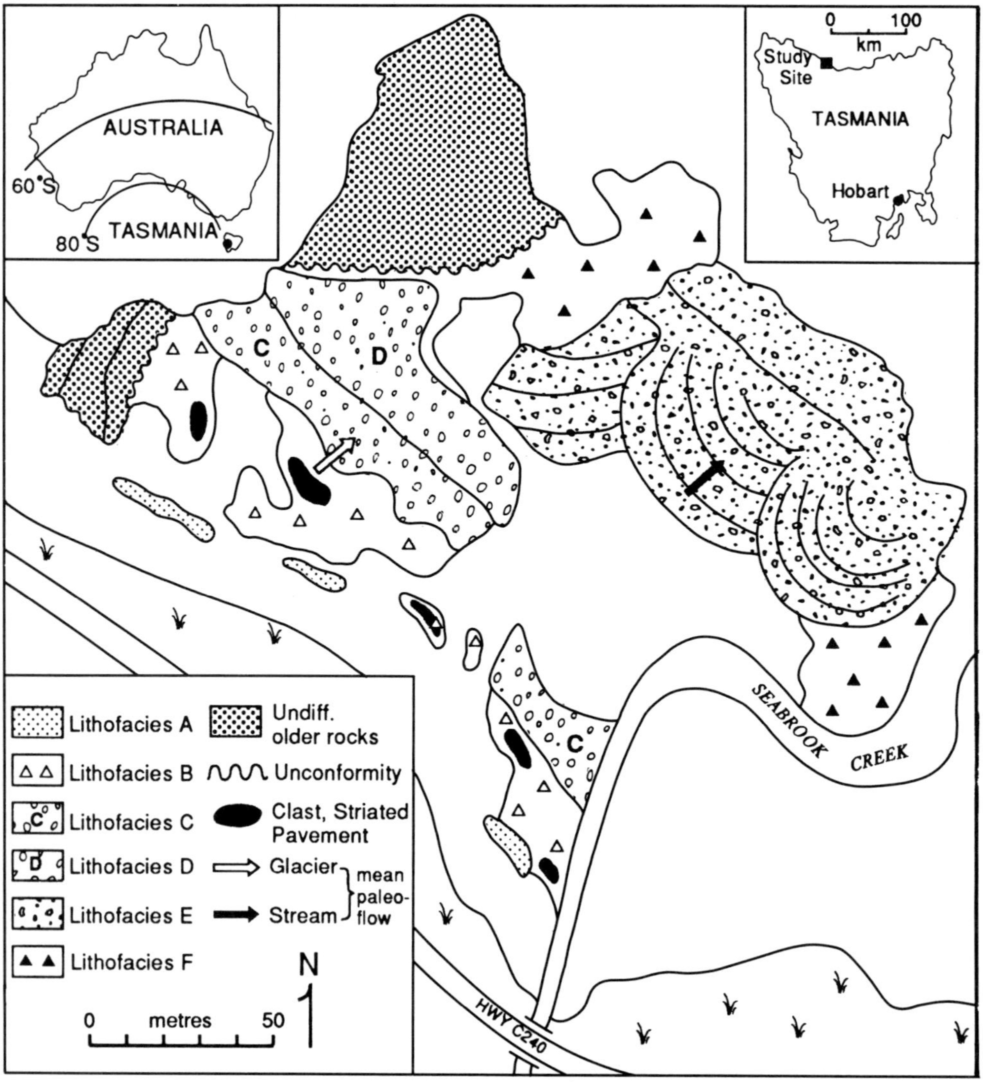

Fig. 6. Lithofacies associations of a late Carboniferous grounding-line fan at Wynyard, Tasmania, Australia. Upper left inset shows palaeomagnetic latitudes for Australia and Tasmania at the time.

Supergroup included in the basal portion of the Wynyard Formation. The sequence is considered to have been deposited by a glacier lobe descending from western Tasmania and flowing northward in a basement valley (Clarke & Banks 1975). The sequence at Seabrook Creek just east of Wynyard is interpreted here as being deposited at the grounding line of the lobe as it flowed into a fjord along a late Carboniferous coastline that roughly paralleled the present coast.

There are five main lithofacies exposed at Seabrook Creek. Lithofacies A is stratified conglomerates with rounded to subrounded clasts. Some beds are unstratified, very poorly sorted, cobble to boulder conglomerate with small to large pebble matrix. Other beds are medium-scale trough cross-bedded, moderately sorted, medium pebble to granule conglomerate, medium to coarse sandstone and rare, rippled fine sandstone in small (100 by 30 cm) channels. Lithofacies A lies stratigraphically above lithofacies B.

Lithofacies B is a stratified diamictite with 25% clasts larger than small pebbles within a sandy mudstone matrix (Fig. 7). Clasts are mainly subangular to subrounded and 15 to 20% of them are faceted or striated. The diamictite contains deformed rafts of fine sandstone with 3–D climbing ripples in beds 10 to 80 cm thick which was originally interbedded with diamictite. Matrix and internal stratification indicate deformation while soft. There are local lenses of poorly sorted medium to coarse sandstone interstratified with poorly sorted small pebble conglomerate. Important constituents of lithofacies B are clast pavements, which at one locality were stacked three high in a 0.5 m thickness of diamictite. Clasts have faceted and striated tops, their a–b planes dip up ice-flow, some are bullet-shaped, and they can make up 5 to 80% of a surface. The matrix between clasts is grooved and striated. Pavements give a palaeoflow direction of 037°. Lithofacies B is laterally equivalent to and just older than lithofacies C and overlies older rocks in the area unconformably.

Lithofacies C is deformed medium to small pebble conglomerate and medium to coarse sandstone (Fig. 8). This lithofacies occurs in blocks up to 8 m long that rest on and are surrounded by diamictite with a sandy mudstone

Fig. 7. Diamictite of lithofacies B, showing stratified nature and a raft of thin rippled fine sandstone (A). Two areas of clast pavements: (B) showing alignment of clasts and grooved/striated matrix, and (C) three successive horizons of pavements (one in foreground, another with scale on, another in background). Scale bar is 10 cm.

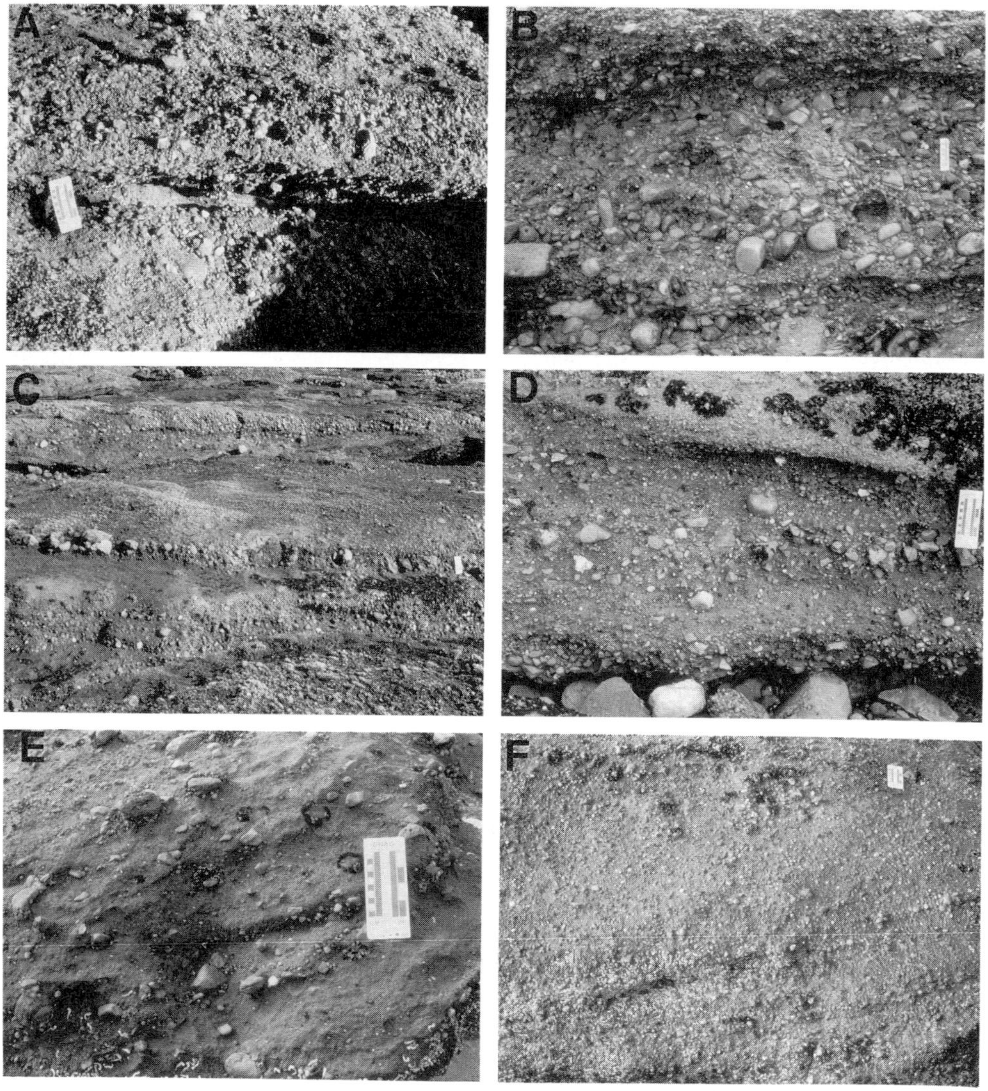

Fig. 8. Examples of beds within lithofacies C, D and E; marine outwash deposits in the runout of the jet's zone of flow establishment. Scale bar is 10 cm in all photos. (A) Diamictite injected upward (at an angle toward the left) along a fracture following a bedding plane in lithofacies C. (B) Poorly sorted conglomerate bed tending matrix-supported locally, reflecting an unsteady flow about a mean flow energy. (C) A rare boulder bed. (D) Welded contacts in cross-beds of poorly sorted, pebbly, coarse to very coarse sandstone and matrix-supported, pebble conglomerate reflecting wide flow-energy fluctuations and pulsating sediment discharge. (E) Weakly cross-bedded, pebbly, coarse to very coarse sandstone; part of a plug-flow deposit. (F) Thickly cross-bedded, well washed granule to small pebble conglomerates with dispersed larger clasts showing very rapid deposition over a period of uniform flow energy. (G) Well washed, cross-bedded, medium to coarse sandstone.

matrix (locally muddy sandstone) and 40% clasts larger than small pebble. The matrix shows flow structure around the conglomerate blocks. This lithofacies is equivalent to and just younger than lithofacies B, and equivalent to lithofacies D. The diamictite could be equivalent to lithofacies F.

Lithofacies D is trough cross-bedded sandstones and conglomerates similar to lithofacies E, but the cross-beds show deformation and some diamictite is injected upward between beds. The diamictite could be equivalent to lithofacies F. Lithofacies D appears to be a less deformed version of lithofacies C and laterally equivalent to lithofacies E.

Lithofacies E is cross-bedded sandstones and conglomerates. Rare clasts have striations. Beds that make up co-sets vary quite widely in character (Figs 8 and 9) and they include the following types: (1) Poorly sorted boulder beds having subrounded clasts from very coarse pebbles up to boulders 80 cm across within a granule to coarse sandstone matrix; (2) medium to large pebble conglomerates occurring as stringers or in moderately sorted, open-work beds; other medium to large pebble conglomerates are poorly sorted with larger clasts dispersed (matrix-supported) in a granule to coarse sandstone matrix; (3) small pebble conglomerates that have similar characteristics to (2); (4) poorly sorted, granule conglomerates that have up to 10% dispersed small to medium pebbles and coarse to medium sandstone matrix, are commonly matrix-supported, but rare beds are open-work and grade into small pebble conglomerates and very coarse to coarse sandstones; (5) very coarse sandstones that have up to 5% pebbles and granules either dispersed or rarely in stringers, and commonly grade up from granule and small pebble conglomerates; (6) coarse sandstones similar to (5); (7) rare, moderately sorted, medium sandstones that have planar laminae of coarse sandstone, and have rare, sorted granules and pebbles; (8) local lenses of unstratified diamictites of 40 to 50% pebbles within a sandy mudstone matrix.

In general, bedding is thicker and not well defined with common welded contacts. Coarser beds are commonly found toward the southwest. The better stratified area is made up of a series of three very large scale trough cross-beds, each up to 45 m across and 1.5 m high (Fig. 9). They show migration of barchanoid bedforms toward 054°. The lower sections of foresets tend to be

Fig. 9. (A) The easternmost barchanoid form in panorama showing sweep of foresets away from viewer. (B) Mudflow diamictite from the centre of this bedform. (C) Nest of pebbles on top of barchanoid surface from an iceberg roll-dump. Scale bar is 10 cm in all photos.

coarser grained, and contacts are commonly graded or welded, or locally sharp. The foreset of one of the bedforms failed, and flowed to form a diamictite (50% clasts in muddy sandstone) with floating sandstone rafts up to 2 m long.

Just before the limit of exposure, the barchanoid forms appear to end and the last deposits in a down-current sense are generally finer grained and less stratified. The beds are mostly matrix-supported, granule to small pebble conglomerate in beds 20 cm to 1 m thick. There are also sandstones with scattered pebbles and rare large pebble to cobble conglomerate beds. The beds appear weakly channelized locally; others are planar and dip up to 24° toward 055°.

The last features of note are rare boulders scattered throughout lithofacies E and clusters of pebbles and cobbles, some in low depressions, on top of the barchanoid forms. This lithofacies is equivalent to lithofacies D and overlies lithofacies F.

Lithofacies F is stratified diamictite in which clast proportions vary areally from 5 to 50%. Clasts are subangular to subrounded and are supported by a sandy mudstone matrix. Rafts of rippled, fine sandstone, medium to coarse sandstone, and small to medium pebble conglomerate, exhibit soft-sediment deformation and apparently were originally interstratified with diamictite.

These lithofacies associations are interpreted as an incipient grounding-line fan. The deposits are subaquatic because of close association of traction current deposits and flow diamictites (debris flow deposits), the presence of turbidites in soft-sediment rafts within the flow diamictites, and the isolated boulders and clast clusters that are interpreted as ice-rafted deposits. In fact, the water body must have been marine because, following the interpretation given below for the large-scale trough cross-beds, the barchanoid bedforms would not be produced in a lake since flows with such high sediment concentrations would not detach from the lake floor. The glacier margin must also have been a vertical tidewater cliff because the pebble-cobble clusters are from iceberg rafting; they cannot be from an ice shelf because it would not have concentrated clasts in such a manner. The clusters are within 100 m of clast pavements formed by an oscillating grounding-line, which necessitates a vertical glacier face. Furthermore, the glacier was most likely sub-polar or temperate because of the volumes of meltwater present.

The glacier is interpreted to have advanced onto the site over older glacial deposits (here undifferentiated). Debris flow deposits from the grounding-line covered the area (lithofacies F) probably before and while the grounding-line occupied the area covered by lithofacies B when the glacier produced the pavements. In fact, lithofacies B is a combination of subglacial and proglacial, ice-contact deposits that were still soft. They deformed as the grounding-line oscillated in the zone represented by the lithofacies. The diamictites are flows from, and/or pushed by, the glacier which then partially overrode them. The rafts of sorted sediment are lateral, finer-grained deposits (commonly turbidites) from the submarine discharge source. All of the diamictites they were deposited with were highly water-saturated. Redepositional slumps and flows in the grounding-line zone were common. Further mixing and deformation occurred as the glacier pushed into the deposits when it oscillated forward.

Whereas the grounding line was quasi-stable in lithofacies B zone, lithofacies C, D and E are thought to have been deposited as a grounding-line fan that prograded over the still soft diamictite of lithofacies F. Most of the fan lithofacies association which is preserved, is marine outwash deposited by a horizontal turbulent jet in contact with the sea floor from hyperconcentrated plug flows. Some other facies are sediment gravity flow deposits resedimented from the marine outwash. Lithofacies C and D are deformed, probably by three processes: loading into the soft diamictite of lithofacies F and also some consequent flowage, the glacier pushing the sediment as it accumulated against the face during oscillations, and flowage as the fan lost support, when the terminus retreated. Once the grounding-line retreated from lithofacies B, lithofacies A was deposited.

The fan deposits are a good basis for evaluating processes of transport and deposition within a plane turbulent jet in the ZFE. The large-scale cross-beds are interpreted as the detachment zone of the jet as it changed to a VTJ. The whole character of deposits indicates very rapid deposition of particles falling out of suspension. Particles as large as small pebbles are barely edge-rounded, indicating their transport after being released from ice was almost entirely in suspension. Rare boulder beds were probably transported periodically by a large surge in the subglacial discharge. Other deposits over the runout distance show welded and graded contacts between beds of different textures that were probably deposited by slugs of sediment from the glacial tunnel. Other beds of matrix-supported conglomerates and diamictites are interpreted as hyperconcentrated and slurry-flow mixtures in transport by the plug flow of the jet.

The detachment zone of the jet is interpreted to have formed a barchanoid bar that migrated along flow with fluctuations in discharge, to produce the large-scale cross-bedding. The foresets would have formed from pulses of bed load deposits moving down the foreslope with more poorly sorted mixtures dropping-out from suspension transport in the jet. Occasionally, the foreslope failed and produced the flow diamictite. During the period when the jet was active it appears to have migrated laterally more than 100 m to produce the overlapping detachment zone bars. Beyond the foresets are varieties of flow deposits and jet (ZEF) deposits that apparently settled rapidly in a sediment veil, and then also occasionally flowed down depositional slope.

A more quantitative description of the jet can be obtained using the empirical and dimensionless relationships presented earlier and using realistic estimates for some parameters. Assuming constant velocity in the ZFE the runout distance can be estimated between the clast pavements and lithofacies E, which is a range of 60 to 130 m. The tunnel diameter was then between 4.5 to 9.7 m (Equation (9)); D = 8 m will be used. The slope was low (≃ 1°) and we can assume a viscosity and density of freshwater at 1°C. If the tunnel had a boulder roughness of k_s = 0.5 m, then the mean velocity of tunnel flow was about 6.4 m s^{-1} (Equation (1)), which could transport particles to small pebble size in suspension (Equation (3); using C = 15 g l^{-1} (0.57%) and particle specific gravity of 2.60). The discharge at the tunnel mouth was a jet (R = 0.034; Equation (4)) that came out at a very low angle near horizontal (F = 52.3; Equation (5)). Assumed parameters here were taken from modern Alaskan conditions: ρ_a = 1.018 kg l^{-1}, ρ_o = 1.008 kg l^{-1} (using C = 15 g l^{-1}), in sea water temperatures of 2 to 3°C. By the end of the runout distance of 100 to 130 m the jet velocity was 0.17 m s^{-1} (Equation (8)), that is capable of transporting at least granule-size particles. The VTJ changed to a plume 14.6 m above the sea floor (Equation (11)) and the plume initially had a velocity of 0.1 m s^{-1} (Equation (14)) which was still capable of transporting granule-size particles. Although crude estimates have been assumed, these descriptors conform well with what was described in the field.

Some special cases

The discussion so far has considered situations in which a grounding line is quasi-stable and grounding-line fans are built from meltwater discharges. However, conditions may differ from those considered. As a grounding line advances, immediate proglacial sediment tends to be reworked by the glacier and consequently, glacial-advance fans are not well preserved. For some advances, they can decrease the relative water depth of the grounding line (Powell 1988, 1990a). During an advance, coarser proglacial sediments can be pushed directly by the glacier with finer sediment squeezed-up in front of the grounding line by loading and consequently pushed (Powell 1988, 1990b).

Fans are not produced during rapid retreat of a grounding line because there is insufficient time for them to build. The marine outwash is spread as a thinner sheet over a larger area (Powell 1981; 1983).

An englacial stream will form an axisymmetric jet incapable of transporting coarse bed load, which will then be dumped as rockfall or grainfall down to the grounding line. A submarine talus cone could accumulate and perhaps build into a fan from which sediment gravity flows may originate. Cyclopsams and cyclopels may accumulate around the efflux. In general, for temperate glaciers at least, submarine englacial streams are inferred to have far lower sediment concentrations than basal and subglacial streams inferred from suspended sediment samples of their HBPs (Cowan & Powell unpublished data). Consequently, englacial streams are thought to be minor contributors to grounding-line fans compared with subglacial streams.

When a grounding line ends at a hanging valley or on the edge of a sea floor high, the fans may lack marine outwash and be made of sediment gravity flow and suspension deposits (Fig. 1). The subglacial stream forms an axisymmetric jet because of the steep slope and deep water, and all the outwash is redeposited from the efflux or rains from the jet and is then redeposited.

Fans may not form if, as suggested by Elverhøi *et al.* (1980), the stream mixes with salt water back behind the grounding line within a melted vertical crevasse. In such cases, eskers would be the most common form of 'marine' outwash.

Alternatively, fans may not form as a discrete entity if the streams do not have a significant coarse-grained load. Fine sediment transported as an overflow can settle and form sheets of glacimarine mud across the grounding line. If water discharge velocity is relatively high, then the ZFE of the jet may scour out a channel within the mud.

Location of an efflux across a grounding line can change over significant distances from one melt season to the next or sometimes within one season. Fans may, therefore, overlap asynchronously as a younger fan progrades toward an older fan. Because of the relatively wide spacing of coexisting basal tunnels, only large fans could interfinger by being active at the same time. Alternatively, at quasi-stable termini some tunnels may be reoccupied for several melt seasons as described above, and the efflux may migrate along the head of a large fan as calving occurs because of slight sinuosity of subglacial and basal tunnels.

A variation of changing efflux positions is having subglacial sheet flow discharging along a grounding line to produce linear wedges of marine outwash. Morainal banks in Maine deposited from the Laurentide Ice Sheet are probably such deposits (cf. Smith 1982). Sheet flow may be initiated if the grounding line becomes slightly buoyant, or if the back-slope of a fan or morainal bank is quite steep. When the glacier bed slopes back up-glacier at about twice the surface slope of the glacier, then freezing in the tunnel can eventually close it and cause a change to sheet flow (cf. Röthlisberger & Lang 1987).

Exactly what happens at effluxes at the grounding lines of modern ice shelves is uncertain at present. It has been suggested that fans could not aggrade very high before being ended and 'planed off' by the ice shelf (Powell 1984). An ice shelf fan may be like the deforming sediment wedge shown by Alley *et al.* (1987; 1989), but that is also speculative.

Fan aggradation and ice-contact delta formation

When a grounding line becomes stable for tens of years or more, fans have the possibility of growing large. In fact, the larger the fan(s) grows the more stable the margin can become at that site.

The fan can grow by direct addition from streams of marine outwash and sediment gravity flow deposits. Mass flows from elsewhere along the grounding line may contribute sediment; other contributions to fan aggradation stem from grounding-line oscillation. Glacier push deforms sediment into folds and produces thrust slices, both of which increase the height of the fan (Fig. 10). If the glacier continues to advance it can ride up the back of the fan to deposit lodgement till and perhaps form morainal bank accumulations (Stemen 1979; Smith 1982; Powell 1983). If the terminus then retreats, basal ice may become buried. This increases the height of the fan for a period of time.

As the fan grows, the subglacial tunnels have an increasingly steep slope up-glacier before reaching the grounding-line. On the slope on the back side of the fan, coarser bed load may accumulate in the tunnel except during high discharge. If the tunnel is maintained by melting up into the glacier, then net accretion

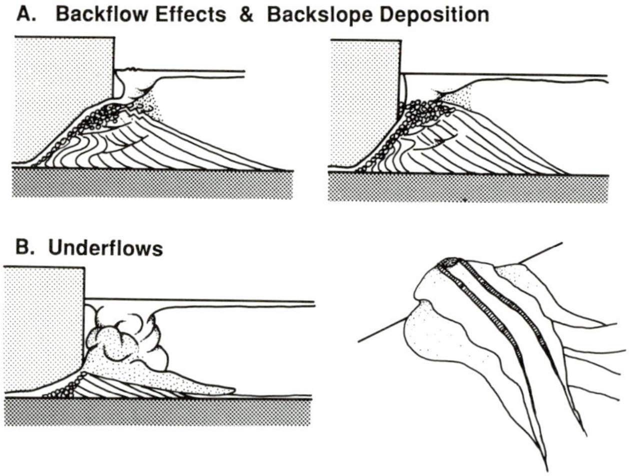

Fig. 10. Two special cases of grounding-line fans (not to scale) where the glacier has fluctuated about its position and pushed and thrust the fan (A). In the left-hand example the jet flows horizontally to cause a backflow effect in the lee of the fan apex and backflow ripples form. In the right-hand example the jet discharges into the back of a fan after the grounding line retreats slightly, and coarse sediment is dammed. An underflow situation may arise (B) when the stream carries a high sediment load that falls back and collapses down the initial part of the VBJ to increase sediment concentrations sufficiently to produce an underflow.

on the tunnel floor may occur. If this deposition enhanced blockage of the tunnel to at least partially dam it, the eventual burst can produce a pulse of highly sediment-charged water to the issuing jet, initiating a quasi-stable underflow.

During terminus retreat much of the coarser load in the jet will be dumped on the back side of the fan, if the jet is directed into the fan up its back slope (Fig. 10). Some back-sliding and back-collapse of sediment could be expected. A half-jet effect may occur if the jet flows over the apex of the fan. As the jet detaches from the apex, a back-flow eddy is produced by the horizontal jet at the fore-slope of the fan (cf. Boersma *et al.* 1968). The result can be a set of back-flow ripples climbing toward the grounding line or fan apex (Fig. 10). This condition has been found recently in a Pleistocene example from Maine (Hunter & Powell 1990).

With continued oscillations of the grounding line, even on a seasonal basis, more sediment can be added to the fan. Larger retreats have the potential of pushing more sediment into the fan area.

Growth rates of Alaskan fans are of the order of 10^6 m^3 a^{-1} (Powell 1988; Powell & Molnia 1989). Consequently, after being a depocentre for a period of time, a grounding-line fan can aggrade to sea level and continue to grow as an ice-contact fan delta. These are good indicators of a grounded tidewater cliff rather than a floating terminus.

Observations at Riggs Glacier, Alaska, of the early stages of delta formation, when the system is transforming from a grounding-line fan into a normal delta, showed that when the plain was fully intertidal, stream channels change position with nearly every tidal cycle (Fig. 11; Heiny 1983). During flood tide the stream is buoyantly lifted off the delta plain, then with ebb tide, the stream is progressively lowered back to the delta plain to potentially erode completely new channels. Consequently, very little sediment is stored on the delta plain and aggradation occurs only by progradation. Resedimentation events from the delta plain are very common and prodeltaic sediment is intensely stratified with sand and mud. A couplet can be generated over a tidal cycle (Powell 1981; 1983), which when combined with random events of sediment slugs from discharge pulses and sediment gravity flows, can lead to several sand laminae being produced per day.

As the delta plain progrades, channels become quasi-stable. Pulses of high sediment concentration are generated during peak discharges. However, major events of prodeltaic deposition occur during tidal draw-downs over monthly spring tide periods where low tide level is at or near the break between the delta plain and delta front. Then, large quantities of sediment temporarily stored on the delta plain, are transported down the front into prodelta areas (Smith *et al.* 1986; 1990). Resedimentation by mass flows is also common from the delta front (Phillips *et al.* 1987). The ice-contact fan delta at Riggs Glacier formed very rapidly and prograded and aggraded in front of the tidewater cliff (Fig. 12). In 1979 the glacier terminated in about 55 m of water (Powell 1983); by 1981 a delta with a fully intertidal plain had prograded 200 m from the glacier. By 1985, the delta plain had aggraded above tidal range and prograded 100 m farther into what had been 32 m of water in 1981 (Powell & Cowan 1986). A total volume of about 10^6 m^3 of sediment had accumulated in four years.

Conclusions

Grounding-line fans form in contact with glaciers ending subaquatically from the discharge of sediment by englacial and subglacial streams. Simple conceptual models have been presented for marine examples, which could differ in detail from lacustrine examples because of ease of underflow formation in lakes, and perhaps flocculation in the marine environment.

Sediment sources are full-tunnel discharges whose flow conditions with sediment loads are not well constrained. However, the likelihood of bed load moving in waves down the tunnel and a tendency for tunnel blockage both enhance the pulsatory nature of sediment discharge.

At the tunnel efflux, discharge is fully turbulent and it forms a jet that is either axisymmetric or plane depending on whether the jet is in contact with the sea floor. Jets with sediment loads have not been studied extensively but, by analogy with pure jets, subglacial discharges will probably remain on the sea floor for some distance beyond the efflux. This distance depends on sediment and water discharge, tunnel size and buoyancy forces. Imbricate boulders can be deposited near the efflux and finer gravels and sands accumulate under rapid sedimentation, along the runout distance of the jet. Characteristics of the flow are a fluctuation about a mean energy, a pulsating, continuous sediment supply, and some hyperconcentrated or slurry flows. Resulting deposits have welded or graded contacts with wide variation in sorting and mean texture. Deposits of marine outwash may be sheets or scour-and-fill forms if the jet

Fig. 11. Sequence of photographs taken during July, 1981 (day is shown on each photo), of an intertidal delta plain at Riggs Glacier, Glacier Bay, Alaska. New channels are commonly cut as the stream comes back into traction when tide is falling from a high stage.

wavers spatially. At the detachment point barchanoid bars may form and major redeposition and sediment gravity flows originate from the foresets. The detachment zone is an area of high sediment concentrations because of particulate fall-back effects as the flow changes to a vertical turbulent jet. The coarsest and thickest cyclopsams can be deposited and underflows have the potential of forming.

Finer particles remain entrained by the jet, which eventually transforms to a plume after buoyancy forces become larger than inertial forces. Rarely, the flow will be an interflow. More commonly a turbid overflow forms in which particles move away as a buoyant flow from the upwelling vertical jet or plume. Particles tend to be held near the base of the flow and released when internal turbulence decreases to form rhythmites of finer-grained cyclopsams and cyclopels on the sea floor.

Grounding-line fans can aggrade rapidly, and at some temperate sites rates are about 10^6 m^3 a^{-1} of sediment. Sediment may additionally be pushed or squeezed as the glacier over-rides the fan to contribute subglacial deposits directly. If the grounding line remains quasi-stable and the subglacial tunnel system maintains a constant position, then fans may grow to sea level and

Fig. 12. Vertical aerial photographs (U.S. Geological Survey, Tacoma, Washington) of Riggs Glacier, Glacier Bay, Alaska. The first was taken on 21 June, 1978 (A) when there was 60 m of water in the embayment at the tidewater cliff. The same scene on 24 August, 1988 (B) showing the delta that has aggraded from a grounding-line fan and prograded, so that the delta plain is above high tide. It grew at about 10^6 m^3 a^{-1}.

form ice-contact deltas. A critical period in delta formation is the transitional stage where the delta plain is intertidal. New stream channels can be eroded quite regularly to move coarse sediment to the prodelta in pulses. Consequently, an intense period of rhythmite formation can be used to determine the transition stage. Recognizing such features may be important in distinguishing fan deposits (not graded to sea level) and deltas. The need for a distinction becomes important in areas where sea level curves depend on locating delta plains, but where grounding-line fans are also present.

I would like to thank John Shaw, Tom Stewart and Ellen Cowan for discussions about grounding-line fans and Norm Smith for work with deltas. Max Banks and his department provided guidance and facilities in Tasmania and Steve Hand contributed to my field knowledge. Thoughtful reviewer comments by J.P.M. Syvitski, R. Gilbert and J.A. Dowdeswell helped improve the quality of the manuscript. Research in Alaska was supported by NSF grants DPP-8420749, DPP-8619153, and DPP-8642364; the National Park Service, James Luthy and Exploration Holidays. Research in Australia was supported by NSF grant INT-8714182. I thank the Geological Society of London for support to present this paper.

Appendix: list of symbols with dimensions

B	Buoyancy flux, [M L T^{-3}]
B_{po}	Initial buoyancy flux of plumes, [M L T^{-3}]
B_p	Final buoyancy flux of plumes, [M L T^{-3}]
C	Volumetric concentration of sediment particles, [M L^{-3}]
D	Tunnel or pipe diameter (circular section) or width (rectangular section), [L]
d	Particle diameter, [L]
F	Densimetric Froude number
f_b	Resistance coefficient between jets and sea floor
f_i	Resistance coefficient between jets and ambient fluid
g	Gravitational acceleration, (L T^{-2}]
g'	Reduced gravity ($= g(\rho_a - \rho_o)\,\rho_o^{-1}$), [L T^{-2}]
h	Thickness of flow, [L]
k_a	Proportionality constant for axisymmetric plumes ($\simeq 4.7$)
k_g	Proportionality constant for gravity flow ($\simeq 1$)
k	Proportionality constant for plane plumes ($\simeq 1.65$)
k_s	Surface roughness, [L]
M	Momentum flux, [M L T^{-2}]
N	Frequency for linear stratification in ambient fluid ($= ((-g/\rho_a)\,(d\rho_a/dz))^{1/2}$), [$T^{-1}$]
Q	Discharge or plume flux, [L^3 T^{-1}]
R	Initial Richardson number
Re	Reynolds number
R_h	Full tunnel or pipe hydraulic radius, [L]
s	Particle specific gravity
S_o	Slope of hydraulic gradient
U_a	Mean ambient fluid velocity, [L T^{-1}]
U_c	Critical mean flow velocity for deposit-free transport tunnels or pipes, [L T^{-1}]
U_g	Mean velocity of plume acting as a gravity current, [L T^{-1}]
U_m	Mean centreline velocity in jets and plumes, [L T^{-1}]
U_{mo}	Mean initial centreline velocity in jets and plumes at tunnel or pipe mouths, [L T^{-1}]
U_p	Mean plume velocity, [L T^{-1}]
U_t	Mean flow velocity in tunnels or pipes, [L T^{-1}]
U_w	Phase velocity of interfacial waves, [L T^{-1}]
x	Axial distance in jets or plumes from centre of tunnel or pipe mouths, [L]
z_c	Critical height where jets terminate and transform to plumes, [L]
z_{ja}	Terminal height of axisymmetric jets in stratified ambient fluids, [L]

z_{jp} Terminal height of plane jets in stratified ambient fluids, [L]
z_{pa} Terminal height of axisymmetric plumes in stratified ambient fluids, [L]
z_{pp} Terminal height of plane plumes in stratified ambient fluids, [L]
β Bulk density of plumes, $[M\ L^{-3}]$
Θ Angle of inclination to the horizontal
λ Wavelength ($= 2\pi U_c/\omega$), [L]
μ Dynamic viscosity, $[M\ L^{-1}\ T^{-1}]$
ν Kinematic viscosity, $[L^2\ T^{-1}]$
ρ Fluid density, $[M\ L^{-3}]$
ρ_a Ambient fluid density, $[M\ L^{-3}]$
ρ_o Initial plume and jet density at tunnel or pipe mouths, $[M\ L^{-3}]$
ρ_p Plume fluid density, $[M\ L^{-3}]$
ω Angular frequency, $[T^{-1}]$

References

ABRAHAM, G. 1965. Horizontal jets in stagnant fluid of other density. *Journal of the Hydraulic Division, Proceedings of the American Society of Civil Engineers*, **91** (HY4), 139–154.

ALBERTSON, M.L., DAI, Y.B., JENSEN, R.A. & ROUSE, H. 1950. Diffusion of submerged jets. *Transactions of the American Society of Civil Engineers*, Paper 2409, 639–664.

ALLEY, R.B., BLANKENSHIP, D.D., BENTLEY, C.R. & ROONEY, S.T. 1987. Till beneath ice stream B. 4. A coupled ice-till flow model. *Journal of Geophysical Research*, **92** (B9), 8931–8940.

——, BLANKENSHIP, D.D., ROONEY, S.T. & BENTLEY, C.R. 1989. Sedimentation beneath ice shelves — the view from ice stream B. *Marine Geology*, **85**, 101–120.

BANKS, M.R. 1981. Late Palaeozoic tillites of Tasmania. *In*: HAMBREY, M.J. & HARLAND, W.B. (eds.) *Earth's pre-Pleistocene Glacial Record*. Cambridge University Press, Cambridge, 495–501.

BATES, C.C. 1950. Rational theory of delta formation. *Bulletin of the American Association of Petroleum Geologists*, **37**, 2119–2162.

BOERSMA, J.R., MEENE, E.A. & VAN DE TJALSMA, R.C. 1968. Intricated cross-stratification due to interaction of a mega ripple with its lee-side system of backflow ripple (upper-pointbar deposits, lower Rhine). *Sedimentology*, **11**, 147–162.

BOULTON, G.S. 1986. Push-moraines and glacier-contact fans in marine and terrestrial environments. *Sedimentology*, **33**, 677–698.

CAREY, S.N., SIGURDSSON, H. & SPARKS, R.S.J. 1988. Experimental studies of particle-laden plumes. *Journal of Geophysical Research*, **93** (B12), 15314–15328.

CARSTENS, T. & RAO, U.M. 1971. Sediment-laden submerged horizontal jet. *International Association for Hydraulic Research, Proceedings of 14th Congress*, 135–144.

CHADWICK, A.J. & MORFETT, J.C. 1986. *Hydraulics in Civil Engineering*. Allen & Unwin, London.

CLARKE, M.F. & BANKS, M.R. 1975. The stratigraphy of the lower (Permo-Carboniferous) parts of the Parmeener Super-group, Tasmania. *In*: CAMPBELL, K.S.W. (ed.) *Gondwana Geology*. Australia National University Press, Canberra, 15–35.

CONDOLIUS, E. & CHAPUS, E.E. 1963. Transporting solid materials in pipelines. *Chemical Engineering*, June, 140–148.

COWAN, E.A. 1988. Sediment transport and deposition in a temperate glacial fjord, Glacier Bay, Alaska. *Ph.D. Thesis*, Northern Illinois University, DeKalb.

—— & POWELL, R.D. 1990. Suspended sediment transport and deposition of cyclically interlaminated sediment in a temperate glacial fjord, Alaska, U.S.A. *In*: DOWDESWELL, J.A. & SCOURSE, J.D. (eds), *Glacimarine Environments: Processes and Sediments*. Geological Society, London, Special Publication, **53**, 75–90.

CROWELL, J.C. & FRAKES, L.A. 1971. Late Paleozoic glaciation in Australia. *Journal of the Geological Society of Australia*, **17**, 115–155.

DURAND, R. 1951. Some observations on the transportation of gravel in pipes. *International Association of Hydraulic Research, 4th Meeting, Bombay, India*, 655–666.

——, 1953. Basic relationships of the transportation of solids in pipes — experimental research. *International Association for Hydraulic Research, Proceedings of the Minnesota International Convention*, 89–103.

ELLISON, T.H. & TURNER, J.S. 1959. Turbulent entrainment in stratified flows. *Journal of Fluid Mechanics*, **6**, 423–448.

ELVERHØI, A., LIESTØL, O. & NAGY, J. 1980. Glacial erosion, sedimentation and microfauna in the inner part of Kongsfjorden, Spitsbergen. *Norsk Polarinstittut Skrifter*, **172**, 38–58.

FISCHER, H.B., LIST, E.J., KOH, R.C.Y., IMBERGER, J & BROOKS, N.H. 1979. *Mixing in Inland and Coastal Waters*. Academic, New York.

FOX, D.G. 1970. Forced plume in a stratified fluid. *Journal of Geophysical Research*, 75, 6818–6835.

GILBERT, R. 1983. Sedimentary processes of Canadian arctic fjords. *Sedimentary Geology*, **36**, 147–175.

GRAF, W.H. 1971. *Hydraulics in Sediment Transport*. McGraw-Hill, New York.

—— & ACAROGLU, E.R. 1968. Sediment transport in conveyance systems. Part I. A physical model for sediment transport in conveyance systems. *Bulletin of the International Association of Scientific Hydrology*, **13**, 20–39.

HEINY, J.S. 1983. Sediment characteristics of a rapidly retreating temperate valley glacier. *Unpublished M.Sc. Thesis*, Northern Illinois University, DeKalb.

HOOKE, R. LeB. 1989. Englacial and subglacial hydrology: a qualitative review. *Arctic and Alpine Research*, **21**, 221–233.

——, WOLD, B. & HAGEN, J.O. 1985. Subglacial hydrology and sediment transport at Bondhusbreen, southwest Norway. *Geological Society of America Bulletin*, **96**, 388–398.

HUNTER L.E. & POWELL, R.D. 1990. Grounding-line fan processes during deglaciation of coastal Maine. *Geological Society of America, Abstracts with Programs*, **22**(2), 25.

LIST, E.J. 1982. Turbulent jets and plumes. *Annual Review of Fluid Mechanics*, **14**, 189–212.

MACKIEWICZ, N.E., POWELL, R.D., CARLSON, P.R. & MOLNIA, B.F. 1984. Interlaminated ice-proximal glacimarine sediments in Muir Inlet, Alaska. *Marine Geology*, **57**, 113–147.

MORTON, B.R. 1959. Forced plumes. *Journal of Fluid Mechanics*, **5**, 151–163.

NEWITT, D.M., RICHARDSON, J.F., ABBOTT, M. & TURTLE, R.D. 1955. Hydraulic conveying of solids in horizontal pipes. *Transactions of the Institution of Chemical Engineers*, **33**, 93–110.

NOVAK, P. & NALLURI, C. 1972. A study into the correlation of sediment motion in pipe and open channel flow. *Hydrotransport 2, 2nd International Conference on the Hydraulic Transport of Solids in Pipes*, Paper D4, 33–51.

PHILLIPS, A.C., SMITH, N.D. & POWELL, R.D. 1987. Marine sediment laminations controlled by tidal fluctuations over an ice-contact delta. *Geological Society of America, Abstracts with Programs*, **19** (4), 238.

POWELL, R.D. 1981. A model for sedimentation by tidewater glaciers. *Annals of Glaciology*, **2**, 129–134.

—— 1983. Glacial marine sedimentation processes and lithofacies of temperate tidewater glaciers, Glacier Bay, Alaska. *In*: MOLNIA, B.F. (ed.) *Glacial-Marine Sedimentation*. Plenum, New York, 195–232.

—— 1984. Glacimarine processes and inductive lithofacies modelling of ice shelf and tidewater glacier sediments based on Quaternary examples. *Marine Geology*, **57**, 1–52.

—— 1988. Processes and facies of temperate and subpolar glaciers with tidewater fronts. *Short Course Notes, Annual Meeting, Centennial Geological Society of America, Denver*.

—— 1990a. Advance of tidewater fronts of temperate glaciers. *2nd Glacier Bay Science Symposium, Proceedings*.

—— 1990b. Grounding-line systems as second order controls on fluctuations of tidewater termini. *In*: ANDERSON, J.B. & ASHLEY, G.M. (eds) *Glacial Marine Sedimentation: Polaeoclimatic Significance*. Geological Society of America, Special Publication.

—— & COWAN, E.A. 1986. Depositional processes at McBride Inlet and Riggs Glacier. *In*: ANDERSON, P.J., GOLDTHWAIT, R.P. & MCKENZIE, G.D. (eds) *Observed Processes of Glacial Deposition in Glacier Bay, Alaska. Institute of Polar Studies, Ohio State University, Columbus. Miscellaneous Publication*, **236**, 140–156.

—— & MOLNIA, B.F. 1989. Glacimarine sedimentary processes, facies and morphology of the south-southeast Alaska shelf and fjords. *Marine Geology*, **85**, 359–390.

RÖTHLISBERGER, H. & LANG, H. 1987. Glacial hydrology. *In*: GURNELL, A.M. & CLARK, M.J. (eds) *Glacio-fluvial Sediment Transfer*. Wiley, New York, 207–284.

RUST, B.R. & ROMANELLI, R. 1975. Late Quaternary subaqueous outwash deposits near Ottawa, Canada. *In*: JOPLING, A.V. & MCDONALD, B.C. (eds) *Glaciofluvial and Glaciolacustrine Sedimentation. Society of Economic Paleontologists and Mineralogists, Special Publication*, **23**, 177–192.

SHREVE, R.L. 1985. Esker characteristics in terms of glacier physics, Katahdin esker system, Maine. *Geological Society of America Bulletin*, **96**, 639–646.

SIKONIA, W.G. & POST, A. 1979. Columbia Glacier, Alaska: recent ice loss and its relationship to seasonal terminal embayments, thinning, and glacier flow. *U.S. Geological Survey, Open-file Report*, 79–1265.

SINGAMSETTI, S.R. 1966. Diffusion of sediment in a submerged jet. *Journal of the Hydraulic Division, Proceeding of the American Society of Civil Engineers*, 92 (HY2), 153–168.

SMITH, G.W. 1982. End moraines and the pattern of last ice retreat from central and south coastal Maine. *In*: LARSON, G.J. & STONE, B.D. (eds) *Late Wisconsinan Glaciation of New England*. Kendall/Hunt Publishers, Dubuque, 195–210.

SMITH, N.D., PHILLIPS, A.C., WOODLAND, S.R. & POWELL, R.D. 1986. Marine sediment laminations formed by tidal effects on a glacier-fed delta. *12th International Congress, International Association of Sedimentologists, Canberra*, 280–281.

——, PHILLIPS, A.C. & POWELL, R.D. 1990. Tidal drawdown: a mechanism for producing cyclic sediment laminations at glacimarine deltas. *Geology*, **18**, 10–13.

STEMEN, K.S. 1979. Glacial stratigraphy of portions of Lincoln and Knox Counties, Maine. *M.Sc. Thesis, Ohio University, Athens*.

SYVITSKI, J.P.M. 1989. On the deposition of sediment within glacier-influenced fjords: oceanographic controls. *Marine Geology*, **85**, 301–329.

——, SMITH, J.M., CALABRESE, E.A. & BOUDREAU, B.P. 1988. Basin sedimentation and the growth of prograding deltas. *Journal of Geophysical Research*, **93** (C6), 6895–6908.

TASK COMMITTEE. 1970. Sediment transport mechanics: J. Transportation of sediment in pipes. *Journal of the Hydraulics Division, Proceedings of the American Society of Civil Engineers*, **96** (HY7), 1503–1538.

TURNER, J.S. 1973. *Buoyancy Effects in Fluids*. Cambridge University Press, Cambridge.

—— 1986. Turbulent entrainment: the development of the entrainment assumption, and its application to geophysical flows. *Journal of Fluid Mechanics*, **173**, 431–471.

WIEDENROTH, W. & KIRCHNER, M. 1972. A summary and comparison of known calculations of critical velocity of solid-water mixtures and some aspects of the optimisation of pipelines. Hydrotransport 2. *2nd International Conference on the Hydraulic Transport of Solids in Pipes, Paper E1*, 1–22.

WILSON, K.C. 1969. Comments on "Sediment transport in conveyance systems." *Bulletin of the International Association of Scientific Hydrology*, **14**, 131–135.

WRIGHT, L.D. 1978. River deltas. *In*: DAVIS, R.A., Jr. (ed.) *Coastal Sedimentary Environments*. Springer, New York, 5–68.

Suspended sediment transport and deposition of cyclically interlaminated sediment in a temperate glacial fjord, Alaska, U.S.A.

ELLEN A. COWAN[1] & ROSS D. POWELL[2]

[1] *Department of Geology, Appalachian State University, Boone, NC 28608, USA*

[2] *Department of Geology, Northern Illinois University, DeKalb, IL 60115, USA*

Abstract: Cyclically interlaminated sediment is a distinctive lithofacies within the sediment package of temperate glacial fjords. The tidewater terminus of McBride Glacier is at the head of a small fjord that receives abundant suspended sediment and is a site of rapid accumulation of cyclically interlaminated sediment. Buoyant sediment-laden meltwater rises from a subglacial stream at the base of the glacier and mixes with fjord water to produce a thick, brackish overflow. Peak suspended sediment concentrations occur beneath the surface of the overflow at 3 to 10 m depth. Particle release from the overflow is controlled by semi-diurnal tidal fluctuations. The major vertical flux of suspended sediment is initiated at low tide because of low horizontal current velocities and reduced vertical eddy velocities. Sorting occurs as the particles settle; sand and coarse silt settle as single grains and finer flocculated particles settle as turbid layers at a rate between 2.5 m/h and 10.9 m/h. Each low water produces a couplet of a coarser grained lamina that is sorted as coarser single grains settle, and a finer grained lamina of flocculated particles in turbid layers. Individual couplet thicknesses and particle size decrease with distance from the discharge source. Semi-diurnal tides produce two couplets each day. Interstratified with these tidal rhythmites are coarser and poorly sorted laminae deposited by sediment gravity flows and coarse laminae contributed by peak daily discharge.

Valley glaciers in temperate climates have meltwater streams that transport large quantities of sediment from basal and lateral debris-rich ice. This results in sediment accumulation rates in temperate glacial fjords that are orders of magnitude higher than in non-glacial or glacial fjords in polar climates (e.g. Eyles *et al.* 1985, Syvitski *et al.* 1987). Fluvially derived sediment forms much of the marine sediment within temperate fjords. One of the most distinctive lithofacies within this package is fine-grained laminated sediment which has been recognized in fjords under both temperate and sub-polar conditions (Elverhøi *et al.* 1980, 1983, Powell 1981, 1983, Gilbert 1983, Görlich 1986). Characteristics of this lithofacies, such as cyclicity, vary but generally laminae become thinner and finer with distance from stream discharges. In glacial fjords, very fine laminae are preserved within the sediment package because high sedimentation rates reduce bioturbation and because bottom currents are absent or of low velocity. Although laminated sediments are recognized as a significant lithofacies, they are currently considered as a problematic glacimarine facies type (Eyles *et al.* 1988). This results from lack of a universally applied classification for types of laminated glacimarine sediments and confusion in distinguishing between fine-grained glacimarine rhythmites and glacilacustrine varves.

Cyclically interlaminated sediments were first described from Glacier Bay, southeastern Alaska, in sediment traps deployed by Hoskin *et al.* (1978). In a 9 m core from Muir Inlet, Molnia (1983) found more than 1000 fine layers in radiographs. Seismic reflection profiles indicated that this core was deposited in only 3 to 4 years (Molnia 1983). These laminae are not, therefore, varves. In a detailed study of cores from Muir Inlet, Mackiewicz *et al.* (1984) described characteristics of these laminated sediments. Cyclic couplets of thin, fine-grained sand or silt lamina that grade normally into a thicker mud lamina were recognized in proximal, medial and distal cores. The term cyclopel was introduced to describe each of these very fine sand/silt and mud couplets. The term cyclopsam describes the coarser, thicker ice-proximal equivalent. These couplets were interpreted to have been deposited by suspension settling when turbid plumes interacted with diurnal tidal currents.

This paper describes quantitatively how sediment plumes and tidal currents interact within a temperate fjord to produce rhythmites. A depositional model is presented for cyclic sedi-

From DOWDESWELL, J. A. & SCOURSE, J. D. (eds), 1990, *Glacimarine Environments: Processes and Sediments*. Geological Society Special Publication No 53, pp 75–89.

mentation within 2 km of a tidewater glacier terminus. The model is constructed using marine suspended sediment data, deposits collected in suspended sediment traps, and salinity and temperature measurements.

Setting

Glacier Bay National Park and Preserve is located in the panhandle of southeastern Alaska (Fig. 1). The Glacier Bay fjord system extends northward, bifurcating into the West Arm fjord system and Muir Inlet, a 50 km long tributary to the east. McBride Inlet, a tributary to Muir Inlet is located at 58° 50′N and 136° 00′ W (Fig. 1).

McBride Glacier extends 24 km from an elevation of 2100 m to sea level, where it terminates as a tidewater front (an ice cliff grounded in the sea with no portion floating). Glacier retreat since 1976 has opened up a 1.9 km long inlet, McBride Inlet that joins Muir Inlet through a narrow channel.

McBride Inlet has two basins (Fig. 2): the eastern basin slopes toward the glacier where it reaches a maximum depth of 100 m adjacent to the terminus. The gently sloping western basin averages 20 m depth. This area acts as a broad, shallow, entrance sill over which tidal exchange occurs between Muir Inlet and the deep basin. A shoal, exposed at low tide, was built-up during slow glacier retreat between 1978 and 1980 and is thought to originate from subglacial streams (Powell & Molnia 1989; Fig. 2).

The climate in Glacier Bay is cool–temperate maritime with very high precipitation. The initiation and length of the melt season is variable but usually begins in early May and extends into September. Diurnal fluctuations in discharge occur on cloudless days but they are damped during overcast periods (N.D. Smith, pers. comm.). Heavy rainfall late in the melt season can initiate high runoff and sediment discharge from meltwater conduits (Cowan *et al.* 1988).

The tide in the Glacier Bay fjord system is semi-diurnal unequal with a period of 12.42 h. The mean tidal range in Muir Inlet is 4.2 m with a maximum spring range of 7.3 m. The period and elevation of tides within McBride Inlet are controlled by exchange of water through the narrow channel. Low tide in McBride Inlet was found to lag behind low tide in Muir Inlet by 10 to 90 minutes in spring and neap tides respectively (Cowan 1988).

Methods

The observations presented here are based on over 1000 water samples collected from McBride Inlet during summers 1984 to 1987. Water samples were collected with a van Dorn bottle to measure suspended sediment concentration and particle size. A known volume (between 400 and 800 ml) of sample was vacuum-filtered onto a pre-weighed 47 mm diameter, 0.8 micron nominal pore diameter membrane filter. Filtered fresh water was washed through after the sample to ensure that no salts precipitated onto the filter. Each filter was air dried and reweighed in the laboratory. The total suspended sediment concentration was then calculated using the sample volume and sediment weight.

Large (>2.5 l) water samples were collected for particle size analysis and determination of

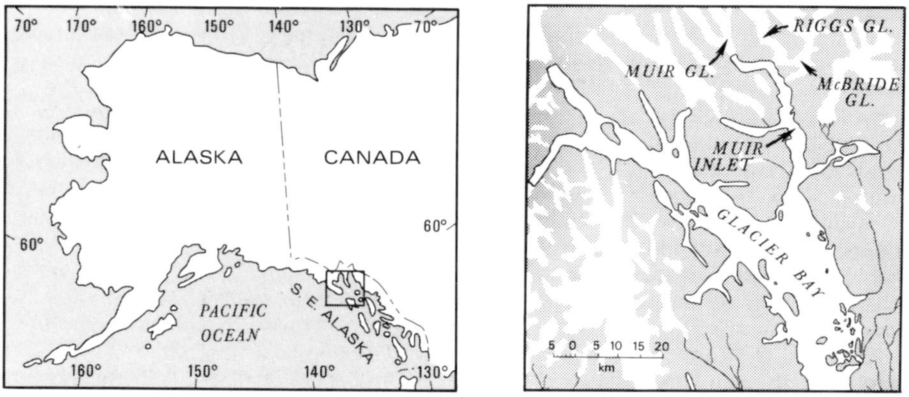

Fig. 1. Location map of Alaska and Glacier Bay National Park and Preserve. Larger scale map shows location of McBride Glacier.

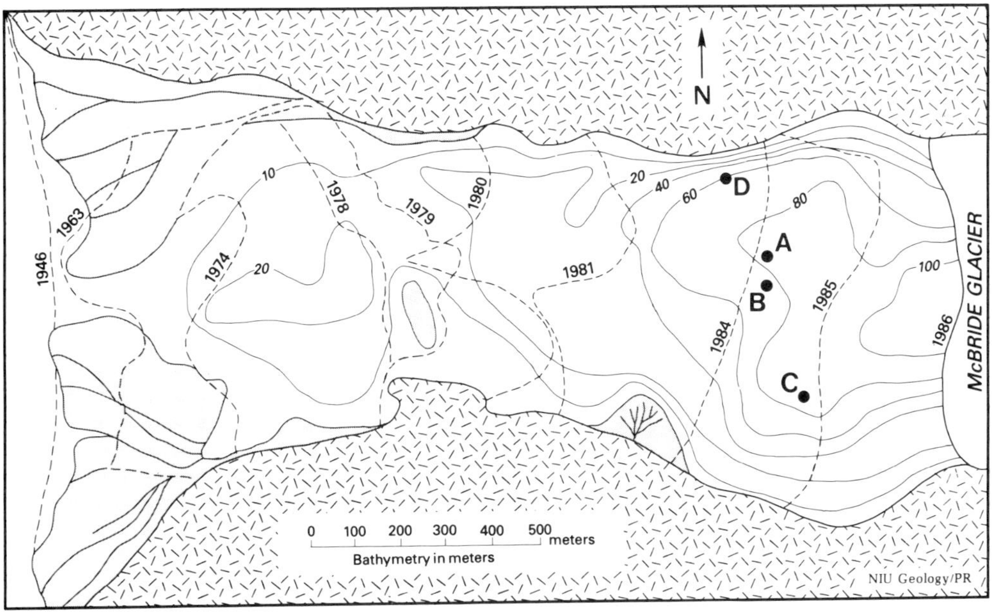

Fig. 2. Map showing position of the McBride Glacier terminus from 1946–1986. Ice front positions in 1946–1981 mapped from aerial photography; 1984–1986 mapped by ground survey. Bathymetry is shown for summer 1986. Positions of station locations, A–D are shown (see Table 1 for dates and types of samples collected).

sediment concentration. These samples were allowed to settle for at least 24 h after collection. Sample volume was noted and clear water siphoned off from above the settled particles. The sediment was then washed into a whirlpack bag. In the laboratory, samples were rinsed with distilled water to remove salts and wet sieved to separate sand (>63 microns) from silt and clay. Each fraction was weighed and both the suspended sand concentration (mg^{-1}) and total suspended sediment concentration were calculated.

Water column properties were measured at anchored stations within McBride Inlet (Table 1). Time-series vertical salinity and temperature (CTD) profiles were collected with an electrodeless induction salinometer (Kahlsico model RS5–3). All measurements were made as the cable was lowered from the surface to within 1 m of the bottom. Salinity measurements were rounded to the nearest 0.1‰ and temperature to the nearest 0.1°C. These measurements are sufficiently precise because of large variability in both salinity and temperature. Current velocities were measured by a Savonius rotor current meter (Interocean Systems Model 135R).

Sediment traps were transparent plastic cylinders 25.4 cm long and 4.7 cm diameter, and closed on one end with a plastic cap. Many field and laboratory studies have shown that this is the most effective trap design (Hargrave & Burns 1979, Gardner 1980, Blomqvist & Kofoed 1981, Butman 1986, Butman *et al.* 1986). Traps were moored from a line stretched vertically between an anchor and a submerged float suspended below wave base. Further description of field and laboratory methods is presented in Cowan (1988).

Subglacial discharge

McBride Glacier is drained by one main meltwater stream that discharges from a subglacial tunnel into the fjord. The high suspended sediment load of the stream indicates that the discharge is subglacial rather than englacial.

The subglacial discharge appears as a turbid upwelling at the water surface near the glacier (Fig. 3). An embayment in the ice cliff usually coincides with an upwelling because meltwater enhances calving of the tidewater front (e.g. Sikonia & Post 1980). A hemispherical cave can form in the ice cliff adjacent to the upwelling but it is unstable and will be calved away over short periods.

Subglacial discharge is buoyant with respect

Table 1. *Time series studies in McBride Inlet referred to in this paper*

Station Location (see Fig. 2)	Date/Time, interval of sampling	Water Column Data	Depths of Samples (m)	Depths of Traps (m)
D	July 19–20, 1985 1000–1000, 2 h	Salinity, Temp.	0, 40	40 m above bottom, 0.5 m above bottom
C	June 6, 1986 0730–1530, 1 h	Salinity, Temp.	4, 20, 1 m above bottom	5 m above bottom
A	June 18, 1986	–	13 samples throughout water column	–
B	June 21–22, 1986 0645–0900, 1 h	Salinity, Temp., Current Vel.	0, 4, 20, 5 m above bottom	5 m above bottom
A	July 12, 1986	Salinity, Temp.	20 samples throughout water column	–
A	June 22, 1987 1300–2000, 1 h	–	0, 2, 4, 6, 10, 15 20, 25, 35, 45, 60	–

to the ambient fjord water. Meltwater is discharged from a submarine tunnel into the base of the fjord water column as a turbulent jet. Because the momentum flux is small for most stream discharges relative to the buoyancy flux, the jet quickly transforms to a vertical jet and plume (Syvitski 1989, Powell 1990). This plume rises until it reaches a point of neutral buoyancy and spreads out laterally at this level, either at the surface or within the water column.

Fig. 3. McBride Glacier terminus showing the turbid upwelling on June 8, 1987. The iceberg-free zone at the terminus at the bottom of the photograph outlines the extent of the upwelling.

During a given sampling interval, the highest suspended sediment concentrations within McBride Inlet were measured from or near the upwelling. Ten suspended sediment concentrations collected from the surface of the upwelling between May and August averaged 730 mg/l. Concentrations from the upwelling increase progressively during summer (Fig. 4). Concentrations as high as 1.4 g/l were collected during late July and early August.

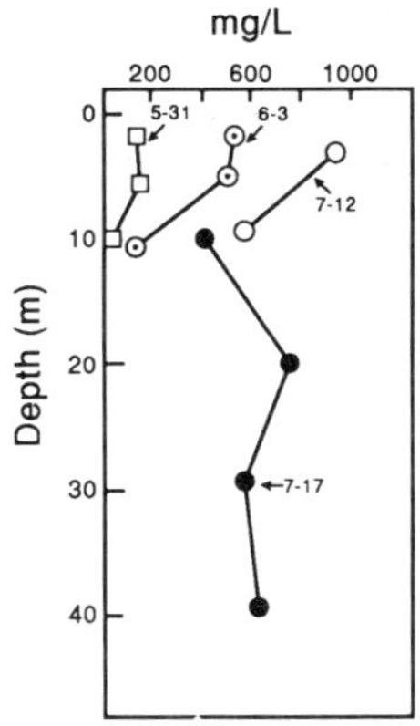

Fig. 4. Suspended sediment concentrations collected from the upwelling between May and July, 1986.

Turbid overflows

Suspended sediment is transported down fjord from the upwelling in a turbid, low-salinity overflow plume. Suspended sediment concentrations and observations of an upwelling near the glacier indicate that under usual conditions of meltwater discharge the plume rises to the surface to form an overflow. Turbid overflows occur above the pycnocline and range in thickness from 8 m to 20 m. Water samples from the overflow (with the exception of those collected from the surface) usually have suspended sediment concentrations greater than 100 mg/l within 1.5 km of McBride Glacier.

The vertical variation in total suspended sediment concentration and the sand concentration is shown in water samples collected from closely spaced depths at Location A (Fig. 2) on June 18, 1986 and July 12, 1986 (Fig. 5). Both profiles show that the maximum sediment concentration in the overflow occurred beneath the water surface, at either 3 or 10 m depth. On June 18, sediment concentrations were less than 450 mg/l and sand concentrations were less than 6 mg/l (Fig. 5). On July 12, concentrations of suspended sediment were greater than 500 mg/l and sand concentrations were as high as 79 mg/l (Fig. 5). The increase in sediment concentrations results from the seasonal increase in stream discharge. However, the presence of much higher suspended sediment concentrations at 0 m and a visible turbid plume on the water surface was caused by increased mixing of

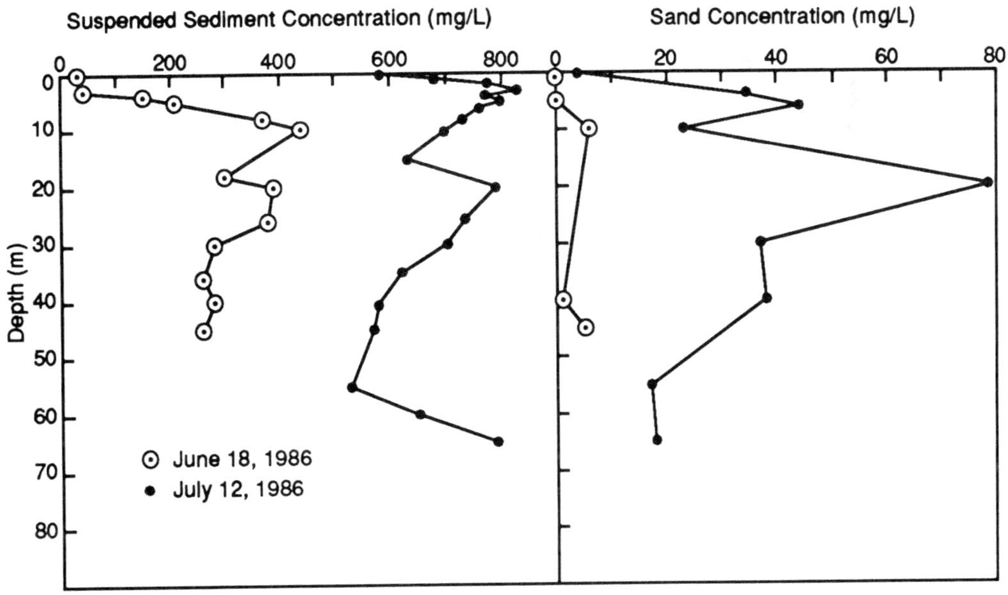

Fig. 5. Profiles of total suspended sediment concentrations (including sand) and sand concentrations from June 18 and July 12, 1986. Both profiles were collected from Location A (see Fig. 2 for station location).

the near surface layer by turbulence from katabatic wind shear. On June 18 the fjord surface was calm when the samples were collected. In contrast, the down fjord wind was strong when the surface sample was collected on July 12. By the late afternoon on windy days, highly turbid water completely covers the surface of McBride Inlet.

Although surface sediment is clearly visible under windy conditions, it is a minor component in comparison with deeper transport in the overflow. Salinity and temperature profiles indicate that the wind-blown surface plume is less than 1 m thick. Although total suspended sediment concentrations are high, the wind-blown surface plume transports little sand (Fig. 5). Much higher sediment concentrations always exist lower in the overflow regardless of wind activity. Edges of the surface plume are sharp and vorticies are shed into the surrounding water. Since surface plumes are subjected to constant mixing from waves and wind they do not maintain their integrity for long distances (Wright 1971).

Horizontal current velocities within the overflow are weak and they respond to tidal frequency (cf. Reed 1988). During flood tide, current velocities in the overflow are highest above 10 m; during ebb tide the highest velocities are deeper in the overflow, between 10 m and 20 m depth (Fig. 6). The horizontal suspended sediment flux (g m^{-2} s^{-1}) was calculated for one 24 h period from measurements of current velocities and suspended sediment concentrations at several depths (Fig. 7).

The horizontal sediment flux in the surface

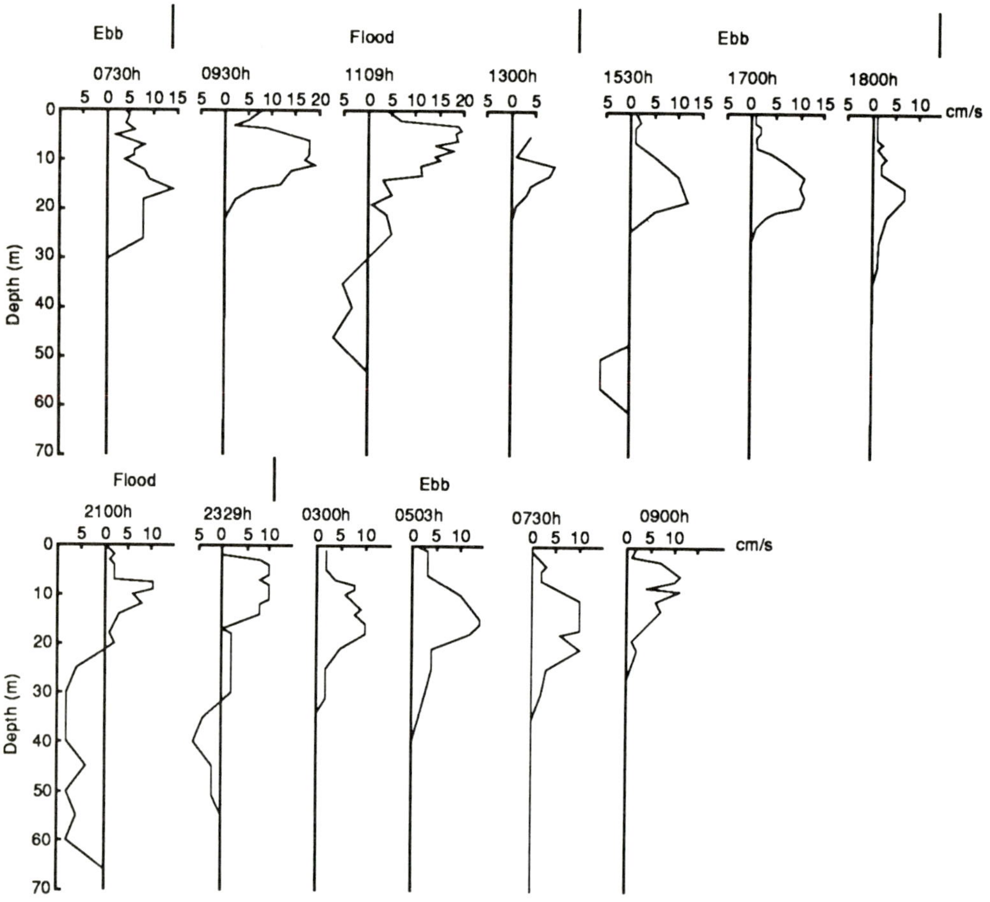

Fig. 6. Current velocity profiles from Location B on June 21–22, 1986. Down-fjord direction is to the right; up-fjord is to the left.

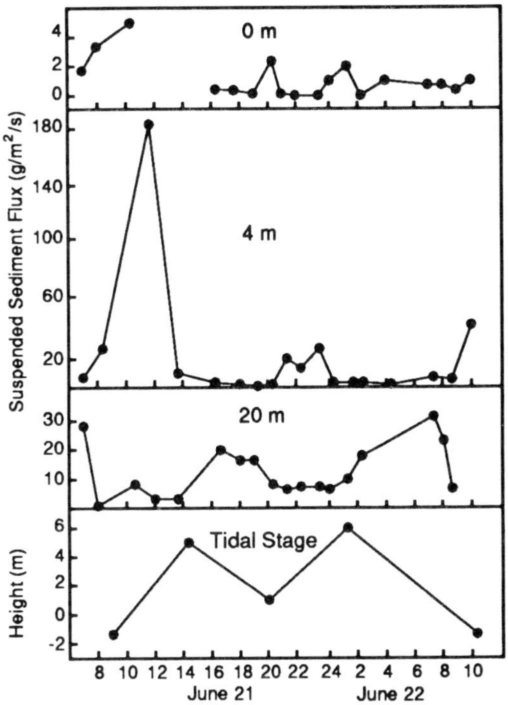

Fig. 7. Horizontal suspended sediment flux ($g\ m^{-2}\ s^{-1}$) at Location B on June 21 and June 22, 1986 during spring tides.

layer was variable. On average the flux at 4 m exceeded that at 20 m depth (Fig. 7). Within the upper 20 m, the horizontal suspended sediment flux varied with tidal frequency. Generally, the horizontal flux at 0 m and 4 m was greater during flood tide than during ebb tide (Fig. 7). At 20 m depth, this pattern was reversed and the horizontal flux was greater during ebb than flood tide. The highest horizontal flux of suspended sediment occurred at depths where current velocities were highest.

The horizontal suspended sediment flux was greater at 0 m and 4 m depth than at 20 m depth, resulting in sediment by-pass of the monitoring station (Location B, Fig. 2). This was especially noticeable when wind influenced suspended sediment transport at the surface. During periods when wind was not present, suspended sediment concentrations in the overflow decreased very rapidly with distance from the glacier. For example, in a transport distance of 1 km concentrations were reduced by more than an order of magnitude. This observation, along with large catches in sediment traps, indicates that the vertical flux was high. Sediment flux beneath 20 m depth is dominantly vertical because of low horizontal current velocities (Fig. 6).

Characteristics of particle settling

Particle release from the overflow plume

Particle transport and release from the overflow plume is controlled by tidal fluctuations. During flood tide, sediment concentrations are highest in the upper part of the surface layer because of relatively high current velocities. Weak horizontal current velocities at slack high water, and reduced vertical eddy velocities, allow even flocs of clay sized particles to settle deeper within the overflow to between 10 and 20 m depth where they are transported during ebb (Fig. 8). Then, at slack low water horizontal velocities decrease below 15 m depth allowing both sand and finer particles to be released from the overflow to settle to the fjord floor.

This model explains the variation of vertical flux with tidal frequency observed in sediment traps and how so much silt and clay are deposited within McBride Inlet rather than being transported for longer distances. Very fine sand can be transported by the surface layer beyond 400 m (Location A). These sand-size particles settle as single grains and finer particles settle in turbid layers.

Settling rates of turbid layers and coarse particles

Particle release from the overflow at slack low water and the subsequent settling of a discrete turbid layer in McBride Inlet is shown by sequential suspended sediment profiles (Fig. 9). During ebb tide on June 22, 1987, the highest near-surface concentrations occurred at 15 m depth (1328 h to 1529 h; Fig. 9). Near slack low water (1629 h profile), the concentrations at 15 m had decreased considerably as this turbid layer settled from the overflow. Its descent can be traced during flood tide by following the high suspended sediment concentrations at 20 m (1740 h), 25 m (1830 h) and 35 m (1928 h) in successive profiles (Fig. 9). Density boundaries do not control the depth of high suspended sediment concentrations in McBride Inlet. These turbid layers are distinct entities with settling velocities determined by Stokes' Law once they are released from the base of the overflow plume.

Time series measurements by Burrell (1971), Hoskin *et al.* (1976), and Hoskin & Burell

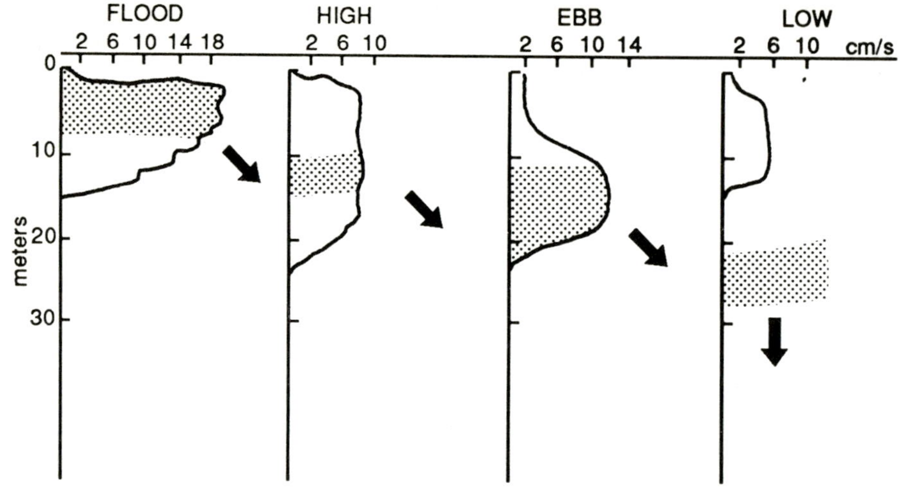

Fig. 8. Sequence of particle transport and release from the overflow in McBride Inlet. Shaded area represents the zone of highest suspended sediment flux. This flux is dominantly horizontal until low tide when settling from the base of the overflow occurs.

(1972) documented turbid layers within the water column of Queen Inlet, a turbid outwash fjord in Glacier Bay. These turbid layers did not correspond to pycnoclines and settled at rates between 3 and 10 m/h. Each layer was interpreted to be an individual flocculation of the surface plume initiated during flood tide (Burrell 1971).

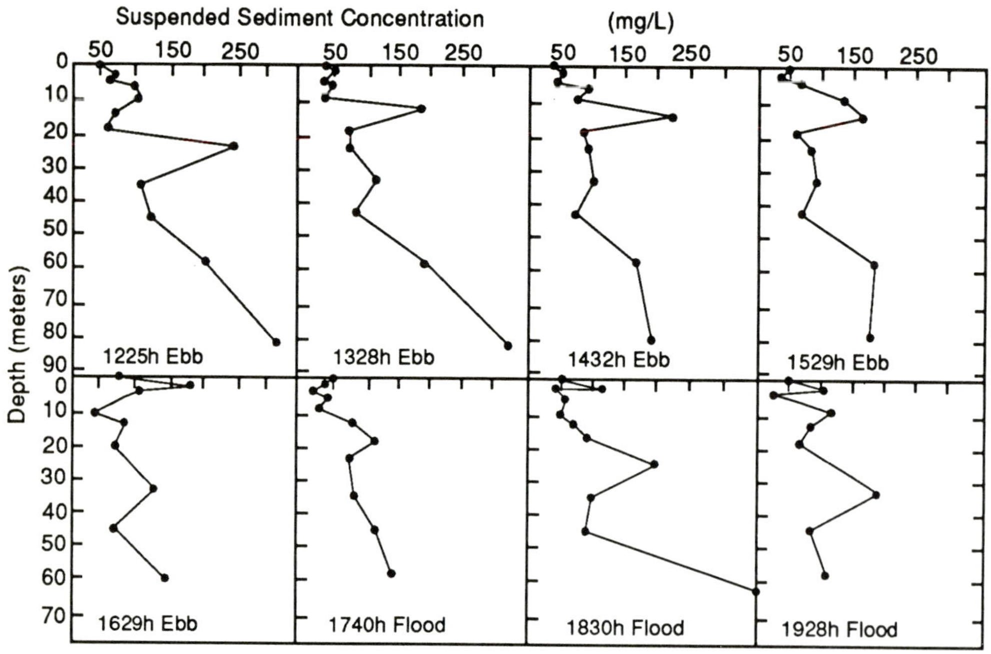

Fig. 9. Profiles of suspended sediment concentration on June 22, 1987. The time that the first water sample was collected in each profile, and its position in the tidal cycle, is listed.

Approximate rates of vertical descent of turbid layers in McBride Inlet were calculated directly from suspended sediment profiles. The average observed settling rate was 5.1 m/h; the range from 2.5 m/h to 10.9 m/h may be due to the resolution of sampling. Comparison of observed turbid layer settling velocities in McBride Inlet and Queen Inlet with theoretical particle settling rates, indicates that turbid layers settled within the range of coarse silt-sized particles (Officer 1981).

Particles are sorted during settling by the behaviour of different particle sizes. The average size distribution of sediment collected in traps 400 m from the glacier (at Location A, Fig. 2) is 10% sand, 75% silt and 15% clay. Sand and coarse silt are released from the overflow with finer particles, but the former settle more rapidly. This is shown by a plot of sand concentration versus total sediment concentration from the samples collected on June 22, 1987 (Fig. 10). Samples with high total sediment concentration usually do not contain a greater proportion of sand than those with lower concentrations. Since turbid layers are identified by high sediment concentrations, this indicates that sand separates from finer particles (the turbid layer) during settling.

Of the finer particles forming the turbid layers, the clays must aggregate and flocculate because of the high suspended sediment concentrations and salinities greater than 3‰ (Syvitski *et al.* 1985). The clays agglomerate and settle coherently as turbid layers with an effective fall diameter in the silt size range. The initiation of floc settling usually results in a decrease in the total sediment concentration in the overlying water column (Kranck 1986), but in McBride Inlet, sediment newly advected into the overflow is released during the next slack low water.

Deposition of cyclically interlaminated sediment

During the summers of 1985 and 1986, data on salinity, temperature, and suspended sediment concentrations in the water column, and samples of suspended sediment from traps were collected simultaneously at anchored stations. The aim was to study the vertical flux of suspended particles under changing conditions of tide and stream discharge (Table 1). This type of time-series sampling has allowed us to describe the characteristics of laminae deposition. The major results of these studies follow.

The sand flux into near bottom sediment traps occurs at discrete intervals, as is predicted by the cyclopsam/cyclopel model for McBride Inlet (Fig. 11). The highest sand flux occurs within 2 h of low tide and the sand flux is lowest about 1 h after high tide. At Location C (Fig. 2), a maximum of 47% sand was collected in a trap deployed between 1004 h and 1103 h and a sand minimum of 12% was observed between 1510 h and 1610 h (Fig. 11). These traps also collected a large amount of very fine sand relative to mud. The high sand/mud ratio probably

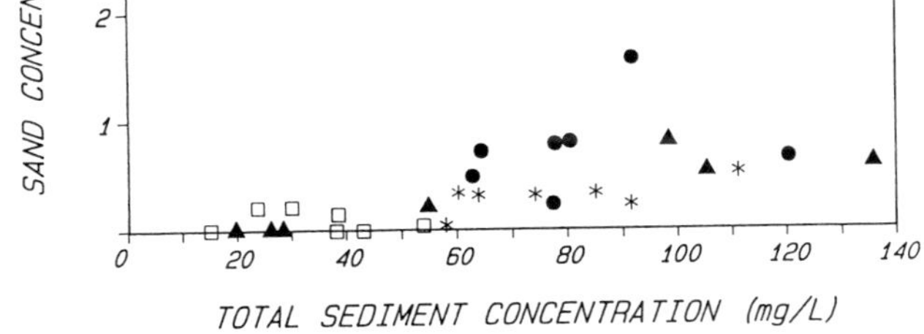

Fig. 10. Total suspended sediment concentration versus sand concentration from June 22, 1987. Symbols designate the depth of collection of the water sample.

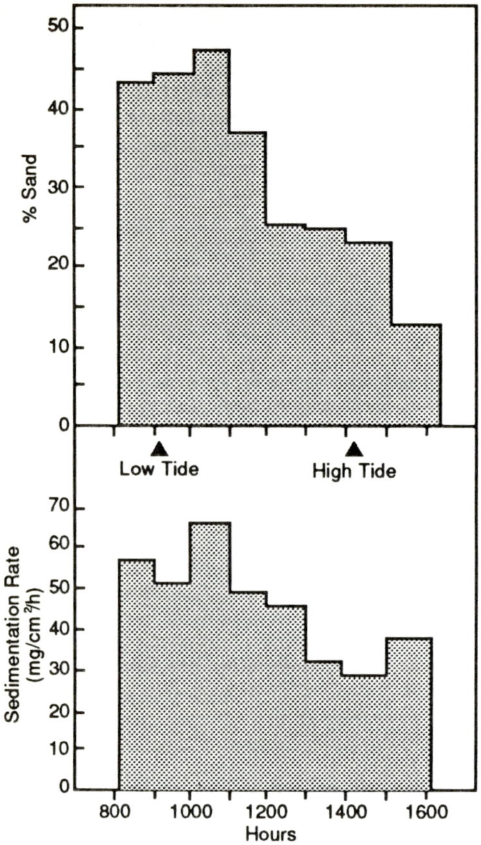

Fig. 11. Percent sand and sedimentation rates in traps from Location C (Fig. 2). Traps were moored 5 m above bottom. The time of high and low tide in McBride Inlet is indicated.

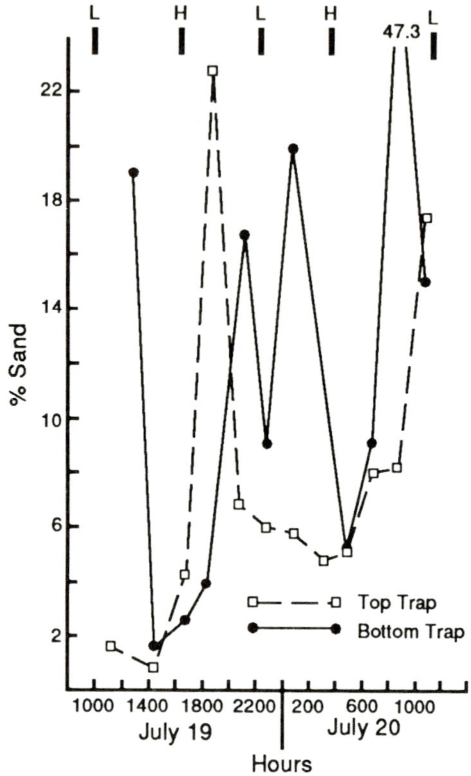

Fig. 12. Percent sand from top and bottom traps from Location D (Fig. 2) on July 19–20, 1985. The top trap (squares) was moored 20 m above bottom; the bottom trap (circles) was 0.5 m above bottom. Points are plotted at the time of trap retrieval and represent an average over the collection interval.

resulted from both the proximal location of the anchored station and large quantities of sand discharged early in the melt season. Sedimentation rates were variable but were highest (66 mg cm^{-2} h^{-1}) during the period of the greatest percent sand capture and were lowest four hours later, near high tide (Fig. 11).

Particles deposited in these laminae originate from the overflow and are deposited from suspension because peaks in the percent sand and sedimentation rate occur earlier in traps suspended in the water column above near bottom traps (Fig. 12). These sets of traps were suspended from the same line and were sampled at the same time. For example, a period of relatively high percent sand was measured (1640 h to 1824 h on July 19, 1985) in a top trap (Fig. 12). A peak in percent sand was then collected in the bottom trap during the next interval (1830 h to 2030 h). Peak sand fluxes in other bottom traps from this period could not be correlated with other top traps probably because of the timing of trap deployment and the length of the study.

Mean particle size of coarse tidally-controlled laminae collected from Location D (Fig. 2) is 6.5 to 6.7 phi (Fig. 13). These laminae are very poorly sorted (Folk 1980). Mean particle size of finer laminae is 7.1 to 7.5 phi and they are very poorly sorted, but comparatively slightly better sorted than the coarser laminae (Fig. 13). Increased sedimentation rates and coarser particle sizes collected in suspended sediment traps deployed closer to the stream discharge confirm the deposition of coarser and thicker laminae in proximal areas.

The particle size distribution of samples collected in sediment traps and the sedimentation

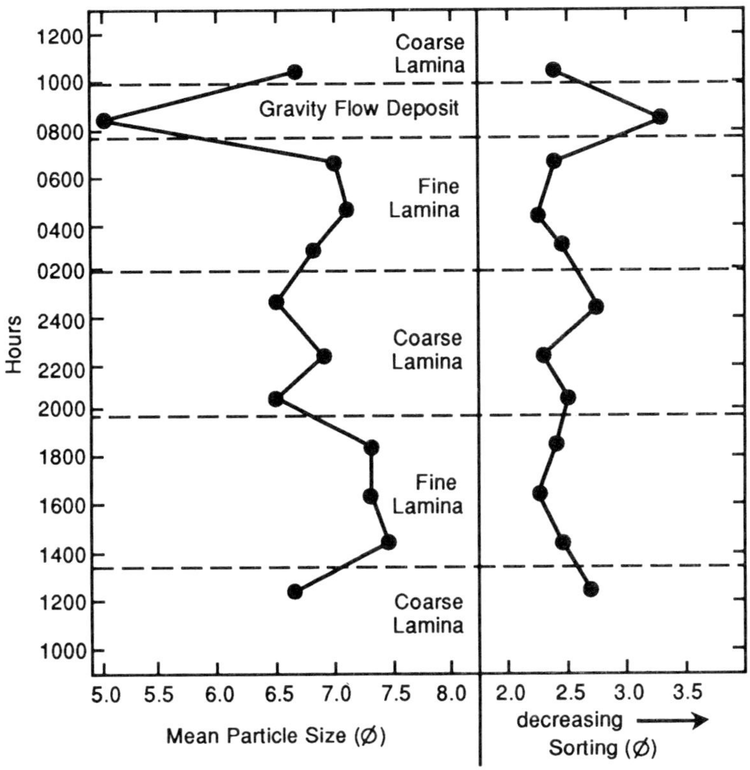

Fig. 13. Changes in mean particle size and sorting of samples collected in the bottom trap from Location D (Fig. 2) on July 19–20, 1985.

rate are related at Location D. For example, the sample in the bottom trap retrieved at 1230 h was skewed towards a coarse distribution and accumulated at a relatively rapid rate (Fig. 14). The next sample (1430 h) collected less sand and coarse silt but more clay at a lower rate. The sample collected at 1633 h had the most silt and the highest sedimentation rate of these three consecutive traps. This sequence documents the rapid transition from deposition of a coarse lamina (at 1230 h) to deposition of a fine lamina (1430 h to 1824 h). A similar sequence of particle size distributions and sedimentation rates was repeated in traps collected at 2434 h, 0320 h, and 0403 h (Fig. 14). These observations are explained by the settling model; the first trap in each sequence collected single grains of sand and coarse silt that settled rapidly from the overflow plume at slack low tide. The next traps in the sequence (1430 h and 0320 h) received less sand and had an uneven distribution of silt. At both 1633 h and 0403 h, turbid layers were collected in the traps. The turbid layer contributes fine particles, as shown by an even distribution of silt and high percent clay relative to sand, and increases the sedimentation rate.

Coarse laminae may be deposited in proximal areas independently of tidal control. On July 19, 1985, two intervals of relatively high percent sand input were measured in the top sediment trap (Fig. 12). The first peak in percent sand (1640 h to 1824 h on July 19) did not coincide with particles settling from slack low tide and could have been introduced by high stream discharge. Peak discharge was measured that day in a proglacial stream at Riggs Glacier, 3 km to the North of McBride Inlet, at about 1700 h (N.D. Smith, pers. comm.). The sand collected in this top trap must have fallen through the surface layer independently of tidal dynamics because it was collected following high tide and not following low tide as is predicted by overflow plume dynamics. This peak in per cent sand was then observed in the bottom trap during the next interval (1830 to 2030 h; Fig. 12). This settling rate agrees with that predicted by Stokes' Law for very fine sand. The dispersed particle size distribution for this sediment trap

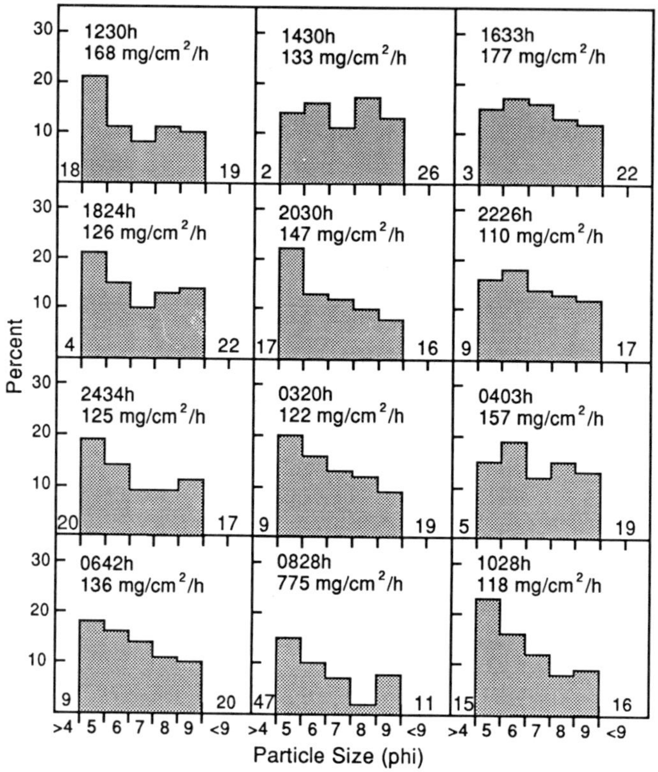

Fig. 14. Histograms of particle size distribution of samples collected from bottom sediment traps from Location D, (Fig. 2) July 19–20, 1985. Numbers above the horizontal axis of each graph indicate the percent greater than 4 phi and percent less than 9 phi, respectively. The time of trap retrieval and sedimentation rate is given for each histogram.

indicates a mode in the very fine sand and coarse silt range (2030 h; Fig. 14).

Sediment gravity flow deposits are also interstratified with rhythmites deposited by suspension settling. A sample collected in the bottom trap at 0820 h on July 20, 1985, is interpreted as a sediment gravity flow deposit because of the high percentage of angular coarse sand (62%) and poor sorting (Figs 12 & 13). This trap was moored 0.5 m above the bottom so it is possible that it intercepted a gravity flow. Ice-rafted debris could also produce coarse laminae but a top trap suspended 19.5 m above the bottom trap did not capture unusually coarse sand during this period.

An evaluation of the model of cyclopsam and cyclopel deposition

The model of suspended sediment transport and release of particles from the overflow plume is based on horizontal sediment fluxes (from June 1986) within 1 km of McBride Glacier. Water column measurements and sediment traps sampled in time-series provide independent verification of this model (Table 1). These studies yield general predictions of both frequency and characteristics of the resulting cyclically interlaminated sediments (Fig. 15).

All particles in one couplet should be released from the bottom of the near-surface layer at each low tide. A smaller percentage of mud than sand plus coarse silt is deposited when single grains settle to the fjord floor at a faster rate than flocs of finer silt and clay. At a later time, turbid layers of flocculated fine silts and clays reach the bottom, depositing mud and increasing the sedimentation rate (Fig. 14).

Particle size histograms of sediment trap samples 350 m from the stream discharge show particle size variations during cyclopsam deposition (Fig. 14). As predicted by particle settling rates, fine to very fine sand settled into near

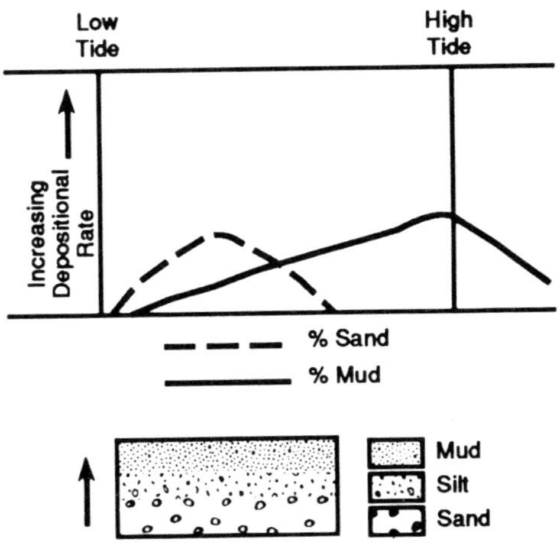

Fig. 15. Deposition of an ideal cyclopsam or cyclopel in McBride Inlet.

bottom traps soon after low tide, and high rates of mud deposition were recorded in traps about six hours later. This timing is relative, however, because deposition on the fjord floor depends on water depth and thickness of the overflow plume. In addition to fine particles settling in layers, these particles also occur at a background level throughout the water column due to their slow settling velocity. Particles within this background are deposited continuously and are, therefore, present within the coarse lamina of a cyclopsam or cyclopel couplet (Fig. 15). This produces poorer sorting than in the fine lamina because of the wider range of particle sizes deposited in coarse lamina (sand through clay sizes). Proximal coarse couplets collected in sediment traps 300 m from the stream discharge (Location C) are better sorted than more distal laminae because the sand/mud ratio is higher. Sand laminae have sharp basal contacts because the coarsest grains settle uniformly onto the fjord floor.

Laminae that are not tide-related may be deposited in areas close to stream discharges. For example, one such lamina, attributed to peak stream discharge, was recorded 350 m from the glacier (Location D; Fig. 13). Most likely, these lamina will be produced only in proximal areas and will contribute an additional coarse lamina to the two tidal laminae when peak discharge does not coincide with low tide. At increasing distances from subglacial stream effluxes, tidal control on the overflow dominates the discharge effects. Coarser sand sizes are deposited close to the efflux and finer sand and silt are transported farther down-fjord in the overflow because the eddy velocity is sufficient to keep them in suspension longer than the coarser particles. As a result, two laminae are deposited each day, if the tidal cycle is diurnal.

Discussion

This study has documented how cyclically interlaminated sediment (rhythmites) can be deposited from suspension in an ice-proximal glacimarine environment. Cyclopels and cyclopsams, first described by Mackiewicz *et al.* (1984) from Muir Inlet, have been collected in sediment traps within McBride Inlet at a rate of two each day. In this macrotidal fjord, suspended sediment transport and settling from the thick overflow is controlled by tidal fluctuations. At each slack low water a pulse of suspended sediment is released from the base of the surface layer and begins its vertical descent. Single grains, such as sand and coarse silt, settle rapidly and the flocculated finer particles settle more slowly as a discrete turbid layer. These layers were identified in water samples from closely spaced depths. Turbid layers settle at rates between 2.5 m/h and 10.9 m/h. Each low water period produces a couplet: a coarser grained lamina that is sorted as coarser single grains settle, and a finer grained lamina of flocculated particles in turbid layers. Individual couplet thickness and particle size decrease with distance from the stream discharge. Horizontal current velocities and turbulence maintain very fine sand in suspension in the surface layer for a distance of approximately 1 km.

Although tidal fluctuations are the major control on rhythmic sedimentation in McBride Inlet, discharge dynamics also play a role. Additional coarse laminae are produced by peak daily meltwater discharges within 350 m of the glacier. The ability of regular discharge events to produce cyclic glacimarine deposits is significant because Pleistocene cyclopsams and cyclopels have been described from areas with tides that may have been smaller than in Glacier Bay. For example, ice-proximal rhythmic glacimarine sediments on Whidbey Island, Washington, were attributed to seasonal fluctuations of meltwater and suspended sediment load (Domack 1984). Cyclopels and cyclopsams were described from Pleistocene deposits on Ellesmere Island, Arctic Canada, where the tidal range was estimated to be less than 0.5 m (Stewart 1988). Although they were deposited from turbid plumes in a setting similar to McBride Inlet (Stewart 1988), the depositional controls were probably not tidally related. Daily and seasonal variations in sediment discharge are most likely the cause of these cyclopsams and cyclopels. Laminated sediments deposited from meltwater plumes have also been described from cores collected in Baffin Island fjords (Andrews 1988).

Several critical factors for rhythmic sediment deposition occur in ice-proximal glacimarine environments, including: (1) very high suspended sediment loads in overflow plumes that are capable of depositing individual laminae thick enough to be observed, (2) high concentrations of rock flour (silt) in subglacial meltwater produce highly concentrated flocculated turbid layers that can settle even in very proximal locations, (3) low fjord water temperatures improve sorting by delaying settling of fine particles without interfering with settling of coarser grain sizes (Fraser 1929), and (4) high ice-proximal turbidity limits infaunal communities which improves laminae preservation by reducing bioturbation.

These observations suggest that cyclopsams and cyclopels should be abundant in ice-proximal temperate and sub-polar glacimarine settings because conditions for their formation are optimal.

Research was supported by National Science Foundation grants DPP-8420749/DPP-8642364 to Powell and by grants from the Geological Society of America, American Association of Petroleum Geologists, and the Goldich Fund of the Department of Geology, Northern Illinois University to Cowan. We thank the Superintendent and Resource Manager of Glacier Bay National Park for allowing us to work there. We are grateful for the essential logistical support that was provided by Captain James Luthy and Brad Carlquist (M/V *Nunatak*), and the crew of the *Thunder Bay*. We are indebted to Keith Seramur, Jim Duncker, and Steve Mattox for assistance in collecting the field data. We also wish to thank Norman Smith, Andrew Phillips and Thomas Stewart for fruitful discussions. We are grateful to E.W. Domack and R. Gilbert for their constructive reviews of the manuscript.

References

ANDREWS, J.T. 1988. Sediment flux into McBeth, Itirbilung, and Tingin Fiords, Home Bay, Baffin Island, N.W.T. over the last glacial/interglacial transition. *American Quaternary Association Program and Abstracts*, tenth biennial meeting, 6–8 June 1988, 103.

BLOMQVIST, S. & KOFOED. C. 1981. Sediment trapping — a subaquatic *in situ* experiment. *Limnology and Oceanography*, **26**, 585–590.

BURRELL, D.C. 1971. Suspended sediment distribution patterns within an active turbid outwash fjord. *Port and Ocean Engineering under Arctic Conditions*, **1**, 227–245.

BUTMAN, C.A. 1986. Sediment trap biases in turbulent flows: results from a laboratory flume study. *Journal of Marine Research*, **44**, 645–693.

——, GRANT, W.D. & STOLZENBACH, K.D. 1986. Predictions of sediment trap biases in turbulent flows: a theoretical analysis based on observations in the literature. *Journal of Marine Research*, **44**, 601–644.

COWAN, E.A. 1988. *Sediment transport and deposition in a temperate glacial fjord, Glacier Bay, Alaska*. Ph.D. Thesis, Northern Illinois University.

——, POWELL, R.D. & SMITH, N.D. 1988. Rainstorm-induced event sedimentation at the tidewater front of a temperate glacier. *Geology*, **16**, 409–412.

DOMACK, E.W. 1984. Rhythmically bedded glaciomarine sediments on Whidbey Island, Washington. *Journal of Sedimentary Petrology*, **54**, 589–602.

ELVERHØI, A., LIESTØL, O. & NAGY, J. 1980. Glacial erosion, sedimentation and microfauna in the inner part of Kongsfjorden, Spitsbergen. *Norsk Polarinstitutt, Skrifter*, **172**, 33–61.

——, LØNNE, O. & SELAND, R. 1983. Glaciomarine sedimentation in a modern fjord environment, Spitsbergen. *Polar Research*, **1** (new series), 127–149.

EYLES, C.H., EYLES, N. & MIALL, A.D. 1985. Models of glaciomarine sedimentation and their application to the interpretation of ancient glacial sequences. *Palaeogeography, Palaeoclimatology, Palaeoecology*, **51**, 15–84.

EYLES, N., EYLES, C.H. & MIALL, A.D. 1988. Glacial facies models, Penrose Conference report. *Geology*, **16**, 374–375.

FOLK, R.L. 1974. *Petrology of Sedimentary Rocks*. Hemphill Publishing Company, Austin, Texas.

FRASER, H.J. 1929. An experimental study of varve deposition. *Transactions of the Royal Society of Canada*, **23**, 49–60.

GARDNER, W. 1980. Sediment trap dynamics and calibration: a laboratory evaluation. *Journal of Marine Research*, **38**, 17–39.

GILBERT, R. 1983. Sedimentary processes of Canadian Arctic fjords. *Sedimentary Geology*, **36**, 147–175.

GÖRLICH, K. 1986. Glacimarine sedimentation of muds in Hornsund Fjord, Spitsbergen. *Annales Societatis Geologorum Poloniae*, **56**, 433–477.

HARGRAVE, B. & BURNS, W. 1979. Assessment of sediment trap collection efficiency. *Limnology and Oceanography*, **24**, 1124–1136.

HOSKIN, C.M. & BURRELL, D.C. 1972. Sediment transport and accumulation in a fjord basin, Glacier Bay, Alaska. *Journal of Geology*, **80**, 539–551.

——, BURRELL, D.C. & FREITAG, G.R. 1976. Suspended sediment dynamics in Queen Inlet, Glacier Bay, Alaska. *Marine Science Communications*, **2**, 95–108.

——, —— & —— 1978. Suspended sediment

dynamics in Blue Fjord, Western Prince William Sound, Alaska. *Estuarine and Coastal Marine Science*, **7**, 1–16.

KRANCK, K. 1986. Settling behavior of cohesive sediment. *In*: Mehta, A.J. (ed.) *Estuarine Cohesive Sediment Dynamics*. Springer, New York, 151–169.

MACKIEWICZ, N.E., POWELL, R.D., CARLSON, P.R. & MOLNIA, B.F. 1984. Interlaminated ice-proximal glacimarine sediments in Muir Inlet, Alaska. *Marine Geology*, **57**, 113–147.

MOLNIA, B.F. 1983. Subarctic glacial-marine sedimentation: a model. *In*: MOLNIA, B.F. (ed.) *Glacial-Marine Sedimentation*. Plenum Press, New York, 95–144.

OFFICER, C.B. 1981. Physical dynamics of estuarine suspended sediments. *Marine Geology*, **40**, 1–14.

POWELL, R.D. 1981. A model for sedimentation by tidewater glaciers. *Annals of Glaciology*, **2**, 129–134.

—— 1983. Glacial-marine sedimentation processes and lithofacies of temperate tidewater glaciers, Glacier Bay, Alaska. *In*: MOLNIA, B.F. (ed.) *Glacial-Marine Sedimentation*. Plenum Press, New York, 185–232.

—— 1990. Glacimarine processes at grounding-line fans and their growth to ice-contact deltas. *In*: DOWDESWELL, J.A. & SCOURSE, J.D. (eds) *Glacimarine Environments: Processes and Sediments*, Geological Society, London, Special Publication, **53**, 53–73.

—— & MOLNIA, B.F. 1989. Glacimarine sedimentary processes, facies and morphology of the south-southeast Alaska shelf and fjords. *Marine Geology*, **85**, 359–390.

REED, D.J. 1988. Tidal currents and glacial discharge, Laguna San Rafael, Southern Chile. *Journal of Coastal Research*, **4**, 93–102.

SIKONIA, G.W. & POST, A. 1980. Columbia Glacier, Alaska: recent ice loss and its relationship to seasonal terminal embayments, thinning, and glacial flow. *U.S. Geological Survey Hydrologic Investigation Atlas* HA-619, 3 sheets.

STEWART, T.G. 1988. *Deglacial-marine sediments from Clements Markham Inlet, Ellesmere Island, N.W.T., Canada*. Ph.D. Thesis, University of Alberta, Edmonton.

SYVITSKI, J.P.M. 1989. On the deposition of sediment within glacier-influenced fjords: oceanographic controls. *Marine Geology*, **85**, 301–329.

—— ASPREY, K.W., Clattenburg, D. & Hodge, O.D. 1985. The prodelta environment of a fjord: suspended particle dynamics. *Sedimentology*, **32**, 83–107.

——, BURRELL, D.C. & SKEI, J.M. 1987. *Fjords: Processes and Products*. Springer, New York.

WRIGHT, F.F. 1971. Suspended transport in Southern Alaska coastal waters. *Offshore Technology Conference*, 235–242.

Laminated terrigenous sediments from the Antarctic Peninsula: the role of subglacial and marine processes

EUGENE W. DOMACK

Hamilton College, Clinton, New York 13323, USA

Abstract: Although laminites from the continental slope and rise are known from a number of regions in Antarctica, laminated muds on the continental shelf have not been widely recognized. During USAP-88 cruise III of the RV *Polar Duke*, a number of piston cores were collected from bays and fjords along the Danco Coast of the Antarctic Peninsula. A fine-grained sedimentary sequence of thin-bedded to randomly laminated mud and sand with dropstones is underlain by a rhythmically laminated sequence of silt and clay. The deposits were collected within a submarine valley (450 m deep) which appears to extend beneath the terminus of a tidewater glacier (Cayley Glacier). The glacial regime of the area today is polar to sub-polar with restricted ice surface melting. Fine-scale horizontal layers are observed in the water column and these transport fine-grained sediment at mid-water depths. Such features are not related to surface meltwater processes, but may be caused by tidal pumping of the basal cavity. The random bedding thickness of the sediments is related to the present depositional environment which includes iceberg rafting, bioturbation, pelagic settling of biogenic detritus, and sediment gravity flows. The rhythmic units likely represent subglacial marine deposition within the valley; possibly under the influence of tidal processes.

Laminated glacimarine sediments are important from a number of perspectives including the fact that the style and nature of lamination often reflects subtle details about the depositional processes and the climatic variables which may control deposition (Gilbert 1982; Domack 1984; Mackiewicz *et al.* 1984; Cowan *et al.* 1988). To date, studies in modern Antarctic settings have recognized two general regimes under which such deposits may be formed. Laminated glacimarine muds have been reported from regions of the continental slope and rise (Anderson *et al.* 1982; Hampton *et al.* 1987) and from ancient slope sediments recovered from the Antarctic margin through deep drilling (Barron, Larsen *et al.* 1989). In the case of the former, laminated silts and clays are thought to be the product of contour currents (of geostrophic origin). However, the stratigraphic facies relationship of such sediments to glacier fluctuations and/or terrigenous sediment supply is not clear. In contrast, laminated muds on the Antarctic continental shelf have been found to consist primarily of biogenic laminae, with a dominance of biosiliceous (diatom) components rather than terrigenous debris (Domack 1988a). Hence, productivity models rather than fluctuating terrigenous sediment supply (in the polar setting) are used to explain these Holocene biogenic laminites.

This paper reports on the occurrence of laminated terrigenous muds from the Antarctic Peninsula and discusses a unique combination of glacial and marine processes which are believed to control deposition in a polar to sub-polar climate. The facies context of these sediments as ice-proximal (in the strictest sense) has important implications regarding the interpretation of ancient glacimarine sequences as well as the glacial stratigraphic record along the Antarctic Peninsula.

Methods

The results presented here are based on field data collected during the 1987–1988 United States Antarctic Program, cruise III of the RV *Polar Duke*. The cruise programme included an oceanographic and marine geological survey of fjords along the Antarctic Peninsula and South Shetland Islands, the details of which are reported in Domack (1988b).

Bathymetry

The bathymetry in Fig. 1 is taken from a combination of U.S. Defense Mapping Agency, Chilean charts, and 12 kHz precision depth records obtained on-board the *Polar Duke*. Resolution of the coastline is consistent with earlier reports of Hattersley-Smith & Thomson (1988). The bathymetry is well resolved due to the crossing of several ship tracks during the investigation of the bay. Glacier flow direction and ice terminus position are determined from LANDSAT imagery acquired in 1978 and 1986.

From DOWDESWELL, J. A. & SCOURSE, J. D. (eds), 1990, *Glacimarine Environments: Processes and Sediments*. Geological Society Special Publication No 53, pp 91–103.

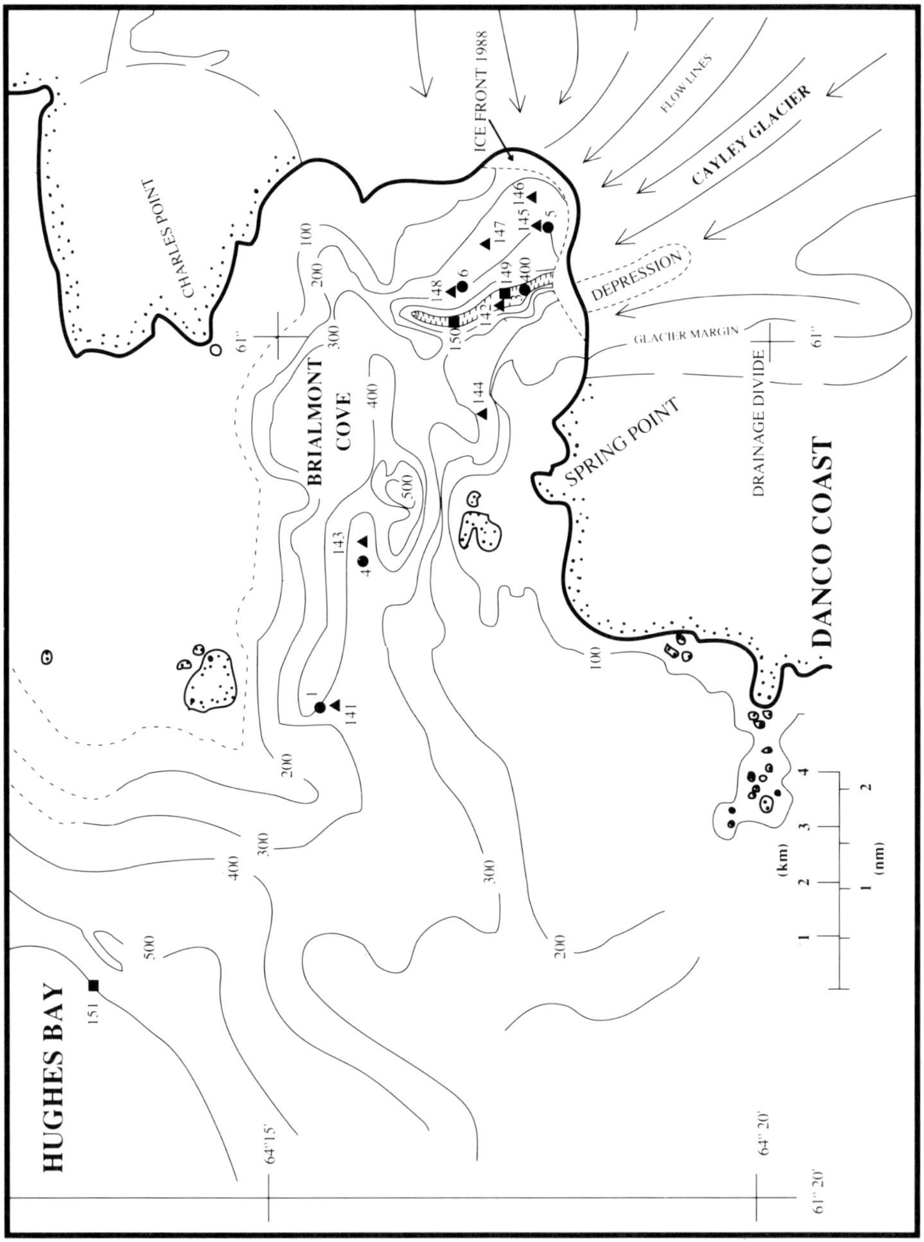

Fig. 1. Bathymetry and coastline features of Hughes Bay in the vicinity of Brialmont Cove, Danco Coast, Antarctic Peninsula. Sample locations are indicated by triangles (surface grabs), squares (piston cores) and dots (hydrographic stations).

Field observations and the LANDSAT data indicate that the terminus is somewhat further seaward than recorded by coastal charts. A linear depression which parallels glacier flow was also observed in both of the LANDSAT images. Thus the depression is a persistent feature of the glacier terminus region (Fig. 1).

Sediment and Water Sampling

Bottom sediment samples were obtained using either Smith McIntyre grabs of the surface and near surface (15 cm) sediment or modified piston cores of up to 6 m in length.

Hydrographic stations consisted of vertical casts for temperature, conductivity (salinity), pressure (depth) and transmissivity using a SEABIRD model SBE-19 profiler with a SEATECH model ST010-A transmissometer. The instrument package was lowered at a minimum rate of 0.75 to 0.5 ms^{-1}. Data processing is discussed further in Domack & Williams (1990). Water bottle samples were collected at selected stations from 10 m and near bottom depths. Water samples were filtered using 0.4 micron polycarbonate filters according to standard procedures. Beam attenuation coefficients were calculated from the transmissometer data according to relationships discussed by Bishop (1986).

Sediment analyses

Sediment samples were analyzed for the overall weight proportions of gravel, sand and mud using standard wet sieving techniques. Total organic carbon was determined, after first measuring the amount of calcium carbonate, by combustion in an induction furnace (LECO).

Observations

Bathymetry

The sea floor within Brialmont Cove is marked by a conspicuous flat-bottomed valley which has a uniform depth of 450 m (Fig. 2). The trend of the valley was traced to within 50 m of the

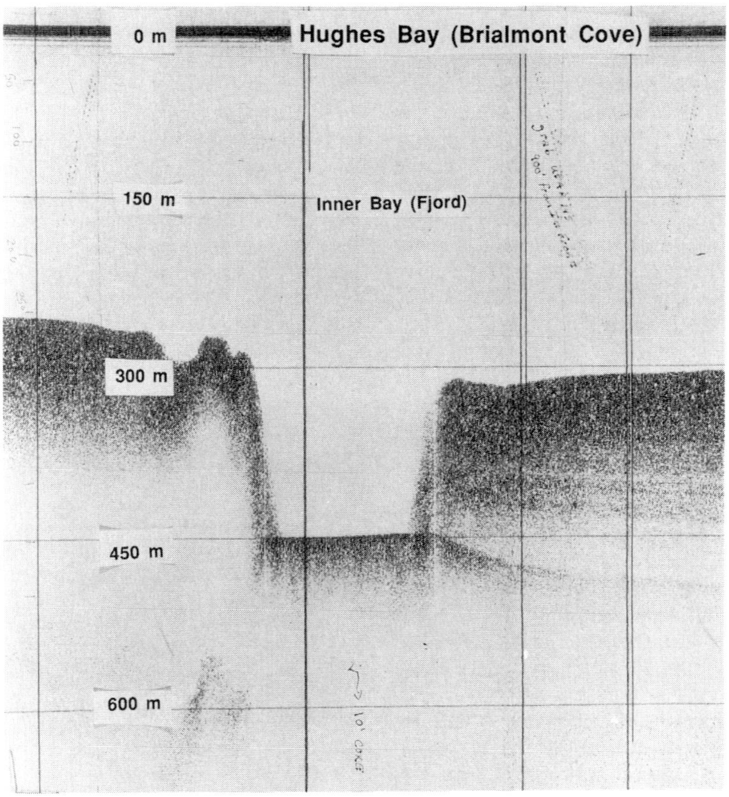

Fig. 2. Photograph of precision depth record across submarine valley system in the inner reaches of Brialmont Cove (see Fig. 1 for location).

glacier terminus (the closest approach by the *Polar Duke*) and, thus, it appears to continue beneath the glacier. Towards the outer bay, a linear bank marks the opposite side of the valley. The valley appears to terminate against this bank to the north. Overall, the valley extends for approximately 5.5 km and is 0.5 km wide. Although the valley bottom appears to be perfectly flat there is a very slight slope of about 1 m km^{-1} away from the ice terminus, down the axis of the valley. Beyond mid-bay, the bottom of Brialmont Cove is marked by a sinuous series of discontinuous lows, some as deep as 500 m, and poorly defined cross-valley banks.

Glacial characteristics and climate

The head of Brialmont Cove receives ice drainage from the Cayley Glacier and an unnamed tributary, both of which drain the western slopes of the northern Antarctic Peninsula. Both systems have a combined drainage area of approximately 820 km^2. Thus, this glacier complex is one of the largest along the western side of the Antarctic Peninsula, as documented by Williams *et al.* (1989). The elevation at the calving front of the Cayley Glacier was determined at a number of stations by measuring angles sighted to the edge of the terminus, horizontal distances (from the *Polar Duke's* radar) and compensating for deck elevation. A height of approximately 48 m above sea level was determined for the calving ice front adjacent to the valley. Calculations of potential freeboard depths suggest the ice front should be grounded in the water depths found along most of the terminus. An exception is the region directly over the submarine valley. Here water depths of 450 m are at least 200 m deeper than the potential freeboard of a freely floating Cayley Glacier. If the Cayley Glacier was grounded in 450 m of water, then it should have a much greater terminus elevation above sea level.

The linear surface depression on the Cayley Glacier, being aligned with the trend of the submarine valley, appears to be the result of buoyant support and sagging of the glacier over the subglacial valley. Hence, the valley system is believed to extend an additional 2 km beneath the terminus of the Cayley Glacier. The expression of the 'tunnel valley' as a depression on the glacier surface contrasts with observations of meltwater-filled systems which actually cause glaciers to rise over the tunnel system due to hydraulic pressure (Pfirman & Solheim 1989).

Climatic conditions of the area are sub-polar to polar with mean summer temperatures reported to be just at 0°C (Fig. 3). Late summer snow lines were observed at elevations of several tens of metres above sea level; hence the glacier terminus lies just within the ablation zone. Because the lower reach of the Cayley Glacier has a relatively gentle slope (45 m km^{-1}), this system may be somewhat susceptible to short term climatic changes, with consequent increases or decreases in surface meltwater production. Basal ice temperatures are unknown but are assumed to be at pressure melting near the terminus since basal temperatures in glaciers further south are known to be at pressure melting (Smith 1972) despite englacial temperatures of less than −4.0°C for the area (Martin & Peel 1978).

Physical oceanography

The waters within Brialmont Cove are characterized by three basic regimes (Figs 4 & 5). Deep saline and clear waters derived from Hughes Bay dominate the outer and middle regions of the cove. At the head of the bay, and directly adjacent to the glacier, such waters are mixed with slightly less saline, colder and more turbid water. Mixing occurs along the boundaries of fine scale layers (5 to 25 m thick) which have a very complex structure discussed further by Domack & Williams (1990). The layers are associated with the submarine valley system where several such features were observed within or directly above the axis of the valley, adjacent to the ice terminus. Suspended sediment concentrations (2 to 3 mg l^{-1}) within the layers are significantly higher than deep outer bay waters (Fig. 6). The layers appear to attenuate rather quickly; within 4–5 km in a down-fjord direction (Fig. 7).

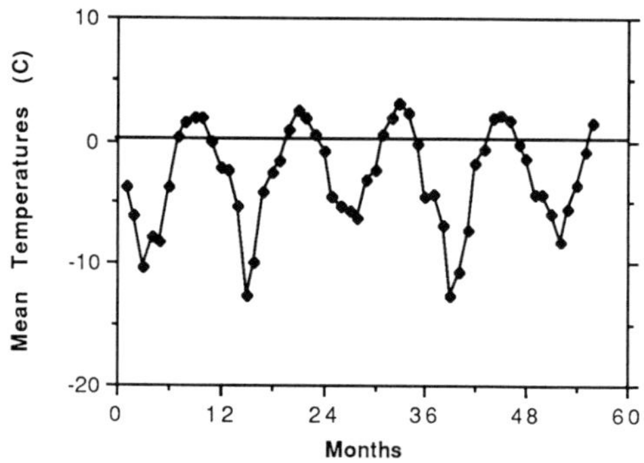

Fig. 3. Temperature trends as observed at the Argentine station *Primavera* located near Cierva Cove (northeastern Hughes Bay).

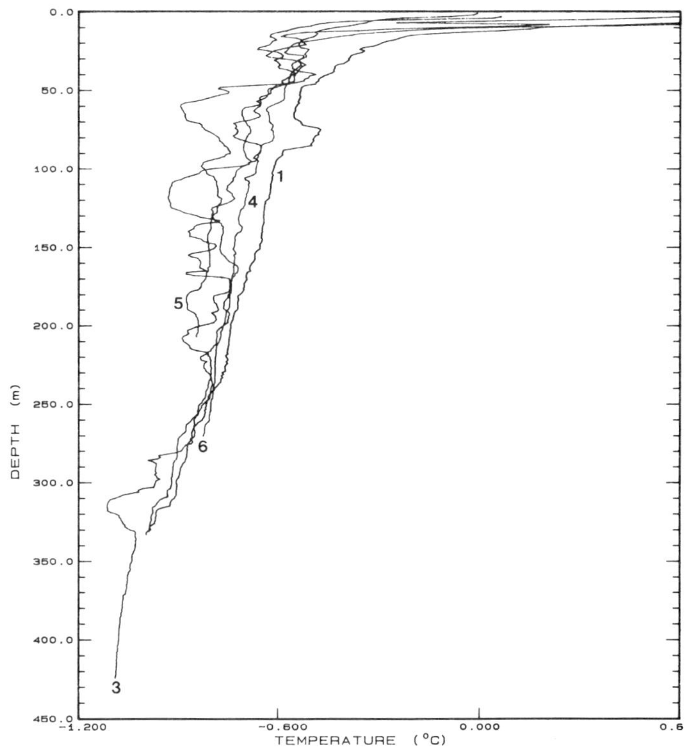

Fig. 4. Temperature casts made at CTDT stations in Brialmont Cove (see Fig. 1 for location).

A warm, low salinity surface layer occurs throughout the bay and is characterized by uniform salinity (Fig. 5) and a strong temperature gradient at depths less than 10 m. This indicates that the temperature gradient is due primarily to surface warming of outer bay waters rather than cooling of inner bay waters. If surface meltwater input was significant there should be a dilution of the surface layer salinity near the ice front. Yet, there is a salinity gradient in the near-surface (between 15 and 100 m depth) suggesting that melting along the glacier front is taking place. This may be coupled with thorough mixing (by wind or ice drift) at depths less than 10 m. The surface layer contains moderate suspended particulate concentrations (2–5 mg l^{-1}) consisting of phytoplankton (diatoms). It should be noted that these observations were made on January 19, 1988, during the peak of the austral summer, and despite thorough examination of the ice front environment by small boat no surface plumes of terrigenous sediment were observed.

Bottom sediments

Surface sediment textures and compositions are quite variable but contain distinctive trends with respect to proximity to the ice front and water depth (Figs 8 & 9). Sediments along the ice front and on the gently sloping floor of the inner (eastern) bay consist of diamicton. Within the submarine valley, silty muds and sandy muds predominate. Outer bay sediments consist of biosiliceous muds with minor amounts of poorly sorted sand and some gravel. Hence there is a trend to finer textures northwest of the ice front and within the submarine valley. Total organic carbon (non-carbonate) contents increase with distance from the glacier front and reflect the dilution effect of terrigenous sedimentation in proximity to the glacier (Fig. 9b).

Subsurface sediments were obtained by three piston cores whose lithology and sedimentary structures are illustrated in Fig. 10. Seismic reflection data (limited to a short segment obtained from a nine element sparker array) demonstrate that the submarine valley is a site of

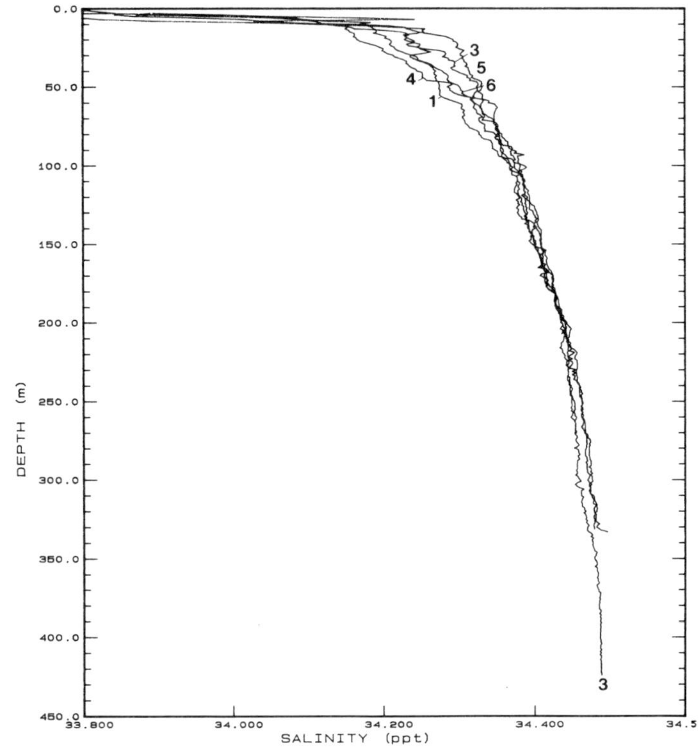

Fig. 5. Salinity casts made at CTDT station in Brialmont Cove. Note slight dilution of salinities below 200 m depth at stations #3 and #6.

ponded sediment accumulation, which reaches thicknesses of at least 20 m. The two cores obtained within the axis of the valley system encountered three dominant units. The uppermost of these consists of a thin bedded to thickly laminated succession of sandy muds and silty muds which are interrupted by thin beds of medium to fine-grained sand. Bed thickness as observed in x-radiographs is random or nonexistent due to bioturbation. A significant component of gravel and poorly sorted sand is also present (Fig. 11b). Based upon the two cores recovered, individual sand units tend to 'thin' and 'fine' distally down the axis of the valley, indicating a source presently tied to the ice-proximal part of the valley system.

At depths of between 4.1 and 4.3 m in core 150, a thin laminated silty mud and sandy silt unit is encountered. The lamination has a pronounced cyclicity or rhythmic character which is best developed between 420 and 450 cm depth (Figs 11a & 12). The thickest laminae are found every 12–13 laminae, as demonstrated between 440 and 450 cm in the core (Fig. 12). This pattern is less pronounced above 440 cm, where the thickest laminae occur every 4–8 laminae (Fig. 12). Small scale bioturbation structures of both horizontal and subvertical character occur above and below the laminated unit. Very little gravel or poorly sorted sand is present within the laminated section which contrasts with the overlying units in both cores 149 and 150. The bottom metre of core 150 consists of a massive (structureless) mud unit which contains no discernable bedding or dropstones.

Deposits of the outer-bay (core 151) drape the underlying topography and consist of biosiliceous pebbly muds (Fig. 11c). Such units contain a persistent component of poorly sorted sand (1–4%) which, together with the gravel, attest to the influence of ice rafting upon deposition of these organic rich muds. The stratification is best described as ill-defined. No distinct laminae are found within the mud; rather the x-radiographs demonstrate diffuse bedding and stratification (Fig. 11c).

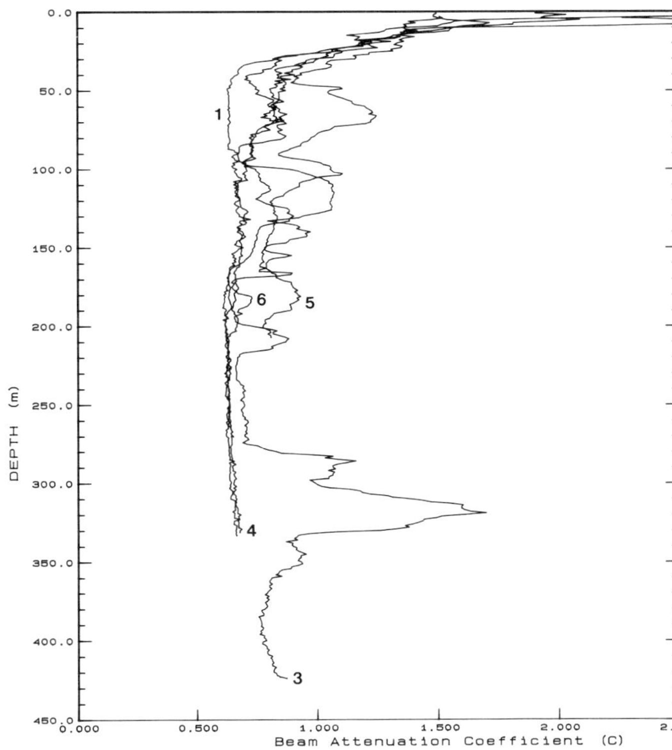

Fig. 6. Beam attenuation coefficient, C (directly proportional to suspended sediment concentration). Note correlation of temperature (Fig. 4) and C for station #3.

Discussion

Depositional processes

The pattern of inner basin ponds of terrigenous sediment and outer bay drapes of biogenic muds is consistent with the observations of Griffith & Anderson (1989) who noted this as a characteristic of fjord sedimentation along the Palmer Archipelago (Fig. 1). They suggested that the pattern was an indication of slightly warmer climates with greater precipitation than colder and drier fjord environments (which lack ponded sediment) found further south along the Danco Coast. Thus, meltwater-derived sediment sources and sediment gravity flows were implicated as the primary contributors to inner-basin ponding of terrigenous deposits (Griffith & Anderson 1989).

Although runoff and surface melting are components of the sediment regime in Brialmont Cove, the primary source of sediment appears to be related to input near the base of the glacier, particularly the 'tunnel valley' system. Basal melting processes and tidal flushing of crevasse systems are possibilities (Domack & Williams 1990). Surface meltwater that is conducted to the base of the glacier and eventually to the terminus, needs a conduit such as a moulin. Conduit systems which reach the glacier base are only possible where the ice temperatures are entirely at pressure melting, though dolines and vertical tunnels have been reported from the Larsen Ice Shelf (Swithinbank *et al.* 1988) and the George VI Ice Shelf (Sugden & Clapperton 1981). Thus, other avenues would have to provide the pathway for drainage to the base, and even these would only be effective to limited depths (i.e. 100 m or less). It seems unlikely that the fine-scale oceanographic layering, especially that observed within the axis of the submarine valley, could have been derived from surface-derived meltwater exiting at the glacier terminus. The character of such features is quite different from typical meltwater plumes observed in melt-dominated (sub-polar) glacimarine settings (Syvitski 1989; Pfirman & Solhiem 1989). The oceanographic data in Figs 5–7 indicate very little dilution of adjacent sea

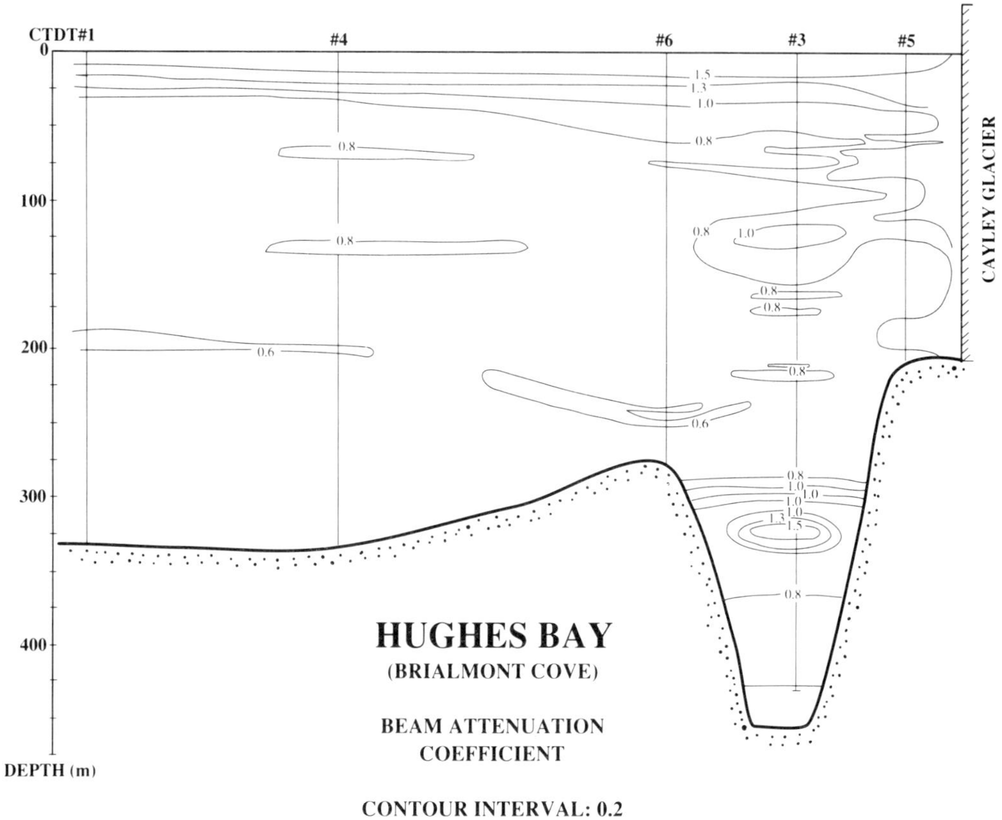

Fig. 7. Contoured values of C for Brialmont Cove obtained on January 19 & 20, 1988.

water in the ice proximal environment and no vertical or rising plume pattern of suspended sediment. True estuarine circulation is not developed. Rather, the fine-scale oceanographic layering and associated sediment load are best explained by either direct ice melting at the sea water boundary or tidal exchange within the sub-ice 'tunnel valley' and surface crevasse system (Domack & Williams 1990).

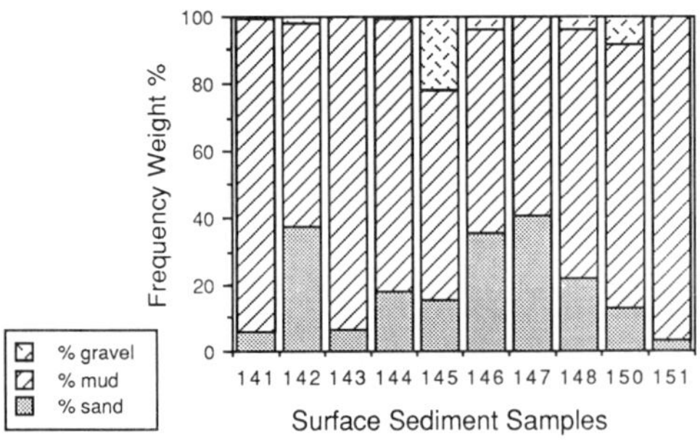

Fig. 8. Cumulative weight percentages of sand, mud and gravel for surface sediments in Brialmont Cove (see Fig. 1 for sample location).

The repetitive nature of the fine structure places constraints on the origin of such phenomena. Direct ice melting caused by heat supplied from sea water can produce horizontal layering within a salinity stratified water body, as demonstrated in laboratory experiments (Josberger & Martin 1981; van Heijst 1987). Thermohaline circulation cells produced by the diffusion differential of heat and salt are characterized by rather unique vertical gradients in temperature, salinity and density. Features such as those observed in the Ross Sea, adjacent to the Erebus Glacier Tongue, are characterized by a stepped vertical profile as suggested by van Heijst (1987) for the data of Jacobs *et al.* (1981). Such profiles reflect the steep salinity/temperature gradients found along the boundary of individual circulation cells. Though the role of such layering in sediment transport has yet to be addressed, it is

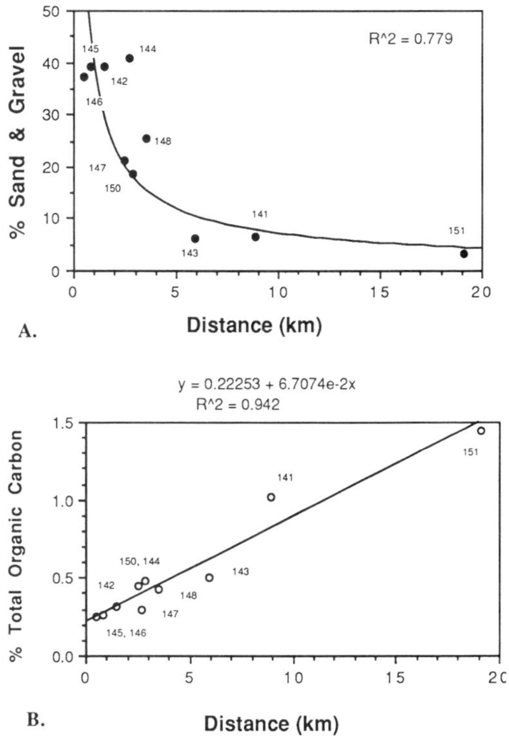

Fig. 9. Down-fjord trends in surface sediment texture and total organic (non-carbonate) carbon. Organic carbon data provided by L. Burkley (Hamilton College), pers. comm.

clear that fine silt and clay released from the ice during melting could be incorporated into the horizontal circulation via suspension. However, the concentrations would reflect the englacial debris content which is generally quite low, especially in Antarctic glacial systems.

Detailed observations of the fine-scale layering as illustrated in Fig. 4 are not explained adequately by the diffusive convection model. The gradients in salinity/temperature are not stepped, but alternate between maximum and minimum values. Thus, the layering is clearly related to alternating zones of saline, clear and relatively warm water with less saline, turbid and cold water. The deep layers are best explained by considering tidal influences upon circulation within the subglacial 'tunnel valley' system. Tidal pumping of the grounding zone within the subglacial valley system could realistically produce alternate pulses of in-flowing waters during the uplift (high tide) cycle and out-flowing waters during the let-down (falling tide) cycle. Alternate pulses would occur close to one another and would be separated by stable or well mixed waters. This is because the maximum tidal effect would occur during and just following high tide; tides for this area are moderate with semi-diurnal ranges of between 1 and 2 metres. Water properties of the out-flowing pulse would then reflect the cooling and slight freshening which occurs in the subglacial cavity. The presence of a basal melt layer or direct melting within the cavity would lead to increased suspended sediment loads, derived from the basal debris zone or bed of the glacier. In-flowing pulses would reflect the relatively warm, saline and clear characteristics of the open bay waters. One would expect this pattern to dominate within the subglacial valley system and be less effective in transporting sediment further out into the bay. This is confirmed by the oceanographic data which demonstrate attenuation of the layers due to mixing beyond the innermost portion of the bay (Fig. 7). If one extrapolates the conditions observed at the ice front to within the subglacial cavity it is reasonable to expect that the cold tongues would contain even more suspended sediment, of coarser texture.

Depositional environments

The silty mud within the axis of the submarine valley is most likely the result of settling from suspension and sediment gravity flows. Transport by and attenuation of the fine-scale plumes is directly implicated in this process and, in part, explains the ponded nature of the fine-grained deposit. The plumes exit the subglacial 'tunnel valley' system and do not migrate upwards. Indeed there is very little difference in density (salinity) properties of the plume and ambient sea water which would induce such circulation (Fig. 5). Rather, horizontal transport and attenuation would limit the distribution of the suspended sediment to the submarine valley. Vertical settling of suspended sediment could, therefore, occur preferentially within the axis of the valley, in contrast to the rest of the bay system. This pattern would accentuate the pond ing of sediment within the valley. Sediment gravity flows also add to deposition; this process most likely produced the sand layers found in both cores 149 and 150. The random bed thickness and structureless sediment found in the near surface is explained by the present position of the glacier terminus. This is because mixing processes limit the direct influence of the fine-scale plumes upon deposition in the ice-distal or open marine environment. In the present setting the direct tidal signal is masked under generally

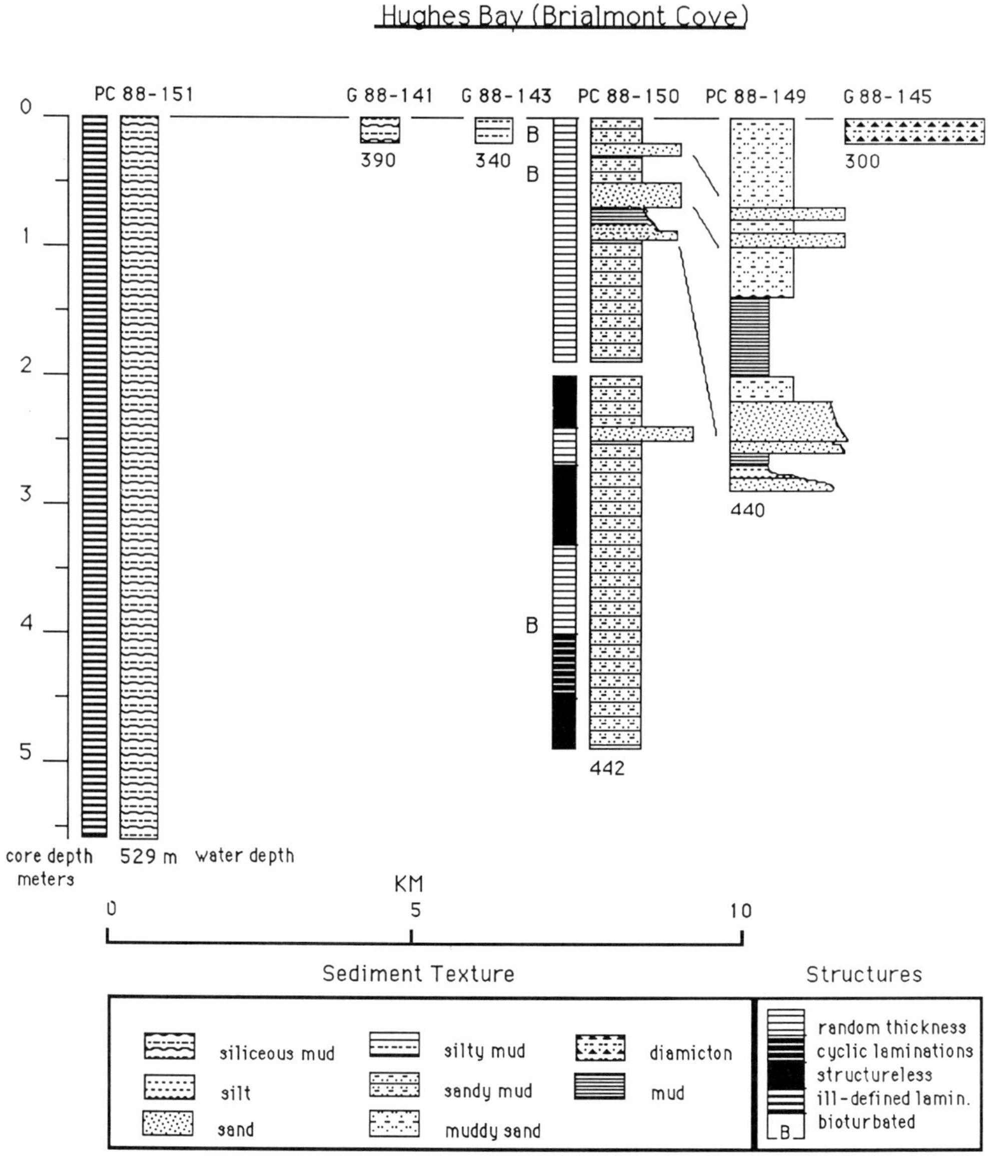

Fig. 10. Texture and sedimentary structures for selected surface sediment samples and piston cores.

lower sedimentation rates, bioturbation, and iceberg rafting. The scarcity of ice rafted debris and low organic carbon content within the laminated intervals also suggest that the depositional setting was rather unique. Present trends of sand, gravel and total organic carbon illustrate that with increasing distance from the ice front the ice rafted and organic carbon contents increase (Figs 8 & 9). The laminated muds lack ice rafted debris and have very low organic carbon contents. Hence, they cannot be explained as ice-distal deposits. Soft sediment disturbance would also be expected in an open marine setting due to ice rafting and bioturbation.

The laminations have to be explained with special conditions that address the lack of ice rafted debris and an absence of bioturbation. It is possible that lowered sea levels could have led to restricted circulation within the inner basin of Brialmont Cove, resulting in lowered

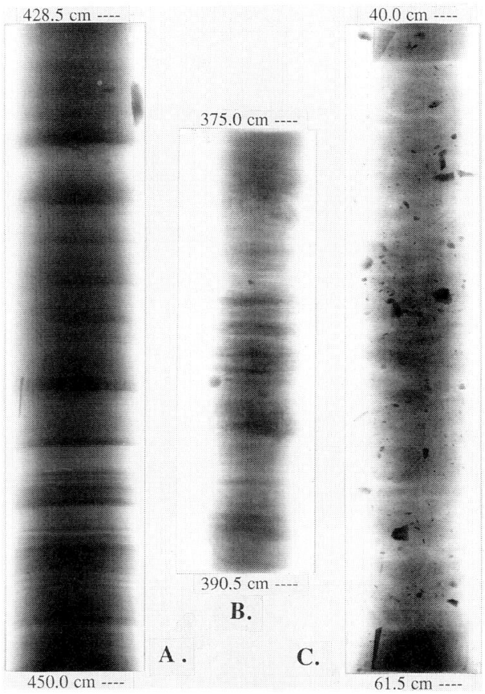

Fig. 11. X-radiographs of piston core PD88–150 428.5–450 cm (a); PD88–150, 375.0–390.5 cm (b); PD88–151, 40–61.5 cm (c).

oxygen levels and minimal bioturbation. However, minimum 'sill' depths are on the order of 280 m (Fig. 1) and so it does not seem realistic that relative sea levels could have played a role in basin isolation. Today the bottom waters are well oxygenated (56% saturation for bottom water at CTDT #3), as is typical for the Antarctic Shelf.

A warmer climatic regime could have led to increased input of meltwater-derived sediment which resulted in rapid sedimentation and pronounced seasonal (varves) or even meteorological (storm) cycles. However, even in temperate oceanic climates seasonal laminations are difficult to preserve and distinguish from storm or outburst flood signals. So far, definite modern glacimarine varves from warmer climates have yet to be recognized (Domack 1984; Mackiewicz *et al.* 1984), although Stevens (1985) suggests they may be present in late Pleistocene deposits in Sweden.

If the ice front was further seaward during the deposition of the rhythmic intervals then the submarine valley system could have been in a subglacial marine setting, similar to the present situation beneath the Cayley Glacier depression (Fig. 1). This environment would have been drastically different from the present open marine setting and would have allowed for three important conditions:

(1) bioturbation would have been reduced under the glacier terminus in the absence of pelagic organic carbon sources; nutrient supply would be limited to material advected laterally beneath the ice by oceanic circulation;

(2) iceberg rafting would be absent and any debris released from the base of the Cayley Glacier would have to be derived from basal debris, the extent of which would be limited by the rate of basal melting and glacier flow velocity; both of these factors need further evaluation as the ice surface topography would suggest that ice flow (basal debris supply) should converge towards the depression;

(3) tidal influences on sedimentation would be considerably amplified as would the sedimentation rate since the sediment source is closer and attenuation of the tidal plumes would be less. A greater sedimentation rate would

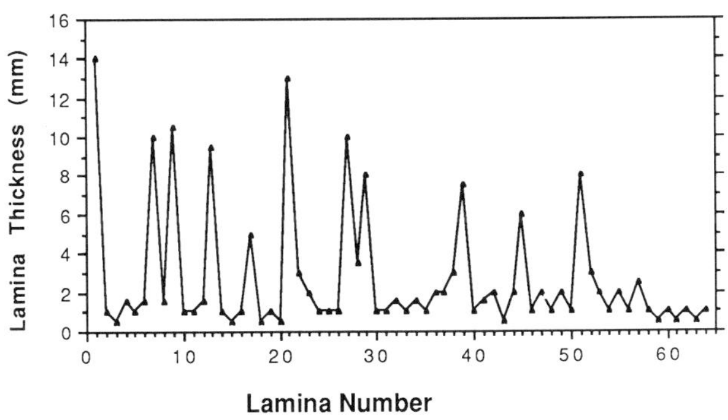

Fig. 12. Lamina thickness within rhythmic laminated intervals of core PD88–150.

also reduce the content of ice rafted debris relative to sites of lower sedimentation rate found in the open marine setting; hence the laminations in core PD88–150 may directly reflect tidal processes which operated within a subglacial marine environment.

Studies based in other areas of Antarctica suggest that dilute sediment gravity flows or bottom currents are generated at the grounding lines of ice tongues and ice shelves (Drewry & Cooper 1981; Barrett *et al.* 1988; Anderson & Molnia 1989). Despite an inadequate understanding or observational database for these processes, similar models have been used to interpret Cenozoic sequences recovered from the Antarctic margin (Hambrey *et al.* 1990). The observations made in the present study suggest that mid-water plumes may also contribute to subglacial sedimentation within marine cavities. The unique bathymetric conditions in Brialmont Cove have provided a natural trap for sediments produced by a combination of subglacial and marine processes. Additional field work to examine the relationship between tidal cycles, vertical glacier movement, and the dynamics of fine-scale structures will provide a more thorough understanding of this system and the controls on deposition. On-going chronological and detailed sedimentological studies should also help clarify the relationship between rhythmic laminations, age and sediment composition.

This study was funded by the U.S. National Science Foundation, Division of Polar Programs (DPP86–13565). I would like to thank Dennis Cassidy at Florida State University, for help in sending cores to Hamilton, and the undergraduate students from Hamilton College who assisted me with great perseverance during cruise III of the *Polar Duke*. The comments of R.D. Powell and A. Solheim on an earlier draft of this manuscript are also appreciated.

References

ANDERSON, J.B., KURTZ, D., WEAVER, F. & WEAVER, M. 1982. Sedimentation on the West Antarctic Continental margin. *In*: CRADDOCK, C. (ed.) *Antarctic Geoscience*. University of Wisconsin Press, Madison, 1003–1012.

—— & MOLNIA, B.F. 1989. Glacial-Marine Sedimentation. *Short Course in Geology: Volume 9*. American Geophysical Union, Washington D.C.

BARRETT, P.J., PYNE, A.R. & MACPHERSON, A.J. 1988. Sedimentation in a polar coastal setting, Antarctica. *Geological Society of America: Abstracts with Programs*, 1988 Annual Meeting, A84–5.

BARRON, J., LARSEN, B. *et al.* 1989. *Proceedings of the Ocean Drilling Program*, Initial Reports, Volume 119A: College Station, Texas.

BISHOP, J.K.B. 1986. The correction and suspended matter calibration of Sea Tech transmissometer data. *Deep-Sea Research*, **89**, 8197–8203.

COWAN, E.A., POWELL, R.D. & SMITH, N.D. 1989. Rainstorm-induced event sedimentation at the tidewater front of a temperate glacier. *Geology*, **16**, 409–412.

DOMACK, E.W. 1984. Rhythmically bedded glaciomarine sediments on Whidbey Island, Washington. *Journal of Sedimentary Petrology*, **54**, 589–602.

—— 1988a. Biogenic facies in the Antarctic glacimarine environment: basis for a Polar glacimarine summary. *Palaeogeography, Palaeoclimatology, Palaeoecology*, **63**, 357–372.

—— 1988b. Depositional environments of the Antarctic continental shelf: fjord studies from the R.V. *Polar Duke. Antarctic Journal of the United States*, 1988 Review Issue, **23**, 96–102.

—— & WILLIAMS, C.R. 1990. Fine-structure and suspended sediment transport in three Antarctic fjords. *In*: BENTLEY, C.R. (ed.) *Antarctic Research Series*, first annual volume, Washington, D.C., American Geophysical Union, in press.

DREWRY, D.J. & COOPER, A.P.R. 1981. Processes and models of Antarctic glaciomarine sedimentation. *Annals of Glaciology*, **2**, 117–122.

GILBERT, R. 1982. Contemporary sedimentary environments on Baffin Island, N.W.T., Canada: glaciomarine processes in fjords of eastern Cumberland Peninsula. *Arctic and Alpine Research*, **14**, 1–12.

GRIFFITH, T.W. & ANDERSON, J.B. 1989. Climatic controls of sedimentation in bays and fjords of the northern Antarctic Peninsula. *Marine Geology*, **85**, 181–204.

HAMBREY, M.J., EHRMANN, W. & LARSEN, B. 1990. The Cenozoic glacial record of the Prydz Bay continental shelf, East Antarctica. *Proceedings of the Ocean Drilling Program*, Scientific Results, Volume 119B. in press.

HAMPTON, M.A., KRAVITZ, J.H. & LUEPKE, G. 1987. Geology of sediment cores from the George V continental margin, Antarctica. *In*: EITTREIM, S.L., & HAMPTON, M.A. (eds) *The Antarctic Continental Margin: Geology and Geophysics of Offshore Wilkes Land*, CPCMR Earth Science Series, Vol 5A: Houston Texas, 151–174.

HATTERSLEY-SMITH, G. & THOMSON, M.R.A. 1988. Confusion of place names; an example from Antarctica. *Polar Record*, **24**, 239–242.

HERMAN, G.F. & JOHNSON, W.T. 1980. Arctic and Antarctic climatology of a GLAS general circulation model. *Monthly Weather Review*, **108**, 1974–1991.

JACOBS, S.S., HUPPERT, H.E., HOLDSWORTH, G. & DREWRY, D.J. 1981. Thermohaline steps induced by melting of the Erebus Glacier Tongue. *Journal of Geophysical Research*, **88**, 6547–6555.

JOSBERGER, E.G. & MARTIN, S. 1981. A laboratory and theoretical study of the boundary layer adjacent to a vertical melting ice wall in salt water. *Journal of Fluid Mechanics*, **111**, 439–473.

MACKIEWICZ, N.E., POWELL, R.D., CARLSON, P.R. &

MOLNIA, B.F. 1984. Interlaminated ice-proximal glacimarine sediments in Muir Inlet, Alaska. *Marine Geology*, **57**, 113–147.

MARTIN, P.J. & PEEL, D.A. 1978. The spatial distribution of 10 m temperatures in the Antarctic Peninsula. *Journal of Glaciology*, **20**, 311–317.

PFIRMAN, S.L. & SOLHEIM, A. 1989. Subglacial meltwater discharge in the open-marine tidewater glacier environment: observations from Nordaustlandet, Svalbard Archipelago. *Marine Geology*, **86**, 265–281.

SMITH, B.M.E. 1972. Airborne radio echo soundings of glaciers in the Antarctic Peninsula. *British Antarctic Survey, Scientific Report*, 72.

STEVENS, R. 1985. Glaciomarine varves in late-Pleistocene clays near Göteborg, southwestern Sweden. *Boreas*, **14**, 127–132.

SUGDEN, D.E. & CLAPPERTON, C.M. 1981. An ice-shelf moraine, George VI Sound, Antarctica *Annals of Glaciology*, **2**, 135–141.

SWITHINBANK, C., CHINN, T.J., WILLIAMS, R.S. & FERRIGNO, J.G. 1988. Satellite Image Atlas of Glaciers of the World, Antarctica. *U.S. Geological Survey, Professional Paper*, 1386-B.

SYVITSKI, J.P.M. 1989. On the deposition of sediment within glacier-influenced fjords: oceanographic controls. *Marine Geology*, **85**, 301–329.

TALJAARD, J.J., VAN LOON, H., CRUTCHER, H.L. & JENNE, R.L. 1969. Climate of the upper air: Southern Hemisphere 1, Temperatures, Dew Points and Heights at selected pressure levels. NAVIR 50-1C-55. Washington, D.C.

VAN HEIJST, G.J.F. 1987. On the oceanic circulation near a shelf-ice edge. *In*: VAN DER VEEN, C.J. & OERLEMANS, J. (eds) *Dynamics of the West Antarctic Ice Sheet*. Reidel, Boston, 37–56.

WILLIAMS, C., BOIES, C. & DOMACK, E.W. 1989. Characteristics of glacial systems along the Graham Land Coast and Palmer Archipelago, Antarctic Peninsula. *Antarctic Journal of the United States*, 1989 Review Issue.

Rafting in glacimarine environments

ROBERT GILBERT

Department of Geography, Queen's University, Kingston, Canada K7L 3N6

Abstract: Sources of rafted sediment include glacial, colluvial, fluvial, aeolian and littoral settings. Processes of incorporation and rafting include active incorporation by freezing and passive loading of icebergs and sea ice, and rafting by vegetation, principally marine algae. Passively loaded sediments, being loose on the ice surface, are more easily lost from the raft and are more likely to form deposits of more than one particle at a time. The effectiveness of passive loading depends on sea ice being in contact with the shore in the spring, when colluvial and fluvial processes are most active, and during winter when aeolian processes occur. Actively incorporated sediment becomes concentrated on the surface of the raft by melting. Other factors that control distribution of rafted sediments include the strength and stability of the raft, its rate of deterioration, its ability to leave the source area, and its movement at sea.

Sediment released from the ice is deposited as single particles (referred to as a 'drop'), agglomerations of more than one particle ('dump'), frozen aggregates, and deposits from the melting of sediment-laden ice in contact with the sea floor. Characteristics that have been used to distinguish the mode of rafting include grain size, sorting, shape, surface features (especially striations), fabric, fauna associated with the environment of origin, distance of transport, and water depth. The most reliable in assessing origin from an iceberg is probably the presence of striations, although none is conclusive, and interpretations must be made cautiously.

Rafted sediment is a major component of the total sediment deposited in the glacimarine environment (Ruddiman (1977) estimates 2×10^{14} m^3 to the deep Atlantic during 3 Ma). Rafted sediment provides information about the oceanic and terrestrial environment. Many workers have used the presence of coarse particles in sediment as evidence of ice rafting and, therefore, of glacial conditions. In this paper 'glacial' refers to those environments influenced by seasonal or permanent ice cover. This definition allows consideration of both glacier and sea ice, the effects of which, in terms of rafting, are often indistinguishable.

The study of rafting of sediment to the marine environment is difficult because (1) the processes dictate unpredictable events especially in the short term, (2) some assumptions about the processes and forms resulting are not straightforward, and (3) observation and measurement of the process is difficult or impossible. This paper is a review of rafting in glacimarine settings; it draws on the established literature and is illustrated with material from studies especially in the eastern Canadian Arctic. It is necessarily qualitative.

Discussion is organized around a conceptual model (Fig. 1) indicating sources of material, processes and deposits (cf. Campbell & Collin (1958) Table 1). Each of glacial, periglacial and paraglacial (in the sense of Church & Ryder 1972) settings is an important source for rafted sediments. The rafting itself is carried out by icebergs, sea ice and vegetation. The role of sea ice and icebergs in rafting may be divided into active and passive processes. In the active role, sediment is incorporated into the ice by freezing or associated with flow. In passive ice rafting, the ice is simply the agent for distributing sediment placed on it by other processes.

The only significant agent of organic rafting in the glacimarine setting is marine algae of the littoral environment, although small amounts may be related to driftwood and marine mammals (Emery 1963), including polar bears (McKinlay 1976). Material rafted by human activity, especially clinker, scale and ash from steamships, is found with naturally rafted sediments (Huggett & Kidd 1983), and in the North Atlantic is the largest modern source of sediment (Kidd & Huggett 1981). The author's students have found clinker in samples from the sea floor near Pangnirtung, Baffin Island probably associated with the annual 'Arctic Patrol' (Marriott 1940), but it is easily recognized and forms a minor, recent component of the rafted sediment record.

Four types of sedimentary deposits are distinguished in Fig. 1. Following the terminology of Thomas & Connell (1985), a 'drop' is a single particle deposited from rafting, and a 'dump' consists of a number of particles laid down in a

From DOWDESWELL, J. A. & SCOURSE, J. D. (eds), 1990, *Glacimarine Environments: Processes and Sediments*. Geological Society Special Publication No 53, pp 105–120.

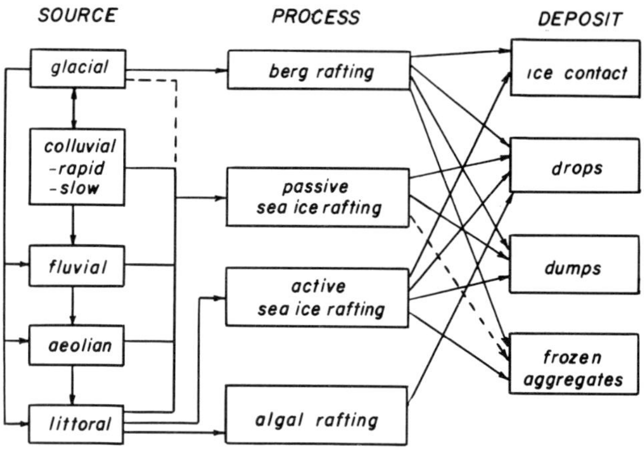

Fig. 1. Conceptual model of sources, processes and forms related to rafting in the glacimarine environment. Extraterrestrial and volcanic sources are not included because these materials are difficult to distinguish as rafted. Floating of particles on surface tension (Hume 1964; Syvitski & van Everdingen 1981) and rafting by trees and mammals (including humans) is not included because these are of minor importance in glacial seas.

single event. 'Frozen aggregates' are held together by interstitial ice and may be released by breaking apart of the ice raft (Huggett & Kidd 1983). 'Ice contact' sediments are deposited by ice in contact with the sea floor. This more general term is used here in preference to 'ice-berg till' (Dreimanis 1979) because the sediment is not associated directly with glacial deposition whether it originates from icebergs or sea ice.

Table 1. *Characteristics of rafted sediments*

Characteristic	Rafting agent		
	Iceberg	Sea Ice	Vegetation
Form	drops, dumps, frozen aggregates, ice contact deposits		mainly drops
Grain size	clay to boulder	clay to boulder maximum size about 4 m	pebbles some sand
Sorting	poor	poor to good	–
Shape	normally angular may be rounded faceted	angular to rounded	normally rounded
Surface features	may be striated	striations less likely	–
Fabric	random except in frozen aggregates and ice contact deposits		random
Fauna	absent	shallow water fauna attached or mixed with rock boring	short-lived holdfasts
Friable and weathered material	absent	may be present	absent
Distance of transport	more distally	more proximally	most proximally
Water depth	deep	shallow to deep	

Sources and modes of rafting

Glaciers and icebergs

Sediment is actively incorporated in glacial ice primarily as basal debris, or passively as a result of colluvial processes on valley sides above the glacier (Anderson *et al.* 1980). The amount of sediment incorporated at the base of the ice varies depending largely on the thermal regime of the glacier (Dowdeswell 1987; Dowdeswell & Dowdeswell 1989). Some cold-based glaciers (Shaw 1977) and those experiencing transition from warm to cold base along their length (Weertman 1961) incorporate a zone of up to several tens of metres of sediment-rich ice. Much of the sediment from warm-based glaciers and from floating ice shelves is lost by melting before icebergs calve (Piper 1976; Elverhøi & Roaldset 1983; Dunbar *et al.* 1985).

Many workers have observed that only a few per cent of icebergs carry visible sediment (Anderson *et al.* 1980), although Kindle (1924) cautioned about the reliability of these observations with the analogy of judging the number of barnacles on a ship by looking at the topsides. The question of the amount of sediment released from bergs is compounded by the near impossibility of direct measurement. However, rates of melt can be calculated (Huppert 1980; Dowdeswell & Murray 1990), and statistics are available on the decline in numbers of bergs away from source areas (Andrews & Matsch 1983), so that rates of deposition may be inferred (Dowdeswell & Murray 1990).

Fluvial and colluvial sources

Timing of delivery of fluvial and colluvial sediments to the shore is of critical importance in determining the potential for passive loading of sea ice. Even at high latitudes, these processes are well established by June. For example, in eastern Baffin Island, rockfalls, landslides and avalanches are most frequent at that time (Church *et al.* 1979; Fig. 2a). Similarly, slow colluvial processes (gelefluction) are most active during about the same period (Washburn 1967; Fig. 2b). In some cases slow colluvial processes may contribute sediment directly to sea ice, but it is more likely that they act as feeders to streams.

Stream flow records (for example, Fig. 2c) indicate that nival melt may begin in May, normally is greatest in June, and in the high Arctic may last into July (Woo 1986). Melt runoff usually lasts only a few weeks (Woo & Steer 1986). Glacial streams peak later in the summer (Church 1974; Gilbert *et al.* 1987).

The breakup of shorefast sea ice in the Arctic normally occurs between mid-June and mid-August after these peaks of sediment production. Except where the inflow of warm water or breakup of dirty ice on intertidal flats has

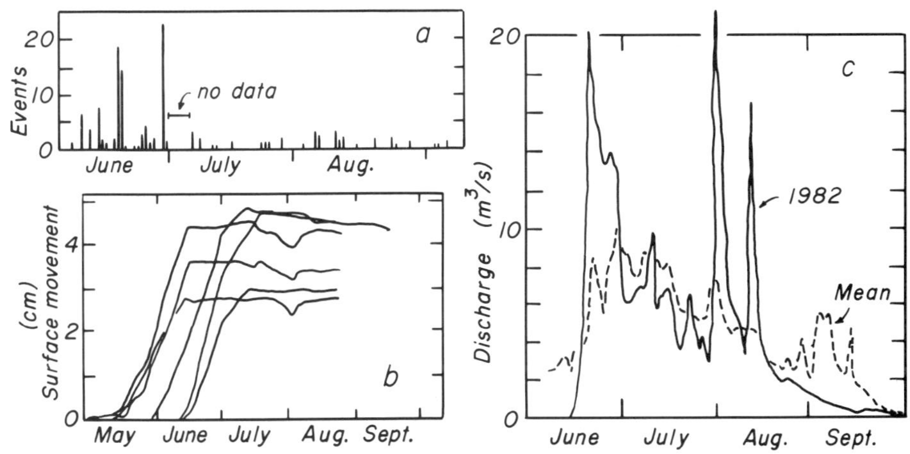

Fig. 2. (a) Number of events of rockfall at Ekalugad Fiord, Baffin Island, June 4 to September 9, 1967 (redrawn from Church *et al.* 1979). (b) Selected measurements of cumulative slope movement near Mesters Vig, northeast Greenland 1957 to 1961 (redrawn from Washburn 1967). (c) Daily mean stream discharge, Duval River at Pangnirtung, Baffin Island 1982 and mean for the years 1974 to 1982 (from the Water Survey of Canada and Gilbert unpublished data).

removed ice from near the shore, there is a reasonable chance in a given year that fluvial and colluvial processes may deliver sediment to the sea ice before breakup. Stefansson (1921) describes an accumulation of sediment ranging from mud to boulders of more than 45 kg on the ice of Peary Channel which he ascribes to landslides. The role of snow in aiding delivery of rock fall to ice is discussed by Ferguson (1970) and Luckman (1975). Howarth & Bones (1972) suggest that sea ice encourages colluvial processes by oversteepening talus slopes at water level as it removes sediment. Kindle (1924) and Keys (1978) describe the delivery of streamborne sediment to sea ice on Greenland and Ellesmere Island. The largest amounts of sediment delivered to sea ice are from the major rivers of the U.S.S.R. (Fuchs & Whittard 1930) and the Alaskan north slope (Reimnitz & Bruder 1972), although most is lost as the ice melts in situ and little is rafted to sea.

Because the process involves coordination of supply of sediment before breakup of sea ice, spatial and temporal variations are great. It is probably impossible to assign a relative importance to these sources in comparison to loading that occurs by other means.

Aeolian sources

Kindle (1924) describes accounts of dust, sand and gravel carried by wind 'far out on the sea ice' (p. 258) (cf. Campbell & Collin 1958). Bentley (1979) and Barrett *et al.* (1983) conclude that aeolian sand deposited initially on sea ice is the major source of sediment on the floor of western McMurdo Sound.

Many surfaces in glacial regions have a cover of sparsely vegetated sediment and there is, therefore, abundant material for aeolian erosion. The major sources are ice-contact sediments exposed by glacial retreat, and fluvial and lacustrine sediments exposed during periods of low water (Molnia 1983; Migala & Sobik 1984). The presence of water, ice and snow is important in determining the availability of these sediments. During winter aeolian processes are probably most effective because winds are generally strongest and source areas are most exposed due to drying (McKenna Neuman & Gilbert 1986). Cold air temperature results in higher air density and, therefore, in lower critical erosion velocities. Except in temperate alpine environments, snow cover is not sufficient to retard erosion significantly, and blowing snow is itself an agent of erosion (McKenna Neuman 1987).

Except for fine grains (less than about 0.02 to 0.05 mm: Reineck & Singh 1975) which are transported in suspension, aeolian transport of sediment to sea requires a solid substrate of sea ice on which saltation, creep and movement by impact from other grains occur (Fig. 3). Thus, only where sea ice is in contact with the shore can coarse-grained aeolian sediment be carried to sea. At Pangnirtung Fiord, Baffin Island in May 1986 patches of silt, sand and gravel were found up to 5 km from their source on the sandur of Weasel River at the head of the fjord (Fig. 4). Much of the sediment (including cobbles to 128 mm diameter) occurred in pockets where it was trapped around sastrugi and snow dunes (cf. Bentley 1979; Nedell *et al.* 1987).

Under ideal conditions of unlimited supply of dry sand, aeolian rates of transport may reach 1 Mg h^{-1} per metre width at wind speeds of 16 m s^{-1} (Bagnold 1941). This may be compared with values in the range about 0.05 to 5 Mg h^{-1} bedload discharge of sand in water discharge of 0.1 to 1 m^3 s^{-1} per metre width (Blench 1969). Thus, at different times during one year especially on sandurs, aeolian transport of sand may rival fluvial transport of sand.

Melting of sea ice involves the establishment of pools, some of which melt through the ice to form conduits which drain the nearby ice (Barry & Jacobs 1978). It is possible that a significant amount of aeolian sediment deposited on the ice surface is released to the water beneath in a short time during this event, and that it may form a distinct annual layer in the glacimarine sediment (Fig. 5). If so, the layers are important stratigraphic markers in sediments in which varves are rare and in which bioturbation often destroys subtle laminations. If annual aeolian/rafted layers can be recognized, they offer the opportunity for study of year-to-year variation in climatic events over long periods.

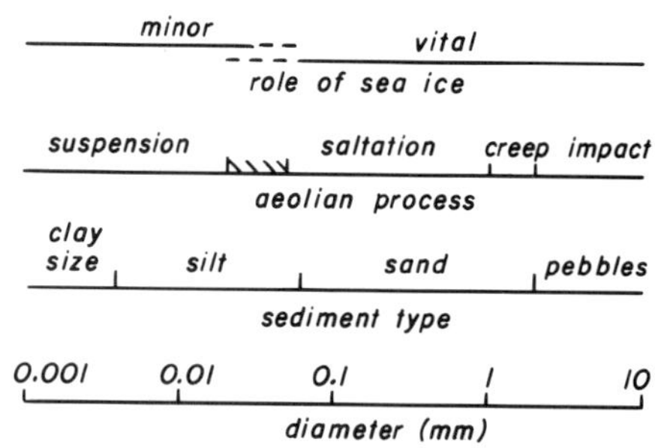

Fig. 3. Relation between grain size and mode of aeolian transport in moderate winds. For transport to sea of particles intermittently or continuously in contact with the surface, the presence of sea ice is vital.

Fig. 4. Aeolian sand and silt on the ice of Pangnirtung Fiord 5 km from the head and about 8 to 15 km from the source on the sandur of Weasel River, June 1986. Photograph by C. McKenna Neuman.

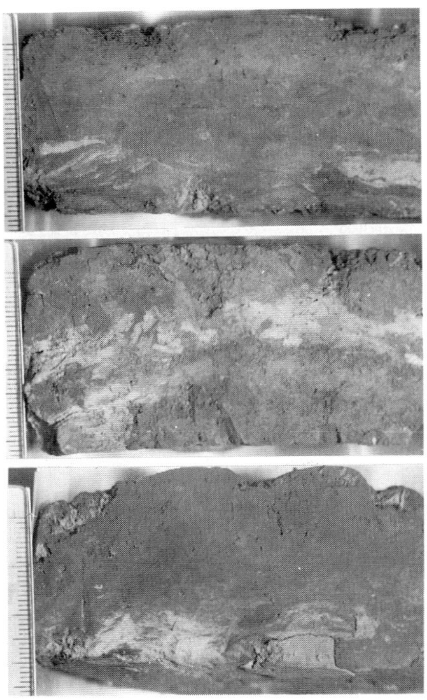

Fig. 5. Ekman grab samples from the floor of Maktak Fiord, eastern Baffin Island at 7 km (upper), 10 km (middle) and 14 km (lower) from a large aeolian deposit (Gilbert & McKenna Neuman 1986) at the head of the fiord. Scale in millimetre divisions. Quartz-rich white sand layers are interpreted as aeolian. Bioturbation following deposition has partially mixed sediments from above and below. Rates of aeolian sedimentation are about 8 to 17 kg m^{-2} a^{-1}.

Littoral environment

The littoral environment is the only one in which both active and passive processes of sediment incorporation into sea ice are important.

Active incorporation occurs in three ways. First, sediment may be frozen onto the bottom of sea ice as it rests on the sea floor. This occurs in shallow water (inshore of the isobath of maximum seasonal ice thickness) and is best developed where there are broad intertidal flats. Commonly, sediment accreted into sea ice by this means is found interlayered with clear ice (Fig. 6). The sediment-rich bands form when the ice is in contact with the sea floor at low tide, and the clear bands when the ice floats off at high tide. In a study of ice on intertidal flats at Pangnirtung, Baffin Island in June 1986 active incorporation accounted for sediment accumulation in the zone within about 200 m of shore of up to 670 g l^{-1} (Fig. 7).

The size distribution of the sediment incorporated by active freezing reflects that on the sea floor because little or no sorting occurs. Commonly, grain size varies from clay size to pebbles. Larger cobbles and boulders are also incorporated into ice as it breaks and refreezes around the stones especially during the highest tides (Fig. 8). The process is described in detail by Rosen (1979) and Lauriol & Gray (1980).

The second active process involves incorporation into the growing sea ice of fine-grained sediment suspended in the sea water. The sediment is scavenged from the water column or sea floor in shallow, turbulent areas by frazil and anchor ice formed in initial stages of freeze-up (Campbell & Collin 1958; Ackley 1982; Barnes *et al.* 1982; Dionne 1984; Reimnitz & Kempema 1987, 1988). The loosely held sediment is later frozen into the ice canopy to form a more-or-less uniformly darkened layer. In the Pangnirtung study (Fig. 7) this ice was observed to occur intermittently in the outer areas of the intertidal flats. Sediment concentrations were variable, but were generally less than several tens of grams per litre. Values of up to 1.6 g l^{-1} from sea ice on the Alaskan north slope have been measured by Barnes *et al.* (1982).

The third process of active incorporation involves the formation of anchor ice on the sea floor (Dayton *et al.* 1969; Reimnitz *et al.* 1987; Reimnitz & Kempena 1988). This is best developed in moderately shallow water where radiative cooling from the sea water and sea floor to a cold sky allows ice (often as frazil) to accumulate especially on rocky bottoms. Areas exposed at low tide may also be cooled sufficiently that, when the tide comes in, ice formation on the

Fig. 6. Ice on intertidal flats at Pangnirtung, Baffin Island during breakup, June 15, 1986, showing actively incorporated sediment by basal freezing and passively loaded sediment on the tops of the pans. Photograph by B.A. Crawford.

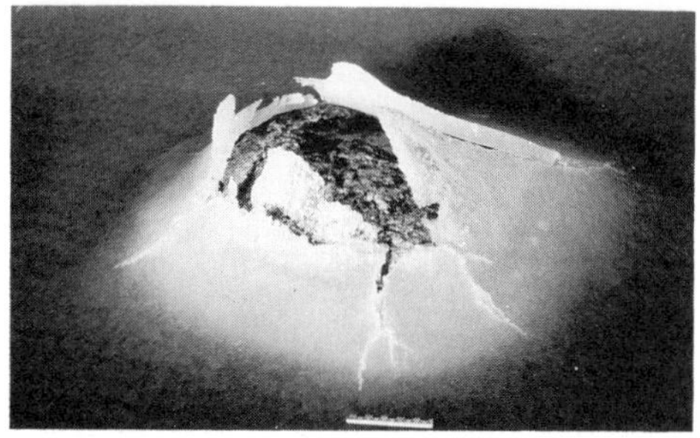

Fig. 8. Photograph of sea ice around a boulder during freeze-up, September 1983 in McBeth Fiord, eastern Baffin Island showing the method of incorporation of cobbles and boulders into sea ice. Two tidal cycles have occurred since the ice began to form. At low tide ice has broken over the top of the boulder, and is freezing around the sides. As the ice thickens, buoyancy increases and the boulder is lifted on each tidal cycle. Scale is 10 cm long.

substrate occurs. Fresh water inflow beneath the surface may also cause anchor ice formation (Sadler & Serson 1981).

As isolated patches of anchor ice grow, their buoyancy may be sufficient to lift sediments and plants from the sea floor. Scour by floating ice may also break a crust of anchor ice allowing it to float upward. These parcels become incorporated in the sea ice cover on the water surface (Reimnitz *et al.* 1987).

Passive loading of sea ice in the littoral environment occurs mainly in spring, as currents generated by rising tides force sediment eroded from the sea floor immediately below through cracks and melt channels onto the ice surface. Wind and waves may be locally important in passive loading at any time during the period of ice cover, but especially when ice is driven ashore, forming gouges and mixing with material from beaches (Andrews & Matsch 1983). However, unless this occurs at low tide, and the tidal range is sufficient to allow refloating, the forces involved usually ground the ice so securely that rafting of its load back to sea is precluded (Kovacs & Sodhi 1980).

In a study at Pangnirtung during breakup in

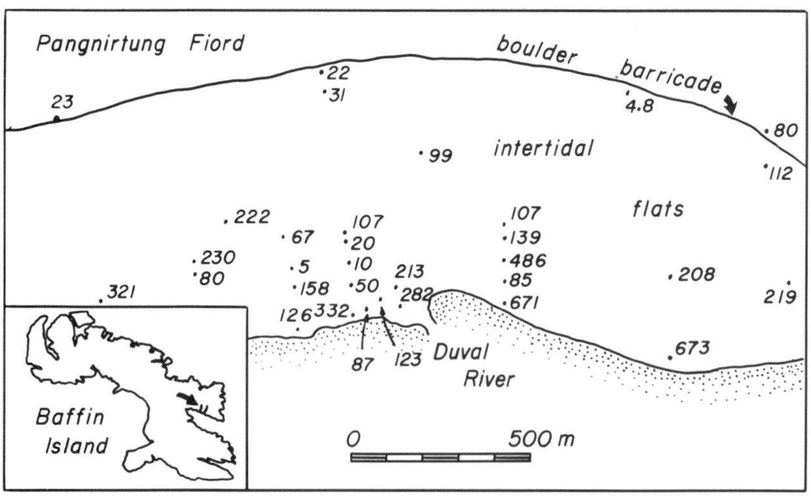

Fig. 7. Intertidal flats at Pangnirtung, June 1986 showing concentrations in g l^{-1} of actively incorporated sediment in sea ice.

June 1986, upwelling through the ice during each rising tide (range 4 to 7 m) allowed sand and small pebbles to be deposited on the ice surface (Fig. 9). During a period of two weeks up to 10 cm of sediment was deposited on the ice, occasionally as much as 1 cm per tidal cycle (cf. Dionne 1984).

Rafting by vegetation

In many glacimarine settings, especially of subpolar regions, benthic macroalgae are an important component of both the littoral and sea floor environments (e.g. Lee 1980). Algae with gas-filled vesicles in the thalli exert lift on the particles to which they are attached. For example in a study of the alga *Fucus vesiculosus* at Pangnirtung Fiord, Gilbert (1984a) showed that when a plant grows to about three times the weight of the stone to which it is attached, it is capable of floating the stone (Fig. 10).

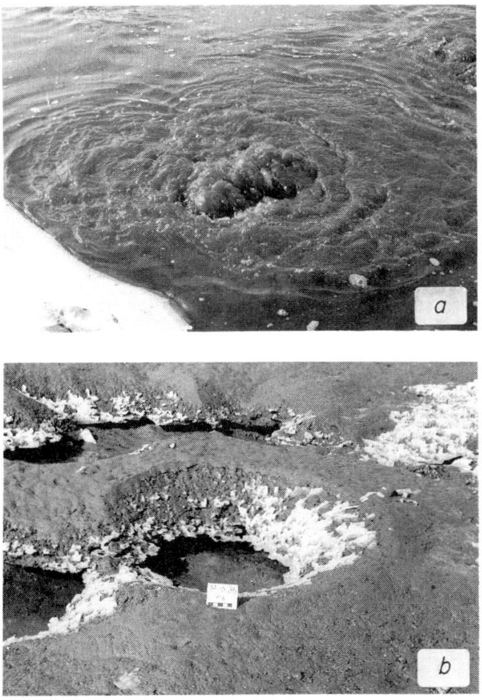

Fig. 9. Melt hole in sea ice on the intertidal flats at Pangnirtung, Baffin Island (a) during rising tide on June 6, 1986 and (b) with sand and gravel deposited as a result of passive loading by June 14. Maximum sediment accumulation is about 10 cm. Scale is 10 cm long. Photographs by B.A. Crawford.

Fig. 10. Stone weighing 72 g with attached alga, *Fucus vesiculosus* sufficient to float it from Pangnirtung Fiord, Baffin Island. Scale is in centimetre divisions.

In addition, the algae are more easily frozen into sea ice around the plants, or as anchor ice accumulates on the plants at some depth below the water surface. In the latter case, the additional buoyancy may be enough to float a stone and attached algae. Beds of algae, some with attached stones and shells, are common in the sea ice of littoral areas.

It is impossible to assess the relative importance of ice rafting and algal rafting in the glacimarine record because both processes are variable in space and time, and are sufficiently unpredictable to make measurement difficult. As well, the evidence of algal transport, namely holdfasts and their marks (Huggett & Kidd 1983), is quickly obliterated by consuming organisms when the plant dies. However, as an indication of algal rafting, rafted particles larger than 16 mm from 49 samples from the floor of Pangnirtung Fiord, Baffin Island were examined. Evidence of *Fucus* holdfasts was found on 5 of 157 particles varying in diameter from 42 to 61 mm.

The presence of coarse particles rafted by algae in glacimarine sediments is important even if the actual amount of sediment moved is small. The occurrence of large particles in a fine-grained sediment far from the source is evidence of rafting and is used to define glacial environments. Emery (1963) shows that the zone of algal rafting overlaps much of the subpolar area of ice rafting in the Northern Hemisphere. Thus, caution must be exercised to distinguish ice rafting and algal rafting when environmental interpretations are being made (Menard 1953).

Distribution of rafted sediment

Rafting depends on a number of factors which must act together. They include:

the availability of material to be rafted;

the amount and distribution of sediment in and on the raft;

the strength and stability of the raft;

the rate of deterioration of the raft;

the freedom of movement of the raft;

patterns and persistence of winds, waves and currents to drive the raft (cf. Piper 1976).

Since most of these factors are independent of each other, their combination to produce a deposit of rafted sediment at any given place and time is normally completely unpredictable, although longer-term average conditions (especially oceanic circulation) produce recognizable patterns over large areas.

One of the critical factors is the relative amount of sediment and ice in the raft, because even a modest concentration of sediment is sufficient to prevent the raft from floating (Fig. 11). Some of the measured concentrations due to active incorporation on the intertidal flats at Pangnirtung (Fig. 7) substantially exceed these values, and ice submerged by its sediment load probably is common in nearshore areas. Single large particles may also be incorporated in sea ice (Fig. 8). Drake & McCann (1982) present calculations that indicate a clean pan of about 35 m diameter and 1 to 2 m thick may float a single boulder of 5 to 6 m diameter. Gilbert & Aitken (1981) show that only boulders smaller than about 4 m diameter are lifted by sea ice. The basal ice of glaciers may also carry concentrations of sediment more than sufficient to sink that part of the ice (Shaw 1977; Dowdeswell 1986; Dowdeswell & Dowdeswell 1989).

Submerged, sediment-laden sea ice or icebergs may be transported by bottom currents in the glacimarine environment. In the absence of direct observation, Josenhans & Woodworth-Lynas (1988) suggest that small furrows in the bottom sediments of the Labrador shelf may originate from movement of submerged ice. Rafting by submerged ice may distribute sediment in very different patterns than occurs in circulation driven by surface currents and winds. Submerged ice that comes to rest probably deposits sediments having the characteristics of frozen aggregates or ice contact deposits.

The distribution of sediment in the raft is important. Sediment actively frozen into bergs or sea ice is usually released by melting out one particle at a time. Thus, particles are much more likely to be deposited as single drop stones over a wider area and a longer time (Dowdeswell & Murray 1990).

Both actively and passively derived sediment may be concentrated on the ice surface by melting during summer. The relation between melt-

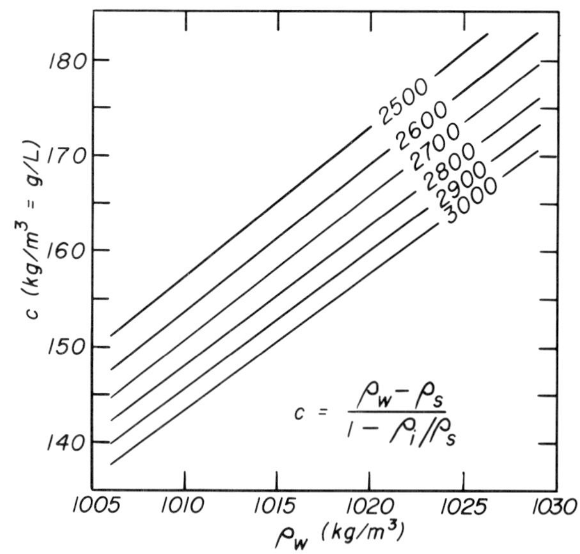

Fig. 11. Concentration, c, of sediment with density, ρ_s, from 2500 to 3000 kg m^{-3} in ice of density, $\rho_i = 910$ kg m^{-3} necessary to sink the ice in sea water having density, ρ_w.

ing of ice and its sediment content or cover is a complex one involving feedback between the energy balance (especially the net radiation component), characteristics of the sediment (thermal conductivity, shear strength, water content, albedo) and morphology of the melting forms (height, shape, roughness) (Drewry 1972). Even a light dust cover greatly reduces albedo (Woo & Dubreuil 1985). Single particles larger than about 40 mm diameter (depending on surface conditions) accelerate melting (Simmons *et al.* 1986). A complete cover of sediment increases the melting rate until its thickness is about 15 to 25 mm (Driedger 1981), with maximum increase when the cover is about 3 mm thick. Thickness of about 100 mm may decrease the melt rate by 4 to 5 times that of clear ice (Nakawo & Young 1981). The most effective particle size in reducing melt is in the range 0.6 to 5 mm (Drewry 1972). Smaller particles flow down slope more easily to expose ice beneath; larger particles allow freer circulation of air and thus heat through the sediment cover.

Sediment concentrated on the surface of sea ice may retard melting of pans in the nearshore until after breakup has occurred offshore. In some cases this may give these sediment-laden, nearshore pans greater opportunity to drift to sea.

New ice may form on the bottom of berg or sea ice, especially during winter in cold polar seas (Fuchs & Whittard 1930). Thus, although sediment does not move up through the ice, the melting above and freezing below eventually leads to a concentration of sediment on the ice surface. Rafts bearing sediment loads on their surface may be more unstable (the more dense sediment is above the centre of gravity). They are more likely to turn over and dump their loads in a single event (Vorren *et al.* 1983).

Ice rafting is most successful when the raft remains intact until it moves some distance from the source. Icebergs are efficient agents of transportation because of their size. Sea ice presents a different picture. In some cases unconsolidated slush ice that forms in early winter may transport large quantities of sediment before a solid landfast ice sheet forms (Reimnitz & Kempema 1987). However, the dispersal of stable pans produced during spring breakup is affected by several factors (Barry & Jacobs 1978). Most of the sediment load is obtained in the nearshore area where ice normally breaks up and melts first. The relatively stable barrier of ice offshore prevents the escape of sediment-laden pans from near shore. This mechanism has been proposed to explain the presence of boulder barricades at the seaward limit of intertidal flats (Rosen 1979). In the study at Pangnirtung, almost none of the ice in the intertidal area (Fig. 7) was observed to escape to sea for this reason. Domack's (1988) model of regions dominated by sea ice rafting near shore and by berg rafting at a much greater distance (especially at the limit of and beyond the sea ice zone) reflects this difference in behaviour of bergs and sea ice.

Deposition from rafts

Sediment is lost from the raft (1) by melting from the ice, (2) by washing or falling especially from unstable rafts, or (3) by sinking as a result of sediment concentration by melting or as a sediment-charged portion breaks off (Campbell & Collin 1958; Ovenshine 1970; Vorren *et al.* 1983). Loss from vegetation rafts occurs as a portion of the raft is damaged so that floatation is decreased or as the holdfast deteriorates.

Once free of the raft, single sediment particles fall through the water at rates that depend on (in decreasing order of importance) the size of the particle, its shape, its submerged weight, and the viscosity of the water which is a function of temperature. Various empirical and theoretical solutions (e.g. Dietrich 1982) produce a range of settling velocities depending on these characteristics (Fig. 12).

There are very few observations on the distance through which acceleration occurs (that is, the distance to reach terminal fall velocity). From values for larger particles given by Herbich (1981) and shown in Fig. 12, it can be seen that within a short distance even large particles reach terminal velocity, and the energy expended by a particle striking the bottom is independent of depth beyond this distance.

The behaviour of clouds of falling sediment as might be released from the surface of a raft during overturn has received some attention because of concern about the disposal of dredge spoil in deep water. Krishnappan (1975) presents a model for settling of a cloud of sediment which involves an entrainment phase during which the cloud spreads laterally and a settling phase below where velocity and size of the cloud remain constant. Results of calculations of these characteristics from Krishnappan's experiments with sand clouds are shown in Fig. 13. The effect of mixed grain-size dumps can be accounted for by a method of partitioning. Although the calculations are extrapolated from small-scale laboratory experiments, they indicate that the spreading may occur to considerable depth and that the area over which the

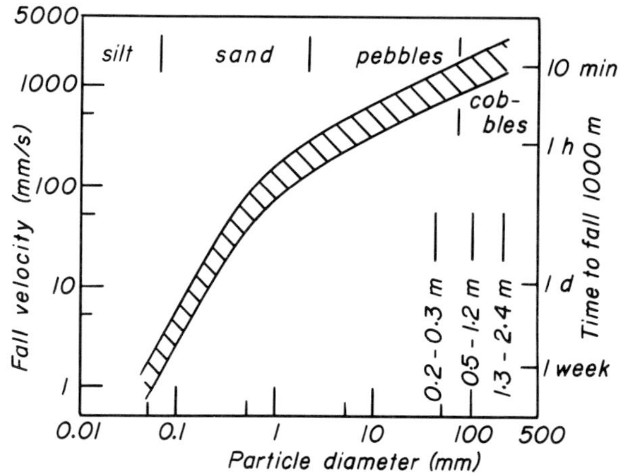

Fig. 12. Range in settling velocities for rounded to angular particles of density 2650 to 2900 kg m^{-3} in sea water at 0°C (after Dietrich 1982). Values in lower right are the distances to reach terminal fall velocity for particles accelerating from rest (Herbich 1981).

sediment spreads in deep water may be rather large.

Rafted deposits

Free-falling particles may significantly turbate soft sediments into which they fall (Thomas & Connell 1985). The amount of disturbance is a function of the kinetic energy of the particle on impact, the angle of impact determined by the slope of the sea floor and the fluttering motion of the particle during descent, the projection of the particle into the sediment based on its shape, and the strength of the host sediment. Impact turbation may rival bioturbation as a mechanism for mixing sediments. For example, measure-

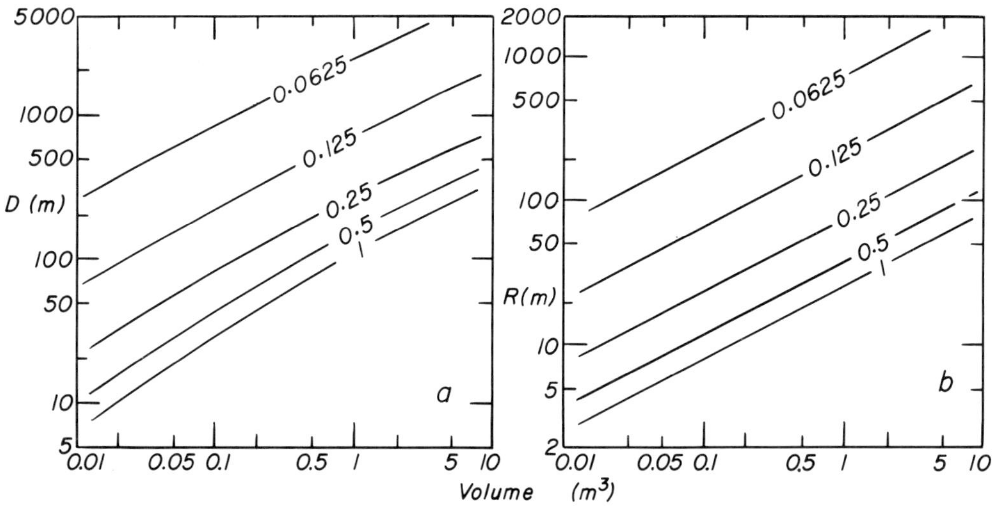

Fig. 13. (a) Depth, D, to which entrainment and spreading of settling clouds occurs and (b) radius, R, of the cloud at that depth for volumes of sediment having grain diameters between 0.0625 and 1 mm. Calculated from analysis by Krishnappan (1975). In water depths greater than D, R, is approximately constant. In water depths less than D, the cloud has radius, $r = Rz/D$ where z is the depth in metres.

ment of the concentration of drop stones larger than 2 mm in Cambridge Fiord, northern Baffin Island, indicates as many as several hundred particles per litre at some sites (Fig. 14). While fine laminations observed in cores from the same area are only slightly disturbed by the smaller stones, particles larger than about 8 mm have sufficiently mixed the fine sediments that laminations may be obliterated.

A dump of sediment of mixed grain size spreads over a wide area (Fig. 13) and separates by grain size to form a graded bed. This deposit might be mistaken for a turbidite, even when seen in an extensive exposure. Lateral separation by grain size also occurs if transport by currents occurs during descent (Krishnappan 1975), thus concentrating coarse sediments near the dump site and fine sediments distally (Hoskin & Valencia 1976; Elverhøi & Roaldset 1983). As a result, a dump may appear similar to the distally fining deposit from a turbid plume issuing from a glacier or stream.

Rafted sediment deposited as a frozen aggregate is difficult to identify because the ice has normally melted long before examination of the feature in a core or section. Indirect evidence includes the occurrence of sedimentary structures preserved when the sediment was incorporated into the raft. For example, active incorporation of beach sands by sea ice may involve little disruption (Medcof & Thomas 1974). Other indicators are the presence of a sharp and irregular boundary between the aggregate and the host sediment, and the presence of an open structure among the particles, previously filled with ice (Thomas & Connell 1985).

The blocks of well sorted medium sand shown in Fig. 15 suggest deposition as frozen aggregates. Dark layers are concentrations of heavy minerals, especially garnet. Similar laminations are common on beaches in the region. The two

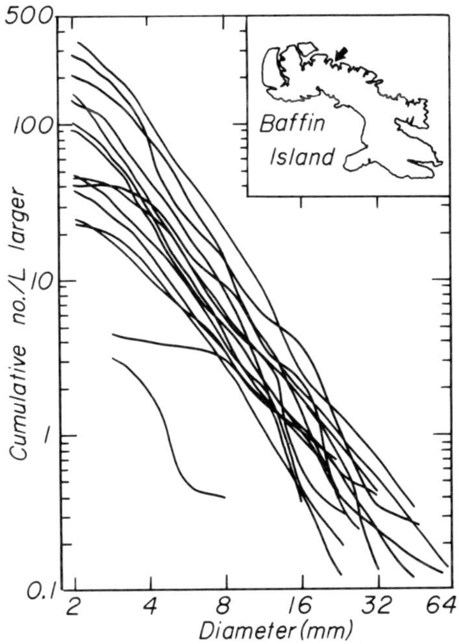

Fig. 14. Number of drop stones in the sediments of Cambridge Fiord, northern Baffin Island. Samples were recovered along the length of the fiord with a vanVeen grab sampler and washed on a 2 mm sieve to recover the drop stones (from Gilbert 1984b).

Fig. 15. Frozen aggregates of beach sand at 5.36 m below the sediment surface (top of photograph) in a piston core recovered 26 km from the head of Cambridge Fiord, northern Baffin Island. (Scale is in 2.5 mm divisions.)

blocks appear to have been deposited at the same time from the same frozen piece.

Distinguishing rafting agents

Because many authors have used the presence of rafted debris as an indicator of glacial history, separation of modes of rafting is important. A list of the major characteristics used to distinguish rafting agents is given in Table 1. Both bergs and sea ice deposit drops, dumps, frozen aggregates and ice-contact sediments, although large dumps may be more characteristic of bergs.

Grain size is a poor indicator of the rafting agent (Fig. 16). It has been suggested that fine-grained sediments are absent in sea-ice rafted debris (Schermerhorn 1974), but both active and passive processes may involve loading with considerable amounts of silt and clay-sized particles (Barnes *et al.* 1982). Very large particles (more than about 4 m diameter) are most likely incorporated and transported only by icebergs (Snelgrove 1984). Vegetation rafting is confined mainly to pebbles, although some sand may be involved. Berg rafted sediments are more likely to be poorly sorted with up to one half or more mud (Hoskin & Valencia 1976; Anderson *et al.* 1980), although sediments of colluvial origin may also include a wide range in a single event (Stefansson 1921).

Shape may be used to suggest origin, but it is not definitive. Well rounded particles are typical of beach and nearshore areas, but in polar environments frost shattering, protection from wave action by ice, and isostatic adjustment to expose new material mean that sedimentary particles may be less rounded (Nichols 1961; McCann & Owens 1969). As well, sediment loaded by colluvial processes is normally angular (Luckman 1975). Angular particles are typical of glaciers, but rounding does occur (Snelgrove 1984; Dowdeswell & Dowdeswell 1989), especially during transport by water within or on the ice.

Fabric is useful for separating dumps and drops from subglacial till (Domack & Lawson 1985). However, it cannot distinguish ice contact sediment from subglacial till, or iceberg rafting from sea ice rafting.

The presence of striations and facets on particles is probably the best evidence of glacial origin (Huggett & Kidd 1983), although sediment transported above the base of the glacier does not become striated (Domack *et al.* 1980). On the other hand, the action of ice push on shores is known to create striations indistinguishable from those made by glaciers (Nichols 1961; McLellan 1971; Dionne 1985).

The presence of shallow water macrofauna within rafted sediments (Reineck 1976) or evidence of shallow water organisms, such as

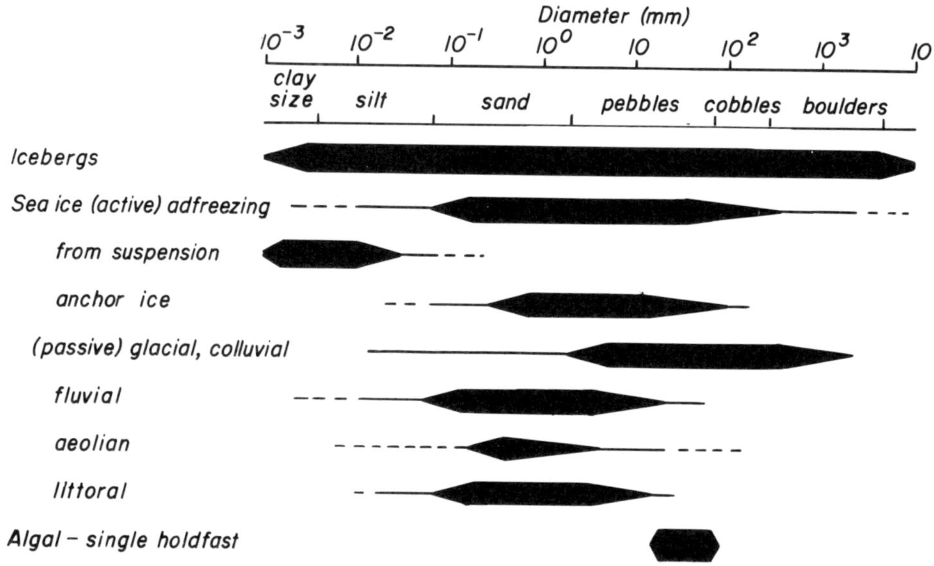

Fig. 16. Approximate range in grain size of rafted sediments.

rock borings (Huggett & Kidd 1983) is an indication of sea ice rafting. However, the absence of these is not a good indication of berg rafting because it is also absent in sea ice loaded by colluvial, aeolian and fluvial processes.

Sea ice is likely to deposit its sediment closer to the point of origin. Because the rafts are smaller, they melt more quickly and are more easily upset (Schermerhorn 1974; Spjeldnaes 1981; Domack 1988). Nevertheless, there are large areas of the sea over which both range together and, except in shallow water where bergs are unable to float, distance of transport may be an unreliable indication of source.

It is clear that sand and gravel particles deposited in deep water at some distance from source areas, and not associated with gravitational processes, are rafted. Even if it is possible to eliminate vegetation rafting by sediment size (Fig. 16) or by the amount of material involved, it is often very difficult to separate sea ice and iceberg rafting conclusively (Spjeldnaes 1981; Reimnitz & Kempema 1988). Conclusions about the extent of glaciation based on interpretation of rafted sediment must be made cautiously.

Conclusions

1. Rafted sediments are a major (in some cases the largest) component of glacimarine sediment. They provide information on the extent of glacial processes and the characteristics of glacial seas.

2. The major agents of rafting are icebergs and sea ice, although algal rafting may be significant locally. Drop stones from algae may not be distinguishable from ice rafted sediment.

3. The major sources of rafted sediments are glacial erosion, colluvial, fluvial and aeolian processes and the littoral environment.

4. Loading occurs by active freezing of sediment to sea ice (as a result of contact with the sea floor, scavenging by frazil ice or formation of anchor ice) or in glacier ice associated with flow, and by passive loading onto the ice surface. Passive loading occurs when sea ice persists unbroken from shore during the period in which loading processes are most active (during nival melt for colluvial and fluvial processes and throughout winter for aeolian processes).

5. Since actively incorporated sediment is distributed through bergs or sea ice, it is more likely to be freed as individual drop stones melting out one at a time, unless it has been concentrated on the ice surface by melting. In this case it behaves as passively loaded sediment, more likely to be dumped *en masse* as a result of slope failure, runoff or upset.

6. Sediment dispersed through berg or sea ice may be concentrated on the ice surface by melting from the top, in some cases while freezing is occurring at the base. Rates of melting depend in part on the ice cover, with accelerated melting caused by sediment load until a thickness of about 15 to 20 mm is reached, after which melting is retarded.

7. Dispersal of sand and mud from a single dumping event may be widespread as a result of the behaviour of the cloud of falling sediment, and if settling velocity is small enough to allow currents to spread the falling material. Resulting deposits may be confused with those created by turbid plumes in underflow or overflow.

8. Deposition of non-buoyant frozen aggregates may occur. As part of this process, the transport of slightly negatively buoyant frozen blocks along the sea floor may raft sediment long distances in patterns different from those at the surface. Preservation of sedimentary structures in frozen aggregates may offer an indication of the origin and mode of transport.

9. Distinguishing the agent of rafting from the nature of the sedimentary deposit is often difficult. The most reliable indication of iceberg rafting is the presence of striated facets on large particles. Other indicators that may be useful in particular circumstances include particle shape, associated shallow water fauna, presence of friable or weathered material, and the depth of water. Just as the presence of rafted debris may not be directly correlated with glacial events, its absence may not be related to non-glacial environments.

Parts of the study were supported by the Natural Sciences and Engineering Research Council of Canada and the Bedford Institute of Oceanography. For the latter, I appreciate the opportunities provided by J.P.M. Syvitski and C.S. Schafer. Thanks to my students, B.A. Crawford, J.E. Dale, and C. McKenna Neuman, for allowing me to use their material to illustrate the paper. The Geological Society of London provided financial assistance to present the paper. Valuable comments on a draft manuscript were provided by D.L. Naldrett, J.T. Andrews and P.W. Barnes.

References

Ackley, S.F. 1982. Ice scavenging and nucleation: two mechanisms for incorporation of an algae into newly forming sea ice. *EOS Transactions of the American Geophysical Union*, **63**, 54.

ANDERSON, J.B., DOMACK, E.W. & KURTZ, D.D. 1980. Observations of sediment-laden icebergs in Antarctic waters: implications to glacial erosion and transport. *Journal of Glaciology*, **25**, 387–396.

ANDREWS, J.T. & MATSCH, C.L. 1983. *Glacial Marine Sediments and Sedimentation. An annotated Bibliography*. Geo Abstracts, Bibliography No. II. Elsevier, Norwich.

BAGNOLD, R.A. 1941. *The Physics of Blown Sand and Desert Dunes*. Methuen, London.

BARNES, P.W., REIMNITZ, E. & FOX, D. 1982. Ice rafting of fine-grained sediment, a sorting and transport mechanism, Beaufort Sea, Alaska. *Journal of Sedimentary Petrology*, **52**, 493–502.

BARRETT, P.J., PYNE, A.R. & WARD, B.L. 1983. Modern sedimentation in McMurdo Sound, Antarctica. *In*: OLIVER, R.L., JAMES, P.R. & JAGO, J.B. (eds) *Antarctic Geoscience*, Australian Academy of Science, Canberra, 550–554.

BARRY, R.G. & JACOBS, J.D. 1978. *Energy Budget Studies in Relation to Fast-Ice Breakup Processes in Davis Strait: Climatological Overview*. University of Colorado, Institute of Arctic and Alpine Research, Occasional Paper No. 26.

BENTLEY, P.N. 1979. *Characteristics and Distribution of Wind Blown Sediment, Western McMurdo Sound, Antarctica*. BSc. Thesis, Victoria University of Wellington, New Zealand.

BLENCH, T.R. 1969. *Mobile-Bed Fluviology*. The University of Alberta Press, Edmonton.

CAMPBELL, N.J. & COLLIN, A.E. 1958. The discoloration of Foxe Basin ice. *Journal of the Fisheries Research Board of Canada*, **15**, 1175–1188.

CHURCH, M. 1974. Hydrology and permafrost with reference to northern North America. *In*: *Proceedings Workshop Seminar on Permafrost Hydrology*, Canadian National Committee, International Hydrological Decade, Ottawa, 7–20.

—— & RYDER, J.M. 1972. Paraglacial sedimentation: a consideration of fluvial processes conditioned by glaciation. *Geological Society of America, Bulletin*, **83**, 3059–3072.

——, STOCK, R.F. & RYDER, J.M. 1979. Contemporary sedimentary environments on Baffin Island, N.W.T.: debris slope accumulation. *Arctic and Alpine Research*, **11**, 371–402.

DAYTON, P.K., ROBBILIARD, G.A. & DE VRIES, A.L. 1969. Anchor ice formation in McMurdo Sound, Antarctica, and its biological effects. *Science*, **163**, 273–274.

DIETRICH, W.E. 1982. Settling velocity of natural particles. *Water Resources Research*, **18**, 1615–1626.

DIONNE, J.-C. 1984. An estimate of ice-drifted sediments based on the mud content of the ice cover at Montmagny, Middle St. Lawrence estuary. *Marine Geology*, **47**, 149–166.

—— 1985. Drift-ice abrasion marks along rocky shores. *Journal of Glaciology*, **31**, 237–241.

DOMACK, E.W. 1988. Biogenic facies in the Antarctic glacimarine environment: basis for a polar glacimarine summary. *Palaeogeography, Palaeoclimatology, Palaeoecology*, **63**, 357–372.

——, ANDERSON, J.B. & KURTZ, D.D. 1980. Clast shape as an indicator of transport and depositional mechanisms in glacial marine sediments: George V continental shelf, Antarctica. *Journal of Sedimentary Petrology*, **50**, 813–820.

—— & LAWSON, D.E. 1985. Pebble fabric in an ice-rafted diamicton. *Journal of Geology*, **93**, 577–591.

DOWDESWELL, J.A. 1986. The distribution and character of sediments in a tidewater glacier, southern Baffin Island, N.W.T., Canada. *Arctic and Alpine Research*, **18**, 45–56.

—— 1987. Processes of glacimarine sedimentation. *Progress in Physical Geography*, **11**, 52–90.

—— & DOWDESWELL, E.K. 1989. Debris in icebergs and rates of glaci-marine sedimentation: observations from Spitsbergen and a simple model. *Journal of Geology*, **97**, 221–231.

—— & MURRAY, T. 1990. Modelling rates of sedimentation from icebergs. *In*: DOWDESWELL, J.A. & SCOURSE, J.D. (eds) *Glacimarine Environments: Processes and Sediments*, Geological Society, London, Special Publication **53**, 121–138.

DRAKE, J.J. & MCCANN, S.B. 1982. The movement of isolated boulders on tidal flats by ice flows. *Canadian Journal of Earth Sciences*, **19**, 748–754.

DREIMANIS, A. 1979. The problem of waterlain tills. *In*: SCHLÜCHTER, Ch, (ed.) *Moraines and Varves* Balkema, Rotterdam, 176–177.

DREWRY, D. 1972. A quantitative assessment of dirt-cone dynamics. *Journal of Glaciology*, **11**, 431–446.

DRIEDGER, C.L. 1981. Effect of ash thickness on snow ablation. *U.S. Geological Survey, Professional Paper*, **1250**, 757–760.

DUNBAR, R.B., ANDERSON, J.B. & DOMACK, E.W. 1985. Oceanographic influences on sedimentation along the Antarctic continental shelf. *In*: *Oceanology of the Antarctic Continental Shelf*, Antarctic Research Series, 291–312.

ELVERHØI, A. & ROALDSET, E. 1983. Glaciomarine sediments and suspended particulate matter, Weddell Sea shelf, Antarctica. *Polar Research*, 1 n.s., 1–21.

EMERY, K.O. 1963. Organic transportation of marine sediments. *In*: HILL, M.N., (ed.) *The Sea*, Wiley Interscience, New York, 776–793.

FERGUSON, L. 1970. "Armored snowballs" and the introduction of coarse terrigenous material into sea-ice. *Journal of Sedimentary Petrology*, **40**, 1057–1060.

FUCHS, V.E. & WHITTARD, W.F. 1930. The east Greenland pack-ice and the significance of its derived shells. *Geographical Journal*, **76**, 419–424.

GILBERT, R. 1984a. The movement of gravel by the alga *Fucus vesiculosus* (L.) on an arctic intertidal flat. *Journal of Sedimentary Petrology*, **54**, 463–468.

—— 1984b. Coarse particles in the sediments of Cambridge, McBeth and Itirbilung fiords. *In*:

Sedimentology of Arctic Fjords Experiment: HU 83–028 Data Report, Volume 2. Canada Data Report of Hydrography and Ocean Sciences, No. 28. Fisheries and Oceans Canada, 9–1 to 9–25.

GILBERT, R. & AITKEN, A.E. 1981. The role of sea ice in biophysical processes on intertidal flats at Pangnirtung (Baffin Island), N.W.T. *In*: DIONNE, J.-C, (ed.) *Workshop on Ice Action on Shores*, National Research Council of Canada, Associate Committee for Research on Shoreline Erosion and Sedimentation, 89–103.

——, ——, MCKENNA NEUMAN, C., DALE, J.E. & GLEW, J.R. 1987. Sedimentary environments: Pangnirtung Fiord. *In*: Cumberland Sound and Frobisher Bay, southeastern Baffin Island, N.W.T. XIIth INQUA Congress Field Excursion C-2, 23–51.

HERBICH, J.B. 1981. *Offshore Pipeline Design Elements*. Marcel Dekker, New York.

HOSKIN, C.M. & VALENCIA, S.M. 1976. Sediment transport by ice-rafting in southcentral Alaska. *In*: HOOD, D.W. & BURRELL, D.E., (eds) *Assessment of the Arctic Marine Environment: Selected Topics*, Institute of Marine Science, University of Alaska, Fairbanks, 173–185.

HOWARTH, P.J. & BONES, J.G. 1972. Relationship between process and geometrical form on High Arctic debris slopes, southwest Devon Island. *In*: PRICE, R.J. & SUGDEN, D.E., (eds) *Polar Geomorphology*, Institute of British Geographers, Special Publication **4**, 139–154.

HUGGETT, Q.J. & KIDD, R.B. 1983. Identification of ice-rafted and other exotic material in deep-sea dredge hauls. *Geo-Marine Letters*, **3**, 23–29.

HUME, J.D. 1964. Floating sand and pebbles near Barrow, Alaska. *Journal of Sedimentary Petrology*, **262**, 267–273.

HUPPERT, H.E. 1980. The physical processes involved in the melting of icebergs. *Annals of Glaciology*, **1**, 97–102.

JOSENHANS, H. & WOODWORTH-LYNAS, C. 1988. Enigmatic linear furrows and pits on the upper continental slope, northwest Labrador Sea: are they sediment furrows or feeding traces. *Maritime Sediments and Atlantic Geology*, **24**, 149–155.

KEYS, J.E. 1978. *Water Regime of Disraeli Fiord, Ellesmere Island, Canada*. Department of National Defence, Defence Research Establishment, Report No. 792.

KIDD, R.B. & HUGGETT, Q.J. 1981. Rock debris on abyssal plains in the Northeast Atlantic: a comparison of epibenthic sledge hauls and photographic surveys. *Oceanological Acta*, **4**, 99–104.

KINDLE, E.M. 1924. Observations on ice-borne sediments by the Canadian and other arctic expeditions. *American Journal of Science*, 5th Series, **7**, 251–286.

KOVACS, A. & SODHI, D.S. 1980. Shore ice pile-up and ride-up: field observations, models, theoretical analysis. *Cold Regions Science and Technology*, **2**, 209–288.

KRISHNAPPAN, B.G. 1975. *Dispersion of Granular Material Dumped in Deep Water*. Inland Waters Directorate, Canada Centre for Inland Waters, Scientific Series No. 55.

LAURIOL, B. & GRAY, J.T. 1980. Processes responsible for the concentration of boulders in the intertidal zone in Leaf Basin, Ungava. *Geological Survey of Canada Paper 80–10*, 281–292.

LEE, R.K.S. 1980. *A Catalogue of the Marine Algae of the Canadian Arctic*. National Museums of Canada, Publications in Botany **9**.

LUCKMAN, B. 1975. Drop stones resulting from snow-avalanche deposition on lake ice. *Journal of Glaciology*, **14**, 186–188.

MARRIOTT, R.S. 1940. Canada's Eastern Arctic Patrol. *Canadian Geographical Journal*, **20**, 156–161.

MCCANN, S.B. & OWENS, E.H. 1969. The size and shape of sediments in three arctic beaches, southwest Devon Island, N.W.T., Canada. *Arctic and Alpine Research*, **1**, 267–278.

MCKENNA NEUMAN, C. 1987. *Aeolian Processes and Landforms in South Pangnirtung Pass, Southeast Baffin Island, N.W.T., Canada*. PhD. Thesis, Queen's University, Kingston.

—— & GILBERT, R. 1986. Aeolian processes and landforms in glaciofluvial environments of southeastern Baffin Island, N.W.T., Canada. *In*: NICKLING, W.H., (ed.) *Aeolian Geomorphology*, Allen and Unwin, Boston, 213–235.

MCKINLAY, W.L. 1976. *Karluk*. Weidenfeld and Nicolson, Great Britain.

MCLELLAN, A.G. 1971. Ambiguous glacial striae formed near water bodies. *Canadian Journal of Earth Sciences*, **8**, 477–479.

MEDCOF, J.C. & THOMAS, M.L.H. 1974. Surfacing on ice of frozen-in marine bottom materials. *Journal of Fisheries Research Board of Canada*, **31**, 1195–1200.

MENARD, H.W. 1953. Pleistocene and Recent sediment from the floor of the northeastern Pacific Ocean. *Geological Society of America, Bulletin*, **64**, 1279–1294.

MIGALA, K. & SOBIK, M. 1984. Deflation and nival eolian phenomena observed under conditions of congelation in the forefield of the Werenskiold Glacier (SW Spitsbergen). *Zeitschrift für Gletscherkunde und Glazialgeologie*, **20**, 197–206.

MOLNIA, B.F. 1983. Subarctic glacial-marine sedimentation: a model. *In*: MOLNIA, B.F. (ed.) *Glacial-Marine Sedimentation,* Plenum, New York, 95–143.

NAKAWO, M. & YOUNG, G.J. 1981. Field experiments to determine the effect of a debris layer on ablation of glacier ice. *Annals of Glaciology*, **2**, 85–91.

NEDELL, S.S., ANDERSON, D.W., SQUYRES, S.W. & LOVE, F.G. 1987. Sedimentation in ice-covered Lake Hoare, Antarctica. *Sedimentology*, **34**, 1093–1106.

NICHOLS, R.L. 1961. Characteristics of beaches formed in polar climates. *American Journal of Science*, **259**, 694–708.

OVENSHINE, A.T. 1970. Observations of iceberg rafting in Glacier Bay, Alaska, and the identification of ancient ice-rafted deposits. *Geological Society of America, Bulletin*, **81**, 891–894.

PIPER, D.J.W. 1976. The use of ice rafted marine sediments in determining glacial conditions. *Revue de Géographie de Montréal*, **30**, 207–212.

REIMNITZ, E. & BRUDER, K.F. 1972. River discharge into an ice-covered ocean and related sediment dispersal, Beaufort Sea, coast of Alaska. *Geological Society of America, Bulletin*, **83**, 861–866.

—— & KEMPEMA, E.W. 1987. Field observations of slush ice generated during freeze-up in arctic coastal waters. *Marine Geology*, **77**, 219–231.

—— & —— 1988. Ice rafting: an indication of glaciation? *Journal of Glaciology*, **34**, 254–255.

REIMNITZ, E., KEMPEMA, E.W. & BARNES, P.W. 1987. Anchor ice, seabed freezing, and sediment dynamics in shallow arctic seas. *Journal of Geophysical Research*, **92**, 14671–14678.

REINECK, H.-E. 1976. Drift ice action on tidal flats, North Sea. *Revue de Géographie de Montréal*, **39**, 197–200.

—— & SINGH, I.B. 1975. *Depositional Sedimentary Environments*. Springer, Berlin.

ROSEN, P.S. 1979. Boulder barricades in central Labrador. *Journal of Sedimentary Petrology*, **49**, 1113–1124.

RUDDIMAN, W.F. 1977. Late Quaternary deposition of ice-rafted sand in the subpolar North Atlantic (lat 40° to 65°N). *Geological Society of America, Bulletin*, **88**, 1813–1827.

SADLER, H.E. & SERSON, H.V. 1981. Fresh water anchor ice along an arctic beach. *Arctic*, **34**, 62–63.

SCHERMERHORN, L.J.G. 1974. Late Precambrian mixtites: glacial and/or nonglacial? *American Journal of Science*, **274**, 673–824.

SHAW, J. 1977. Till body morphology and structure related to glacier flow. *Boreas*, **6**, 189–201.

SIMMONS, G.M., Jr., WHARTON, R.A., Jr., MCKAY, C.P., NEDELL, S. & CLOW, G. 1986. Sand/ice interactions and sediment deposition in perennially ice-covered Antarctic lakes. *Antarctic Journal of the United States*, **21**, 217–220.

SNELGROVE, T. 1984. Where's that iceberg from? *C-CORE News*, Centre for Cold Ocean Research and Engineering, Memorial Univ. **9(2)**, 4.

SPJELDNAES, N. 1981. Ice-rafting, an indication of glaciation? *Science*, **214**, 687–688.

STEFANSSON, V. 1921. *The Friendly Arctic*. MacMillan, New York.

SYVITSKI, J.P.M. & VAN EVERDINGEN, D.A. 1981. A re-evaluation of the geologic phenomenon of sand flotation: a field and experimental approach. *Journal of Sedimentary Petrology*, **51**, 1315–1322.

THOMAS, G.S.P. & CONNELL, R.J. 1985. Iceberg drop, dump, and grounding structures from Pleistocene glacio-lacustrine sediments, Scotland. *Journal of Sedimentary Petrology*, **55**, 243–249.

VORREN, T.O., HALD, M., EDVARDSEN, M. & LIND-HANSEN, O.-W. 1983. Glacigenic sediments and sedimentary environments on continental shelves: general principles with a case study from the Norwegian shelf. *In*: EHLERS, J., (ed.) *Glacial Deposits in North-West Europe*, Balkema, Rotterdam, 61–73.

WASHBURN, A.L. 1967. Instrumental observations of mass wasting in the Mesters Vig District, Northeast Greenland. *Meddelelser om Grønland*, **166**, Nr. 4.

WEERTMAN, J. 1961. Mechanism for the formation of inner moraines found near the edge of cold ice caps and ice sheets. *Journal of Glaciology*, **3**, 965–978.

WOO, M.-K. 1986. Permafrost hydrology in North America. *Atmosphere-Ocean*, **24**, 201–234.

—— & DUBREUIL, M. 1985. Empirical relationship between dust content and arctic snow albedo. *Cold Regions Science and Technology*, **10**, 125–132.

—— & STEER, P. 1986. Runoff regime of slopes in continuous permafrost areas. *Canadian Water Resources Journal*, **11**, 58–68.

Modelling rates of sedimentation from icebergs

JULIAN A. DOWDESWELL & TAVI MURRAY

Scott Polar Research Institute, University of Cambridge, Cambridge CB2 1ER, UK

Abstract: Icebergs transport debris from ice masses into the glacimarine environment, but iceberg production, debris content and sedimentation rate will vary with glacier dynamics and the nature of marine waters. A model incorporating glaciological and oceanographic parameters calculates the flux of debris calved in icebergs, iceberg melt rates and sedimentation rates with distance from calving tidewater ice fronts. The flux of basal debris calved through a 1 m width of tidewater margin is $(h_d\, c/100)\, v_c$, where h_d and c are basal debris-rich ice thickness and sediment content (%), and v_c is iceberg calving rate. Iceberg basal melting for any horizontal interval (dx) away from the calving glacier is $dy = (\dot{M}_b\, dx)/v_b$, where $\dot{M}_b$ is iceberg melt rate and v_b is horizontal iceberg velocity. Sediment release for any x increment depends on iceberg debris concentration and previous berg melt history. Model parameters vary with the nature of climatic conditions, on a continuum from cold polar environments (e.g. High Canadian Arctic, Antarctic Peninsula), through sub-polar environments (e.g. Baffin Island, Svalbard) to more temperate environments (e.g. SE Alaska). These parameters include: glacier flow velocity, glacier thermal regime, basal debris-rich ice thickness, and marine water temperature. Model results suggest that iceberg sedimentation rates are lowest for high polar settings, and most rapid in warm maritime and some sub-polar areas. However, icebergs in cold waters transport debris to significantly greater distances from the parent glacier. The model provides calculations of order of magnitude rates of iceberg sedimentation in varying glacimarine settings.

Icebergs transport into the marine environment debris derived from glaciers and ice sheets. The diagnostic environmental significance of sedimentary features such as ice rafted dropstones has been recognized for some time (e.g. Ovenshine 1970), but much less is known concerning the rates at which icebergs deliver material to marine waters. A better understanding of this problem is important, for example, to our interpretation of the depositional and environmental history of the Arctic Ocean (e.g. Clark & Hanson 1983; Sejrup *et al.* 1984). The lack of information on rates of sedimentation from icebergs stems largely from the difficulties involved in making precise field observations of parameters such as the rate of release of material from bergs. An alternative approach is to model the processes which influence iceberg sedimentation, and to derive sedimentation rates from these calculations (Dowdeswell & Dowdeswell 1989). Field data on glaciological, sedimentological and oceanographic parameters are still required, but the problematic direct observations of iceberg debris release over time are no longer necessary. Modelling also gives opportunities for the effects of changes in key parameters to be examined, allowing iceberg sedimentation rates under different glaciological and oceanographic conditions to be assessed.

In this paper, a model is presented for the calculation of sedimentation rates from icebergs. Each major parameter in the model is discussed and examined using a sensitivity analysis. The model is then applied to several environmental settings for which glaciological and oceanographic data are available. Model results are then compared with observed rates of total glacimarine sedimentation derived from dated core material.

A model for iceberg sedimentation

The model presented here is two-dimensional, implying that a longitudinal transect through the parent glacier and the ocean is considered. The tidewater glacier front, icebergs, and the ocean are specified as a two-dimensional grid of variable horizontal and vertical resolution (Fig. 1). A value for debris content is given to each grid element which is designated as an ice cliff or iceberg. A temporally-averaged water temperature is assigned to each grid element within the ocean, with finer resolution possible towards the top of the water column to take account of the seasonal layer. The model is applicable to any grounded ice mass ending in tidewater from which icebergs are calved (Fig. 2). It is not immediately appropriate to the study of icebergs calved from the large floating ice shelves of Antarctica, where undermelt may take place prior to calving. This may not be an important restriction, however, because Drewry & Cooper

From DOWDESWELL, J. A. & SCOURSE, J. D. (eds), 1990, *Glacimarine Environments: Processes and Sediments*. Geological Society Special Publication No 53, pp 121–137.

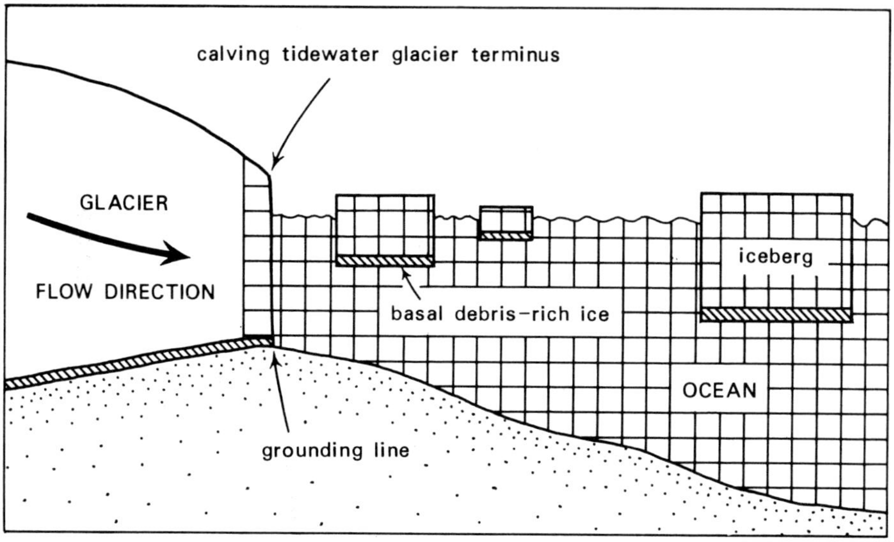

Fig. 1. Schematic diagram of the two-dimensional model of a glacier–ocean system with icebergs calved from the tidewater glacier ice front. The glacier terminus, icebergs and ocean can be divided into a grid of varying resolution. The diagram and grid are not to scale.

(1981) showed that bergs derived from large ice shelves will contain little basal debris-rich ice.

The major factors controlling sedimentation from icebergs include: (i) the concentration and distribution of debris in the parent ice mass; (ii) the rate of iceberg calving; (iii) the iceberg melt rate; (iv) the temperature structure of the ocean; and (v) iceberg velocity, stability and drift track.

Glacier and iceberg debris content

To establish the rate at which debris is released from tidewater glaciers to the marine environment, both the distribution of debris within the tidewater ice margin and the rate of iceberg calving must be specified. The flux of basal debris (f_s) through a 1 m width of tidewater ice margin is:

$$f_s = (h_d \, c/100)v_c \tag{1}$$

where h_d and c are the thickness and percentage debris concentration of basal debris-rich ice, and v_c is the rate of iceberg calving.

Equation (1) holds if we assume that the bulk of clastic material in tidewater glaciers is in a basal debris-rich zone, and that calving events include this basal ice. The significance of basal debris relative to material held at higher, englacial levels can be demonstrated. Let us take englacial debris content as 0.001% by volume, a typical observed value for a Baffin Island tidewater glacier (Dowdeswell 1986). For a 200 m deep iceberg with a 1 m basal layer of 20% debris content, englacial debris will contribute 1% of total iceberg debris load. Our model can take account of more complex debris distributions, however, because the ice margin, the icebergs and the ocean are defined as a two-dimensional grid. The vertical distribution of debris within a tidewater glacier or an iceberg calved from it can, therefore, be specified in the model.

The two parameters describing iceberg debris content have rather different effects on calculated sedimentation rates. This is demonstrated in Fig. 3, where calculations assume that basal melt rate, water temperature and iceberg drift velocity are held constant. Variations in basal debris layer thickness influence the distance over which an iceberg continues to transport sediment before that material is exhausted (Fig. 3). They do not, however, affect the rate at which sedimentation takes place. Conversely, the sedimentation rate from icebergs varies with the concentration of debris within the ice, but this does not affect the distance to berg debris exhaustion greatly (Fig. 3). The effects on sedimentation of variations in debris content within basal ice are shown in Fig. 4.

Iceberg calving rate

Once glacier debris content and distribution are defined, the rate of iceberg calving from a

a b c d

Fig. 2. Tidewater glacier termini where icebergs are calved from grounded ice into marine waters. (a) A calving tidewater valley glacier entering a northwest Spitsbergen fjord. (b) A small ice cap outlet glacier entering Watts Bay, an inlet in Frobisher Bay, southern Baffin Island. (c) Tidewater terminus of a large ice cap on Edgeøya, Svalbard. Large tabular icebergs are calved from here as well as small irregular bergs. (d) A tidewater glacier in eastern Svalbard, with winter shorefast sea ice.

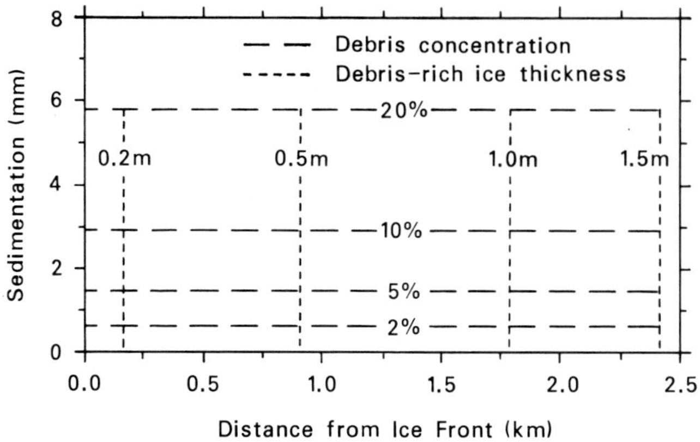

Fig. 3. The effect on sedimentation from a single iceberg of variations in the thickness and sediment concentration of basal debris-rich ice. Calculations use the melt rate equation of Russell-Head and assume an ocean temperature of 4°C and iceberg velocity of 0.005 m s^{-1}.

tidewater ice front (v_c) controls the amount of glacier-derived debris input to the marine environment in unit time. Iceberg production is calculated in one of two ways, depending on the type of field data available. Where direct observations on tidewater glacier velocity and terminal stability are made, or aerial photographic data are available for analysis:

$$v_c = u - m - \dot{M}_c \qquad (2)$$

where u is velocity at a tidewater glacier terminus, m is the rate of advance or retreat of the terminus, and $\dot{M}_c$ is the rate of direct melting of the tidewater ice cliffs. Thus, for a stable tidewater ice margin, calving rate is related mainly to glacier velocity, with a generally small subtraction for direct melting of the terminal ice cliffs (see Fig. 6). For a constant velocity, terminal advance will reduce the rate of iceberg production, whereas retreat will increase it.

If field data on iceberg melt rates are unavailable, an alternative approach may be used, based on the link between calving rate and water depth. Brown *et al.* (1982) proposed an empirical calving relation, based on observations of calving rate and water depth at the margins of 14 Alaskan tidewater glaciers:

$$v_c = b\, h_w \qquad (3)$$

where b is an empirical calving coefficient and

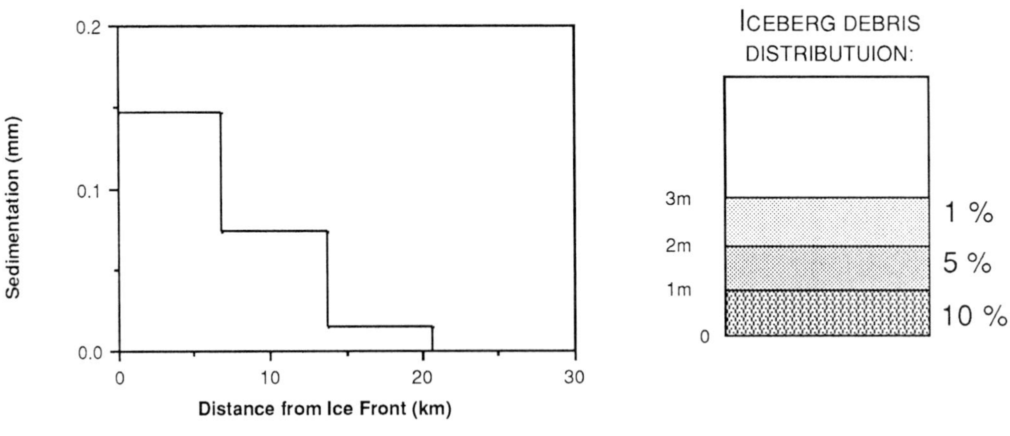

Fig. 4. Sedimentation from a single iceberg with an idealized basal sediment layer of varying debris concentration. The melt equation of Russell-Head is used, with water temperature of 4°C and berg velocity of 0.02 m s^{-1}.

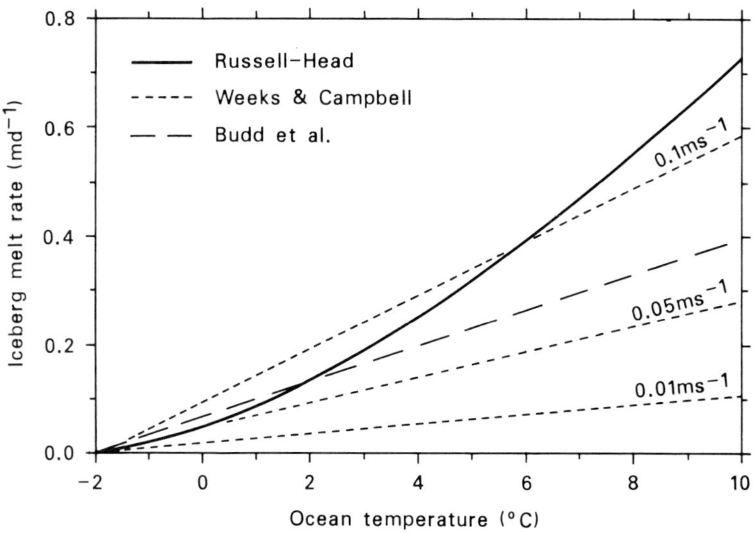

Fig. 5. The relationship between ocean temperature and iceberg basal melt rates from the predictive equations of Russell-Head (1980) and Weeks & Campbell (1973) and the observations of Budd *et al.* (1980). An iceberg of length 50 m is assumed for calculations using Weeks & Campbell's equation.

h_w is water depth at a tidewater glacier terminus. The best estimate of b is 27 a^{-1}, and the goodness of fit to data from the 14 glaciers is 0.91. Where direct observations of a given tidewater glacier are available they should be used instead of the calving relation in equation (3), because it is based on a small subset of temperate glaciers from Alaska only.

Iceberg basal melt rates

The melt rate of icebergs ($\dot{M}$) is crucial to the release of sediment into the water column. Several equations are available for the calculation of melting at the base of an iceberg ($\dot{M}_b$). They show marked differences in their results at higher water temperatures, and two are selected

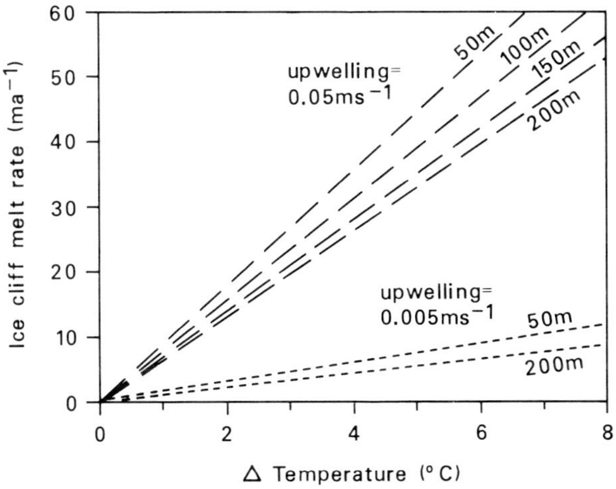

Fig. 6. The rate of melting of tidewater glacier ice cliffs predicted from the equation of Weeks & Campbell (1973) for varying velocities of upwelling and vertical ice cliff heights. Δ Temperature is the difference between ice cliff and water temperatures.

which provide relatively high and relatively low values for $\dot{M}_b$. The work of Russell-Head (1980) is based on laboratory experiments on small ice blocks (<1 m long) floating in saline water of varying temperature, whereas the calculations of Weeks & Campbell (1973) are based largely on theoretical considerations.

The empirical equation provided by Russell-Head (1980) is:

$$\dot{M}_b = 2.08 \times 10^{-7} (T_s + 1.8)^{1.5} \quad (4)$$

where T_s is sea water temperature (°C). That presented by Weeks & Campbell (1973) involves external parameters in addition to temperature and is:

$$\dot{M}_b = 6.74 \times 10^{-6} v_w^{0.8} \Delta T / X^{0.2} \quad (5)$$

where v_w is free stream relative horizontal water velocity past a berg, ΔT is temperature difference between an iceberg and the surrounding water, and X is the length of an iceberg side. Both melt rates are in units of m s^{-1}. Melting is more rapid at higher free stream water velocities and water temperatures, and it is relatively insensitive to changes in iceberg size. However, the melt rate and, therefore, the rate of sedimentation, is enhanced for relatively small bergs (Weeks & Campbell 1973). Both equations are based on pure ice. Debris rich ice has a lower thermal conductivity than pure ice, and this will reduce the rate of iceberg melting (Drewry 1986; Dowdeswell & Dowdeswell 1989). However, the presence of debris in basal ice may also make the ice–water interface more rough. This would increase melting.

These two equations provide an envelope of predicted iceberg melt rates (Fig. 5). Note that the curves have their origins at a temperature of −1.8°C. This is the freezing point of ocean water at a salinity of 35‰. At the relatively low water temperatures usually encountered in polar environments the equations predict fairly similar rates of melting (Fig. 5). Divergence in predicted melt rates becomes significant only at ocean temperatures above approximately 4 to 5°C and when iceberg velocity is either extremely high or very low (Fig. 5). Iceberg melt rates have also been inferred from observations of berg size variations over space around Antarctica (Budd *et al.* 1980). The relationship Budd *et al.* derive for melt rates falls between the estimates from equations (4) and (5) (Fig. 5). Combining iceberg basal melt rates with values for berg debris content and drift velocity, the choice of melt rate equation does not affect calculated sedimentation rates greatly at the temperatures typically encountered in polar waters.

Melt rates at tidewater glacier ice cliffs

We showed in equation (2) that direct melting of tidewater glacier terminal ice cliffs ($\dot{M}_c$) will affect the rate of iceberg calving. The ice cliff melt rate can be calculated using a modified version of equation (5), in which the free stream water velocity and length terms are now measured in the vertical rather than the horizontal direction (Syvitski 1989). This is because at tidewater ice cliffs upwelling of fresh water derived from ice wall melting and subglacial stream inputs (where present) dominates over horizontal flow across the terminus. As a consequence of this, length is also expressed as the thickness of the tidewater glacier below water. Using realistic values for submerged ice cliff height and velocity of upwelling, direct ice wall melting is seldom likely to be an important influence on iceberg calving rates, or indeed on meltwater release to fjord waters (Fig. 6). Even at a relatively high velocity of upwelling of 0.05 m s^{-1}, ice cliff melt rate only exceeds 20 m a^{-1} at water temperatures above 2–3°C (Fig. 6).

Iceberg debris release through time

We have now defined the major parameters influencing glacimarine sedimentation from icebergs, and it remains to calculate rates of melting and debris release with time since calving from a tidewater glacier front. The model assumes that time elapsed from calving is proportional to distance from the ice front. Debris release by basal melting ($\dot{M}_b$) in any small horizontal element ($\mathrm{d}x$) away from the ice front is calculated from:

$$\mathrm{d}y = (\dot{M}_b \, \mathrm{d}x)/v_b \quad (6)$$

where v_b is iceberg horizontal velocity. The amount of sediment released at any time after calving will depend on the initial debris concentration in the iceberg, and its history of melting and overturn. This approach is similar to that used to model sedimentation from ice shelves by Drewry & Cooper (1981). The effect of varying iceberg drift velocity on the rate of sedimentation is shown in Fig. 7. As iceberg velocity rises, the rate of sedimentation decreases but the maximum distance to which sediment is transported away from the ice front increases.

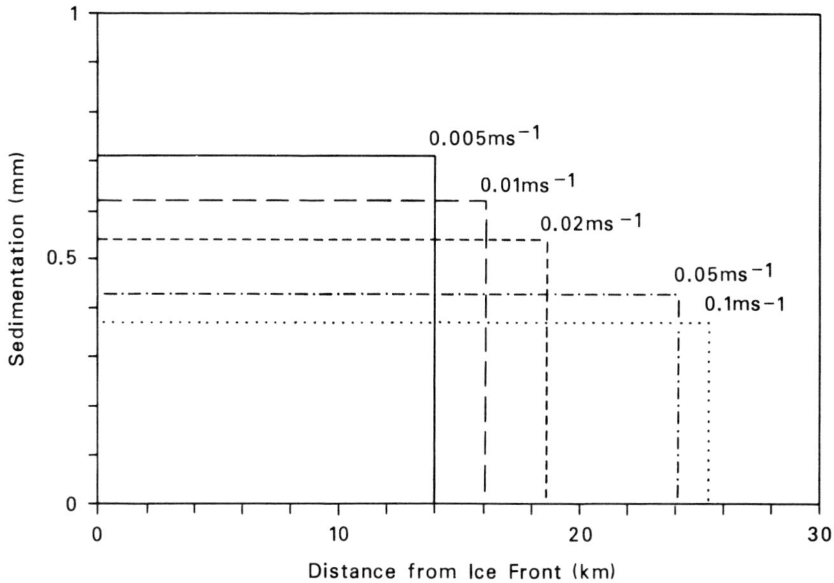

Fig. 7. The effect on sedimentation from a single iceberg of variations in the velocity of iceberg drift. Calculations use the basal melt rate equation of Weeks & Campbell (1973). They assume an iceberg 100 m in length, with a 1 m thick layer of sediment-laden ice of 10% debris content and an ocean temperature of 4°C.

Complications to the model

Iceberg drift tracks

The model does not consider complex berg drift tracks due, for example, to eddies or gyres in the offshore waters or to the effects of bathymetric highs in trapping bergs. In its simplest form it is, therefore, most suited to fjord environments where bergs move directly down-fjord and away from the calving ice front. In specific situations where icebergs are known to be held for long periods within a certain distance of the ice front, this can be simulated in the model by reducing the berg velocity. This will enhance the sedimentation rate in the area where bergs are being held. This simple solution is appropriate when it is considered that detailed data on iceberg drift tracks through time are seldom available.

Iceberg stability, overturn and grounding

Iceberg stability is difficult to model, because rigorous calculations of stability criteria depend on a detailed knowledge of iceberg morphology (e.g. Bass 1980). We are modelling individual icebergs as simple cuboids in order to conform to the grid we define. Because of this, we use a simple matrix incorporating iceberg morphology and the degree of exposure to wave and wind fields to provide a general indicator of iceberg stability (Table 1). Large, tabular icebergs with high width to depth ratios are more stable than generally smaller, irregular-shaped bergs with low width–depth ratios. Icebergs calved along open coastlines, where strong winds, storm and swell waves are more frequent, are also likely to overturn more often. Iceberg melting and fragmentation over time will also tend to make bergs less stable (Table 1). Table 1 can be used to make qualitative estimates of the frequency of iceberg overturn events, which can be incorporated into the model.

There are several ways in which we can deal with the complication of iceberg overturn and debris release, storage and dumping from the berg surface. First, where we are considering tidewater glaciers which calve mainly tabular icebergs, whose shape is relatively stable (Table 1), we may ignore the problem. Such icebergs are usually depleted in basal sediment before they fragment into smaller, less stable bergs. The second course is to state simply that our calculated sedimentation rates are maximum values because they take no account of times when dirty ice is above the water line. We could, in addition, adjust iceberg sedimentation rates by a constant related to the stability of the

Table 1. *Variations in iceberg stability with size, shape and the exposure of the marine environment.*

Berg shape & exposure	Berg size (long-axis) Large (> 1 km)	Medium (0.05–1 km)	Small (< .05 km)
Tabular	*	*/**	—
Irregular (fjord)	—	**	***
Irregular (open ocean)	**	***	***

* Stable; ** Intermediate; *** Unstable.

bergs calved from a particular glacier on the basis of typical berg shapes and degree of exposure (Table 1).

Icebergs may also ground in water that is shallower than their maximum keel depth. Two points are important (Syvitski *et al.* 1983). First, bergs continue to melt and, therefore, to release sediment, for the period they are at rest, increasing sedimentation at that point and depleting berg debris load. Secondly, icebergs may incorporate sediment from the sea floor during grounding events, thus increasing their debris load. Neither of these complications is addressed in the model.

Melting at the subaerial iceberg surface

We have already stated that a large proportion of the debris held in icebergs will be at the base when they are calved from the parent ice mass. However, unstable bergs will undergo multiple overturn events before final melting. Thus, material that is held at the iceberg base on calving may later appear above water at the berg surface when overturn occurs. Repeated overturn events may, in turn, re-submerge the dirty basal ice and dump any debris that has melted out subaerially.

The model can simulate iceberg overturn events by a rotation of the matrix defining the debris distribution within the berg. This is clearly a simplified treatment, in that bergs do not always turn through 180°. The amount of sediment released at a subaerially exposed berg surface any time after calving depends on the initial debris concentration in the iceberg, and its history of melting and overturn. Rates of subaerial debris melt-out are derived from empirical results obtained by Østrem (1959), allowing the changing thickness of melted out sediment to be calculated over time. The melt rate is dependent on the thickness of the sediment layer (Østrem 1959). A surface debris cover less than 0.5 cm in thickness enhances melting. At greater thicknesses the melt rate of the underlying ice is progressively retarded. The following polynomial approximates Østrem's data on the relationship between melt rate (y in cm d^{-1} and surface debris thickness (x in cm d^{-1}):

$$y = 5.1 - 0.37x + 7.5x^2. \qquad (7)$$

At the iceberg surface, debris that is melted out is not released directly to the water column. Instead, much of it remains in storage on the iceberg until an overturn event occurs, when the accumulated material is dumped from the berg into the water column. The model assumption that all debris released at the berg surface is stored is a simplification. Some material will slump and flow off the berg sides prior to overturn (Clark & Hanson 1983). Thus, in contrast to the quasi-continuous sedimentation from basal iceberg melting, surface melting leads to debris release as a series of single point inputs occurring when overturn takes place. The modelled sedimentation pattern for a small iceberg undergoing regular overturn events is shown in Fig. 8. However, when this pattern of sedimentation is applied to, for example, the annual flux of icebergs through a fjord, the random element in individual iceberg overturn events will provide a smoothing effect, spreading the sites of dumping over wide areas of the fjord floor.

Modelling different glaciological and oceanographic settings

The aim of this section is to investigate the trends in rates of iceberg sedimentation as we move from warmer to colder environmental settings. Thus, the fjords and tidewater glaciers of Glacier Bay in Alaska represent one end of a continuum of environments where iceberg calving takes place. The other end member includes the Antarctic continental margin and the high Canadian Arctic, exemplified by the fjords of Ellesmere Island. Between the two fall the more intermediate settings of, for example,

Baffin Island and Svalbard tidewater glacier systems.

Tidewater glacier debris content and calving rates

Where detailed observations of either tidewater glacier or iceberg debris content exist, these values can be assigned to individual grid elements within the model ice front or bergs (e.g. Anderson *et al.* 1980; Dowdeswell 1986; Dowdeswell & Dowdeswell 1989). However, field measurements of these parameters are for the most part absent or qualitative in nature.

A much greater amount of systematic field evidence is available for more easily accessible terrestrial glacier termini. In many glaciers, and all large ice caps and ice sheets, the bulk of debris is transported in a basal debris-rich zone of limited vertical extent. Above this, englacial debris is often very sparse and mainly of aeolian origin on large ice sheets. Where valley walls confine glaciers, rockfalls and avalanches can supply debris to the glacier surface and, where rates of subaerial weathering are particularly high, significant quantities of debris are also transported as a surface debris cover. Debris can also be transported upward from the basal zone to a high level within a glacier where strong compression and glacitectonic deformation occur. However, this is much less common in tidewater glaciers than in their terrestrial counterparts, because tidewater glacier margins are usually relatively unconstrained and extensional in stress regime. Debris derived from subaerial and basal sources can be distinguished through its shape characteristics, with more rounded material linked to processes of abrasion and crushing at the ice–bed interface (e.g. Boulton 1978; Dowdeswell 1986). The origins of debris in icebergs and glacimarine sediments can also be analysed using clast shape criteria (Domack *et al.* 1980; Dowdeswell & Dowdeswell 1989).

The thermal regime of the parent ice mass is likely to be a dominant control on its debris content, in that it influences debris entrainment processes at the glacier bed (Boulton 1972). Studies of a number of terrestrial glacier termini suggest the following generalizations:

(i) *Temperate Glaciers*, at the pressure melting point throughout, will have basal debris-rich layers on the order of centimeters to tens of centimeters in thickness (e.g. Kamb & LaChapelle 1964). This is related to the pressure melting-regelation process of glacier motion. Where predominantly temperate ice masses are thermally complex at a more detailed scale, or in strongly compressive tectonic settings, greater thicknesses of basal debris may build up. This is the situation at the margins of a number of Alaskan glaciers (e.g. Powell & Molnia 1989). Values of between 3 and 15 m of debris-rich ice have been reported, with debris concentrations varying from 3 to over 40% by volume.

(ii) *Sub-Polar Glaciers*, where part of the glacier bed is frozen (often the terminus region) and part remains at the melting point, have basal debris-rich zones of one to several meters in thickness. This is associated with the bulk freezing-on of debris-rich meltwater and sediment at the transition from a warm to a cold-based thermal regime (e.g. Weertman 1961; Lawson 1979). Debris-rich basal ice between 0.8 and 2.9 m in thickness, with debris content of up to 50%, has been observed at the margins of a Baffin Island tidewater glacier (Dowdeswell 1986). In Svalbard, basal debris-rich ice has been studied in both glaciers and icebergs (Boulton 1970; Dowdeswell & Dowdeswell 1989). Layer thicknesses of up to 4 m have been observed. Griffith & Anderson (1989) also report observations from the northern Antarctic Peninsula suggesting that tidewater glaciers may contain significant thicknesses of basal debris-rich ice; values of up to 3 m are given.

(iii) *High Polar Glaciers*, where meltwater is scarce or absent, will not tend to entrain basal debris by any of the above processes. Instead, significant thicknesses of debris-rich ice can be entrained where glaciers over-ride and incorporate a frontal apron composed of ice and debris (Shaw 1977a, b). Taylor Glacier, in the McMurdo Dry Valleys, Antarctica, has a basal debris layer at least 5 m thick of this origin. This debris is stratified, and reaches concentrations of 40–50% by volume (Shaw 1977b). Similar thicknesses of basal debris may be typical of many Ellesmere Island ice masses (T. Stewart, pers. comm.).

(iv) *Fast Flowing and Surge-Type Glaciers* may also contain basal debris several metres in thickness, because their high flow rates are associated with water lubrication and sometimes the deformation of basal sediments under high pore water pressure. Compressive glacitectonic activity during surges, and the overriding of marginal debris aprons, are also important factors in the buildup of large thicknesses of dirty basal ice in surge-type glaciers. Basal ice of significant debris content and thickness up to about 10 m has been observed at the margins of surge-type glaciers in Iceland and Svalbard (Clapperton 1975).

The values for iceberg debris content given in Table 2 are generalized from the sources dis-

Table 2. *Variations in calculated iceberg sedimentation rate and distance to berg debris exhaustion in different environmental settings*

Tidewater Glacier Location	Input Parameters			Calculated results for differing basal melt rates & berg velocities					
				Russell-Head at 0.05 m s^{-1}		Weeks & Campbell at 0.05 m s^{-1}		Weeks & Campbell at 0.01 m s^{-1}	
	Basal ice thickness (m)	Calving rate (m a^{-1})	Ocean temp. (°C)	sed.rate (mm a^{-1})	distance (km)	sed.rate (mm a^{-1})	distance (km)	sed.rate (mm a^{-1})	distance (km)
S.E. Alaska	10	2000	5.0	14.8	130	11.0	190	14.7	140
Svalbard	3	300	2.5	1.1	80	1.0	90	1.4	65
High Arctic	3	50	−1.0	0.0015	1000	0.0005*	3000*	0.0005*	3000*
Antarctic pen	2	100	−1.0	0.004	670	0.001*	2000*	0.001*	2000*

Calculations include the basal melt rates of Russell-Head (1980) and Weeks & Campbell (1973). Iceberg horizontal velocity is also varied. A debris concentration of 10% by volume is assumed throughout, and iceberg length is 100 m (50 m for the High Arctic).
* Rounding errors at very low sedimentation rates make these values essentially the same.

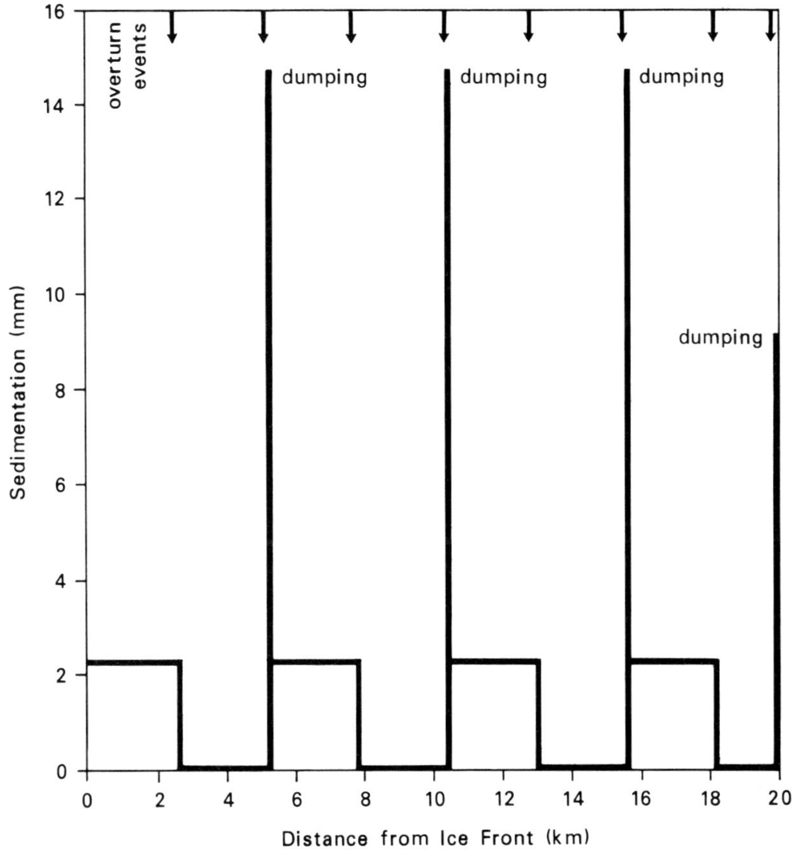

Fig. 8. The predicted pattern of iceberg sediment deposition from dumping and basal melting for a single 50 m long berg moving at 0.01 m s^{-1} and undergoing 180° overturn events at three-day intervals. Iceberg debris content is assumed to be a 1 m thick layer of debris-rich ice of 10% sediment concentration. The ocean temperature is set at 4°C and basal melting is calculated from the equation of Weeks & Campbell (1973). Note that dumping would also take place at the calving line if significant supraglacial sediment was present on the glacier surface.

cussed above to provide input data to calculations of iceberg sedimentation in several environmental settings. The iceberg calving rates in Table 2 are also derived, through equation (2), from a number of sources concerning glacier velocities, activity indices, and ice front advance and retreat (e.g. Koerner 1964; Elverhøi *et al.* 1980; Brown *et al.* 1982; England 1986; Powell 1988; Dowdeswell unpublished). Data on iceberg calving rates are particularly sparse for Ellesmere Island and the Antarctic Peninsula but, unless tidewater ice cliffs are grounded in very deep water, rates are likely to be low because of the low mass inputs and activity indices in these areas (England 1986).

Ocean temperature fields

The two-dimensional model is well suited to accommodating observations on the ocean temperature field. This is because temperature variations with depth are often acquired at a number of stations along a transect perpendicular to tidewater glacier termini. However, the aim of this paper is to examine general trends in iceberg sedimentation rate with changing environmental conditions. In this case we will use temperatures averaged through the water column. For model investigations of iceberg sedimentation in specific localities a more detailed grid of observations can be used. An idealized example of model results in a variable

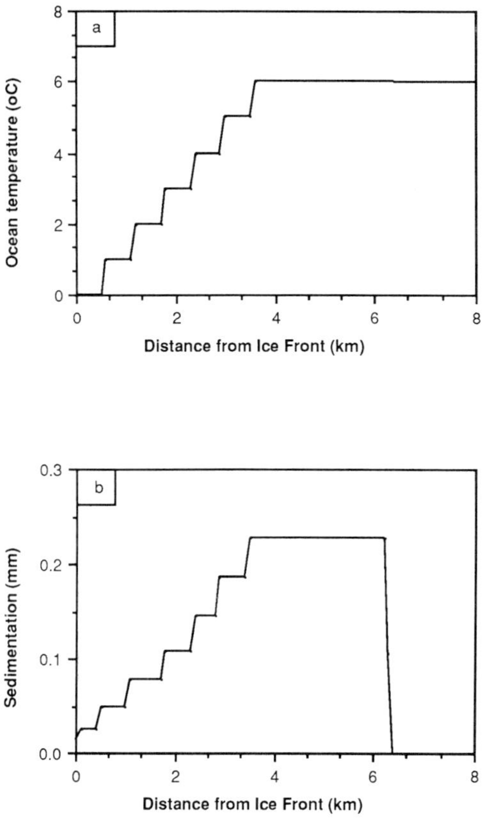

Fig. 9. Sedimentation from a single iceberg (a) moving through ocean waters of varying temperature (b). The berg has a basal debris layer 1 m thick with a concentration of 10%. The Russell-Head melt equation is used with a berg velocity of 0.02 m s^{-1}.

ocean temperature field is shown in Fig. 9.

Year-round observations of offshore water temperature are unavailable in many parts of the polar regions, but seasonal variations can be significant in some areas (e.g. Matthews & Quinlan 1975). However, this is not a severe problem in our modelling work for two reasons. First, in high Arctic and Antarctic waters, temperatures are often as low as −1°C or less even in summer (e.g. Lake & Walker 1976), although the intrusion of Circumpolar Deep Water onto the Antarctic continental shelf may yield somewhat higher temperatures (Jacobs 1989). Thus, there is little opportunity for marked winter cooling in many high latitude locations. At lower latitudes, the seasonal change in water temperature below the thermocline can be on the order of several degrees. In a southeast Alaskan fjord, for example, variations of over 3°C have been observed (Matthews & Quinlan 1975). However, the lack of winter temperature data for many other intermediate polar areas is not critical for a second reason. This is the tendency for iceberg calving to be a largely seasonal phenomenon. Calving at the margins of many Svalbard tidewater glaciers is restricted mainly to the summer (Dowdeswell 1989). This is because the buildup of winter fast ice, combined with a characteristic slowing of glacier velocity, reduces winter calving greatly. The relatively few icebergs that are produced in winter are held close to the ice front by the shorefast sea ice, and only begin to move down-fjord on spring breakup (Dowdeswell 1989). Thus, the use of water temperature data collected during summer is an acceptable modelling strategy.

The ocean temperature values in Table 2, used in calculations of iceberg sedimentation rates, are averaged over the upper 100 to 200 m of the water column. This encompasses the likely thicknesses of all but the largest tabular icebergs. Averaged temperatures do not include the very high near-surface values found above the pycnocline in some areas during summer. Water temperature data for southeast Alaska in Table 2 are from Matthews & Quinlan (1975). Information from the fjords of Spitsbergen, Svalbard, is from Elverhøi *et al.* (1980) and our own unpublished oceanographic data. Measurements from the high Arctic are from d'Iberville Fjord in Ellesmere Island (Lake & Walker 1976). Water temperature values for the Antarctic Peninsula come from an unpublished 1988 cruise report by J.B. Anderson & E.W. Domack concerning coves and bays on the west side of the Peninsula.

Variations in iceberg sedimentation with environmental setting

The results of model calculations using typical parameter values for several different environmental settings are shown in Table 2. The table incorporates calculations based on the basal melt rate equations of both Russell-Head (1980) and Weeks & Campbell (1973). It is important to note that the predicted sedimentation rates and transport distances in Table 2 are quite similar for the two melt rate equations, indicating the relative robustness of the results. Data on iceberg drift velocities are very sparse (Vinje 1980), but the effect of different drift velocities is also demonstrated in Table 2. In each case it is further assumed that basal debris is the sole contributor to iceberg sedimentation.

The trapping of icebergs close to the tidewater ice front would increase sedimentation rates but clearly decrease the distance to berg exhaustion. Iceberg overturn events are also ignored, but would increase local variability in the rate of sedimentation. Thus the changing rates shown in the table are a function of variations in basal debris-rich ice thickness, iceberg calving rates and ocean temperatures.

In terms of both rates of iceberg sedimentation and travel distances to iceberg debris exhaustion, the model predicts wide variations between the four selected areas: southeast Alaska, Svalbard, Ellesmere Island in the high Arctic, and the Antarctic Peninsula. However, only order of magnitude differences in calculated values should be regarded as significant.

Predicted iceberg sedimentation rates are highest for the fjords of southeast Alaska, where ocean waters are warmest and the calving of icebergs most rapid. It is in this area that the most detailed data on both calving rates and temperatures are available (Brown *et al.* 1982; Matthews & Quinlan 1975). Calculated sedimentation rates are 1 mm a^{-1} for Svalbard; an order of magnitude less than the 11 to 15 mm a^{-1} predicted for Alaskan fjords. This is a result of lower water temperatures combined with less rapid iceberg calving. In the high Arctic and the Antarctic Peninsula region, predicted iceberg sedimentation of approximately 0.004 to 0.0005 mm a^{-1} is three orders of magnitude less rapid than in Svalbard, and four orders of magnitude less than in Alaska (Table 2). Again, both rates of iceberg calving and water temperatures are very much lower than elsewhere. The calculated rates for the high Arctic and the Antarctic Peninsula are essentially identical at the model's level of resolution. The model results in Table 2 show, therefore, that there is a marked reduction in the rate of sedimentation from icebergs as the environmental gradient from relatively warm maritime to colder and drier tidewater glacier settings is traversed.

Iceberg sedimentation rates for fast flowing outlet glaciers and surge-type glaciers in the active phase of the surge cycle are not included in Table 2. This is because relatively few data exist on calving rates from such ice masses. We can demonstrate the possible effect of a surge on the rate of iceberg rafting, using as an example one of the large number of tidewater glaciers in Svalbard that surge (Liestøl in press). Assume that calving during a surge is 2000 m over an active phase lasting a year. This is an order of magnitude greater than calving rates from Svalbard glaciers not known to surge (Table 2). It is probably two orders of magnitude greater than iceberg calving from surge-type glaciers during their quiescent period. Using water temperature, debris content and iceberg velocities as in Table 2, the rate of iceberg sedimentation rises from about 1 to between 6 and 9 mm a^{-1} during the surge event. Thus, the effect of a tidewater glacier surge is to inject a pulse of poorly sorted iceberg rafted material into the marine environment. However, because glacier surging is linked to water lubrication of the ice-bed interface, inputs of meltwater-derived sediment are also likely to increase, in part as a pulse of highly turbid water released in association with surge termination (Kamb *et al.* 1985).

The large variations in iceberg production associated with a glacier terminus retreating through alternately deep and shallow water, for example Columbia Glacier, Alaska (Brown *et al.* 1982), might also be expected to place spikes of ice rafted debris in the sedimentary record. In this case the ice rafted debris would be masked less by large inputs of meltwater-derived fine material than would spikes produced during surges.

The predicted distances over which icebergs will transport sediment are also given in Table 2. Here the input parameters of importance are the ocean temperature field and the thickness of basal debris-rich ice contained in the iceberg. The latter is less well known. Again there is a marked contrast between icebergs derived from Alaskan and Svalbard glaciers on the one hand, and Antarctic Peninsula and high Arctic glaciers on the other. Icebergs offshore of Alaska and Svalbard may carry debris on the order of 100 km from the tidewater ice front (Table 2). By contrast, iceberg rafting of debris may continue for one or more thousands of kilometres of distance travelled from glacier sources in the high Arctic and Antarctic. It is the order of magnitude difference in distance to berg exhaustion that is of interest, rather than the more detailed values given in Table 2. Thus icebergs in cold, high polar waters will transport debris for considerably greater distances than bergs calved into warmer waters. The implication of this is that in the colder waters associated with full glacial conditions in the North Atlantic, iceberg debris would be transported to lower latitudes. This has been inferred through analysis of the pattern of North Atlantic ice rafted sand deposition during the late Quaternary (e.g. Ruddiman 1977).

To develop greater resolution for the model, more information is required on basal debris thickness and rates of iceberg drift in particular. The amount and distribution of debris will clearly vary for individual icebergs, and so the predicted distances to berg debris exhaustion in Table 2 are approximations only.

Comparison with observed rates of glacimarine sedimentation

A problem with many models is that it is difficult to test them against observational datasets. In our case, this can be overcome at least in part by comparing model predictions of iceberg sedimentation in different environments with field data on rates of overall glacimarine sedimentation (Table 3). Errors will certainly have occurred in model specification if calculated iceberg sedimentation rates exceed total observed sedimentation in the glacimarine environment.

In the fjords of Svalbard, overall glacimarine sedimentation rates have been derived from radiocarbon and ^{210}Lead dating of core material (Elverhøi *et al.* 1983; Pers. comm. D.J. Sexton). Rates from two fjords in northwest Spitsbergen fall between 30 and 100 mm a^{-1} at distances up to about 10 km from tidewater glacier margins. These rates are over an order of magnitude greater than those predicted in our modelling exercise (Table 3). This is a reasonable finding, considering the predominantly fine-grained nature of the offshore sediments and the characteristic presence of turbid suspended sediment plumes emanating from Spitsbergen tidewater glaciers and indicating the importance of sedimentation from meltwater (Elverhøi *et al.* 1980, 1983; Dowdeswell & Dowdeswell 1989).

In the fjords of Glacier Bay, Alaska, overall rates of glacimarine sedimentation are very rapid, ranging from 2000 up to 9000 mm a^{-1} (Powell 1983). Calculated iceberg sedimentation rates of 11 to 15 mm a^{-1} are relatively small compared with these very high total sedimentation rates (Table 3). However, the higher figures for total glacimarine sedimentation are from sites within only a few hundred metres of calving ice fronts and sedimentation will decline in more distal locations. Nonetheless, the build-up of material from meltwater sedimentation, often associated with the development of proximal fans, is clearly a dominant process in this area.

A more specific calculation of iceberg sedimentation can be made using the observations of Powell (1988) from McBride Inlet in southeast Alaska, where rates of total glacimarine sedimentation are comparatively well known. This example also illustrates the importance that constraints on iceberg movement can have on calculated sedimentation rates. Taking a 10 m thick basal debris-rich zone (10% debris content by volume), a calving rate of 540 m a^{-1}, and 5°C ocean temperature, an iceberg sedimentation rate of close to 4 mm a^{-1} is calculated. However, almost all calved icebergs are trapped by a sill within 1 km of the tidewater ice front. This revises the iceberg sedimentation rate upward to approximately 500 mm a^{-1}. The rate of total sedimentation in McBride Inlet is about 2000 mm a^{-1} (Powell 1988). The model predicts, therefore, that about 25% of total sedimentation in this Alaskan Inlet may be assigned to the process of iceberg rafting.

For tidewater glaciers in southeast Alaska and Svalbard, modelled rates of iceberg sedimentation fall an order of magnitude or more below rates of overall glacimarine sedimentation (Table 3). Two points can be made. First, predicted rates of iceberg sedimentation are lower than overall sedimentation rates, providing an indication that the model is not overestimating the importance of iceberg rafting vastly. Secondly, in these relatively maritime polar settings, where significant summer melting takes place, sediment delivery to the marine environment appears to be dominated by meltwater transport rather than by iceberg rafting (e.g. Syvitski 1989).

Information on rates of glacimarine sedimentation in high Arctic fjords and off the Antarctic Peninsula is largely absent. In Ellesmere Island, and the high Arctic in general, it is likely that sedimentation from icebergs, although low in absolute terms, provides a higher relative proportion of total glacimarine sedimentation than further south. This is because in such high-latitude, arid polar areas, meltwater inputs are relatively less important. For the Arctic Basin there is considerable controversy over the rate at which overall sedimentation has taken place, and of the relative roles of sediment delivery by icebergs and sea ice (e.g. Clark & Hanson 1983; Sejrup *et al.* 1984). The model for iceberg sedimentation presented here offers some scope for furthering our understanding of this problem, but detailed consideration of iceberg drift tracks and calving rates both today and during late Cenozoic glacial maxima is required before detailed calculations can be presented.

Summary and conclusions

The model outlined above allows calculation of both the rate of sedimentation from icebergs and the maximum distance to which they will transport a given debris load. The model is applicable to any ice mass ending in tidewater

Table 3. *Comparing modelled rates of iceberg sedimentation from Table 2 with rates of overall glacimarine sedimentation derived from dated core material.*

Environmental setting	Model iceberg sedimentation rate (mm a^{-1})	Observed total glacimarine sed. rate (mm a^{-1})
Alaska	14	2000–9000
Svalbard	1	30–100

where ice is grounded rather than floating (Fig. 2). Where ice floats to form an ice tongue or ice shelf, undermelt can take place prior to calving. The model does not take account of this. The model provides data on iceberg debris deposition which is very difficult to acquire by direct observation, due to the logistical problems involved. The model focuses attention on the main parameters affecting the rate of sedimentation from icebergs. These parameters are:

(i) the distribution and concentration of debris in the parent ice mass;

(ii) the rate of iceberg calving and, therefore, the rate of iceberg debris introduction to the marine environment;

(iii) the rate of melting at the base of floating icebergs;

(iv) the temperature structure of the offshore ocean waters;

(v) the velocity, stability and drift track of icebergs.

Iceberg sedimentation is linked to both the glaciological regime of the parent glacier or ice sheet, and to the oceanography of the offshore marine environment. Ice masses ending in tidewater are found in many parts of the Arctic and Antarctic, and both glaciological and oceanographic parameters will vary in different polar environmental settings. The model has been used to assess iceberg sedimentation rates in a number of these settings on a continuum from warm southeast Alaska, through Svalbard, to high polar environments such as the Antarctic Peninsula and Ellesmere Island (Table 2). Calculated rates of sedimentation from icebergs are most rapid in southeast Alaska, falling by an order of magnitude in Svalbard waters. This is a result of lower water temperatures combined with less rapid iceberg calving. In the high Arctic and the Antarctic Peninsula regions, predicted iceberg sedimentation rates are three orders of magnitude less rapid than in Svalbard, and four orders of magnitude less than in Alaska. These rates of iceberg sedimentation are considerably less than rates of overall glacimarine sedimentation derived from dated core material in Alaska and Svalbard fjords (Table 3). In these environments, debris inputs from glacial meltwater appear to be the dominant source for offshore deposition. Sedimentation from icebergs may become proportionally more important as meltwater availability declines in high polar settings, but few datasets on rates of total glacimarine sedimentation are available here. Icebergs will transport debris over much greater distances in cold, high polar regions, because basal debris-rich ice is melted off less quickly. Icebergs may raft sediment over distances up to two orders of magnitude greater in high polar areas than in the warmest tidewater glacier settings.

In general, model results give a guide to the rate of iceberg sedimentation in different glaciological and oceanographic environments within the polar regions. The calculated iceberg sedimentation rates should be seen as order of magnitude estimates. The resolving power of the model is restricted by the relatively sparse observational datasets for several parameters. However, predictions concerning iceberg sedimentation rates and debris transport distances are important in assessing the contribution of material rafted by bergs to overall glacimarine deposition and how this contribution may vary over time in association with changing climate. The nature of late Cenozoic sedimentation within the Arctic Basin is an example of a potential future model application.

We thank Eugene Domack, Evelyn Dowdeswell, David Drewry, Robert Gilbert, Martin Sharp and Anders Solheim for their comments on this paper. The Nuffield Foundation kindly provided financial support. Discussions with Anders Solheim and other colleagues at the Norsk Polarinstitutt, Oslo, were funded through NATO Collaborative Research Grant 0747/87. T. Murray is currently in receipt of a NERC studentship.

References

Anderson, J.B., Domack, E.W. & Kurtz, D.D. 1980. Observations of sediment-laden icebergs in Antarctic waters: implications to glacial erosion and transport. *Journal of Glaciology*, **25**, 387–396.

Bass, D.W. 1980. Stability of icebergs. *Annals of Glaciology*, **1**, 43–47.

Boulton, G.S. 1970. On the origin and transport of englacial debris in Svalbard glaciers. *Journal of Glaciology*, **9**, 213–229.

—— 1972. The role of thermal regime in glacier sedimentation. *Institute of British Geographers, Special Publication*, **4**, 1–19.

—— 1978. Boulder shapes and grain-size distributions of debris as indicators of transport paths through a glacier and till genesis. *Sedimentology*, **25**, 773–799.

Brown, C.S., Meier, M.F. & Post, A. 1982. Calving speed of Alaska tidewater glaciers, with application to Columbia Glacier. *U.S. Geological Survey Professional Paper*, 1258C.

Budd, W.F., Jacka, T.H. & Morgan, V.I. 1980. Antarctic iceberg melt rates derived from size distributions and movement rates. *Annals of Glaciology*, **1**, 103–112.

Clapperton, C.M. 1975. The debris content of surging glaciers in Svalbard and Iceland. *Journal of Glaciology*, **14**, 395–406.

Clark, D.L. & Hanson, A. 1983. Central Arctic Ocean sediment texture: a key to ice transport mechanisms. *In*: Molnia, B.F. (ed), *Glacial-Marine Sedimentation*. Plenum, New York, 301–330.

Domack, E.W., Anderson, J.B. & Kurtz, D.D. 1980. Clast shape as an indicator of transport and depositional mechanisms in glacial marine sediments: George V continental shelf, Antarctica. *Journal of Sedimentary Petrology*, **50**, 813–820.

Dowdeswell, J.A. 1986. The distribution and character of sediments in a tidewater glacier, southern Baffin Island, N.W.T., Canada. *Arctic and Alpine Research*, **18**, 45–56.

—— 1989. On the nature of Svalbard icebergs. *Journal of Glaciology*, **35**, 224–234.

—— & Dowdeswell, E.K. 1989. Debris in icebergs and rates of glaci-marine sedimentation: observations from Spitsbergen and a simple model. *Journal of Geology*, **97**, 221–231.

Drewry, D.J. 1986. *Glacial Geologic Processes*. Arnold, London.

—— & Cooper, A.P.R. 1981. Processes and models of Antarctic glaciomarine sedimentation. *Annals of Glaciology*, **2**, 117–122.

Elverhøi, A., Liestøl, O. & Nagy, J. 1980. Glacial erosion, sedimentation and microfauna in the inner part of Kongsfjorden, Spitsbergen. *Norsk Polarinstitutt Skrifter*, **172**, 33–58.

——, Lønne, Ø. & Seland. R. 1983. Glaciomarine sedimentation in a modern fjord environment, Spitsbergen. *Polar Research*, **1** (n.s.), 127–149.

England, J. 1986. A paleoglaciation level for north-central Ellesmere Island, N.W.T., Canada. *Arctic and Alpine Research*, **18**, 217–222.

Griffith, T.W. & Anderson, J.B. 1989. Climatic control of sedimentation in bays and fjords of the northern Antarctic Peninsula. *Marine Geology*, **85**, 181–204.

Jacobs, S.S. 1989. Marine controls on modern sedimentation on the Antarctic continental shelf. *Marine Geology*, **85**, 121–153.

Kamb, B. & LaChapelle, E.R. 1964. Direct observation of the mechanism of glacier sliding over bedrock. *Journal of Glaciology*, **5**, 159–172.

——, Raymond, C.F., Harrison, W.D., Engelhardt, H., Echelmeyer, K.A., Humphrey, N., Brugman, M.M. & Pfeffer, T. 1985. Glacier surge mechanism: 1982–1983 surge of Variegated Glacier, Alaska. *Science*, **227**, 468–479.

Koerner, R.M. 1964. Glaciological observations in Trinity Peninsula and the islands of the Prince Gustav Channel, Graham Land. *British Antarctic Survey Scientific Report*, **42**.

Lake, R.A. & Walker, E.R. 1976. A Canadian Arctic fjord with some comparisons to fjords of the western Americas. *Journal of the Fisheries Research Board, Canada*, **33**, 2272–2285.

Lawson, D.E. 1979. Sedimentological analysis of the western terminus region of the Matanuska Glacier, Alaska. *Cold Regions Research and Engineering Laboratory Report*, 79–9. Hanover, New Hampshire.

Liestøl, O. in press. Glaciers of Svalbard, Norway. *U.S. Geological Survey Professional Paper*, 1386–5E.

Matthews, J.B. & Quinlan, A.V. 1975. Seasonal characteristics of water masses in Muir Inlet, a fjord with tidewater glaciers. *Journal of the Fisheries Research Board, Canada*, **32**, 1693–1703.

Østrem, G. 1959. Ice melting under a thin layer of moraine, and the existence of ice cores in moraine ridges. *Geografiska Annaler*, **41**, 228–230.

Ovenshine, A.T. 1970. Observations of iceberg rafting in Glacier Bay, Alaska, and the identification of ancient ice-rafted deposits. *Geological Society of America, Bulletin*, **81**, 891–894.

Powell, R.D. 1983. Glacial-marine sedimentation processes and lithofacies of temperate tidewater glaciers, Glacier Bay, Alaska. *In*: Molnia, B.F. (ed), *Glacial-Marine Sedimentation*. Plenum, New York, 185–232.

Powell, R.D. 1988. Processes and facies of temperate and sub-polar glaciers with tidewater fronts. *Geological Society of America*, Short Course Notes, GSA Centennial Meeting, Denver, 114 pp.

Powell, R.D. & Molnia, B.F. 1989. Glacimarine sedimentary processes, facies and morphology of the south-southeast Alaska shelf and fjords. *Marine Geology*, **85**, 359–390.

Ruddiman, W.F. 1977. Late Quaternary deposition of ice-rafted sand in the subpolar North Atlantic (lat 40°–65°N). *Bulletin, Geological Society of America*, **88**, 1813–1827.

Russell-Head, D.S. 1980. The melting of free-drifting icebergs. *Annals of Glaciology*, **1**, 119–122.

Sejrup, H.P., Miller, G.H., Brigham-Grette, J., Løvlie, R. & Hopkins, D. 1984. Amino acid epimerization implies rapid sedimentation rates in Arctic Ocean cores. *Nature*, **310**, 772–775.

SHAW, J. 1977a. Tills deposited in arid polar environments. *Canadian Journal of Earth Sciences*, **14**, 1239–1245.

—— 1977b. Till body morphology and structure related to glacier flow. *Boreas*, **6**, 189–201.

SYVITSKI, J.P.M. 1989. On the deposition of sediments within glacier-influenced fjords: oceanographic controls. *Marine Geology*, **85**, 301–329.

——, FADER, G.B., JOSENHANS, H.W., MACLEAN, B. & PIPER, D.J.W. 1983. Seabed investigations of the Canadian east coast and Arctic using Pisces IV. *Geoscience Canada*, **10**, 59–68.

VINJE, T.E. 1980. Some satellite-tracked iceberg drifts in the Antarctic. *Annals of Glaciology*, **1**, 83–87.

WEEKS, W.F. & CAMPBELL, W.J. 1973. Icebergs as a fresh water source: an appraisal. *Journal of Glaciology*, **12**, 207–233.

WEERTMAN, J. 1961. Mechanism for the formation of inner moraines found near the edge of cold ice caps and ice sheets. *Journal of Glaciology*, **3**, 665–978.

Development of slope valleys in the glacimarine environment of a complex subduction zone, Northern Gulf of Alaska

PAUL R. CARLSON, TERRY R. BRUNS & MICHAEL A. FISHER

U.S. Geological Survey, Menlo Park, CA 94025, USA

Abstract: Morphological, seismic-reflection, and sedimentological evidence indicates that glacial ice tongues cut large sea valleys into the Gulf of Alaska continental shelf during the Pleistocene. During the Holocene, glacially-derived sediments from the Copper River and other meltwater streams have been prograding seaward across the shelf, covering the glacial and glacimarine upper Yakataga diamicts that blanketed the shelf and accumulated on the upper slope seaward of the sea valleys during the Pleistocene. GLORIA imagery near Middleton Island provides a new perspective on the glacimarine depositional environment on the continental slope in a collision zone between the Pacific and North American plates. Southwest of Middleton Island, along the subduction margin, sinuous valleys funnel sediment around shelf-edge-parallel, subduction-created, anticlinal ridges that have deflected and locally trapped glacimarine sediment. The slope south and southeast of Middleton Island where oblique convergence occurs, is incised by dendritic, erosional gulley systems, contains no compressional ridges, and thus, the apparently active sediment pathways to the trench are unrestricted. However, below the sea valley mouths, apparently both glacial and glacimarine sediments blanket the upper slope, covering any dendritic gulley systems that may have formed during or since the Pleistocene low stands of sea level.

The northern Gulf of Alaska continental margin is a region greatly affected by both tectonics and glaciation as expressed by bordering high mountains and large glaciers. Late Miocene and younger glaciation has deposited as much as 5.5 km of glacimarine strata along the margin (Plafker & Addicott 1976). At the same time, plate motions have created a subduction zone between the Pacific and North American plates, resulting in extensive uplift and deformation (Bruns 1985; Plafker 1987). The intertwining of these events has created complex sea floor morphology and sediment distribution patterns for the rapidly deposited glacimarine sediments.

The purpose of this paper is to examine the glacimarine depositional environment of part of this tectonically active region. Recently acquired GLORIA (Geological Long Range Inclined Asdic) wide-range side-scan sonar data gives a perspective of the sea floor morphology, structure, and sediment distribution pathways substantially better than that derived from previously available seismic reflection and bathymetric data. The area covered by the GLORIA data lies below the shelf break in the northern extreme of the Gulf of Alaska, between Pamplona Spur (143°W) and Amatuli Trough (149°W) (Fig. 1).

Data collection

GLORIA data discussed in this paper were collected in 1986 as part of the U.S. Geological Survey EEZ (Exclusive Economic Zone) program. The nominal width of high quality GLORIA imagery is generally 30 km in this area. Data collected simultaneously with GLORIA imagery include 160 in^3 air gun two-channel seismic-reflection profiles, 3.5 kHz and 10 kHz high-resolution acoustic profiles, and gravity and magnetic measurements. The one-week cruise to the Gulf of Alaska, which recorded imagery of 35 500 km^2 of the continental slope and Aleutian Trench (Fig. 1), was planned to obtain morphological coverage of the continental margin seaward of Middleton Island in support of the USGS Trans Alaska Crustal Transects program. The remainder of the Gulf of Alaska EEZ coverage will be completed in the summer of 1989. Seabeam bathymetry was collected by the NOAA ship SURVEYOR in the study area in 1987 and 1988. The bathymetry provided valuable additional information to verify and aid interpretation of the GLORIA imagery, but security classification prohibits publication at this time (EOS 1988).

Regional setting

The northern Gulf of Alaska continental margin and southern Alaska are underlain by rocks of Paleozoic through Cenozoic age that form fault-bounded tectonostratigraphic terranes (Fig. 2) that were accreted to North America in late Mesozoic and Cenozoic times

From Dowdeswell, J. A. & Scourse, J. D. (eds), 1990, *Glacimarine Environments: Processes and Sediments*. Geological Society Special Publication No 53, pp 139–153.

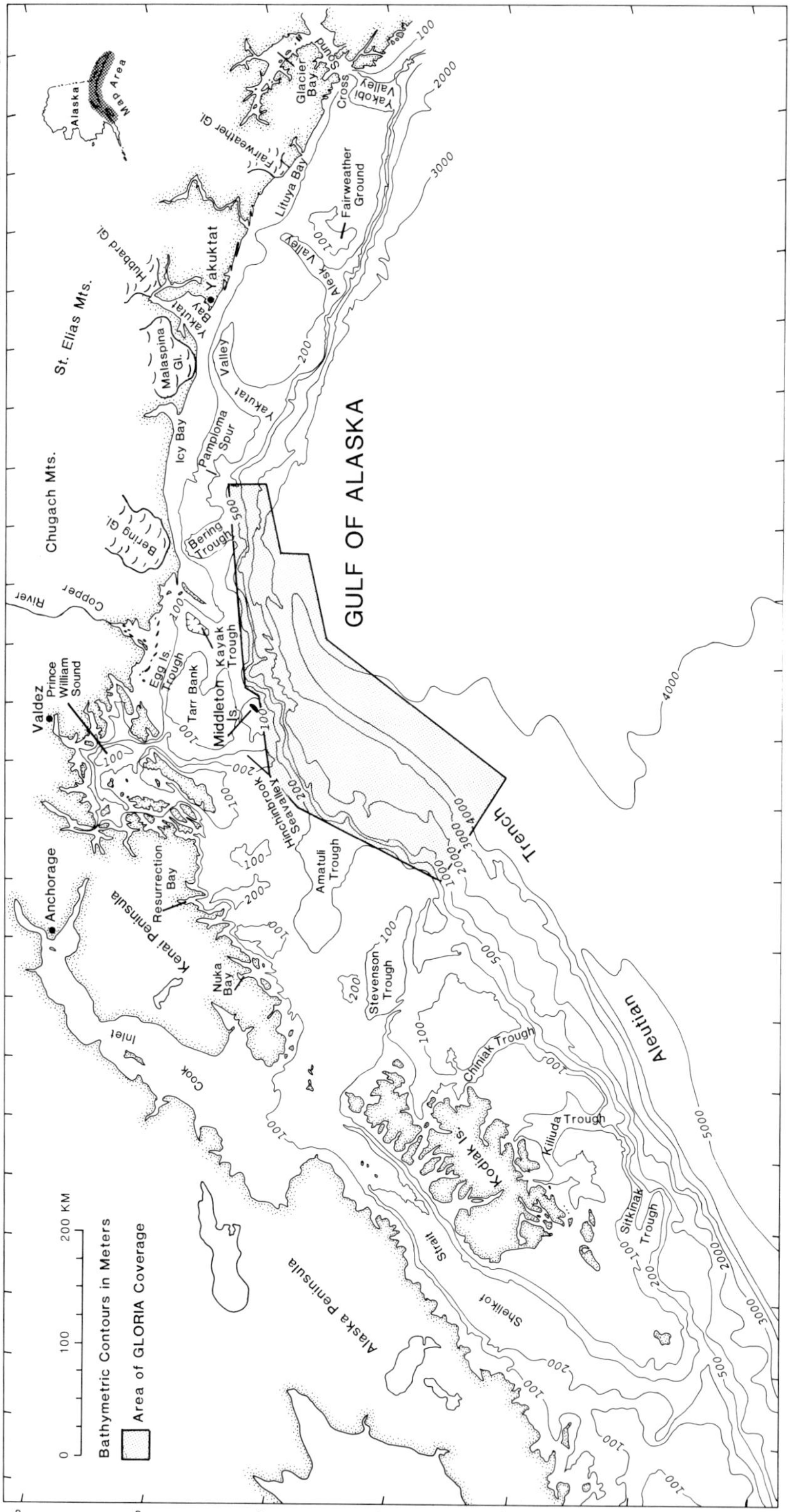

Fig. 1. Location map of Gulf of Alaska showing area of GLORIA mosaic. Bathymetry from Atwood *et al.* (1981).

(Jones *et al.* 1986). At present, plate motion between the Pacific and North American plates is taken up by strike-slip motion along the Queen Charlotte–Fairweather fault zone, and by subduction along the Aleutian Trench (Fig. 2). In the transition region, between the strike-slip and subduction regimes, the small Yakutat terrane is moving with the Pacific plate and colliding with and subducting beneath southern Alaska (Bruns 1983, 1985; Plafker 1987). The northern Gulf of Alaska is structurally complex, with rapid uplift of the coastal mountains and with the development of a fold and thrust belt (the Pamplona zone) across the western part of the Yakutat terrane and adjacent slope areas (Fig. 3).

The structure of the continental shelf in the Middleton Island to Kayak Island region consists of small basins separated by tightly folded, extensively faulted anticlines (Fig. 3) (Bruns 1985). The basins are underlain by Eocene and Oligocene strata and are filled with up to 5 km of lower Miocene through Quaternary strata of the Yakataga Formation, a unit strongly affected by glacial processes and consisting mostly of glacimarine pebbly mudstone, till-like diamictite, and marine mudstone (Plafker & Addicott 1976).

Structure along the Middleton slope region is markedly different from west to east (Bruns 1985; Bruns & Schwab 1983). Southwest of Middleton Island, the slope is characterized by large, laterally extensive, shelf-break parallel anticlines and synclines. At the base of the slope, an accretionary wedge becomes more pronounced northeast to southwest. Between Middleton Island and Pamplona Spur, a large anticlinal fold forms a ridge (Khitrov Ridge; Fig. 3) at the base of the slope which splays northeast into multiple anticlines and continues northeastward as the Pamplona zone fold and thrust belt.

The continental slope directly seaward of Middleton Island shows little structure. Bruns (1985) suggests that this variation in structure from the large folds and accretionary structures southwest of Middleton Island to a region of little structure adjacent to Middleton Island, to structures of the Pamplona zone fold and thrust belt at Khitrov Ridge, may be a result of the Yakutat terrane passing through the subduction zone. Subduction of the terrane may remove

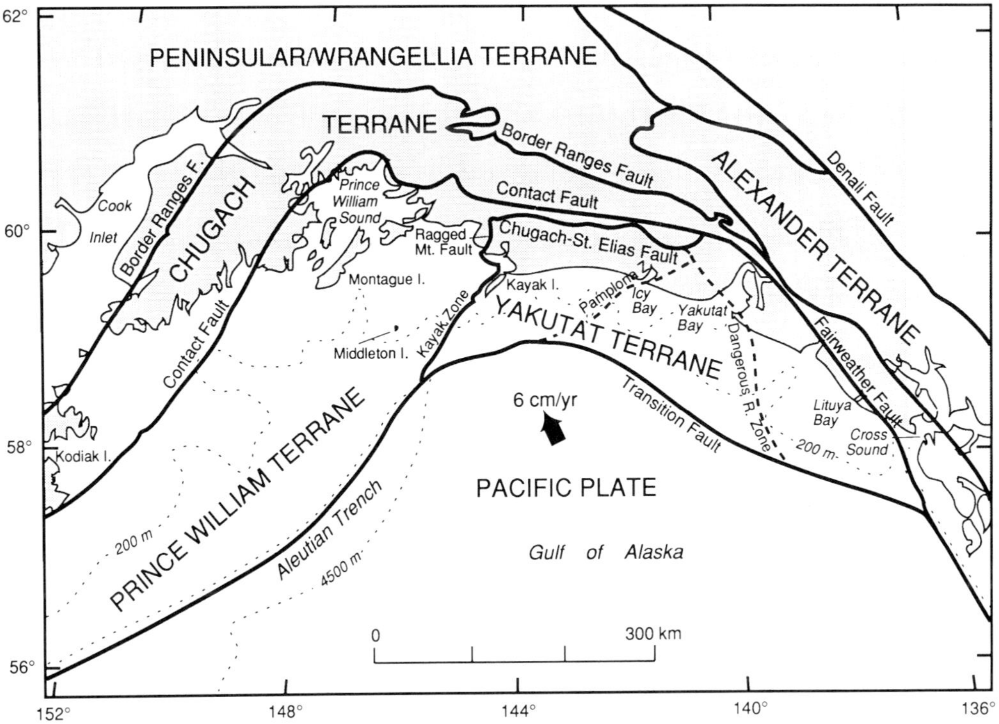

Fig. 2. Tectonostratigraphic terrane map and related major fault system of southern Alaska and Gulf of Alaska modified from Jones *et al.* (1986).

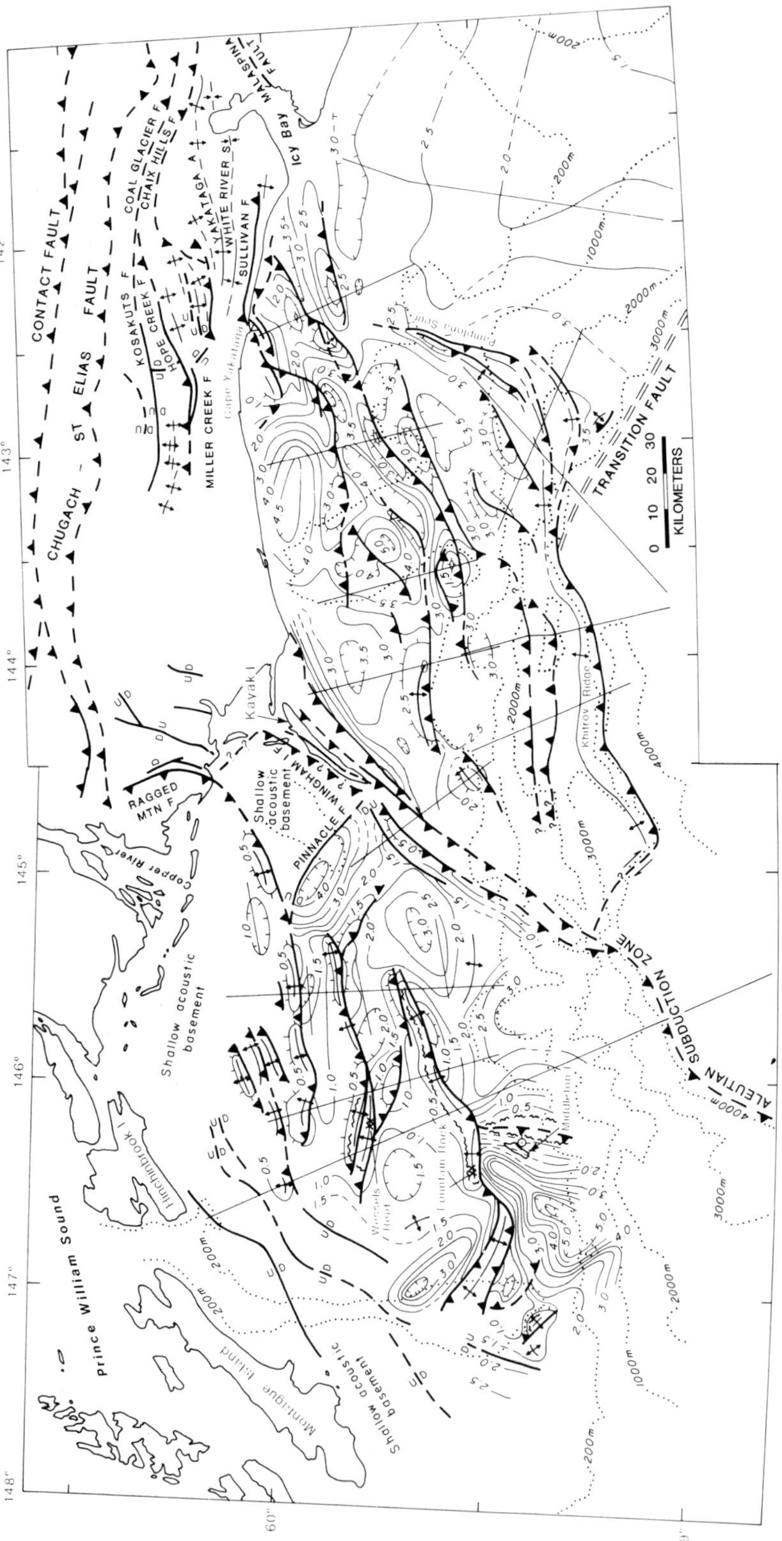

Fig. 3. Structure map of northern Gulf of Alaska from Bruns (1985).

previously developed structures, and, after passage of the Yakutat terrane, new subduction and accretion-related structures may develop as the Pacific plate continues to collide with the North American plate.

Regional morphology

The shelf, which ranges in width from 20 to 80 km, has an average gradient of about 0.25° from shore to the shelf break which lies at a depth of about 200 m (Atwood *et al.* 1981). Six prominent valleys cut the shelf between Cross Sound and Prince William Sound and five valleys or troughs cut the shelf from Middleton Island to the southwest of Kodiak Island (Fig. 1). All are attributed primarily to the advance of tongues of glacier ice across the shelf during low stands of sea level (Carlson *et al.* 1982; Thrasher 1979; Hampton *et al.* 1986). Characteristics of the submarine valleys, many of which are located seaward of large glaciers, include: a sub-bottom erosional surface incised into the lithified strata of the shelf; U-shaped cross sections; concave longitudinal sections that commonly shoal at their seaward end; till-like sediment along the walls of the valleys and at the outer shelf and upper slope adjacent to the troughs; seismic stratigraphy and bottom samples indicative of glacially-derived strata; and disconnected depressions on the modern valley floor.

The continental slope in the northern Gulf of Alaska varies in width from 25 to 100 km and varies in gradient from about 5° to 10°. In the area investigated (Fig. 1), the slope is about 100 km wide off Amatuli Trough and less than 30 km wide off Bering Trough. The base of the slope coincides with the landward wall of the Aleutian Trench, the morphological expression of which terminates in the study area. Farther to the southwest, along the Aleutian arc, the trench floor exceeds water depths of 7000 m, however, in the northern Gulf of Alaska the trench floor shoals to less than 4000 m. The trench floor has an axial gradient of 0.21° at its northeastern end, a slightly lower gradient than the nearby continental shelf.

Quaternary sediment

The distribution of surficial sediment on the northern Gulf of Alaska shelf (Fig. 4) combined with high-resolution seismic-reflection profiles (Fig. 5) provides a useful picture of the Quaternary depositional environment (Carlson *et al.* 1977; Molnia & Carlson 1978, 1980). In the simplest case, a graded shelf sedimentary profile appears to be developing today (Fig. 6). Sand accumulating in the nearshore zone grades offshore into silty sand and finally to clayey silt by mid-shelf depths. This Holocene glacimarine deposit wedges out near the outer shelf where an underlying thin Pleistocene(?) diamict consisting of gravelly sand, muddy gravel, and pebbly mud crops out (Figs. 5 & 6). The diamict appears to mantle the outer shelf and upper slope in much of the Northern Gulf of Alaska (Figs. 4, 5, 6).

Complicated sediment distribution patterns emerge where tectonic activity causes positive relief features such as banks (i.e. Tarr Bank) and islands (i.e. Middleton and Kayak Island) and glacial advances created negative relief features such as sea valleys (i.e. Hinchinbrook Sea Valley and Amatuli Trough) (Fig. 4). Winnowing of the fine sediment from the tops of banks and islands creates cobble and boulder lag deposits such as those present today on Tarr Bank (Carlson 1989). Lag deposits have formed in the past as seen in outcrops of the Yakataga Formation exposed in the uplifted terraces on Middleton Island (Eyles *et al.* 1985). Eyles (1987) interprets some of the Yakataga Formation sand and gravel facies on Middleton Island as fill deposited by various kinds of sediment-gravity flows within a submarine-channel complex. She suggests that this channel system was cut in an outer shelf-upper slope environment by sediment gravity flows generated by glacial meltwater streams which drained from a large ice sheet that covered much of the shelf. The modern glacially-derived sea valleys have concentrations of relict coarse sediment adjacent to their mouths on the outer shelf and upper slope (Carlson *et al.* 1982; Sharma 1979). These accumulations of coarse sediment are possibly end morainal deposits (Fig. 4). The sea valleys are pathways for currents and sediments to traverse the shelf and can also be sediment sinks. Seismic-reflection profiles show accumulations of modern sediment tens to hundreds of metres thick in some of the sea valleys (Carlson *et al.* 1982; Hampton 1983).

Although much is known about the distribution and physical properties of the shelf sediment, the information about slope sediment in this area is sparse. Sharma (1979) shows data from 12 sample stations in the area of the GLORIA mosaic. These shelf edge and upper slope samples contain quantities of gravel ranging from a trace to more than 25% and all samples contain less than 40% clay. These samples are primarily classified as gravelly sandy silts and silty sands (Fig. 4). Such sediment, whether glacimarine pebbly mud or basal till or

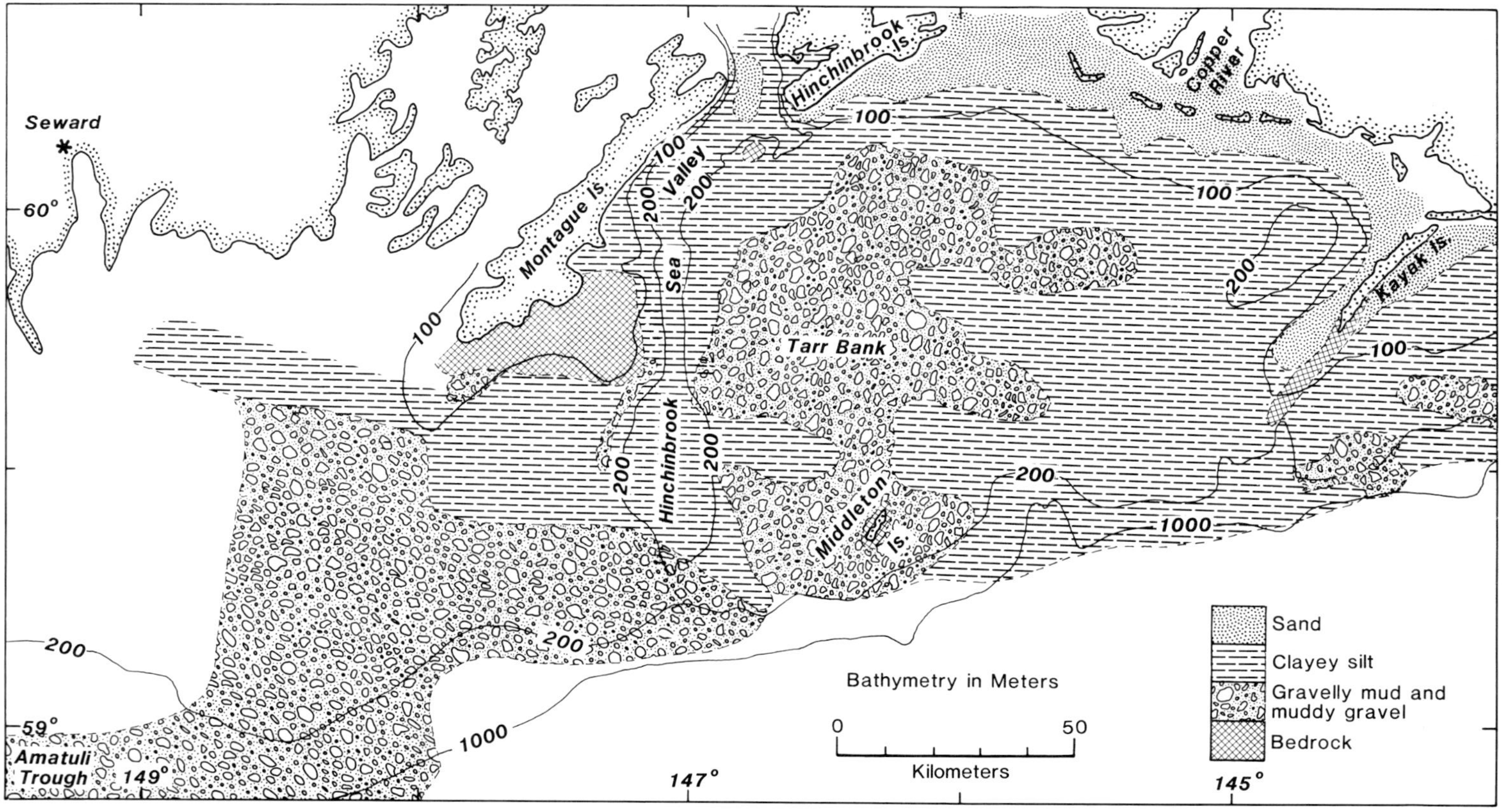

Fig. 4. Sediment distribution map of shelf and upper slope in northern Gulf of Alaska. Modified after Carlson *et al.* (1977). Data west of Hinchinbrook Sea Valley from Sharma (1979) and Hampton *et al.* (1986).

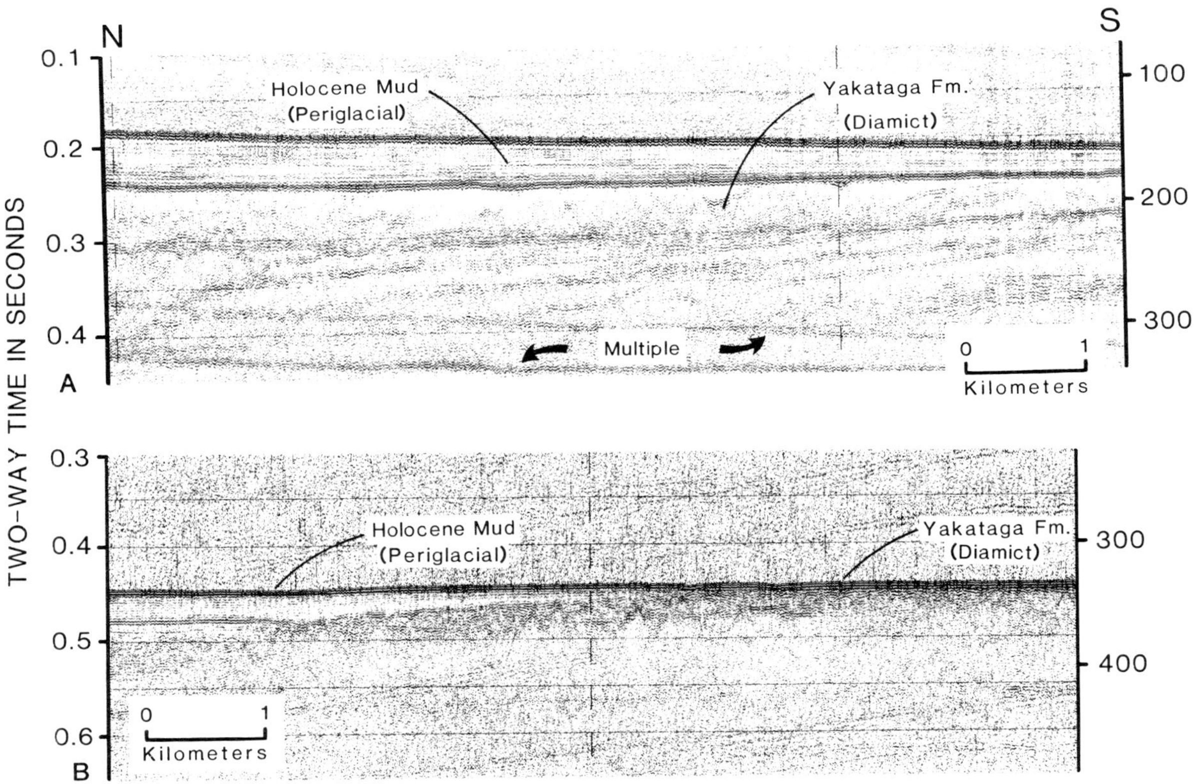

Fig. 5. Minisparker (600 J) profiles showing diamict unit of Yakataga Formation overlain by seaward-thinning wedge of Holocene mud. (V.E.~10.5x). Profile B shows diamict cropping out on outer shelf.

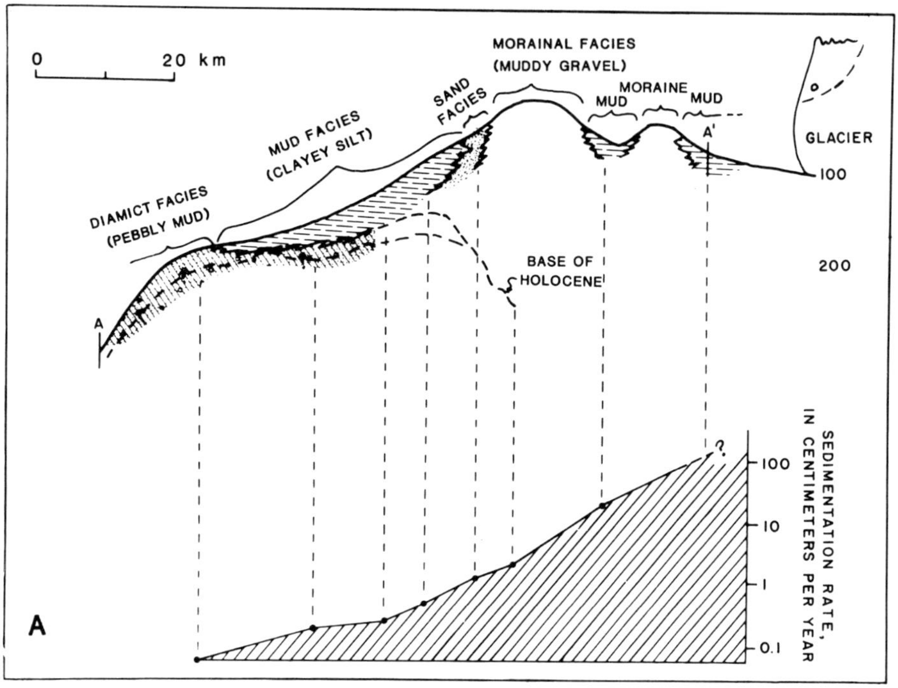

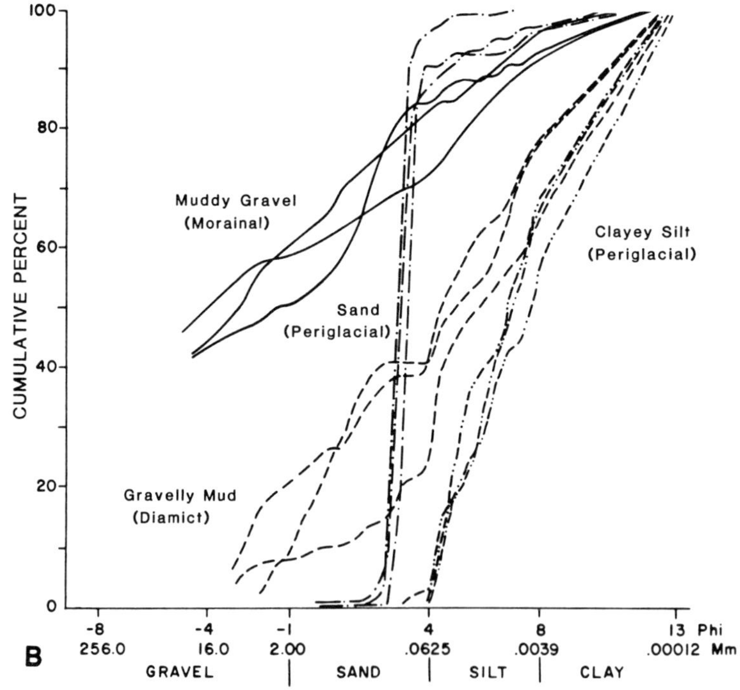

Fig. 6. (A) Diagrammatic profile of northern Gulf of Alaska shelf and upper slope sedimentary facies. (B) Typical grain-size curves of various glacial and glacimarine facies.

outwash delta sediment, owes its origin to some glacially-related processes. The sediment was probably deposited at the shelf edge during one of the most recent glacial stages when sea level was near the shelf break. Two USGS cruises in 1978 provided samples (18 cores and 24 dredges from the slope in the northern Gulf. Several gravity cores cut through lenses of coarse clastic sediment or in some cases bottomed in gravelly sand and sandy or silty mud. Some samples suggest deposition of glacimarine mud with dropstones. Others samples may be the result of mass movement of basal tills that were deposited by ice tongues at the mouths of sea valleys and were then carried to the base of the slope by earthquake or gravity-triggered slides and slumps.

Morphology interpreted from GLORIA and bathymetric data

The GLORIA mosaic constructed for this small segment of the Gulf of Alaska EEZ clearly shows marked changes in morphology throughout this transitional subduction zone (Fig. 7). Large ridges dominate the slope to the southwest of Hinchinbrook Sea Valley (Fig. 7). These ridges vary in length from less than 5 km to greater than 35 km and in width from 2 to 9 km and vary in relief from 200 m to more than 1000 m. The ridges are asymmetrical in transverse section with slopes that vary from 10° to 30°. The slopes of the ridges are generally steeper on the seaward side, which is a result of both fold vergence and of ponded sediment, often several hundred metres thick, onlapping the ridge from the landward side. South of the area between Kayak Island and Bering Trough the large Khitrov Ridge is greater than 50 km in length and 1 to 5 km in width, slopes 17° to 21° seaward and 7° to 12° landward. The main ridge is cut at about 144.5°W longitude by a low pass or saddle with relief of about 25 m. This pass could be overtopped by large, thick turbidity currents that are thought to flow down the adjacent valley (Fig. 7). A shorter ridge that appears to be aligned with Khitrov Ridge is located southwest of and separated from the main ridge by a slope valley that has a relief of 400 to 600 m and a width of 2.5 to 3.5 km. The shorter ridge is 13 km long, 2 km wide, and has declivities of 23° seaward and 7° landward. Between the west end of Khitrov Ridge and the group of ridges developing on the slope southwest of Middleton Island, a distance of about 90 km, there are no shelf-break parallel ridges, only erosional ridges that separate the numerous gullies that are incised into the slope (Fig. 7).

Slope valleys

The continental slope in the northern Gulf of Alaska is different to many other slopes where large submarine canyons dominate the morphology. Here there are no large canyons. The shelf valleys, rather than emptying into large slope canyons, end abruptly. Drainage systems on the slope are more likely to be a group of gullies cut as a shallow dendritic pattern into the upper slope (Fig. 7). However, immediately seaward of Bering Trough, and to a lesser degree Hinchinbrook Sea Valley, the pervasive dendritic patterns are not as evident. The smoothness of the slope suggests a covering by sediment, perhaps dumped there when ice tongues occupied the shelf valleys. Well developed dendritic drainage patterns are present southeast of Middleton Island and Tarr Bank where, in each case, the drainage gullies merge to form small canyons that are incised into the lower slope (Fig. 7). These canyons empty into the Aleutian Trench. The most prominent and largest dendritic gulley system heads at the shelf break near Tarr Bank southeast of Middleton Island (Fig. 8) and is herein named the Tarr Canyon system. Numerous gullies merge into tributary canyons and eventually at 2900 m into one canyon that extends to the floor of the Aleutian Trench at a depth of about 4400 m. The overall length of Tarr Canyon is about 48 km with a thalweg gradient of about 4.6°. The width of the canyon floor varies from 0.5 to 1 km, and the width from rim to rim is as wide as 5 km (Fig. 9).

West of longitude 146° the slope trends to the southwest and is dominated by compressional ridge morphology. The ridges are especially prevalent below mid-slope depths (2000 to 2600 m). The drainage patterns reflect the increased ridge dominance. East of Middleton Island the canyons and gulleys are dendritic in pattern and extend directly down slope to the trench but, to the west of Middleton Island, the ridges obstruct the down slope drainage causing detours of the flow patterns parallel to the ridges (Fig. 7). This obstruction results in ponding of sediment behind the ridges. In some places the channels can be traced around the ridges and the flow may extend in a circuitous pattern from the shelf break all the way to the Aleutian Trench (Fig. 10). A well developed winding channel pattern, which we have named the Amatuli Channel system, has formed seaward of Amatuli Trough. Similar patterns can be seen on the GLORIA mosaic further to the southwest (Fig. 7). Amatuli Channel is about 105 km long from the shelf break at 200 m to

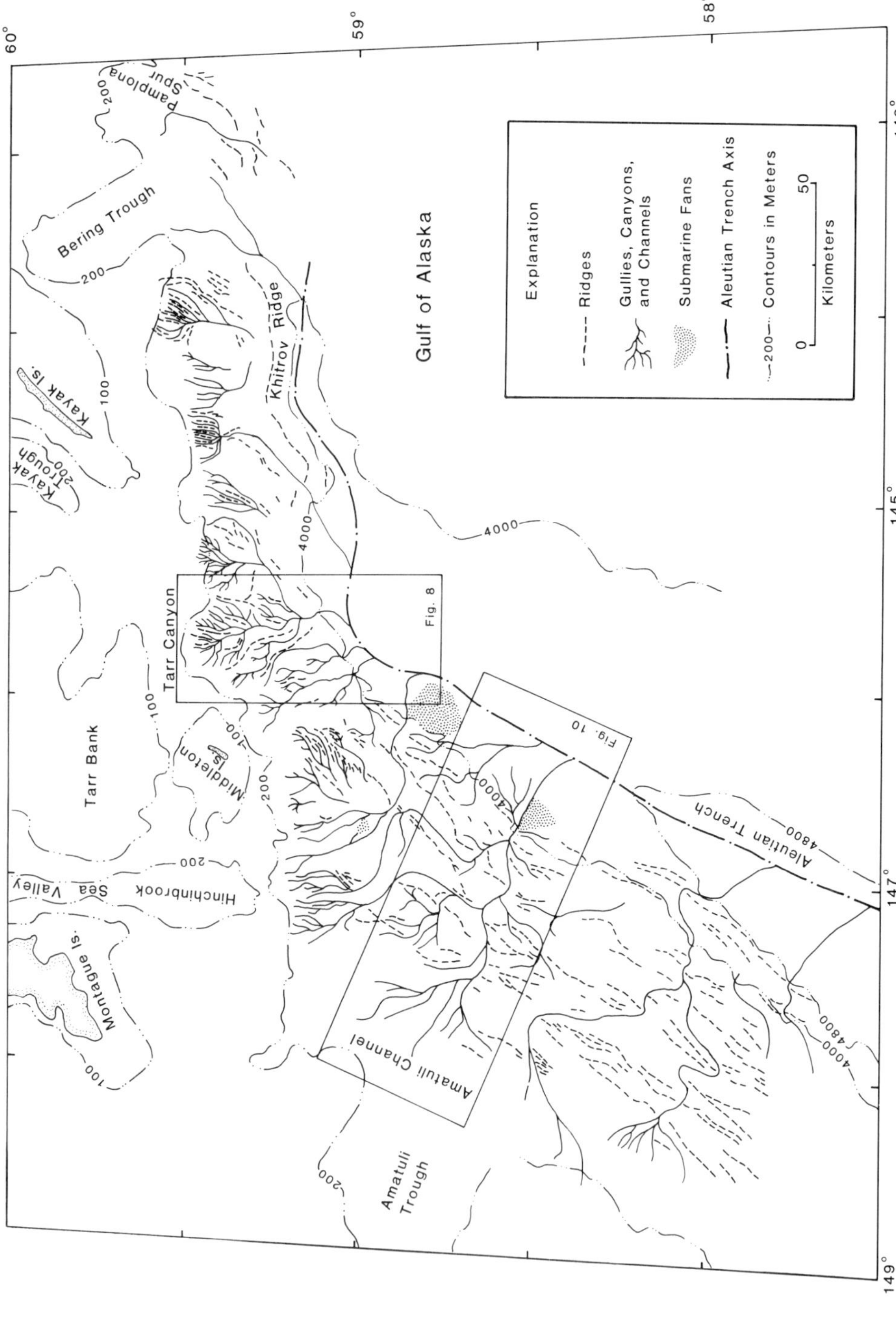

Fig. 7. Interpretive sketch of morphological features shown on GLORIA mosaic of northern Gulf of Alaska. Bathymetry taken from Atwood *et al.* (1981).

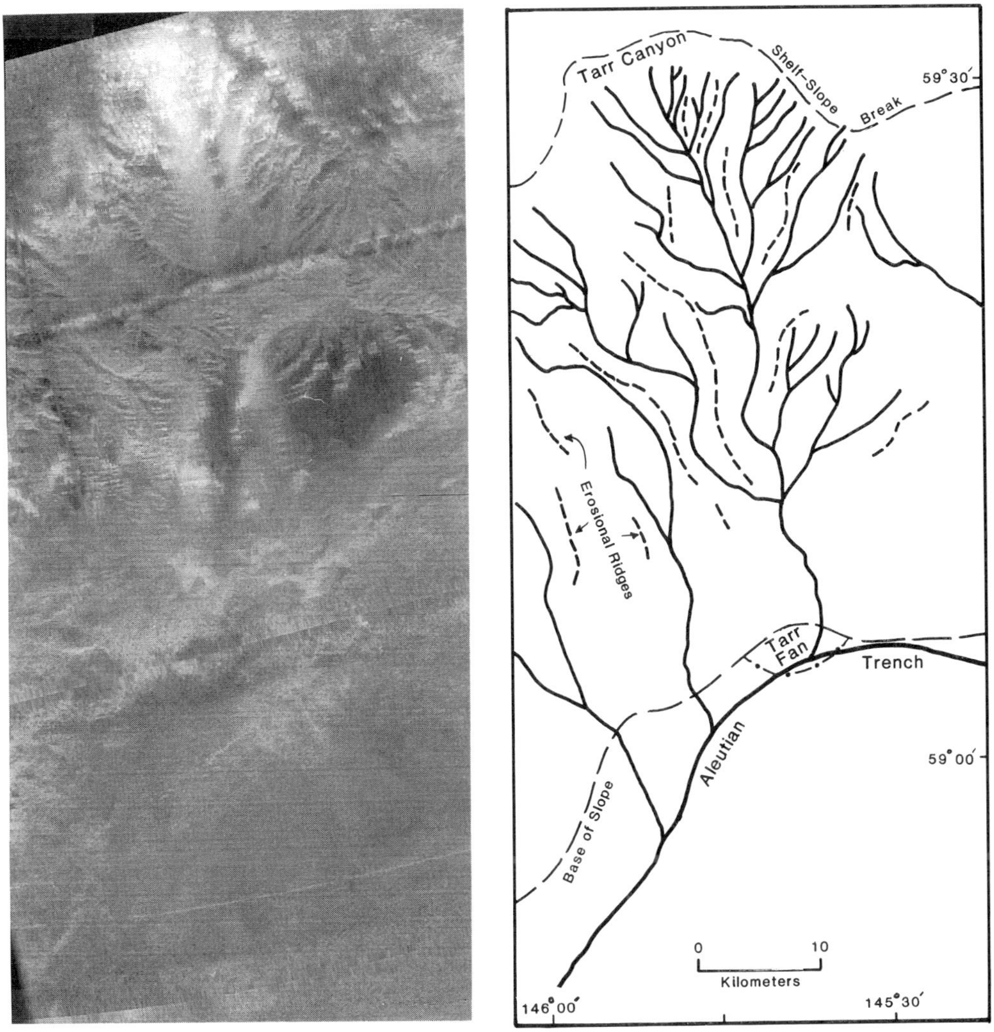

Fig. 8. Digitally mosaiced segment of GLORIA imagery and interpretive sketch showing intensely gullied dendritic head of Tarr Canyon. See Fig. 7 for location.

the base of the slope at 4200 m where it empties onto Amatuli Fan, a small deposit on the trench floor. The gradient of the main thalweg of the channel is about 2.2°. The width of the channel floor varies from 1 to 7 km and the width from rim to rim is as much as 12 km (Fig. 9).

Submarine fans

At the base of the slope in the inner part of the Aleutian Trench, small submarine fans are forming at the outlets of at least two of the slope channel systems. The larger fan is located at the inner wall of the trench below the channel system that heads near Hinchinbrook Sea Valley (Fig. 7). This fan has a radius of 15 km and a width of about 20 km. The gradient of the fan is about 1.9°. A more prominent fan on the GLORIA mosaic is located at the mouth of the Amatuli Channel at a depth of about 4200 m (Fig. 10). This fan is portrayed on the side-scan imagery as consisting of a series of ridges that are similar in appearance to bedforms, but are more likely slide or slump structures that have formed a fan-like deposit of debris elements. The fan has a radius of 10 km and a width of about 13 km. The gradient down the fan is 0.6°. Cores collected from the trench floor sediment

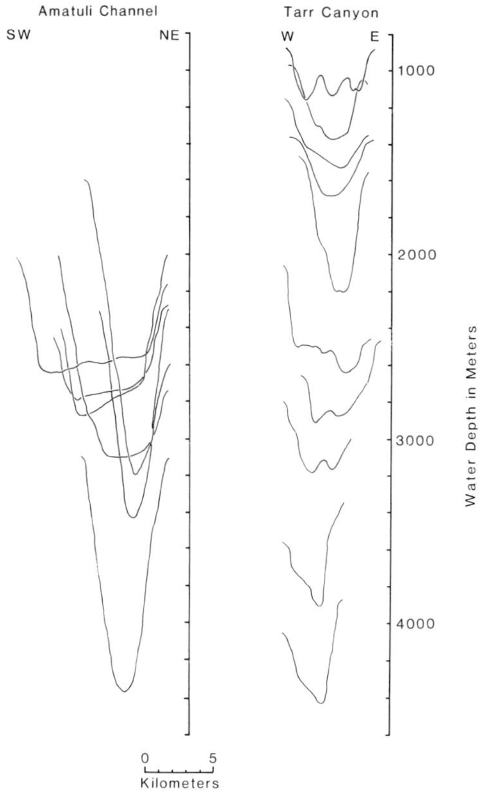

Fig. 9. Selected bathymetric profiles of traverses across Amatuli Channel (left) and Tarr Canyon (right).

contain graded sand layers attributed to turbidity current deposition (Johnson 1986; Underwood 1986).

Development of the glacimarine sedimentary environment

GLORIA imagery, seismic-reflection profiles, and sediment samples provide a preliminary picture of the Pleistocene and Holocene depositional environments on the northern Gulf of Alaska continental margin. During the glacial stages of the Pleistocene and earlier times, the shoreline was probably near the present shelf edge and large glaciers that fringe the modern Gulf of Alaska advanced far out across the shelf. When lobes of these massive glaciers reached the outer shelf, they were actively attacked by the Gulf waters so that large blocks, calved from the ice front, carried coarse sediment further into the Gulf, where, upon melting, the debris would be deposited on the slope and the abyssal plain as was seen at DSDP sites 178, 181, and 182 (Von Huene *et al.* 1973). The sediment transported by the glaciers was also deposited at the face of the glaciers in the form of terminal moraines. If these moraines were deposited near the edge of the shelf, storm waves or earthquakes could trigger sliding and slumping of the coarse glacial debris. The GLORIA imagery shows dendritic drainage gullies along the upper slope. These gullies were probably eroded by sediment gravity flows during the Pleistocene low stands of sea level. Gullies appear to be sparse or lacking beneath the mouths of some of the large sea valleys where a smooth slope suggests abundant sediment accumulation. We think that much of this sediment was dumped there when the latest ice tongues occupied the sea valleys, thus effectively covering some of the erosional gullies. Evidence for glaciation of much of the shelf includes: the morphology of the large shelf valleys that extend to the shelf break, gravelly sand and sandy gravel at the outer edge of the shelf and upper slope, especially adjacent to the shelf valleys, and a thin layer of discordant, irregular reflections that have been interpreted as diamicton, including till and outwash (Molnia & Carlson 1980).

As the climate ameliorated, sea level rose and wave action smoothed out the surviving glacial deposits. Accompanying the glacial advances and retreats and attendant sea-level fluctuations, was the continuing tectonic activity as the Pacific plate pushed slowly beneath the North American plate. Subduction of the Pacific plate formed numerous discontinuous compressional ridges on the slope southwest of Middleton Island. These ridges disrupted the downslope transport of glacially-derived sediment deposited at the shelf edge by the large ice tongues. Instead of the gravity flows and slides carrying the debris directly down slope to the Aleutian Trench, the ridges ponded much of the sediment. Sediment that reached the abyssal depths arrived there by a circuitous route that is clearly evident on the GLORIA sonar imagery. Disruption of the downslope transport also occurred on the transpressional margin to the southeast of Middleton Island, but not to the extent that disruption occurred to the southwest. As a result of these tectonic differences, the drainage patterns of the two areas are very different. Dendritic patterns prevail on the southeastern margin and near-trellised patterns dominate the southwestern margin.

Another result of subduction is periodic uplift of shelf areas such as the Middleton Island

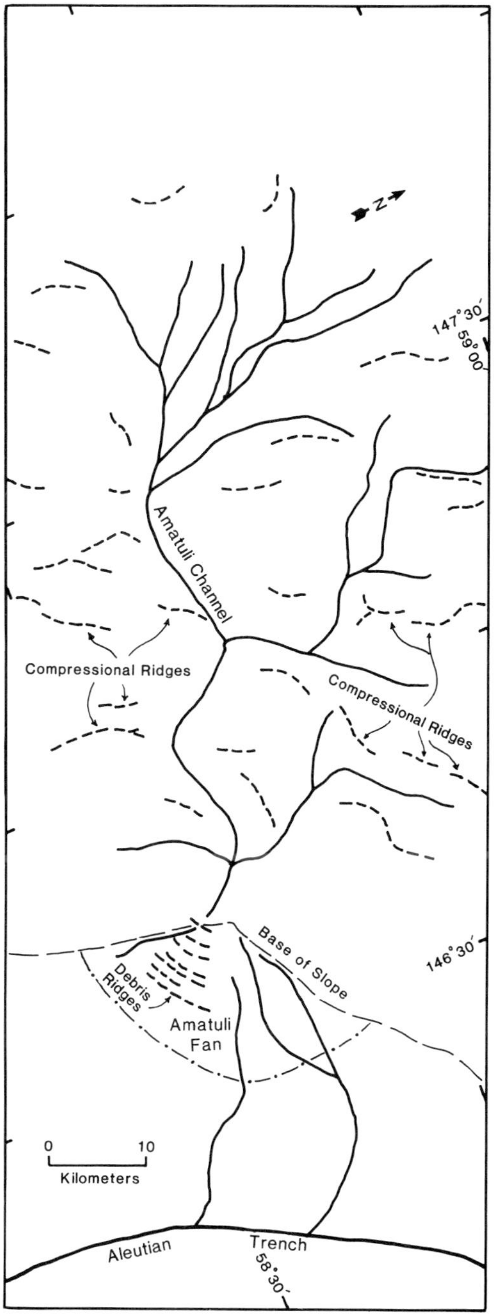

Fig. 10. Digitally mosaiced segment of GLORIA imagery and interpretive sketch showing winding course of Amatuli Channel as controlled by compressional ridges. Debris fan at centre of lower quarter of image. See Fig. 7 for location.

platform which has created several raised terraces (Plafker & Rubin 1967), Tarr Bank and other islands and banks. Thus the uplift has, in some cases, brought features to within reach of the frequent storm waves that sweep through the area.

During the Holocene, glacimarine sediment has been accumulating on the shelf. This sediment is prograding across the shelf and, in places, covers all but the outermost shelf with a wedge-shaped deposit that grades from sand in the littoral zone to clayey silt beyond a water depth of about 50 metres. Exceptions to this sedimentary environment include the mouths of some of the fjord-like bays such as Icy and Yakutat Bays where glaciers extended to and beyond the mouths of the bays, leaving large terminal moraines. These glaciers filled the bays in historic time, less than 100 years ago in Icy Bay (Molnia 1977) and less than 1000 years ago in Yakutat Bay (Plafker & Miller 1958).

Most of the sediment by-passing the littoral zone today is suspended sediment from streams and rivers that drain from the massive ice fields of south central and southeast Alaska. The large visible surface plumes of suspended matter become incorporated in the coastal current that sweeps from east to west along the shelf (Burbank 1974; Reimnitz & Carlson 1975). Complex gyres move some particulate suspended matter across the shelf and over the slope (Sharma 1979; Molnia & Carlson 1980; Feely *et al.* 1979) where the glacimarine sediment is deposited on the slope and the abyssal areas. Another mode of sediment transport to the slope is the occasional katabatic wind storm that moves down the Copper River gorge and across the delta and barrier islands at speeds of 75 mph (120 km/hr) and more, carrying clouds of silt-size sediment more than 50 km into the Gulf (Post 1976; Swift *et al.* 1978; Molina 1983). Some of the silt-size sediment has accumulated on the upper slope, blanketing the glacial debris overlying the dendritic drainage systems cut into the upper slope.

Conclusions

(1) GLORIA side-scan sonar imagery has added significantly to our knowledge of the morphology and sediment dispersal pathways on the continental margin in the tectonically complex northern Gulf of Alaska.

(2) Southwest of Middleton Island, the slope is characterized by discontinuous compressional ridges that are barriers to downslope transport of glacial and glacimarine sediment. However, circuitous pathways have developed that channel some sediment around the subduction ridges to the Aleutian Trench.

(3) South of Middleton Island, the slope lacks the compressional ridges and is dominated by several intensively gullied dendritic drainage systems that can be traced directly downslope to the trench.

(4) On the slope east of Middleton Island, between Kayak Island and Pamplona Spur, one large ridge (Khitrov), present on the lower slope, has deflected and ponded Pleistocene and Holocene sediment.

(5) On the upper slope, off several large glacially scoured sea valleys, a blanket of glacial and glacimarine sediment apparently covers any dedritic gulley systems that may have formed during the Pleistocene low stands of sea level.

We are grateful to the crew and scientific party of the R/V FARNELLA for their enthusiastic cooperation during the collection of the GLORIA imagery. We appreciate the assistance provided by Douglas Rearic and Chip Stevens in producing enhanced segments of the GLORIA mosaic. Thanks to Carolyn Eyles, Herman Karl, Bruce Molnia and Erk Reimnitz for their constructive criticisms of the manuscript.

References

Atwood, T.A., Bruns, T.R., Carlson, P.R., Molnia, B.F. & Plafker, G. 1981. Bathymetric map of the eastern Gulf of Alaska. U.S. Geological Survey Miscellaneous Field Study Map MF-859, 3 sheets, scale 1:250 000.

Bruns, T.R. 1983. A model for the origin of the Yakutat block, an accreting terrane in the northern Gulf of Alaska. *Geology*, **11**, 718–721.

—— 1985. Tectonics of the Yakutat block, an allochthonous terrane in the northern Gulf of Alaska. U.S. Geological Survey Open-File Report 85–13, 112.

—— & Schwab, W.C. 1983. Structure maps and seismic stratigraphy of the Yakataga segment of the continental margin, northern Gulf of Alaska. U.S Geological Survey Miscellaneous Field Studies Map MF-1424, 20 p., 4 sheets, scale 1:250 000.

Burbank, D.C. 1974. *Suspended sediment transport and deposition in Alaskan coastal waters*. MS Thesis, University of Alaska, Fairbanks, AK.

Carlson, P.R. 1989. Seismic reflection characteristics of glacial and glacimarine sediment in the Gulf of Alaska and adjacent fjords. *In*: Powell, R.D. & Elverhoi, A. (eds.) Modern Glacimarine Environments: Glacial and Marine Controls on Modern Lithofacies and Biofacies. *Marine Geology*, **85**, 391–416.

——, Molnia, B.F., Kittleson, S.C. & Hampson, J.C. 1977. Distribution of bottom sediments on the continental shelf, northern Gulf of Alaska. U.S. Geological Survey Miscellaneous Field

Study Map MF-876, 13 p., 2 sheets, scale 1:500 000.

CARLSON, P.R., BRUNS, T.R., MOLNIA, B.F., & SCHWAB, W.C. 1982. Submarine valleys in the northeastern Gulf of Alaska: characteristics and probable origin. *Marine Geology*, **47**, 217–242.

EOS. 1988. Navy ponders a total ban on bathymetry. *Transactions, American Geophysical Union*, **69**, 1571–1573.

EYLES, C.H. 1987. Glacially influenced submarine-channel sedimentation in the Yakataga Formation, Middleton Island, Alaska. *Journal of Sedimentary Petrology*, **57**, 1004–1017.

——, EYLES, N., & MIALL, A.D. 1985. Models of glaciomarine sedimentation and their application to the interpretation of ancient glacial sequences. *Palaeogeography, Palaeoclimatology, and Palaeocology*, **51**, 15–84.

FEELY, R.A., BAKER, E.T., SCHUMACHER, J.D., MASSOTH, G.J. & LANDING, W.M. 1979. Processes affecting the distribution and transport of suspended matter in the northeastern Gulf of Alaska. *Deep-Sea Research*, **26**, 445–464.

HAMPTON, M.A. 1983. Geology of the Kodiak shelf, Alaska: environmental considerations for resource development. *Continental Shelf Research*, *1*, 253–281.

——, CARLSON, P.R., LEE, H.J. & FEELY, R.A. 1986. Geomorphology, sediment, and sedimentary processes. *In*: HOOD, D.W. & ZIMMERMAN, S.T. (eds.) *The Gulf of Alaska: Physical Environment and Biological Resources*. National Oceanic and Atmospheric Administration, 93–143.

JOHNSON, K.H. 1986. Recent sedimentation in the central and western Aleutian Trench. M.S. Thesis, Calif. St. Univ., Hayward, Calif.

JONES, D.L., SILBERLING, N.J., CONEY, P.J. & PLAFKER, G. 1986. Lithotectonic terrane map of Alaska (west of 141st meridan). U.S. Geological Survey Miscellaneous Field Studies Map MF-1874, scale 1:2 500 000.

MOLNIA, B.F. 1977. Rapid shoreline erosion at Icy Bay, Alaska–A staging area for offshore petroleum development. *Proceedings 1977 Offshore Technology Conference*, IV, 115–126.

—— 1983. Subarctic glacial-marine sedimentation: a model *In*: MOLNIA, B.F. (ed.) *Glacial-marine sedimentation*. Plenum, New York, 95–144.

—— & CARLSON, P.R. 1978. Surface sedimentary units of the northern Gulf of Alaska continental shelf. *American Association of Petroleum Geologists Bulletin*, **62**, 633–643.

—— & —— 1980. Quaternary sedimentary facies on the continental shelf of the northeast Gulf of Alaska. *In*: FIELD, M.E., BOUMA, A.H., COLBURN, I.P., DOUGLAS, R.G. & INGLE, J.C. (eds.) *Quaternary Depositional Environments of the Pacific Coast*, Pacific Coast Paleogeography Symposium 4, Society of Economic Paleontologists and Mineralogists, Los Angeles, Calif. p. 157–168.

PLAFKER, G. 1987. Regional geology and petroleum potential of the northern Gulf of Alaska continental margin. *In*: SCHOLL, D.W., GRANTZ, A & VEDDER, J.G. (eds.) *Geology and Resource Potential of the Continental Margin of Western North America and adjacent ocean basins–Beaufort Sea to Baja California*. Circum-Pacific Council for Energy and Mineral Resources Earth Science Series, 6, Houston, Texas, 229–268.

—— & ADDICOTT, W.O. 1976. Glaciomarine deposits of Miocene through Holocene age in the Yakataga Formation along the Gulf of Alaska margin, Alaska. *In*: MILLER, T.P (ed.) *Recent and Ancient Sedimentary Environments in Alaska*. Proceedings of Alaska Geological Society Symposium, Anchorage, AK, Q1–Q23.

—— & MILLER, D.J. 1958. Glacial features and surficial deposits of the Malaspina district, Alaska. U.S Geological Survey Miscellaneous Geological Investigation Map I-271, scale 1:25 000.

—— & RUBIN, M. 1967. Vertical tectonic displacement in south-central Alaska during and prior to the great 1964 earthquake. *Journal of Geoscience*, Osaka City University, **10**, 53–66.

POST, A. 1976. Environmental geology of the central Gulf of Alaska coast. *In*: WILLIAMS, R.S., Jr. & CARTER, W.D. (eds.) *ERTS-1: a new window to our planet*. U.S. Geological Survey Professional Paper 929, 117–119.

REIMNITZ, E. & CARLSON, P.R. 1975. Circulation of nearshore surface water in the Gulf of Alaska. *In*: CARLSON, P.R., CONOMOS, T.J., JANDA, R.J. & PETERSON, D.H. (eds.) Principal sources and dispersal patterns of suspended particulate matter in nearshore waters of the northeast Pacific Ocean. Resources Technology Satellite final report to NASA. National Technical Information Service, E75-10266, Springfield VA. 145 p.

SHARMA, G.D. 1979. *The Alaskan shelf: Hydrographic, Sedimentary, and Geochemical Environment*. Springer, New York, NY.

SWIFT, D.J., MOLNIA, B.F. & JACKSON, R.G., II. 1978. Intermittent structure in the atmospheric boundary layer made visible by entrained sediment: example from the Copper River delta, Alaska. *Journal of Sedimentary Petrology*, **48**, 897–900.

THRASHER, G.P. 1979. Geologic map of the Kodiak outer continental shelf, western Gulf of Alaska. U.S. Geological Survey Open-File Report 79–1267, 2 sheets, scale 1:250 000.

UUNDERWOOD M.B. 1986. Transverse infilling of the central Aleutian Trench by unconfined turbidity currents. *Geo-Marine Letters*, **6**, 7–13.

VON HUENE, R., LARSON, E. & CROUCH, J. 1973. Preliminary study of ice-rafted erratics as indicators of glacial advances in the Gulf of Alaska. *In*: KULM, L.D., VON HUENE, R. *et al*. 1973. *Initial Reports of the Deep Sea Drilling Project*, **18**, Washington (U.S. Government Printing Office) 835–842.

Fossilization potential of Arctic fjord and continental shelf benthic macrofaunas

ALEC E. AITKEN

Fisheries and Oceans Canada, Institut Maurice Lamontagne, C.P. 1000, 850 Route de la Mer, Mont-Joli, Quebec, Canada G5H 3Z4 (Present Address: Department of Geography, Queen's University, Kingston, Ontario, Canada K7L 3N6)

Abstract: The organic remains of many organisms do not survive the post-mortem period. This study was undertaken to identify patterns of preservation in arctic marine macrofaunas with the goal of aiding Quaternary researchers in the interpretation of fossiliferous marine sediments. The probable fate of 310 genera of marine invertebrates which inhabit the fjords and continental shelf in the eastern Canadian Arctic was investigated. The preservation potential of marine invertebrates was assessed in three ways: by morphology (presence of macroscopic hardparts); by the previous existence of a fossil record (data for genera only taken from the *Treatise on Invertebrate Paleontology*); and by the previous record in Quaternary marine sediments on Baffin Island. Considering the arctic benthos as a whole, up to 86% of the genera have not been preserved and are not likely to be preserved.

The fossil record preferentially preserves infaunal suspension feeding and epifaunal browsing organisms. Infaunal suspension feeding bivalves frequently contribute more than 50% of the macroscopic hardparts in Quaternary fossil assemblages. The lack of preservation of soft-bodied carnivorous and deposit feeding organisms suggests that little understanding of community trophic dynamics based on the study of trophic group composition is likely to be derived from Quaternary fossil assemblages. Bivalves and gastropods inhabiting fjords and nearshore waters exhibit the greatest preservation potential of all the arctic marine benthos. The excellent preservation potential of arctic marine molluscs offers palaeoecologists the potential to reconstruct ancient marine biocoenoses in glacimarine environments.

Marine invertebrate fossils, notably molluscs and foraminifera, have been collected extensively from Quaternary sediments that crop out along the eastern coast of Baffin Island to provide the materials needed for geochronological research (Miller *et al.* 1977; Andrews 1978; Miller 1985). Quaternary marine invertebrate assemblages from eastern Baffin Island are described by Matthews (1967), Andrews (1972), Feyling-Hanssen (1976, 1977, 1980). Andrews *et al.* (1981), Mode (1985) and Aitken & Risk (1988). The study of the palaeoecology of these invertebrate assemblages has focused on their utility in inferring past marine climates in Baffin Bay. The repeated association of marine sediments containing subarctic marine faunas with glacimarine diamictons suggests a marine climate warmer than the present during Quaternary glaciations of eastern Baffin Island (Andrews *et al.* 1981; Aksu 1985; Mode 1985). Quaternary research conducted in the eastern Canadian arctic has largely ignored the study of other fundamental aspects of community palaeoecology such as community trophic structure and population dynamics. The works of Andrews (1972, 1973) on mollusc growth and Aitken & Risk (1988) on predator–prey interactions are notable exceptions in the study of Quaternary marine invertebrate palaeoecology in the eastern Canadian arctic.

Comparison of living faunas with fossil assemblages may enable palaeoecologists to reconstruct ancient biocoenoses. An understanding of the degree to which fossil assemblages differ from the living faunas from which they are derived is required if palaeoecological reconstructions are to be successful. The purpose of this study is to identify patterns of preservation of macrobenthic organisms inhabiting the fjords of eastern Baffin Island and the adjacent continental shelf with the goal of aiding Quaternary researchers in the interpretation of fossiliferous marine sediments. I have tried to assess how much of the macrobenthos is potentially preservable, and what if any differences in preservation potential exist amongst benthic habitats and amongst life habits. It is obvious that the organic remains of many organisms do not survive the post-mortem period (Lawrence 1968; Petersen 1976; Schopf 1978;

From DOWDESWELL, J. A. & SCOURSE, J. D. (eds), 1990, *Glacimarine Environments: Processes and Sediments*. Geological Society Special Publication No 53, pp 155–176.

Cummins *et al.* 1986). An assessment of how taphonomy has altered the fossil assemblage from the living fauna should aid us in understanding what information on life habitats and life habits can reasonably be expected in the fossil record.

Methods and materials

This study is based on data collected by a number of agencies as part of multidisciplinary studies devised to investigate the role of benthic organisms in arctic marine ecosystems. Zoobenthos were collected by grab sampling in upper Frobisher Bay (Wacasey *et al.* 1979, 1980) (Fig. 1A), Brevoort Harbour (Arctic Biological Station 1980) (Fig. 1B), southern Davis Strait and Ungava Bay (Stewart *et al.* 1985) (Fig. 2), Pangnirtung Fjord (Dale *et al.* 1989) (Fig. 2) and Cambridge, Itirbilung and McBeth Fjords (Syvitski *et al.* 1989) (Fig. 2). The study of the fjord macrobenthos was supplemented by numerous underwater photographs (Dale *et al.* 1989; Syvitski *et al.* 1989). Methods for the collection of the zoobenthos are presented in the publications cited above. The stations occupied during the various studies and data pertaining to the collections are listed in Appendix 1.

A species list for the study area (available from the author upon request) was compiled from the publications cited above and the list was verified by professional systematists. Organisms occurring at two or more stations within the study area were included in the list. The species list consists of representatives of 310 genera of marine invertebrates. Organisms on the list represent the modern equivalent of the benthic macrofauna which palaeontologists might expect to find in an outcrop if the organisms were preserved.

The conceptual framework established by Schopf (1978) to assess the preservation potential of individual taxa has been adopted in this study with minor modification. Preservation potential was judged in three ways: by morphology (presence of macroscopic hard parts); by the previous existence of a fossil record (genera only); and by the recorded presence in Quaternary marine sediments on Baffin Island.

Morphology

All organisms possessing durable shells or tubes composed of calcium carbonate are placed in Class 1: bivalves, gastropods, scaphopods, brachiopods, polychaetes inhabiting calcareous tubes (*Circeis, Chitinopoma*) and an infaunal

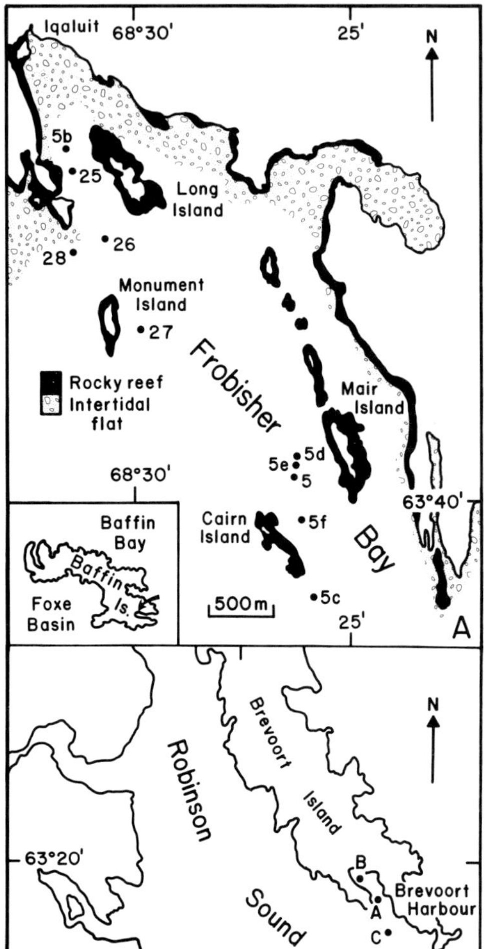

Fig. 1. Location of benthic sampling sites in (A) upper Frobisher Bay and at (B) Brevoort Island. The inset maps show the location of the sampling localities, indicated by arrows, with respect to Baffin Island.

sea urchin *Brisaster*. These organisms are expected to yield *many* easily identifiable fossils. *Brisaster* was placed in Class 1, while *Strongylocentrotus*, an epifaunal sea urchin, was placed in Class 2 based on the information presented in Kier (1977).

Organisms possessing largely uncalcified skeletons and those whose skeletons or other hard parts disaggregate upon death into plates,

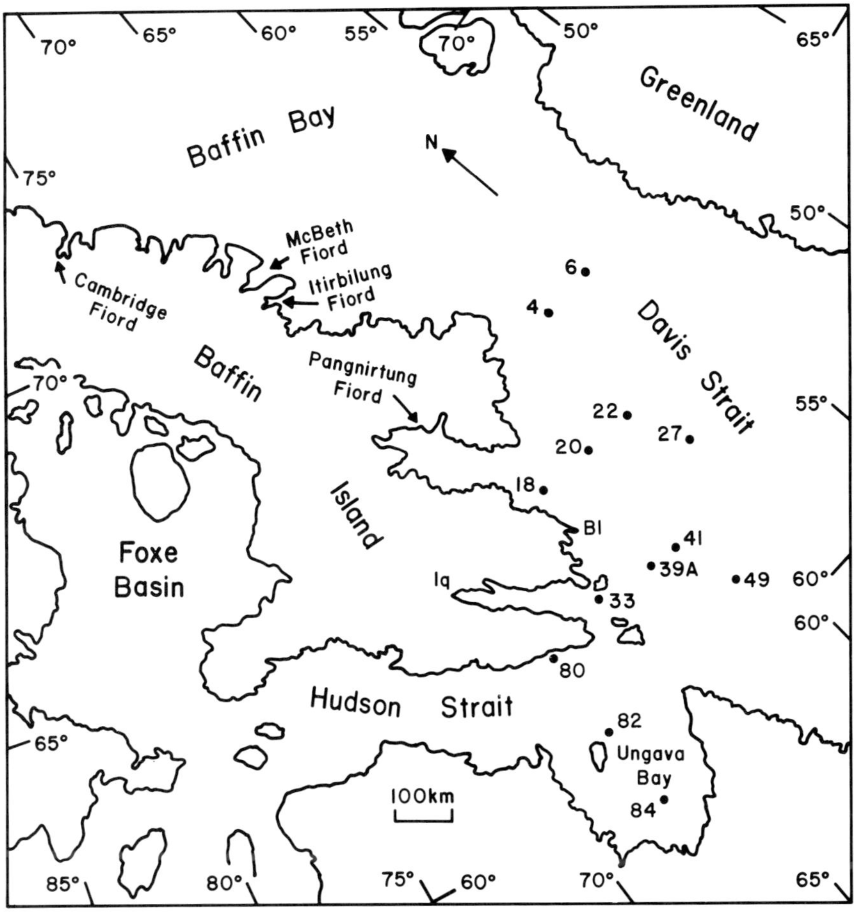

Fig. 2. Location of benthic sampling sites on the continental shelf in Davis Strait, Hudson Strait and Ungava Bay and the fjords referred to in the text; Iq = Iqaluit, BI = Brevoort Island.

spicules, jaws, etc. are placed in Class 2: crustaceans (amphipods, barnacles, cumaceans, isopods, tanaidaceans), chitons, echinoderms (asteroids, ophiurians, crinoids, holothurians and the echinoid *Strongylocentrotus*), sponges and polychaetes possessing jaws (e.g. nephtyids). These organisms are expected to yield *few* easily identified fossils.

Organisms lacking macroscopic hardparts are assigned to Class 3: most polychaetes, octocorals, pennatulaceans, sipunculans, priapulids, nemerteans, ascidians, bryozoans and tunicates. These organisms are expected to yield *no* easily identifiable fossils.

Previous fossil record

Each of the 310 genera was checked with the appropriate volume of the *Treatise on Invertebrate Paleontology*, with the exception of some Caenogastropoda. The fossil record of these organisms was verified by professional systematists.

Previous record in Quaternary marine sediments

A list of the subfossil organisms recovered from Quaternary marine sediments on Baffin Island was compiled from the works of Matthews (1967), Andrews (1972), Miller *et al.* (1977), Andrews *et al.* (1981), Mode (1985) and Aitken & Risk (1988).

Trophic groups

Information on the feeding habits of marine invertebrates is generally available only at the family level. The predominant feeding method employed by each family of organisms represented in this study was determined from information presented by Barnes (1980), Enequist (1949), Fauchald & Jumars (1979), Kohn (1983), and Morton (1983). The taxa within each family were designated as carnivores/scavengers, suspension feeders, deposit feeders, browsers or commensals using the classification scheme developed by Walker & Bambach (1974) based on this information.

Modern biota and habitats

The biology of the fjords and continental shelf of eastern Baffin Island has been the subject of several recent studies (Thomson 1982; Stewart *et al.* 1985; Thomson *et al.* 1986; Aitken *et al.* 1988; Dale *et al.* 1989; Syvitski *et al.* 1989). Four groups of organisms: bivalves, polychaetes, echinoderms and crustaceans contribute significantly to the biomass of continental shelf benthic communities and are the dominant organisms of the arctic marine benthos (Table 1). Data on the biomass of benthic organisms inhabiting eastern Baffin Island fjords is not available. Common organisms collected from Pangnirtung, Cambridge, McBeth and Itirbilung Fjords are listed in Table 2. I will focus my attention on the molluscs, particularly bivalves, throughout this paper as these organisms possess the greatest preservation potential and are therefore most likely to yield useful information relating to life habits and habitats.

Molluscan associations occurring within eastern Baffin Island fjords and on the Baffin Island continental shelf are presented in Table 3 and illustrated in Figs 3 and 4. The distribution of water masses within the fjords and over the continental shelf of eastern Baffin Island influence directly the distribution of these marine invertebrates (Macpherson 1971; Lubinsky 1980).

Several distinct water masses occur along the eastern Baffin Island coast. A seasonally variable, shallow water mass develops in July and August from the mixing of Baffin Current water with freshwater from rivers and melting sea ice. The temperature of this water mass ranges from 0°C to 8°C and the salinity ranges from 20‰ to 30‰ (Ellis 1960; Gilbert 1978; Trites 1985). This shallow water mass develops to depths of 5 to 60 m within the fjords (Gilbert 1978; Trites 1985) and 50 to 100 m depth across the continental shelf (Ellis 1960). The nearshore habitat is defined to be those areas influenced by this seasonally variable, shallow water mass on the continental shelf. Two molluscan associations are known to inhabit the nearshore habitat: a *Portlandia* association on muddy sand substrates in proximity to river mouths, and an *Astarte* association occurring on mud and sand substrates (Table 3). These associations are equivalent, with respect to their species composition and habitat, to the *Portlandia arctica* association (Thorson 1934) and the *Astarte borealis* zone of the Arctic *Macoma* association (Thorson 1933) described from east Greenland fjords.

The Baffin Island Current lies below the shallow surface water mass. It is formed by cold, low salinity water (−1.8°C to 1.8°C, 32.0‰ to 34.0‰; Dunbar 1951) originating in the Arctic Ocean that flows southward at depths of 50 to 300 m across the continental shelf of eastern Baffin Island. Trites (1985) reported Baffin Island Current water at depths as great as 325 m in Baffin Island fjords. The inner shelf habitat is defined to be those areas influenced by the Baffin Island Current on the continental shelf. A *Macoma–Astarte* association, occurring on mud, sand and gravel substrates, inhabits the inner shelf habitat (Table 3). The characteristic species of this association are similar to those of the Arctic *Macoma* association described from east Greenland fjords (Ockelmann 1958). The Arctic *Macoma* association, however, inhabits depths influenced by seasonally variable 'fjord' surface waters (5 to 50 m) in east Greenland fjords.

The Greenland Current flows below the Baffin Island Current. It is a water mass of Atlantic origin (1.0°C to 4.0°C, 34.5‰ to 35‰; Bailey 1957) that flows southward at depths of 200 to 900 m across the outer continental shelf and slope of eastern Baffin Island. Greenland Current waters enter Baffin Island fjords where sill depths permit, forming a warmer bottom layer at the greatest depths in the fjords (Trites 1985). The outer shelf is defined to be those areas influenced by the Greenland Current on the continental shelf. A *Bathyarca–Astarte* association, occurring on mud substrates, inhabits the outer shelf habitat (Table 3). This association is equivalent to the *Bathyarca–Astarte crenata* association which inhabits depths of 50 to 550 m in east Greenland fjords (Thorson 1933, 1934; Ockelmann 1958).

Molluscan associations within eastern Baffin Island fjords inhabit broader depth ranges than equivalent macrofaunas on the Baffin Island continental shelf. The names of the fjord macrofauna associations employed by Syvitski *et al.* (1989) are retained in this paper, although two of the associations: the Onuphid association and the Maldanid association, are characterized by tubiculous polychaetes rather than molluscs.

Table 1. *Dominant organisms based on biomass ($g\ m^{-2}$) for stations on the continental shelf and slope of Baffin Island*

Region	Station	(D/M/Y)	Depth (m)	Molluscs	Polychaetes	Echinoderms	Crustaceans
Upper Frobisher Bay	69–5c	23/8/69	31	*Hiatella arctica* *Mya truncata* *Serripes groenlandicus*	*Nichomache lumbricalis* *Nephthys paradoxa* *Pista flexuosa*	*Ophiocantha bidentata* *Ophiopus arcticus*	
	69–5d	23/8/69	43	*Hiatella arctica* *Clinocardium ciliatum* *Mya truncata*	*Nichomache lumbricalis* *Nephthys paradoxa* *Branchiomma infarcta*	*Ophiocantha bidentata* *Ophiopus arcticus*	*Balanus balanus* *Nymphon hirtipes* *Cumella* sp.
	69–5e	23/8/69	58	*Hiatella arctica* *Clinocardium ciliatum* *Musculus discors*	*Nichomache lumbricalis* *Nephthys paradoxa* *Phyllodoce groenlandica*	*Ophiocantha bidentata* *Ophiopus arcticus*	*Balanus balanus* *Socarnes* sp.
	69–5f	23/8/69	73	*Hiatella arctica* *Musculus niger* *Macoma calcarea*	*Nichomache lumbricalis* *Nephthys paradoxa* *Enipo gracilis*	*Ophiocantha bidentata* *Ophiopus arcticus*	*Balanus crenatus*
	70–5	6/8/70	72	*Hiatella arctica* *Clinocardium ciliatum* *Musculus discors*	*Nichomache lumbricalis* *Enipo gracilis* *Maldane sarsi*	*Ophiocantha bidentata* *Ophiopus arcticus*	*Balanus crenatus* *Orchomene serrata* *Unciola leucopis*
	73–5b	8/8/73	14	*Macoma calcarea* *Cylichna alba* *Axinopsida orbiculata*	*Praxiellella praetermissa* *Heteromastus* sp. *Nephthys ciliata*		*Paroediceros lynceus*
	73–25	8/8/73	30	*Serripes groenlandicus* *Astarte borealis* *Nucula belloti*	*Nephthys ciliata* *Praxiellella praetermissa* *Phyllodoce groenlandica*		
	73–26	8/8/73	40	*Hiatella arctica* *Colus tortuosus* *Margarites umbilicalis*	*Nichomache lumbricalis* *Nephthys paradoxa* *Ampharete finmarchica*	*Ophiocantha bidentata* *Ophiopus arcticus*	*Paroediceros lynceus* *Diastylis rathkei*
	73–27	8/8/73	43	*Hiatella arctica* *Colus tortuosus* *Musculus discors*	*Pista flexuosa* *Nichomache lumbricalis* *Nephthys paradoxa*	*Ophiocantha bidentata* *Ophiopus arcticus*	*Balanus balanus*
	73–28	13/8/73	30	*Hiatella arctica* *Astarte borealis* *Macoma moesta*	*Nephthys ciliata* *Clymenella* sp. *Phyllodoce groenlandica*	*Myriotrochus rinki*	*Ampelisca eschrictii*
Davis Strait	4	16/8/77	622	*Bathyarca pectunculoides* *Nucula delphinodonta* *Cylichna alba*	*Lumbrineris impatiens*	*Ophiura robusta*	*Haploops tubicola* *Haploops setosa* *Diastylis rathkei*
	6	19/8/77	658	*Astarte crenata* *Cylichna alba*	*Terebellides stroemi* *Glycera capitata* *Lumbrineris impatiens*	*Ophiura robusta*	*Haploops tubicola* *Diastylis rathkei*
	18	22/9/77	292	*Thyasira gouldi* *Nuculana pernula*	*Galathowenia oculata* *Prionospio cirrifera* *Chaetozone setosa*	*Ophiocten sericeum*	*Haploops tubicola*
	20	22/9/77	283	*Astarte crenata*	*Onuphis conchylega* *Chaetozone setosa* *Prionospio cirrifera*	*Ophiura sarsi*	*Byblis gaimardi*

Table 1. (*contd*)

Region	Station	(D/M/Y)	Depth (m)	Molluscs	Polychaetes	Echinoderms	Crustaceans
	22	21/9/77	920	*Astarte crenata* *Bathyarca pectunculoides*	*Terebellides stroemi* *Euchone* sp.	*Strongylocentrotus droebachiensis*	*Haploops tubicola*
	27	13/8/77	970	*Nucula delphinodonta*	*Glycera capitata* *Lumbrineris fragilis* *Aricidea suecica*	*Ophiomusium lymani*	*Harpinia mucronata* *Harpinia propinqua* *Diastylis rathkei*
	33	9/9/77	177	*Lepeta caeca* *Macoma moesta*	*Chaetozone setosa*	*Ophiura robusta* *Ophiopholis aculeata*	*Ischyrocerus latipes* *Unciola leucopsis* *Gammaropsis melanops*
	39a	9/8/77	507	*Bathyarca pectunculoides* *Nucula delphinodonta* *Dentalium* sp.	*Onuphis conchylega* *Glycera capitata*	*Ophiura robusta*	*Ischyocerus megacheir*
	41	1/8/77	532	*Astarte crenata* *Bathyarca pectunculoides*	*Glycera capitata* *Syllis cornuta*	*Amphiopholus squamata*	*Ampelisca declivitatus* *Diastylis rathkei*
	49	6/9/77	621	*Yoldiella* sp. *Thyasira gouldi* *Dentalium occidentale*	*Glycera capitata* *Ophelina cylindricaudatus* *Lumbrineris impatiens*	*Amphiopholus squamata* *Amphiura* sp.	*Ischyrocerus megacheir* *Ampelisca gibba* *Ampelisca ambylops*
Hudson Strait/ Ungava Bay	80	3/9/77	179	*Lepeta caeca*	*Onuphis conchylega* *Ophelina aulogaster*	*Ophiura robusta*	*Phoxocephalus holbolli* *Unciola leucopis*
	82	6/8/77	106	*Lophyrochiton albus* *Lepeta caeca* *Astarte striata*	*Spaerodorum gracilis* *Lumbrineris fragilis* *Terebellides stroemi*	*Ophiura robusta* *Ophiopholis aculeata*	*Ampelisca eschrictii* *Byblis gaimardi* *Syrrhoe crenulata*
	84	5/8/77	145	*Lepeta caeca* *Nuculana pernula* *Macoma calcarea*	*Aglaophamus malmgreni* *Euchone analis* *Chaetozone setosa*	*Stegophiura nodosa* *Ophiopholis aculetata*	*Eudorellopsis integra* *Ischyrocerus latipes* *Ampelisca eschrictii*
Brevoort Harbour	79-a	8,13/8/79	35–37	*Astarte montagui* *Macoma calcarea* *Astarte borealis*	*Pectinaria granulata* *Pectinaria hyperborea*	*Stegophiura nodosa* *Myriotrochus rinki*	*Diastylis rathkei*
	79-b	6,10/8/79	13		*Travisia* sp. *Microspio* sp. *Scoloplos armiger*		*Amphipods*
	79-c	12/8/79	77	*Macoma calcarea* *Nucula belloti*	*Thelepus cincinnatus* *Harmothoe oerstedi* *Pista flexuosa*	*Ophiopholis aculeata* *Ophiocantha bidentata* *Ophiocten sericeum*	
	79-d	5,11/8/79	146	*Astarte montagui* *Macoma calcarea* *Nuculana minuta*	*Onuphis conchylega* *Chaetozone setosa*	*Ophiura sarsi* *Ophiura robusta* *Ophiocten sericeum*	Amphipods
	79-e	11/8/79	245	*Macoma calcarea* *Astarte montagui* *Astarte borealis*	*Onuphis conchylega* *Chaetozone setosa* *Galathowenia oculata*	*Ophiura robusta* *Ophiocten sericeum* *Pteraster* sp.	

Data are taken from Wacasey *et al.* (1979, 1980), Arctic Biological Station (1980) and Stewart *et al.* (1985).

Table 2. *Common organisms recovered in grab samples or viewed in underwater photographs from Pangnirtung Fjord (Dale et al., 1989) and Cambridge, McBeth and Itirbilung Fjords (Syvitski et al., 1989; Aitken, unpublished)*

	Pangnirtung Fjord			Cambridge Fjord, McBeth Fjord, Itirbilung Fjord		
Benthic Association	Soupground	Softground	Shell/gravel Lag	*Portlandia*	Onuphid	Maldanid
Bivalvia	*Portlandia arctica* *Nucula belloti* *Nuculana pernula*		*Chlamys islandica* *Hiatella arctica*	*Portlandia arctica* *Musculus discors* *Hiatella arctica*	*Macoma calcarea* *Nucula belloti* *Yodiella lenticula* *Axinopsida orbiculata*	*Astarte crenata* *Bathyarca glacialis* *Yoldiella intermedia* *Axinopsida orbiculata*
Gastropoda		*Buccinum hydrophanum*	*Buccinum hydrophanum*	*Buccinum* sp.	*Buccinum* sp.	*Buccinum* sp.
Polychaeta	*Nereis pelagica* *Thelepus cincinnatus*	tubiculous polychaetes	*Nothria conchylega* *Circeis* cf. *spirillum*	Maldanidae Sabellidae	*Nothria* cf. *conchylega* Maldanidae	*Asychis biceps* Maldanidae Onuphidae Sabellidae
Echinodermata	*Amphiophiura* cf. *bullata convexa* *Stegophiura nodosa*	*Asterias* sp. *Henricia eschricti* *Heliometra glacialis* *Gorgonocephalus arcticus* *Strongylocentrotus droebachiensis* *Ophiura sarsi*	*Henricia eschricti* *Heliometra glacialis* *Gorgonocephalus arcticus* *Strongylocentrotus droebachiensis* *Ophiura sarsi*	Ophiuroides	*Gorgonocephalus arctica* *Ophiacantha bidentata* *Ophiura robusta*	*Gorgonodephalus arcticus* *Ophiacantha bidentata* *Ophiopleura borealis* Holothurians
Anthozoans		*Actinauge* sp. *Alcyonium* cf. *glomeratum* *Cerianthus* sp. *Gersemia* cf. *rubiformis*	*Actinauge* sp. *Alcyonium* cf. glomeratum *Cerianthus* sp. *Gersemia* cf. *rubiformis*		Actinaria (anemones) Agglutinated foraminifera	Actinaria (anemones) Agglutinated foraminifera

Table 3. *Molluscan associations present within Baffin Island fjords and on the Baffin Island continental shelf*

1.	*Fjords*		*Nearshore associations*	
	McBeth Fjord *Portlandia* association *Portlandia arctica* *Musculus discors* *Mya truncata* *Hiatella arctica* *Macoma calcarea*	Frobisher Bay *Astarte borealis* *Hiatella arctica* *Mya truncata* *Macoma calcarea* *Musculus discors*	Scott Inlet *Portlandia* association *Portlandia arctica* *Thyasira* spp.	Eclipse Sound, Lancaster Sound, NW Baffin Bay *Astarte* association *Astarte borealis* *Astarte montagui* *Macoma moesta* *Mya truncata* *Musculus niger* *Nuculana minuta*
Depth (m):	5–40	30–75	5–20	15–50
Substrate:	sandy mud	sandy mud	muddy sand	silty sand
Source:	Syvitski *et al*. (1989) Aitken (unpublished)	Wacasey *et al*. (1979, 1980)	Thomson *et al*. (1986)	Thomson *et al*. (1986)
2.	*Fjords*	*Inner Shelf Associations*		
	Cambridge Fjord, Itirbilung Fjord Onuphid association *Macoma calcarea* *Nucula belloti* *Yoldiella lenticula* *Yoldiella intermedia* *Axinopsida orbiculata*	Hudson Strait, Davis Strait *Astarte striata* *Macoma calcarea* *Thyasira gouldi* *Nuculana pernula* *Yoldiella lucida* *Lepeta caeca*	Eclipse Sound, Lancaster Sound, NW Baffin Bay *Macoma-Astarte* association *Macoma calcarea* *Astarte montagui* *Hiatella arctica* *Margarites* sp. *Lepeta caeca*	
Depth (m):	100–560	100–300	50–250	
Substrate:	sandy mud, gravel	sandy mud, gravelly sand	muddy sand	
Source:	Syvitski *et al*. (1989) Aitken (unpublished)	Stewart *et al*. (1985)	Thomson (1982)	
3.	*Fjords*	*Outer Shelf Associations*		
	Cambridge Fjord, McBeth Fjord, Itirbilung Fjord Maldanid association *Astarte crenata* *Bathyarca glacialis* *Axinopsida orbiculata* *Yoldiella intermedia* *Yoldiella lenticula*	Davis Strait *Astarte crenata* *Bathyarca pectunculoides* *Nucula delphinodonta*	Lancaster Sound, NW Baffin Bay *Bathyarca* association *Bathyarca raridentata*	
Depth (m):	255–750	300–700	250–500	
Substrate:	sandy mud	mud	mud	
Source:	Syvitski *et al*. (1989) Aitken (unpublished)	Stewart *et al*. (1985)	Thomson (1982)	

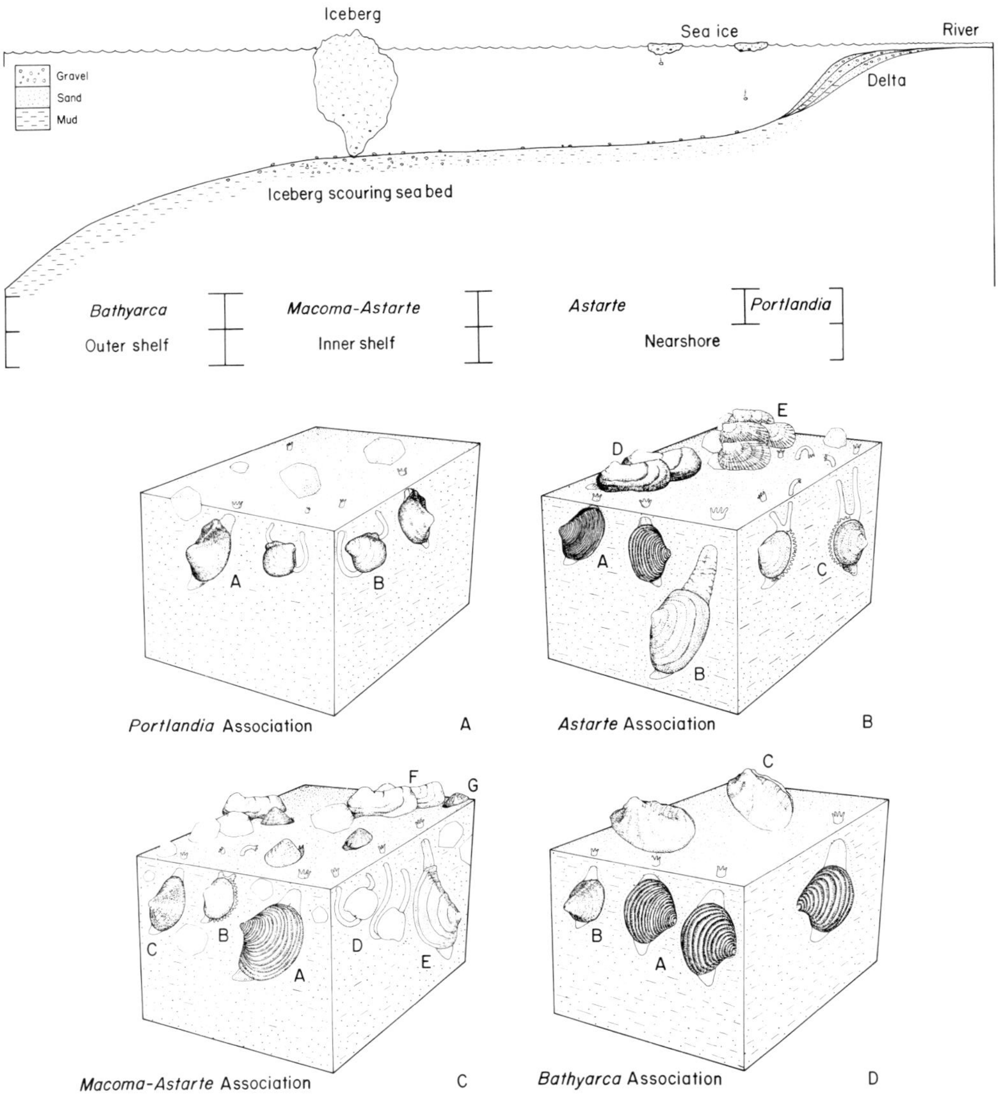

Fig. 3. (Top) Schematic representation of the Baffin Island continental shelf and the distribution of molluscan associations on the continental shelf. (Bottom) Block diagrams illustrating the characteristic species of Baffin Island shelf molluscan faunas. A: *Portlandia* Association; *Portlandia arctica* (A), *Thyasira gouldi* (B). B: *Astarte* Association; *Astarte borealis* (A), *Mya truncata* (B), *Macoma calcarea* (C), *Hiatella arctica* (D), *Musculus discors* (E). C: *Macoma-Astarte* Association; *Astarte striata* (A), *Macoma calcarea* (B), *Yoldiella lucida* (C), *Thyasira gouldi* (D), *Nuculana pernula* (E), *Hiatella arctica* (F), *Lepeta caeca* (G). D: *Bathyarca* Association; *Astarte crenata* (A), *Nucula delphinodonta* (B), *Bathyarca pectunculoides* (C).

The *Portlandia* association, which occurs on sandy mud substrates in proximity to river mouths, is composed of species from both the *Portlandia* and *Astarte* associations of the nearshore habitat on the shelf (Table 3). The Onuphid association is similar, with respect to its species composition and habitat, to an association characterized by *Macoma calcarea* and *Astarte striata* on the inner shelf of Hudson Strait and Davis Strait (Table 3). The Maldanid association is equivalent to associations characterized by *Astarte crenata* and *Bathyarca* spp. on the outer shelf.

The fjords included in this study consist of several basins separated by sills composed of bedrock or glacial till (Gilbert 1985). The sills

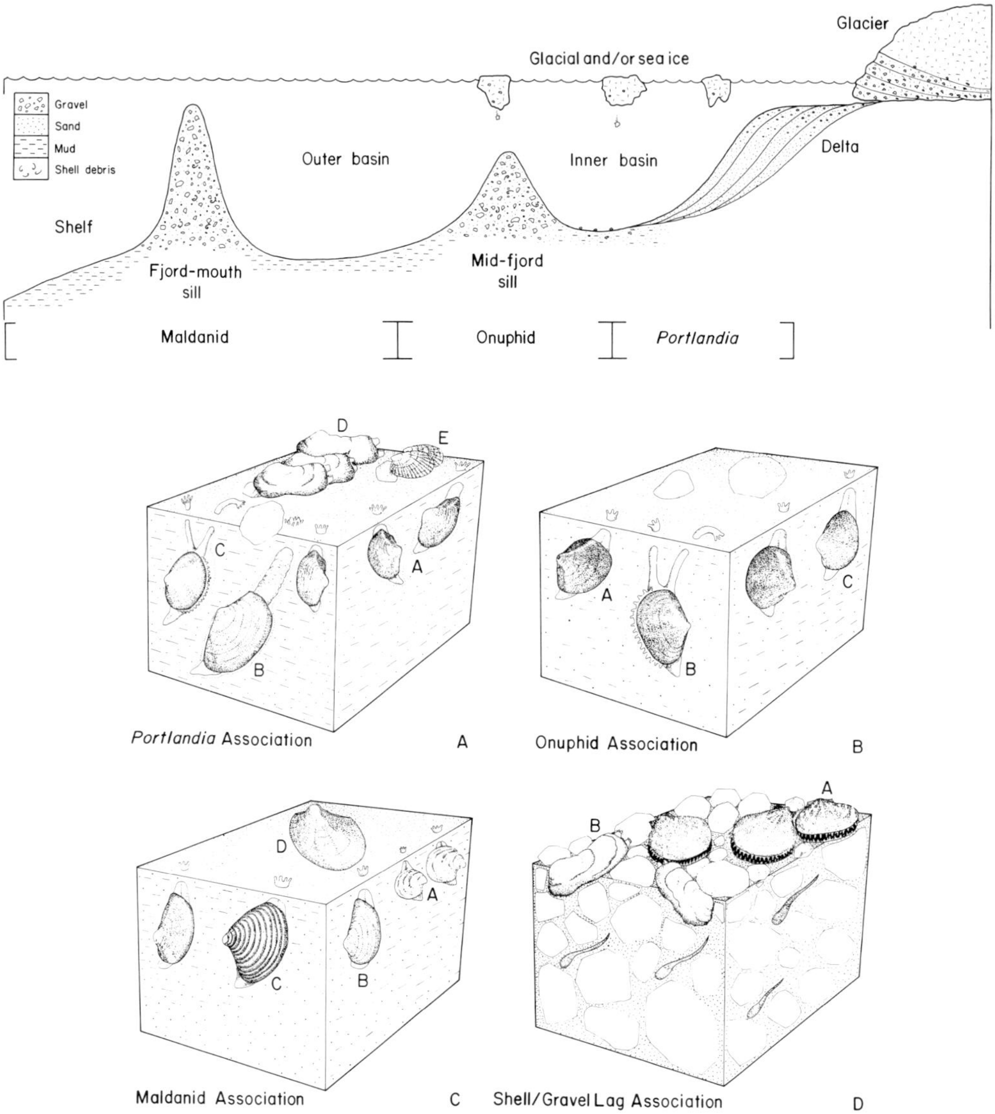

Fig. 4. (Top) Schematic representation of a Baffin Island fjord and the distribution of molluscan associations within the fjord. (Bottom) block diagrams illustrating the characteristic species of Baffin Island fjord molluscan faunas. A: *Portlandia* Association; *Portlandia arctica* (A), *Mya truncata* (B), *Macoma calcarea* (C), *Hiatella arctica* (D), *Musculus discors* (E). B: Onuphid Association; *Nucula belloti* (A), *Macoma calcarea* (B), *Yoldiella lenticula* (C). C: Maldanid Association; *Axinopsida orbiculata* (A), *Yoldiella intermedia* (B), *Astarte crenata* (C), *Bathyarca glacialis* (D). D: Shell/gravel lag Association; *Chlamys islandica* and/or *Delectopecten greenlandicus* (A), *Hiatella arctica* (B).

are subjected to strong currents which produce winnowed surface lags of boulders, cobbles, gravel, sand and shell debris (Syvitski 1989). Bivalves inhabiting these coarse substrates are *Chlamys islandica*, *Delectopecten greenlandicus* and *Hiatella arctica*, notably at fiord mouth sills (Dale *et al.* 1989).

Results

The three methods of estimating preservation potential yield similar results with regards to life habitats. The macrofauna inhabiting fjords and nearshore waters exhibit the greatest preservation potential, followed in succession by macrofaunas inhabiting the inner and outer continental shelf (Table 4). The morphological method indicates that the proportion of genera expected to yield *many* identifiable fossils decreases from 32% in the fjords, to 21% in the nearshore, to 18% on the inner shelf to 14% on the outer shelf. The proportion of genera actually known from the fossil record decreases from 43% in the fjords, to 25% in the nearshore, to 22% on the inner shelf to 21% on the outer shelf, reflecting the trend in the morphological data. The method based on morphology does not account for the preservational loss due to biostratinomic processes and diagenesis. It is noteworthy, therefore, that the morphological method yields results similar to those based on the actual fossil record. In each habitat, preservation potential assessed by the fossil record exceeds that anticipated from the genera which yield *many* identifiable fossils. The difference can be attributed to the genera expected to yield *few* identifiable fossils (Type 2), notably asteroid and ophiuroid echinoderms.

The third method, based on the presence of organisms in Quaternary marine sediments, accounts for the influence of biostratinomic and diagenetic processes on the formation of fossil assemblages. The proportion of genera which presently inhabit the fjords and are recorded from Quaternary marine sediments is 24%, in comparison to 19% which presently inhabit the nearshore, 13% which presently inhabit the inner shelf and 8% which presently inhabit the outer shelf (Table 4). The Quaternary fossil assemblages consist entirely of genera expected to yield *many* identifiable fossils, with the exception of the sea urchin *Strongylocentrotus* and the barnacle *Balanus* (Table 5).

Quaternary fossil assemblages represent the end products of all taphonomic processes which operated on contemporary fjord and continental shelf macrofaunas: hence, they provide the most reliable measure of preservation potential for modern arctic fjord and continental shelf macrofaunas. Considering the benthos as a whole, the data presented in Table 4 indicate that 68% to 86% of the genera have not been preserved, as indicated by the percentage of Type 1 organisms, and are unlikely to be preserved.

The trophic group composition of the potentially preservable macrofauna differs noticeably from that in the modern macrofauna. Suspension feeders and browsers exhibit the greatest preservation potential. Data presented in Table 6 indicate that 32 of 74 (43%) of the suspension feeding genera and 9 of 9 (100%) of the browsing genera are represented in the fossil record, in contrast to 21% (22 of 103) of the carnivorous genera and 9% (11 of 123) of the deposit feeding genera. The preferential preservation of suspension feeding and browsing genera is also reflected in the Quaternary fossil assemblages. Data presented in Table 6 indicate that 16 of 74 (22%) of the suspension feeding genera and 3 of 9 (33%) of the browsing genera are represented in the Quaternary fossil record, in contrast to 8% (8 of 103) of the carnivorous genera and 7% (9 of 123) of the deposit feeding genera.

Discussion

The present study indicates that Quaternary fossil assemblages exhibit a pronounced taphonomic bias with regards to information on the

Table 4. *Preservation potential of macrofauna inhabiting eastern Baffin Island fjords and the Baffin Island continental shelf*

	Fjords	Nearshore (< 100 m)	Inner Shelf (100–300 m)	Outer Shelf (300–900 m)
		Number of genera		
Morphology				
Type 1 (many fossils)	36 (32.1)	45 (20.7)	35 (17.8)	17 (13.5)
Type 2 (few fossils)	25 (22.3)	74 (34.1)	75 (38.1)	50 (39.7)
Type 3 (no fossils)	51 (45.6)	98 (45.2)	87 (44.1)	59 (46.8)
Total number of genera	112	217	197	126
Number of genera from fossil record	48 (42.9)	54 (24.9)	44 (22.3)	26 (20.6)
Number of genera from Quaternary sediments	24 (21.4)	30 (13.8)	22 (11.2)	11 (8.7)

Numbers in brackets indicate the relative abundance (%) of genera in each category. The relative abundance of genera from the fossil record (*Treatise* and Quaternary sediments) is expressed as a percentage of the total number of genera in each category.

Table 5. *List of macrofauna recovered from Quaternary marine sediments on the eastern coast of Baffin Island and their present distribution in eastern Baffin Island fjords and on the Baffin Island continental shelf*

	Cambridge Fjord	McBeth Fjord	Itirbilung Fjord	Pangnirtung Fjord	Nearshore (< 100 m)	Inner Shelf (100–300 m)	Outer Shelf (300–900 m)	Trophic Group	Organism Group	Morphological Type
Bathyarca glacialis (Gray)	X			X				Deposit feeder	Bivalve	1
Axinopsida orbiculata (G.O. Sars)	X	X	X	X	X			Deposit feeder	Bivalve	1
Macoma balthica (L.)				X	X			Deposit feeder	Bivalve	1
Mya truncata (L.)		X		X	X			Suspension feeder	Bivalve	1
Portlandia arctica (Gray)		X	X	X	X			Deposit feeder	Bivalve	1
Serripes groenlandicus (Brugière)				X	X			Suspension feeder	Bivalve	1
Acmaea testudinalis (Müller)				X	X			Browser	Gastropod	1
Hemithiris psittacae (Gmelin)				X	X			Suspension feeder	Brachiopod	1
Astarte montagui (Dillwyn)	X			X	X	X		Suspension feeder	Bivalve	1
Astarte striata (Leach)			X	X	X	X		Suspension feeder	Bivalve	1
Chlamys islandica (Müller)	X			X		X		Suspension feeder	Bivalve	1
Clinocardium ciliatum (Fabricius)				X	X	X		Suspension feeder	Bivalve	1
Hiatella arctica (L.)		X		X	X	X		Suspension feeder	Bivalve	1
Macoma calcarea (Gmelin)	X	X	X	X	X	X		Deposit feeder	Bivalve	1
Musculus discors (L.)	X	X		X	X	X		Suspension feeder	Bivalve	1
Yoldia hyperborea (Löven) Torell				X	X	X		Deposit feeder	Bivalve	1

Boreotrophon truncatus (Ström)				X	X	X		Carnivore	Gastropod	1
Margarites sp.				X	X	X		Carnivore	Gastropod	1
Trichotrophis borealis Broderip & Sowerby				X	X	X		Suspension feeder	Gastropod	1
Circeis sp.				X	X	X		Suspension feeder	Polychaete	1
Astarte crenata (Gray)	X		X	X		X	X	Suspension feeder	Bivalve	1
Delectopecten greenlandicus (Sowerby)	X					X	X	Suspension feeder	Bivalve	1
Nuculana minuta (Fabricius)				X	X	X	X	Deposit feeder	Bivalve	1
Nuculana pernula (Müller)	X		X	X	X	X	X	Deposit feeder	Bivalve	*1*
Thyasira gouldi (Phillipi)	X	X	X	X	X	X	X	Deposit feeder	Bivalve	1
Yoldiella fraterna Verrill & Bush				X	X	X	X	Deposit feeder	Bivalve	1
Lepeta caeca (Müller)	X			X	X	X	X	Browser	Gastropod	1
Strongylocentrotus droebachiensis (O.F. Müller)				X	X	X	X	Scavenger	Echinoderm	2
Astarte elliptica (Brown)					X			Suspension feeder	Bivalve	1
Cyrtodaria sp.					X			Suspension feeder	Bivalve	1
Mya pseudoarenaria Schlesch					X			Suspension feeder	Bivalve	1
Mytilus edulis (L.)					X			Suspension feeder	Bivalve	1
Amauropsis sp.					X			Carnivore	Gastropod	1
Cylichna gouldi (Couthouy)					X			Carnivore	Gastropod	1
Natica clausa Broderip & Sowerby					X			Carnivore	Gastropod	1
Oenopota decussata (Couthouy)					X			Carnivore	Gastropod	1
Balanus balanus (L.)					X			Suspension feeder	Crustacean	2

Table 5. (*contd*)

	Cambridge Fjord	McBeth Fjord	Itirbilung Fjord	Pangnirtung Fjord	Nearshore (< 100 m)	Inner Shelf (100–300 m)	Outer Shelf (300–900 m)	Trophic Group	Organism Group	Morphological Type
Balanus crenatus Brugière					X			Suspension feeder	Crustacean	2
Balanus sp.					X			Suspension feeder	Crustacean	2
Musculus niger (Gray)					X	X		Suspension feeder	Bivalve	1
Pandora glacialis Leach					X	X		Deposit feeder	Bivalve	1
Boreotrophon clathratus (L.)					X	X		Carnivore	Gastropod	1
Colus islandicus (Gmelin)					X	X		Carnivore	Gastropod	1
Colus spitzbergensis (Reeve)					X	X		Carnivore	Gastropod	1
Nucula delphinodonta Mighels & Adams						X	X	Deposit feeder	Bivalve	1
Bathyarca pectunculoides Scacchi						X	X	Deposit feeder	Bivalve	1
Dentalium entale (L.)						X	X	Carnivore	Scaphopod	1
Cyclocardia borealis (Conrad)							X	Suspension feeder	Bivalve	1

Data are taken from the following sources: subfossil; Matthews (1967), Andrews (1972), Miller *et al.* (1977), Andrews *et al.* (1981), Mode (1985), Aitken and Risk (1988): modern; Wacasey *et al.* (1979, 1980), Arctic Biological Station (1980), Stewart *et al.* (1985), Dale *et al.* (1989), Syvitski *et al.* (1989) and Aitken (unpublished).

Table 6. *Trophic categories for the benthic macrofauna inhabiting eastern Baffin Island fjords and the Baffin Island continental shelf*

	Trophic Category				
	Carnivores/ Scavengers	Suspension Feeders	Deposit Feeders	Browsers	Commensals
	Number of Genera				
Morphology					
Type 1 (many fossils)	17	27	11	7	0
Type 2 (few fossils)	56	15	56	2	1
Type 3 (no fossils)	30	32	56	0	0
Total number of genera	103 (33.2)	74 (23.9)	123 (39.7)	9 (2.9)	1 (0.3)
Number of genera from fossil record	22 (21.4)	32 (43.2)	11 (8.9)	9 (100)	0 (0)
Number of genera from Quaternary sediments	8 (7.8)	16 (21.6)	9 (7.3)	3 (33.3)	0 (0)

The trophic terminology follows Walker and Bambach (1974). Numbers in brackets indicate the relative abundance (%) of genera in each category. The relative abundance of genera from the fossil record (*Treatise* and Quaternary sediments) is expressed as a percentage of the total number of genera in each category.

life habits and habitats of the arctic marine benthos from which they are derived. This discussion will examine the nature of the information loss and the implications for reconstructing ancient marine biocoenoses.

Life habits

The Quaternary fossil record preferentially preserves suspension feeding and epifaunal browsing organisms. Infaunal suspension feeding bivalves frequently contribute more than 50% of the macroscopic hardparts in Quaternary fossil assemblages (Matthews 1967; Simonarson 1981; Petersen 1982; Aitken & Risk 1988). The preferential preservation of suspension feeding organisms in fossil assemblages is not restricted to arctic marine environments. Fossil assemblages derived from temperate and subtropical marine faunas also exhibit this trend (Cadée 1968, 1984; Stanton & Dodd 1976; Bosence 1979; Staff *et al.* 1986). The trophic group composition of the Quaternary fossil assemblage differs greatly from that of the living arctic marine benthos as a result of the loss of soft-bodied carnivorous and deposit feeding organisms. The lack of preservation of these organisms suggests that little understanding of community trophic dynamics based on the study of trophic group composition is likely to be derived from Quaternary macrofossil assemblages.

The presence of organisms whose bodies have low preservation potential may be deduced by searching for traces produced by them. Biogenic structures recorded from Quaternary marine environments are presented in Table 7. The biogenic structures reveal the presence of predatory gastropods, soft-bodied, suspension feeding and deposit feeding organisms, and epifaunal browsing organisms. The recognition of biogenic structures is of considerable importance to palaeoecologists as they facilitate the reconstruction of the trophic structure of arctic Quaternary marine macrofaunas (see Thomsen & Vorren 1986; Aitken & Risk 1988).

Life habitats

Bivalves, ophiuroid echinoderms and polychaetes dominate the benthic biomass over most of the arctic continental shelves (Curtis 1975). In terms of biomass, bivalves dominate the benthos inhabiting the inner shelf, ophiuroid echinoderms dominate the benthos inhabiting the outer shelf and polychaetes dominate the benthos inhabiting the continental slope (Curtis 1975; Thomson 1982). Data on the biomass of benthic organisms inhabiting eastern Baffin Island fjords is not available. Abundance data indicate, however, that the shallow water *Portlandia* association is dominated by bivalves and the deeper water Onuphid and Maldanid associations are dominated by polychaetes (Syvitski *et al.* 1989). The preservation potential of these organisms decreases in order from bivalves (Type 1, *many* fossils) to ophiuroid echinoderms (Type 2, *few* fossils) to polychaetes (Type 3, *no* fossils). The preservation potential of macrofaunas, as indicated by the relative abundance of Type 1 organisms, decreases from the fjords across the continental shelf (Table 4),

Table 7. *Biogenic structures (trace fossils) recorded from Quaternary marine sediments*

Biogenic structure	Organism group	Function	Source
Skolithos	soft-bodied, infaunal suspension feeder	dwelling burrow	Petersen (1985), Aitken & Gilbert (1989)
Monocraterion	soft-bodied, infaunal suspension feeder	dwelling burrow	Aitken & Gilbert (1989)
?*Chondrites*	soft-bodied, infaunal deposit feeder	feeding trace	Petersen (1985)
Lockeia	infaunal bivalves	dwelling burrow	Aitken & Gilbert (1989)
pyritized burrow/tube fillings	tubiculous polychaetes	dwelling burrows	Thomsen & Vorren (1984), Neilsen & Holthe (1985)
Oichnus	carnivorous gastropods	predation	Petersen (1982), Thomsen & Vorren (1986), Aitken & Risk (1988)
Caulostrepsis	soft-bodied, epifaunal suspension feeder	dwelling burrow in mollusc shells	Aitken & Risk (1988)
Gnathicnus	gastropods, chitons, sea urchins	grazing traces on hard substrates	Thomsen & Vorren (1986)

reflecting the composition of the benthos within each habitat.

The macrofauna inhabiting Baffin Island fjords apparently exhibits the greatest preservation potential as assessed by all three methods employed in this study. The data presented for the fjords in Table 4 are undoubtedly inflated and reflect the fact that the faunal lists available for the fjords consist largely of molluscs and echinoderms, the most thoroughly studied marine organisms in these environments (Grainger 1955, 1966; D'yakonov 1954; Macpherson 1971; Lubinsky 1980). These animal groups are expected to yield *few* to *many* easily identifiable fossils and have extended geologic histories. Our present knowledge of several important animal groups, notably polychaetes and crustaceans, inhabiting Baffin Island fjords is poor.

The data presented in this study indicate that 14% to 32% of the genera, as indicated by the percentage of Type 1 organisms, which constitute the arctic marine benthos are preserved in Quaternary fossil assemblages (Table 4). These fossil assemblages retain useful information on the life habitats of the arctic benthos despite the obvious bias in preservation. The data presented in Table 5 indicate that 41 of 48 species recorded from Quaternary fossil assemblages inhabit the nearshore, 28 of 48 species inhabit the fjords and/or inner continental shelf and 12 of 48 species inhabit the outer shelf. Glacimarine environments are defined to be marine environments where sediment is deposited after being released from grounded or floating, glacial or sea ice (Powell 1984). Among the four habitats examined in this study (fjords, nearshore, inner and outer shelf) only the fjords and nearshore waters are considered to be glacimarine environments as defined above. Bivalves and gastropods inhabiting the fjords and nearshore waters exhibit the greatest preservation potential of all the arctic marine benthos.

The excellent preservation potential of arctic marine molluscs offers a palaeoecologists the potential to reconstruct ancient marine biocoenoses in glacimarine environments. Syvitski *et al.* (1989) have developed a model that describes the macrofaunal succession during glacier retreat in Baffin Island fjords based on their observations of fjord-head macrofaunas. The model is presented in Table 8. A macrofauna association consisting of nuculanid bivalves, notably *Portlandia arctica*, is replaced by a macrofauna consisting of *Macoma calcarea, Mya truncata* and *Hiatella arctica* which, in turn, is replaced by a macrofauna association consisting of *Macoma calcarea, Astarte borealis, Astarte striata* and *Musculus discors*. The changes in the composition of fjord-head macrofaunas are related to the diminution of the rate of sedimentation and increasing salinity of the bottom water.

Equivalent macrofaunas are recorded from marine sediments of the postglacial Champlain Sea (Table 8). Deep waters of the Champlain Sea (> 50 m, ≤ 0°C, 30‰ to 34‰) were inhabited by *Portlandia arctica, Hiatella arctica, Nuculana* sp., *Nucula* sp. and *Balanus hameri*. Nearshore waters of the Champlain Sea (5 to 50 m, 0°C to 3°C, 25 to 30‰) were inhabited by *Macoma calcarea, Mya truncata, Hiatella arctica, Mytilus edulis, Astarte* sp. and *Balanus crenatus*. Surface waters of the Champlain Sea (0 to 5 m, ≥ 5°C, 15‰ to 25‰) were inhabited by *Macoma balthica, Mya arenaria, Mytilus edulis*, and *Balanus balanus*.

Table 8. *Comparison of Champlain Sea and Baffin Island fjord-head macrofauna associations*

Western Champlain Sea		Eastern Champlain Sea		Baffin Island Fjords	
Epifauna on coarse substrates	Infauna in fine substrates	Epifauna on coarse substrates	Infauna in fine substrates	Epifauna on coarse substrates	Infauna in fine substrates
	Estuarine (0–5 m, ≥ 5°C, 15‰–25‰)			Ice Proximal (high sedimentation rates, 20‰–30‰)	
Mytilus edulis	*Macoma balthica*	*Mytilus edulis*	*Macoma balthica*		*Portlandia arctica*
	Mya arenaria	*Balanus balanus*	*Mya arenaria*		*Nuculana pernula*
	Nearshore (5–50 m, 0°C–3°C, 25‰–30‰				*Nucula belloti*
Hiatella arctica	*Macoma calcarea*	*Hiatella arctica*	*Macoma calcarea*	Ice Distal (moderate sedimentation rates, ≥ 30‰)	
Mytilus edulis	*Mya truncata*	*Mytilus edulis*	*Astarte* spp.	*Hiatella arctica*	*Macoma calcarea*
Balanus crenatus		*Balanus crenatus*	*Chlamys islandica*		*Mya truncata*
		Mya truncata	*Clinocardium ciliatum*	Ice Distal (low sedimentation rates, ≥ 30‰)	
			Serripes groenlandicus	*Musculus discors*	*Macoma calcarea*
	Offshore (50–150 m, ≤ 0°C, 30‰–34‰)				*Astarte borealis*
Hiatella arctica	*Portlandia arctica*	*Hiatella arctica*	*Portlandia arctica*		*Astarte striata*
Balanus hameri	Nuculanid	*Mya truncata*	*Nuculana* spp.		
Lepeta caeca	bivalves	*Balanus hameri*	*Nucula* spp.		

Ecological data for the Champlain Sea macrofauna are taken from Hillaire-Marcel (1980, 1981), Rodrigues & Richard (1983) and Rodrigues (1988).

Appendix 1. *Station data for the stations examined in this study*

Region	Station	Date (D/M/Y)	North Latitude	West Longitude	Depth (m)	Substrate	Bottom Temperature (°C)	Bottom Salinity (‰)	Benthic Association
Upper Frobisher Bay	69–5c	23/8/69	63° 39.0′	68° 25.9′	31	Sandy mud	0.22	32.38	Nearshore
	69–5d	23/8/69	63° 40.4′	68° 26.2′	43	Sandy clay	0.22	32.38	Nearshore
	69–5e	23/8/69	63° 40.3′	68° 26.2′	58	Sandy mud	0.11	32.42	Nearshore
	69–5f	23/8/69	63° 39.8′	68° 26.1′	73	Sandy mud	0.11	32.42	Nearshore
	70–5	6/8/70	63° 40.2′	68° 26.3′	72	Sandy mud	−0.64	32.38	Nearshore
	73–5b	8/8/73	63° 43.5′	68° 31.7′	14	Muddy sand	N/A	32.41	Nearshore
	73–25	8/8/73	63° 43.3′	68° 31.4′	30	Sandy mud	N/A	32.40	Nearshore
	73–26	8/8/73	63° 42.6′	68° 30.7′	40	Sandy mud	N/A	N/A	Nearshore
	73–27	8/8/73	63° 41.7′	68° 29.9′	43	Sandy mud	N/A	N/A	Nearshore
	73–28	13/8/73	63° 42.5′	68° 31.4′	30	Muddy sand	N/A	N/A	Nearshore
Davis Strait	4	16/8/77	66° 01′	59° 04′	622	N/A	1.4	N/A	Outer Shelf
	6	19/8/77	65° 56′	56° 59′	658	Sandy mud	1.4	N/A	Outer Shelf
	18	22/9/77	64° 00′	64° 00′	292	N/A	0.2	N/A	Inner Shelf
	20	22/9/77	64° 00′	62° 00′	283	N/A	2.2	N/A	Inner Shelf
	22	21/9/77	64° 00′	60° 00′	920	N/A	2.9	N/A	Outer Shelf
	27	13/8/77	62° 59′	58° 59′	970	Sandy silt	3.0	N/A	Outer Shelf
	33	9/9/77	62° 14′	65° 01′	177	N/A	0.4	N/A	Inner Shelf
	39a	9/8/77	62° 00′	62° 38′	507	Silty sand	4.3	N/A	Outer Shelf
	41	1/8/77	62° 02′	61° 09′	532	Sandy gravel	4.3	N/A	Outer Shelf
	49	6/9/77	61° 00′	61° 50′	621	Sandy silt	4.0	N/A	Outer Shelf
Hudson Strait/ Ungava Bay	80	3/9/77	61° 55′	67° 32′	179	Gravely sand	0.7	N/A	Inner Shelf
	82	6/8/77	60° 34′	67° 35′	106	Sandy gravel	0.0	N/A	Inner Shelf
	84	5/8/77	59° 11′	67° 29′	145	Sandy silt	−0.7	N/A	Inner Shelf
Brevoort Harbour	79-a	8,13/8/79	63° 18.5′	64° 08.2′	35–37	Silty sand	−0.5	33.05	Nearshore
	79-b	6,10/8/79	63° 19.4′	64° 10.1′	13	Sand	0.15	33.00	Nearshore
	79-c	12/8/79	63° 17.0′	64° 07.3′	77	Silty sand	−0.8	33.16	Nearshore
	79-d	5,11/8/79	63° 15.0′	64° 07.4′	146	Muddy sand	−1.2	33.38	Inner Shelf
	79-e	11/8/79	63° 13.0′	64° 08.3′	245	Silty sand	−1.2	33.38	Inner Shelf
Cambridge Fiord	CA0.2	22/9/83	71° 11.5′	75° 02.5′	125	Silty sand	−1.30	33.05	Onuphid
	CA0.3	22/9/83	71° 12.6′	75° 02.0′	200	Sandy mud	N/A	N/A	Onuphid
	CA1.0	22/9/83	71° 18.0′	75° 02.0′	181	Sandy mud	−1.27	33.24	Onuphid
	CA1.1	22/9/83	71° 12.7′	74° 59.0′	183	N/A	−1.21	33.26	Onuphid
	CA1.2	22/9/83	71° 12.6′	75° 01.0′	190	Sandy mud	−1.21	33.36	Onuphid
	CA1.3	22/9/83	71° 13.4′	74° 59.0′	240	N/A	N/A	N/A	Onuphid
	CA1.4	22/9/83	71° 13.4′	74° 59.0′	218	Sandy mud	N/A	N/A	Onuphid
	CA1.5	22/9/83	71° 14.0′	74° 57.0′	262	Sandy mud	N/A	N/A	Onuphid

	CA1.7	22/9/83	71° 15.0′	74° 54.0′	310	Gravelly sandy mud	N/A	N/A	Onuphid
	CA2.2	22/9/83	71° 19.4′	74° 46.2′	292	N/A	−1.26	33.44	Onuphid
	CA3.0	22/9/83	71° 23.5′	74° 38.0′	365	Gravelly sandy mud	−0.44	33.93	Maldanid
	CA4.1	21/9/83	71° 25.5′	74° 45.7′	520	Sandy muddy gravel	0.47	34.23	Onuphid
	CA4.2	21/9/83	71° 25.5′	74° 48.4′	513	Muddy sand	0.41	34.22	Onuphid
	CA4.3	21/9/83	71° 32.0′	74° 50.5′	560	Sandy mud	N/A	N/A	Onuphid
	CA6.0	23/9/83	71° 35.5′	74° 38.4′	665	N/A	0.53	34.25	Maldanid
	CA6.1	20/9/83	71° 43.2′	74° 36.5′	750	Gravelly mud	0.60	34.28	Maldanid
	CASILL3.1/8.1	22/9/83	71° 41.5′	74° 15.0′	397	N/A	N/A	N/A	Maldanid
	CASILL3.2/8.2	22/9/83	71° 41.8′	74° 24.0′	327	N/A	N/A	N/A	Maldanid
	CASILL3.3/8.3	22/9/83	71° 42.0′	74° 24.0′	322	N/A	N/A	N/A	Maldanid
	CASILL3.4	22/9/83	71° 41.9′	74° 24.0′	292	N/A	N/A	N/A	Maldanid
	CASILL3.5	22/9/83	71° 42.1′	74° 23.6′	255	N/A	N/A	N/A	Maldanid
	CASILL7.1	20/9/83	71° 46.2′	74° 24.5′	660	Mud	0.58	34.27	Maldanid
Itirbilung Fiord	IT0.1	26/9/83	69° 16.4′	69° 15.1′	55	Sandy mud	−0.73	32.64	*Portlandia*
	IT0.2	26/9/83	69° 16.4′	69° 15.0′	88	Sandy silt	−0.79	32.65	*Portlandia*
	IT0.3	26/9/83	69° 17.5′	69° 11.1′	155	Mud	−1.31	32.95	*Portlandia*
	IT0.4	26/9/83	69° 17.9′	69° 12.1′	148	Sandy mud	−1.31	32.98	*Portlandia*
	IT1.1	26/9/83	69° 20.0′	69° 03.8′	256	Sandy mud	−1.00	33.59	Maldanid
	IT1.2	27/9/83	69° 01.5′	69° 20.7′	293	Sandy mud	−0.98	33.60	Maldanid
	IT2.1	28/9/83	69° 25.3′	68° 51.2′	310	Muddy sand	−1.01	33.68	Maldanid
	IT2.2	27/9/83	69° 19.3′	68° 45.5′	402	Sandy mud	−0.93	33.65	Maldanid
	IT2.3	28/9/83	69° 17.5′	68° 27.0′	424	Sandy mud	−0.98	33.72	Maldanid
	IT3.1	28/9/83	69° 17.6′	68° 12.3′	356	N/A	−1.00	33.69	Onuphid
	IT5.0	28/9/83	69° 04.2′	67° 10.0′	175	Silty sand	−1.28	33.45	Onuphid
	IT6.0	25/9/83	68° 52.7′	66° 24.5′	502	Sandy silt	1.01	34.33	Maldanid
McBeth Fiord	MC0.1	30/9/83	69° 31.0′	69° 57.0′	152	Mud	−1.24	32.84	Maldanid
	MC2.0	30/9/83	69° 33.5′	69° 40.2′	320	Mud	−1.28	33.45	Maldanid
	MC2.1	30/9/83	69° 32.1′	69° 27.3′	320	Mud	−1.28	33.54	Maldanid
	MC4.1	1/10/83	69° 36.6′	69° 44.5′	549	Silty sand	−1.32	33.63	Maldanid

Data are taken from Wacasey *et al.* (1979, 1980), Arctic Biological Station (1980), Stewart *et al.* (1985), Schafer *et al.* (1984), Syvitski & Schafer (1984) and Petrie & Trites (1984).

Macrofauna associations consisting of *Portlandia arctica, Hiatella arctica* and *Balanus hameri* are found at the base of raised Champlain Sea marine sections, and macrofauna associations consisting of *Macoma balthica* and *Mya arenaria* occur at the top of the raised marine sections. The changes in the composition of macrofauna associations within the raised marine sections are related to increasing temperature and decreasing salinity accompanying marine regression within the Champlain Sea basin.

The presence of mollusc associations in Champlain Sea marine sediments comparable in composition to modern arctic mollusc associations demonstrates clearly the excellent preservation potential of molluscs in glacimarine sediments and their potential in the reconstruction of ancient marine biocoenoses. Thomsen & Vorren (1986) provide further evidence of the utility of marine molluscs in reconstructing Quaternary marine environments on the Norwegian continental shelf. Quantitative studies of mollusc assemblages from Quaternary marine sediments in the eastern Canadian arctic are required to test the utility of the model presented by Syvitski *et al.* (1989). Piston and Lehigh cores recovered from eastern Baffin Island fjords are remarkably unfossiliferous with regards to their macrofossil content, in contrast to the fossiliferous core material recovered from the fjords and continental shelf of Norway described by Thomsen & Vorren (1986). Future quantitative studies of Quaternary mollusc assemblages, therefore, must focus on the fossiliferous marine sediments exposed on land. The combined study of lithofacies (see Mode *et al.* 1983) and molluscan assemblages represents a powerful approach to the reconstruction of Quaternary marine environments.

Funds in support of this research were provided through the Visiting Scientist in Government Laboratories program administered by the Natural Sciences and Engineering Research Council of Canada and by Fisheries and Oceans Canada. Luci Bossé, Bernard Ste.-Marie, Edward H. Grainger and E. Gary Atkinson of Fisheries and Oceans Canada, and Judith Fournier and Diane Laubitz of the National Museums of Canada provided valuable information on the systematics and natural history of arctic marine invertebrates. Annemarie Plint assisted in the preparation of the figures and Lionel Corriveau handled the photography. The content of the paper has been improved through discussions with Janis Dale (Queen's University, Kingston, Ontario, Canada) and Michael Risk (McMaster University, Hamilton, Ontario, Canada), and the constructive reviews provided by the referees, Eric Robinson and William Mode.

References

AITKEN, A.E. & RISK, M.J. 1988. Biotic interactions revealed by macroborings in arctic bivalve molluscs. *Lethaia*, **21**, 339–350.

——, —— & HOWARD, J.D. 1988. Animal-sediment relationships on a subarctic intertidal flat, Pangnirtung Fiord, Baffin Island, Canada. *Journal of Sedimentary Petrology*, **58**, 969–978.

—— & GILBERT, R. 1989. Holocene nearshore environments and sea-level history in Pangnirtung Fiord, Baffin Island, N.W.T., Canada. *Arctic and Alpine Research*, **21**, 34–44.

AKSU, A.E. 1985. Climatic and oceanographic changes over the past 400 000 years: evidence from deep-sea cores in Baffin Bay and Davis Strait. *In*: ANDREWS, J.T. (ed.) *Quaternary Environments: Eastern Canadian Arctic, Baffin Bay and Western Greenland*. Allen & Unwin, Boston, 181–209.

ANDREWS, J.T. 1972. Recent and fossil growth rates of marine bivalves, Canadian Arctic, and Late-Quaternary arctic marine environments. *Palaeogeography, Palaeoclimatology, Palaeoecology*, **11**, 157–176.

—— 1973. Late Quaternary variations in oxygen and carbon isotopic compositions in Canadian Arctic marine bivalves. *Palaeogeography, Palaeoclimatology, Palaeoecology*, **14**, 187–192.

—— 1978. Sea level history of arctic coasts during the Upper Quaternary: dating, sedimentary sequences, and history. *Progress in Physical Geography*, **2**, 375–407.

——, MILLER, G.H., NELSON, A.R., MODE, W.N. & LOCKE, W.W., III. 1981. Quaternary near-shore environments on Eastern Baffin Island. *In*: Mahaney, W.C. (ed.) *Quaternary Paleoclimate*. Geo Abstracts, Norwich, UK, 13–44.

ARCTIC BIOLOGICAL STATION. 1980. A marine biological study of Brevoort Harbour and nearby waters of Baffin Island. *Canadian Manuscript Report of Fisheries and Aquatic Sciences*, No. 1557.

BAILEY, W.B. 1957. Oceanographic features of the Canadian Archipelago. *Journal of the Fisheries Research Board of Canada*, **14**, 731–769.

BARNES, R.D. 1980. *Invertebrate Zoology*. Saunders College/Holt, Rinehart and Winston, Philadelphia.

BOSENCE, D.W.J. 1979. Trophic analyses of communities and death assemblages. *Lethaia*, **12**, 120.

CADÉE, G.C. 1968. Molluscan biocoenoses and thanatocoenoses in the Ria de Arosa, Galicia, Spain. *Zoologische Verhandelingen*, No. 95.

—— 1984. Macrobenthos and macrobenthic remains on the Oyster Ground, North Sea. *Netherlands Journal of Sea Research*, **18**, 160–178.

CUMMINS, H., POWELL, E.N., STANTON, R.J., Jr. & STAFF, G. 1986. The rate of taphonomic loss in modern benthic habitats: How much of the potentially preservable community is preserved? *Palaeogeography, Palaeoclimatology, Palaeoecology*, **52**, 291–320.

CURTIS, M.A. 1975. The marine benthos of arctic and subarctic continental shelves. *Polar Record*, **17**,

595–626.

Dale, J.E., Aitken, A.E., Gilbert, R. & Risk, M.J. 1989. Macrofauna of Canadian Arctic fjords. *In*: Powell, R.D. & Elverhøi, A. (eds) Modern glacimarine environments: Glacial and marine controls of modern lithofacies and biofacies. *Marine Geology*, **85**, 331–358.

Dunbar, M.J. 1951. Eastern arctic waters. *Bulletin of the Fisheries Research Board of Canada*, No. 88.

D'Yakonov, A.M. 1954. *Ophiuroids of the USSR Seas*. Keys to the fauna of the USSR. Zoological Institute of the Academy of Sciences of the USSR, No. 55. (English translation by the Israel Program for Scientific Translations, Jerusalem, 1967).

Ellis, D.V. 1960. Marine infaunal benthos in Arctic North America. *Arctic Institute of North America, Technical Paper*, No. 5.

Enequist, P. 1949. Studies on the soft-bottom amphipods of the Skagerrak. *Zoologiska Bidrag frän Uppsala*, **18**, 297–492.

Fauchald, K. & Jumars, P. 1979. The diet of worms: a study of polychaete feeding guilds. *Oceanography and Marine Biology, Annual Review*, **17**, 193–284.

Feyling-Hanssen, R.W. 1976. The stratigraphy of the Quaternary Clyde Foreland Formation, Baffin Island, illustrated by the distribution of benthic foraminifera. *Boreas*, **5**, 57–94.

—— 1977. The Clyde Foreland Formation, a micropaleontological study of Quaternary stratigraphy. *Maritime Sediments Special Publication*, **10**, 315–377.

—— 1980. Microbiostratigraphy of young Cenozoic marine deposits of the Qivituq Peninsula, Baffin Island. *Marine Micropaleontology*, **5**, 153–184.

Gilbert, R. 1978. Observations on oceanography and sedimentation at Pangnirtung Fiord, Baffin Island. *Maritime Sediments*, **14**, 1–9.

—— 1985. Quaternary glaciomarine sedimentation interpreted from seismic surveys of fiords on Baffin Island, N.W.T. *Arctic*, **38**, 271–280.

Grainger, E.H. 1955. Echinoderms of Ungava Bay, Hudson Strait, Frobisher Bay and Cumberland Sound. *Journal of the Fisheries Research Board of Canada*, **12**, 899–916.

—— 1966. Sea stars (Echinodermata: Asteroidea) of arctic North America. *Fisheries Research Board of Canada Bulletin*, **152**.

Hillaire-Marcel, C. 1980. Les faunes des mers postglaciaires du Québec: quelques considérations paléóecologiques. *Géographie physique et Quaternaire*, **1**, 3–59.

Hillaire-Marcel, C. 1981. Paleo-océanographie isotopique des mers post-glaciaires du Québec. *Palaeogeography, Palaeoclimatology, Palaeoecology*, **35**, 63–119.

Kier, P.M. 1977. The poor fossil record of the regular echinoid. *Paleobiology*, **3**, 168–174.

Kohn, A.J. 1983. Feeding biology of gastropods. *In*: Saleuddin, A.S.M. and Wilbur, K.M. (eds.) *The Mollusca, Physiology*, vol 5, Part 2, Academic, New York, 2–63.

Lawrence, D.R. 1968. Taphonomy and information losses in fossil communities. *Geological Society of America Bulletin*, **79**, 1315–1330.

Lubinsky, I. 1980. Marine bivalve molluscs of the Canadian central and eastern Arctic: Faunal composition and zoogeography. *Canadian Bulletin of Fisheries and Aquatic Sciences*, No. 207.

Macpherson, E. 1971. The marine molluscs of Arctic Canada. *National Museums of Canada, Publications in Biological Oceanography*, **3**.

Matthews, B. 1967. Late Quaternary marine fossils from Frobisher Bay (Baffin Island, N.W.T., Canada). *Palaeogeography, Palaeoclimatology, Palaeoecology*, **3**, 343–363.

Miller, G.H. 1985. Aminostratigraphy of Baffin Island shell-bearing deposits. *In*: Andrews, J.T. (ed.) *Quaternary Environments: Eastern Baffin Island, Baffin Bay and Western Greenland*. Allen & Unwin, Boston, 394–427.

——, Andrews, J.T. & Short, S.K. 1977. The last interglacial-glacial cycle, Clyde Foreland, Baffin Island, N.W.T.: Stratigraphy, biostratigraphy, and chronology. *Canadian Journal of Earth Sciences*, **14**, 2824–2857.

Mode, W.N. 1985. Pre-Holocene pollen and molluscan records from Eastern Baffin Island. *In*: Andrews, J.T. (ed.) *Quaternary Environments: Eastern Baffin Island, Baffin Bay and Western Greenland*. Allen & Unwin, Boston, 502–519.

——, Nelson, A.R. & Brigham, J.K. 1983. A facies model of glacial-marine cyclic sedimentation along eastern Baffin Island, Canada. *In*: Molnia, B.F. (ed.) *Glacial-Marine Sedimentation*. Plenum Press, New York, 495–533.

Morton, B. 1983. Feeding and digestion in the Bivalvia. *In*: Saleuddin, A.S.M. & Wilbur, K.M. (eds.) *The Mollusca, Physiology*, Vol 5, Part 2. Academic, New York, 65–147.

Neilsen, R. & Holthe, T. 1985. Arctic and Scandinavian Oweniidae (Polychaeta) with a description of *Myriochele fragilis* sp. n., and comments on the phylogeny of the family. *Sarsia*, **70**, 17–32.

Ockelmann, W.K. 1958. The zoology of east Greenland. Marine lamellibranchiata. *Meddelelser om Grönland*, **122** (4), 1–256.

Petersen, C.H. 1976. Relative abundances of living and dead molluscs in two California lagoons. *Lethaia*, **9**, 642–653.

Petersen, K.S. 1982. Attack by predatory gastropods recognized in an interglacial marine molluscan fauna from Jameson Land, east Greenland. *Malacologia*, **22**, 721–726.

—— 1985. Marine macrofaunaelementer i Kap Kobenhavn Formationen og Independence Fjord, Nordgrönland. *Dansk geologisk Forening, Arsskrift for 1985*, 35–37.

Petrie, W.M. & Trites, R.W. 1984. Synoptic oceanography. *In*: Syvitski, J.P.M. (compiler). Sedimentology of Arctic Fjords Experiment: HU 83–028 and HU 82–031 Data Report, Volume 2. *Canadian Data Report of Hydrography and Ocean Sciences*, No. 28, 2–1 to 2–133.

POWELL, R.D. 1984. Glacimarine processes and inductive lithofacies modelling of ice shelf and tidewater glacier sediments based on Quaternary examples. *In*: BORNHOLD, B.D. & GUILCHER, A. (eds) Sedimentation on high-latitude continental shelves. *Marine Geology*, **51**, 1–52.

RODRIGUES, C.G. 1988. Late Quaternary invertebrate faunal associations of the western Champlain Sea basin. *In*: GADD, N.R. (ed.) The Quaternary development of the Champlain Sea basin. *Geological Association of Canada Special Paper* **35**, 155–176.

—— & RICHARD, S.H. 1983. Late glacial and postglacial macrofossils from the Ottawa-St. Lawrence Lowlands, Ontario and Québec. *Geological Survey of Canada, Paper* 83–1A, 371–379.

SCHAFER, C.T., CLATTENBURG, D., COLE, F.E., LEBLANC, W. & SYVITSKI, J.P.M. 1984. SAFE: 1983 *Hudson* bottom grab samples. *In*: SYVITSKI, J.P.M. (compiler). Sedimentology of Arctic Fjords Experiment: HU 83–028 and HU 82–031 Data Report, Volume 2. *Canadian Data Report of Hydrography and Ocean Sciences*, No. 28, 7–1 to 7–73.

SCHOPF, T.J.M. 1978. Fossilization potential of an intertidal fauna: Friday Harbour, Washington. *Paleobiology*, **4**, 261–270.

SIMONARSON, L.A. 1981. Upper Pleistocene and Holocene marine deposits and faunas on the north coast of Nugssuaq, West Greenland. *Grönlands Geologiske Undersögelse Bulletin*, No. 140.

STAFF, G.M., STANTON, Jr., R.J., POWELL, E.N. & CUMMINS, H. 1986. Time-averaging, taphonomy and their impact on paleocommunity reconstruction: Death assemblages in Texas bays. *Geological Society of America Bulletin*, **97**, 428–443.

STANTON, Jr., R.J. & DODD, J.R. 1976. The application of trophic structure of fossil communities in paleoenvironmental reconstruction. *Lethaia*, **9**, 327–342.

STEWART, P.L., POCKLINGTON, P. & CUNJAK, R.A. 1985. Distribution, abundance and diversity of benthic macroinvertebrates on the Canadian Continental Shelf and Slope of southern Davis Strait and Ungava Bay. *Arctic*, 38, 281–291.

SYVITSKI, J.P.M. 1989. On the deposition of sediment within glacier-influenced fjords: oceanographic controls. *In*: POWELL, R.D. & ELVERHØI, A. (eds) Modern glacimarine environments: Glacial and marine controls of modern lithofacies and biofacies. *Marine Geology*, **85**, 301–330.

——, FARROW, G.E., ATKINSON, R.G.A., MOORE, P.G. & ANDREWS, J.T. 1989. Baffin Island fjord macrobenthos: Bottom communities and environmental significance. *Arctic*, **42**, 232–247.

THOMSEN, E. & VORREN, T.O. 1984. Pyritization of tubes and burrows from Late Pleistocene continental shelf sediments off North Norway. *Sedimentology*, **31**, 481–492.

—— & —— 1986. Macrofaunal palaeoecology and stratigraphy in Late Quaternary shelf sediments off northern Norway. *Palaeogeography, Palaeoclimatology, Palaeoecology*, **56**, 103–150.

—— & SCHAFER, C.T. 1984. Introduction. *In*: SYVITSKI, J.P.M. (compiler) Sedimentology of Arctic Fjords Experiment: HU83–028 and HU 82–031 Data Report, volume 2. *Canadian Data Report of Hydrography and Ocean Sciences*, No 28, 1–1 and 1–8.

THOMSON, D.H. 1982. Marine benthos in the eastern Canadian High Arctic: multivariate analyses of standing crop and community structure. *Arctic*, **35**, 61–74.

——, MARTIN, C.M. and CROSS, W.E. 1986. Identification and characterization of arctic nearshore benthic habitats. *Canadian Technical Report of Fisheries and Aquatic Sciences*, No. 1434.

THORSON, G. 1933. Investigations on shallow water animal communities in the Franz Joseph Fjord (east Greenland) and adjacent waters. *Meddelelser om Grönland*, **100** (2), 1–70.

—— 1934. Contributions to the animal ecology of Scoresby Sound Fjord complex (east Greenland). *Meddelelser on Grönland*, **100** (3), 1–67.

TRITES, R.W. 1985. Oceanographic reconnaissance of selected Baffin Island fjords. *Fourteenth Arctic Workshop, Arctic Land-Sea Interaction, Program with Abstracts*, Bedford Institute of Oceanography, Dartmouth, Nova Scotia, Canada, November 6–8, 1985, 54–58.

WACASEY, J.W., ATKINSON, E.G. & GLASSPOLE, L. 1979. Zoobenthos data from upper Frobisher Bay, 1967–1973. *Canadian Data Report of Fisheries and Aquatic Sciences*, No. 164.

——, —— & —— 1980. Zoobenthos data from inshore stations of upper Frobisher Bay, 1969–1976. *Canadian Data report of Fisheries and Aquatic Sciences*, No. 205.

WALKER, K.R. & BAMBACH, R.K. 1974. Feeding by benthic invertebrates: classification and terminology for paleoecological analysis. *Lethaia*, **7**, 67–78.

The flux and preservation of organic carbon in Baffin Island fjords

JAMES P.M. SYVITSKI, K. WILLIAM G. LEBLANC & R.E. CRANSTON

Geological Survey of Canada, Bedford Institute of Oceanography, Dartmouth, N.S., Canada

Abstract: Except for shell hash lags found along minor portions of fjord walls and on current swept sills, organic carbon dominates over inorganic carbon, accounting for between 95 and 100% of the total carbon composition within the fjords. The deposition of terrestrial organic carbon is of minor significance except proximal to river mouths; i.e. the atomic carbon/nitrogen ratio, both within the water column and within the fjord sediments, fluctuates around 6.4. Organic carbon is an important constituent of the suspended particulate matter throughout most of the year when the suspended sediment concentration is $< 2\ g\ m^{-3}$. During the summer meltwater season, the organic fraction of the suspended sediment is overwhelmed by inorganic mineral matter. In the autumn season, the organic carbon content of the suspended particulate matter can exceed 30%, especially in the seasonal layer where productivity is highest. Bacterial biomass accounts for more than 50% of the total biomass found at the sediment–water interface.

Microbial respiration of organic carbon is highly effective, consuming > 90% of the annual deposition of carbon to the sea floor. The remaining and mostly refractory carbon preserved within the sediment column is < 2% by weight of the sediment and, in high sedimentation environments, such as proximal to tidewater glaciers, organic carbon is < 0.1%. The organic carbon content of sea floor samples is found inversely proportional to the modern suspended load delivered to these fjords. Anoxic degradation of this buried carbon appears to be a slow and ineffective process over the time period of 10000 years. The abundance of organic carbon buried within Arctic fjord sediments is therefore considered as an inverse indicator of deglacial and post-glacial sedimentation rates. A consistent organic carbon stratigraphy relating to the deglacial and Hypsithermal period, the Neoglacial period culminating in the Little Ice Age, and the present-day warm conditions has been identified in all ten Baffin Island fjords investigated. A simple technique is presented to convert the content of organic carbon observed down core to relative or absolute sedimentation rates, depending on the availability of ^{14}C dates. Instantaneous deposition of sediment from slides or gravity flows remains problematic.

Deep fjord basins, by their nature of being overdeepened, provide an ideal locale for the uninterrupted accumulation of sediment. Sediments delivered to a fjord are derived principally from drainage of the hinterland, and from water–mass exchanges between the continental shelf and the fjord. Sedimentary particles are also generated in situ within the fjord, through primary productivity and biogeochemical processes. The resulting sedimentary deposits can vary widely, even between neighbouring fjords. This variability reflects local conditions, including: (1) the topography and size of the drainage basins; (2) the hinterland distribution of lakes and glaciers; (3) the hydrography of the fjord-valley rivers; (4) the sediment load carried by these rivers; (5) the sill depth. The availability of wind-transportable sediment can also be important locally.

Together, these factors influence the main mechanisms of sediment deposition within a fjord: (1) bedload dumping at river mouths, (2) hemipelagic sedimentation from the seasonal surface layer, (3) turbidity currents generated from prodelta failure, (4) diffusive processes, such as slides, wave action and tides, that transport nearshore sediment into deeper water, and (5) ice rafting of sediment by sea ice and icebergs.

Our objective is to discuss the accumulation of carbon within the Baffin Island fjords (Fig. 1). We wish to demonstrate that down-core trends of organic carbon content can provide an excellent means for understanding sedimentation history, especially in discerning climatic trends through the Holocene. This study is particularly timely considering the recent interest in the carbon cycle and how the impact of humans has contributed to an apparent global warming trend (i.e. the International Geosphere–Biosphere Program on Global Change), and the role of the oceans in ameliorating the build-up of atmospheric carbon gases (i.e. the Joint Global Ocean Flux Study). It has been suggested (Walsh *et al.* 1981) that burial of carbon in shallow shelf and coastal marine areas

From DOWDESWELL, J. A. & SCOURSE, J. D. (eds), 1990, *Glacimarine Environments: Processes and Sediments*. Geological Society Special Publication No 53, pp 177–199.

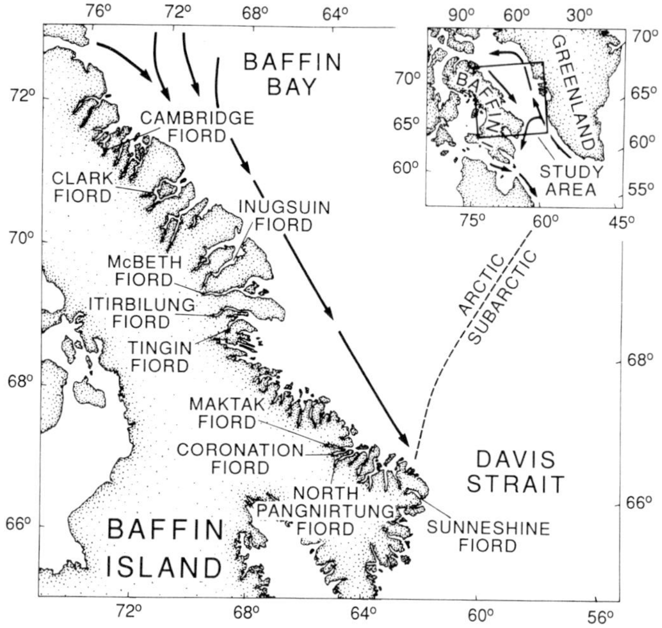

Fig. 1. Location of the ten Baffin Island fjords surveyed in the Sedimentology of Arctic Fjords Experiment (SAFE). Also indicated is the direction of the Baffin Island current.

may constitute an important sink for anthropogenic increases in atmospheric carbon dioxide. Our study synthesizes a large data set on organic carbon, collected as part of the international Sedimentology of Arctic Fjords Experiment (SAFE: Syvitski & Schafer 1985; Syvitski & Blakeney 1983; Syvitski 1984; Syvitski & Praeg 1987).

Our approach is to identify and discuss the significance of the various sources of organic carbon that contribute to Arctic fjord sediments, to understand the preservation potential of this buried carbon, and to discern and interpret vertical trends in organic carbon within the stable sediment column of the deep fjord basins. Unfortunately, we were not able to collect synoptic data relating to the seasonal cycling of carbon — Baffin Island fjords are very hostile and remote environments (Syvitski & Schafer 1985) — and inferences must be made from better studied fjords.

Baffin Island fjords can be described as high energy environments, although they are deep (400 to 700 m) and many have sills. Much of their water column is very cold (< -1.0°C) year-round and always well oxygenated (> 3.5 mL/L). Fjords such as Sunneshine, North Pangnirtung, McBeth and Itirbilung contain strongly reducing sediments (pE < 3), others such as Tingin and Inugsuin fjords contain moderately reducing sediments ($3 \leq$ pE ≤ 5), and Cambridge, Coronation and Maktak fjords contain mildly reducing sediments ($5 \leq$ pE ≤ 8) (Fitzgerald & LeBlanc 1983; Fitzgerald 1984). The open water season can vary from weeks to four months with an average of three months. The fluvial discharge season is also short and except for the larger rivers, many streams flow for only a few weeks. Except for Sunneshine Fiord, which is macrotidal (> 4.5 m large tide), most of the fjords are microtidal (< 1.5 m large tide).

Methodology

Field

We have measured the particulate organic carbon (POC) content in 250 water samples, the detrital organic carbon (OC) content of 140 grab samples of the sea floor, 3 samples of glacier ice, and 1 380 subsamples of 27 Lehigh and 20 piston cores (cores were ≤ 12 m in length) obtained using an array of oceanographic instruments. Details of our field and analytical methods are given elsewhere (Asprey *et al.* 1983; Clattenburg *et al.* 1983; Fitzgerald & LeBlanc 1983; Andrews *et al.* 1984; Fitzgerald 1984; Schafer *et al.* 1984; Syvitski *et al.* 1984a; Mothersill & Tabrez 1987; Andrews 1987). The samples were collected within the late summer and fall seasons (August through October) principally in 1982 and 1983 with additional data from mid to late summer in 1985, 1986 and 1987. Deep water samples were collected from C.S.S. Hudson, the M/V Pandora II and the submersible Pisces IV. Nearshore samples were collected from scientific launches. Terrestrial samples were collected from modern and early to mid-Holocene raised deposits using both ship-based shore parties and land-based helicopter-supported surveys. Reported water temperatures were obtained through 130 calibrated Guildline CTD profiles.

Laboratory

The concentration of suspended particulate matter from 680 fjord and river water samples was determined by volumetric filtration of samples onto preweighed Nucleopore and Selas filters. Suspensates deposited on Selas Ag-filters were subjected to CHN (carbon:hydrogen: nitrogen) elemental analysis using a Perkin-Elmer model 240B elemental analyzer equipped with a Perkin–Elmer AD-2z autobalance. Details and error analysis are reported in Asprey *et al.* (1983), and are typically $\leq$ 10% of the reported value.

Total carbon determinations on 1 to 2 grams of grab or core subsamples were performed with a Leco induction furnace with an attached Leco WR-12 Carbon Determinator (Clattenburg *et al.* 1983). Organic carbon (OC) analysis was similar, but conducted only after any carbonates were first removed using 1N hydrochloric acid washings. Details and error analysis are reported in Clattenburg *et al.* (1983), and are typically $\leq$ 5% of the reported value. Total nitrogen determinations of the grab samples were analyzed using the 'Kieldahl' method.

Sources of carbon

Syvitski *et al.* (1987) have identified four principal sources of organic carbon for any given fjord: (1) freshwater and terrigenous detritus, (2) macrophytic algae and marsh detrital production, i.e. both along rocky shorelines and tidal flats, (3) marine plankton production and its associated by-products, i.e. marine snow, phytodetritus and faecal pellets, derived both from within the fjord and imported from the continental shelf, and (4) epilithic production. Locally, OC may be remobilized from bottom sediment (slides, turbidity currents, bioturbation) and added to basinal sediments.

Freshwater and terrigenous POC

In temperate fjords, where watersheds are covered in lush vegetation, freshwater and terrigenous POC can be an extremely important source. The vertical flux of POC is highest near the outflow of fjord-valley rivers (Fig. 2a: Farrow *et al.* 1983), although the percentage of OC incorporated within the sediment column is very low because of the rapid deposition of carbon-poor sedimentary minerals. In high-latitude or polar fjords, there is little terrestrial vegetation so that even streams not draining glaciers carry little POC. Most of the terrestrially-discharged OC is derived from fluvial cannibalism of raised early Holocene glacimarine and mid- to late Holocene marine deposits, which in themselves contain little carbon (< 0.3% by weight, n=35). Rivers that drain glaciated hinterlands volumetrically carry little POC – these rivers transport primarily a bedload of gravel and sand. An important excep-

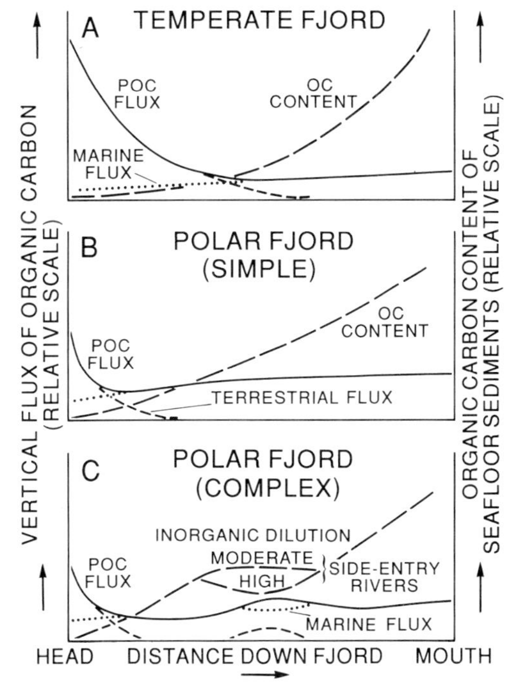

Fig. 2. The vertical flux (relative) of particulate organic carbon (POC) and the amount of POC incorporated into bottom sediments, as a function of distance downfjord. (A) Temperate fjord (after Farrow *et al.* 1983), (B) Polar fjord – simple (based on no side-entry inputs), and (C) Polar fjord – complex (with significant side-entry discharges, mid fjord).

tion, not applicable to our Baffin Island study, is for rivers that drain hinterlands containing organic carbon-bearing rocks (Armentrout 1983; Elverhøi *et al.* 1989).

Terrestrial POC, initially transported seaward within the river plume, tends to be hydrodynamically heavy, and usually settles out close to river mouths (Syvitski *et al.* 1985a). Thus the larger the discharge, the stronger the seaward-flowing surface current and the further these particles will be transported. Fluvial discharge into the fjords of Baffin Island is approximately one order-of-magnitude smaller than for comparable sized temperate fjord–valley rivers (Syvitski *et al.* 1987). Thus the contribution of land-derived POC is considerably lower, as is the distance that this POC can be transported seaward (Fig. 2b).

Most freshwater is discharged into continental fjords through the head of the fjord, but fjords located on islands may have a significant proportion of incoming water and sediment discharged laterally from marginal streams. For the ten Baffin Island fjords investigated, side-entry discharge contributes from 20% (e.g. Coronation and Maktak Fjords) to 90% (e.g. Inugsuin Fiord) of the annual fluvial discharge. Furthermore, present day Baffin Island ice fields are largely situated along the fjord margins, where the mountains are at their highest elevation (Syvitski & Schafer 1985). Thus these side-entry streams supply POC-deficient sediments that act as a mid-fjord dilutant to the deposition of POC (Fig. 2c). Exceptions include those fjords that drain the Penny Ice Cap (Coronation, Maktak and North Pangnirtung Fjords).

Land-derived organic carbon from polar regions is largely refractory, i.e. resistant to chemical and thermal weathering and diagenesis (Syvitski *et al.* 1987). This contrasts with temperate regions, especially agricultural regions, where the proportion of refractory POC is much lower and the volume of dissolved organic matter (i.e. humic acids) is much higher.

Littoral production

The production of POC from tidal flats is considered low: (1) there is little POC entering the flats from the land (see above), (2) the intertidal area is small; most of the fjords investigated have a mean tidal range < 1 m and the thalwegs of the streams are quite steep, (3) the tidal flats are high energy environments with limited habitat for stable POC-producing flora and fauna, and (4) the flats are ice-covered nine months of the year. One exception is a small lagoon somewhat isolated from the advancing delta at the head of McBeth Fiord. There, strongly reduced sediment was found to contain up to 2% organic carbon.

The subtidal (0 to 10 m) regions along the rocky fjord margins are more productive and dense beds of kelp have been observed (Syvitski *et al.* 1985b; Syvitski 1987; Gilbert pers. comm.). We have no data on this production of OC, although it could be substantial at least compared with intertidal portions of the SAFE-studied fjords.

Marine production

Phytoplankton in fjords are generally the largest source of POC (Syvitski *et al.* 1987). Marine snow and phytodetritus account for much of the POC observed within the water column of Baffin fjords (LeBlanc *et al.* 1988). To understand this production, we describe the oceanographic character of the Baffin fjords, and include axial cross-sections of the distributions of six carbon-related properties from three representative Baffin fjords (Coronation Fiord: Fig. 3; Inugsuin Fiord: Fig. 4; McBeth Fiord: Fig. 5).

Baffin fjords experience three oceanographic seasons. The first is the winter season (about Oct. 1 to June 15), a time of sea ice growth wherein salt is rejected during the crystallization process and super-saline 'salt fingers' sink through the water column. This brine rejection is believed to establish isohaline circulation that initiates the break down in water stratification and the formation of large circulation cells (Syvitski 1989). In terms of carbon flux, little is known for this season, although we suggest that the vertical flux of POC would be low, corresponding to seasonally low nutrient levels. Ice algae may make a significant contribution to the primary production in ice-covered waters (Horner 1976), although there are no data specifically relating to Arctic fjords (cf. Cross 1982).

The second oceanographic season occurs during the spring to early summer (about June 15 to July 15), when the seasonal surface layer is re-established through a combination of nival melt, sea ice melt and warming of the surface waters. Wind mixing is relatively minor at this time, and water stratification becomes well developed. Although phytoplankton production would begin in earnest during this period, with surface radiation high and nutrients at their maximum, turbid river plumes may initially delay phytoplankton growth. Phytoplankton transported from the shelf to the more nutrient-rich waters of fjords may trigger plankton

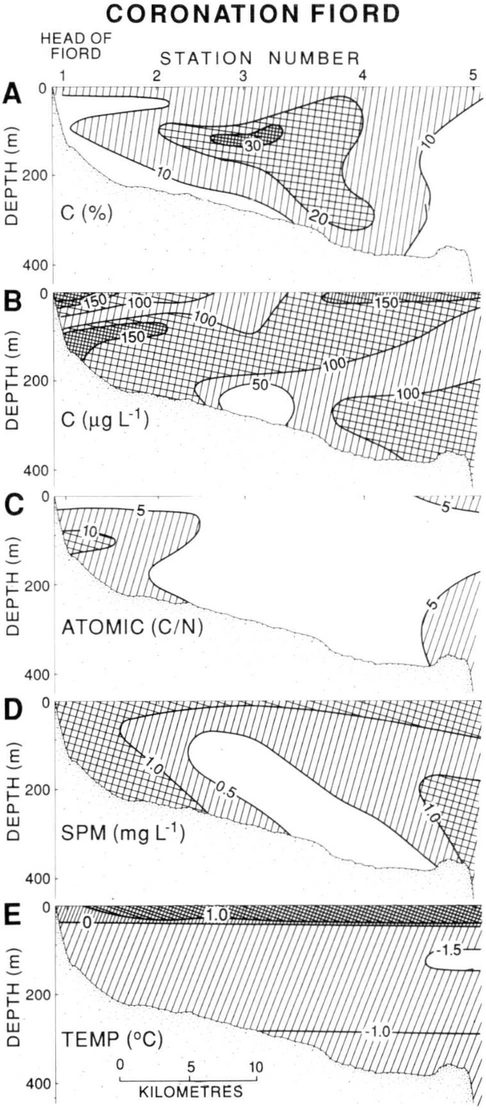

Fig. 3. Station locations and oceanographic properties (based on continuous profiles or up to 12 water samples per station) within Coronation Fiord, Baffin Island, measured on September 13, 14, 1982 (data after Asprey *et al.* 1983). (A) Carbon content of suspended particulate matter. (B) Particulate organic carbon concentration. (C) Atomic carbon nitrogen ratios of suspended particulate matter. (D) Suspended particulate matter concentration. (E) Water temperature.

blooms and a substantial flux of organic matter to the sediments (Syvitski *et al.* 1987). According to Petersen (1978), west Greenland fjords act as a sink for organic matter, most of which originates from shelf waters. Baffin Island fjords similarly contain higher nutrient levels than observed on the continental shelf (Trites 1985).

The third oceanographic season extends through summer into fall (July 15 to Oct. 1), a time of surface layer thickening through wind mixing. In the case of a Greenland fjord, wind dominates the energy budget during this period (Nielsen & Ottesen-Hansen 1980). During this season, the Baffin Island fjords are comprised of three thermal layers (Figs 3e, 4e, 5e, after Trites 1985): a relatively warm surface (seasonal) layer (0–3°C) of varying thickness (5–60 m), a cold intermediate layer (< −1°C) of varying thickness (0–325 m) which forms part of the Baffin Island current, and a warmer deep layer (to > −1°C), where depth permits. The depth boundaries of these layers are quite variable. Bacterial biomass accounts for < 15% of the total biomass within the Baffin fjord surface waters (Albright & Stroh 1983). Suspended particulate matter sampled during this period was comprised largely of diatoms, dino- and silico-flagellates (Andrews *et al.* 1984; LeBlanc *et al.* 1988). Maximum phytoplankton abundance was observed just below or within the seasonal surface layer. In the deeper water, intact specimens were replaced largely by zooplankton-egested faecal pellets, organic mucoid matter (marine snow) and other phytodetritus. A large part of the mucoid matter appears to relate to the large concentrations of discarded *Oikopleura* housings observed floating between 50–300 m within the water column (Syvitski *et al.* 1985b).

Coronation Fiord is an example of a simple polar basin (i.e. Fig. 2b), and is presently occupied by a large tidewater glacier at its head through which most freshwater and sediment is discharged. Both the concentration of suspended particulate matter (SPM: Fig. 3d) and POC (Fig. 3b) demonstrate the competing sediment sources from the turbid continental shelf and the tidewater glacier. The carbon content of the SPM is highest (up to 30%) between these two sources (Fig. 3a), where the concentration of SPM is low. The proportion of POC drops below 10% when the SPM exceeds 2 mg/L (Fig. 6a).

Organic carbon to nitrogen ratios (atomic C/N ratio) can indicate the contribution of terrigenous carbon to marine carbon within the marine environment (Bornhold 1978; Vilks *et al.* 1979; Pocklington & Leonard 1979). Most marine organic matter has atomic C/N of ≤ 10, whereas most terrigenous organic matter has much higher atomic C/N values. Examination of the atomic C/N ratio of Baffin Island fjord SPM indicates a strong marine source of POC.

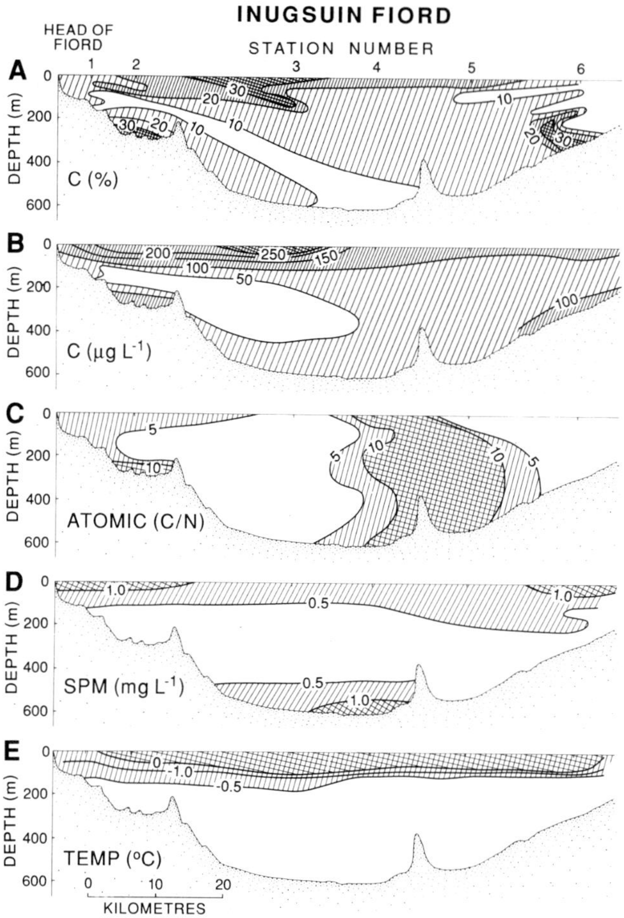

Fig. 4. Station locations and oceanographic properties (based on continuous profiles or up to 12 water samples per station) within Inugsuin Fiord, Baffin Island, measured on September 21, 1982 (data after Asprey *et al.* 1983). (A) Carbon content of suspended particulate matter. (B) Particulate organic carbon concentration. (C) Atomic carbon:nitrogen ratios of suspended particulate matter. (D) Suspended particulate matter concentration. (E) Water temperature.

Slightly elevated values of atomic C/N (10–12) were observed near the tidewater glacier at the head of Coronation Fiord (Fig. 3c), which compare well with the predicted carbon flux scenario of a simple polar marine basin (Fig. 2b). Maktak and North Pangnirtung fjords also conform to this model of a simple polar fjord.

Inugsuin Fiord represents the extreme opposite in conditions to Coronation Fiord. First, it has a pronounced sill at its mouth (Fig. 4). Thus the import of shelf sediment is much less (cf. Fig. 4d & 3d). Second, only 2.3% of the total sediment delivery to the fjord enters through its head (compared to 99.9% for Coronation Fiord: LeBlanc *et al.* 1988). Most of the sediment delivered to the fjord is derived from numerous glaciers whose meltwater enters the fjord along its margins. Inugsuin Fiord represents a complex polar fjord in the carbon flux scheme of Fig. 2c. Slightly elevated values of atomic C/N (10–12) were thus observed mid fjord (Fig. 4c). POC was highest in the surface waters, an indication of ongoing productivity (Fig. 4a). Tingin, Clark, Cambridge and Sunneshine fjords also conform to this complex polar fjord scenario of carbon flux (Fig. 2c).

McBeth and Itirbilung fjords lie between the above (Coronation and Inugsuin) end-member basins, in terms of sedimentation and carbon flux. McBeth Fiord drains the largest hinterland (4070 km^2) of the ten fjords investigated, with an estimated total annual freshwater runoff of

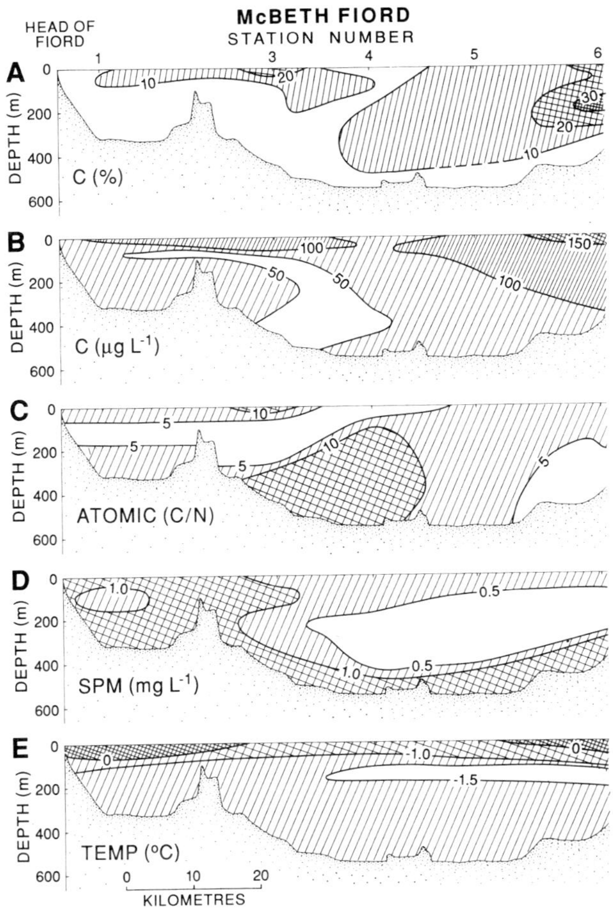

Fig. 5. Station locations and oceanographic properties (based on continuous profiles or up to 12 water samples per station) within McBeth Fiord, Baffin Island, measured on September 18, 19, 1982 (data after Asprey *et al.* 1983). (A) Carbon content of suspended particulate matter. (B) Particulate organic carbon concentration. (C) Atomic carbon:nitrogen ratios of suspended particulate matter. (D) Suspended particulate matter concentration. (E) Water temperature.

0.77 km^3. It has a fjord-head sandur that receives very little contribution from ice meltwater, yet this sandur accounts for 68% of the estimated 220 000 tonnes of suspended sediment that enters the fjord annually. McBeth Fiord is surrounded by terrestrial glaciers whose meltwater enters along the margins of the fjord and may account for the slightly elevated atomic C/N values observed mid fjord (Fig. 5c). POC is highest in the surface waters, especially in the outer and shelf-influenced regions of the fjord (Winters *et al.* 1985). During the time of sampling, an on-going deep-water exchange of fjord and shelf water may account for the higher than average SPM concentrations in the bottom waters of the fjord (Fig. 5d). Unlike the concentration of total SPM, POC concentration decreases with depth (Fig. 5b), supporting the notion of organic carbon utilization through respiration and grazing.

Epilithic production

Baffin Island fjords have between 30 to 60% of their sea floor in the form of steep (30° on average) side wall slopes. Except near river mouths, many of these side wall slopes are barren of sediment, especially in the outer parts of the fjords (Syvitski 1987). Benthos found on these side walls are dominated by sea urchins, anemones, sponges, serpulids, Asteroedea, shrimp, and shellfish. Shell hash was ubiquitous.

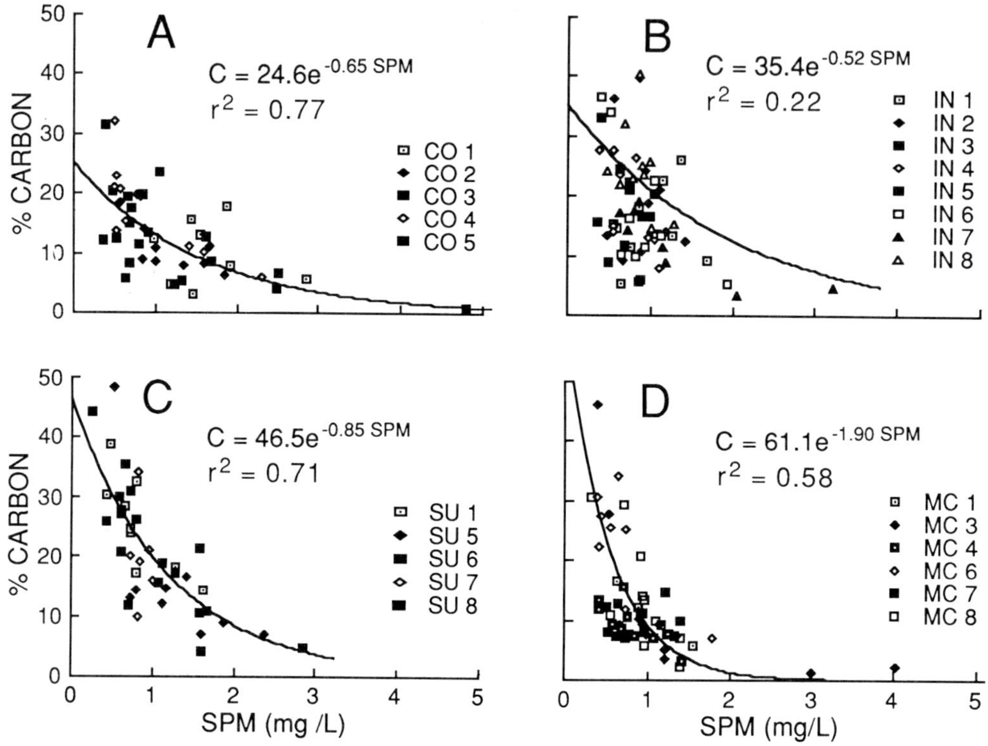

Fig. 6. Exponential decrease in carbon content (%) with increasing levels of suspended particulate matter (SPM in mg/L) for four representative Baffin Island fjords: (A) Coronation Fiord, (B) Inugsuin Fiord, (C) Sunneshine Fiord, and (D) McBeth Fiord. The samples are numbered from near the head of a fjord, i.e. CO1 or MCl, to outside the mouth of a fjord, i.e. CO5 or MC8. The equations given are best fit exponential regressions between carbon content (C) and SPM; r is the correlation coefficient of regression.

Considering the large portion of the water column that overlies the fjord margins, a large proportion of the settling phytodetritus must pass through and be consumed by the epilithic community.

Remobilized bottom sediment

Bottom sediments in Baffin Island fjords may be remobilized by three processes, which in descending order of importance are: (1) subaqueous slope failures, especially within the proximal prodelta environment (Syvitski 1985; Syvitski & Farrow 1988); (2) resuspension through the foraging action of crustaceans, fish and mammals (Hein & Syvitski 1989) and the bioturbating mechanisms of benthic infauna (Syvitski *et al.* 1989); and (3) current resuspension, especially along the shallower reaches of the side wall environment (Syvitski *et al.* 1985b). Through these processes there exists the possibility of introducing low carbon sediment en masse to a basin accumulating higher levels of carbon, or through affecting the steady-state rates of sediment accumulation.

We have already stressed that a high vertical flux of POC does not necessarily result in high carbon content on the sea floor (Fig. 2). This concept of inorganic dilution of POC forms the fundamental rule for understanding and predicting hydrocarbon-generating source rocks (Magara 1986). With this concept is the geologic axiom that areas receiving high rates of sedimentation (and therefore deficient in carbon) are more prone to subaqueous slope failures. These failed sediment masses may be transported downslope by turbidity currents, slides and debris flows. For slides and debris flows, little sorting of the sediment mass takes place during transport and, in most cases, carbon-deficient sediment is deposited on top of carbon-rich (relative) basinal sediment. Turbidity currents do sort sedimentary particles during their transport, and the upper mud layers of a distal turbidite may become enriched in carbon relative to the initial failed sediment mass.

Foraging by bottom-dwellers and sea floor scavengers can also mobilize the bottom sediment and, in the case of whales, can resuspend an amount equivalent to between 1% and 30% of the annual delivery of sediment by Baffin rivers (Hein & Syvitski 1989). However, due to weak bottom currents, the effect is one of mixing rather than removal.

Currents within the fjords of Baffin Island are similar to fjords found elsewhere, especially during the POC-producing summer/fall season when a two-layer estuarine circulation is best established. The strongest currents are found within the seasonal surface layer and over sills (Syvitski 1989). If the fjords have sills, then the current energy within the basin decreases rapidly with depth. Currents are driven largely by meltwater discharge during the summer, wind-generated drift during the late summer and fall, deep-water exchanges during the fall, isohaline-generated circulation of winter, and tidal currents which operate over the entire year. We expect estuarine surface currents to have little effect on the vertical flux of POC although, as mentioned above, the distribution of terrigenous-derived POC can be influenced strongly. Tidal currents within silled basins simply oscillate back and forth with the tide producing little residual flow. They are, however, capable of sweeping the steeper side walls of the fjord and sills clean of sediment (Syvitski 1989), in effect increasing the vertical flux of POC to the basin floors. The relative proportion of POC to lithogenic particles would, however, remain largely unchanged.

Summary

Most of the carbon in the water column is marine in origin, mainly from the primary production of phytoplankton within the seasonal surface layer. Some of this carbon is derived from land sources and deposited close to the river mouths, but much of this terrigenous POC is from cannibalized marine terraces. Much of the primary production of POC is consumed during its passage through the water column, especially along the steep margins of the fjords where a significant epilithic benthos exists. Finally, mass failures may play a significant role in redistributing POC. Below we discuss sedimentation dynamics and the character of the POC sampled from the sea floor.

Deposition of carbon

To understand the deposition of POC we must, for lack of seasonal synoptic data, focus on two important data sets. The first is the antipathetic (inverse exponential) relationship between the concentration of SPM and the proportion of carbon contained within the suspended matter (Fig. 6). The relationship is significant as turbidity levels are too low to be a limiting factor on primary production (Farrow *et al.* 1983). During the height of fluvial discharge in early summer, we expect the suspended POC to be even further diluted by silt and sand grains.

The second data set involves comparing the estimated annual input of suspended sediment, both from fjord head and for the entire fjord, i.e. including marginal side-entry contributions, to the fjord-averaged organic carbon content of sea floor sediments (Fig. 7). The potential use of organic carbon content as an inverse indicator of sedimentation rates in fjords is suggested (cf. Andrews 1987; Syvitski *et al.* 1989), although the relationship is not linear and fjords that receive low sediment input, such as Cambridge and Inugsuin fjords, are overpredicted (Fig. 7). Together, Figs 6 & 7 suggest the possibility of interpreting changes in the carbon content of marine cores as indications of changes in lithogenic sedimentation rates.

Character and origin of POC

Most of the carbon observed within the sediments of Baffin Island fjords is organic (Fig. 8a). There is only a minor contribution of inorganic carbon from carbonate debris and mostly within continental shelf samples where the total carbon content exceeds 1%. Submersible observations from the fjord side-walls do suggest more local enrichment of carbon from carbonate hash (Syvitski 1987). For the linear relationship between total carbon (which based on Fig. 8a can be considered organic) and total nitrogen within sea floor sediments of Baffin Island fjords (Fig. 8b), the slope of the regression line, when corrected for the atomic weights of carbon and nitrogen, indicates an atomic C/N ratio of 6.4, typical of marine organic matter. Bacterial biomass accounts for less than 15% of the total biomass of SPM but exceeds 50% of the total biomass within sea floor sediments (Albright & Stroh 1983).

Nature of POC sedimentation

Based on SEM examination of a large number of Baffin Island water samples (LeBlanc *et al.* 1988), POC settles largely as part of agglomerates and faecal debris. This is similar to suspended particles settling in more temperate fjords (Syvitski *et al.* 1985a). As no data exist

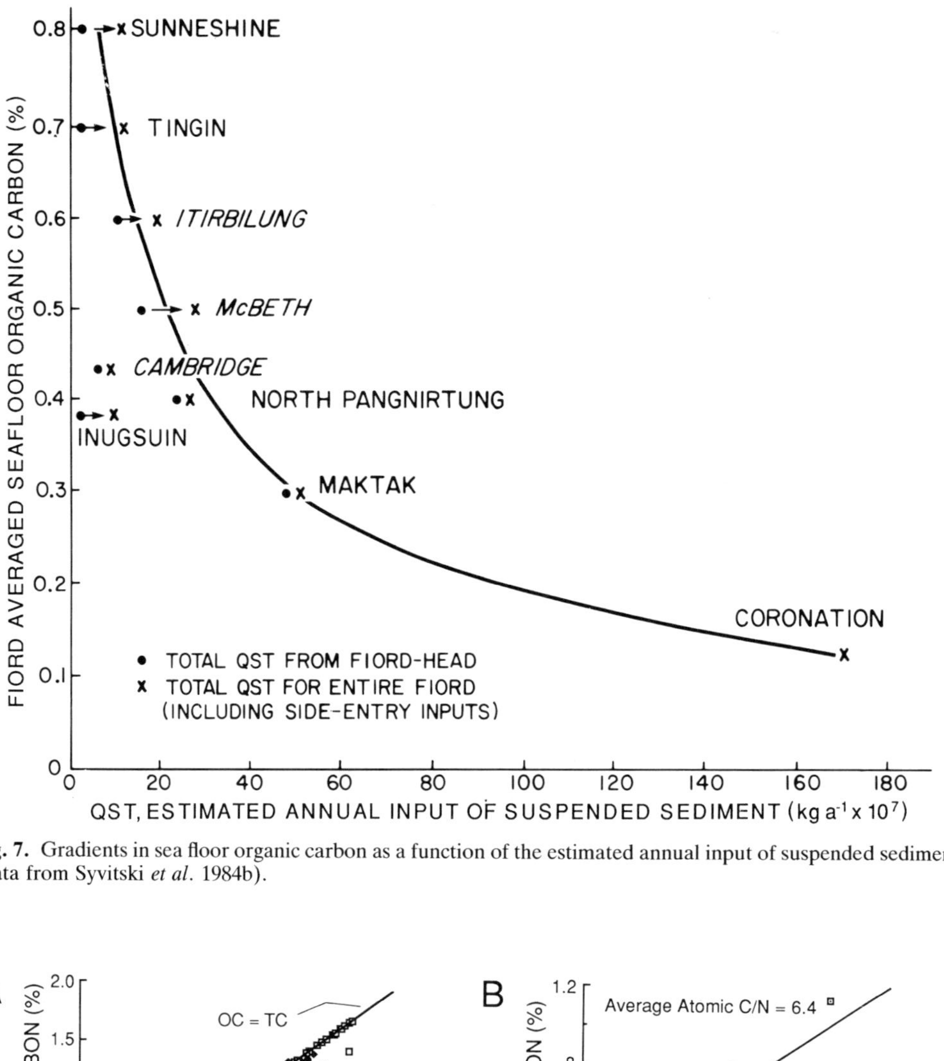

Fig. 7. Gradients in sea floor organic carbon as a function of the estimated annual input of suspended sediment (data from Syvitski *et al.* 1984b).

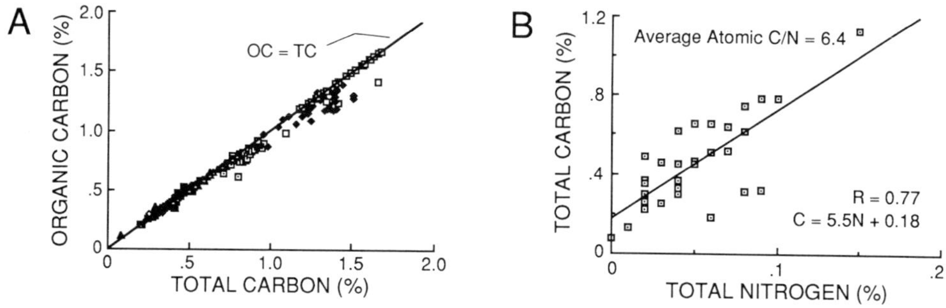

Fig. 8. (A) Almost 99% of the carbon observed within the sediments of Baffin fjords is organic in nature. Minor scattering within the plot relates to inorganic carbon contribution from shells and shell debris which is of minor importance from continental shelf samples where the carbon content was greater than 1%. (B) The relationship between total carbon and total nitrogen in sea floor sediments from Baffin Island fjords. The regression line indicates an atomic C/N ratio of 6.4, typical of marine organic matter (data after Clattenburg *et al.* 1983).

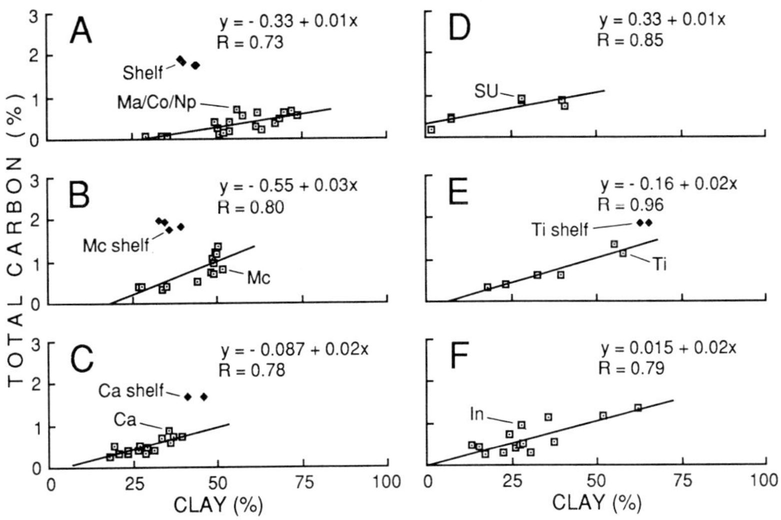

Fig. 9. Relationship between total carbon and clay content of sea floor sediment samples from eight Baffin Island fjords: (A) Maktak, Coronation and North Pangnirtung fjord and shelf environment, (B) McBeth fjord and shelf environment, (C) Cambridge fjord and shelf environment, (D) Sunneshine fjord, (E) Tingin fjord and shelf environment, and (F) Inugsuin fjord. The regression equations are only for fjord samples: shelf data are excluded.

on the settling rates of Baffin Island fjord SPM, we assume we may use values reported from British Columbia fjords where particles settle out at rates of 30 to 100 m/d. The Baffin Island fjords are between 400 and 700 metres deep, thus the residence time of POC would range from 4–23 days. We also note that the carbon content of Baffin fjord sea floor sediments increases with the clay-size content (Fig. 9), similar to observations in the Barents Sea (Elverhøi *et al.* 1989). This is understandable as coarse silt and sand-sized particles introduced from a river largely settle out with little attached carbon (Syvitski *et al.* 1985a). In the more distal portions of the fjord basin, where sedimentation of lithogenic particles is greatly reduced and dominated by clays and fine silt-sized agglomerates, the carbon content of the sea floor is high. Therefore, the low carbon content of the sea floor (typically $< 1.5\%$ OC) relative to carbon content of the sampled suspended matter (typically $> 10\%$) is a result of both carbon utilization during a relatively short residence time and seasonal inorganic dilution. Diagenesis of OC in the sediment is discussed below in *Burial of Carbon* section. Sea floor samples collected from the continental shelf invariably contain a higher proportion of carbon for a given clay content when compared to fjord samples (Fig. 9).

Sea floor distribution of organic carbon

With the above relationships in mind, the distribution of sea floor carbon sampled along the thalweg of a given fjord can be understood (Fig. 10). For Cambridge Fiord, the carbon content of the sea floor initially increases seaward of the fjord head deltas as expected (cf. Figs 2b & 10a). Seaward of the inner Cambridge sill, however, the diluting effect of lithogenic sedimentation from side-entry glacial streams is apparent, both in the decrease in carbon content and in the increase of the sand content (cf. Figs 2c & 10a). In the outer fjord environment, outside the influence of the glacial meltwater plumes, the carbon content (and clay content) increases seaward (Fig. 10a). McBeth Fiord similarly shows the effect of mid-fjord sedimentation of lithogenic particles on the distribution of sea floor carbon (cf. Figs 2c & 10b), but to a lesser degree.

Itirbilung and Coronation fjords both demonstrate the decrease in lithogenic sedimentation rates found seaward of a fluvial discharge point source, and as a result sea floor carbon increases systematically out of the fjord (cf. Figs 2b & 10b). The decrease in sedimentation rates with distance out of the fjord has been established previously for these Baffin Island fjords (Andrews *et al.* 1985).

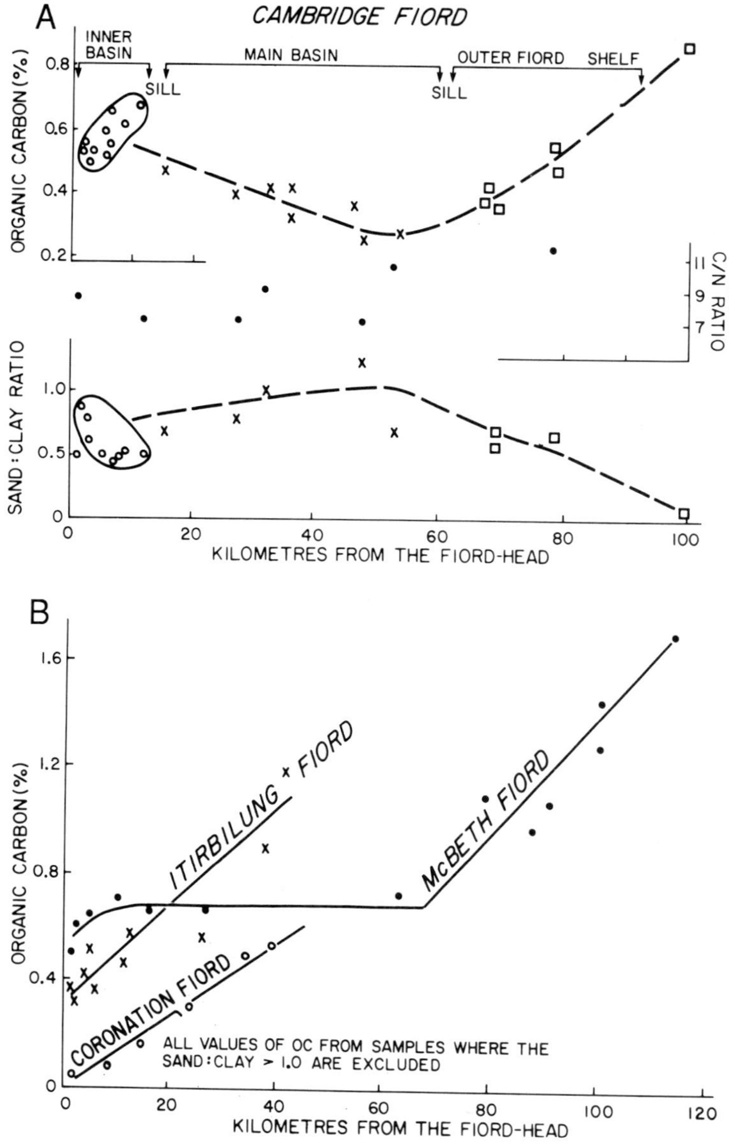

Fig. 10. (A) Gradients in sea floor sediment parameters (sand:clay ratio, percent organic carbon, atomic C/N ratio) for Cambridge Fiord. (B) Gradients in organic carbon for Itirbilung, McBeth and Coronation fjords.

Summary

The carbon content of the sea floor of Baffin Island fjords is derived largely from marine phytoplankton-generated POC that settles quickly through the water column and in association with agglomerates of clays and fine-silt-sized particles. The low carbon content of the sea floor relates both to high trophic utilization during the sedimentation process and lithogenic dilution, especially near stream mouths. As the rate of lithogenic sedimentation increases, the carbon content demonstrates a concomitant decrease.

Burial of carbon

The steady state relationship of OC at the sediment-water interface can be expressed as:

Deposition + deep-water production
= resuspension + burial + grazing + microbial oxidation

Deposition of OC in a nearby West Greenland fjord was measured at 90 gC m^{-2} a^{-1} (Anderson 1977). If we assume an average sediment accumulation rate of 1 mm a^{-1}, a dry bulk density of sea floor sediment of 700 kg m^{-3} and an average POC value of 13% (from data associated with Figs 3a, 4a, 5a and other SAFE-fjords), then the deposition rate for OC for a typical Baffin Island fjord is similarly 91 gC m^{-2} a^{-1}. Sea floor production of POC in fjords from dissolved forms of organic carbon is very low (≪ 10%: Syvitski *et al.* 1987). Although deep living algae have been identified (LeBlanc *et al.* 1988), they form only a minor constituent of the benthic biomass. We assume a deep-water production of 5 gC m^{-2} a^{-1}. The resuspension of sediment term can be largely ignored for silled fjord basins (Syvitski *et al.* 1987). Assuming a total sediment accumulation rate of 1 mm a^{-1}, a dry bulk density of sea floor sediment of 700 kg m^{-3} and from Figure 7 an average OC value of 0.6%, then for a typical Baffin Island fjord the burial rate is 4.2 gC m^{-2} a^{-1}. Although Syvitski (1987), Dale *et al.* (1989) and Syvitski *et al.* (1989) have described the benthic community within Baffin Island fjords, they provide no data on the rate of benthic grazing on the deposited OC. We assume a 10% trophic transfer efficiency, i.e. a benthic grazing value of 9.1 gC m^{-2} a^{-1}. Thus microbial oxidation dominates in the consumption of deposited OC:

$$91 + 5 \text{ gC m}^{-2} \text{ a}^{-1} = 0 + 4.2 + 9.1 + ? \text{ gC m}^{-2} \text{ a}^{-1} \text{ (microbial oxidation)} \quad (1)$$

$$\text{microbial oxidation} = 82.7 \text{ gC m}^{-2} \text{ a}^{-1}.$$

To understand the preservation of organic carbon buried within the sediment column we must consider three factors: (1) the OC flux into the sediment column (i.e. the burial term shown above), (2) the accumulation rate of non-OC material, and (3) the degradation rate of labile OC within the sediment column. Changes to any of these terms over time can effect a change in the observed carbon content preserved at depth. If oceanographic conditions changed, say through an increase of solar radiation into the water column or the presence of warmer waters, primary production could increase and the flux of carbon to the sediment column might increase. If lesser amounts of lithogenic particles were deposited at the sea floor, then the organic carbon content within the sediment column might increase. If organic carbon was once more labile or the sediments were more oxygenated, then it is possible that the rate of OC degradation would increase and the amount of buried carbon would decrease. Other possible scenarios are obvious.

Degradation of OC

The amount of OC buried in a sediment represents the difference between that deposited at the sediment surface and the amount oxidized to CO_2 (Berner 1989). This oxidation process occurs within the first few centimetres in most marine sediments and within the first three centimetres in the Baffin fjords, after which reduced sediments are found (Fitzgerald & LeBlanc 1983; Fitzgerald 1984). To determine the rate of OC degradation below the oxic layer, one must either ascertain the decomposition kinetics operating within the sediment column, or define the change in organic carbon with depth within an environment of invariant oceanographic conditions and over an appropriate time scale.

Using stable isotope ratios ($^{15}N/^{14}N$ and $^{13}C/^{12}C$) and the composition of bound amino acids, Macko *et al.* (1987) attempted the first approach for the sediments of Baffin fjords. They found some enrichment of $\delta^{15}N$ in the deeper portion of a core from North Pangnirtung Fiord suggesting at least limited degradation of organics with depth. They similarly found a depletion of $\delta^{13}C$ with depth in other Baffin cores, an indication of selective degradation of the more labile carbon compounds, i.e. amino acids and carbohydrates which are enriched in $\delta^{13}C$.

For the second approach, we examine the change in OC within the upper few centimetres of samples collected from all of the fjords investigated. In this manner we average out local variations in sedimentation, geochemistry and benthic community. The deviation between OC content of the surface oxidized skin (upper 2 to 3 cm of the sea floor) and the OC content of the sediment sampled just below this layer (5 to 10 cm layer) is shown in Fig. 11. The local sedimentation rates vary considerably between sample locations and the elapsed time between these surface and subsurface samples is in the range of 15 to 150 years in the deep, more distal environments (data after Andrews 1987), and 2 to 5 years in the proximal deltaic environment of the fjord head (data after Syvitski & Praeg 1987; Syvitski 1989). In all cases, we observe very little depletion of organic carbon in the subsurface samples (Fig. 11).

To support this notion further, we examine organic carbon versus depth profiles from subsamples collected from Lehigh cores. Lehigh cores are large diameter (10 cm), capable of returning an undisturbed sediment-water interface followed by an uncompressed 2 to 3 m sediment column. OC values from five McBeth cores collected along the length of the fjord are presented in Fig. 12.

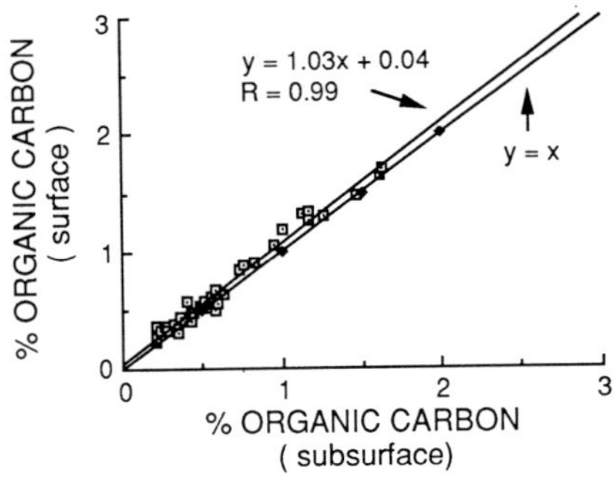

Fig. 11. The deviation between POC content of the surface skin (upper 2 to 3 cm of the sea floor) and the POC content of the sediment sampled just below this layer (5 to 10 cm layer). Also shown is the one-to-one regression between surface and subsurface POC with which to compare. There appears to be little POC consumption with depth.

The three inner fjord cores (Mc 0.1, 2.0 and 2.1) have roughly the same OC content. Mc 4.1, situated near a mid-fjord glacial sediment source, is very low in carbon (cf. Fig. 10b), and Mc 7.0 from the most distal portion of the fjord basin contains relatively high carbon values (Fig. 12). In all five cores, OC either remains constant with depth or increases with depth, based on linear regression analysis (Fig. 12). Organic carbon analysis of seven Lehigh cores collected from Itirbilung Fiord provides similar results (Fig. 13). Generally, the carbon content increases with distance from the main sediment source at the head of the fjord (cf. Fig. 10b). The OC content in three cores is found to increase with core depth (It 0.1, 2.2 and 2.3), and for four cores (It 0.3, 0.4, 1.1 and 3.1) OC is found to decrease with depth. These short Lehigh cores, both from Itirbilung and McBeth fjords, contain strongly reducing conditions wherein buried carbon can be considered geochemically stable.

The lack of a strong decay in the content of OC with depth and the stable reducing environment found in our other Baffin fjord cores suggests that, in itself, degradation of buried carbon is of minor influence on the fluctuation of carbon with depth. We emphasize here that we do not rule out the possibility of some increased diagenetic reduction of OC locally, for instance in the 15 to 25 cm depth interval in Mc 7.0 (Fig. 12). However, the effect appears to be minor.

Variable carbon production and deposition with time

There are no unambiguous methods to determine the fluctuations of OC production during the Holocene (last 10 000 years). We have previously demonstrated that modern POC production and deposition is almost entirely from marine primary production. Has this remained the case during the Holocene period and at deposition levels presently observed?

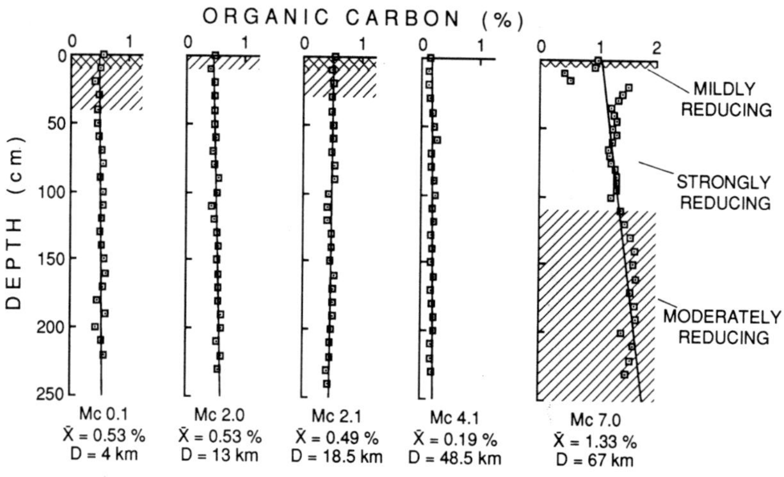

Fig. 12. The down-core variation of organic carbon from five Lehigh cores collected from McBeth Fiord. Also shown is the averaged organic carbon value ($\bar{X}$) and the distance the core was collected from the head the fjord (D). Three redox zones are also indicated based on ship-board electrochemical measurements (data after Fitzgerald 1984): mildly reducing ($5 \leqslant pE \leqslant 8$), moderately reducing ($3 \leqslant pE \leqslant 5$), and strongly reducing ($pE < 3$).

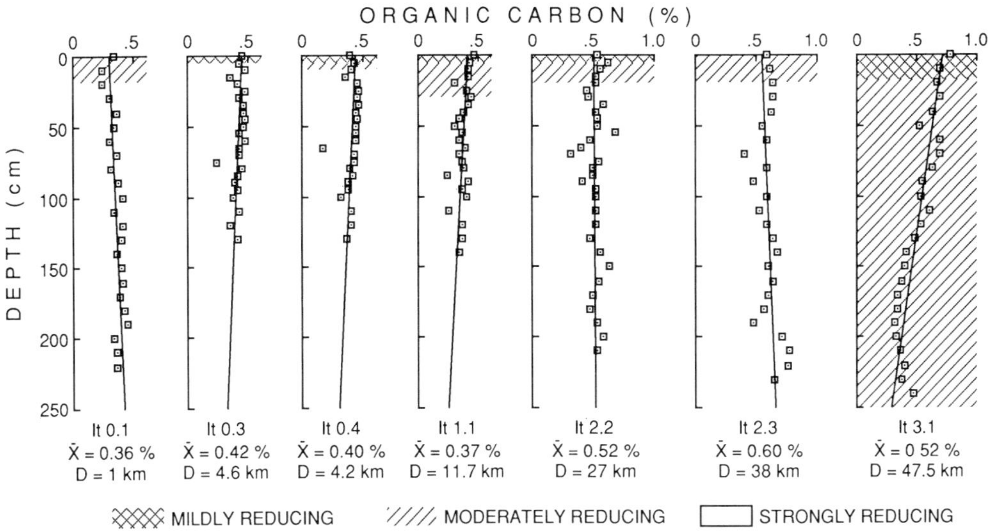

Fig. 13. The down-core variation of organic carbon from seven Lehigh cores collected from Itirbilung Fiord. Also shown is the averaged organic carbon value ($\bar{X}$) and the distance the core was collected from the head of the fjord (D). Three redox zones are also indicated based on ship-board electrochemical measurements (data after Fitzgerald 1984): mildly reducing ($5 \leqslant pE \leqslant 8$), moderately reducing ($3 \leqslant pE \leqslant 5$), and strongly reducing ($pE < 3$).

This very difficult question involves tenuous estimates on possible changes of factors affecting marine primary production and terrestrial plant community changes. As a clue we know that the mean annual phytoplankton primary production in modern fjords ranges from 90 gC m^{-2} a^{-1} for west Greenland fjords to 200 gC m^{-2} a^{-1} for fjords draining the western North American rain forests, with those of Scotland and Norway falling in between (see Syvitski *et al*. 1987, p. 216, for full references). Even the fjords of the Antarctic experience significant primary production (80 gC M^{-2} a^{-1}: Dunbar *et al*. 1989). Thus we have a factor of two range in possible primary production levels which is surprisingly small considering the wide variation in latitude (solar radiation), nutrient and turbidity levels, water temperature, sea ice conditions, and levels of dissolved organic carbon. Research has shown the existence of a warm Hypsithermal period between 7000 to 3000 years ago in Baffin Island, and a colder Neoglacial period between 3000 to 100 years ago and the present warm conditions of today (Miller 1973; Church 1978; Short *et al*. 1985). Yet these predicted variations in Baffin Island climate are smaller than the range in climate observed between fjords located in different parts the Earth, and there is little reason to expect that Arctic waters within Baffin fjords were much warmer (i.e. 10°C) than today (cf. Andrews 1972), and they cannot get much colder.

Similarly, there is little evidence of substantial increases in the vegetation cover of Baffin Island during the Hypsithermal as compared to today. By 'substantial' we refer to a radical change in vegetation cover, i.e. from rain forests to desert conditions. Short *et al*. (1985) do provide evidence for the degradation of vegetation and a decline in pollen productivity since the Hypsithermal for the eastern Baffin region. We suggest that the terrestrial flux of carbon has only played a minor role in the flux of carbon to the sea floor during the whole of the Holocene. Andrews (1987) and Short *et al*. (1989) have noted that a portion of the refractory carbon within the sediment column of the Baffin fjords is old, based on the presence of pre-Quaternary pollen and differences in age from ^{14}C dated marine mud and accelerator-dated shells. The former are always older. Andrews (1987) also feels that the rate of carbon production (or preservation in the cores) has not changed drastically during the last 10 000 years. Syvitski *et al*. (1984b) suggested the possibility of a permanent sea ice cover over Cambridge Fiord between 2000 and 3000 years ago. That conceivably could substantially, and locally, reduce the flux of POC to the sea floor, relying completely on imported carbon from the continental shelf waters. So in the extreme we might expect a factor of two change in productivity within Baffin fjords.

Lithogenic dilution

We have already demonstrated the importance of lithogenic dilution on the organic carbon content of the modern sea floor (Figs 7 & 11). The leading question then is 'have the inorganic sedimentation rates changed over the Holocene in Baffin fjords?' Church (1978) has calculated for a medium sized Baffin Island river flowing into Ekalugad Fiord that the sediment flux to the sea was in excess of one order-of-magnitude larger between 5700 to 4300 years ago than the period 4300 years ago to the present. Syvitski *et al.* (1984b) similarly calculated for a river flowing into Cambridge Fiord, that the annual rate of sediment delivery decreased exponentially from 6000 years ago, with little sediment transported seaward over the last 4750 years. Together these data support the notion of paraglacial sedimentation from the period of ice withdrawal at the heads of these fjords between 9 and 5 ka and reaching steady state conditions at around 4.5 ka. Prior to 8 ka the inner portion of the northern fjords may have been occupied by glacial ice draining the Foxe ice dome (Andrews & Miller 1984). The outer fjord basins, however, would have experienced very high proglacial sedimentation rates. The Little Ice Age maximum in the Baffin fjord region was between 0.35 to 0.1 ka (Locke & Locke 1977). Since that period, ice terminus retreat has occurred along much of the Baffin coast (Syvitski & Praeg 1987) and a new phase of increased (paraglacial) sedimentation has begun.

Summary

There is little evidence of post-oxic degradation of organic carbon within the Baffin fjord sediment column. This might reflect the low temperature biokinetics operating within these cold polar sediments (e.g. Dunbar *et al.* 1989) and the near-surface location of the anoxic zone. Although we have limited direct data, we argue that carbon production, either on land or in the sea, has changed little during the Holocene. Even extreme changes in natural conditions could only provide a factor of two variation in primary phytoplankton productivity. Finally, there is good evidence that non-carbon sedimentation rates have changed by an order-of-magnitude over the Holocene of Baffin Island, with peak rates occurring in the Late Foxe to Early Holocene and then again in the last 50 to 100 years.

The buried carbon signal: climate and sedimentation trends

Given the above information on operating conditions during the Holocene, we suggest four units in the up-core trends in organic carbon. Unit A reflects the portion of the sediment column deposited during the retreat of the Late Foxe ice tongues (Cockburn retreat): – i.e. during the local thermal maximum period (after Andrews pers. comm.) of high sediment accumulation rates. OC should be very low, diluted by high vertical flux of lithogenic particles. This unit should be overlain by a Unit B, one of gradual increasing OC that reflects the onset of the cold Neoglacial period and its association of decreased fluvial discharges and increased glacial ice storage. Above this is Unit C, where the content of OC should decrease as the core mirrors the decline of the Little Ice Age with the increased melt of glaciers and concomitant increase in sedimentation rates. Finally, there might be a small increase in carbon within the surface skin of the core (upper few centimetres) that reflects the increased content of labile OC that is associated with a surficial biodegradation zone of oxygenated to mildly reducing conditions. This upper zone might include a portion or all of unit C depending on the local rates of sediment accumulation and the local biogeochemical environment. The thickness of any unit would depend both on the local rate of hemipelagic sedimentation and the sediment accumulation from non-steady state events such as the mass emplacement of sediments from slides and sediment gravity flows discussed previously.

To test this hypothesis, OC content of piston cores was investigated. Conventional piston cores are capable of retrieving upwards of 10 m of the sediment column. However, they may not collect the surface of the sediment column (upper 1 m). Therefore we have adjusted OC piston core data relative to OC data from Lehigh or equivalent gravity cores collected from the same location so that the entire stratigraphic sequence is represented. We have also double checked the surficial OC values of the Lehigh cores with our subsampled grab data (cf. Fig. 11). Without these corrections, the biodegradation zone (unit D), the period following the Little Ice Age (unit C) and, from basins that have experienced low sedimentation rates, much of the Neoglacial interval (unit B) may be missing.

The similarity between the hypothesized up-core trend in OC outlined above and those observed from two cores collected from Maktak Fiord (Fig. 14) is unmistakable. Maktak Fiord

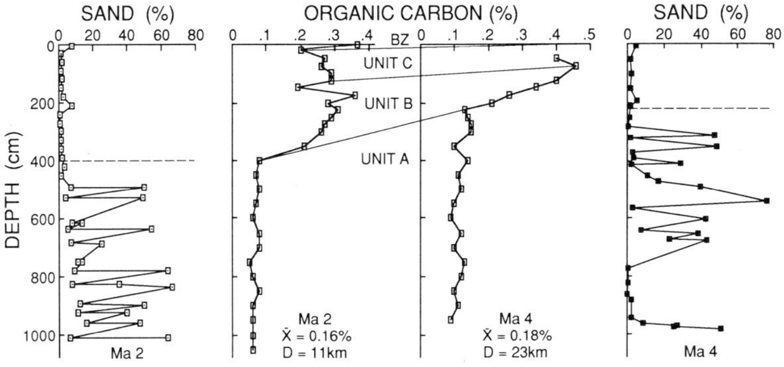

Fig. 14. Down-core variation in percent sand and organic carbon for two cores collected at variable distance (D) from the head of Maktak Fiord. Also indicated are the interpreted units based on OC trends (see text for details) and the averaged organic carbon value ($\bar{X}$).

presently drains part of the Penny Ice Cap and experiences high rates of sediment delivery to its waters (cf. Fig. 7), and as a result has presently low OC values. It is an example of a simple polar basin (cf. Fig. 2b) with nearly all of the sediment delivery through the head of the fjord. The two cores are located 11 km and 23 km from the present river mouth (Ma2 and Ma4, respectively). All four units or zones are represented. The top of unit A is located twice as deep (4 m) in Ma2 compared to Ma4 (2 m). The tops of the other units are similarly more deeply buried in Ma2, as would be expected from a core collected closer to the sediment source. Furthermore, Ma4 being a more distal core has higher OC values in all units. As unit A is hypothesized to reflect a period of high sedimentation rates, it follows with little surprise that the unit, in both cores, is associated with sandy turbidites (Fig. 14). The turbidites would have been generated from the rapidly aggrading sea floor at the head of the fjord. What is remarkable is the lack of signal noise in the OC data associated with these turbidites. The onset of increased carbon levels within unit B is associated with the establishment of a mature polychaete community as recorded from bioturbation intensity (Farrow *et al.* 1984).

Figure 15 is a collation of down-core (piston & Lehigh) profiles from each of the fjords investigated. The same general trend as observed in the Maktak cores can be seen to varying degrees in all of these cores. Gross variations will reflect the variability in sediment accumulation at a particular site, particularly the influence of fjord marginal rivers. If we examine the available ^{14}C dates (Andrews *et al.* 1989) on these cores, we see that the boundary between Unit A and Unit B is around 6 ± 1 ka (Fig. 15). The timing of this transition should not, however, be considered formalized.

Interestingly, Tan & Vilks (1987) found a similar OC stratigraphy for sediment cores collected in Lake Melville, a marine inlet on the coast of Labrador, although none of their three cores analyzed recorded unit C or D. The lack of unit D simply reflects field sampling problems (piston versus Lehigh corer: G. Vilks pers. comm.). The lack of a strong unit C appears to differentiate Baffin fjords that have experienced a rapid ablation of their ice fields since the time of the Little Ice Age, and more temperate fjords, such as Lake Melville, that have not had glacial ice within their hinterlands since the early Holocene period of Late Wisconsinan ice ablation. Other differences between the two systems include the greater role of terrestrial carbon input and subsequent diagenesis in the more temperate basins (Tan & Vilks 1987).

Figure 16 idealizes the basic time-history trend in organic carbon accumulation in the Baffin Island fjord region (cf. Figs 14 & 15). Three scenarios for the development of this down-core trend are provided. In the first case (Fig. 16a), and the case we feel is most reasonable, carbon burial is time invariant, biodegradation of OC affects only the most recently deposited material, and the inorganic sedimentation rates vary considerably with time. In the second case (Fig. 16b), the flux of POC to the sea floor has increased through the Holocene, but so has the rate of OC biodegradation which in effect neutralizes the increased flux of POC. However, our data have shown the rate of anoxic biodegradation to be low. The third scenario (Fig. 16c) sees the inorganic sedimentation rate to be time-invariant, and the down-core variability in OC is a function of marked changes in the production rate during the Neoglacial period. This can be discounted since we know that carbon production during periods of cooler summers should be even smaller both on

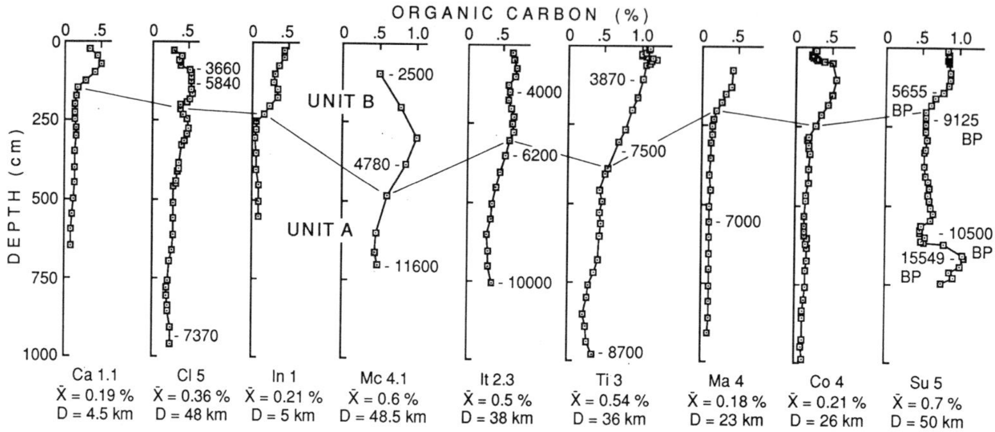

Fig. 15. An example of the down-core organic carbon trend from each of the nine Baffin Island fjords investigated (Ca = Cambridge, Cl = Clark, In = Inugsuin, Mc = McBeth, It = Itirbilung, Ti = Tingin, Ma = Maktak, Co = Coronation, and Su = Sunneshine Fiord). The fjords are given in order of decreasing latitude from the upper left to the lower right. Also shown is the averaged organic carbon value ($\bar{X}$) and the distance the core was collected from the head of the fjord (D). ^{14}C dated intervals are from Andrews *et al.* (1984), Andrews (1987) and Andrews and Jennings (1987). The boundary between units A and B is also shown (cf. Fig. 14).

land and in the water, and POC flux rates would need to be more than currently occurring in warm temperate fjords of Scotland, Norway and British Columbia.

Based on the first scenario (Fig. 16a), it is possible to convert OC values observed in marine cores to sedimentation rates with the explicit assumption that the buried OC signal is derived wholly from a constant flux of marine carbon deposited uniformly throughout the fjord. However, if a small fraction of the fluvially-transported sediment consists of organic carbon, then, from the calculations shown in Table 1, the method will provide minimum estimates of sedimentation rates. For example, if the accumulation rate of fluvial sediment contains 0.1% OC (case 1, Table 1), then a change in the OC content within a marine core from 0.2 to 0.5% is associated with a 4-fold decrease in the sedimentation rate. If, however, the accumulation of OC is strictly from marine production (case 2, Table 1), then a change in the OC content from 0.2 to 0.5% is associated with a 2.5-fold decrease in the sedimentation rate. The difference between the two cases, in the magnitude of predicted sedimentation rates, increases in high sedimentation environments.

To predict sediment accumulation rates depends on the availability of absolute age determinations for intervals within a core. For instance, Unit A of Ma4 has an average OC content of 0.1% (Fig. 14). Unit B sees OC

Table 1. *Calculations of the organic carbon (OC) content of sea floor sediments based on varied assumptions of the initial input parameters*

Distance (km)	ISR (mm a^{-1})	Case 1 OSR$_T$ (mm a^{-1})	OC content %	Case 2 OSR$_T$ (mm a^{-1})	OC content %
1	100	0.11	0.1	0.01	0.01
5	5	0.015	0.3	0.01	0.2
10	4	0.014	0.35	0.01	0.25
50	3	0.013	0.4	0.01	0.33
100	2	0.012	0.6	0.01	0.5

Given are predicted accumulation rates of inorganic sediment (ISR) along a fjord basin in terms of distance from a fjordhead fluvial source. The sedimentation rate of marine-produced organic carbon (OSR$_M$) is assumed constant at 0.01 mm a^{-1} throughout the fjord basin. For case 1, the sedimentation rate of fluvial-transported organic carbon (OSR$_F$) is set at 0.1% of ISR. For case 2, the OSR$_F$ is set to nil. The total sedimentation rate of organic carbon (OSR$_T$) = (OSR$_M$) + (OSR$_F$).

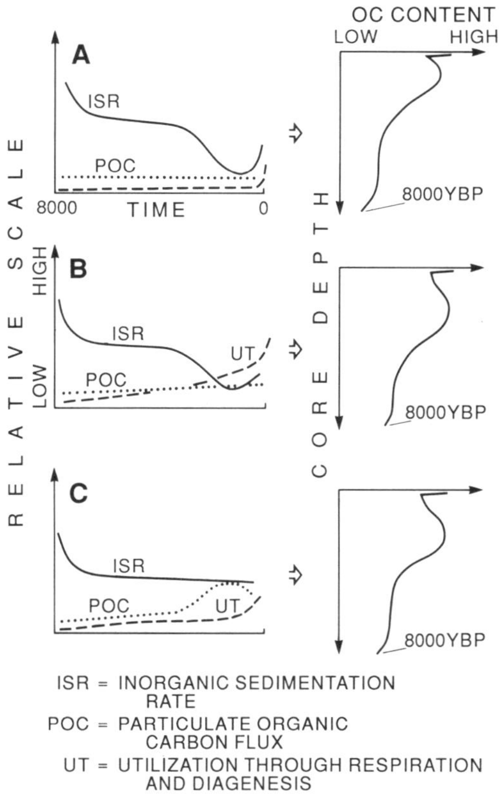

Fig. 16. Three scenarios that see fluctuations in inorganic sedimentation rates (ISR), the vertical flux of particulate organic carbon (POC), and the utilization or degradation of buried organic carbon (UT) varied to produce the given idealized down-core trend in organic carbon found in Baffin Island fjord cores (cf. Figs 14 & 15).

increase to 0.47%. Therefore we can say that sediment accumulation rates have decreased by some unknown factor. Based on Fig. 7, OC apparently increases exponentially with decreasing sedimentation rates. In support of a non-linear relationship, we also know that the rate of OC increases linearly with distance away from a sediment source (cf Fig. 10b), and sedimentation rates decrease exponentially out from these sediment sources within Baffin fjords (Andrews & Syvitski 1990). Finally, with two or more dated intervals within a core, an average accumulation rate can be determined and compared to the averaged OC content between these intervals. Figure 17a shows this relationship of two Baffin fjord cores, and again the inverse exponential relationship between the OC-content and the bulk sedimentation rates is apparent.

As an example of converting an OC stratigraphy to sediment accumulation rates, we choose Su5 from Sunneshine Fiord as it is well dated using various ^{14}C techniques (Andrews *et al.* 1985; Andrews *et al.* 1989): dates are illustrated on Figure 15. By using the regression curve shown on Figure 17a, i.e. log Sed. Rate (in cm/a) = 0.73−2.54 OC (in %), we can convert our OC data (cf. Fig. 15) to sedimentation rates. To view the appropriateness of this method we have also plotted against core depth the known changes in sedimentation rate as given by the ^{14}C determined rates (Fig. 17b). Our method

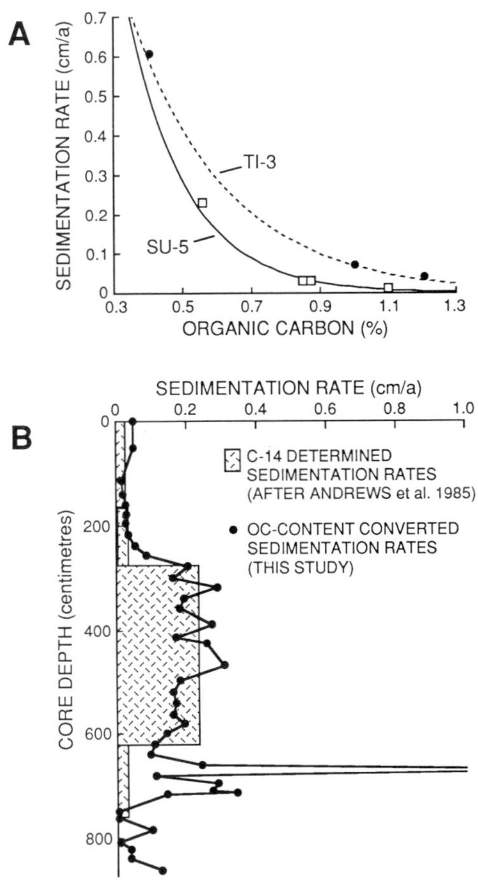

Fig. 17. (A) Inverse exponential relationship between ^{14}C determined sedimentation rates (after Andrews *et al.* 1989) and the interval-averaged organic carbon content from two Baffin fjord cores (cf. Fig. 15). (B) The down-core trend in organic carbon from a core collected in Sunneshine Fiord (Su5: Fig.15) has been converted to sedimentation rates according to the method outlined in the text and plotted as a function of core depth. The interval sedimentation rates as determined from ^{14}C dates are also plotted for comparison.

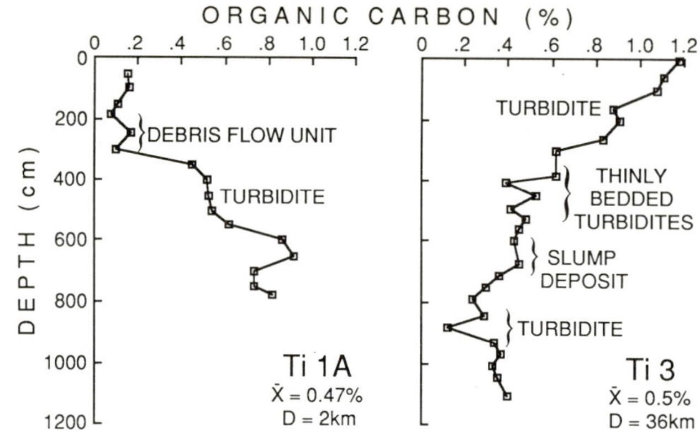

Fig. 18. Down-core variation in organic carbon for two cores collected at variable distances (D) from the head of Tingin Fiord. The down-core averaged OC value ($\bar{X}$) is also given. High-frequency fluctuations in OC appear to relate to mass flows and turbidites.

allows for more precise location of unit boundaries and delineation of other subtle changes.

Our final example of OC stratigraphy is from two cores collected from the head and mouth of Tingin Fiord (Ti1A and Ti3, respectively). The boundary between units A and B, for Ti3, is situated at 4 m below the sea floor (cf. Fig. 15). Although coring problems may account for the missing unit C, it may also reflect the distal position of the core site, whereby the site is outside the influence of sediment plumes issuing from margins of the fjord during the ablation period following the Little Ice Age. Ti1A is one of our most nearshore sites, collected just less than 2 km from the fjordhead delta. It has an expanded OC stratigraphy that reflects significantly higher sedimentation rates than Ti3 and apparently only covers the period defined by unit C and the upper portion of unit B. Based on detailed core and X-radiography description, many of the high-frequency fluctuations in the carbon signal of both cores can be related to debris flow deposits, slump deposits and turbidites (Fig. 18). It is remarkable that emplacements of mass flow deposits do not alter significantly the major OC content trends downcore.

Conclusions

Of the five sources of carbon identified, most of the flux of carbon to the sea floor of Baffin Island fjords is a result of primary productivity of phytoplankton within the seasonal surface layer during the summer and fall season. This marine POC settles quickly through the water column in association with agglomerates of fine-grained mineral matter, in the form of phytodetritus and marine snow. The distribution of OC within the sea floor sediments is very much dependent on the distance from river mouths, which are the main point sources of sediment delivery to these fjords. As the rate of lithogenic sedimentation increases, the carbon content demonstrates a concomitant decrease.

The preservation of OC buried within the sediment column is considered in terms of the vertical flux of POC to the sea floor, the dilution effect of sedimenting lithogenic particles and the degradation of labile carbon. The time-history of OC burial appears to depend on the fluctuation of sediment delivery to these fjords during the Holocene. Analysis of long cores shows a consistent carbon stratigraphy that in many respects is more diagnostic and easier to interpret than lithostratigraphic data that are sensitive to the effects of sediment gravity flows. We present a simple method to interpret the carbon stratigraphy in terms of relative and, with additional data, absolute sedimentation rates through time. We propose to further elaborate the hypothesis developed here by a more detailed analysis of an expanded suite of ^{14}C dated piston cores.

We thank the many SAFE participants that helped collect the water and sediment samples, including the officers and crew of CSS Hudson and M/V Pandora II, and the Polar Continental Shelf Project for helicopter field support. D.C. Clattenburg and K.W. Asprey performed some of the grain size analysis given in this report. We particularly thank J.T. Andrews for his support and encouragement for this project. J.T. Andrews, G. Vilks, R. Gilbert and I.J. Fairchild are thanked for their conscientious and critical comments. This manuscript forms GSC contribution no 31489 and SAFE contribution no 26.

References

ALBRIGHT, L.J. & STROH, E.M.D. 1983. Heterotrophic bacterial dynamics within waters and surficial sediments of several Canadian eastern Arctic fjords. *In*: SYVITSKI, J.P.M. & BLAKENEY, C.P. (compilers) *Sedimentology of Arctic Fjords Experiment: HU82–031 Data Report, Vol. 1*. Canadian Data Report of Hydrography and Ocean Sciences, **12**, 6–1 to 6–13.

ANDERSON, J.S. 1977. Primary production associated with sea ice at Godhaven Disko, West Greenland. *Ophelia*, **16**, 205–220.

ANDREWS, J.T. 1972. Recent and fossil growth rates of marine bivalves, Canadian Arctic, and Late-Quaternary Arctic marine environments. *Palaeogeography, Palaeoclimatology, Palaeoecology*, **11**, 137–156.

—— 1987. Downcore variations in the carbon content of fjord piston cores and association with sedimentation rates. *In*: SYVITSKI, J.P.M. & PRAEG, D.B. (compilers) *Sedimentology of Arctic Fjords Experiment: Data Report, Vol. 3*. Canadian Data Report of Hydrography and Ocean Sciences, **54**, 12–1 to 12–19.

—— & JENNINGS, A.E. 1987. Influence of sediment source and type on the magnetic susceptibility of fiord and shelf deposits, Baffin Island and Baffin Bay, N.W.T. *Canadian Journal of Earth Sciences*, **24**, 1386–1401.

—— & MILLER, G. 1984. Quaternary glacial and nonglacial correlations, in the eastern Canadian Arctic. *In*: R.J. FOULTON (ed.) *Quaternary Stratigraphy of Canada – a Canadian Contribution to IGCP project 24*. Geological Survey of Canada, Paper **84–10**, 101–116.

—— & SYVITSKI, J.P.M. 19⁰0. Sediment fluxes along high latitude glaciated continental margins. National Research Council and National Academy of Science Press.

——, JULL, A.J.T., DONAHUE, D.J., SHORT, S.K. & OSTERMAN, L.E. 1985. Sedimentation rates in Baffin Island fiord cores from comparative radiocarbon dates. *Canadian Journal of Earth Sciences*, **22**, 1827–1834.

——, LAYMON, C.A. & BRIGGS, W.M. 1989. Radiocarbon date list III: Labrador & Northern Quebec, Canada, and Radiocarbon date list IV: Baffin Island, N.W.T., Canada. *Institute of Arctic and Alpine Research, Occasional Paper*, **46**.

——, OSTERMAN, L.E. & KRAVITZ, J. 1984. Quaternary studies on Baffin Island fjord cores. *In*: SYVITSKI, J.P.M. (compiler) *Sedimentology of Arctic Fjords Experiment: HU83–028 Data Report, Vol. 2*. Canadian Data Report of Hydrography and Ocean Sciences, **28**, Chapter 5.

ARMENTROUT, J.M. 1983. Distribution of organic carbon in the Gulf of Alaska Neogene and Holocene sedimentary record. *In*: THIEDE, J. & SUESS, E. (eds.) *Coastal Upwelling, Its Sedimentary Record, Part B: Sedimentary Records of Ancient Coastal Upwelling*. Plenum, New York, 421–449.

ASPREY, K.W., BISHOP, P., BLAKENEY, C., LEBLANC, K.W., SYVITSKI, J.P.M. & WINTERS, G.V. 1983. SAFE: 1982 suspended particulate matter data. *In*: SYVITSKI, J.P.M. & BLAKENEY, C.P. (compilers) *Sedimentology of Arctic Fjords Experiment: HU82–031 Data Report, Vol. 1*. Canadian Data Report of Hydrography and Ocean Sciences, **12**, 5–1 to 5–30.

BERNER, R.A. 1989. Biogeochemical cycles of carbon and sulphur and their effect on atmospheric oxygen over Phanerozoic time. *Palaeogeography, Palaeoclimatology, Palaeoecology*, **75**, 97–122.

BORNHOLD, B.D. 1978. Carbon/Nitrogen (C/N) ratios in surficial marine sediments of British Columbia. *In*: Current Research Part C, *Geological Survey of Canada Paper*, **1978–1C**, 108–112.

BURRELL, D.C. 1983. The biogeochemistry of Boca de Quadra and Smeaton Bay, southeast Alaska: A summary report on investigations 1980–1983. *Report Inst. Marine Sciences Univ. of Alaska*, Fairbanks.

CHURCH, M. 1978. Paleohydrological reconstructions from a Holocene valley fill. *In*: MIALL, A.D. (ed.) *Fluvial Sedimentology*. Canadian Society of Petroleum Geologists Memoir **5**, 743–772.

CLATTENBURG, D.D., COLE, F., KELLY, B., LEBLANC, K.W., BISHOP, K., RASHID, M., SCHAFER, C.T., & SYVITSKI, J.P.M. 1983. SAFE 1982 bottom grab samples. *In*: SYVITSKI, J.P.M. & BLAKENEY, C.P. (compilers) *Sedimentology of Arctic Fjords Experiment: HU82–031 Data Report, Vol. 1*. Canadian Data Report of Hydrography and Ocean Sciences, **12**, 8–1 to 8–94.

CROSS, W.E. 1982. Under-ice biota at the Pond Inlet ice edge and in adjacent fast ice areas during spring. *Arctic*, **35**, 13–27.

DALE, J.E., AITKEN, A.E., GILBERT, R. & RISK, M.J. 1989. Macrofauna of Canadian Arctic fjords. *Marine Geology*, **85**, 331–358.

DUNBAR, R.B., LEVENTER, A.R. & STOCKTON W.L. 1989. Biogenic sedimentation in McMurdo Sound, Antarctica. *Marine Geology*, **85**, 155–179.

ELVERHØI, A., PFIRMAN, S.L., SOLHEIM, A. & LARSSEN, B.B. 1989. Glaciomarine sedimentation in epicontinental seas exemplified by the northern Barents Sea. *Marine Geology*, **85**, 225–250.

FARROW, G.E., ATKINSON, J. & MOORE, G. 1984. SAFE 1983 macrobenthos and bioturbation. *In*: SYVITSKI, J.P.M. (compiler) *Sedimentology of Arctic Fjords Experiment: HU83–028 Data Report, Vol. 2*. Canadian Data Report of Hydrography and Ocean Sciences, **28**, Chapter 6.

——, SYVITSKI, J.P.M. & TUNNICLIFFE, V. 1983. Suspended particulate loading on the macrobenthos in a highly turbid fjord; Knight Inlet, British Columbia. *Canadian Journal of Fisheries and Aquatic Sciences*, **40**, 100–116.

FITZGERALD, R.A. 1984. Geochemical data for marine sediments from eastern Baffin Island including Cambridge Fiord, Itirbilung Fiord and McBeth Fiord. *In*: SYVITSKI, J.P.M. (compiler) *Sedimentology of Arctic Fjords Experiment: HU82–028 Data Report, Vol. 2*. Canadian Data Report of Hydrography and Ocean Sciences, **28**,

Chapter 10.

—— & LeBlanc, K.W.G. 1983. Geochemical properties of Baffin Island marine sediments (Lehigh cores). *In*: Syvitski, J.P.M. & Blakeney, C.P. (compilers) *Sedimentology of Arctic Fjords Experiment: HU82–031 Data Report, Vol. 1*. Canadian Data Report of Hydrography and Ocean Sciences, **12**, 10–1 to 10–21.

LeBlanc, K.W.G., Syvitski, J.P.M. & Maillet, L. 1988. Examination of the suspended particulate matter within Arctic fjords. *Geological Survey of Canada Open File Report* **1733**.

Locke, C.W. & Locke, W.W. III 1977. Little Ice Age snow cover extent and paleoglaciation thresholds: north-central Baffin Island, N.W.T., Canada. *Arctic and Alpine Research*, **9**, 291–300.

Hein, F.J. & Syvitski, J.P.M. 1989. Seafloor gouges and pits in deep fjords, Baffin Island: Possible mammalian feeding traces. *Geo-Marine Letters*, **9**, 91–94.

Horner, R.A. 1976. Sea ice organisms. *Oceanographic Marine Biology Annual Review*, **14**, 167–182.

Magara, K. 1986. *Geological Models of Petroleum Entrapment*. Elsevier Applied Science Publishers, New York. Chapter 2: Organic carbon concentrations in source rocks, 25–48.

Macko, S.A., Pulchan, K. & Ivany, D.E. 1987. Organic geochemistry of Baffin Island Fjords. *In*: Syvitski, J.P.M. & Praeg, D.B. (compilers) *Sedimentology of Arctic Fjords Experiment: Data Report, Volume 3*. Canadian Data Report of Hydrography and Ocean Sciences, **54**, 13–1 to 13–34.

Miller, G.H. 1973. Late Quaternary glacial and climatic history of northern Cumberland Peninsula, Baffin Island, N.W.T., Canada. *Quaternary Research*, **3**, 561–583.

Mothersill, J.S. & Tabrez, A.R. 1987. Sedimentologic data from 82–031 S.A.F.E. cores MC-1, MC-7, IT-1 and TI-1A. *In*: Syvitski, J.P.M. & Praeg, D.B. (compilers) *Sedimentology of Arctic Fjords Experiment: Data Report, Volume 3*. Canadian Data Report of Hydrography and Ocean Sciences, **54**, 11–1 to 11–8.

Nielsen, T.K. & Ottesen-Hansen, N.-E. 1980. Mixing and exchange processes in a small Greenland sill fjord. *In*: Freeland, H.J., Farmer, D.M. & Levings, C.D. (eds.) *Fjord Oceanography*, Plenum Press, New York, 219–225.

Petersen, G.H. 1978. Life cycles and population dynamics of marine benthic bivalves from the Disko Bugt area of West Greenland. *Ophelia*, **17**, 95–120.

Pocklington, R. & Leonard, J.D. 1979. Terrigenous organic matter in sediments of the St. Lawrence Estuary and the Saguenay Fjord. *Journal of the Fisheries Research Board of Canada*, **36**, 1250–1255.

Schafer, C.T., Clattenburg, D., Cole, F.E., Le Blanc, W. & Syvitski, J.P.M. 1984. SAFE 1983 Hudson bottom grab samples. *In*: Syvitski, J.P.M. (compilers) *Sedimentology of Arctic Fjords Experiment: HU83–028 Data Report, Vol. 2*. Canadian Data Report of Hydrography and Ocean Sciences, **28**, Chapter 7.

Short, S.K., Mode, W.N. & Andrews, J.T. 1985. Palynology of nearshore marine and terrestrial sediments, east coast, Baffin Island. *In*: Vilks, G. & Syvitski, J.P.M. (eds.) *Arctic Land–Sea Interactions*. Geological Survey of Canada Open File Report **1223**, 205–206.

——, Andrews, J.T. & Mode, W.N. 1989. Modern and late Quaternary pollen spectra of fiord sediments, eastern Baffin Island, Artic Canada. *Marine Micropaleontology* (in press).

Syvitski, J.P.M. 1984. Sedimentology of Arctic Fjords Experiment: HU83–028 Data Report, Vol. 2. *Canadian Data Report of Hydrography and Ocean Sciences*, **28**.

—— 1985. Subaqueous slope failures within seismically active fjords. *In*: Vilks, G. & Syvitski, J.P.M. (eds.) *Arctic Land–Sea Interactions*. Geological Survey of Canada Open File Report **1223**, 60–63.

—— 1987. Submersible observations and other analytical results from the third SAFE cruise. *In*: Syvitski, J.P.M. & Praeg, D.B. (compilers) *Sedimentology of Arctic Fjords Experiment: Data Report, Volume 3*. Canadian Date Report of Hydrography and Ocean Sciences, **54**, 7–1 to 7–10.

—— 1989. On the deposition of sediment within glacier-influenced fjords: oceanographic controls. *Marine Geology*, **85**, 301–329.

—— & Blakeney, C.P. 1983. Sedimentology of Arctic Fjords Experiment: HU82–031 Data Report, Vol. 1. *Canadian Data Report of Hydrography and Ocean Sciences*, 12, 935 pp.

—— & Schafer, C.T. 1985. Sedimentology of Arctic Fjords Experiment (SAFE): 1. Project Introduction. *Arctic*, **38**, 264–270.

—— & Praeg, D. 1987. Sedimentology of Arctic Fjords Experiment: Data Report, Volume 3. *Canadian Data Report of Hydrography and Ocean Sciences*, **54**.

—— & Farrow, G.E. 1988. Fjord sedimentation as an analogue for small hydrocarbon-bearing fan deltas. *In*: Whateley, M.K.G. & Pickering, K.T. (eds) *Deltas: Sites and Traps for Fossil Fuels*. Geological Society Special Publication, **41**, 21–43.

——, Hay, A.E., Schafer, C.T. & Asprey, K.W. 1984a. SAFE 1983 bayhead prodelta investigations. *In*: Syvitski, J.P.M. (compilers) *Sedimentology of Arctic Fjords Experiment: HU83–028 Data Report, Vol. 2*. Canadian Data Report of Hydrography and Ocean Sciences **28**, Chapter 17.

——, ——, Taylor, R., Gilbert, R. & Emery-Moore, M. 1984b. SAFE 1983 delta survey report. *In*: Syvitski, J.P.M. (compiler) *Sedimentology of Arctic Fjords Experiment: HU83–028 Data Report, Vol. 2*. Canadian Data Report of Hydrography and Ocean Sciences, **28**, Chapter 18.

——, Asprey, K.W., Clattenburg, D.A., Hodge, G.D. 1985a. The prodelta environment of a fjord: suspended particulate dynamics. *Sedimentology*, **32**, 83–107.

——, ——, HEIN, F.J., HODGE, G.D. & GILBERT, R. 1985b. Sedimentology of Arctic fjords experiment: PA-85–062 expedition report. *Geological Survey of Canada Open File Report,* **1234**.

——, BURRELL, D.C. & SKEI, J.M. 1987. *Fjords: Processess and Products*. Springer, New York.

——, FARROW, G.E., ATKINSON, R.J.A., MOORE, P.G. & ANDREWS, J.T. 1989. Baffin Island fjord macrobenthos: bottom communities and environmental significance. *Arctic*, **42**, 232–247

TAN, F.C. & VILKS, G. 1987. Organic carbon isotope ratios and paleoenvironmental implications for Holocene sediments in Lake Melville, southeastern Labrador. *Canadian Journal of Earth Sciences*, **24**, 1994–2003.

TRITES, R.W. 1985. Oceanographic reconnaissance of selected Baffin Island fjords. *In*: VILKS, G. & SYVITSKI, J.P.M. (eds.) *Arctic Land–Sea Interactions*. Geological Survey of Canada Open File Report **1223**, 54–57.

VILKS, G., RASHID, M.A. & DELGADO, C. 1979. Interpretation of paleoceonographic conditions on the Peru Shelf using paleoecology and geochemistry. *In: Proceedings of the ICANE Project Workshop*, Lima, Peru, November 1979, 269–273.

WALSH, J.J., ROWE, G.T., IVERSON, R.L. & MCROY, C.P. 1981. Biological export of shelf carbon is a neglected sink of the global CO_2 cycles. *Nature*, **291**, 196–201.

WINTERS, G.V., SYVITSKI, J.P.M. & MAILLET, L. 1985. Distribution and dynamics of suspended particulate matter in Baffin Island fjords. *In*: VILKS, G. & SYVITSKI, J.P.M. (eds.) *Arctic Land–Sea Interactions*. Geological Survey of Canada Open File Report, **1223**, 73–77.

Carbonate minerals in glacial sediments: geochemical clues to palaeoenvironment

IAN J. FAIRCHILD[1] & BARUCH SPIRO[2]

[1] *School of Earth Sciences, University of Birmingham, PO Box 363, Birmingham B15 2TT, UK*

[2] *NERC Isotope Geoscience Laboratory, British Geological Survey, Keyworth, Nottingham NG12 5GG, UK*

Abstract: Various subglacial, proglacial, periglacial and glacimarine parageneses of carbonates are known, but have not been systematically compared. Precipitation-triggering mechanisms include subglacial regelation, lacustrine photosynthesis and evaporation, freezing in permafrost, skeletal biomineralization, and bacterial oxidation of organic matter in organic-rich sediments. New data from late Proterozoic carbonates suggest another, potentially important mechanism: recrystallization of glacially transported (and hence chemically unstable) carbonate.

Unlike the bulk of warm-water carbonate sediments, most glacial carbonates have been precipitated as stable phases which therefore preserve a good memory of environmental parameters. Carbon stable isotopic data have been used to constrain precipitation mechanisms whilst oxygen isotopes together with minor element chemistry can characterize salinity and distinguish the effects of evaporative from freezing processes. Elevated Fe and Mn contents point to anoxic precipitational conditions. Petrographic data on crystal size, shape and internal zonation are vital to avoid misinterpretations of diagenetically-altered material.

Carbonate geochemistry has proved valuable in interpretation of Quaternary and older palaeoenvironments, and could even yield information to aid Proterozoic palaeogeographic reconstructions.

Whole-rock geochemical studies of glacial sediments aimed at interpreting palaeoenvironment have had mixed success (Eyles *et al.* 1985). This is not surprising since there is a philosophical difficulty in interpreting analyses of sediments with several genetically distinct components (Fairchild *et al.* 1988). Several investigators have explored the possibilities of palaeoenvironmental interpretation from the geochemistry of carbonates in glacial settings, but we aim to demonstrate that the potential may be very much greater than had been supposed. There are many geochemical data on some deposits, but few have been subjected to the combined petrographic–geochemical approach that is now standard for carbonates, and which is most likely to yield meaningful palaeoenvironmental results. Late Quaternary carbonates forming as subglacial precipitates, concretions in glacimarine and glaciolacustrine sediments and precipitates in saline lakes have been intensively studied in a few study areas by isotope geochemists (e.g. Hendy *et al.* 1979, Hillaire-Marcel *et al.* 1979, Aharon 1988). By a combination of radiometric dating and stable isotopic analysis they have elucidated the timing and nature of former glacial events with some elegance and precision. In the marine realm, the nature and distribution of Quaternary skeletal carbonate has been studied from a facies viewpoint and analogies have been made with the widespread carbonate facies associated with glacial deposits in upper Proterozoic sections (Bjørlykke *et al.* 1978, Domack 1988).

Conversely, studies on the Proterozoic rocks indicate that these carbonates are polygenetic, and implicate, in part, processes which have not yet been looked for in Quaternary environments (Fairchild & Hambrey 1984, Fairchild & Spiro 1987, Fairchild *et al.* 1989). The crux of the late Proterozoic carbonate–tillite association is the glacial erosion and transport of underlying carbonate platform sediments. Dissolved carbonate in glacial lakes was reprecipitated in varves, stromatolites and evaporitic facies. A new discovery is that the mud fraction of the glacially-transported carbonate of one Proterozoic Formation is geochemically distinct from coarser detritus. This fine carbonate seems to reflect the chemistry of the depositional waters, and hence was recrystallized in early diagenesis. This is inferred to be because glacial transport rendered it unstable. Thus palaeoenvironmental information may be obtainable from a very

From DOWDESWELL, J. A. & SCOURSE, J. D. (eds), 1990, *Glacimarine Environments: Processes and Sediments*. Geological Society Special Publication No 53, pp 201–216.

common form of carbonate in glacial rocks: the hitherto neglected matrix of clastic lithologies.

Interpretation of carbonate mineral geochemistry

This discussion primarily concerns stable isotopic and elemental chemistry, although there are implications for other aspects, e.g. ^{14}C dating. The chemistry of any precipitated carbonate will depend on the solution composition. Equilibrium compositions are well known for several parameters, but deviations related to kinetics or biological fractionation may be important. Many ancient carbonates are diagenetically-altered: their chemistry will be more difficult to interpret. It is therefore vital to use textural criteria derived from conventional optical and cathodoluminescence or electron microscopy to distinguish diagenetically modified or entirely secondary phases (e.g. Sandberg 1985, Swirydczuk 1988).

The ideal analytical technique would involve isotopic and elemental analyses on the same homogeneous portion of sample minerals. Microscopically inhomogenous rocks would require a combination of microanalytical techniques such as laser extraction for isotopic analysis and ion probe for cations: methods not yet in routine use. A suitable macroscopic technique for components adequately homogeneous on the mm scale is a combination of conventional stable isotopic analysis and ICP cation analysis: this requires total sample weights in the order of 2–20 mg (Fairchild *et al.* 1988). Selective dissolution of the carbonate phase is desirable in impure carbonates.

In this paper we report isotopic compositions of carbonates as δ values on the PDB scale and waters on the SMOW scale. At equilibrium at 0°C, calcite would have δ values about 4‰ greater than water expressed in this way and dolomite very roughly 7‰ greater. Meteoric waters have a lower $^{18}O/^{16}O$ ratio than seawater because of evaporation and condensation effects. These effects are more pronounced in colder settings and at high altitudes and high latitudes; of these parameters temperature seems to be the dominant control (Anderson & Arthur 1983, Gregory *et al.* 1989). The wide spread of $\delta^{18}O$ values of subglacial carbonates shown on Fig. 1 primarily reflects temperature (and latitudinal) effects on meltwater composition. Apart from mixing of waters which yields simple averages of isotopic compositions, the most important modifying processes are evaporation and freezing. Fortunately, these two mechanisms of increasing salinity and carbonate supersaturation have opposite effects on isotopic composition. Freezing leads to a slight depletion in ^{18}O in remnant water (Souchez & Jouzel 1984), whereas evaporation leads to enrichment, most pronounced in low humidity conditions (Matsubaya *et al.* 1979).

Carbon isotopic composition depends on the source of carbon and the nature of the precipitating mechanism. Marine precipitates today have $\delta^{13}C$ values typically of 0 to $+4‰_{PDB}$. This parameter shows variation with time in the geological record of marine carbonates. The dissolved carboxy species of freshwater bodies tend to be isotopically lighter because of the derivation of carbon also by oxidation of isotopically light organic matter and by dissolution of atmospheric CO_2. Precipitates *equilibrated with* gaseous CO_2 at low temperatures could be heavier, up to +6‰ at 0°C, although up to 2‰ lower if precipitation was rapid. Precipitation induced by photosynthesis can lead to ^{13}C-enrichment because of the local selective removal of light carbon into organic matter. In diagenetic porewaters, extreme variations in $\delta^{13}C$ can arise as a result of bacterial decay reactions; most products are light (down to −55‰ for oxidized methane), except CO_2 accompanying methanogenesis which is heavy (+10 to +15‰).

Fig. 1. Summary of stable carbon and oxygen isotopic data on glacial carbonates. Elongation and vertical displacement of trends is explicable by the arrowed processes: see text for explanation. Note Quaternary samples are surrounded by solid lines, Permian ones by broken lines and Proterozoic ones by chain lines. Sample suites: 1, subglacial, Antarctica, Aharon (1988); 2, subglacial, Quebec (Hillaire-Marcel *et al.* 1979); 3, subglacial, Alps (Souchez & Lemmens 1985); 4a glaciolacustrine concretions, 4b glacimarine concretions (Lamothe *et al.* 1983); 5, till concretions, Northern Ireland, (unpublished); 6, upland evaporative precipitates, Alps and Israel (Magaritz 1973); 7, lacustrine carbonates from Taylor Valley, Antarctica (Hendy *et al.* 1979); 8a 20000 years old, 8b 10000 years old and 8c Recent lacustrine carbonates, Lake Fryxell, Taylor Valley, Antarctica (Lawrence & Hendy 1989); 9a modern cold-water skeletal, 9b Permian, Tasmania, skeletal, 9c cements of skeletal limestones, Tasmania (Rao & Green 1982; Rao 1988); 10 Permian, Antarctica authigenic lacustrine (Lord *et al.* 1988) and micrite and coarsely-crystalline cements in clastic glacigenic sediments (Schmidt & Friedman 1974); 11a glaciolacustrine precipitates (see Figs 2 & 3) 11b recrystallized detrital dolomite (Figs 3 & 4), from late Proterozoic of NE Spitsbergen (Fairchild *et al.* 1989).

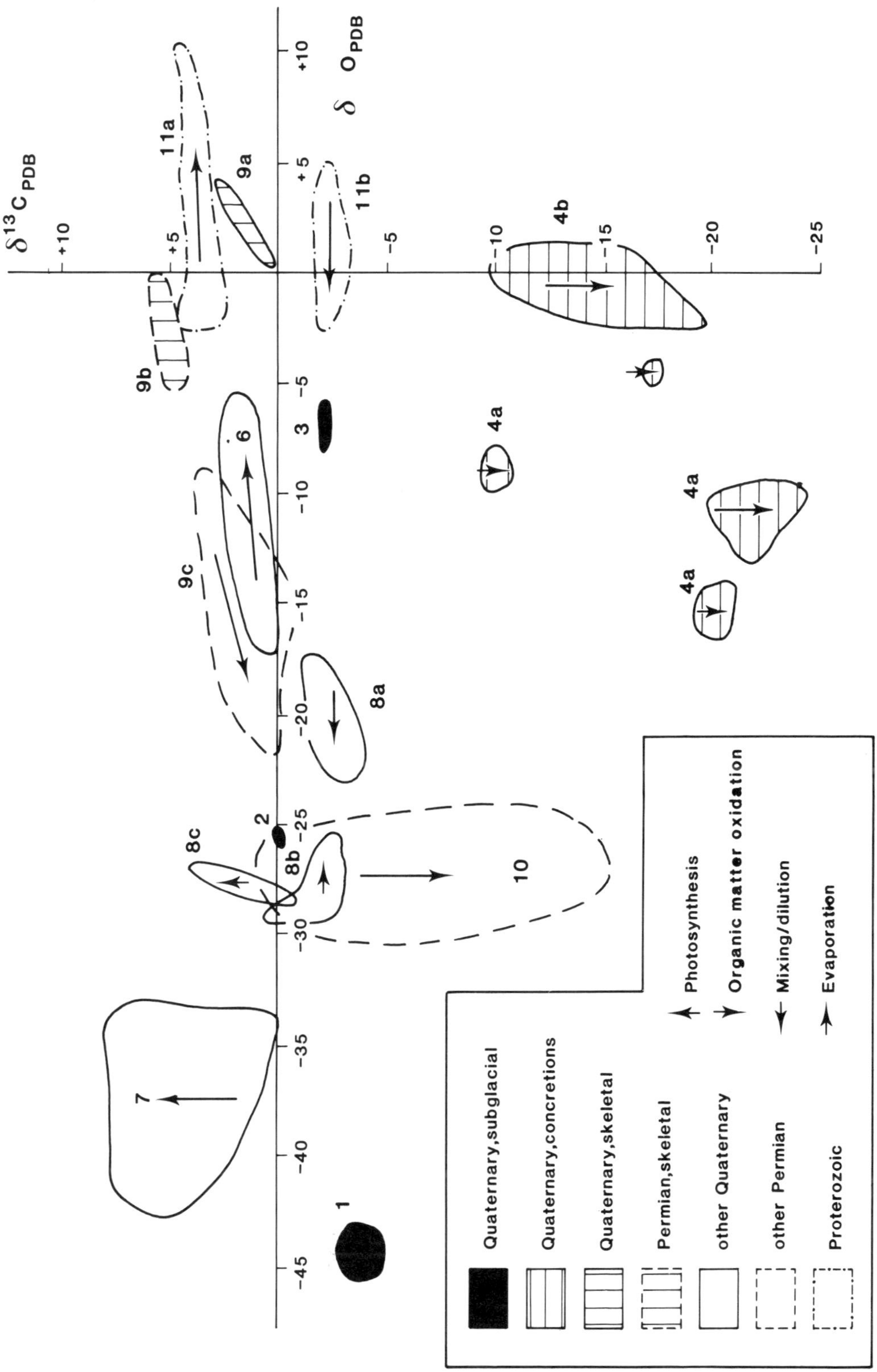

$\delta^{13}C_{PDB}$
$\delta\ O_{PDB}$
+10
+5
-5
-10
-15
-20
-25
-30
-35
-40
-45
1
2
3
4a
4b
6
7
8a
8b
8c
9a
9b
9c
10
11a
11b
Quaternary,subglacial
Quaternary,concretions
Quaternary,skeletal
Permian,skeletal
other Quaternary
other Permian
Proterozoic
Photosynthesis
Organic matter oxidation
Mixing/dilution
Evaporation

Absolute abundances of minor elements in carbonates relate to (1) their abundance in the solution relative to the major element for which they substitute in the mineral lattice, (2) a constant (the partition coefficient) related to their ease of incorporation in the lattice, (3) temperature and (4) kinetic and biological factors. In some cases partition coefficients are not known (e.g. Fe and Mn in dolomite, Veizer 1983), although a general contrast between elements scavenged from solution into carbonates (Fe, Mn in calcite and dolomite, Sr in aragonite) and those relatively depleted in carbonates compared with the carrier element in solution (Mg in calcite, Sr in calcite and dolomite) can be discerned. Na appears to substitute interstitially, and therefore increases in abundance where there are many lattice defects; it also resides in fluid inclusions. Because of the complexity of the chemical systems involved it is often difficult to use absolute values to interpret depositional water chemistry. However comparative study of sample suites may reveal chemical trends which can be interpreted in process terms. For example, a covariation of increasing Sr, Na and $\delta^{18}O$ up a stratigraphic section can be interpreted as reflecting salinity changes resulting from evaporation, whereas there should be a weak negative correlation of $\delta^{18}O$ with Sr and Na if freezing were involved. Fe and Mn will be enriched in precipitates from anoxic solutions, and their relationships with each other and $\delta^{13}C$ can sometimes be interpreted in respect of the timing of carbonate precipitation in relation to the sequence of bacterially mediated redox reactions in organic-rich sediments (e.g. Burns & Baker 1987).

Most carbonate-saturated meteoric waters in glacial settings will favour calcite precipitation. In theory, aragonite could occur where rapid precipitation from highly supersaturated solutions occurs or where calcite precipitation is inhibited by very high solution Mg/Ca ratios in saline lakes. Any calcite formed from such solutions would be Mg-rich, but normally such high-magnesium calcite would be absent because of the decreased incorporation of Mg in calcite at low temperatures (Burton & Walter 1987). Dolomite could form instead of calcite where Mg/Ca in solution was somewhat greater than unity, but this has only been documented from Proterozoic glacigenic facies.

In ancient examples, calcites replacing the metastable polymorphs aragonite or Mg-calcite may preserve some aspects of the original mineral chemistry (e.g. $\delta^{13}C$) relatively unaltered, but there would probably not be enough information preserved to constrain other palaeoenvironmental parameters. Originally low-Mg calcite is not prone to chemical alteration, except by cementation of micropores, or strain-induced recrystallization. Although some finely crystalline dolomites are prone to diagenetic modification of Mg/Ca ratio and $\delta^{18}O$ values (Land 1980), this was not found to be the case for the Proterozoic glacigenic dolomites described by Fairchild *et al.* (1989). Overall it will be argued that a high proportion of glacigenic carbonates are originally stable or stabilized in contact with depositional waters and hence likely to preserve their original chemical (including isotopic) characteristics.

Modes of occurrence

A summary of isotopic data in given in Fig. 1. A composite diagram summarizing the distribution of carbonate precipitates is given at the end of the paper (Fig. 5).

Detrital carbonate

Macroscopic carbonate clasts are a feature of tills in many areas. Variation in abundance of carbonate has been utilized for stratigraphic purposes (e.g. Murray 1953, Baermann *et al.* 1983), or to demonstrate the degree of dispersal of detritus by glaciers (e.g. Dredge 1988). It has also been studied in relation to engineering properties of tills (Baermann & Hübner 1984). No detailed studies on abrasion seem to have been made, but it seems likely that the relatively soft and cleavable nature of calcites and dolomites will lead to a comminuted matrix somewhat finer than for silicates. Murray (1953) found significant dolomite in the $< 2\mu m$ fraction of Wisconsin tills. In some Proterozoic tillites with detrital carbonate, there is abundant carbonate 'rock flour' (Fairchild 1983), whereas in other terrestrial ones, the fine fraction has apparently been dissolved (Deynoux 1985). There is a pressing need for studies on the physical and chemical properties of glacially transported carbonate, particularly given the evidence (see later) for its reactivity.

Subglacial precipitates

In this zone liquid water exists as films within basal ice, or at the ice–bedrock interface, as well as in discrete drainage channels. Cations in this fluid may be derived from chemical weathering of rock matrix or, in coastal regions, from the marine aerosol component of the precipitation that was the source of the glacier ice. Carbonate alkalinity is generated by reaction of atmospheric CO_2 with silicates, but will be particularly high where a carbonate rock substrate is available, which will also buffer its $\delta^{13}C$ content (Souchez & Lemmens 1985). Precipitation of carbonate occurs in response to regelation processes whereby pressure melting occurs on the upflow side of protuberances and refreezing on their lee side. As has been confirmed experimentally (Hallet 1976), dissolved chemicals are concentrated in the fluid phase as freezing proceeds leading to supersaturation and the precipitation of the observed fibrous or micritic banded $CaCO_3$ as thin crusts on clasts or bed-rock. The mineralogy is usually calcitic, although aragonitic examples are also known (Aharon 1988 and references therein).

Temporal variations in pressure are also thought to give rise to the carbonate-coating of pebbles and shear surfaces within debris-rich ice (Lavrushin & Rengarten 1974). There are well-documented examples of such phenomena in Late Proterozoic tillites of Mauritania (Deynoux 1985). The oxygen isotopic composition of these carbonates is largely dependent on that of local glacier ice (Fig. 1, fields 1 to 3). Subglacial carbonates may form from water somewhat depleted in ^{18}O compared with bulk ice composition as a result of fractionation during freezing (there is no fractionation on melting), but since a steady-state system is likely to develop, the effect is an isotopic shift of about 3‰ (Lamothe *et al.* 1979, Aharon 1988). Even allowing for this, discrepancies between the isotopic composition of modern glacier ice and that inferred from the carbonates have been found. In the Antarctic examples of Aharon (1988), the minerals were formed during a previous cold phase when atmospheric precipitation was isotopically-lighter than today. In the case of carbonates formed by the Swiss Tsanfleuron Glacier (Souchez & Lemmens 1985), they formed from glacier-marginal basal ice, isotopically heavier than the bulk glacier.

Although these complications would reduce the precision of interpretations in older deposits, such data on Pre-Cainozoic examples could provide interesting constraints on the glacial setting. There is also much scope for study of cation chemistry of the precipitates.

Periglacial and surficial environments

Freezing of interstitial water in sediments at an advancing permafrost front could yield carbonate cement. Some examples with isotopic compositions consistent with this proposition were found in one of the Dry Valley boreholes in Antarctica (DVDP11, Stuiver *et al.* 1981), although Hendy *et al.* (1990) disputed this interpretation. Texturally carbonate of such an origin could be confused with interglacial ground-water precipitates, although vadose examples of the latter may show a characteristic non-uniform cement distribution (Aber 1979).

A perched water table in the soil zone above permafrost in summer can allow the formation of carbonate (calcrete) crusts around clasts (Swett 1974). Dissolution of carbonate clasts at the top of the permafrost and reprecipitation by evaporation in the warm surface layer is implicated. Texturally there are similarities with both stromatolites and laminated subglacial crusts, but the stratigraphic context should make confusion less likely. Morikawa & Ossaka (1974) and Nishiyama & Kurasawa (1975) reported that various efflorescences and mineral aggregates (dominantly halite and thenardite, but also including widespread calcite and some aragonite and monohydrocalcite) occur under ventifacts and in soils in the Dry Valleys region.

Magaritz (1973) reported on the isotopic composition of veinlets and irregular masses of calcite in karstic cavities of carbonate rocks of two alpine glacial localities, and found isotope characteristics (Fig. 1, field 6) indicative of evaporative precipitation.

There is clearly much scope for petrographic and geochemical studies of permafrost-related and surficial carbonates in comparison with meteoric, lacustrine and subglacial types.

Lacustrine precipitates

Glacial lakes are normally dilute and repositories of purely clastic sediments. In certain climatic-physiographic regimes however (particularly arid closed drainage basins) saline lakes precipitating carbonates and evaporite minerals occur. Although widespread in both polar regions, the Antarctic examples, especially those in the Dry Valleys of the McMurdo Sound area of South Victoria Land, have become the classic examples and are reviewed by Matsubaya *et al.* (1979), Burton (1981) and Hendy (1990) among others.

In this region, individual lakes have a complex Quaternary history, which has been elucidated by a combination of radiometric dating on carbonates and organic carbon, and oxygen isotope measurements on carbonates and water of hydration in gypsum (Hendy 1990). During periods of high sea level, as today, lakes were relatively small and associated with either local alpine glaciers or outlet glaciers from the East Antarctic ice sheet. Conversely, when sea level was low, larger lakes developed associated with an ice sheet in the Ross Sea area. The oxygen isotope composition of waters and hence of carbonate precipitates exhibits wide variations primarily because of differing proportions of water supplied from the continental ice sheet ($\delta^{18}O = -50‰_{SMOW}$), local alpine glaciers (*c.* −30) and undiluted seawater (0), but also because of evaporation, which in extreme cases can lead to enrichments of as much as 20‰. Dissolved ions for carbonate formation were derived variably from rock weathering and sea-water.

Study of present-day lakes coupled with data from cores and outcrops implicates three mechanisms of carbonate precipitation: benthic microbial photosynthesis producing stromatolites, algal photosynthesis in the water column, and evaporation. Freezing out of $CaCO_3$ also seems likely to have occurred to explain the composition of some brine bodies (Hendy 1990), but particular carbonate beds formed by this process have apparently yet to be identified.

Microbial mats are well developed in some lakes (Parker *et al.* 1981) and are mineralized by calcite to yield rigid stromatolite columns in shallow areas of Lake Fryxell in Taylor Valley (Wharton *et al.* 1982). No fabric or geochemical studies of the carbonate appear to have been done.

Platy carbonate beds thought to be related to algal photosynthesis are widespread (Hendy 1990). Studies of water chemistry in Lake Fryxell clearly demonstrate the nucleation of calcite crystals in the water column at around 7–9 m depth, a level characterized by ^{13}C enrichment in dissolved inorganic carbon as a consequence of intense photosynthesis and preferential sequestration of ^{12}C into organic matter (Lawrence & Hendy 1985, 1989). The low to medium-Mg calcite accumulates in the sediment, but is partly corroded by undersaturated bottom waters. In the 20 000-year old sediments under Lake Fryxell (Lawrence & Hendy 1989) there occur aragonitic carbonates thought to be of similar origin. Although varying in $^{13}C/^{12}C$ ratio in the different occurrences (Fig. 1, fields 7, 8a, 8c), $\delta^{13}C$ values are considerably enriched compared with present-day input streams.

Evaporitic carbonates associated with sulphate or chloride evaporites are also known in the young (< 1000 year) sequences of Lake Bonney (Hendy *et al.* 1979) and Lake Vanda (Hendy 1990), and a distinct progressive evaporative episode (10 000 years ago) is recorded in Lake Fryxell by varved laminae of sands and aragonite with increasingly heavy $\delta^{18}O$ values up-section (Lawrence & Hendy 1989; Fig. 1, field 8b).

Most of the carbonates appear to be diagenetically unaltered. In mixed aragonitic-calcitic deposits of the Wright Valley, SEM studies show calcite typically enclosing uncorroded aragonite (Hendy & Utting 1990, Dagel *et al.* 1990). Manganese incorporation in calcites in some Lake Fryxell core material suggested (Lawrence & Hendy 1989) that some aragonite to calcite neomorphism could have occurred, but primary calcite precipitation beneath an anoxic water body (many of the lakes are partly anoxic) is an alternative.

Although intensively investigated in certain respects, there are several aspects of carbonate geochemistry that remain little studied. For instance, the occurrence of aragonite rather than calcite in some instances can be related to solution Mg/Ca ratio (Lawrence & Hendy 1989), but the possible role of varying supersaturation requires investigation. Also little studied is the cation chemistry of individual mineral phases and relation to isotopic chemistry and possible diagenetic alteration, and the relationship of crystal fabrics to inferred modes of origin. A comparison with other modern lacustrine carbonates and ancient analogues would also be instructive. Ancient examples of such phenomena should be represented in the deposits of the Late Palaeozoic Gondwana glaciations and the widespread Late Proterozoic glacials.

Non-marine carbonate beds and lenses occur in Permian formations of the Transantarctic mountains; Sr and carbon isotope data point to an authigenic origin (Lord *et al.* 1988). Petrographic and elemental data on such occurrences would be valuable.

Walter & Bauld (1983) reviewed much of the literature on the Dry Valley regions and suggested Proterozoic analogues from Australia. However, the stromatolitic carbonates they described do not occur within definite glacigenic rocks, but immediately above them, or inferentially laterally adjacent to them. Geochemical data could help provide a test of their ideas.

New data (Fairchild *et al.* 1989) from the upper of two late Proterozoic glacial formations in NE Spitsbergen, the Wilsonbreen Formation (Hambrey 1982, Fairchild & Hambrey 1984), point to a range of carbonates formed in glaciolacustrine conditions from fresh to hypersaline in the central member W2. Some carbonate facies compare with the modern Dry Valley region, but others are apparently not represented there. Here, we emphasize the relationship of geochemistry to mineralogy and facies.

Calcitic facies include stromatolites and rhythmites. No geochemical data are available on the rhythmites at present, but they are probably varves similar to examples from the modern Dry Valleys. The stromatolites incorporate some detrital siliciclastic and dolomitic sediment, but are dominantly composed of manganoan calcite with well preserved textures including radiaxial-fibrous crusts. Low-Mg calcite appears to be the primary mineral phase. Cathodoluminescence reveals that brightly luminescing calcite is cut by veins and vug-fillings of calcite, probably of burial origin, and highly depleted in ^{18}O. Fairchild & Spiro (1987) noted the heaviness of the oxygen isotopes of these precipitates (Fig. 2) in relation to modern high latitude meltwater and hence assumed they

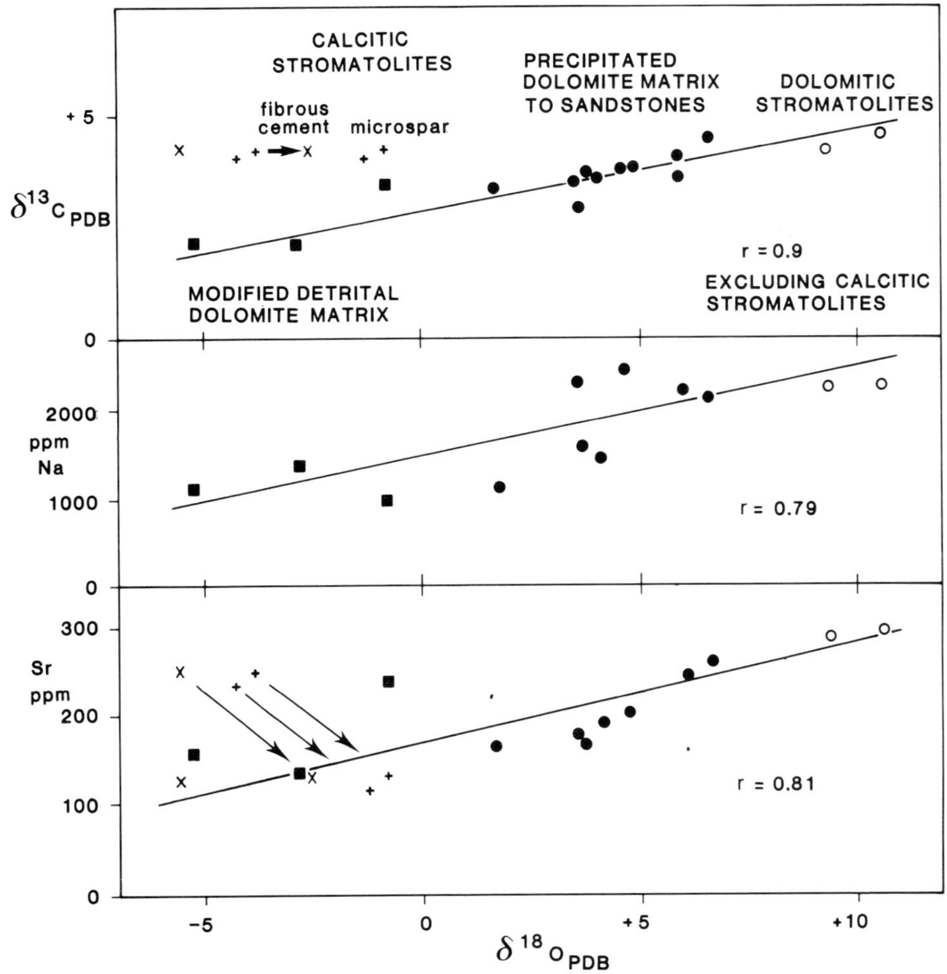

Fig. 2. Covariations of $\delta^{13}C$, Na and Sr with $\delta^{18}O$ in glaciolacustrine precipitates and associated originally detrital carbonates, member W2, Wilsonbreen Formation (late Proterozoic), NE Spitsbergen (combination of data of Fairchild & Spiro 1987 and Fairchild *et al.* 1989). The calcite data are plotted twice: firstly as analysed, and secondly corrected for comparison with the dolomites (the $\delta^{18}O$ values have been increased by three parts per thousand and the Sr values halved, since Sr substitutes for Ca, not Mg, Veizer 1983). The calcitic stromatolites are slightly enriched in ^{13}C in relation to the evaporative trend of the dolomitic samples. High Na values are now described from other lacustrine carbonates with arguably preserved original isotopic composition (Janaway & Parnell 1989).

formed in a seawater-meltwater mixture. A fuller data set leads to the conclusion that the glaciation was not high latitude (in agreement with Harland 1964) and so the glacial meltwaters were not greatly depleted in ^{18}O (perhaps no more negative than −12‰, Fairchild *et al.* 1989). Carbon isotopes are slightly heavy in relation to the trend of dolomitic samples (Fig. 2) which is consistent with local high rates of photosynthesis as in Quaternary African lacustrine stromatolites (Hillaire-Marcel & Casanova 1987) and the glaciolacustrine rhythmites described above, or to partial equilibration with atmospheric CO_2 of the lake waters.

Dolomitic facies are more restricted in distribution and include rhythmites, highly dolomitic sandstones in which displacive growth has occurred, and stromatolites. The sandstones with displacive dolomite growth exhibit generally brightly luminescing crystals with a constant zonation in a given specimen indicating preservation of the products of a single episode of carbonate authigenesis. The dolomitic facies display a distinct trend on a $\delta^{13}C-\delta^{18}O$ plot (Fig. 2), the stromatolites attaining $\delta^{18}O$ values which compare with the heaviest carbonate parageneses, ancient or modern. They and the matrix of diamictites share certain chemical features, particularly Mn-enrichment (Fig. 3) which distinguish them from detrital dolomite and point to formation under suboxic to anoxic conditions. Evidence of vanished evaporites in the presence of crystal pseudomorphs and collapse breccias is present in the lithologies with enriched ^{18}O. A covariance of $\delta^{18}O$ with Na and Sr is found, with Na values extremely high in absolute terms (Fig. 2). These observations support an interpretation of the trend as an evaporative one, the samples heaviest in oxygen being formed from solutions considerably enriched in ^{18}O compared with modern seawater. An environmental continuum is thus recognizable. At one extreme there are relatively fresh, possibly anoxic lakes precipitating calcite and near to or beneath which detrital dolomite in tills is recrystallizing. At the other extreme are saline, probably anoxic lakes with Mg/Ca elevated by precipitation of Ca minerals, in which dolomite formed. The preservation of arguably primary mineral chemistry points to the initial precipitation of stable species within the water column or within the sediment during early diagenesis. The formation of stable dolomite in contrast to the unstable aragonite of modern settings of inferred high salinity and Mg/Ca ratio can tentatively be related to the abundance of detrital dolomite nuclei in the depositional environments. A more detailed comparison with modern glaciolacustrine facies, studied using similar techniques, is now required.

Skeletal carbonates

Skeletal carbonates occur in both lacustrine and marine settings. Lacustrine examples are widely used for ^{14}C-dating purposes; oxygen isotope data can help distinguish proglacial from interglacial types.

In the marine realm, chemical analyses of material from foraminiferal oozes have of course played a major part in our present-day understanding of Quaternary events, but they are not our concern here. Shelf bioclastic sediments of characteristic cold-water organisms ('foramol' assemblage) are widespread at high-latitudes in areas of low terrigenous input, although most do not occur in true glacial conditions. Domack (1988) developed a glacimarine model based on the East Antarctica continental shelf. Here, seasonal melting of sea ice stimulates high plankton productivity leading to organic-rich laminites accumulating in basinal areas. Skeletal deposits occur both in offshore banks, and in the nearshore zone where terrigenous input is limited by the subfreezing base of nearby glaciers, and carbonate productivity may be stimulated by rain out of plankton. This model was applied by Domack (1988) to the Late Permian Tasmanian succession which accumulated in proximity to the major Gondwana ice sheet. Analogies in terms of both basinal and shallow skeletal facies can be found. Rao (1981a, b) demonstrated that the types of organisms, the petrography of early marine cements and the elemental geochemistry of the rocks all pointed to low-Mg calcite formation in contrast to low-latitude carbonates. Isotopic data were presented by Rao & Green (1982) and Rao (1988) for whole rocks, fossils and pore-filling cements (Fig. 1, fields 9b, 9c) which they used to deduce compositions of Permian marine water and meltwater. Their conclusions are in some doubt because of the indirect line of arguments used. The data for fossils show some spread in $\delta^{18}O$ which they chose to interpret as representing recrystallization rather than slight dilution of seawater by meltwater. Later cements trend towards the −24 to -30_{PDB} values found in calcites from continental situations at the time (Fig. 1, field 10; Schmidt & Friedman 1974, Lord *et al.* 1988). A combination of data from cathodoluminescence indicating whether recrystallization has occurred in fossils, together with more detailed sub-sampling of cement zones and combined isotope and elemental data on the same samples could resolve some of the uncertainties and provide more precise estimates of terrestrial and marine fluid chemistry. In turn this could allow a better interpretation of the palaeo-oceanographic conditions (e.g. palaeosalinities) and a test of Domack's (1988) model. There is also potential for isotopic studies on *Quaternary* high latitude shallow marine skeletal deposits.

No definite evidence has yet been found to support the analogy (Bjørlykke *et al.* 1978; Domack 1988) of modern polar skeletal carbonates with carbonates associated with Proterozoic glacials. Carey & Ahmad (1961) proposed carbonate precipitation from marine waters which had salinity increased by sea-ice formation and which were subsequently warmed. However, modern skeletal carbonates exist *despite* the marked undersaturation of glacimarine waters for carbonate. The only organisms capable of being implicated in carbonate formation in late Proterozoic times are the cyanobacteria, but they are never calcified in undersaturated waters today. It is conceivable that Proterozoic glacimarine waters could have been oversaturated: if so isotopically heavy carbonates would have formed because of the low temperatures. Isotopically heavy stromatolites have now been found (Fig. 2), but as described elsewhere in this paper, their associations are typically glaciolacustrine. Therefore, chemical data on apparently marine columnar stromatolites interbedded with tillites (e.g. Young & Gostin 1988) would be valuable.

Diagenetic carbonate concretions

A rare, but characteristic occurrence are calcareous concretions in rhythmites of glaciolacustrine or glacimarine origin. Examples in Quebec have been found to be highly depleted in ^{13}C (Fig. 1, fields 4a and 4b) indicating the derivation of carbonate carbon from organic matter and making the concretions highly suitable for ^{14}C-dating (Lamothe *et al.* 1983). Lacustrine examples are lighter in $\delta^{18}O$ than would be expected for modern meteoric water which points to their origin under glacial conditions. Marine examples are readily distinguished by their oxygen isotope chemistry (Fig. 1, field 4b).

Concretions related to organic matter diagenesis may also occur in till. Unpublished data on calcite-cemented matrix and pore-lining cements in septarian concretions in lignite-bearing till from Northern Ireland (Fig. 1, field 5, specimens of G.R. Coope) indicate an interglacial origin with carbon largely supplied from organic matter. Unusually high Mn and Fe concentrations (both up to 3 wt% carbonate) are also characteristic of the anoxic paragenesis. Other till concretions have been ascribed to subglacial pressure changes (Lavrushin & Golubev 1981); such an hypothesis is testable by chemical analysis.

A surprise outcome of drilling Neogene glacigenic sections in West McMurdo Sound (offshore of the Dry Valleys Region, Antarctica) has been the discovery of extensive carbonate cementation and concretionary carbonate formation in both terrestrial and marine facies (Hambrey *et al.* 1990; Barrett *et al.* 1990). Petrographic studies by Bridle & Robinson (1990) on materials from the CIROS-1 borehole indicate the occurrence of siderite, ankerite and calcite. Although concretionary forms do occur near the present-day sea floor, the carbonates analyzed so far are rather depleted in ^{18}O (Wada & Okada 1990). Combined with the evidence for former excess burial of 400–1100 m this suggests a burial origin for much of the carbonate. Nevertheless if the source and precipitational mechanisms of the carbonate were to be clarified, this would aid interpretation of Proterozoic glacial sediments.

Recrystallized detrital carbonate

In addition to the carbonate concretions just described, the possibility of a precipitated component of carbonate in tills has sometimes been suggested (e.g. Harland & Herod 1975), but remains little investigated.

Evidence for an authigenic till matrix has been found in the lower of the two late Proterozoic glacial units in NE Spitsbergen, member E2 (Fig. 1, field 11b, Figs 3 & 4). Sample locations and full tabulated data are given by Fairchild *et al.* (1989); we emphasize here the chemical aspects of the data interpretation.

The chemistry of the pre-glacial carbonate platform sediments in general and dolomite clasts of diamictites in particular was found to be distinct from the matrix of diamictites and the material composing rhythmites and other mud-grade clastic dolostones. The latter were found to be depleted in ^{13}C compared with actual and potential detritus (Knoll *et al.* 1986) and exhibiting a range of $\delta^{18}O$ values up to +4.7‰ (Fig. 3). This is considerably heavier than any non-glacigenic Precambrian carbonate yet known and consistent with a cool-marine origin (cf. Mullins *et al.* 1985). The matrix also exhibits Fe and Mn enrichment and a weak

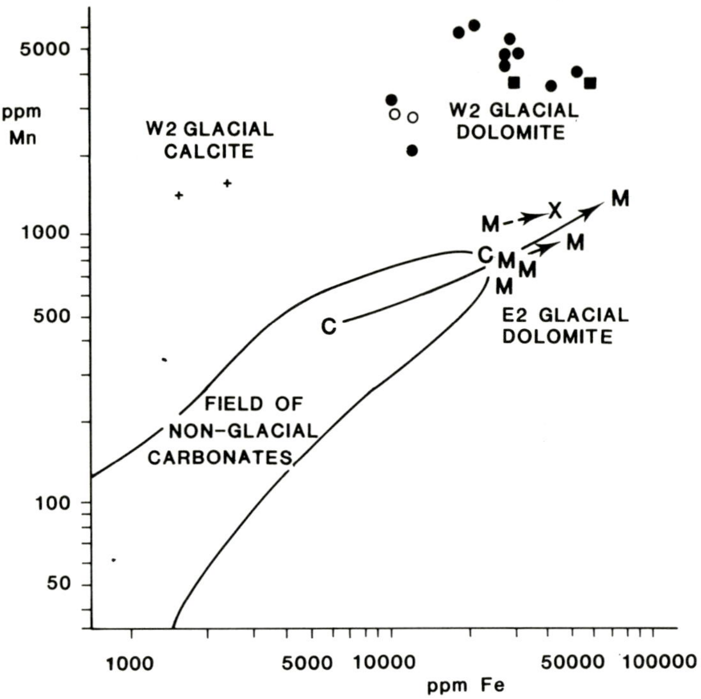

Fig. 3. Mn–Fe chemistry of some late Proterozoic carbonates (key on Figs 2 & 4). The field of non-glacial carbonates includes all data for carbonates below the lower glacial horizon, and between the two horizons in NE Spitsbergen (Fairchild & Spiro 1987) and unpublished analyses on clasts in diamictites from the equivalent of the upper horizon in E Greenland. The recrystallized clastic dolomites from the lower horizon (E2) are notable for their Fe and slight Mn enrichment. The glaciolacustrine facies of the upper horizon show a distinct Mn enrichment.

positive correlation of Sr and Na with $\delta^{18}O$ (Fig. 3).

In the absence of magmatic fluids there does not seem to be any mechanism by which heavy-oxygen dolomite could be produced by secondary processes. The high absolute values of Na are also consistent with a preservation of primary chemistry for the heavy oxygen samples. Given this, three hypotheses for the spread of oxygen isotope values can now be evaluated.

One hypothesis would be that the variation in sample chemistry reflects recrystallization during burial. Such a process would inevitably lead to ^{18}O-depletion of initially heavy carbonates. However the lighter samples all have the same fine crystal size with median crystal size of < 3 μm, found by Fairchild (1983) to be much finer than non-glacial dolostones and to indicate the formation of rock flour by glacial transport. A genuine burial recrystallization effect is shown by a sub-sample of diffuse slightly more coarsely crystalline areas in a fine-grained dolostone (point x on Figs 3 & 4). Preservation of primary chemistry across the range of oxygen isotope values is thus indicated.

A second hypothesis is that the oxygen isotope spread from positive to negative values reflects the progressive freezing of seawater. This is contradicted because there is a positive correlation of $\delta^{18}O$ with Na and Sr, not a negative one.

The third hypothesis is that the oxygen isotope

Fig. 4. Covariations of $\delta^{13}C$, Na and Sr with $\delta^{18}O$ in originally glacidetrital dolostones from the lower glacial unit in the later Proterozoic succession of NE Spitsbergen (Fairchild *et al.* 1989). The source rocks have distinctively high $\delta^{13}C$ and variable Na and Sr. The fine matrix recrystallized in the depositional environment (isotopically distinct from detritus) in variable salinities. Sodium and Sr incorporation was related to palaeosalinity as expressed by $\delta^{18}O$; Na values in more saline samples approach that of modern marine dolomite (Land & Hoops 1973); Sr values partly depend on local variations in supply from detritus.

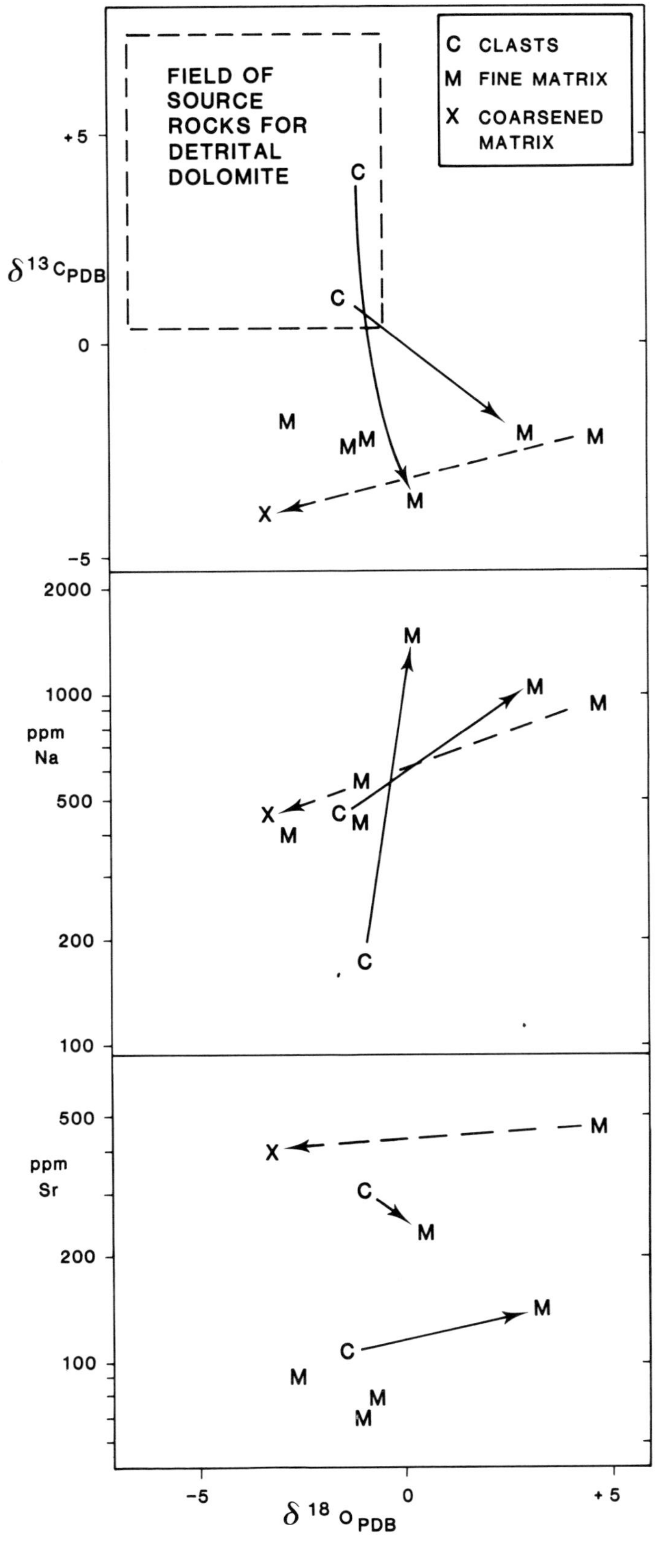

C CLASTS
M FINE MATRIX
X COARSENED MATRIX
FIELD OF SOURCE ROCKS FOR DETRITAL DOLOMITE
$\delta^{13}C_{PDB}$
+5
0
-5
2000
1000
ppm Na
500
200
100
500
ppm Sr
200
100
-5
0
+5
$\delta^{18}O_{PDB}$

spread reflects partial dilution of glacimarine waters by meltwater. This is consistent with the Na and Sr data.

The carbon isotopic data point to the incorporation of some organically-derived carbon. This reveals the influence of bacterially-mediated oxidation of organic material in these sediments, consistent also with the co-occurrence of pyrite and the Fe and Mn-enrichment of the carbonate matrix (derived by reduction of oxides in the sediment). Since the matrix has the macroscopic appearance of representing the fine-fraction of clastic carbonate in these detrital lithologies, it is logical to suppose that this is essentially a recrystallization process where the bulk of the carbon and major cations originated in the clastic dolomite. The driving force for recrystallization is unlikely to be provided by bacterial processes alone, otherwise all fine-grained marine carbonates would show bulk recrystallization after deposition. Also, matrix dolomite samples of similar fine crystal size in the upper glacial unit (Wilsonbreen Formation) in NE Spitsbergen show elevated Mn values (Fig. 3) suggestive of recrystallization, but do not show any carbon isotope deviations (Fig. 2). Bacterial activity must have been more limited here.

The unique feature, which is likely to drive recrystallization in these lithologies, is that the carbonate matrix has experienced the rigours of subglacial transport abrasion. This would inevitably give rise to particles with a high specific surface area and lattice defects. Such particles would be prone to extensive dissolution-reprecipitation by the mechanism known to chemists as Ostwald's ripening (Morse & Casey 1988).

The significance of this discovery is that such a process could automatically cause recrystallization of detrital carbonate in glacigenic rocks. The products of recrystallization would be expected to be of the same bulk mineralogy as the detrital precursor, but with minor element and carbon isotopic chemistry modified by, and oxygen isotope chemistry largely determined by, the porewater chemistry during early diagenesis. Thus a record of the depositional environment may be left which could be particularly valuable in distinguishing the depositional setting of coarse diamictons, whose origin is often controversial.

Investigations are clearly required in Quaternary sediments to determine the rates of such a process and its dependence on the physical state of the detrital material. The result could be a new technique for palaeo-environmental analysis available for both Quaternary and Proterozoic glacigenic successions.

Future work

We have shown that represented in glacial environments are diverse carbonate sediments produced by a variety of precipitation mechanisms (Fig. 5). The principles of carbonate geochemistry are sufficiently well understood that precipitation mechanisms can often be inferred from sets of chemical data which should include both elemental and isotopic data. It is recommended here that such data should be based on materials whose field relations are well documented and whose petrography is well described. In ancient examples in particular, this should include techniques such as cathodoluminescence to characterize the degree of chemical inhomogeneities and diagenetic alteration of the material being analysed.

Applying such methods will complement the excellent existing body of literature on Quaternary glacial carbonates which has been aimed primarily at establishing glacial chronology. An urgent objective is the investigation of the chemical stability of glacially-transported carbonate in Quaternary settings, and to apply the results to palaeoenvironmental analysis. Other needs are to clarify the relationship between precipitating process and texture, and to identify the extent of carbonate precipitation in response to bacterial decomposition of organic matter, and freezing.

Improved knowledge of Quaternary processes will aid the interpretation of Palaeozoic and Proterozoic glacial carbonates. In particular, development of geochemical criteria for distinguishing terrestrial from marine diamictites, and to derive isotopic compositions of meltwater from the former, would be a great bonus in palaeogeographic reconstructions and interpreting the nature and origin of Proterozoic ice ages.

Proterozoic samples were collected on expeditions funded by NERC grant GR3/4342 to expedition director Brian Harland (Cambridge University) and isotopic work was carried out at the NIGC. IJF's work on stromatolites is supported by NERC Grant GR3/6533. Isotopic data on Irish till concretions were provided by Roger Williams (University of Birmingham). Dr. Nick Walsh (London University) is thanked for access to ICP facilities. Mike Hambrey (Cambridge University) is thanked for his advice and encouragement. This is a contribution to IGCP Project 260: Earth's Glacial Record.

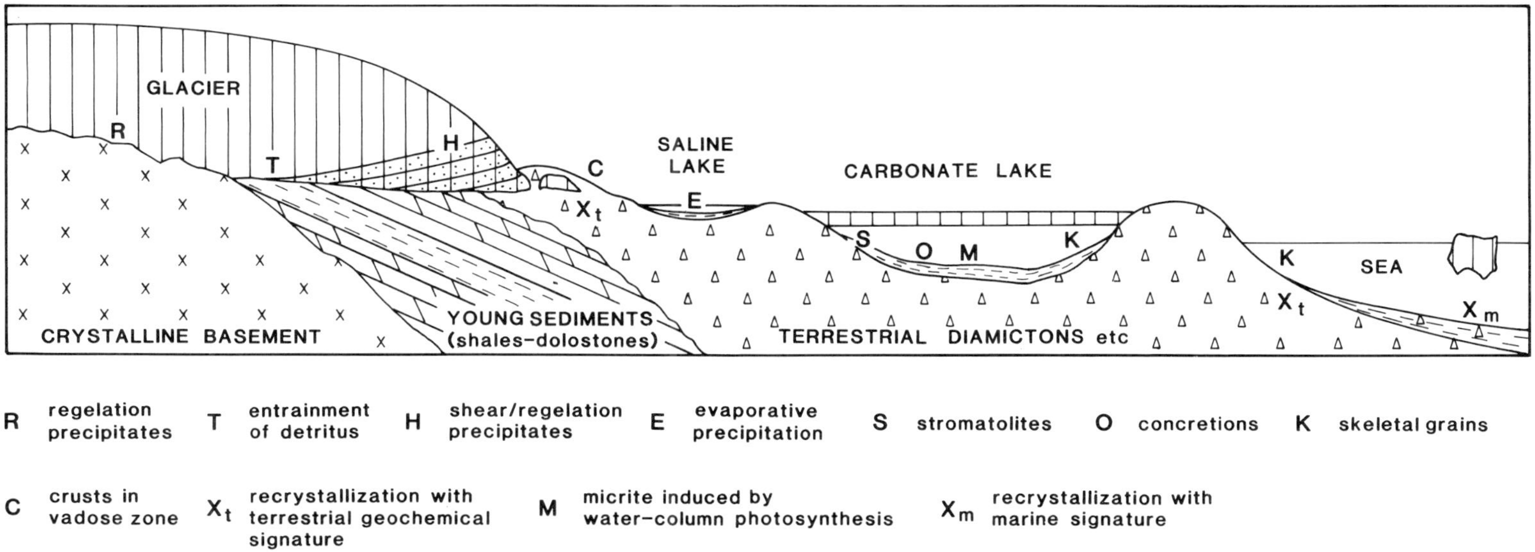

Fig. 5. Cartoon summarizing the occurrence of glacigenic carbonates. The diagram is composite, including both the characteristic dolomitic detritus of Proterozoic glacial successions and skeletal precipitates of Phanerozoic time. See text for further explanation.

References

ABER, J.S. 1979. Glacial conglomerates of the Appalachian Plateau, New York. *Quaternary Research*, **11**, 185–196.

AHARON, P. 1988. Oxygen, carbon and U-series isotopes of aragonites from Vestfold Hills, Antarctica: clues to geochemical processes in subglacial environments. *Geochimica Cosmochimica Acta*, **52**, 2321–2331.

ANDERSON, T.F. & ARTHUR, M.A. 1983. Stable isotopes of oxygen and carbon and their application to sedimentologic and palaeoenvironmental problems. *In*: *Stable Isotopes in Sedimentary Geology. Short Course of the Society of Economic Palaeontologists and Mineralogists, Tulsa*, **10**, Chapter 1, 1–151.

BAERMANN, A. & HÜBNER, S. 1984. Ingenieurgeologische und geophysikälische Untersuchungen an Geschiebemergeln im Norddeutschen Raum. *Geologisches Jahrbuch*, **C37**, 17–57.

——, IWANOFF, A. & WILKE, H. 1983. The calcium carbonate content of North German tills. *In*: EHLERS, J. (ed.) *Glacial Deposits in North-west Europe*. Balkema, Rotterdam 259–262.

BARRETT, P.J., HAMBREY, M.J. & ROBINSON, P.R. 1990. Cenozoic glacial and tectonic history from CIROS-1, McMurdo Sound. *In*: *5th International Symposium on Antarctic Earth Sciences* Cambridge University Press, Cambridge.

BJØRLYKKE, K., BUE, B. & ELVERHØI, A. 1978. Quaternary sediments in the northwestern part of the Barents Sea and their relation to the underlying Mesozoic bedrock. *Sedimentology*, **25**, 227–246.

BRIDLE, I.M. & ROBINSON, P.H. 1990. Diagenesis. *In*: BARRETT, P.J. (ed.) Antarctic Cenozoic history from the CIROS-1 drillhole McMurdo Sound. *DSIR (NZ) Bulletin*, **245**.

BURNS, S.J. & BAKER, P.A. 1987. A geochemical study of dolomite in the Monterey Formation, California. *Journal of Sedimentary Petrology*, **57**, 128–139.

BURTON, E.A. & WALTER, L.M. 1987. Relative precipitation rates of aragonite and Mg calcite from seawater: temperature or carbonate ion control. *Geology*, **15**, 111–114.

BURTON, R. 1981. Chemistry, physics and evolution of Antarctic saline lakes. *Hydrobiologica*, **82**, 339–362.

CAREY, S.W. & AHMAD, N. 1961. Glacial marine sedimentation. *In*: *Geology of the Arctic, 2* (ed. G.O. Raasch), Toronto University Press, Toronto, 865–894.

DAGEL, M.A., HENDY, C.H., DENTON, G.H., STUIVER, M. & JUDD, F.M. 1990. Stratigraphy and chronology of isotope stage 6 and 2 drift, Marshall Valley, Antarctica. *Boreas*,

DEYNOUX, M. 1985. Terrestrial or waterlain glacial diamictites? Three case studies from the Late Precambrian and Late Ordovician glacial drifts in West Africa. *Palaeogeography, Palaeoclimatology, Palaeoecology*, **51**, 97–141.

DOMACK, E.W. 1988. Biogenic facies in the Antarctic glacimarine environment: basis for a polar glacimarine summary. *Palaeogeography, Palaeoclimatology, Palaeoecology*, **63**, 357–362.

DREDGE, L.A. 1988. Drift carbonate on the Canadian Shield. II: carbonate dispersal and ice-flow patterns in northern Manitoba. *Canadian Journal of Earth Sciences*, **25**, 783–787.

EYLES, C.H., EYLES, N. & MIALL, A.D. 1985. Models of glaciomarine sedimentation and their application to the interpretation of ancient glacial sequences. *Palaeogeography, Palaeoclimatology, Palaeoecology*, **51**, 15–84.

FAIRCHILD, I.J. 1983. Effects of glacial transport and neomorphism on Precambrian dolomite crystal sizes. *Nature*, **304**, 714–716.

—— & HAMBREY, M.J. 1984. The Vendian of NE Spitsbergen: petrogenesis of a dolomite-tillite association. *Precambrian Research*, **26**, 111–167.

——, ——, SPIRO, B. & JEFFERSON, T.H. 1989. Late Proterozoic glacial carbonates in NE Spitsbergen: new insights into the carbonate-tillite association. *Geological Magazine*, **126**.

——, HENDRY, G.L., QUEST, M. & TUCKER, M.E. 1988. Chemical analysis of sedimentary rocks. *In*: TUCKER, M.E. (ed.) *Techniques in Sedimentology* Blackwells, Oxford, 271–352.

—— & SPIRO, B. 1987. Petrological and isotopic implications of some contrasting Late Precambrian carbonates, NE Spitsbergen. *Sedimentology*, **34**, 973–989.

GREGORY, R.T., DOUTHITT, C.B., DUDDY, I.R., RICH, P.V. & RICH, T.H. 1989. Oxygen isotopic composition of carbonate concretions from the lower Cretaceous of Victoria, Australia: implications for the evolution of meteoric waters on the Australian continent in a paleopolar environment. *Earth and Planetary Science Letters*, **92**, 27–42.

HALLET, B. 1976. Deposits formed by subglacial precipitation of $CaCO_3$. *Geological Society of America Bulletin*, **87**, 1003–1015.

HAMBREY, M.J. 1982. Late Precambrian diamictites of northeastern Svalbard. *Geological Magazine*, **119**, 527–551.

——, BARRETT, P.J. & ROBINSON, P.H. 1990. Stratigraphy and sedimentology. *In*: BARRETT, P.J. (ed.) Antarctic Cenozoic glacial history from CIROS-1 drillhole, McMurdo Sound, Antarctica. *DSIR (NZ) Bulletin*, **245**

HARLAND, W.B. 1964. Evidence of Late Precambrian glaciation and its significance. *In*: NAIRN, A.E.M. (ed.) *Problems in Palaeoclimatology* Wiley, Chichester. 119–149 and 179–184.

—— & HEROD, K.N. 1975. Glaciation through time. *In*: WRIGHT, A.E. & MOSELEY, F. (eds.) *Ice Ages: Ancient and Modern*. Seel House Press, Liverpool, 189–216.

HENDY, C.H. 1990. The history of Quaternary lakes in the McMurdo Sound region of Antarctica. *Quaternary Reviews*.

——, HEALY, T.R., RAYNER, E.M., SHAW, J. & WILSON, A.T. 1979. Late Pleistocene glacial chronology of the Taylor Valley, Antarctica, and the global climate. *Quaternary Research*, **11**, 172–184.

—— & UTTING, A.J. 1990. The occurrence of mixed calcite and aragonite beds in ice free valleys of the Transantarctic mountains. *Journal of Sedimentary Petrology*.

HILLAIRE-MARCEL, C. & CASANOVA, J. 1987. Isotopic hydrology and paleohydrology of the Magadi (Kenya)- Natron (Tanzania) Basin during the late Quaternary. *Palaeogeography, Palaeoclimatology, Palaeoecology*, **58**, 155–181.

——, SOUCY, J.-M. & CAILLEUX, A. 1979. Analyse isotopique de concrétions sous-glacaires de l'inlandsis laurentidien et teneur en oxygen 18 de la glace. *Canadian Journal of Earth Sciences*, **16**, 1494–1498.

JANAWAY, T.M. & PARNELL, J. 1989. Carbonate production within the Orcadian Basin, northern Scotland: a petrographic and geochemical study. *Palaeogeography, Palaeoclimatology, Palaeoecology*, **70**, 89–105.

KNOLL, A.H., HAYES, J.M., KAUFMAN, A.J., SWETT, K. & LAMBERT, I.B. 1986. Secular variation in carbon isotope ratios from Upper Proterozoic successions of Svalbard and East Greenland. *Nature*, **321**, 832–839.

LAMOTHE, M., HILLAIRE-MARCEL, C. & PAGÉ, P. 1983. Découverte de concrétions calcaires striées dans le till de Gentilly, basses-terres du Saint-Laurent, Québec. *Canadian Journal of Earth Sciences*, **20**, 500–505.

LAND, L.S. 1980. The isotopic and trace element geochemistry of dolomite: the state of the art. *In*: ZENGER, D.H., DUNHAM, J.B. & ETHINGTON, R.L. (eds.) *Concepts and models of dolomitization Special Publications of the Society of Economic Palaeontologists and Mineralogists*, **28**, 87–110.

—— & HOOPS, G.K. 1973. Sodium in carbonate sediments and rocks: a possible index to the salinity of diagenetic solutions. *Journal of Sedimentary Petrology*, **43**, 614–617.

LAVRUSHIN, Y.A. & GOLUBEV. Y.K. 1981. Diagenetic carbonate concretions in till. *Doklady Akademii Nauk SSSR*, **245**, 131–133.

—— & RENGARTEN, N.V. 1974. Main features of the glacial type of lithogenesis. Plenum, New York, 658–666 (translation).

LAWRENCE, M.J.F. & HENDY, C.H. 1985. Water column and sediment characteristics of Lake Fryxell, Taylor Valley, Antarctica. *New Zealand Journal of Geology and Geophysics*, **28**, 543–552.

—— & HENDY, C.H. 1989. Carbonate deposition and Ross Sea ice advance, Fryxell Basin, Taylor Valley, Antarctica. *New Zealand Journal of Geology and Geophysics*, **32**, 267–277.

LORD, B.K., JONES, L.M. & FAURE, G. 1988. Evidence for the existence of the Gondwana ice sheet in the ^{18}O depletion of carbonate rocks in the Permian formations of the Transantarctic Mountains. *Chemical Geology*, **72**, 163–171.

MAGARITZ, M. 1973. Precipitation of secondary calcite in glacier areas: carbon and oxygen isotopic composition of calcites from Mt. Hermon, Israel, and the European Alps. *Earth and Planetary Science Letters*, **17**, 385–390.

MATSUBAYA, O., SAKAI, H., TORII, T., BURTON, H. & KERRY, K. 1979. Antarctic saline lakes-stable isotope ratios, chemical compositions and evolution. *Geochimica et Cosmochimica Acta*, **43**, 7–25.

MORIKAWA, H. & OSSAKA, J. 1974. The distribution of secondary minerals at Lake Vanda. *Dry Valley Drilling Project Bulletin*, **4**, 39–40.

MORSE, J.W. & CASEY, W.H. 1988. Ostwald processes and mineral parageneses in sediments. *American Journal of Science*, **288**, 537–560.

MULLINS, H.T., WISE, S.W., LAND, L.S., SIEGEL, D.I., MASTERS, P.M., HINCHEY, E.J. & PRICE, K.R. 1985. Authigenic dolomite in Bahamian peri-platform slope sediment. *Geology*, **13**, 292–295.

MURRAY, R.C. 1953. The petrology of the Cary and Valders Tills of northeastern Wisconsin. *American Journal of Science*, **251**, 140–155.

NISHIYAMA, T. & KURASAMA, H. 1975. Distribution of secondary minerals from Taylor Valley. *Dry Valley Drilling Project Bulletin*, **5**, 120–133.

PARKER, B.C., SIMMONS, G.M., LOVE, F.G., WHARTON, R.A. & SEABURG, K.G. 1981. Modern stromatolites in Antarctic Dry Valley lakes. *Bioscience*, **31**, 656–661.

RAO, C.P. 1981a. Criteria for recognition of cold-water carbonate sedimentation: Berriadale Limestone (Lower Permian), Tasmania, Australia. *Journal of Sedimentary Petrology*, **51**, 491–506.

—— 1981b. Geochemical differences between tropical (Ordovician) and subpolar (Permian) carbonates, Tasmania, Australia. *Geology*, **9**, 205–209.

—— 1988. Oxygen and carbon isotope composition of cold-water Berriadale Limestone (Lower Permian), Tasmania, Australia. *Sedimentary Geology*, **60**, 221–231.

—— & GREEN, D.C. 1982. Oxygen and carbon isotopes of Early Permian cold-water carbonates, Tasmania, Australia: *Journal of Sedimentary Petrology*, **52**, 1111–1125.

SANDBERG, P. 1985. Aragonite cements and their occurrence in ancient limestones. *In*: SCHNEIDERMANN, N. & HARRIS, P.M. (eds.), *Carbonate Cements*, SEPM (Special Publication 36), Tulsa, 33–57.

SCHLÜCHTER, C. (ed.) 1979. *Moraines and Varves*. Balkema, Rotterdam, 441.

SCHMIDT, D.L. & FRIEDMAN, I. 1974. Continental deposition of Antarctic tillite indicated by carbon and oxygen isotopes. *Journal of Research of the US Geological Survey*, **2**, 711–715.

SOUCHEZ, R. & JOUZEL, J. 1984. On the isotopic composition of δD and $\delta^{18}O$ of water and ice during freezing. *Journal of Glaciology*, **30**, 369–372.

—— & LEMMENS, M. 1985. Subglacial carbonate deposition: an isotopic study of a present-day case. *Palaeogeography, Palaeoclimatology, Palaeo-*

ecology, **51**, 357–364.

STUIVER, M., YANG, I.C., DENTON, G.H. & KELLOGG, T.B. 1981. Oxygen isotope ratios of Antarctic permafrost and glacier ice. *In*: MCGINNIS, L.D. (ed.) *Dry Valley Drilling Project Antarctic Research Series of the American Geophysical Union*, **33**, 131–140.

SWETT, K. 1974. Calcrete crusts in an Arctic permafrost environment. *American Journal of Science*, **274**, 1059–1063.

SWIRYDCZUK, K. 1988. Mineralogical control on porosity types in Upper Jurassic Smackover oolitic grainstones, southern Arkansas and northern Louisiana. *Journal of Sedimentary Petrology*, **58**, 339–347.

VEIZER, J. 1983. Chemical diagenesis of carbonates: theory and application of trace element technique. *In*: *Stable Isotopes in Sedimentary Geology. Short Course of the Society of Economic Palaeontologists and Mineralogists, Tulsa*, **10**, Chapter 4, 1–100.

WADA, H. & OKADA, H. 1990. Carbonate isotopes. *In*: BARRETT, P.J. (ed.) Antarctic Cenozoic history from the CIROS-1 drillhole McMurdo Sound. *DSIR (NZ) Bulletin*, **245**.

WALTER, M.R. & BAULD, J. 1983. The association of sulphate evaporites, stromatolitic carbonates and glacial sediments: examples from the Proterozoic of Australia and the Cainozoic of Antarctica. *Precambrian Research*, **21**, 129–148.

WHARTON, R.A., PARKER, B.C., SIMMONS, G.M., SEABURG, K.G. & LOVE, F.G. 1982. Biogenic calcite structures forming in Lake Fryxell, Antarctica. *Nature*, **295**, 403–405.

YOUNG, G.M. & GOSTIN, V.A. 1988. Stratigraphy and sedimentology of Sturtian glacigenic deposits in the sestern part of the North Flinders Basin, South Australia. *Precambrian Research*, **39**, 151–170.

Iceberg scours in the geological record: examples from glacial Lake Agassiz

C.M.T. WOODWORTH-LYNAS & J.Y. GUIGNÉ

C-CORE, Memorial University of Newfoundland, St. John's, Newfoundland, Canada A1B 3X5

Abstract: Observation of sub-surface deformation structures beneath modern oceanic iceberg scours is impossible, but Quaternary scours exposed above sea or lake level on land provide this opportunity. Well developed iceberg scours which are clearly visible on the former bed of glacial Lake Agassiz in southeastern Manitoba were chosen for detailed studies of sub-scour deformation.

Partial burial of the scour troughs and the absence of bioturbation of the clay sediments have preserved the surface morphology and sub-scour deformation structures in pristine condition. Shallow-angle faults and sub-horizontal thrust faults are developed below the scour trough, the former extending to at least 3 m beneath the deepest part of the scour incision surface. These faults formed in response to vertical and horizontal loading of the lake bed by the encroaching keel. High-angle normal faults developed outside the scour margin represent gravity collapse of sediment blocks that were piled up on the scour berms as a result of lateral displacement of material away from the axis of the scour. Definition of deeply penetrating faults has significant engineering implications for the burial depths required to protect oil and gas production pipelines in areas of active scouring today.

Remnant pods and lenses of tightly folded laminated sediments, in places carrying a well developed fracture cleavage, are all that remain of original primary layering in the severely reworked sub-scour sediments. Based on our findings, criteria for the recognition of scours and of ice keel turbates exposed in argillaceous sediments are proposed.

The keels of drifting icebergs which touch, penetrate and plough through sea floor or lacustrine sediments generate characteristic curvilinear furrows referred to as iceberg scours. A great deal is known about the surface morphology of Quaternary scours, mostly from descriptions of side scan sonograms and from visual examination of fresh scours from submarines. However, virtually no data exist on the nature and extent of sediment deformation beneath iceberg scours. As a result, discrete scour features and sediments reworked by multiple scour events have not been recognized in the pre-Quaternary glacimarine or glacilacustrine rock record.

Reworking of the sea floor by scouring and grounded icebergs and sea ice pressure ridges has been and continues to be a major glacimarine process affecting enormous areas of high latitude continental shelves, yet recognition of the importance of the process has not been fully realized. Severe reworking by multiple iceberg scour events may obliterate primary sedimentary structures and result in the generation of iceberg turbates, a facies defined by Vorren *et al.* (1983). The less specific term ice keel turbate, proposed by Barnes & Lien (1988), is preferred because deformation structures associated with icebergs, sea ice and lake ice are likely to be similar.

Along the eastern Canadian seaboard relict Pleistocene scours, still visible on the sea bed, may have incision depths as great as 20 m (Fader *et al.* 1982), widths of 250 m and lengths of several kilometres in present day water depths up to 700 m (Josenhans & Woodworth-Lynas 1988; Pereira *et al.* 1985). Modern scouring occurs in water depths up to 235 m in the Arctic and sub-Arctic (Hodgson *et al.* 1988), and possibly to 500 m in the Antarctic (Barnes & Lien 1988).

Presently on the eastern Canadian continental shelf curvilinear scours are formed by continuously moving icebergs which excavate characteristically flat-bottomed troughs. In the troughs of individual scours small-scale, axis-parallel ridge and groove microtopography (up to 30 cm relief) is formed at the trailing edge of the scouring keel by boulders and cobbles embedded in the ice (Hodgson *et al.* 1988). Most of this is probably clastic sea bed material mechanically incorporated into the keel. The presence of small sand volcanoes (less than 0.5 m high) on top of the ridge and groove microtopography suggests that sub-scour liquefaction and dewatering occurs after, and possibly during, the scour event (Hodgson *et al.* 1988). Similar 'extrusion mounds' have been described in association with ice scour berms in the Beaufort Sea

From DOWDESWELL, J. A. & SCOURSE, J. D. (eds), 1990, *Glacimarine Environments: Processes and Sediments*. Geological Society Special Publication No 53, pp 217–223.

(e.g. Barnes & Reimnitz 1979). Linear scour berms, which consist of 'blocky' material up to 6 m high on either side of the trough, are characterized by a network of vertical columnar or radial fractures which extend downwards at least 1 m (Hodgson *et al.* 1988). Similar structures have been observed along berm crests of fresh scours in the Beaufort Sea (e.g. Barnes & Reimnitz 1979). Small irregular pits up to 1 m deep abruptly truncate ridge and groove microtopography in the scour trough. The pits are formed by blocks of ice which are broken out of the keel, overridden and embedded in the sea floor (Hodgson *et al.* 1988). Subsequent melting leaves a void in the scour trough.

Observations of surface features clearly suggest that deformation affects sediments to some depth beneath the sea bed. However, documentation of these effects in terms of deformation structures is sparse (Fischbein 1987; Thomas & Connell 1985; Eyles & Clark 1988).

This paper describes in detail the cross-sectional morphology and subscour deformation structures associated with relict iceberg scours exposed subaerially on the floor of glacial Lake Agassiz in southern Manitoba, Canada. Based on these descriptions and interpretations, we propose criteria by which iceberg scours in pre-Quaternary sediments may be recognized.

Lake Agassiz iceberg scours

Because of difficulties experienced with permafrost while studying Quaternary marine scours exposed on King William Island in the Canadian Arctic (Woodworth-Lynas *et al.* 1986) we chose an area in southern Manitoba where access is easier and where permafrost is not a problem (Fig. 1). A region immediately southeast of Winnipeg was selected where prolific, well preserved exposed scours are found in thick clays of former glacial Lake Agassiz (Fig. 2). The scours are partially buried beneath a 2–3 m thick layer of silty clay, thick enough to protect scoured sediments from the effects of farming activities and from deformation related to seasonal freeze/thaw of the ground. Two scours (A and B) were chosen for study from analysis of aerial photographs of the area (Fig. 2). In the regions chosen for excavation each scour is relatively free from the influence of other scours.

Methods

In 1987 a backhoe was used to expose cross sections of the two scours to maximum depths of 4 m. A geotechnical programme was conducted and involved coring and the use of a piezocone, pressuremeter, dilatometer, and hand held shear vane and penetrometer. Results of that study will be reported elsewhere. The contact between the scour surface and overlying infilling sediment was well defined in both scours (Fig. 3). In 1988 another cross section of scour B was made because this scour contained the best structures. In both years block samples were taken from the sub-scour sediments for the analysis of microstructures.

General geology

At its maximum, Lake Agassiz was the largest lake in North America, covering a total area of 950 000 km^2, although it never exceeded more than 350 000 km^2 at any one time (Teller & Clayton 1983). It was as deep as 213 m in the Winnipeg area (Elson 1967) and was supplied with most of its water from the melting of the Wisconsinan ice sheet via meltwater channels or directly from the glacier (Teller 1976). Sediment-laden meltwater probably travelled as slow-moving density currents that may have flowed the entire length of the Lake Agassiz Basin (Teller 1976). The majority of sediments in the study region averages 79% clay, 20% silt and 1% sand (Last 1974).

In southern Manitoba up to three stratigraphic clay-rich units overlie till (Teller 1976). The lowermost unit (Unit 1) was correlated with the Brenna Formation in North Dakota and Minnesota by Teller (1976) and Fenton *et al.* (1983). The unit has an average thickness of about 15 m (Harris *et al.* 1974). It is characterized by a general absence of laminae, the presence of iceberg-rafted clasts of white calcareous silt, carbonate and crystalline rocks, and the presence of slickensided surfaces. Fenton *et al.* (1983) interpret deposition of this unit (in the study area) between about 11 200 and 10 000 years ago as glacial ice retreated north in the Red River Valley.

A final ice advance from the north caused lake levels to rise, allowing deposition of Unit 3 between about 10 000 and 9000 years ago after several hundred years of low water levels (Fenton *et al.* 1983). Unit 2 of Teller (1976) may not be present in the study area, and Unit 3, which was correlated with the Sherack Formation, rests on the Brenna Formation. Unit 3 is characterized by the presence of laminae and absence of clasts (Teller 1976).

Water depth and age of scours

Prior to scouring a period of subaerial exposure

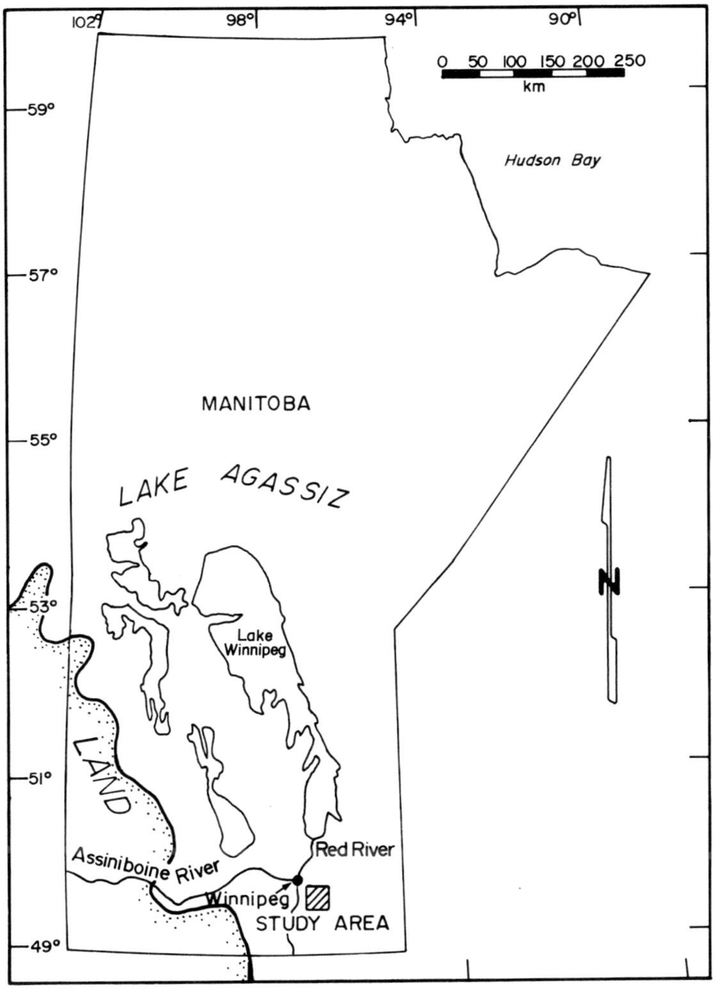

Fig. 1. Location map of the study area in southeastern Manitoba.

existed between about 10 300 and 9900 years ago. By using estimated lake levels in the Lake Manitoba basin (Teller & Last 1981) and from the eastern lake outlets (Teller & Thorleifson 1983) it is likely that the study area was exposed at least 7 m above lake level during this period. The interpretation of exposure and possible desiccation is supported by the results of the 1987 in situ geotechnical testing at the two scours, which indicates overconsolidation of the clays down to a consistent depth of 8 m below the present ground surface (Woodworth-Lynas & Landva 1988).

About 9900 years ago after the eastern outlets were re-dammed by glacial ice, the lake depth increased to about 110 m (Teller & Thorleifson 1983) and deposition of Unit 3 began. Woodworth-Lynas & Matile (1988) present evidence that the rise in lake levels at this time triggered a southeast-directed surge of ice from north of the Winnipeg area terminating east of the study area against a significant (unnamed) north-south oriented moraine. After the surge the ice sheet would have begun to disintegrate, producing numerous icebergs which scoured the previously exposed lake bed sediments, beginning shortly after 9900 years ago when water depths rose to about 110 m. In the study area deposition of a tan-coloured unit (Unit 3?) which fills the scour troughs was under way during the period of scouring, because reworked sediments of the lowermost part of the unit are incorporated within deformed clays of the underlying unit (Unit 1?) at scour B (see below). As water

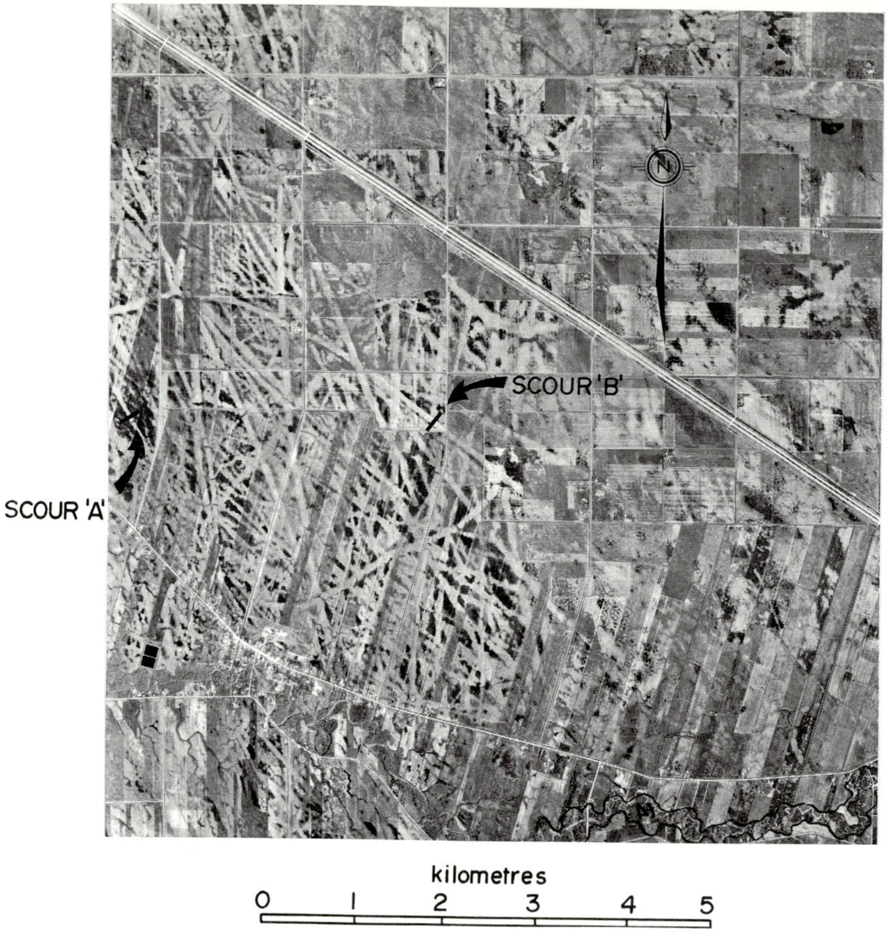

Fig. 2. Aerial photograph of the study region southeast of Winnipeg. Iceberg scours show up as prominent white criss-crossing lines. Two scours, A and B, are described in this paper. The straight diagonal line in the upper right of the photograph is the Trans-Canada Highway.

depth increased beyond 60 m scouring by lake ice pressure ridge keels would not have occurred (e.g. Reimnitz *et al.* 1984).

Geology of the scours

Scour A is about 50 m wide and can be traced for at least 6 km, both ends probably being obliterated beneath thickening sediments of the tan-coloured unit (Unit 3?) (Fig. 2). Scour B is about 50 m wide as measured between berm crests exposed in the excavations, and can be traced for at least 8.5 km before dying out beneath overburden deposits at both ends.

The scours are incised into brown and grey mottled clays (of Unit 1?) (Fig. 3), and sub-scour sediments at scour B also contain zones of reworked tan-coloured unit silty clay. The brown and grey clay in the study area is characterized by the presence of plate-like clasts of white calcareous silt. The clasts are commonly concentrated along the scour incision surface, which is a contact between the brown and grey clay and tan-coloured units. Clasts of carbonate and friable crystalline rock are also present. The clay is generally plastic in both scours with notable exceptions at scour A where discrete zones of brittle clay occur immediately below the incision surface. In these zones, which are 50 to 75 cm thick, the clay fractures and disaggregates when handled, in contrast to the plastic clay which is cohesive and easily sampled.

Except where it has been affected by scouring, the tan-coloured unit occurs as a well laminated deposit of silty clay filling the scour troughs. At scour B discrete horizons, generally less than

Fig. 3. View from the northeast berm looking towards the centre of the trough at scour B, showing the incision surface (abrupt contact) separating dark, scoured massive sediments of the brown and grey clay, from overlying, well laminated silty clays of the tan-coloured unit. The incision surface dips away from the camera (white cards on the trench wall are spaced 1 m apart and define a horizontal line).

10 cm thick, of yellow/white finely cross-bedded silty clay occur.

Sub-scour deformation structures

Scour A

Small-scale kink folds are seen immediately below the incision surface on the southwest edge of the scour. The folds are sharp-crested and symmetric with flat limbs and vertical axial surfaces, and have amplitudes and wavelengths no greater than 5 cm. It is possible that the folds formed in response to horizontal pressures generated by the lateral displacement of sediment during the passage of the scouring keel past this region. The scour incision surface has not been affected by the folds, implying that the iceberg keel may have been in contact with the clay either during fold generation or immediately after they were formed so that they were truncated by the passing keel.

Unlike scour B, faults were not seen in any part of the section. This may in part be the result of the general absence of bedding, truncation and offset of which is an important criterion for identifying faults.

Scour B

The concave incision surface has apparent dips of between about 4° and 6° towards the centre of the scour trough. The absence of ridge and groove microtopography on the scour surface, characteristic of modern oceanic scours, may be because coarse material was not available on the lake bed for incorporation into the scouring keel.

Faults. A suite of symmetrically-arranged faults affects the sub-scour clays (Fig. 4). Faults are sometimes defined by offset bedding, but where bedding is absent they are hard to discern. After excavating the trenches minor settling sometimes caused small displacements (1–

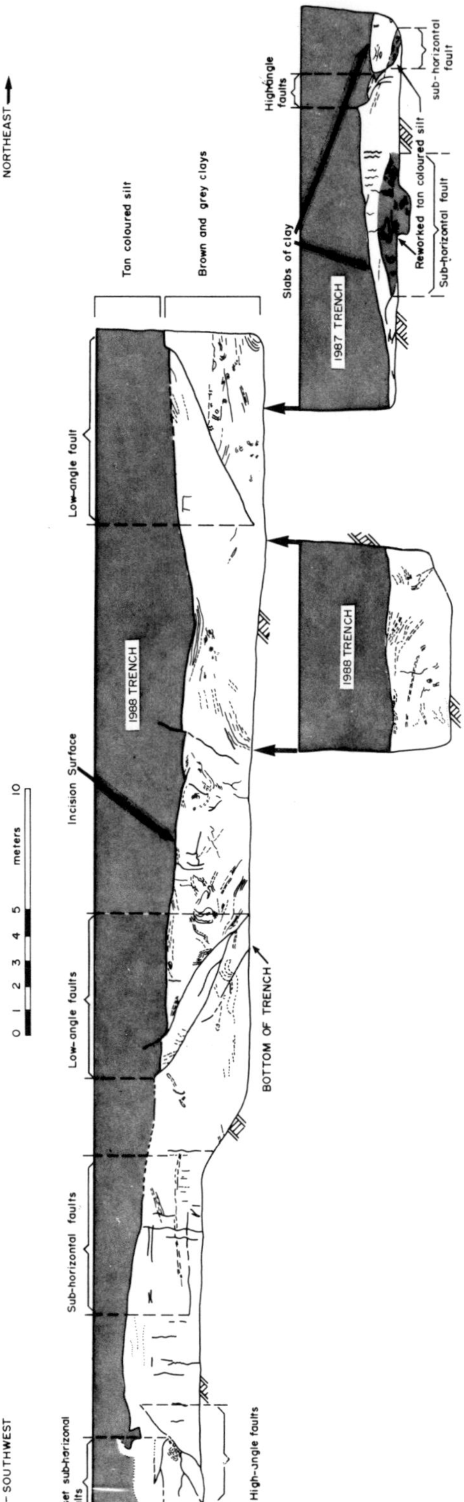

Fig. 4. Composite cross section through scour B showing the scour incision surface, sub-scour faults (heavy lines) and deformed bedding. No vertical exaggeration.

2 mm) of the trench walls along fault surfaces causing shadows to be thrown on the trench wall, thereby allowing their recognition in areas of structureless clay. Three types of fault are distinguished, each associated with different parts of the scour, and each related to different mechanisms of displacement during the scouring event. Fault types include: (1) deeply penetrating, low-angle faults underneath the central part of the scour trough, (2) sub-horizontal faults beneath the scour berms, and (3) high angle faults located below the outer berm regions.

Low angle faults. On both sides of the scour, between the inner berm flanks and the centre of the trough, a series of low-angle faults with dips ranging from 18° to 65° extend from the scour incision surface to at least 3 m below the deepest part of the scour trough. Excavation of the faults revealed well developed highly polished slickensided slip surfaces (Fig. 5). It is possible that faults such as these are equivalent to the slickensided fracture surfaces described by Teller (1976) as one of the diagnostic features of Unit 1.

Along the single low-angle fault on the northeast side of the scour (Fig. 4) a band of severely distorted clay laminae is truncated at the hanging wall (Fig. 6). Continuity of this band is not seen on the hanging wall side, suggesting dip slip displacement of at least 3.5 m along the fault. There is no distinct marker horizon associated with the anastomosing shallow angle faults on the southwest side of the scour so that relative amounts of displacement cannot be inferred.

The sense of fault motion (normal or reversed) is not clear although apparent drag folds associated with the southwest fault set suggest reverse movement. Stereoplots show that the low-angle faults are symmetrically disposed about the scour axis within two distinct domains (Fig. 7). The spread of points parallel to the scour axis suggests an overall cusp-shaped fault geometry, the general attitude of the faults being slightly oblique to the axis. Slickenside data, also falling into two domains on either side of the scour axis (Fig. 7), are consistent with a sense of displacement symmetrically oblique to the scour axis.

We are uncertain about the sense of movement along the fault surfaces, but Poorooshasb *et al.* (1989) in their analysis of model scour tests and of data presented in this paper, suggest that loading of the lake bed by the scouring keel caused shallow foundation failure in the subkeel sediments whereby a triangular-shaped wedge of material, bounded by the low angle faults and the scour incision surface, is displaced downwards. This interpretation indicates a normal sense of fault motion (Fig. 8) and implies the continuance of the faults well below the depth of our excavations. Extrapolation of the faults at a constant angle shows that they would

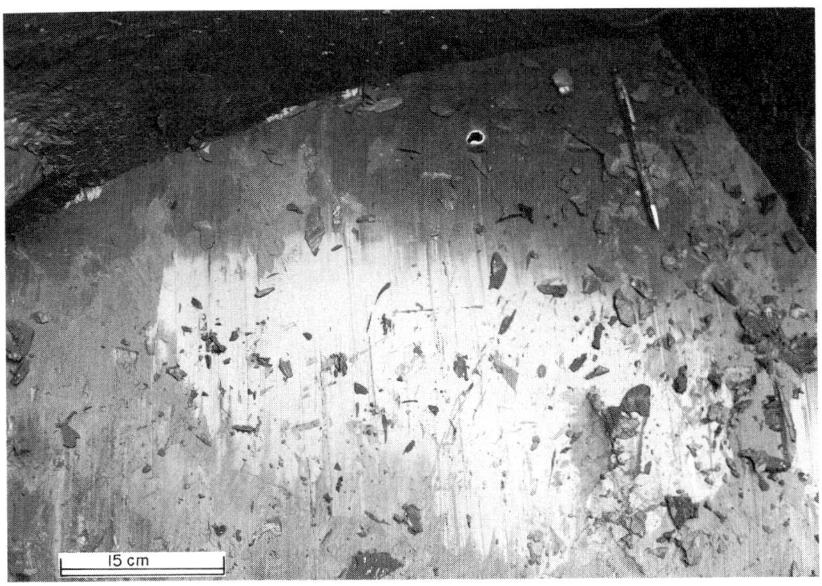

Fig. 5. View of the highly polished slickensided surface of a low angle fault plane from scour B.

Fig. 6. Low-angle fault on the northeast side of scour B showing truncation of deformed bedding. Dip slip offset along this fault of at least 3.5 m can be inferred.

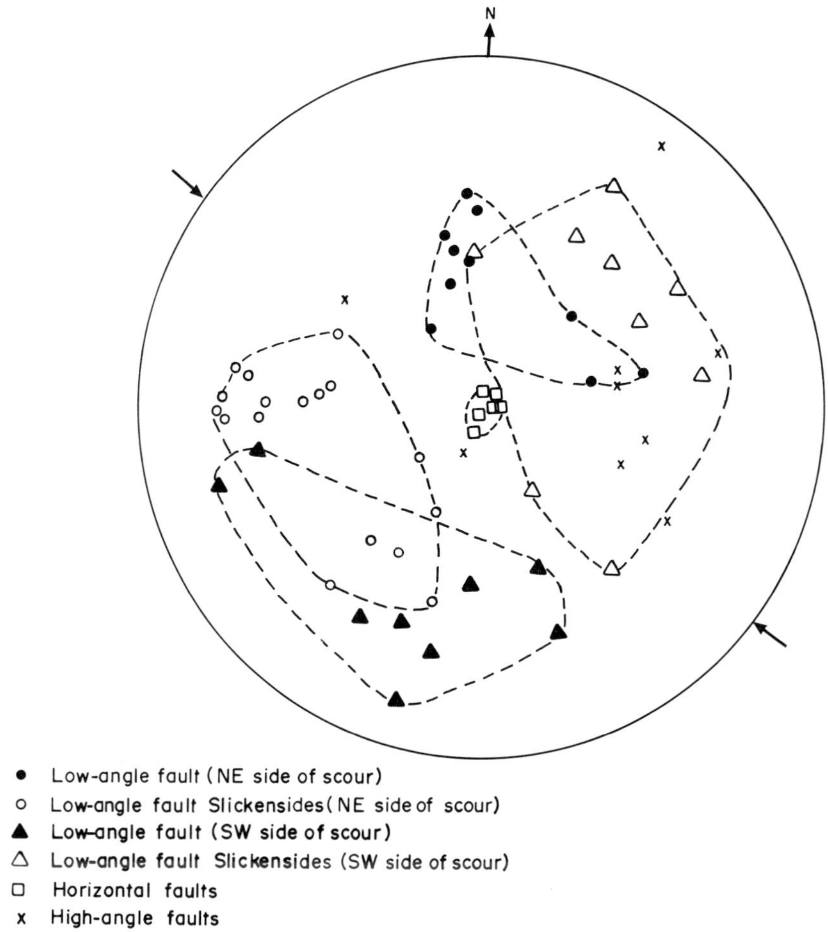

Fig. 7. Stereonet of fault (plotted as poles to planes) and slickenside orientation, scour B. Note how the low-angle fault and associated slickenside orientations, plotted with respect to the side of the scour axis on which they occur, fall into distinct structural domains arranged symmetrically about the axis (the orientation of which is shown as two heavy black arrows). Sub-horizontal fault data for both sides of the scour are tightly clustered in the centre of the diagram. High-angle fault orientations (measured only from the southwest side) are widely dispersed, reflecting unconstrained failure.

intersect at a point about 6 m below the deepest part of the scour incision surface. For the purposes of discussion we have adopted this tentative interpretation.

In places the low-angle faults offset the scour incision surface by about 20 cm of normal displacement (Fig. 9). Because the faults also extend into, and offset, laminated sediments of the overlying tan-coloured unit by the same amount, the movement is clearly a fault reactivation phenomenon of post-scour and post-early tan-coloured unit age. Scanning electron microscopy of slickensided surfaces of freeze-dried samples of brown and grey clays shows a dominant set of micro-grooves interpreted to correlate with the large offsets which occurred during the scouring event (Fig. 10). A less well developed subordinate set of micro-grooves cross-cuts the dominant set at a slight angle. The subordinate set is interpreted to have formed during the later phase of normal faulting which may have been induced by increasing overburden pressures as the tan-coloured unit increased in thickness.

Sub-horizontal faults. Sub-horizontal faults are developed at depths between 1 to 3 m below the scour berms (Fig. 4). Tight clustering of fault plane data (collected solely from the southwest

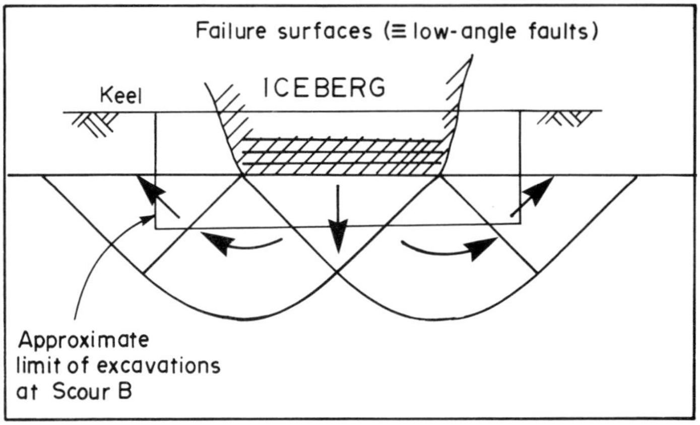

Fig. 8. The kinematic model for two-dimensional shallow foundation failure beneath a scouring keel adapted from Poorooshasb *et al.* (1989). The triangular-shaped area beneath the keel represents the area bounded by the shallow-angle faults and the scour incision surface. A normal sense of fault motion is implied. The area represented by the excavations of scour B is shown. Note that this model implies the continuance of the faults to at least 6 m below the deepest part of the scour trough.

Fig. 9. View of the excavated incision surface (scour B) offset by about 20 cm along two low-angle faults (arrows). Offset is the result of fault reactivation possibly caused by overburden pressures. Note the large dolomite dropstone on the extreme right. The stone (21 kg) has negligible penetration into the sub-scour sediments lending support to the interpretation that the clays were overconsolidated before the scour event.

margin of scour B) indicates consistent horizontal fault attitude (Fig. 7). There are no physical links between these faults, and the low-angle faults and slickensides were not seen on most of the excavated fault surface. Excavations of the northeast berm in 1987 exposed a slab of brown and grey clay bounded on the top by the scour incision surface and on the bottom by a sub-horizontal fault (Fig. 4). The block rests on severely contorted laminated silty clays of the

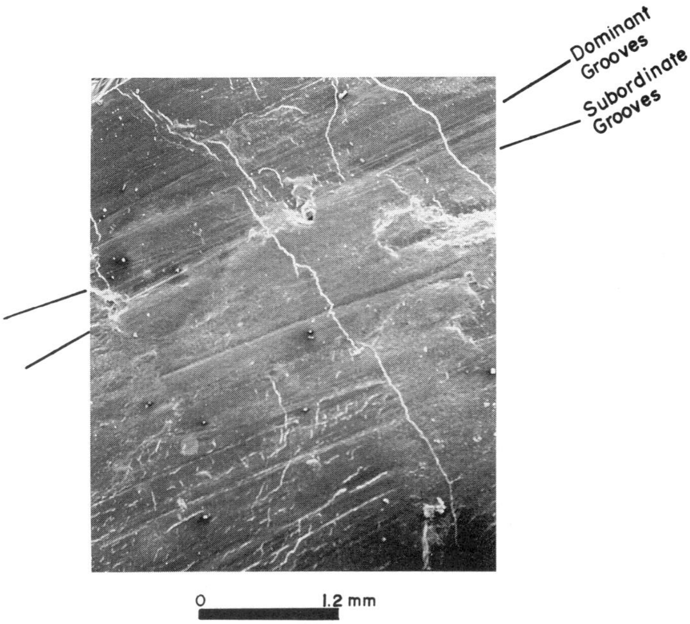

Fig. 10. Scanning electron micrograph of a slickensided surface from low-angle fault on the southwest side of scour B (low-angle faults). See text for details.

tan-coloured unit. Structural stacking of older above younger sediments implies both vertical and horizontal components of movement.

We interpret the block of brown and grey clay to have originated from a structurally lower region near the centre of the scour. Horizontal forces of the encroaching iceberg keel caused decoupling of the block along a decollement within the brown and grey clay. Upward and outward translation of the 'rigid' block occurred as the keel passed. The block overrode reworked sediments of the tan-coloured unit during final emplacement.

Three sub-horizontal faults, arranged one above the other, occur on the southwest scour margin, but there are no tan-coloured unit sediments beneath them. Although there is no direct evidence to indicate the direction of thrusting, the development of these faults beneath the scour margin suggests the sequential structural stacking of horizontal blocks of brown and grey clay displaced, as in the other scour margin, from the central part of the scour by the scouring keel. Such stacking is envisaged as being responsible for generating the positive relief of the scour berms.

High-angle faults. Sediments beneath both outer berms contain high-angle normal faults. Slickensides were developed on some fault surfaces. On the northeast scour margin offsets of the top surface of the berm clearly indicate normal faulting. Here the horizontal base of a down-faulted block of brown and grey clay rests above distorted tan-coloured unit sediments on the outer berm (Fig. 4). Although faults affect the upper berm surface they do not continue into the overlying tan-coloured unit: bedding clearly drapes or onlaps local berm topography indicating the development of berm relief at the time of scouring and prior to burial by continued deposition of the tan-coloured unit.

High-angle faults with low angle offshoots are found below the outer berm on the southwest scour margin. The high-angle faults are probably related to normal faulting although offset cannot be determined. The low-angle offshoots possibly represent sub-horizontal fault contacts between clay blocks. High-angle fault plane orientations are widely spread (Fig. 7) suggesting that fault formation is not related directly to stresses caused by the passing of the scouring iceberg keel.

We interpret the high-angle faults to have formed in response to gravity collapse of the outer scour berms which had an initial relief greater than present. The low-angle offshoot faults associated with high-angle faults on the southwest berm and the horizontal base of the clay block on the northeast berm were probably originally sub-horizontal faults related to the

Fig. 11. Photographs of typical fold geometries (defined by silt laminae) in the brown and grey clay below the incision surface of scour B.

lateral thrusting of clay slabs now preserved below the inner scour margin.

Folds and cleavage. Pre-scour primary bedding below scour B has been largely obliterated by mechanical disruption during scouring. At scour A, bedding in the scoured sediments is absent except where preserved in small scale kink folds (described earlier) in a very small region (few tens of cm^2) immediately below the incision surface.

Below scour B, discontinuous laminations can be traced over small distances but are generally confined to fold structures. Dislocated small scale (5–20 cm), disharmonic, noncylindrical folds typify the brown and grey clay and incorporated tan-coloured unit sediments (Fig. 11). Folds range from open, tight to isoclinal and there is no consistency in fold symmetry even between adjacent structures (Fig. 11).

Some folds are characterized by the presence of a well developed but non-fissile fracture cleavage. Cleavage planes are spaced about 1 mm apart with average displacements of 1 mm along the planes (Fig. 12). Because they are associated with disharmonic folds there is no consistency in apparent cleavage orientation, although it may be axial planar to specific folds. There is evidence that the cleavage is also folded (Fig. 13). The cleavage is seen only in association with the folds and cannot be traced into the structureless ground-mass. In places where cleavage is not developed, micro-faulting may affect the silt/clay laminae (Fig. 14).

Discussion

Sub-scour deformation structures below scour A are sparse whilst scour-related deformation structures are prolific below scour B. We believe that this contrast is due in part to the almost total lack of bedding structures below scour A. As a result folding and faulting, which may well be present, cannot be defined in the unstratified ground-mass. It is also possible that deformation structures, particularly faults, may develop as discontinuous features below scour troughs and that our excavations at scour A did not intersect a faulted region.

Both scours exhibit a curvilinear planform shape with widths and lengths typical of iceberg scours being formed today on Canada's eastern continental shelf area. However, the smooth-bottomed, slightly concave troughs are different from typical modern iceberg scours which generally have flat-bottomed troughs characterized by ridge and groove microtopography.

Original berm topography was observed in scour B, and bears striking resemblance (although seen only in two dimensions) to the 'blocky', fractured berms observed on the flanks of some modern scours (e.g. Hodgson *et al.* 1988). Meltout pits were not seen in either of the scours, but because they are non-linear features it is unlikely that they will be exposed in random cross sections.

Geological model

We present a general geological model to show the progressive development of fault structures during a typical scouring event in very fine-grained sediments, based on our observations and interpretations (Fig. 15). Sediment compacted in front of the scouring keel responds by bulging of the lake bed (or sea floor) surface and the initiation of horizontal faults and thrusting of older sediments (brown and grey clay) over younger (tan-coloured unit) (Fig. 15, Stage 1). Dominant motion of the sediment is directed radially away from the scour axis. Horizontally-displaced slabs of clay generate initial berm topography (Fig. 15, Stage 2). Final emplacement of the stacked assemblages of clay

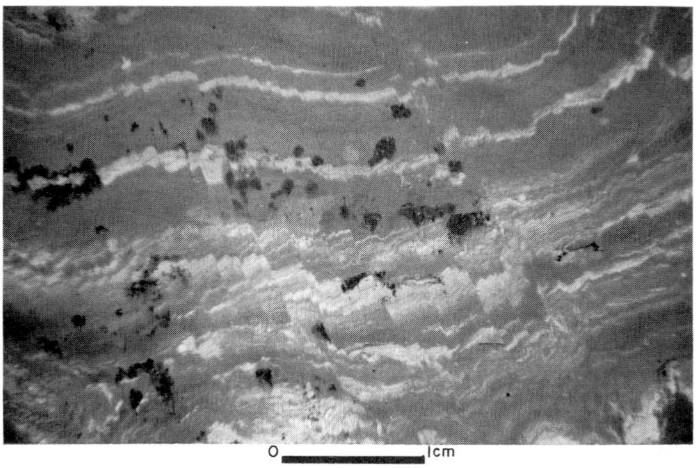

Fig. 12. Fracture cleavage developed in part of a small-scale open fold from the brown and grey clay below scour B. Note the larger displacements associated with a band in the centre of the photograph where several cleavage planes are bounded by regularly spaced (approx. 4 mm) slip planes defined by larger offsets (approx. 2 mm).

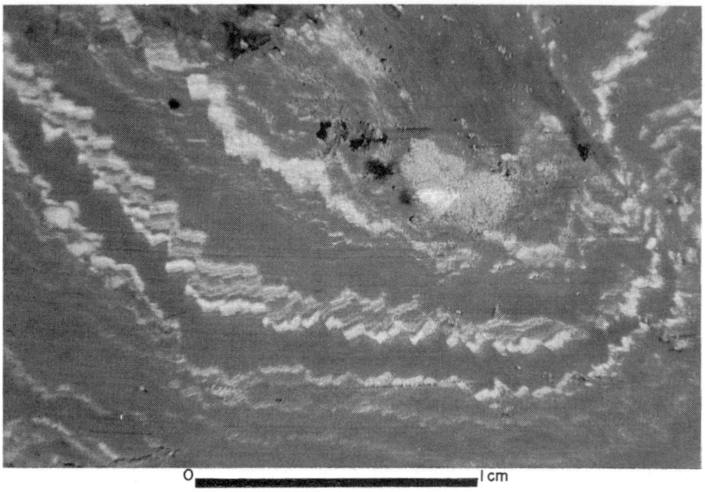

Fig. 13. Folded fracture cleavage in the limb of an earlier fold. Note the change in cleavage orientation from nearly vertical (left) to nearly horizontal (right).

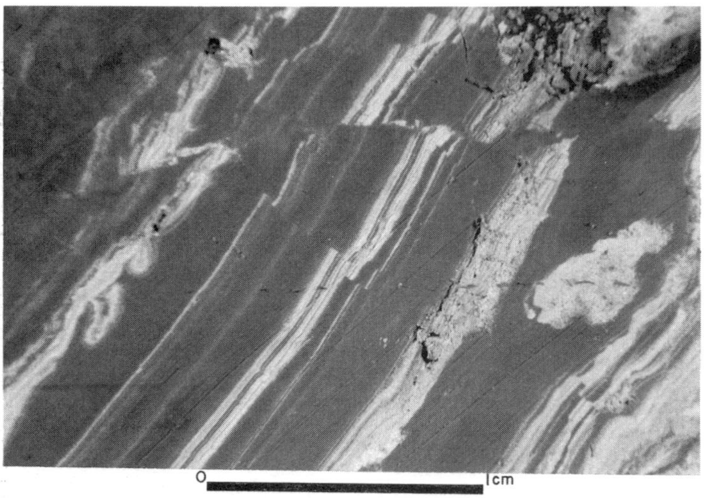

Fig. 14. Micro-faults affecting clay/silt laminae within sub-scour brown and grey clay at scour B. Note load structures beneath the lamina on the left.

slabs at the scour margins occurs as the keel passes through the plane of section (Fig. 15, Stage 3). The keel is shown in contact with the berm slabs to create the characteristically smooth incision surface. Vertical loading by the scouring keel causes possible shallow foundation failure resulting in the formation of shallow angle faults beneath the scour trough to depths possibly as great as 6 m. As the keel moves past, stress release causes high-angle normal faulting of the horizontally-stacked clay slabs on the outer berm margins, creating 'blocky' berm topography commonly seen in fresh iceberg scours (Fig. 15, Stage 4). Continuing (tan-coloured unit) sedimentation fills the scour trough, draping or onlapping against scour topography. As overburden pressures increase in the deeper part of the trough, low-angle faults are reactivated causing small (20 cm) offsets of the scour incision surface and of immediately overlying sediments.

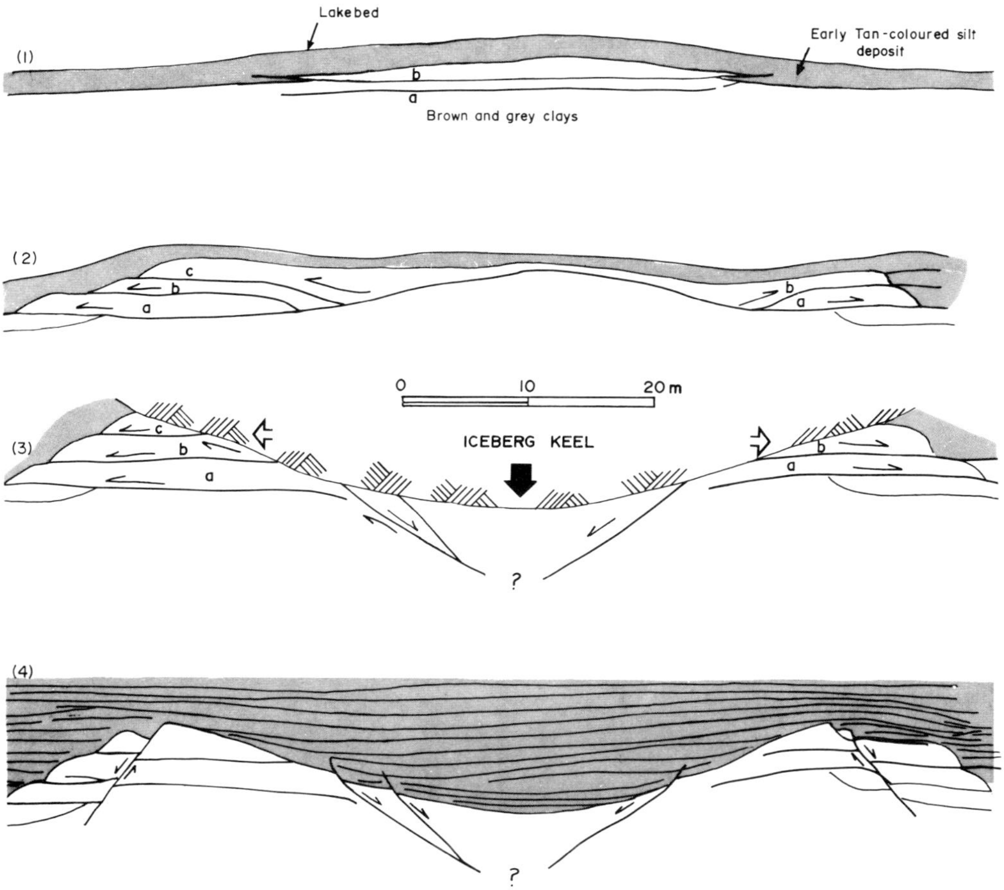

Fig. 15. A general geological model to show the progressive development of fault structures during a typical scouring event in very fine-grained semi-consolidated sediments.

Ice keel turbate

Based on the published observations of Thomas & Connell (1985) and Eyles & Clark (1988) fold symmetry in the sub-scour sediments had been expected. However, these authors describe isolated features preserved in sediments clearly unaffected by previous or subsequent scouring events. Below scour B the chaotic arrangement of disharmonic folds and the patchy development of an associated fracture cleavage, which is also deformed in places, suggests that these features may be the product of more than one scouring event. We therefore tentatively interpret the sub-scour brown and grey clay and incorporated components of the tan-coloured unit as an ice keel turbate. This interpretation is corroborated by the high concentration of intersecting scours visible on the ground surface, multiple scouring obviously having occurred at the cross over points. We speculate that if repeated reworking of the lake bed by scouring icebergs had continued, all traces of sedimentary structures would eventually have been erased, generating a massive, structureless diamicton characterized by the presence of matrix-supported clasts (mostly dropstones). In the pre-Pleistocene rock record such a unit, bounded below (as in this part of the Lake Agassiz basin) and above by glacial till might be described as a tillite or tilloid, its genesis being either misinterpreted or not understood.

Recognition of ice scours and ice keel turbates

The scour model we present is restricted to fine-grained sediments but we feel that it has application to the recognition of individual scours in the pre-Pleistocene record. In exposures where single events are unaffected by previous or subsequent scours (e.g. Thomas & Connell 1985; Eyles & Clark 1988), scour should be relatively easy to define. Savage (1972) described grooves, made when sediments were unconsolidated, on bedding plane exposures from sandstones of the Carboniferous Dwyka glaciation in South Africa. These features could perhaps be re-interpreted as ridge and groove microtopography, formed by drifting ice in wet sediment. Savage (1972) acknowledged difficulty in explaining how the marks were preserved if they were formed beneath an ice sheet.

Ice keel turbates may be harder to interpret, but remnants of sub-scour faults and disharmonic folds, possibly carrying randomly oriented fracture cleavage, may be diagnostic criteria if the degree of reworking is not too severe. Where the base or top surface of an ice keel turbate is exposed, morphologies of the scours and sub-scour structures may be preserved, in which case features of our model may be recognized.

Conclusions

Deformation structures found in sediments beneath iceberg scours preserved in clays on the bed of glacial Lake Agassiz are related to disturbance by floating ice. Three types of sub-scour faults are defined. Low-angle faults penetrate to at least 3 m below the deepest part of the scour incision surface and may extend to depths as great as 6 m. Large (3.5 m) displacements can be inferred from offset of remnant bedding. This finding has significant engineering implications for the protection by burial of oil and gas pipelines and electrical transmission cables in areas of modern scouring by both sea ice and icebergs. Sub-horizontal thrust faults, related to lateral displacement and stacking of clay slabs, occur within the berms. Structural stacking is thought to be an important berm-building process. High-angle faults on the outer berm flanks relate to outward collapse of the stacked slabs, and are responsible for creating 'blocky' berm topography such as is commonly seen in fresh oceanic iceberg scours.

Disharmonic, disarticulated folds characterize the sub-scour sediments, and in places carry a well developed fracture cleavage which may also be folded. We suggest that these features are the product of previous scour events which reworked the sediment prior to final scouring. Accordingly, this facies is interpreted as an ice keel turbate.

We suggest that the scour model can be used to aid the reinterpretation and recognition of individual scours and of ice keel turbates in fine-grained tillites or tilloids exposed in Pleistocene and pre-Pleistocene glacial successions.

This study is supported in part by a contract from the Offshore Geotechnics Program of the Canada Panel on Energy Research and Development and by a Natural Sciences and Engineering Research Council Strategic Grant (No 0032716) entitled 'An integrated investigation into the processes of ice keel/soil interaction'. We are grateful to G. Manaigre and R. LeBrun for permission to excavate scours on their land. We extend sincere thanks to our colleague J. Landva who participated in data collection in the two years of field work. We thank A. Solheim, E. Reimnitz, J.-C. Dionne and J. Teller for critical reviews of the manuscript. C. Emerson and H. Williams provided assistance with electron microscopy and light microscopy, respectively. Drafting of the figures was by D. King.

References

Barnes, P. & Reimnitz, E. 1979. Ice gouge obliteration and sediment redistribution event; 1977–1978, Beaufort Sea, Alaska. *U.S. Geological Survey Open File report* 79–848.

Barnes, P.W. & Lien, R. 1988. Icebergs rework shelf sediments to 500 m off Antarctica. *Geology*, **16**, 1130–1133.

Elson, J.A. 1967. Geology of glacial Lake Agassiz. *In*: Mayer-Oakes, W. (ed.) *Life, Land and Water* University of Manitoba Press, Winnipeg, 36–95.

Eyles, N. & Clark, B.M. 1988. Storm-dominated deltas and ice-scouring in a Late Pleistocene glacial lake. *Geological Society of America Bulletin*, **100**, 793–809.

Fader, G.B., King, L.H. & Josenhans, H.W. 1982. Superficial geology of the Laurentian channel and the western Grand Banks of Newfoundland. *Geological Survey of Canada Paper* 81–22.

Fenton, M.M., Moran, S.R., Teller, J.T. & Clayton, L. 1983. Quaternary stratigraphy and history in the southern part of the Lake Agassiz basin. *In*: Teller, J.T. & Clayton, L. (eds) *Glacial Lake Agassiz*, Geological Association of Canada Special Paper **26**, 49–74.

Fischbein, S.A. 1987. *Analysis and interpretation of ice-deformed sediments from Harrison Bay, Alaska*. MSc. thesis, California State University, Hayward.

Harris, K., Moran, S. & Clayton, L. 1974. *Late Quaternary stratigraphic nomenclature, Red River Valley, North Dakota, and Minnesota*. North Dakota Geological Survey, Miscellaneous Series 52.

Hodgson, G.J., Lever, J.H., Woodworth-Lynas, C.M.T. & Lewis, C.F.M. (eds) 1988. *The dynamics of iceberg grounding and scouring (DIGS) experiment and repetitive mapping of the eastern Canadian continental shelf*. Environmental

Studies Research Funds Report No. 094, Ottawa.

JOSENHANS, H. & WOODWORTH-LYNAS, C.M.T. 1988. Enigmatic linear furrows and pits on the upper continental slope, northwest Labrador Sea: are they sediment furrows or feeding traces? *Maritime Sediments and Atlantic Geology*, **24**, 149–155.

LAST, W. 1974. *Clay mineralogy and stratigraphy of offshore Lake Agassiz sediments in southern Manitoba*. MSc. thesis, University of Manitoba.

PEREIRA, C.P.G., PIPER, D.J.W. & SHOR, A.N. 1985. *SeaMARC I midrange sidescan sonar survey of Flemish Pass, east of the Grand Banks of Newfoundland*. Geological Survey of Canada Open File Report No. 1161.

POOROOSHASB, F., CLARK, J.I. & WOODWORTH-LYNAS, C.M.T. 1989. Small scale modelling of iceberg scouring of the seabed. *In*: *Tenth International Conference on Port and Ocean Engineering under Arctic Conditions (POAC '89), Lulea, Sweden*, **1**, 133–145.

REIMNITZ, E., BARNES, P.W. & PHILLIPS, R.L. 1984. *Geological evidence for 60 meter deep pressure-ridge keels in the Arctic Ocean*. Proceedings, International Association of Hydraulic Research, 7th International Symposium on Ice, Hamburg, Vol. 2, 189–206.

SAVAGE, N.M. 1972. Soft-sediment glacial grooving of Dwyka age in South Africa. *Journal of Sedimentary Petrology*, **42**, 307–308.

TELLER, J.T. 1976. Lake Agassiz deposits in the main offshore basin of southern Manitoba. *Canadian Journal of Earth Sciences*, **13**, 27–43.

—— & CLAYTON, L. 1983. An introduction to Glacial Lake Agassiz. *In*: TELLER, J.T. & CLAYTON, L. (eds), *Glacial Lake Agassiz* Geological Association of Canada Special Paper **26**, 3–5.

—— & LAST, W.M. 1981. Late Quaternary history of Lake Manitoba, Canada. *Quaternary Research*, **16**, 97–116.

TELLER, J.T. & THORLEIFSON, H. 1983. The Lake Agassiz – Superior connection. *In: Glacial Lake Agassiz,* TELLER, J.T. & CLAYTON, L. (eds), Geological Association of Canada Special Paper **26**, 261–290.

THOMAS, G.S.P. & CONNELL, R.J. 1985. Iceberg drop, dump and grounding structures from Pleistocene glacio-lacustrine sediments, Scotland. *Journal of Sedimentary Petrology*, **55**, 243–249.

VORREN, T.O., HALD, M., EDVARDSEN, M. & LIND-HANSEN, O.W. 1983. Glacigenic sediments and sedimentary environments on continental shelves: general principles with a case study from the Norwegian shelf. *In*: EHLERS, J. (ed.) *Glacial Deposits in North-West Europe*, Balkema, Rotterdam, 61–73.

WOODWORTH-LYNAS, C.M.T., CHRISTIAN, D., SEIDEL, M. & DAY, T. 1986. Relict iceberg scours on King William Island, N.W.T. *In*: LEWIS, C.F.M., PARROTT, R., SIMPKIN, P.G. & BUCKLEY, J.T. (eds), *Ice Scour and Seabed Engineering, Environmental Studies Revolving Funds Report No. 049, 64–70.*

—— & MATILE, G.L.D. 1988. Large scale linear ridge and groove topography from glacial Lake Agassiz and Hudson Strait: possible evidence for buoyant, surging ice sheets. *In*: Program with abstracts, Joint Annual Meeting, Geological Association of Canada, Mineralogical Association of Canada and the Canadian Society of Petroleum Geologists, Memorial University of Newfoundland, St. John's, May 23–25, A137.

—— & LANDVA, J. 1988. Sediment deformation by ice scour. Contract Report for Supply and Services Canada.

Nordic Seas surface ice drift reconstructions: evidence from ice rafted coal fragments during oxygen isotope stage 6

JENS BISCHOF[1], JOACHIM KOCH[2], MICHAELA KUBISCH[3], ROBERT F. SPIELHAGEN[3] & JÖRN THIEDE[1,3]

[1] *Sonderforshungsbereich 313, Universität Kiel, Olshausenstr. 40, D-2300 Kiel 1, West Germany*

[2] *Bundesanstalt für Geowissenschaften und Rohstoffe, Stilleweg 2, D-3000 Hannover 51, West Germany*

[3] *Geomar, Forschungszentrum für Marine Geowissenschaften, Wischhofstr. 1–3, D-2300 Kiel 14, West Germany*

Abstract: Sixteen long sediment cores from the eastern Arctic Ocean, the Fram Strait and the Norwegian–Greenland Sea, documenting 200 000 years of sedimentation, were studied for their qualitative dropstone composition (>500 μm-fraction). In sediments from oxygen isotope stages 1–5, coal particles are usually subordinate components of the coarse fraction. In contrast to younger deposits, coal content in oxygen isotope stage 6 (186–128 ka) varies between 20% and 65% in the eastern Arctic Ocean and the Fram Strait and between 5% and 20% in the Norwegian Sea. Southward decreasing coal content and similarities in maturity and petrography of the coals indicate that the coal was transported by iceberg or sea ice rafting more than 1000 km to the south. It is suggested that during intervals of oxygen-isotope stage 6 drifting ice carried abundant coal fragments from the eastern Arctic Ocean southward through the Fram Strait into the eastern Norwegian Sea. Thus, surface circulation was then opposite to that of today.

The set of palaeoceanographic data for the northern oceans in Pleistocene times has strongly increased from research of the past decades (Thiede *et al.* 1990). However, not very much is known about surface currents during the ice ages. Some attempts at reconstruction were made using the quantitative dispersal patterns of ice rafted sand (Ruddiman 1977a, b) or by computer modelling of the estimated behaviour of a suggested permanent ice covered ocean (Lindström & MacAyeal 1986). Other studies deal with surface currents and ice transport in the Norwegian–Greenland Sea and Fram Strait (Kellogg *et al.* 1978; Gard 1988a, b; Gard & Backman 1989; Henrich *et al.* 1989; Köhler & Spielhagen 1990).

Dropstones as an ice drift indicator

During glacial times, vast areas of the northern hemisphere continents were covered by huge ice sheets (Andersen 1981). Due to the movement of the ice and its erosional capability, ice streams strongly modified the underlying land masses. Eroded rock fragments were either deposited as moraines or glacial outwash in valleys, fjords and shallow marine areas or incorporated in debris-rich zones in the ice (Drewry 1986). Glacimarine processes involving iceberg rafting of debris can be observed in presently glaciated nearshore areas (Dowdeswell 1989, Dowdeswell & Dowdeswell 1989). When ice streams terminate in the sea, icebergs are released and drift in the ocean, forced by surface currents and winds. Due to continuous meltout, enclosed rock fragments, termed 'dropstones', are released and sink to the sea floor, leaving behind a carpet of specific lithologies in marine sediments that marks the drift path of the ice.

Sea ice is also an important agent for sediment transport in the Arctic Ocean. Recent investigations of sea ice in the eastern Arctic Ocean, the Barents Sea and the Fram Strait (Pfirman *et al.* 1989a, b) revealed that sea ice strongly contributes to the sediment transport in ice-covered seas. Sediment incorporation occurs during sea ice formation in shallow seas (e.g. Siberian shelves). Only very rare coarse-grained material (> 300μm) was observed in open ocean sea ice of the investigated areas (Wollenburg, pers. comm.). Off the coast of Northeast Greenland, pebble-sized rock fragments were found in landfast ice. The role of sea ice as a sediment transport agent during glacial times is still under discussion.

From DOWDESWELL, J. A. & SCOURSE, J. D. (eds), 1990, *Glacimarine Environments: Processes and Sediments*. Geological Society Special Publication No 53, pp 235–251.

In this study, a method for surface current reconstructions during glacial times will be presented. It is based on the idea that dropstones can be used as tracers for the pathways of drifting ice if the area of provenance of the dropstone rock type is known. This can be converted into a simple model of surface currents for various times during glacial stages. Detailed knowledge of the dropstone source regions is required for palaeocurrent reconstructions. If these locations are known, the ice drift tracks can be reconstructed by following the dispersal patterns of the dropstones in deep-sea sediments. Easily recognizable, characteristic rocks from restricted, well known outcrops are the best tracers. In the Nordic Seas, appropriate lithologies for this purpose are rhomb porphyries from the Oslo–Graben area, spiculites from Spitsbergen or chalk from northern central Europe. Unfortunately, such specific rocks are very rare among dropstones of the Norwegian–Greenland Sea. In the Arctic Ocean and the Fram Strait, it is even more difficult to find rocks serving as a tracer, because detailed descriptions of the geology of the circum-Arctic regions are rare.

Thus, other lithologies with widespread exposures had to be considered. This study is focussed on unusually frequent coal particles that were found in deep sea sediments from the Norwegian Sea, the Fram Strait and the eastern Arctic Ocean. Multiple lines of evidence are used to outline possible source regions for the coal. Regional variations in coal abundance, the stratigraphical extension of coal occurrences in cores from different regions and possible extensions of ice sheets are used in a stepwise process to exclude less likely source areas. In addition, petrographic analyses of coal samples (vitrinite reflectance measurements and description of the maceral composition) were carried out to characterize the coal and compare it with possible sources.

Coal as an indicator of ice drift direction

Coal fragments are a good tracer, because they are easy to characterize by their composition, and important coal occurrences are relatively rare on continental areas adjacent to the Norwegian–Greenland Sea. Exceptions are Spitsbergen (Carboniferous to Tertiary coals, Harland *et al.* 1976) and Franz Josef Land (Carboniferous to Cretaceous coals, Dibner & Krylova 1963). The bedrock geology of the Barents Sea is similar to Svalbard and Franz Josef Land, except that no Tertiary rocks are found (Ulmishek 1985, Elverhøi *et al.* 1988). Minor coal exposures are present on the Barents Shelf sea floor (Elverhøi, pers. comm.).

Recent currents and ice drift

The present water exchange between the Arctic Ocean and the Norwegian–Greenland Sea, connected by the Fram Strait and the Barents Sea, is a significant factor for the climate in the North American–European–Arctic domain. Knowledge of the recent current system in this area (Fig. 1) has greatly improved from oceanographic research in the past decades (Aagaard 1970, Coachman & Aagaard 1974, Swift 1986, Anderson *et al.* 1989) but is still incomplete for the Arctic Ocean. Today, two main current systems are dominating in the Arctic Ocean: the Beaufort Gyre in the west and the Transpolar Drift stream in the east (Gordienko & Laktionov 1969). The Beaufort Gyre is an anticyclonic current in the Amerasian Basin. In the Eurasian Basin, the Transpolar Drift stream transports sea ice from north of Siberia through the Eurasian Basin to the Fram Strait. Running through Fram Strait and along the Greenland coast to north of Iceland (Fig. 1), this cold ice-transporting current is called the East Greenland Drift (Vinje 1982). The Norwegian Current and the West Spitsbergen Current are northern extensions of the North Atlantic Current and transport relatively warm water masses as far as about 81°N (Pickard & Emery 1982, Johannessen 1986, Swift 1986).

South of Spitsbergen, cold water masses from the western Barents Sea are deflected to the north. This feature is of significant importance for the present sea ice distribution. The Barents Sea receives considerable amounts of ice from the Kara Sea and the Arctic Ocean (Vinje 1978, 1982). Buoy experiments (MIZEX '87 Group 1989, Pfirman *et al.* 1989a) showed that this ice may drift with the West Spitsbergen Current to the north, where it loops back to the southwest into the East Greenland Current. Due to the pronounced thermal gradient between the temperate Norwegian Current and the cold East Greenland Current, the western half of the Norwegian–Greenland Sea can be covered by sea ice (Vinje 1985).

Regional setting

The Norwegian Sea sediments were studied in five long gravity cores, recovered during cruise Me 2/2 of RV METEOR (Gerlach *et al.* 1986) between Jan Mayen and the Vøring Plateau (Fig. 2). Two long cores were obtained from the Iceland Sea (RV POLARSTERN cruise ARK

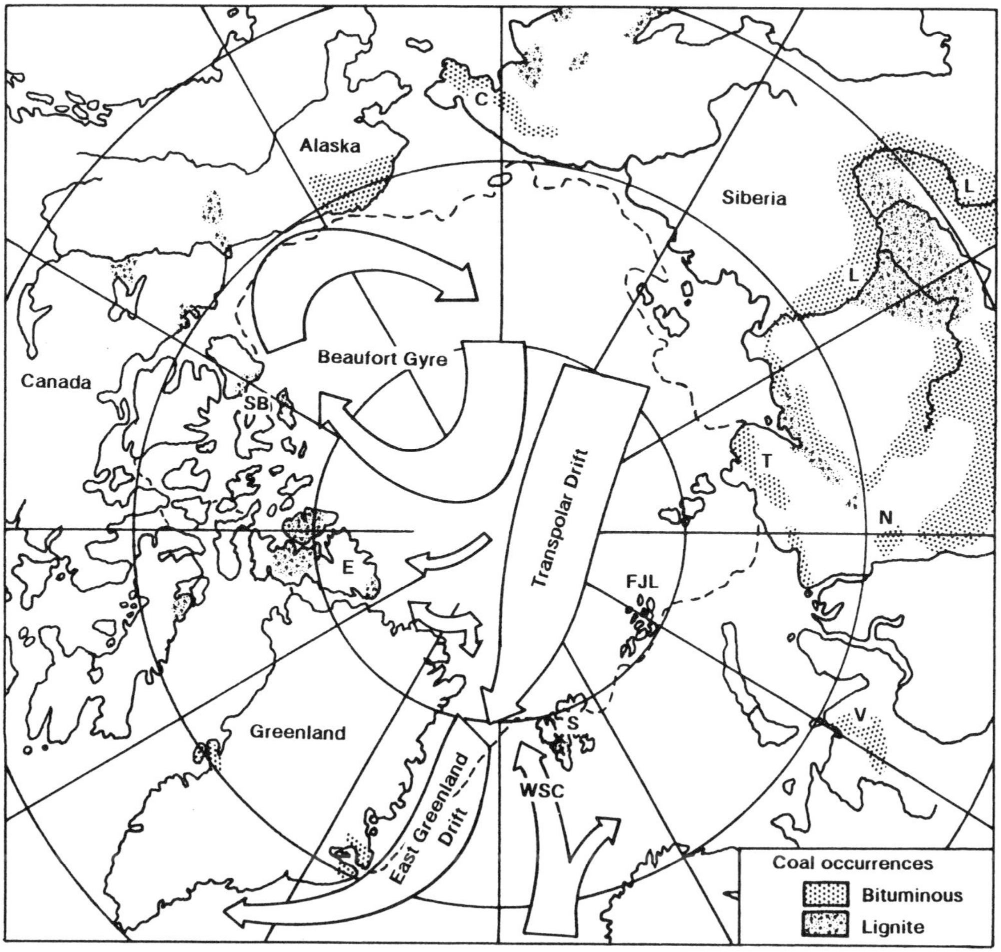

Fig. 1. Ocean surface circulation pattern, geography and coal occurrences in the North Polar area. In addition to stippled areas, important coal exposures discussed in the text are also found on Spitsbergen and Franz Josef Land. Hatched line marks average minimum extent of sea ice. C, = Chukchi Peninsula; E, = Ellesmere Island; FJL, = Franz Josef Land; L, = Lena River; N, = Norilsk; S, = Spitsbergen; SB, = Sverdrup Basin; T, = Taymyr Peninsula; V, = Vorkuta; WSC, = West Spitsbergen Current. Redrawn from Polar Regions Atlas (Anon 1978).

II/5, Augstein 1984b). Two box cores were studied from the southern Greenland Sea (RV POSEIDON cruise 142, von Bodungen 1988).

In the Fram Strait, six long cores of glacimarine sediments were taken from RV POLARSTERN during the cruises ARK I-IV in 1983–1987 (Augstein 1984a, b, Spielhagen *et al.* 1988). Data from five long sediment cores from the eastern Arctic Ocean, obtained during cruise ARK IV/3 of RV POLARSTERN in 1987 (Spielhagen *et al.* 1988) between 21°–33°E and 82°–87°N, are presented in this paper. The core locations are aligned on a transect between the northern Spitsbergen/Barents Sea continental margin and the eastern Arctic Nansen–Gakkel Ridge. All cores (Table 1) were composed of undisturbed, well-stratified sediments.

Methods

The analysed sediments

The cores were mainly composed of brownish, olive grey and dark grey glacimarine muds (Fig. 3). In the Norwegian-Greenland Sea, glacimarine sediments can be distinguished from interglacial deposits by common occurrence of

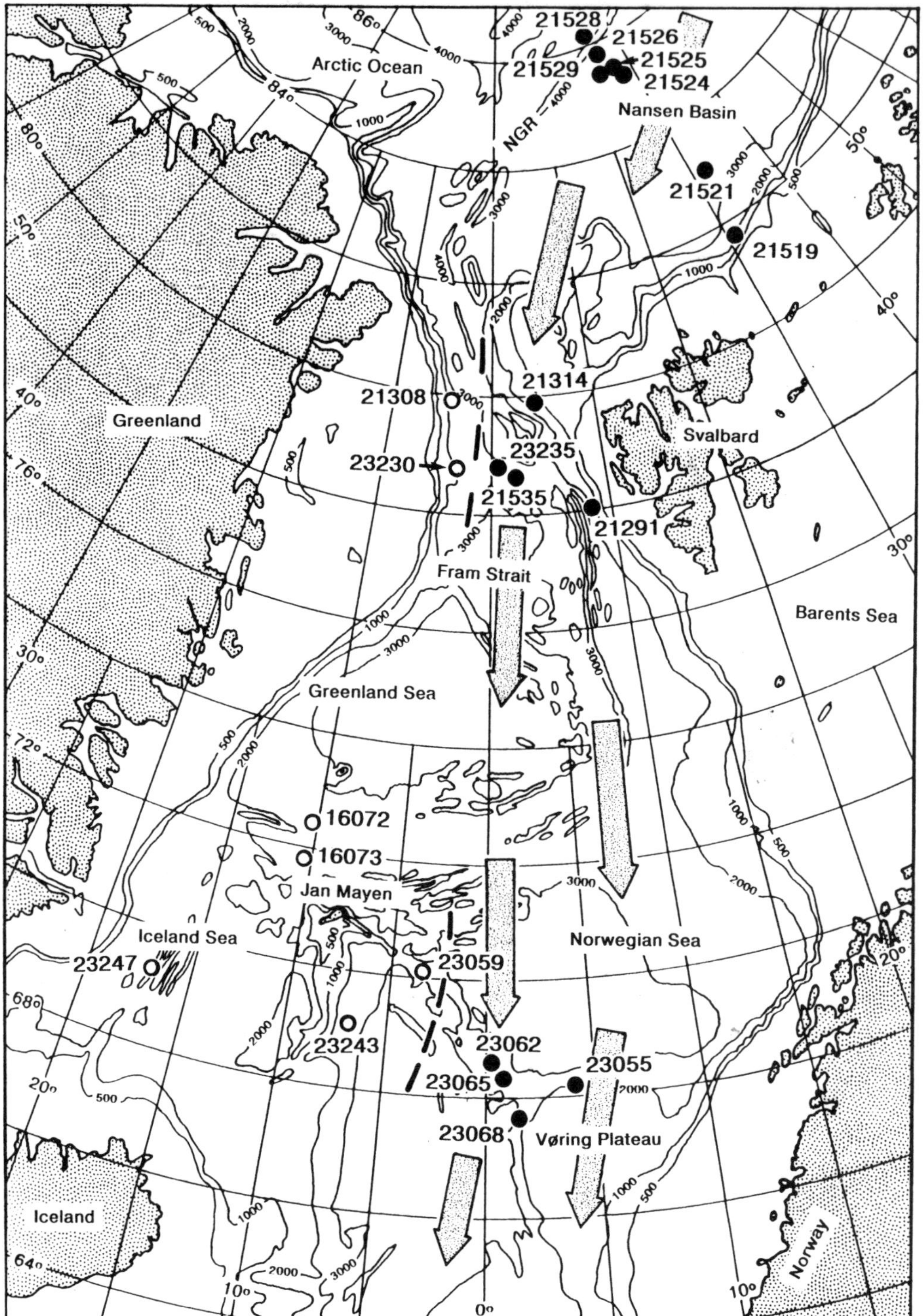

Fig. 2. Location map of analysed cores. Dark dots mark core sites with coal particles in sediments from oxygen-isotope stage 6, open circles mark core sites with no coal. Large arrows indicate proposed ice drift with coal particles during early and late oxygen isotope stage 6.

Table 1. *Location data and length of cores. At site 21526 two cores were obtained from different water depths on the Nansen–Gakkel Ridge*

Site	Area	Latitude	Longitude	Water depth (m)	Core length (m)
16072	Greenl. Sea	N 72°34.1′	W 10°28.0′	2176	0.43
16073	Greenl. Sea	N 71°55.2′	W 11°33.4′	2307	0.48
21291	Fram Strait	N 78°00.6′	E 08°03.1′	2400	4.92
21308	Fram Strait	N 80°01.0′	W 04°49.8′	1444	3.49
21314	Fram Strait	N 80°00.1′	E 04°29.7′	1382	4.79
21519	Fram Strait	N 81°47.4′	E 31°30.1′	2998	5.10
21521	E. Arct. Oc.	N 82°56.5′	E 32°05.2′	3752	4.98
21524	E. Arct. Oc.	N 85°21.8′	E 26°12.9′	3646	4.23
21525	E. Arct. Oc.	N 85°32.4′	E 25°19.4′	3366	3.65
21526	E. Arct. Oc.	N 85°55′	E 22°43.9′	4676/5211	4.50/3.52
21528	E. Arct. Oc.	N 86°08.0′	E 23°07.0′	3990	2.42
21529	E. Arct. Oc.	N 85°23.1′	E 21°44.1′	2917	3.70
21535	Fram Strait	N 78°44.8′	E 01°52.8′	2557	4.96
23055	Norw. Sea	N 68°25.4′	E 04°01.3′	2311	6.67
23059	Norw. Sea	N 70°18.3′	E 03°07.4′	2281	6.22
23062	Norw. Sea	N 68°43.7′	E 00°10.1′	2244	7.00
23065	Norw. Sea	N 68°29.7′	E 00°49.1′	2802	7.85
23068	Norw. Sea	N 67°50.0′	E 01°30.3′	2230	7.60
23230	Fram Strait	N 78°51.5′	E 04°46.8′	1235	5.12
23235	Fram Strait	N 78°51.9′	E 01°19.0′	2456	8.18
23243	Iceland Sea	N 69°22.3′	W 06°32.9′	2673	7.67
23247	Iceland Sea	N 69°29.5′	W 17°06.8′	1355	8.75

dropstones, dark grey colour and a minor proportion of biogenic components (Henrich *et al.* 1989).

In the Arctic Ocean, glacial and interglacial sediments resemble each other. Sediments from the eastern Arctic Ocean and the Fram Strait contain common dropstones throughout the entire core and the content of biogenic components seems to be strongly influenced by variations in carbonate solution. Samples were taken as slices of 2–5 cm at 6–12 cm intervals. A representative sample split (usually 300–500 grains) of the >500μm-fraction was analysed under a binocular microscope for the qualitative dropstone composition using a classification scheme of about 70 different lithologies. Coal particles were easy to identify by their dark colour, high light reflection and relics of plant structures. In cases of doubt, combustibility of particles was an additional indicator.

Stratigraphy

A reliable oxygen isotope stratigraphy has been established for the Norwegian Sea cores (Henrich *et al.* 1989, Fig. 3). The isotopic stage

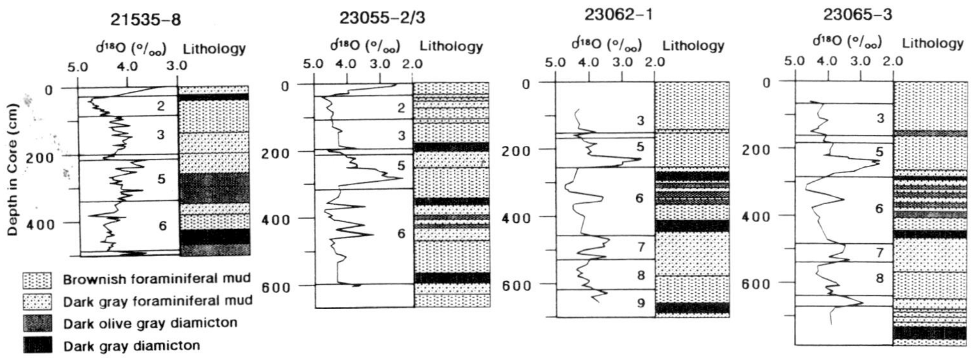

Fig. 3. Lithostratigraphy and oxygen isotope records of cores 21535–8 (Central Fram Strait), 23055–2/3, 23062–1 and 23065–3 (Norwegian Sea). After Henrich *et al.* (1989) and Köhler & Spielhagen (1990).

boundaries and ages were correlated according to the SPECMAP time scale (Imbrie *et al.* 1984). Time periods of high coal accumulation were calculated by interpolation, assuming constant average sedimentation rates between consecutive stage boundaries.

Good stratigraphic control for the cores from the Fram Strait is based on oxygen-isotopes (21535–8, Köhler & Spielhagen 1990, cf. Fig. 3), on the U–Th method (23235–2, Eisenhauer *et al.* 1990), and on nannofossil stratigraphy (21291–4, 21314–4, 23235–2, Gard 1987, 1988a). Preliminary oxygen isotope data from core 23230–2 (Spielhagen & Köhler, pers. comm.) and lithostratigraphic correlation with other cores indicate that oxygen isotope stage 6 may be reached in the lower part of the core. Core 21308–4 can be roughly correlated to 23230–2, but the presence of stage 6-sediments in this core is uncertain. Two cores from the eastern Arctic Ocean have been analysed for nannofossil stratigraphy (21521–15 and 21529–8, Baumann 1990). Cores 21524–2, 21525–7 and 21528–3 were correlated to these cores by using palaeomagnetic and physical properties data (Nowaczyk, pers. comm.). All assigned ages of the analysed cores from the eastern Arctic Ocean should be regarded as preliminary.

Coal petrographic analyses

Sediment samples and separated coal fragments were available for organopetrographic investigations. 21 samples of coal fragments and dispersed organic matter were analysed. Random vitrinite reflectance measurements (R_r, %) and descriptions of the maceral composition were carried out according to standardized procedures using granular polished sections of <1 mm grain size. For microscopic fluorescence investigations, blue light excitation with a high pressure mercury bulb, an excitation filter BG 12/3 mm, and a 530 nm barrier filter were used.

Results

General dropstone composition

The analysed dropstone composition varies with age in the cores. With respect to glacial interglacial cycles, characteristic compositions can be defined for the Arctic Ocean, the Fram Strait and the Norwegian–Greenland Sea. Apart from coal particles, which will be discussed in the next section, dropstones consist of clastic sedimentary rocks (sandstones and grey or dark shales), magmatic and metamorphic rocks and quartz and feldspar grains. In the eastern Arctic Ocean, the dropstone composition is generally dominated by silt- and sandstones, quartz and feldspar grains and quartzitic rocks. In the central and eastern Fram Strait, sediments from oxygen isotope stages 1, 3, 5, and 7 are dominated by dropstones of mostly clastic sedimentary rocks. Stages 2, 4, and 6 are dominated by magmatic and metamorphic rocks, quartz and feldspar. This composition is also present in all cores of the western Fram Strait.

In the Norwegian Sea, magmatic and metamorphic rocks dominate in the east towards the Fennoscandian shield whereas grey and black siltstones and shales dominate in the west. Downcore, a general increase of the silt- and sandstone proportion can be observed in the Norwegian Sea. Near the Norwegian continental margin and on the southern Vøring Plateau, the proportion of magmatic and metamorphic rocks increases significantly.

Coal particles in sediments of the Norwegian Sea, the Fram Strait, and the eastern Arctic Ocean

Coal content and regional distribution. In Norwegian Sea sediments, coal fragments are usually absent or present in minor amounts of 1–2% of the analysed > 500µm-fraction in sediments from oxygen isotope stages 1–5. This indicates that coal must be regarded as a minor dropstone constituent in the Norwegian Sea. The only region where coal percentages are significantly enhanced is northwest of Jan Mayen (sites 16072 and 16073). In this area (Fig. 2), near the recent East Greenland Drift, up to 5% coal fragments can be detected in Holocene sediments. In sediments from oxygen isotope stage 6, however, coal is an important dropstone component with peak values of up to 30% (site 23068, Fig. 4). The high coal content is restricted to the eastern Norwegian Sea where it is registered in a zone approximately 250 km wide. Coal content decreases to the northwest (sites 23062 and 23065) and to the east (site 23055) and can be observed only during early and late oxygen isotope stage 6. West of site 23062 (e.g. at site 23059), the coal content in stage 6 sediments is not significantly higher than during following periods. Other analysed cores from the central Vøring Plateau did not penetrate as deep as oxygen isotope stage 6 (Vogelsang, pers. comm.). No coal particles were found in stage 6 sediments of the Iceland Sea (sites 23243 and 23247).

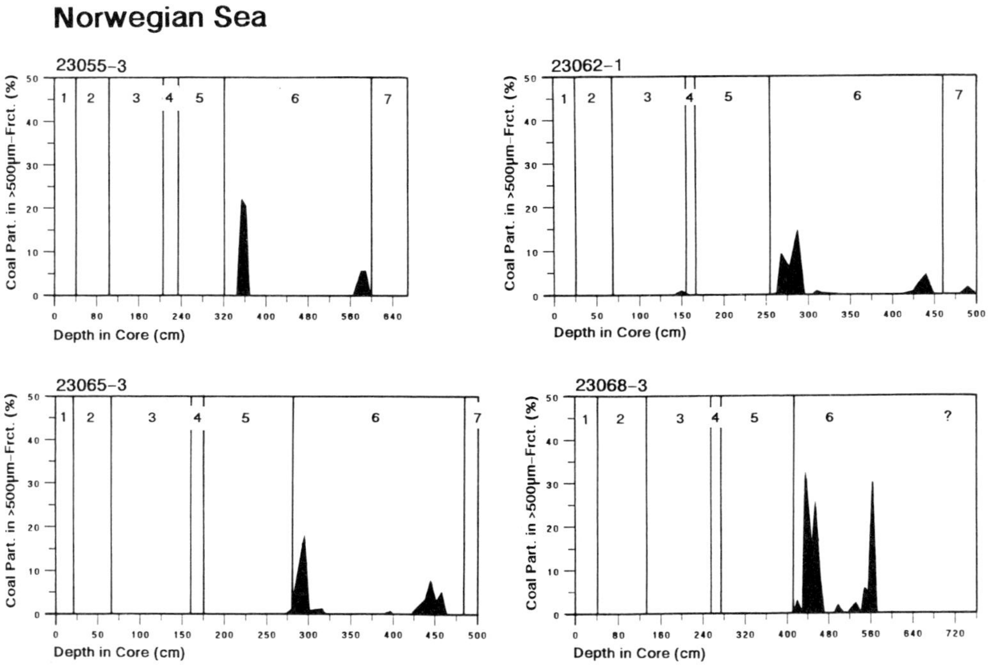

Fig. 4. Coal content in > 500 μm-fraction (grain-%) and oxygen–isotope stage boundaries in analysed cores from the Norwegian Sea.

In the Fram Strait, the pattern of coal occurrence is similar to the Norwegian Sea, with very low values throughout the entire upper succession and a high content in sediments from oxygen isotope stage 6 (Fig. 5). Two peaks in coal abundance can be determined in core 21535–8 from the central Fram Strait. Total amounts reach up to 66% coal particles within the > 500μm-fraction, which is considerably more than in the Norwegian Sea and the Arctic Ocean. The lower one of two peak-complexes shows higher values than the upper one. Coal content is highest in cores from the central Fram Strait (sites 21535 and 23235) and decreases towards the Spitsbergen continental margin (sites 21291 and 21314). No coal was found in cores from the Greenland continental margin (sites 21308 and 23230). At site 21314, coal particles are also present in sediments tentatively dated as oxygen isotope stage 8.

In eastern Arctic Ocean sediments, coal particles can be found nearly continuously during oxygen isotope stage 6. Here, coal content varies between 20 and 50% of the analysed fraction (Fig. 6). Highest coal concentrations are observed in sediments from the Nansen–Gakkel Ridge (86°N) and decrease towards the Barents Sea continental margin. In contrast to the Norwegian Sea and the Fram Strait, coal is also present in minor amounts in younger sediments at sites 21521 and 21528. No coal particles were reported from sediments of the central Arctic Ocean (Morris *et al.* 1985, Mudie & Blasco 1985, Scott *et al.* 1989).

Coal-associated dropstones In Norwegian Sea core 23068–3, silt- and sandstones together with abundant quartzites dominate the dropstones that are associated with the upper coal peak, whereas the lower one is associated with quartzites only. In eastern Norwegian Sea sediments, dropstones from between the coal-bearing sequences consist of 70% silt- and sandstones. Most abundant dropstone lithologies associated with coal in the Fram Strait are quartz and feldspar grains, gneisses, and mica-schists. The coal particles in sediments from the Nansen-Gakkel Ridge (eastern Arctic Ocean) are found together with siltstones, quartz and feldspar grains and quartzitic rocks. In the Nansen Basin, siltstones and quartzites are most abundant.

Analyses of coal particles Extracted coal particles are mostly angular or only slightly rounded and of deep black colour. Surface light reflection varies between mat and shiny. Often relics of fibrous plant structures can be observed. Evi-

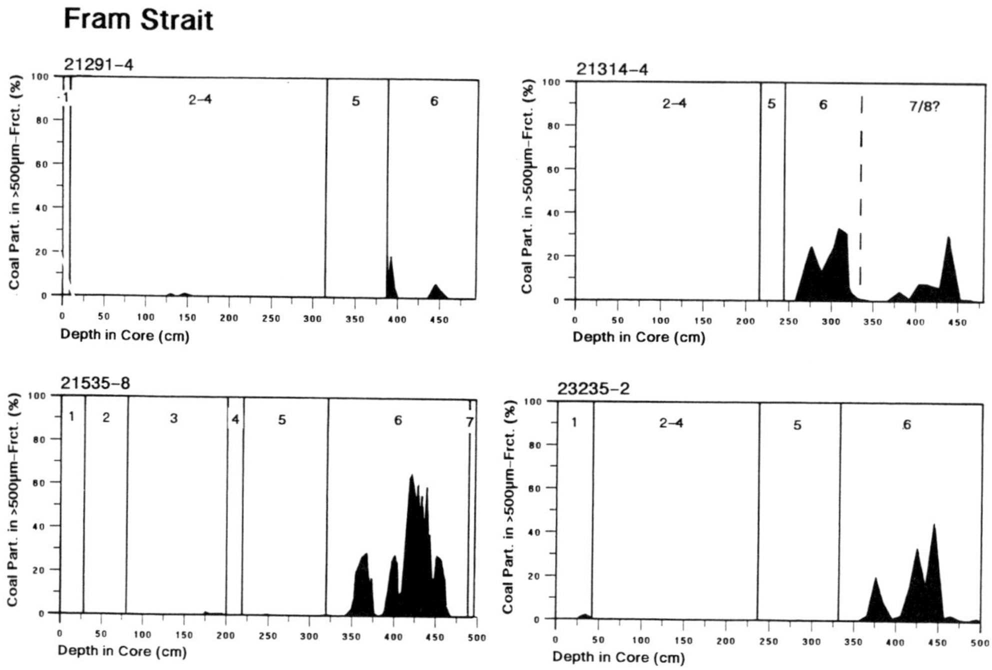

Fig. 5. Coal content in > 500 μm-fraction (grain-%) and oxygen-isotope stage boundaries in analysed cores from the Fram Strait.

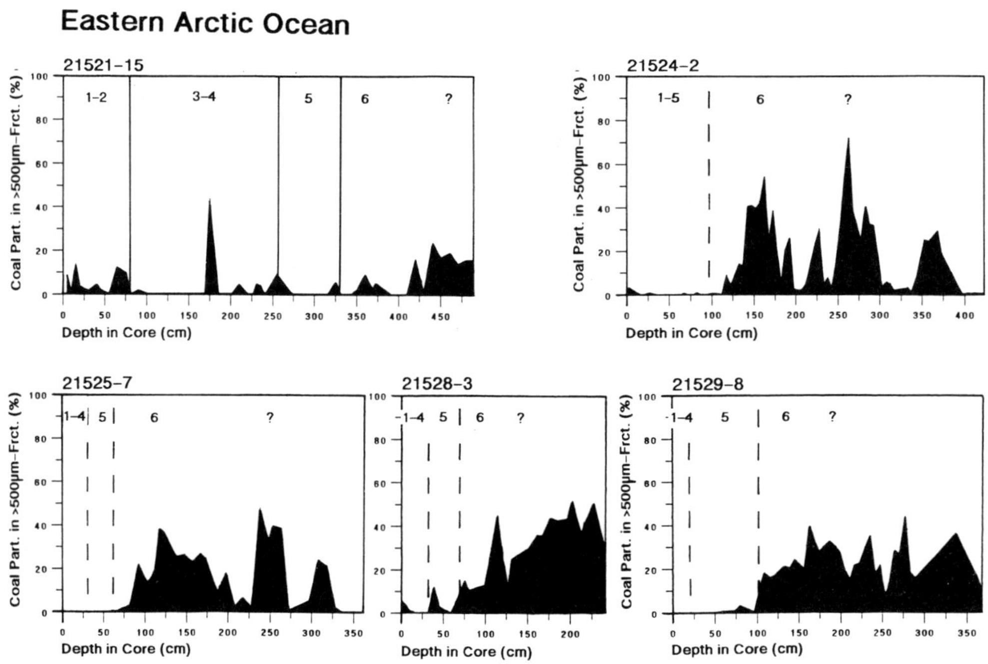

Fig. 6. Coal content in > 500 μm-fraction (grain-%) and oxygen-isotope stage boundaries in analysed cores from the eastern Arctic Ocean.

dence for weathering of the coal particles is rare. Some particles have visible cracks.

Present results of petrographic investigations on dispersed sediment organic matter (OM) and coal fragments have to be regarded as preliminary because the total organic fraction of core sections has yet to be analysed. Sample types and results of vitrinite reflectance measurements are given in Table 2.

Organopetrographic analyses reveal that both coal fragments and nearly the total OM from the analysed sediments (sites 21524, 21526, 23055 and 23235) are allochthonous. Only greenish fluorescing liptodetrinite, sporadically occurring in some samples, may be autochthonous.

With a few exceptions which will be discussed later, the coal or coaly shale fragments can be divided in two classes. The first type is composed mainly of vitrinite (tellocollinite with more or less recognizable plant structures); it is in part rich in pyrite. The second type is composed of vitrinite with liptinite (sporinite, cutinite, liptodetrinite and some resinite) as a minor constituent. Initernite is rare.

The dispersed OM from the analysed sediments is dominated by the very resistant inertinite (as inertodetrinite), while vitrinite and liptinite are rare. In samples from core 23055–3, small heterogeneous coal particles could be identified microscopically. Graphite sporadically occurs as a residue of reworked metamorphic rocks.

The random vitrinite reflectance (R_r) of most of the analysed coal fragments and OM from the different sites is rather uniform. R_r-values range mainly between 0.31% and 0.44%, with an average of 0.38%. This means that most of the coal is of sub-bituminous rank. Lowest values were measured on coalified (vitrinitic) wood remains with well preserved cell structures which often have low reflectance.

These preliminary results show no discernible trends in rank and composition of the coal particles in terms of geographic direction or depth in cores. It can be assumed that most coal grains and dispersed OM stem from a single source of sub-bituminous coal. However, this should be verified by further systematic investigations.

Some rare coal particles from the Arctic Ocean sites 21521 (368–370 cm) and 21528 (113 cm) show exceptionally high inertinite contents (semifusinite, fusinite, and some inertodetrinite) and distinctly higher R_r-values of 0.51% and 0.49%. These values mark the approximate border between sub–bituminous and bituminous coal. Another exception (although not reliably verifiable) comes from the samples from site 23235 (Fram Strait). Measured R_r-values are only 0.28% and 0.29%, which are characteristic for soft brown coals to lignites.

In some samples from the eastern Arctic Ocean, fragments of quartzitic schist were found, containing dispersed OM of anthracitic

Table 2. *Results of random vitrinite reflectance measurements of coal fragments (CF) and dispersed organic matter (OM)*

Core	Sample Depth (cm)	Sample Type	R_r (%)
21519–12	194	CF	0.39
21521–15	369	CF	0.35
21521–15	475	CF	0.42
21524–2	20	OM	0.39
21524–2	29	OM	0.37
21524–2	37	OM	0.35
21524–2	48	OM	0.35
21525–7	80	CF	0.44
21526–12	258	OM	0.34
21526–20	345	OM	0.44
21528–3	113	CF	0.49
21529–8	133	CF	0.31
21529–8	165	CF	0.36
21529–8	276	CF	0.41
23055–3	203	OM	0.39
23055–3	297	OM	0.36
23055–3	360	OM	0.38
23055–3	387	OM	0.40
23055–3	582	OM	0.39
23235–2	384	OM	0.28
23235–2	392	OM	0.29

rank. Particles were big enough for reflectance measurements in only one sample from 21521 (460–470 cm). The R_r–value is 5.49%, with maximum reflectance of 6.38% (meta–anthracite).

Coal occurrences in circum-Arctic areas

Terrestrial and shallow marine occurrences of coal in the circum-Arctic region (Fig. 1) can be discussed as potential source areas of coal particles in the analysed sediments. On Spitsbergen, several coal bearing sequences developed from Devonian to Tertiary times: in the Devonian, Carboniferous, Triassic, Lower Cretaceous and Palaeogene (Cutbill *et al.* 1976, Harland *et al.* 1976). Two of them (Billefjorden Group/Lower Carboniferous and Firkanten Fm./Palaeocene) are economically mined. Investigations of the gravel fraction from the Barents Sea floor showed that sedimentary strata similar to those on Svalbard continue below the bottom of the sea floor (Dibner *et al.* 1970, Edwards 1975, Elverhøi *et al.* 1988).

According to this similarity, coal seams are likely to occur in Barents Sea rocks. Coal particles can be found in minor amounts in surface samples from the central Northern Barents Sea (Elverhøi, pers. comm.). Coal on Franz Josef Land is known from Carboniferous shales, Jurassic sandstones, and lower Cretaceous sandstones (Dibner & Krylova 1963).

Various coal exposures are known from the Soviet Arctic regions. In Siberia coal seams are found in the Vorkuta area in Silurian shales and Cretaceous sandstones (Churkin *et al.* 1981, Ziegler 1988), near Norilsk in the upper Carboniferous and Permian (Wagner & Berthold 1979), on the Taymyr Peninsula in Permian sandstones, middle Triassic shales, and Cretaceous sandstones (Churkin *et al.* 1981), along the Lena River in upper Triassic/lower Jurassic, Cretaceous and Miocene rocks (Wagner & Berthold 1979), and in central Siberia in Permian and middle/upper Triassic silt- and sandstones (Churkin *et al.* 1981).

In northern Alaska coal seams developed in upper Cretaceous marine and non-marine sandstones (Dutro 1981). Early Cretaceous rocks from Ellesmere Island contain coaly sandstones (Churkin *et al.* 1981). In the eastern Sverdrup Basin coal deposits are reported from upper Jurassic to lower Cretaceous and lower Tertiary rocks (Kerr 1981). Lower Carboniferous and Jurassic coal was found in Northeast Greenland (Ziegler 1988).

Coal fragments investigated in this study are characterized by average random vitrinite reflectance values of 0.38% (Table 2, Fig. 7). On Spitsbergen, only the youngest, not very widespread and not economically viable coals (Aspelintoppen Fm./Lower Tertiary) and very few older coals have low R_r-values of about 0.4%. Usually, vitrinite reflectance exceeds 0.5% and is around 0.7% or higher for the minable coals of Carboniferous and earliest Tertiary age (Harland *et al.* 1976, Manum & Throndsen 1978). Even if these coal occurrences extend to the bottom of the Barents Sea, it is doubtful that they are the source of ice rafted coal because of the differences in vitrinite reflectance (Fig. 7).

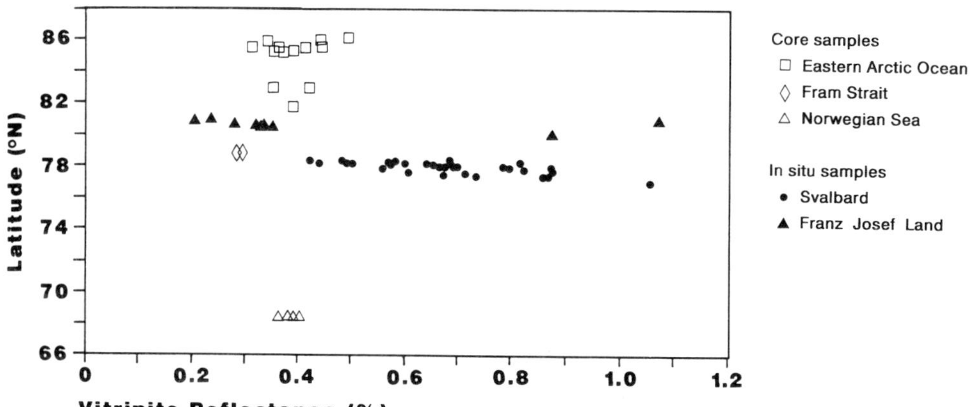

Fig. 7. Vitrinite reflectance (R_r) of coal measures on Spitsbergen and Franz Josef Land, and of analysed coal particles. Spitsbergen data from Manum & Throndsen (1978), Franz Josef Land data on vitrain refraction from Dibner & Krylova (1963), converted to R_r-values after van Krevelen (1961).

Considering the R_r-values of the analysed coal fragments, the Cretaceous to Palaeogene low-rank coals from Franz Josef Land could be a possible source, because they have similar R_r-values (Dibner & Krylova 1963, Fig. 7), as well as coals from northern Siberia. In particular, the Cretaceous coal occurrences on the Taymyr Peninsula and possibly other brown coal deposits within the drainage basins of the Jenessei and Lena Rivers represent potential sources. A provenance from the American continent is unlikely because no coal has been reported from the deep-sea sediments of the Amerasian Basin of the Arctic Ocean (Morris *et al.* 1985, Mudie & Blasco 1985, Scott *et al.* 1989).

Discussion

Towards a palaeoceanographic model

As demonstrated above, the sedimentological analyses of the cores from the eastern Arctic Ocean, the Fram Strait and the Norwegian Sea show that coal particles are a rather unusual sediment constituent in sediments younger than oxygen isotope stage 6. However, in sediments from oxygen isotope stage 6 (approx. 186–128 ka), they are abundant in the eastern Arctic Ocean and the Fram Strait and common in the Norwegian Sea in some sequences. The stratigraphical distribution of coal within the stage 6 sediment column gives evidence that the sedimentation of coal particles occurred almost at the same time in both areas. If the same source is assumed for coals from both areas, a southward ice-drift from the Arctic Ocean through the Fram Strait to the Norwegian Sea can be proposed. A surface current most likely transported ice with incorporated coal fragments from the Arctic to the Norwegian Sea, losing part of this ice by melting on the way south and releasing the enclosed sediment to the sea floor.

Time-correlation of coal occurrences

Selected core segments can be dated using the established oxygen isotope stratigraphy for cores from the Norwegian Sea and the Fram Strait (Vogelsang, in: Henrich *et al.* 1989; Köhler & Spielhagen 1990). A shortening of the time periods with high coal input is observed between the Fram Strait and the eastern Norwegian Sea. Whereas coal influx in the Fram Strait occurred at a relatively high level nearly continuously from early to late stage 6 with only one short interruption, this is not the case in the Norwegian Sea.

In the Fram Strait, the first coal input pulse lasted from early to middle stage 6. Assuming an average linear sedimentation rate (LSR) of approximately 2.6 cm/ka during stage 6 (Köhler & Spielhagen 1990), a duration of about 33 000 years for this first coal input interval can be calculated (Fig. 8). It is characterized by highest coal percentages of 50–66% between approxi-

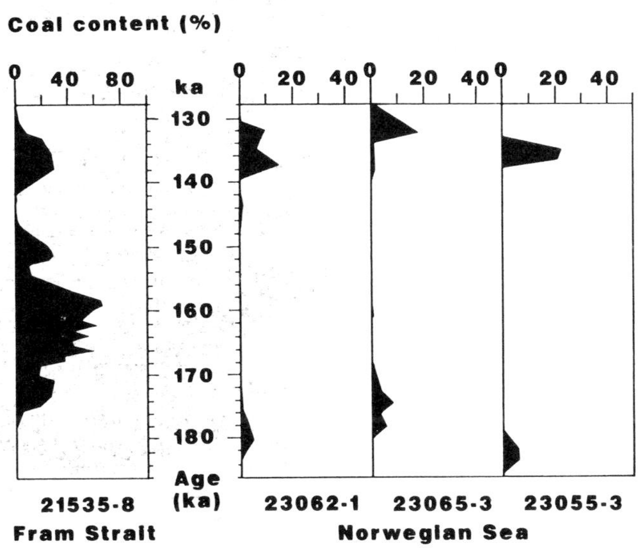

Fig. 8. Coal content and correlation of core sections from oxygen isotope stage 6. Coal content is given as grain-percent in > 500 μm-fraction. Ages are calculated from linear sedimentation rates (Henrich *et al.* 1989; Köhler & Spielhagen 1990).

mately 165–158 ka, followed by a minor peak with values of 25%, culminating at approximately 150 ka. After a short period with no coal deposition, the third, upper peak with a coal content of 30% culminated at approximately 138 ka.

In the Norwegian Sea, two distinct and separate, individual coal input peaks of limited duration can be defined (Fig. 8). Assuming an LSR of 4.8 cm/ka (Henrich *et al.* 1989), the early stage 6 event at site 23055 on the northern Vøring Plateau can be dated as approximately 182 ka, whereas the second coal input pulse occurred at about 136 ka. Each interval of coal deposition at site 23055 is calculated to last less than 5000 years. This indicates that the coal input reflects peak events of short duration in the Norwegian Sea. In the western Norwegian Sea (site 23065), two main coal input events can be recognized at 176 ka and 132 ka. At site 23062, which is the westernmost extension of the area with enhanced coal input, the coal-bearing levels are at 180 ka and 135 ka. These dates were calculated with average LSR of 3.55 cm/ka (23062–1) and 3.53 cm/ka (23065–3) (Henrich *et al.* 1989). According to E. Vogelsang (unpubl. data), the correlation of isotopic events within stage 6 would date the coal inputs at site 23068 as 175 ka and 134 ka.

Although the correlation of most coal input peaks in the Norwegian Sea and the Fram Strait is striking, there appears to be a time lag of about 1000–5000 years between the correlatable events. This may be an artifact caused by the sampling intervals and by the calculations of the dates, which are based on linear sedimentation rates. It is unlikely that sedimentation was constant throughout oxygen isotope stage 6 and in fact the coal input events may well have been isochronous.

If approximately 3000–7000 years are considered to represent the time span of uncertainty, the early (182, 180, 176 and 175 ka) and the late events (136, 135, 132 and 134 ka) may be correlated between cores 23055–3, 23062–1, 23065–3 and 23068–3. The late event can be linked also with the upper coal peak from core 21535–8 (Fram Strait, 136 ka).

In summary, only the late coal input pulse in the Fram Strait can be correlated with an equivalent event in the western and eastern Norwegian Sea. It occurred between 130 and 138 ka. The lowest and strongest Fram Strait peak is apparently not reflected in the Norwegian Sea. This may be explained by melting of most the coal–transporting ice in areas north and/or west of the core sites in the Norwegian Sea or by transport of this ice closer to Greenland. On the other hand, sedimentation may have been unusually high in earliest oxygen isotope stage 6 in the Fram Strait. This would allow correlation of the maximum in coal content around 430 cm core depth in core 21535–8 with the lower coal peak of the cores 23055–3, 23062–1 and 23065–3.

Sources and transport of coal particles

Size (up to 5 cm) and amount of coal particles in the analysed sediments provide convincing evidence for ice rafting. Surface currents and winds are not capable of keeping particles of this size in suspension, but icebergs can carry even big boulders over large distances. However, only fine-grained rock fragments of ≤ 100 μm were observed in three icebergs that were sampled in the Barents Sea during cruise ARK VI/1 of RV POLARSTERN in June 1989. On the other hand, high amounts of fine-grained sediments (≤ 300 μm) were found on sea ice in the eastern Arctic Ocean (Pfirman *et al.* 1989a, b).

When sea ice is formed in shallow waters (< 20–30 m), resuspended sediment can be incorporated from freezing, turbid waters (Barnes *et al.* 1982). Thus, the source of sea ice transported coal fragments in the Arctic Ocean must be located in or near shallow seas, such as the East Siberian Sea, the Laptev Sea, the Kara Sea and the Barents Sea. Other possible sources are coal exposures that are cut by rivers which discharge winter ice into the Arctic Ocean. The transport of coal particles by icebergs requires surficially exposed coal measures that were covered by glacier ice, terminating in the sea.

Reconstructions of ice sheets in circum-Arctic areas are mainly confined to the Weichselian glaciation (Grosswald 1980, Andersen 1981, CLIMAP Project Members 1981, Elverhøi & Solheim 1983, Biryukov *et al.* 1988). In the northern hemisphere, five large ice sheets existed at that time: the Laurentide/Cordilleran ice sheet in North America, the Greenland ice sheet, the Fennoscandian and the Barents Sea ice sheet in northern Europe and a Siberian ice sheet in northern Asia (Fig. 9).

The finds of coal particles and the two possible transport modes allow two hypotheses for the Arctic environment during oxygen isotope stage 6.

(i) Coal transport by icebergs requires an ice sheet near the source area that reached almost to the shelf edge.

(ii) Coal transport by sea ice requires coal exposures near the area of sea ice formation or fluvial transport to that area.

A wider extension of the Siberian ice sheet

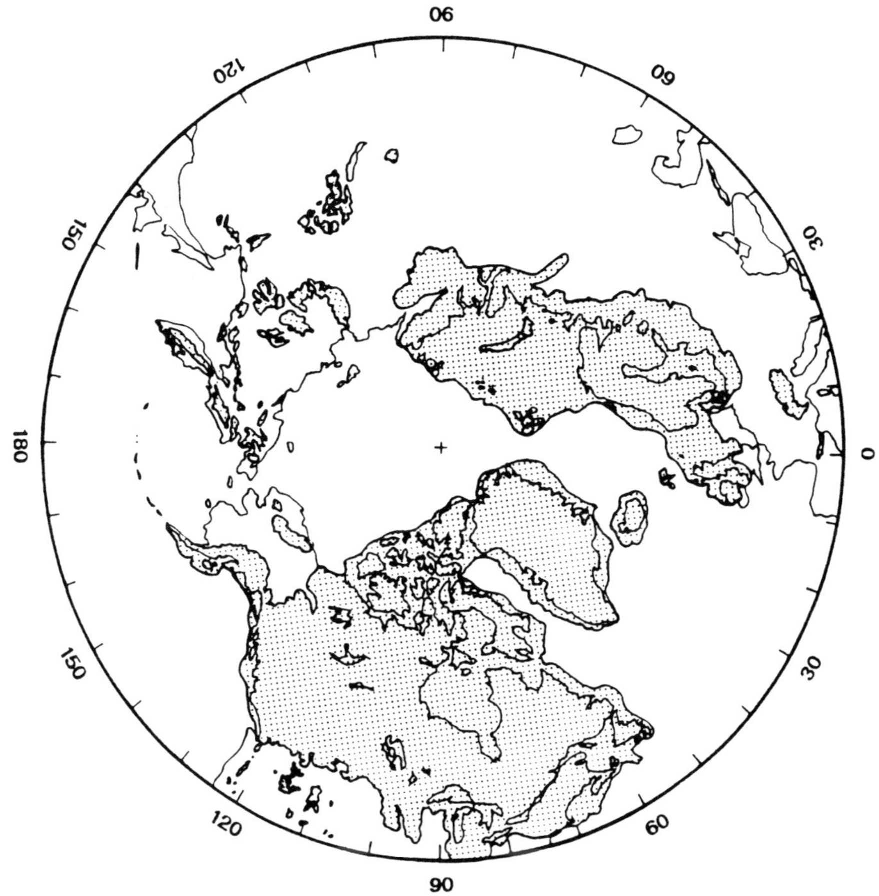

Fig. 9. Extent of Weichselian ice sheets in the northern hemisphere (after CLIMAP Project Members 1981).

during the Saalian (roughly corresponding to oxygen isotope stage 6) than during the Weichselian has been suggested by Flint (1971) and Woldstedt (1969).

The mode of ice transport (iceberg or sea ice rafting) cannot be determined from the size and shape of the coal fragments. However, palaeocurrents can be reconstructed by dispersal patterns of a sedimentary tracer (coal fragments), regardless of whether the tracer was transported by sea ice or icebergs. From the following observations it can be concluded that during oxygen isotope stage 6 the coal particles were transported from the eastern Arctic Ocean through the Fram Strait to the Norwegian Sea.

Coal-bearing stratigraphic intervals shorten southwards. In eastern Arctic Ocean sediments, coal particles are present in the entire oxygen isotope stage 6 sequence. In the Fram Strait, short periods with no coal input are recorded. In the Norwegian Sea, coal input is restricted to short intervals within stage 6.

Coal content of the sediments generally decreases southwards. The only exception is the central Fram Strait where the enrichment in coal may be explained by a process discussed below.

Composition and maturity analyses data of the coal particles are similar to some Siberian coals but different from northern European coals. In the Fram Strait, coal percentages decrease towards the Spitsbergen continental margin, excluding Spitsbergen's coal exposures as a possible source area.

Additional evidence for an Arctic origin of the coals is supplied by the qualitative composition of dropstones from sediments deposited near to the East Greenland Drift. Samples from northwest of Jan Mayen contain four to five times more coal fragments than equivalent samples from easterly locations (Bischof 1990). The East Greenland Drift transports mainly ice out of the Arctic Ocean southwards. Thus, some shelf regions adjacent to the Arctic Ocean may be the most likely source of the coal. The precise origin of the coal fragments and dispersed organic matter in the analyzed sediments is debatable. Although northern Siberia is a likely source area, it remains enigmatic as to why Fram Strait sediments show higher coal values than sediments from the eastern Arctic Ocean. It is possible that the area of maximum coal deposition was not sampled in the eastern Arctic Ocean. On the other hand, this phenomenon may result from a sorting process during ice transport. Coal particles have less than half the density of other sediment particles and may even float for short periods because of enclosed air. It is possible that the ice that carried coal lost many of the other sediment particles to the sea floor by melting, turnover and washoff on its way from the source area to the Fram Strait. On the way further south, sedimentation of coal and other particles from the ice continued and was diluted from ice that was admixed from other areas (Barents Shelf, Fennoscandian Ice Sheet; Bischof 1990). Finally, this resulted in glacimarine sediments in the Norwegian Sea in which only two significant coal peaks can be defined.

Surface current directions during early and late oxygen isotope stage 6

In terms of a general southward ice drift in the Nordic Seas during oxygen isotope stage 6, additional observations allow speculation about regional variations in the origin of ice and the distribution of coal particles. In the Fram Strait, the amount of coal in stage 6 sediments is highest in cores from the central part (21535–8 and 23235–2) and decreases towards the Spitsbergen margin (cores 21291–4 and 21314–4), whereas coal is absent in cores near to the Greenland margin (21308–4 and 23230–2). It can be concluded that during stage 6 the ice east of the Greenland margin did not have the same origin as the ice that covered the central and eastern part of the Fram Strait. Highest amounts of coal in the central Fram Strait indicate that the admixture of coal from Spitsbergen was limited or non-existant. This is supported by the coal analyses. Similarly to the Fram Strait, in the Norwegian Sea coal content in sediments from oxygen isotope stage 6 is highest in the eastern and central parts and decreases towards the Fennoscandian and Greenland continental margins.

The correlation of coal input events allows speculations about tracing the coal path from the Fram Strait to the Norwegian Sea. If the early inputs at sites 21535 (155–175 ka) can be correlated with the early inputs at sites 23055, 23062, 23065 and 23068 (182, 180, 176 and 175 ka), the coal-transporting ice mass covered at least the western Vøring Plateau and the Norwegian Sea from 0°–5°E. No data are available from the southeastern Vøring Plateau because the recovered sediments were younger than oxygen isotope stage 6 (Vogelsang, pers. comm.).

During the second input phase around 134 ka, recorded at all the above mentioned sites, the ice cover may have been wider and covering an even greater area of the Norwegian Sea than in early oxygen isotope stage 6. This assumption is based on the relatively high coal content (> 15%) in all cores east of 0°E in late stage 6, whereas this value was reached only in core 23068–3 in early stage 6. The reasons for the differing width of the coal-transporting ice mass are speculative. The distinct boundary between coal occurrences in sediments from stage 6 in sites each side of O°E and the lack of coal at site 23059 and in the Iceland Sea implies that mixing of ice from different sources was limited in the western Norwegian Sea. A strong ice zonation may have existed in the Norwegian–Greenland Sea, probably with ice from Greenland in the west, followed to the east by ice from the central Arctic Ocean, and further east by coal-transporting ice from the eastern Arctic Ocean. Finally, near the Norwegian continental margin, ice from the Fennoscandian ice sheet dominated. The degree of ice coverage in the Norwegian Sea and the motion of ice near to Fennoscandia remain speculative. It is possible that a northward current of variable intensity and width may have existed in the easternmost Norwegian Sea during the times of coal deposition. This is supported by findings of Cretaceous *inoceramus* prisms and chalk in some coal-bearing horizons from oxygen isotope stage 6 in the Norwegian Sea (Henrich 1990). Exposures of these lithologies are found in northern central Europe and Great Britain. Recent observations from the Fram Strait (MIZEX '87 Group 1989; Pfirman *et al.* 1989a) show that in contact areas of antiparallel ice-transporting currents, drifting ice and buoys can move oblique or even perpendicular to the mean currents. Circulation conditions may have been similar in the eastern Norwegian Sea during early and late oxygen isotope stage 6. Most of the ice probably came from the north (carrying coal from the Arctic Ocean), with admixtures from the east (Fennoscandia?) and subordinate contributions from the south (northern central Europe or Great Britain, carrying Cretaceous rocks).

Conclusions

Coal particles in sediments from the Fram Strait and the Norwegian Sea occur in probably isochronous layers from oxygen-isotope stage 6 (at about 179 ka and 134 ka), whereas in the eastern Arctic Ocean coal is present in sediments from the entire stage 6.

The coal particles are in most cases alike concerning maturity and petrographic composition. It is most likely that they are derived from the same source. The provenance of the coal particles cannot be definitely assigned but the coal exposures in northern Siberia are a possible source.

After the coal particles were incorporated in the ice, they were transported by floating sea ice or icebergs out of the Arctic Ocean through the Fram Strait to the Norwegian Sea at least twice during early and late oxygen-isotope stage 6. The ice was carried by a southward surface current to an area that was a minimum of 250 km wide at 68°N, 0°–5°E.

We wish to thank K.-H. Baumann, R. Botz and I. Wollenburg for valuable comments and critical discussions on coal occurrences and transport mechanisms. L. Birgisdottir and U. Struck provided additional samples. N. Nowaczyk made geophysical data available for intercore correlations. B. Pörksen and S. Rumohr helped with handling of the samples and drawing of the figures. H. Cordt and H. Erlenkeuser operated the mass spectrometer. R. Henrich, H. Kassens and E. Vogelsang are gratefully acknowledged for providing additional samples and unpublished data and for invaluable discussions on the proposed palaeoceanographic model. We thank one anonymous reviewer and especially A. Elverhøi for constructive reviews of the manuscript.

The Alfred-Wegener Institut für Polar- und Meeresforschung (Bremerhaven/FRG) supplied RV POLARSTERN for the ARK I–IV cruises. This study was supported financially by the Bundesministerium für Forschung und Technologie, the Deutsche Forschungsgemeinschaft, and the U.S. Office of Naval Research.

References

AAGAARD, K. 1970. The wind-driven circulation in the Greenland-Norwegian Seas. *Deep-Sea Research*, **17**, 281–291.

ANDERSEN, B. 1981. Late Weichselian Ice Sheets in Eurasia and Greenland. *In*: DENTON, G.H. & HUGHES, T.J., (eds), *The Last Great Ice Sheets*, Wiley, New York, 1–65.

ANDERSON, L.G., JONES, E.P., KOLTERMANN, K.P., SCHLOSSER, P., SWIFT, J.H., & WALLACE, D.W.R. 1989. The first oceanographic section across the Nansen Basin in the Arctic Ocean. *Deep-Sea Research*, **36**, 475–482.

ANON 1978. *Polar Regions Atlas*. Produced by the Foreign Assessment Center, Central Intelligence Agency.

AUGSTEIN, E., HEMPEL, G. & THIEDE, J. 1984a. Fahrtbericht (cruise report) der Polarstern-Reise ARKTIS I, 1983. *Berichte zur Polarforschung*, **17**, Bremerhaven.

——, HEMPEL, G., SCHWARZ, J., THIEDE, J. & WEIGEL, W. 1984b. Die Expedition ARKTIS II des FS Polarstern 1984. *Berichte zur Polarforschung*, **20**, Bremerhaven.

BARNES, P.W., REIMNITZ, E. & FOX, D. 1982. Ice rafting of fine-grained sediment, a sorting and transport mechanism, Beaufort Sea, Alaska. *Journal of Sedimentary Petrology*, **52**, 493–502.

BAUMANN, M. 1990. Coccoliths in Sediments of the Eastern Arctic Basin. *In*: BLEIL, U. & THIEDE, J. (eds) *Geologic History of the Polar Oceans: Arctic versus Antarctic*, NATO ASI Series C, Kluwer, Dordrecht, 437–445.

BIRYUKOV, V.Y., FAUSTOVA, M.A., KAPLIN, P.A., PAVLIDIS, Y.A., ROMANOVA, E.A. & VELICHKO, A.A. 1988. The paleogeography of Arctic shelf and coastal zone of Eurasia at the time of the last glaciation (18000 yr B.P.). *Palaeogeography, Palaeoclimatology, Palaeoecology*, **68**, 117–125.

BISCHOF, J. 1990. Dropstones in the Norwegian-Greenland Sea — Indications of a Late Quaternary Circulation Pattern?. *In*: BLEIL, U. & THIEDE, J. (eds), *Geologic History of the Polar Oceans: Arctic versus Antarctic*, NATO ASI Series C, Kluwer, Dordrecht, 499–518.

CHURKIN, M. JR., SOLEIMANI, G., CARTER, C. & ROBINSON, R. 1981. Geology of the Soviet Arctic: Kola Peninsula to Lena river. *In*: NAIRN, A.E.M., & CHURKIN, M. JR. & STEHLI, F.G. (eds) *The Ocean Basins and Margins, 5, The Arctic Ocean* 331–375.

CLIMAP Project Members 1981. Maps of northern and southern hemisphere continental ice, sea ice, and sea surface temperatures in August for the modern and the last glacial maximum. Geological Society Map and Chart Series, MC-36, maps 7A and 7B.

COACHMAN, L.K. & AAGAARD, K. 1974. Physical Oceanography of Arctic and Subarctic Seas. *In*: HERMAN, Y. (ed) *Marine Geology and Oceanography of the Arctic Seas*, Springer, New York, 1–72.

CUTBILL, J.L., HENDERSON, W.G. & WRIGHT, N.J.R. 1976. The Billefjorden Group (Early Carboniferous) of Central Spitsbergen. *Norsk Polarinstitutt Skrifter*, **164**, 57–89.

DIBNER, V.D., BASOV, V.A., GERKE, A.A., SOLOV'YEVA, M.F., SISOPATROVA, G.P. & SHUL'GINA, N.I. 1970. The age of the Pre-Quaternary deposits on the bottom of the Barents Sea. *Oceanology*, **10**, 520–529.

—— & KRYLOVA, N.M. 1963. Stratigraphic position and material composition of coal measures in Franz Josef Land. *Sovetskaya Geologiya*, **7**, 77–89 [in Russian] (transl.: *International Geology*

Reviews, **7**, (1965), 1030–1038).

DOWDESWELL, J.A. 1989. On the nature of Svalbard icebergs. *Journal of Glaciology*, **35**, 224–234.

—— & DOWDESWELL, E.K. 1989. Debris in icebergs and rates of glacimarine sedimentation: observations from Spitsbergen and a simple model. *Journal of Geology*, **97**, 221–231.

DREWRY, D. 1986. *Glacial Geologic Processes*. Arnold, London.

DUTRO, J.T. 1981. Geology of Alaska bordering the Arctic Ocean. *In*: NAIRN, A.E.M., CHURKIN, M. JR. & STEHLI, F.G. (eds) *The Ocean Basins and Margins*, **5**, *The Arctic Ocean*, 331–375.

EDWARDS, M.B. 1975. Gravel fraction on the Spitsbergen Bank, NW Barents Shelf. *Norges Geologiske Undersøkelse*, **316**, 205–217.

EISENHAUER, A., MANGINI, A., BOTZ, R., WALTER, P., BEER, J., BONANI, G., SUTER, M., HOFMAN, H.J., & WÖLFLI, W. 1990. High resolution ^{230}Th and ^{10}Be stratigraphy of core M23235–2 (Fram Strait). *In*: BLEIL, U. & THIEDE, J. (eds) *Geologic History of the Polar Oceans: Arctic versus Antarctic* NATO ASI Series C, Kluwer, Dordrecht, 475–487.

ELVERHØI, A. & SOLHEIM, A. 1983. The Barents Sea ice sheet, a sedimentological discussion. *Polar Research*, **1**, 23–42.

——, ANTONSEN, P., FLOOD, S.B., SOLHEIM, A. & VULLSTAD, A.A. 1988. The Physical Environment, Western Barents Sea, Shallow Bedrock Geology. *Norsk Polarinstitutt Skrifter*, **179D**.

FLINT, R.F. 1971. *Glacial and Quaternary Geology*. Wiley, New York.

GARD, G. 1987. Late Quaternary calcareous biostratigraphy and sedimentation patterns: Fram Strait, Arctic. *Palaeoceanography*, **2**, 519–529.

—— 1988a. Late Quaternary calcareous nannofossil biozonation, chronology and paleooceanography in areas north of the Faeroe-Iceland Ridge. *Quaternary Science Reviews*, **7**, 65–73.

—— 1988b. Late Quaternary Calcareous Nannofossil Biochronology and Paleo-oceanography of Arctic and Subarctic Seas. *Meddelanden från Stockholms Universitets Geologiska Institution*, **275**, 8–45.

—— & BACKMAN, J. 1990. Synthesis of Arctic and Subarctic Coccolith Biochronology and History of North Atlantic Drift Water Influx during the last 500 000 Years. *In*: BLEIL, U. (ed.) *Geologic History of the Polar Oceans: Arctic versus Antarctic*, NATO ASI Series C, Kluwer, Dordrecht 417–436.

GERLACH, S.A., THIEDE, J., GRAF, G. & WERNER, F. 1986. Forschungsschiff Meteor, Reise 2, Forschungsschiff Poseidon Reise 128, Berichte der Fahrtleiter. Berichte SFB 313, 4, Kiel.

GORDIENKO, P.A. & LAKTIONOV, A.F. 1969. Circulation and physics of the Arctic Basin waters. *In*: *Annals of the International Geophysical Year, XLVI Oceanography*, Pergamon, New York, 94–112.

GROSSWALD, M.G. 1980. Late Weichselian ice sheet of northern Eurasia. *Quaternary Research*, **13**, 1–32.

HARLAND, W.B., PICKTON, C.A.G., WRIGHT, N.J.R., CROXTON, C.A., SMITH, D.G., CUTBILL, J.L. & HENDERSEN, W.G. 1976. Some coal-bearing strata in Svalbard. *Norsk Polarinstitutt Skrifter* **164**.

HENRICH, R., KASSENS, H., VOGELSANG, E. & THIEDE, J. 1989. The sedimentary facies record of glacial–interglacial cycles in the Norwegian Sea during the last 350 ka. *Marine Geology*, **86**, 282–319.

——, 1990. Cycles, rhythms and events in Quaternary Arctic and Antarctic glaciomarine deposits (a review). *In*: BLEIL, U. (ed.) *Geologic History of the Polar Oceans: Arctic versus Antarctic*, NATO ASI Series C, Kluwer, Dordrecht 213–244.

IMBRIE, J., HAYS, J.D., MARTINSON, D.G., MCINTYRE, A., MIX, A.C., MORLEY, J.J., PISIAS, N.G., PRELL, W.L. & SHACKLETON, N.J. 1984. The orbital theory of Pleistocene climate: support from a revised chronology of the marine δ^{18}O record. *In*: BERGER, A.L., IMBRIE, J., HAYS, J.D., KUKLA, G. & SALTZMAN, B. (eds), *Milankovitch and Climate*, **1**. Reidel, Dordrecht, 269–305.

JOHANNESSEN, O.M. 1986. Brief Overview of the Physical Oceanography. *In*: HURDLE, B.G. (ed.) *The Nordic Seas*. Springer, New York, 103–127.

KELLOGG, T.B., DUPLESSY, J.C. & SHACKLETON, N.J. 1978. Planktonic foraminiferal and oxygen isotope stratigraphy and palaeoclimatology of Norwegian deep-sea cores. *Boreas*, **7**, 61–73.

KERR, J.W. 1981. Evolution of the Canadian Arctic islands: a transition between the Atlantic and Arctic Ocean. *In*: NAIRN, A.E.M. & CHURKIN, M. JR. & STEHLI, F.G. (eds) *The Ocean Basins and* Margins, **5**, *The Arctic Ocean*, 105–199.

KÖHLER, S.E.I. & SPIELHAGEN, R.F. 1990. The enigma of Oxygen Isotope Stage 5 in the Central Fram Strait. *In*: BLEIL, U. & THIEDE, J. (eds) *Geologic History of the Polar Oceans: Arctic versus Antarctic*, NATO ASI Series C, Kluwer, Dordrecht 489–497.

LINDSTRØM, D.R. & MACAYEAL, D.R. 1986. Paleoclimatic constraints on the possible ice shelf cover in the Norwegian-Greenland Sea. *Palaeoceanography* **1**, 313–337.

MANUM, S.B. & THRONDSEN, T. 1978. Rank of coal and dispersed organic matter and its geological bearing in the Spitsbergen Tertiary. *Norsk Polarinstitutt Årbok 1977*, 160–177.

MIZEX '87 Group 1989. MIZEX East 1987, Winter Marginal Ice Zone Program in the Fram Strait and Greenland Sea. *EOS Transactions, American Geophysical Union*, **70**, 545–555.

MORRIS, T.H., CLARK, D.L. & BLASCO, S.M. 1985. Sediments of the Lomonosov Ridge and Makarov Basin: a Pleistocene stratigraphy for the North Pole. *Geological Society of America Bulletin*, **96**, 901–910.

MUDIE, P.J. & BLASCO, S.M. 1985. Lithostratigraphy of the CESAR cores. *Geological Survey of Canada Paper*, 84–22, 59–99.

PfIRMAN, S., GASCARD, J.-C., WOLLENBURG, I., MUDIE, P. & ABELMANN, A. 1989a. Particle-laden Eurasian Arctic sea ice: observations from

July and August 1987. *Polar Research*, **7**, 59–66.

——, WOLLENBURG, I., THIEDE, J. & LANGE, M.A. 1989b. Lithogenic sediment on Arctic pack ice: potential aeolian flux and contribution to deep sea sediments. *In*: LEINEN, M. & SARNTHEIN, M. (eds) *Palaeoclimatology and Palaeometeorology: Modern and Past Patterns of Global Atmospheric Transport* NATO ASI Series C, Kluwer Academic Publishers, Dordrecht, **282**, 463–493.

PICKARD, G.L. & EMERY, W.J. 1982. *Descriptive Physical Oceanography, An Introduction*, 4th Enlarged Edition, Oxford.

RUDDIMAN, W.F. 1977a. Late Quaternary deposition of ice-rafted sand in the subpolar North Atlantic (lat 40° to 65°N). *Geological Society of America Bulletin*, **88**, 1813–1827.

—— 1977b. North Atlantic ice-rafting: a major change at 75 000 years before the present. *Science*, **196**, 1208–1211.

SCOTT, D.B., MUDIE, P.J., BAKI, V., MACKINNON, K.D. & COLE, F.E. 1989. Biostratigraphy and late Cenozoic paleoceanography of the Arctic Ocean: foraminiferal, lithostratigraphic, and isotopic evidence. *Geological Society of America Bulletin*, **101**, 260–277.

SPIELHAGEN, R., PFIRMAN, S. & THIEDE, J., (eds) 1988. Geoscientific report on the ARK IV/3 Expedition of the PFVS POLARSTERN Summer 1987 to the Central Eastern Arctic Basin. *Reports Geologisch-Paläontologisches Institut und Museum*, Universität Kiel, **24**.

SWIFT, J.H. 1986. The Arctic Waters. *In*: HURDLE, B.G. *The Nordic Seas*, Springer, New York, 103–127.

THIEDE, J., CLARK, D.L. & HERMAN, Y. 1990. Late Mesozoic and Cenozoic Paleoceanography of the Northern Polar Oceans. The Geology of North America *In*: GRANTZ, A., JOHNSON, L. & SWEENEY, J.F. (eds), *The Arctic Ocean Region*, vol. L, Geological Society of America, Boulder.

ULMISHEK, G. 1985. Geology and Petroleum Resources of the Barents — Northern Kara Shelf in Light of New Geologic Data. Argonne National laboratory, Report ANL/ES-148.

VAN KREVELEN, D.W. 1961. *Coal*. Elsevier, Amsterdam.

VINJE, T. 1978. Sea ice conditions and drift of Nimbus-6 buoys in 1977. *Norsk Polarinstitutt Arbok 1977*, 283–292.

—— 1982: The drift pattern of sea ice in the Arctic with particular reference to the Atlantic approach. *In: The Arctic Ocean*, Rey, L. (ed.), 82–96.

—— 1985. Drift, composition, morphology and distribution of the sea ice fields in the Barents Sea. *Norsk Polarinstitutt Skrifter*, **179C**.

VON BODUNGEN, B. 1988. Poseidon Reise 141–2/142 vom 17.10.87 bis 18.11.87, Fahrtbericht. *Berichte Sonderforschungsbereich 313 der Universität Kiel*, **11**, 1–20.

WAGNER, H. & BERTHOLD, G. 1979. COMECON. Rohstoffwirtschaft-licher Uberblick. *Rohstoffwirtschaftliche Länderberichte der Bundesanstalt für Geowissenschaften und Rohstoffe*, **23**, Hannover.

WOLDSTEDT, P. 1969. *Quartär*. Enke, Stuttgart.

ZIEGLER, P.A. 1988. *Evolution of the Arctic-North Atlantic and the Western Tethys*. American Association of Petroleum Geologists Memoir, **43**.

Glacial geomorphic features in the northern Barents Sea: direct evidence for grounded ice and implications for the pattern of deglaciation and late glacial sedimentation

ANDERS SOLHEIM, LARS RUSSWURM, ANDERS ELVERHØI & MONA NYLAND BERG

Norwegian Polar Research Institute, P.O. Box 158, N-1330 Oslo Lufthavn, Norway

Abstract: Glacial landforms, in particular flutes and associated transverse ridges, have been identified in an area of roughly 4000 km^2 and between approximately 150 and 340 m water depth in the north-central Barents Sea. The transverse ridges are interpreted to be analogous to De Geer moraines and are observed only in the southern part of the fluted area. The flutes, trending roughly north–south, have a relief of less than one metre, widths on the order of 4–8 m and lengths usually 100–500 m. Although the present data do not allow firm conclusions on the exact genesis of these features, they are compared to analogous terrestrial features.

The existence of these glacigenic morphological features has major implications concerning the late glacial history of the Barents Sea. They give the first direct evidence of a grounded ice sheet in the northern Barents Sea. Ice flow directions during the latest phase before final recession were mainly southwards, but a partly topographically influenced flow pattern indicates that the main banks may have acted as local ice centres. The ice sheet was relatively thin and rested on a deformable bed. It retreated rapidly from the fluted area, as evidenced by the lack of sediment masking the low-relief flutes. The general retreat from the region took place in a stepwise manner, apparently without surging and with deposition of locally thicker glacier-proximal deposits during halts in the recession.

The existence of a Late Weichselian Barents Sea ice sheet has been the subject of debate for more than two decades (e.g. Schytt *et al.* 1968, Denton & Hughes 1981, Boulton 1979, Matishov 1980, Grosswald 1980, Andersen 1981, Hughes *et al.* 1981, Salvigsen 1981, Elverhøi & Solheim 1983, Solheim & Kristoffersen 1984, Vorren & Kristoffersen 1986, Elverhøi & Solheim 1987, Solheim *et al.* 1988b). Presently, most views favour the existence of a grounded ice sheet, and the debate is concentrated on the extension and the exact timing and pattern of the deglaciation (Forman *et al.* 1987, Jones & Keigwin 1988, Vorren *et al.* 1988, Elverhøi *et al.* 1990). Evidence cited in the debate includes the uplift curves for Svalbard, the Quaternary sediment distribution and lithostratigraphy and the submarine morphology and oxygen isotope data from the Greenland Sea. However, direct evidence for Late Weichselian grounded glacier ice over extensive areas of the Barents Sea has never been presented.

This study is mainly based on high resolution acoustic data and sediment cores acquired in the northern part of the Bjørnøyrenna trough in 1987 (Fig. 1). In the paper we describe and discuss morphological features that are formed subglacially or near the grounding line of a marine based ice sheet. Hence, this is the first firm evidence presented for the existence of such an ice sheet over extensive parts of the northern Barents Sea. Although the problems of the maximum extent and exact timing of the deglaciation are not solved, a major objective of the study is to discuss the implications these new data have for our understanding of the pattern and character of Late Weichselian deglaciation of the Barents Sea. Combined with data on the late and post glacial sediment distribution in the area, the morphological features provide new insight regarding glacier flow directions during the final stages of glaciation, glacier temperature regime, the nature of the glacier bed, whether the deglaciation took place in one event or stepwise and whether surges occurred.

Materials and methods

The study area (Fig. 1) is situated between the northeastern part of Spitsbergenbanken to the west, and Sentralbanken and Storbanken to the east. Acoustical profiles were run mainly in a northwest–southeast direction (Fig. 2) (Solheim *et al.* 1988a). The profile locations and directions were chosen to fill in a pre-existing

From DOWDESWELL, J.A. & SCOURSE, J.D. (eds), 1990, *Glacimarine Environments: Processes and Sediments*. Geological Society Special Publication No 53, pp 253–268.

Fig. 1. The Barents Sea, with bathymetry and place names used later in the text. Heavy frame indicates the main study area. Note the arrow to the south of the framed area, marking glacial flutes and flute direction in a small area at 74° 55′N.

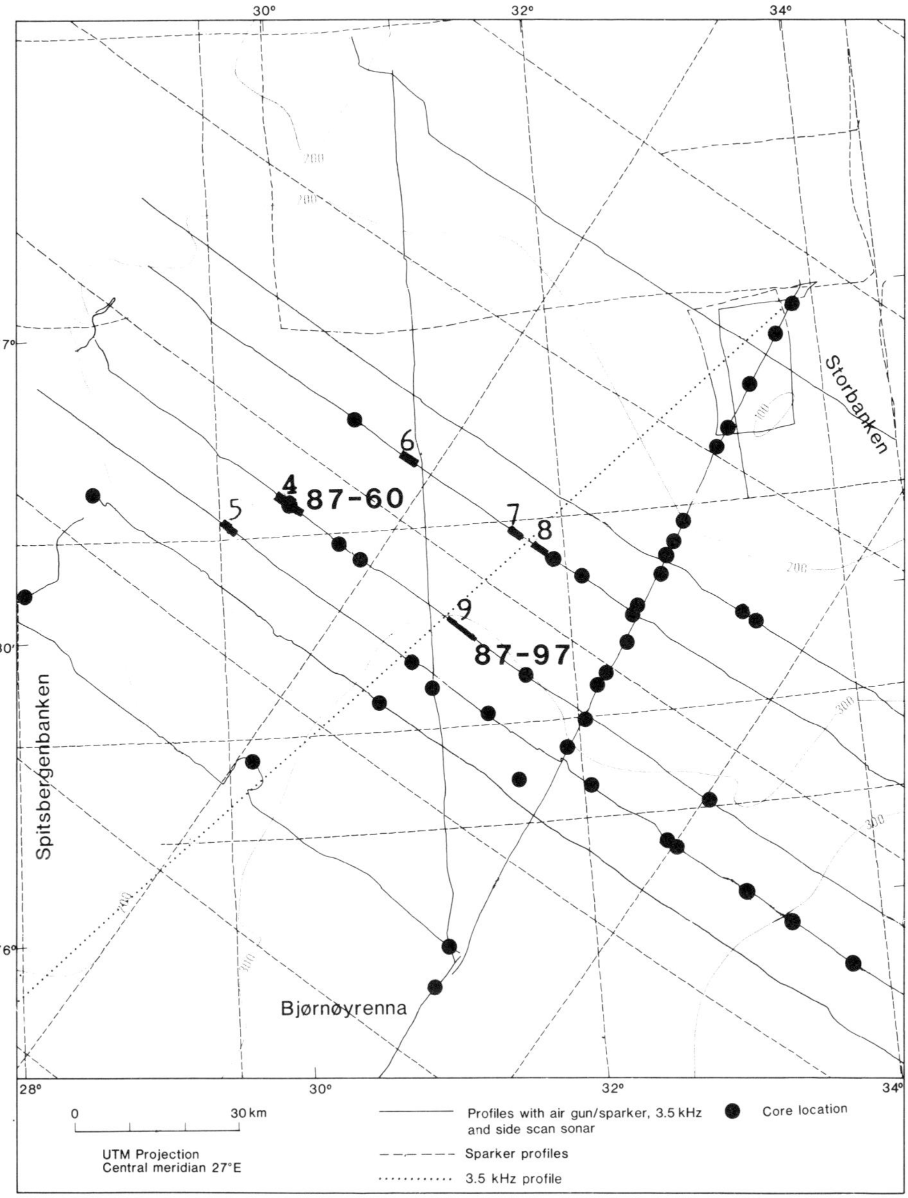

Fig. 2. Map showing the acoustic profiles and core locations in the study area. Of the lines which include side scan sonar data, only the rectangular pattern on Storbanken is from 1983, the rest is a 1987 cruise. Contours are bathymetry in metres. Heavy lines with numbers mark figures referred to in the text. Cores 87–60 and 87–97, shown in Fig. 10, are indicated.

grid of shallow seismic (sparker) profiles. The northwest–southeast alignment was chosen because it is essentially perpendicular to the structural strike of the region.

The geophysical equipment used consisted of: (1) Side scan sonar, operated with 50 kHz transducers and with a range of 300 m to each side. (2) A 30 in^3 air gun equipped with a wave shape kit, with analogue recording via a single channel, 50 element, 7.5 m active length streamer. The signals were band pass filtered in a range of 60 to 500 Hz and TVG amplified before recording. Shot interval was generally 4 s, with 2 s used for short periods. Tow depths were 0.8 to 1.0 m for the air gun and 1 to 2 m for the streamer. (3) 9-electrode sparker-array, with energy levels 3.6 and 4.5 kJ and the same recording facilities and parameters as for (2). The sparker was mainly used as back-up during periods of air gun maintenance. (4) Hull mounted 3.5 kHz echo sounder (PDR).

Geological sampling equipment used for this study included: (1) A 3.5 m vibrocorer with barrel diameter of 90 mm. (2) 3 and 6 m gravity corers with barrel diameter of 110 mm. Navigation with an accuracy on the order of 100 m was achieved with a system combining GPS satellite navigation, Loran C and a rubidium oscillator.

The cores were cut into one metre sections on board. Core tops, bottoms and section ends were described and measured for undrained shear strengths with a hand held vane. Subsequent laboratory work included X-radiography of unsplit cores, core splitting, description and photography, undrained shear strength measurements by fall-cone penetrometer, measurements of water content in per cent of wet weight, bulk density measurements by weighing a sample in a cylinder of known volume, grain size analysis by wet sieving for fractions > 0.063 mm and sedigraph for fractions < 0.063 mm.

Results

Sediment distribution

In general, the northern Barents Sea has a thin (less than 10 to 15 milliseconds (ms), two-way reflection time) cover of Quaternary sediments over the Mesozoic sedimentary bedrock (Elverhøi & Solheim 1983, Solheim & Kristoffersen 1984, Elverhøi *et al.* 1988). However, in troughs and along the margins of bank areas, accumulations may locally exceed 50 ms thickness. The general stratigraphy consists of firm diamicton, interpreted to be a till, overlain by softer sediments interpreted to consist of ice proximal glacimarine sediments and a thin (0.2 to 1.0 m) cover of postglacial mud (Elverhøi *et al.* 1990). The till may be continuous over large areas, but is often patchy and may fill depressions in the bedrock surface (Solheim *et al.* 1988b).

Because of increased data coverage and the use of 3.5 kHz echo sounder, the present study is able to add new information on till composition, distribution and origin, as well as on the general sediment distribution. To emphasize the more significant accumulations, only thicknesses exceeding 10 ms have been contoured (Fig. 3). Only four accumulations significantly exceed 20 ms and have somewhat greater areal extent. The accumulation on the southern part of Storbanken (Fig. 3), forms a low relief ridge of acoustically homogeneous and semi-transparent sediments (Solheim *et al.* 1988b). The two accumulations to the northwest both appear as positive features of low relief, but the western one has the most transparent character and also shows some internal reflections. The main part of the extensive accumulation to the south appears as an acoustically homogeneous and transparent infill to bedrock depressions and continues as a thin drape up the lower part of the slope to the west (Fig. 3). On the basis of comparison with present-day analogues off the Austfonna ice cap on Nordaustlandet, Svalbard (Fig. 1), a likely explanation for the acoustically transparent sediment lenses is that they are deposited by turbid meltwater plumes off an ice front (Pfirman & Solheim 1989).

Sea floor morphology

Morphological features in the study area include pockmarks, iceberg plough marks and a system of straight, parallel grooves and ridges which form a striated pattern. In several places the ridge and groove system is associated with short, straight to arcuate ridges which run subperpendicular to the striated pattern. The distribution of pockmarks is closely related to the distribution of soft sediments. Their dimensions and shape are similar to those described from other parts of the northern Barents Sea (Solheim & Elverhøi 1985, Solheim 1990). Iceberg plough marks are found throughout the study area, but they occur in greatest concentrations in water depths of < 200 m. The average relief of the plough marks is 2–5 m, and their widths generally vary between 20 and 80 m, although wider and deeper examples have been observed. As recent iceberg ploughing only rarely occurs below 120–130 m waterdepth (Solheim 1990), most of the plough marks in the study area must be relict features. The general shapes and dimensions of plough marks from the northern Barents Sea are further described by Solheim *et al.* (1988b).

Of principal importance for this study is the striated pattern of ridges and grooves (Figs 4–9). By comparison with similar features associated with terrestrial glaciers, this pattern is interpreted to represent glacial flutes. Dynamic agents other than grounded glacier ice, such as grounding icebergs or sea ice and current activity, are excluded as possible causes for the pattern, because of the regular, parallel

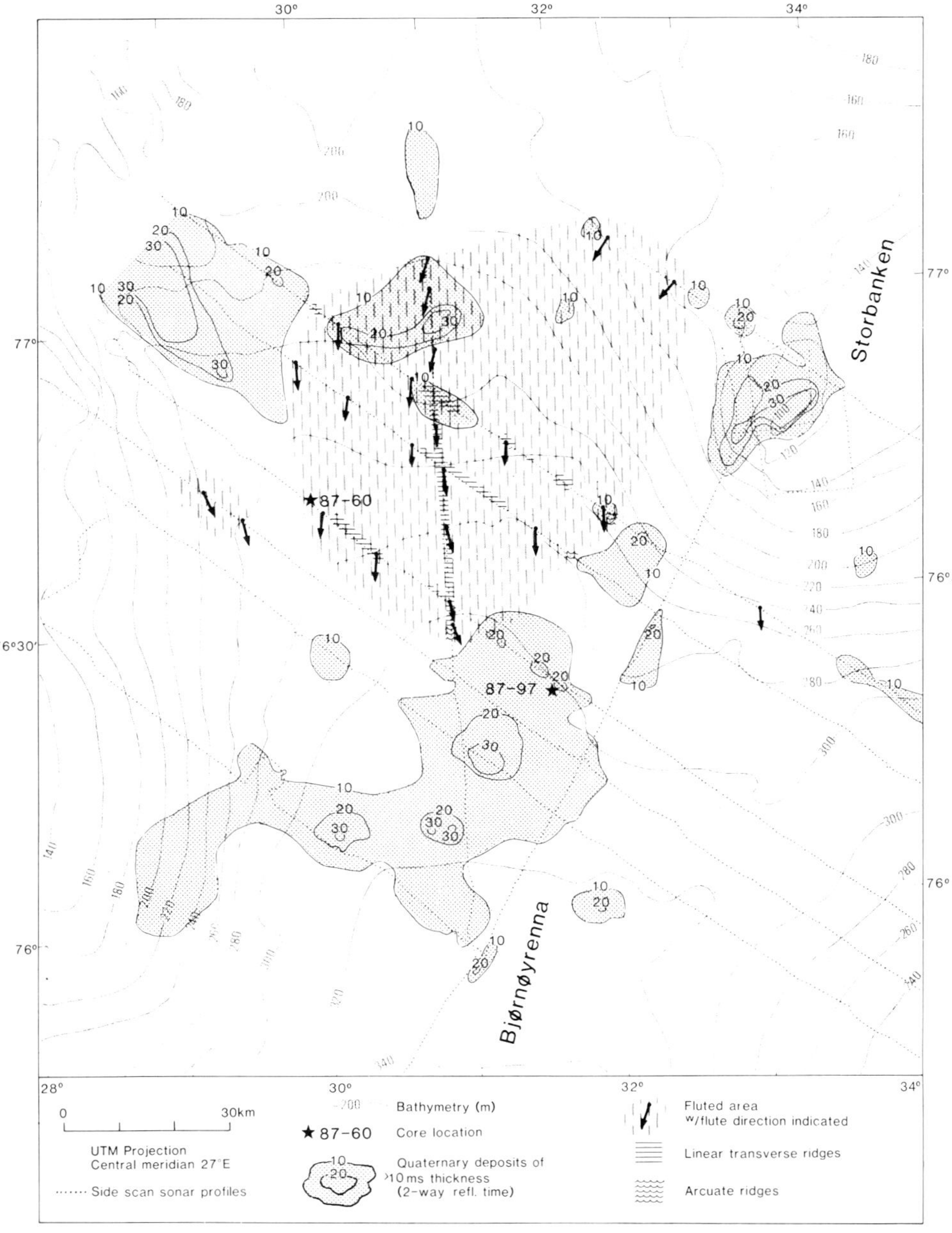

Fig. 3. Map showing the distribution of Quaternary sediments and sea floor morphological features. Only sediment thickness > 10 ms has been contoured. Location of cores shown in Fig. 10 is indicated.

geometry of the pattern and the general physical setting of the region. All iceberg plough marks identified within the fluted area are seen to obliterate the flutes and clearly postdate them (Fig. 4).

The fluted sea floor is recognized over an area of roughly 4000 km^2 at water depths between 160 and 300 m in northernmost Bjørnøyrenna, between Hopen and Storbanken (Figs 1 & 3). Additionally, flutes are also identified in a local area at 340 m water depth further south in Bjørnøyrenna, at approximately 74° 55′ N (Fig. 1). Individual flutes are straight, and their orientations are very consistent on a local scale, although they do show slight regional changes (Fig. 3). The main trend is

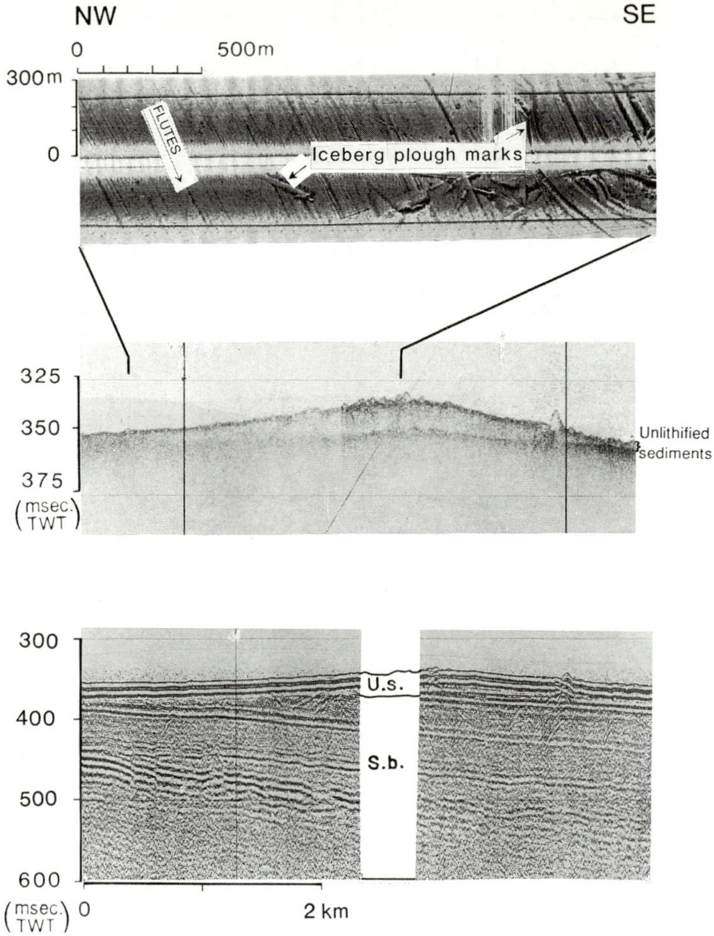

Fig. 4. Side scan sonar (upper), 3.5 kHz (middle) and air gun (lower) records from an area where well developed glacial flutes cover a small local sediment accumulation. In the shallowest part, the flutes are obliterated by iceberg ploughmarks. The parallel, dipping reflectors in the air gun record are sedimentary bedrock layers. U.s.: unlithified sediments, S.b.: sedimentary bedrock. For location, see Fig. 2.

roughly north–south.

The main part of the fluted sea floor is developed in an area where less than 10 ms of Quaternary sediments overlie the sedimentary bedrock. Exceptions to this are small patches of approximately 10 to 15 ms of sediment, and a 35 ms thick accumulation in the northwestern part, all of which have clearly fluted surfaces (Fig. 3). The boundaries of the fluted area are defined by either a significant increase in the abundance of iceberg plough marks or an increased thickness of acoustically transparent sediments, which may obliterate or bury the flutes, respectively. The latter boundary occurs where the fluted area approaches the 10 ms isopach in the south and the northwest (Fig. 3).

The dimensions of individual flutes and their appearance on the side scan sonar records vary slightly. The relief shows the least variation, and seems to be less than one metre throughout the area. Widths range between 1 m and 15 m, with 4 to 8 m as the most typical (Fig. 4). Lengths are most often between 100 and 500 m, but values up to 1200 m are recorded. In the small, restricted area on the southern flank of Storbanken (Fig. 3), the flutes appear more discontinuous, and can only rarely be followed for more than 100 m. A similar pattern is observed to the west, at the flank of Spitsbergenbanken (Fig. 5). In most of the area where the fluted pattern is recognized, the entire sea floor seems to be fluted. Restricted areas with a lower density of flutes appear to be associated with adjacent sediment accumulations. There-

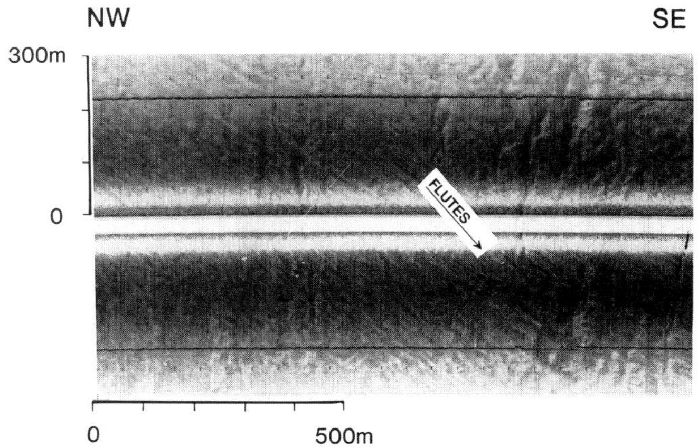

Fig. 5. Side scan sonar record showing discontinuous and weakly defined glacial flutes from the slope east of Spitsbergenbanken. For location, see Fig. 2.

fore, a density lower than full coverage may be apparent and associated with soft sediments covering the lowest relief features.

Associated with the fluted pattern is a system of ridges aligned approximately perpendicular to the flute direction. These transverse ridges are only recognized in the central and southern parts of the fluted area (Fig. 3), where they occur on both flat and undulating sea floor. Their surfaces are fluted, and individual flutes can be followed across ridges, as well as through the gaps between ridge segments (Fig. 6). Typically, the ridges are straight to weakly lunate, with lengths varying from 100 to 500 m, although ridges more than 700 m long have been observed. Widths are most often between

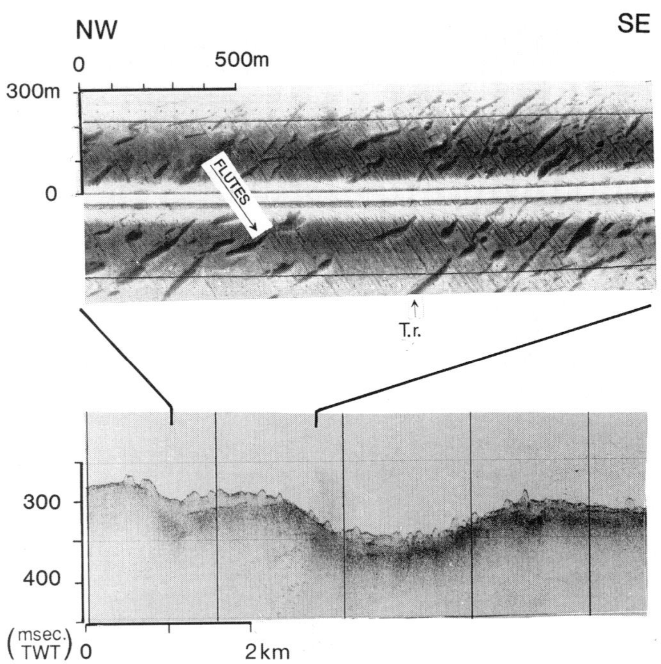

Fig. 6. Side scan sonar (upper) and 3.5 kHz (lower) records showing well developed transverse ridges (T.r.). Note continuations of flutes across the ridges and through gaps between ridge segments, particularly in the southeast. For location, see Fig. 2.

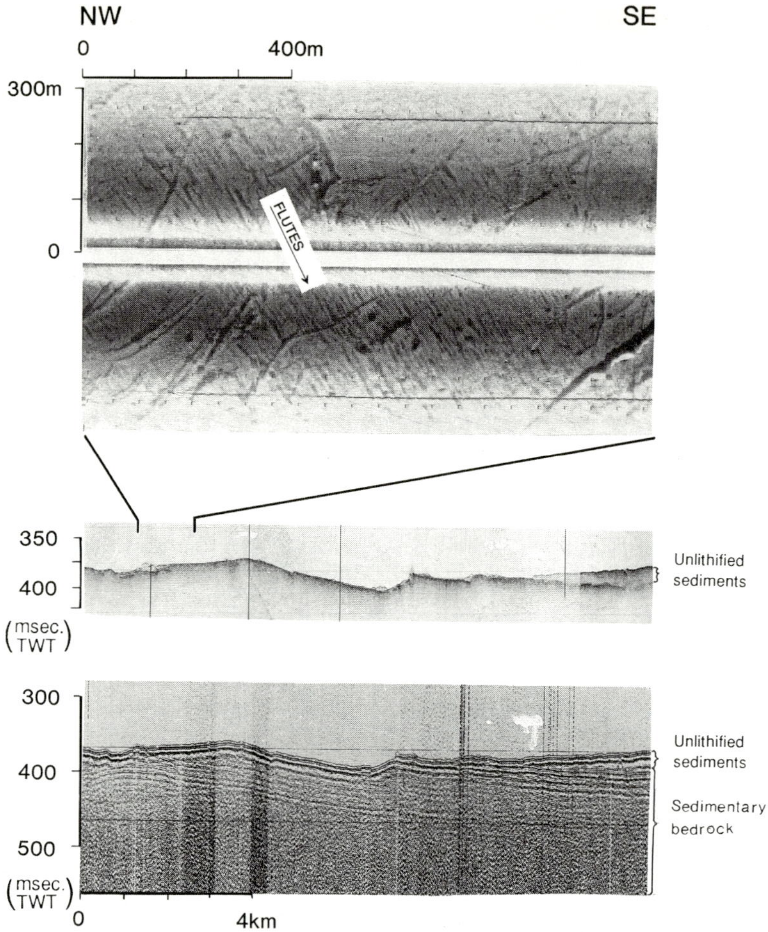

Fig. 7. Side scan sonar (upper), 3.5 kHz (middle) and air gun (lower) records showing glacial flutes with narrow, low relief transverse ridges. Note the more distinct, 'normal' ridge in the lower right of the side scan sonar record. The 3.5 kHz and air gun records also exemplify the typical thin sediment cover outside the local accumulations, one of which can be seen towards the southeast. The parallel, dipping reflectors in the air gun record are sedimentary bedrock layers. For location, see Fig. 2.

15 and 30 m, and relief varies between 1 and 8 m, with 2 to 4 m as an average. Locally, narrower ($\approx$10 m) ridges with relief of < 1 m are present (Fig. 7). The density of transverse ridges may be as high as one per 100 m on a transect (Fig. 6), with the highest density generally found in the central part of the fluted area.

In three areas, approaching the thicker sediment lenses in the south and southeast (Fig. 3), a set of highly arcuate ridges is observed, with limbs which recurve into alignment with the flutes. Individual ridges are 10 to 20 m wide and have a relief of less than 1 to 2 m (Fig. 8).

Due to the difference in relief between the flutes and most of the transverse ridges, the flutes are more vulnerable to sediment masking, and are in some places buried while the transverse ridges remain visible. Particularly in approaching the southern and southeastern sediment accumulations, the flutes become less distinct and eventually disappear. In the southernmost part, a weak and partly buried flute pattern can be followed well onto the sediment accumulation proper (Fig. 9).

Sediment composition and physical properties

Vibrocores recovered from the fluted area showed a lower, stiff layer of diamicton with high bulk density, low water content and shear

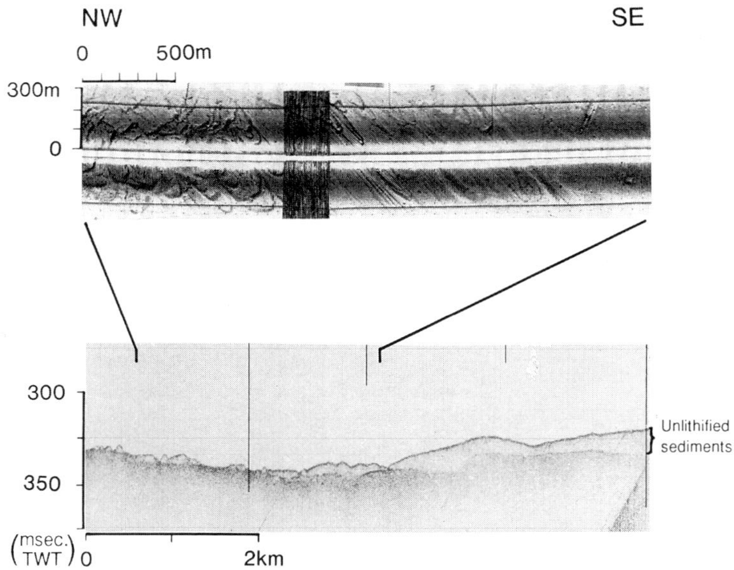

Fig. 8. Side scan sonar (upper) and 3.5 kHz (lower) records showing arcuate ridges associated with glacial flutes, which becomes gradually weaker towards the acoustically transparent sediment lense in the southeast of the profile. For location, see Fig. 2.

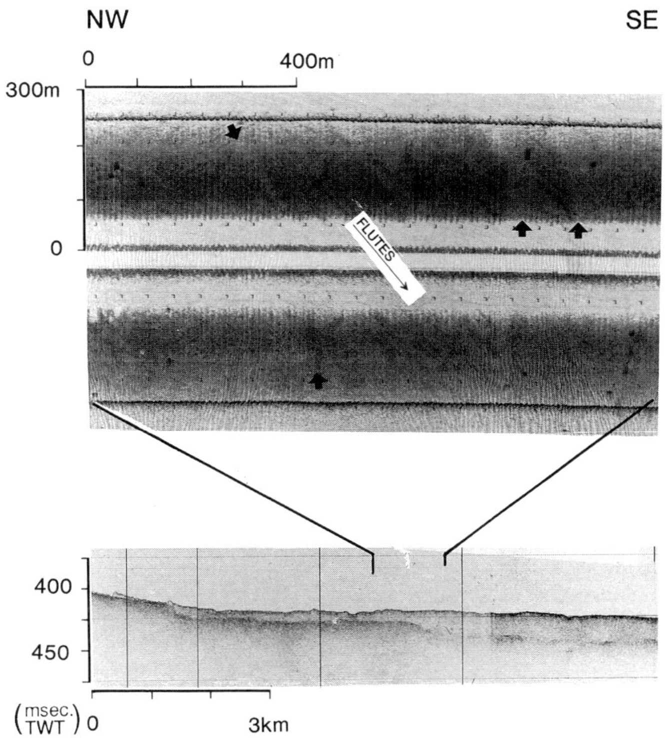

Fig. 9. Side scan sonar (upper) and 3.5 kHz (lower) records showing weak, partly masked glacial flutes in the marginal zone of the acoustically transparent sediment lense that bounds the fluted area to the south. Arrows mark possible iceberg plough marks. For location, see Fig. 2.

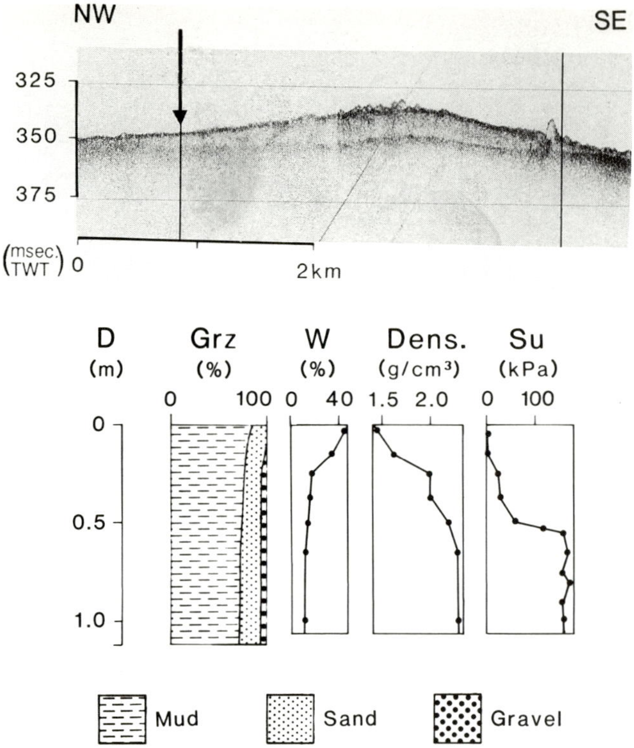

Fig. 10. Vibrocore 87–60 sediment core from the fluted area with corresponding 3.5 kHz records and physical properties indicated. Grz: grain size distribution, W: water content, Dens: wet bulk density, Su: undrained shear strength. For location, see Figs 2 & 3.

strengths exceeding 150 kPa (Fig. 10a). This stiff sediment was only recovered in the lower parts of a few cores, after vibrating times on the order of 20 to 30 minutes. A layer of intermediately firm sediment is often found above the lower, stiff diamicton. This sediment has shear strengths in the range of 20 to 80 kPa, bulk densities of 1.9 to 2.1 g cm^{-3} and water contents of between 15 and 25% (Fig. 10). In previous studies, based only on gravity cores (Elverhøi & Solheim 1983), this has been taken as the basal layer. A thin surface cover of soft sediment has shear strengths of less than 10 kPa, bulk densities of around 1.5 to 1.7 g cm^{-3} and water contents above 30%.

A core recovered from one of the thicker accumulations of acoustically transparent sediment in the deeper parts of the study area (Fig. 11) contained only soft diamicton, which includes both olive grey mud, interpreted as being primarily of Holocene age, and dark grey material inferred to represent Late Weichselian ice proximal glacimarine deposits (Elverhøi & Solheim 1983). Note, however, that based on lithology alone, a distinction between sediments from the fluted area and the non-fluted sediment accumulations is difficult. Neither can the lower, stiff diamicton be distinguished from the intermediately firm diamicton in the fluted area from lithology alone (Fig. 10).

On the basis of its acoustic character and the stratigraphy of adjacent cores, the sediment accumulation at the southern part of Storbanken (Fig. 3) was interpreted by Solheim *et al.* (1988b) to consist of relatively sandy, normally consolidated mud. Cores obtained during the present study confirm this sediment character for the upper two metres (Solheim *et al.* 1988a). No core information exists from the northwestern sediment accumulations. Further sedimentological and stratigraphical analyses are presently being carried out for selected cores from the entire study area.

Formation of the morphological features

Fluted surfaces on till have been described from several regions in front of present-day terrestrial glaciers, where they are most often found in newly exposed lodgement till (Dyson 1952,

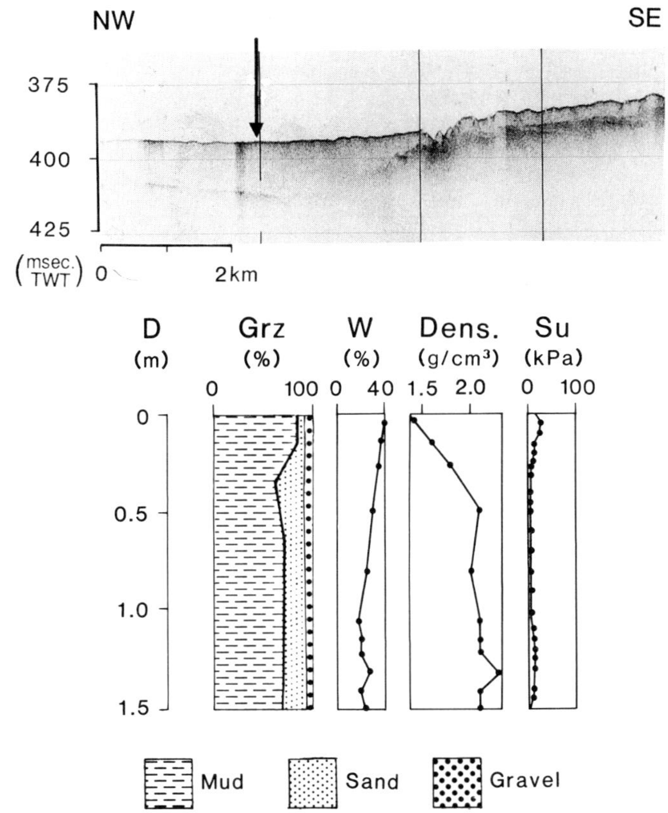

Fig. 11. Gravity core 87–97, from the acoustically transparent lense to the south of the fluted area, with corresponding 3.5 kHz records and physical properties indicated. Grz: grain size distribution, W: water content, Dens: wet bulk density, Su: undrained shear strength. For location, see Figs 2 & 3.

Hoppe & Schytt 1953, Baranowski 1970, Paul & Evans 1974, Clayton & Moran 1974, Boulton 1976, Boulton 1982, Van der Meer 1983). Glacial flutes are scarcely reported from marine areas but, in a recent paper, Josenhans & Zevenhuizen (1990) described flutes from the sea floor of Hudson Bay, Canada.

Boulton (1976) argues that flutes may form when till is intruded into subglacial cavities that open on the lee side of obstacles as basal ice flows over them. The obstacles are most often boulders that are deeply embedded in the till and therefore relatively immobile in response to the drag forces exerted by the glacier, but bedrock irregularities have also been shown to form lee-side cavities. Frost heave mounds have been proposed as cavity-forming obstacles beneath glaciers with cold based margins (Baranowski 1970). Other mechanisms, such as erosion (Price 1973) and transverse flow in basal ice (Shaw & Freschauf 1973, Aario 1977, 1987) have also been proposed as causes for fluted surfaces, but neither of these seem to describe the phenomenon as well as the lee-side cavity mechanism. The dimensions of flutes reported in the literature vary, but they are generally less than 2 m high and 3 m wide. They can often be quite extensive in the flow direction, and individual flutes have been followed for up to one kilometre (Boulton 1976).

On the basis of the apparent thickness of the uppermost soft sediment and the flute relief, the flutes are most likely to be formed in the veneer of intermediately firm sediment (Su=20–80 kPa), but the present data are not considered adequate to resolve the exact formative mechanism. However, regardless of which view is favoured, the flutes are formed beneath actively sliding glaciers and, therefore, indicate the former existence of a grounded ice mass in the study area.

Because of the close association between the flutes and the transverse ridges (Fig. 6), the formation of the latter also seems to be related to glacial processes. Post-glacial formation, as bed forms related to current action, is excluded both because of the fluted surfaces and also because of the irregular shape and spacing of the features. Furthermore, present mean current velocities in the region are less than 10 cm s^{-1} (Eide 1983). The relatively constant orientation, nearly perpendicular to the flutes, also suggests an origin related to the action of moving glacier ice.

Two thoroughly described types of glacial landforms are Rogen moraines (Lundqvist 1969, 1989) and De Geer moraines (De Geer 1889, Zilliacus 1976, 1981, 1989). The typical Rogen moraines are of somewhat greater dimensions (height 10–20 m and width 50–100 m) than the transverse ridges described in this study, and usually they are also concave in the down-glacier direction. Most typical is also their gradual transition into drumlins (Lundqvist 1989). Different views on formational mechanisms for Rogen moraines include differential subglacial deformation in response to laterally changing sediment physical properties (Boulton 1987), folding of sediment rich basal ice in zones of longitudinal compression (Minell 1977, Shaw 1979) and separation of basal ice into zones of debris-rich ice and clean zones in areas of longitudinal extension (Lundqvist 1969). Both the latter theories require passive downwasting of stagnant ice during final deglaciation to preserve the features.

De Geer moraines, on the other hand, show dimensions and morphologies more similar to the transverse ridges of the present study (Sugden & John 1976, Zilliacus 1989). In Finland, De Geer moraines are typically 1–3 m high, 10–20 m wide and 100–200 m long (Zilliacus 1989). The features are most likely related to marginal parts of glaciers ending in water and are formed near the grounding line. Proposed mechanisms include seasonal ice push to form annual moraines (De Geer 1889), sediment transport across the grounding line in response to differences in hydrostatic pressure (Zilliacus 1981) and squeeze-up of deformable sediment into basal crevasses of the marginal zone (Zilliacus 1989).

Based on morphological considerations, we favour De Geer moraines as the most relevant analogue to the transverse ridges in the northern Barents Sea. The present data do not give firm evidence in favour of any exact formational mechanism. However, the facts that the ridges have fluted surfaces and that individual flutes can be followed through gaps in the ridges, imply that the flutes and the transverse ridges are formed contemporaneously. As the study area has a relatively flat sea floor (a maximum slope of 0.10 in the north–south direction), the grounding line may easily fluctuate laterally, e.g. in response to seasonal variations. Hence, transverse ridges formed at or near the grounding line may be fluted and/or degraded by subsequent readvances. Simultaneously, new transverse ridges may be formed at or near the new grounding line position. Indications of fluted iceberg plough marks (Fig. 9) give further support to the idea of a fluctuating grounding line/ice front. Variation in the general recession rate may explain why the transverse ridges are not found over the entire study area.

The low-relief, acurate ridges observed in three areas close to the thicker sediment lenses (Fig. 3), appear to be closely associated with the flutes (Fig. 8), and an origin related to flute formation seems most likely. This implies that the accurate ridges are also formed in the frontal zone, and a possible mode of formation could be transport of mobile till material to the low (hydrostatic) pressure zone immediately off the grounding line. Lobe-shaped irregularities in the grounding line, related to ice flow, may explain the association with the flutes. This mechanism corresponds to one of the proposed mechanisms for formation of De Geer moraines (Zilliacus 1981). The difference in plan form between the transverse and the arcuate ridges may be attributed to variations in grounding line morphology linked to the local stress regime. Alternatively it may imply somewhat different modes of formation, e.g. that the transverse ridges were formed in basal crevasses near the grounding line, as proposed by Zilliacus (1989) for the De Geer moraines in Finland.

Discussion

Implications for sedimentation, ice movement and deglaciation history

The glacial flutes clearly demonstrate a southerly direction of ice flow during the last phase of deglaciation from this part of the northern Barents Sea (Fig. 3). However, regional variations from SSE to SSW are seen. The southwesterly directions observed in the northeastern part and the southeasterly directions in the western part of the fluted area, respectively, indicate ice movement from the banks (Fig. 3). This favours the idea that the ice sheet was fed from several source areas, at least during its later stages, as proposed by Solheim *et al.* (1988b). Hence, Storbanken may have been such a local source area from which flow directions were channelized southwards when ice reached the deeper parts of northern Bjørnøyrenna. The same effect was probably experienced by ice flowing from the west, but directions here indicate the island of Edgeøya and surrounding areas as a likely source area, rather than Spitsbergenbanken (Fig. 1). The latter bank area may, however, have had more influence on the flow directions further southwest. The channelizing effect of Bjørnøyrenna is further exemplified by the NNW-SSE direction of the flutes in the small area at 74° 55′N (Fig. 1).

Despite recent documentation of subglacial deformation of frozen sediments (Echelmeyer & Zhongxiang 1987), extensive formation of glacial geomorphic features like those in the present study area most likely implies that the glacier base was at the pressure melting point during deglaciation. However, this was probably an area of enhanced flow as a result of the channelizing effects of the Bjørnøyrenna trough, and frozen bed conditions cannot be excluded in other parts of the ice sheet, where frictional heating may have been less.

The present data do not resolve the questions of maximum extent of the ice sheet and the exact timing of deglaciation. The fluted area is generally bounded towards the north, west and east by intensive iceberg ploughing in which any pre-existing morphological pattern would have been obliterated. To the south, the boundary is associated with the acoustically transparent, thicker sediment lenses (Fig. 3). Due to the relatively restricted areal extent of the sediment lenses, we find it most probable that they represent glacimarine deposition during a halt in the general ice retreat from the Barents Sea. The existence of flutes further south (Fig. 1) also shows that grounded ice has had an extension beyond the position of the sediment lenses. The general pattern of sediment distribution and sea floor morphology in this part of the Barents Sea (Fig. 3) therefore indicates stepwise retreat rather than recession in one event.

Preservation of the low-relief glacial flutes shows that little late and post glacial sedimentation has occurred and therefore indicates a rapid ice recession from the fluted areas. Rates of deposition in ice-proximal environments may be high, ranging from a few cm/year to more than 1 m/year (Powell 1981, Elverhøi *et al.* 1983, Görlich 1986, Pfirman & Solheim 1989). The acoustically transparent and homogeneous character of the lenses immediately south of the fluted area (Figs 9 & 10) also indicates rapid sedimentation in this region. As shown both from the open, marine environment (Pfirman & Solheim 1989) and in fjords (Elverhøi *et al.* 1983; Syvitski 1989), the majority of sediments brought out by meltwater, fall out from suspension within a few kilometres from the meltwater efflux. Small, local sediment accumulations may represent shorter halts or deposition from smaller, short-lived meltwater outlets.

As some of the local accumulations have fluted surfaces, they probably represent accumulations of Late Weichselian till, or older deposits reworked and deformed by Late Weichselian grounded ice. In particular, this is so for the 35 ms thick accumulation in the northern part of the fluted area (Fig. 3).

Variations in the amount of ice marginal sedimentation may, in addition to stillstands in ice retreat, also be related to temporal and spatial variations in sediment supply to the margin. Significant temporal variations in the availability of erosional products are considered unlikely due to the short time intervals probably involved and the uniform geological character of the region (Elverhøi *et al.* 1988). Variations in the position of main meltwater outlets is another possibility. However, since the study area occupies the deepest parts of the inner Bjørnøyrenna trough (Fig. 1), the most likely positions of the main outlets would be within the study area. Furthermore, because of the low relief of the flutes, only small amounts of sedimentation would be required to mask the features.

From the above discussion, the following recession history can tentatively be inferred: The ice front retreated rapidly through iceberg calving in northern Bjørnøyrenna, leaving subglacially formed flutes exposed on the sea floor at least as far south as 74° 55′N, at 340 m water depth. Then the ice front was stationary for some time, depositing the acoustically transparent lenses, now found at approximately 300 m water depth. After a subsequent phase of rapid retreat which left the main area of flutes and transverse ridges exposed, a second halt occurred when the grounding line was at approximately 150 m water depth or slightly deeper. During this halt the ridge-type deposits at Storbanken and northwest of the fluted area were deposited. The time frame for these events is uncertain, but based on recent depositional rates on the order of 15 cm/year from off meltwater outlets of the Austfonna ice cap on Nordaustlandet, Svalbard (Pfirman & Solheim 1989), all the accumulations described may have been emplaced within periods as short as 150 to 200 years. Further southwest, along the flank of the Spitsbergenbanken area (Fig. 1), locally thicker accumulations have been mapped in approximately the same depth intervals (Elverhøi *et al.* 1990).

The rapid recession is most likely linked to rises in relative sea level. Due to the short time window available, the ice sheet must have been relatively thin in the frontal zone to respond rapidly. The concept of a relatively thin ice sheet is further supported by the fact that most of it, at least in its outer parts, rested on a deformable bed. According to models put forward by Boulton & Jones (1979), this may allow glaciers to develop lower surface profiles than ones resting on bedrock, since a part of the flow takes place by deformation of the subglacial sediments. Formation of the glacial flutes shows the deformable nature of the bed.

From the data off large (> 1000 km^2), surging drainage basins on Nordaustlandet (Fig. 1), a distinct suite of sea floor morphologies, in particular including a set of linear ridges forming a rhombohedral pattern, have been interpreted as diagnostic for surges of marine based glaciers (Solheim 1990). Typical for these surges is post-surge stagnation, during which rapid retreat of the thinned and intensively fractured glacier takes place through calving. Flutes were not observed off the surging glaciers in Nordaustlandet, but are observed off terrestrial surging glaciers (Sharp 1985). However, the near total coverage of the sea floor by flutes in the present study area, combined with a lack of rhombohedral ridge patterns, indicates that the ice sheet had normal movement during deglaciation, and that no surges took place.

Sediment types and physical properties

The geotechnical data are too sparse for a detailed discussion of compaction beneath an overriding glacier or ice sheet. However, a few observations are of importance for the development of physical properties of sediments in such an environment.

1. Glacial flutes are formed in subglacial diamictons which, after retreat, have undrained shear strengths of less than 80 kPa. Diamictons of similar shear strengths have also been sampled in other parts of the Barents Sea, and the action of grounded icebergs, consolidating soft sediments and/or disturbing firm sediments, has been discussed as a possible cause for the strength (Elverhøi *et al.* 1990). The present observations show that these sediments also form subglacially. As full consolidation below even a relatively thin ice sheet would cause higher than the observed shear strengths, this implies poor pore water drainage and low effective pressures under the outer zones of the ice sheet. This is in accordance with Boulton & Paul's (1976) and Boulton's (1979) discussions of a layer of subglacial 'deformation till' of low strength and relatively high water content.

2. The fluted surface can be followed into the marginal parts of the southern sediment lense (Fig. 9), but the 3.5 kHz records show no difference in the acoustic character relative to the main parts of the accumulation, where flutes are non-existent. This observation indicates that at the margins of a marine based ice sheet, there may be a transitional zone where the division between subglacial sediments and ice proximal glacimarine sediments is gradational and not acoustically identifiable with the equipment used in this study. This may be caused by small readvances reworking the glacimarine deposits, or indicate that the till in an outer zone in which the ice may have been near-buoyant, grades laterally into the proximal sediments. The similarities in lithology (Fig. 10) further support the latter suggestion.

Conclusions

The identification of glacial flutes and transverse ridges in the northern Barents Sea provides evidence of the former existence of grounded glacier ice at least to present-day 340 m water depth and as far south as 74° 55′N in central Bjørnøyrenna.

The main fluted area is bounded to the south by accumulations of acoustically transparent and homogeneous sediments, deposited in front of a stationary ice front. This represents a halt during its recession, which took place in a stepwise manner. A second halt took place at approximately 150 m present-day water depth, resulting in the deposition of a new set of thicker sediments. The retreat from the fluted area took place rapidly, not allowing enough ice proximal deposition to mask the low-relief flutes.

Relatively soft sediments (undrained shear strengths less than 80 kPa) formed subglacially in the study area and reflect poor pore water drainage and subglacial deformation of the till. These sediments appear acoustically indistinct from proximal glacimarine deposits, and also have similar lithologies.

The ice had a predominantly southerly flow direction during the stages before final retreat, but deviations due to the effects of regional topography are observed, and the banks may have formed local ice centres.

The base of the ice sheet was most likely at the pressure melting point in the study area. It rested on a deformable till bed and probably had a relatively low surface profile.

The ice sheet apparently had normal activity until retreat took place by destabilization of the front and iceberg calving. There are no indications of surges.

The crew members of M/S Lance are acknowledged for their valuable cooperation in the field. The Norwegian Petroleum Directorate provided older sparker data used for mapping the sediment distribution. Jon Landvik, Heiner W. Josenhans and Weston Blake Jr. gave critical comments on the manuscript. The Norwegian Research Council for Science and the Humanities (NAVF) provided funds for laboratory analyses. Meetings and discussions with Julian A. Dowdeswell were possible due to funding by NATO under grant no. 0747/87. This is the Norwegian Polar Research Institute contribution no. 260.

References

ANDERSEN, B.G. 1981. Late Weichselian ice sheets in Eurasia and Greenland. *In*: DENTON, G.H. & HUGHES, T.J. (eds.) *The Last Great Ice Sheets*. Wiley, Chichester, 1–65.

AARIO, R. 1977. Classification and terminology of morainic landforms in Finland. *Boreas*, **6**, 87–100.

—— 1987. Drumlins of Kuusamo and Rogen-ridges of Ranua, northeast Finland. *In*: MENZIES, J. & ROSE, J. (eds.) *Drumlin Symposium*. Balkema, Rotterdam, 87–101.

BARANOWSKI, S. 1970. The origin of fluted moraine at the fronts of contemporary glaciers. *Geografiska Annaler*, **52A**, 68–75.

BOULTON, G.S. 1976. The origin of glacially fluted surfaces, observations and theory. *Journal of Glaciology*, **17**, 287–309.

—— 1979. Glacial history of the Spitsbergen archipelago and the problem of a Barents Shelf ice sheet. *Boreas*, **8**, 31–57.

—— 1982. Subglacial processes and the development of glacial bedforms. *In*: DAVIDSON-ARNOTT, R., NICKLING, W. & FAHEY, B.D. (eds.) *Research in Glacial, Glaciofluvial and Glaciolacustrine Systems*, 6th Guelph Symposium on Geomorphology, 1–31.

—— 1987. A theory of drumlin formation by subglacial sediment deformation. *In*: MENZIES, J. & ROSE, J. (eds.) *Drumlin Symposium*. Balkema, Rotterdam, 25–80.

—— & JONES, A.S. 1979. Stability of temperate ice caps and ice sheets resting on beds of deformable sediment. *Journal of Glaciology*, **24**, 29–42.

—— & PAUL, M.A. 1976. The influence of genetic processes on some geotechnical properties of glacial tills. *Quarterly Journal of Engineering Geology*, **9**, 159–194.

CLAYTON, L. & MORAN, S.R. 1974. A glacial process-form model. *In*: COATES, D.R. (ed.) *Glacial Geomorphology*. Binghamton, State University of New York, 89–119.

DE GEER, G. 1889. Andmorener i trakten mellan Spånga och Sundbyberg. *Geologiska Foreningen i Stockholm, Forhandlingar*, **11**, 395–397.

DENTON, G.H. & HUGHES, T.J. 1981. The Arctic Ice sheet: An outrageous hypothesis. *In*: DENTON, G.H. & HUGHES, T.J. (eds) *The Last Great Ice Sheets*. Wiley, Chichester, 437–467.

DYSON, J.L. 1952. Ice-ridged moraines and their relation to glaciers. *American Journal of Science*, **250**, 204–211.

ECHELMEYER, K. & ZHONGXIANG, W. 1987. Direct observations of basal sliding and deformation of basal drift at sub-freezing temperatures. *Journal of Glaciology*, **33**, 83–98.

EIDE, L.I. 1983. Environmental conditions in the Barents Sea and near Jan Mayen. Unpublished report of The Norwegian Meteorological Institute, 1–139.

ELVERHØI, A., ANTONSEN, P., FLOOD, S.B., SOLHEIM, A. & VULLSTAD, A.A. 1988. The physical environment, western Barents Sea, 1:1 500 000: Shallow bedrock geology – structure, litho- and biostratigraphy. *Norsk Polarinstitutt Skrifter*, **179(D)**.

——, LØNNE, Ø. & SELAND, R. 1983. Glaciomarine sedimentation in a modern fjord environment, Spitsbergen. *Polar Reseach* **1** (new series) 127–149.

——, NYLAND BERG, M., RUSSWURM, L.F. & SOLHEIM, A. 1990. Late Weichselian ice recession in the central Barents Sea. *Proceedings of the Advanced Research Workshop on the "Geologic History of the Polar Oceans: Arctic versus Antarctic"*, Bremen, F.R. Germany October 10–14, 1988.

—— & SOLHEIM, A. 1983. The Barents Sea ice sheet – a sedimentological discussion. *Polar Research* **1**, 23–42.

—— & —— 1987. Late Weichselian glaciation of the northern Barents Sea – a discussion. *Polar Research*, **5** (new series), 285–287.

FORMAN, S.L., MANN, D.H. & MILLER, G.H. 1987. Late Weichselian and Holocene relative sea-level history of Brøggerhalvøya, Spitsbergen. *Quaternary Research*, **27**, 41–50.

GROSSWALD, M.G. 1980. Late Weichselian ice sheet of northern Eurasia. *Quaternary Research*, **13**, 1–32.

GÖRLICH, K. 1986. Glacimarine sedimentation of muds in Hornsund fjord, Spitsbergen. *Annals of the Geological Society of Poland*, **56**, 433–477.

HOPPE, G. & SCHYTT, V. 1953. Some observations on fluted moraine surfaces. *Geografiska Annaler*, **35**, 105–115.

HUGHES, T.J., DENTON, G.H., ANDERSEN, B.G., SCHILLING, D.H., FASTOOK, J.L. & LINGLE, C.S. 1981. The last great ice sheets: A global view. *In*: Denton, G.H. & Hughes, T.J. (eds.) *The Last Great Ice Sheets*. Wiley, Chichester, 263–317.

JONES, G.A. & KEIGWIN, L.D. 1988. Evidence from Fram Strait (78°N) for early deglaciation. *Nature*, **336**, 56–59.

JOSENHANS, H.W. & ZEVENHUIZEN, J. 1990. Dynamics of the Laurentide ice sheet in Hudson Bay, Canada. *Marine Geology*.

LUNDQVIST, J. 1969. Problems of the so-called Rogen moraine. *Sveriges Geologiska Undersøkning*, **C648**, 1–33.

—— 1989. Rogen (ribbed) moraine – identification and possible origin. *Sedimentary Geology*, **62**, 281–292.

MATISOV, G.G. 1980. Geomorphological indications of the impact of the Scandinavian, Novaja Zemlja, and the Spitsbergen ice cover upon the bottom of the Barents Sea. (Geomorfologiceskie priznaki vozdejstvija Skandivavskogo, Novozemelskogo, Spicbergenskogo lednikovych pokrovov na poverchnostdna Barenceva morja.) *Okeanologija*, **20** (4), 669–680.

MINELL, H. 1977. Transverse moraine ridges of basal origin in Harjedalen. *Geologiska Foreningens Forhandlingar*, **99**, 271–277.

PAUL, M.A. & EVANS, H. 1974. Observations on the internal structure and origin of some flutes in glacio-fluvial sediments, Blomstrandbreen, northwest Spitsbergen. *Journal of Glaciology*, **13**, 393–400.

PFIRMAN, S.L. & SOLHEIM, A. 1989. Subglacial meltwater discarge in the open-marine tidewater glacier environment: observations from Nordaustlandet, Svalbard archipelago. *Marine Geology*, **86**, 265–281.

PRICE, R.J. 1973. *Glacial and Fluvioglacial Landforms*. Oliver & Boyd, Edinburgh.

POWELL, R.D. 1981. A model for sedimentation by tidewater glaciers. *Annals of Glaciology*, **2**, 129–134.

SALVIGSEN, O. 1981. Radiocarbon dated raised beaches in Kong Karls Land, Svalbard, and their consequences for the glacial history of the Barents Sea area. *Geografiska Annaler*, **63**, 283–291.

SCHYTT, V., HOPPE, G., BLAKE, W. Jr. & GROSSWALD, M.G. 1968. The extent of the Wurm glaciation in the European Arctic. *International Association of Scientific Hydrology, I.U.G.G., General Assembly of Bern, 1967, Commission of Snow and Ice*, Publication No. 79, 207–216.

SHARP, M. 1985. "Crevasse-fill" ridges – a landform type characteristic of surging glaciers? *Geografiska Annaler*, **67A**, 213–220.

SHAW, J. 1979. Genesis of the Sveg tills and Rogen moraines of central Sweden: a model of basal melt out. *Boreas*, **8**, 409–426.

—— & FRESCHAUF, R.C. 1973. A kinematic discussion of the formation of glacial flutings. *Canadian Geographer*, **17**, 19–35.

SOLHEIM, A. 1990. The depositional environment of surging subpolar tidewater glaciers: a case study of the morphology, sedimentation and sediment properties in a surge affected marine basin outside Nordaustlandet, northern Barents Sea. *Norsk Polarinstitutt Skrifter*.

—— & ELVERHØI, A. 1985. A pockmark field in the central Barents Sea; gas from a petrogenic source? *Polar Research*, **3** (new series), 11–19.

——, —— & FINNEKÅSA, Ø. 1988a. Marine geophysical/geological cruise in the northern Barents Sea 1987 – Cruise report. *Norsk Polarinstitutt Rapportserie*, **43**.

—— & KRISTOFFERSEN, Y. 1984. The physical environment, western Barents Sea, 1:1 500 000, sheet B; Sediments above the upper regional unconfromity: thickness, seismic stratigraphy and outline of the glacial history. *Norsk Polarinstitutt Skrifter*, **179B**.

——, MILLIMAN, J.D. & ELVERHØI, A. 1988b. Sediment distribution and sea-floor morphology of Storbanken: implications for the glacial history of the northern Barents Sea. *Canadian Journal of Earth Sciences*, **25**, 547–556.

SUGDEN, D.E. & JOHN, B.S. 1976. *Glaciers and Landscape*. Edward Arnold, London.

SYVITSKI, J.P.M. 1989. On the deposition of sediment within glacier-influenced fjords: oceanographic controls. *Marine Geology*, **85**, 301–329.

VAN DER MEER, J.J.M. 1982. A recent drumlin with fluted surface in the Swiss Alps. *In*: EVERNSON, E.B., SCHLUCHTER, Ch. & RABASSA, J. (eds.) *Tills and related deposits*, Balkema, Rotterdam, 105–109.

VORREN, T.O., HALD, M. & LEBESBYE, E. 1988. Late Cenozoic environments in the Barents Sea. *Paleoceanography*, **3**, 601–612.

—— & KRISTOFFERSEN, Y. 1986. Late Quaternary glaciation in the south-western Barents Sea. *Boreas*, **15**, 51–59.

ZILLIACUS, H. 1976. De Geer moräner och isrecessionen i södra Finlands östra delar. *Terra*, **88**, 176–184.

—— 1981. De Geer morenerna på Replot och Björkön i Vasa skärgård. *Terra*, **93**, 12–24.

—— 1989. Genesis of De Geer moraines in Finland. *Sedimentary Geology*, **62**, 309–317.

Geometry and genesis of the glacigenic sediments in the southern Barents Sea

TORE O. VORREN, ERLAND LEBESBYE & KJELL B. LARSEN

Institute of Biology and Geology, University of Tromsø, N–9000 Tromsø, Norway

Abstract: The glacigenic sequence on the continental shelf and shelf break in the Barents Sea has been mapped seismostratigraphically. Four and six major units have been distinguished on the outer and inner shelf, respectively. The outer shelf units increase in thickness towards the shelf break, except for the youngest one. The geometry of these units is determined partly by the morphology of the basement (infill), but to a larger extent by postdepositional erosion. The older, inner shelf units are located in a coast parallel trough. The two youngest units also have a blanketing character. The geometry of the younger units gives indications of source areas, as well as of ice movement direction of subsequent eroding ice sheets. The palaeoenvironment and palaeogeography during the mid-late Weichselian in the Barents Sea is reconstructed.

Recently, Vorren *et al.* (1989) gave an account of the glacigenic sediments in the southern Barents Sea (Fig. 1). We distinguished three depositional systems, each having a distinct geometry and character: the continental slope, the continental shelf and fjord systems. In this paper we will present some new data from the continental shelf system. The aim is: (1) to establish a seismostratigraphy for the glacigenic units on the shelf, (2) to explain the lateral relationship between units on the outer and inner shelf, (3) to explain the geometry and origin of the units, and (4) to reconstruct the palaeoenvironment and palaeoceanography during the mid/late Weichselian.

In this study we have mainly used analogue as well as digital shallow-seismic data (Fig. 2). This includes high-resolution airgun data and sparker data. The seismic data have been collected by the University of Tromsø, the Continental Shelf and Petroleum Technology Research Institute A/S (IKU), the Norwegian Petroleum Directorate and the Norwegian Polar Institute.

The southern Barents Sea is underlain by sedimentary rocks, mainly of Mesozoic age in the inner areas and a Cenozoic wedge at the margin (Spencer *et al.* 1984). The bathymetry of the continental shelf is characterized by a broad E – W channel, the Bear Island Trough which reaches depths of 500 m, and shallower banks to the south and north (Fig. 1). At the mouth of the trough is a huge fan, the Bear Island Trough Mouth Fan (TMF) (Fig. 1). The water masses are dominated, at present, by Atlantic water entering the area from the southwest and Arctic water from the north (Mosby 1968).

Extent and seismostratigraphy of the glacigenic sequence

An upper regional unconformity (URU) separates the variously dipping pre-glacial sedimentary rocks from the overlying glacigenic sequence (Fig. 3B). This unconformity can be followed across most of the Norwegian continental shelf (Dekko 1975; Bugge & Rokoengen 1976; Lien 1976; Bugge *et al.* 1978; Rokoengen 1980; Rokoengen & Rønningsland 1983; Solheim & Kristoffersen 1984; Vorren *et al.* 1986). The morphology of the URU in the southern Barents Sea comprises a wide channel underlying the present Bear Island Trough, and an overdeepened trough just off the coast of northernmost Norway. A glacigenic sequence of varying thickness overlies this unconformity. Its glacigenic origin is documented by several core and drill samples (e.g. Vorren *et al.* 1984, 1989; Sættem & Hamborg 1987; Elverhøi *et al.* 1989). At the shelf break, the glacigenic sequence is between 900 and 1000 milliseconds TWT (two-way travel time) thick (Fig. 3A). On the shelf proper the thickness varies between 0 and approximately 300 milliseconds TWT. A depocentre occurs as an infill in the western part of the trough off the coast.

The glacigenic sequences, called the Barents Synthem (ISSC, 1987) by Vorren *et al.* (1989), can be divided into several seismostratigraphic units. The boundaries between the units are generally defined by semi-regional, continuous, high amplitude reflectors. In most cases the boundaries are level and sub-horizontal, but they may occasionally be quite irregular with steep gradients. In mapping the seismostrati-

From DOWDESWELL, J.A. & SCOURSE, J.D. (eds), 1990, *Glacimarine Environments: Processes and Sediments*. Geological Society Special Publication No 53, pp 269–288.

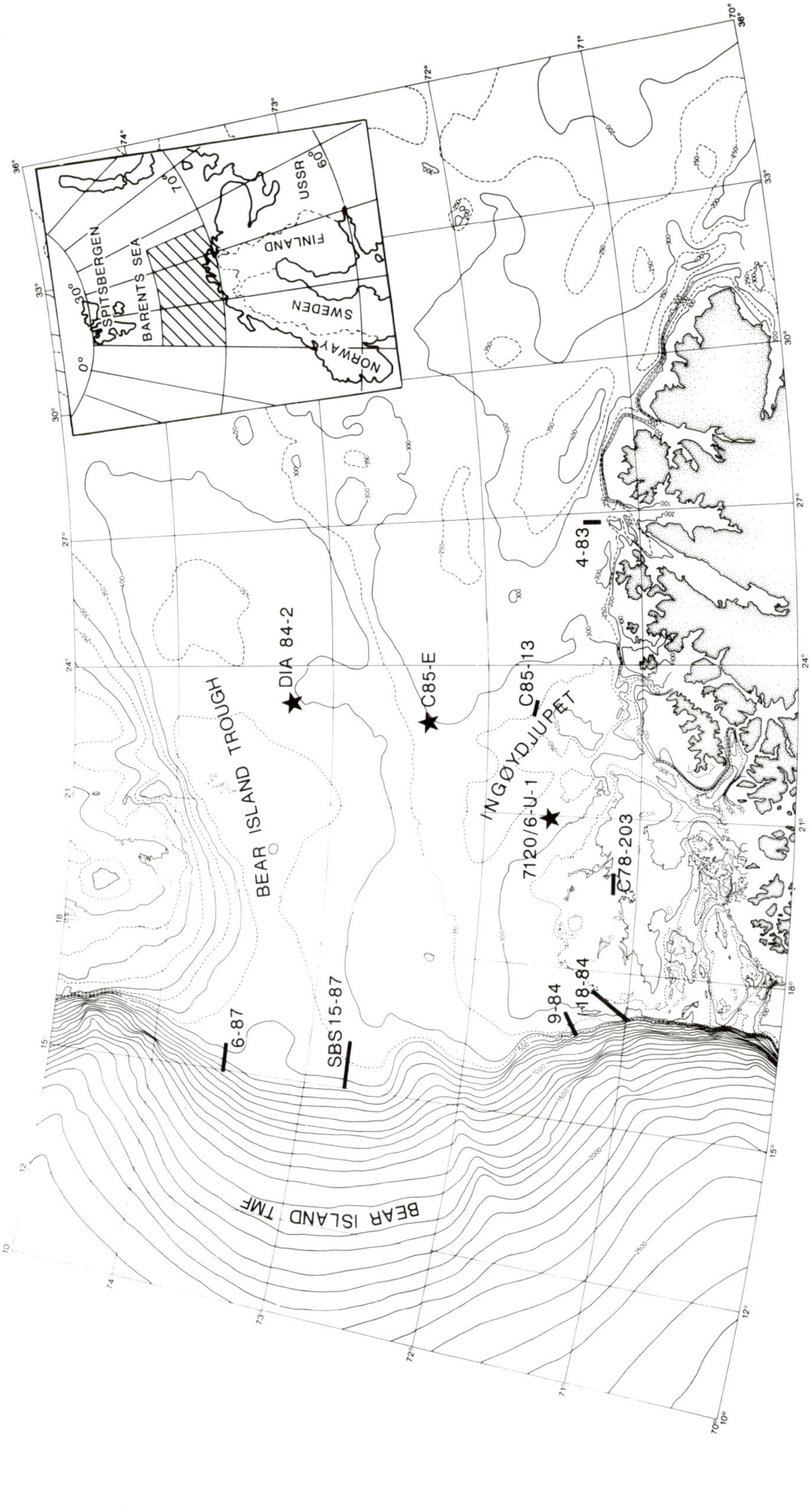

Fig. 1. Location of study area and bathymetry of the southern Barents Sea. Contour interval is 100 m on the continental slope and 50 m on the continental shelf. Seismic profiles are shown by bold lines and borehole localities with dated samples by stars.

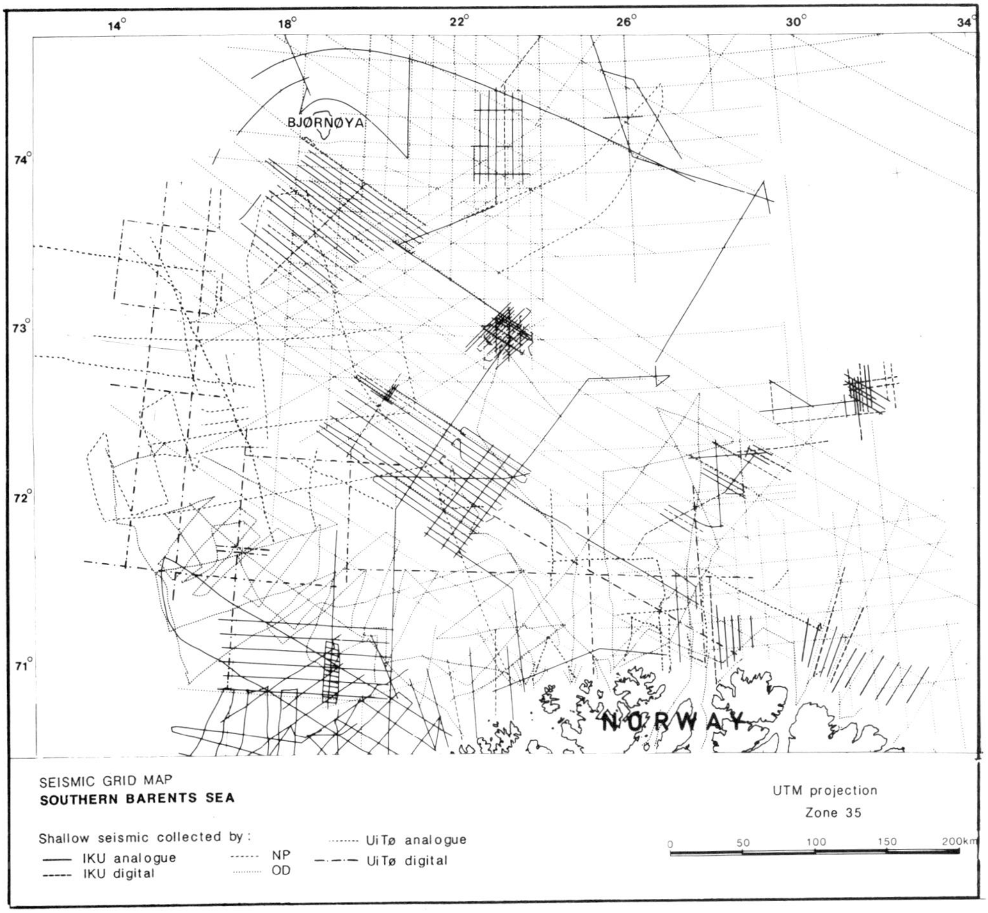

Fig. 2. Shallow seismic grid south of 74°N in the Barents Sea showing lines recorded from 1970 to 1987. Compiled by IKU in 1986. 1987 lines were added by the authors.

graphy we have divided the southern Barents Sea into an outer (western) and inner (eastern) part, due to considerable difficulties in correlating the two areas. The correlation problem is partly due to lack of appropriate data but, even more so, to the fact that most of the units belong either to a western or to an eastern province and wedge out in the central part (about 23° – 25°E) of the southwestern Barents Sea (Fig. 4).

In the west we have, in accordance with Solheim & Kristoffersen (1984), divided the Barents Synthem into four informal major regional units from 1W (oldest) to 4W (W denotes west). These units can be followed down the continental slope, and each unit can be divided into two or more sub-units (except for unit 3W). In addition to these four major units, we have distinguished a unit comprising moraine ridges (Vorren & Kristoffersen 1986). In the east, six major units (1E oldest to 6E) are distinguished. A brief overview of this stratigraphy is given by Vorren *et al.* (1989). The inner shelf deltas (Vorren *et al.* 1986), earlier related to 1W, are correlated here to 4E. The stratigraphic relation between the outer and inner province is that units 5E and 6E onlap unit 4W (Fig. 4). There is a slight possibility that unit 4E also onlaps 4W and may thus be younger.

Temporal framework

The process of dating the glacigenic sequence on the Barents Sea shelf has just begun. Based on results from deep-sea stratigraphy (Jansen *et al.* 1988), Vorren *et al.* (1988a) have speculated that the glaciations of the Barents Sea proper

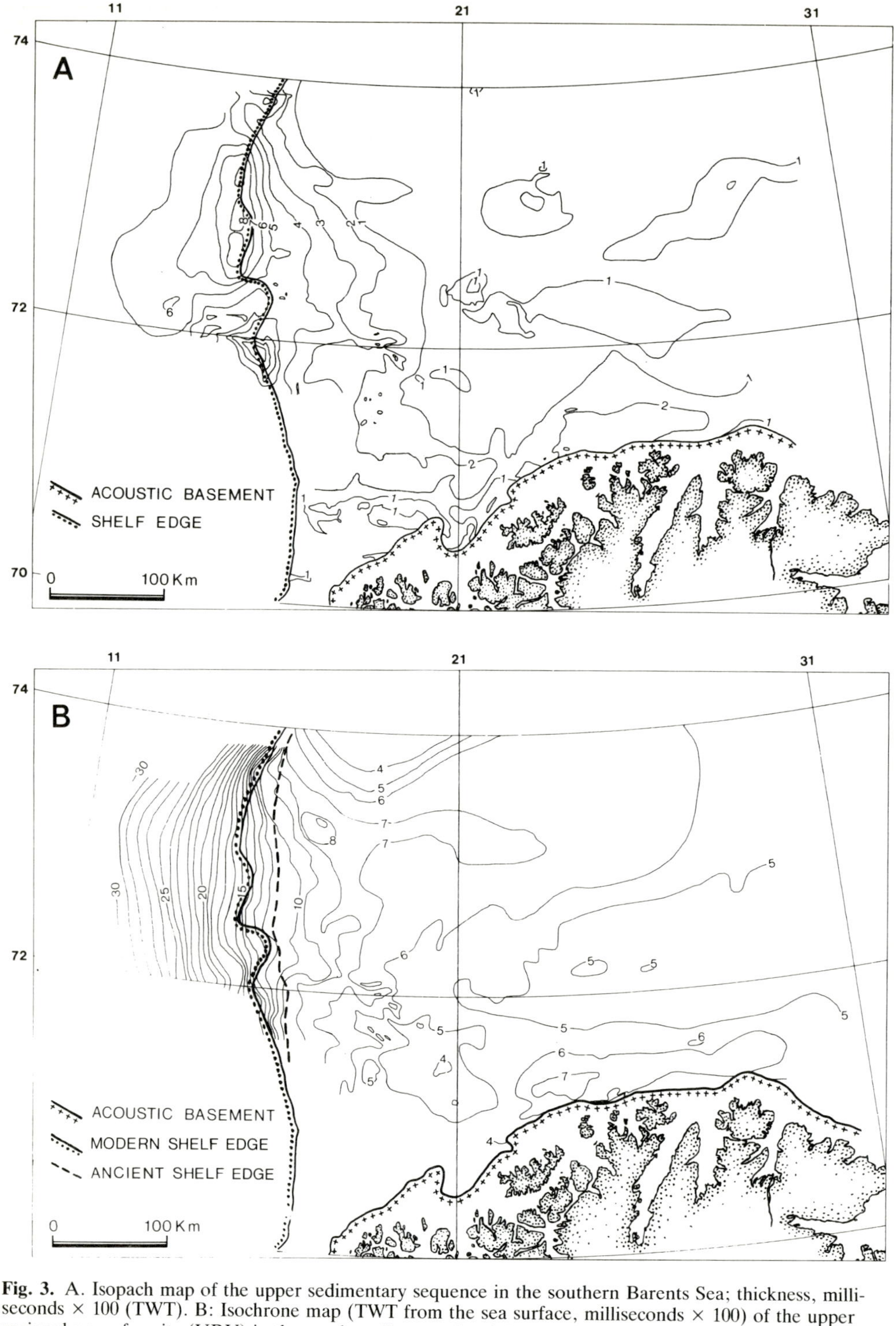

Fig. 3. A. Isopach map of the upper sedimentary sequence in the southern Barents Sea; thickness, milliseconds × 100 (TWT). B: Isochrone map (TWT from the sea surface, milliseconds × 100) of the upper regional unconformity (URU) in the southern Barents Sea (after Vorren *et al.* 1989).

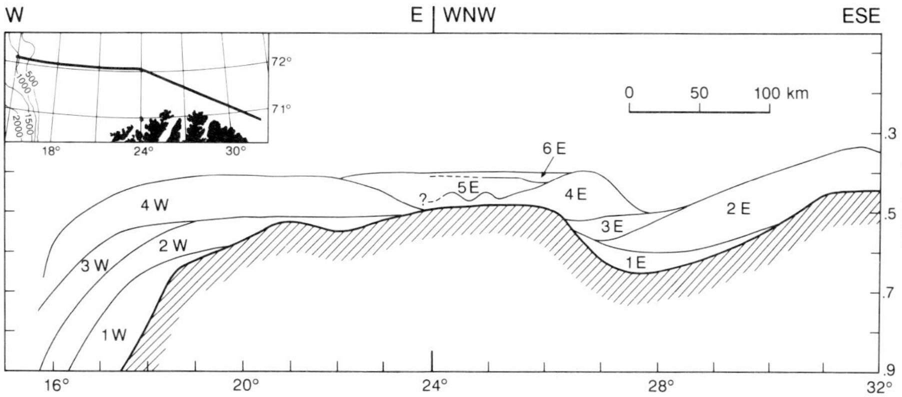

Fig. 4. Generalized geoseismic profile from the upper slope to the inner shelf (inset) showing the stratigraphical and lateral relationship of the different seismostratigraphic units in the upper sedimentary sequence.

commenced approximately at 0.8 Ma ago.

This implies that the glacial shaping of the unconformity (URU), which the glacial sediments superpose, is of that age and younger. However, this does not exclude the possibility that older glacimarine sediments on the continental slope exist, being derived from previous glaciers on the Fennoscandian mainland and from the Svalbard archipelago.

Radiocarbon dates have been obtained (Hald pers. comm) from unit 4W (Table 1, Fig. 1). The dates were made on shell fragments from borecores recovered by the Continental Shalf Institute and studied with respect to lighostratigraphy by T.O. Vorren and biostratigraphy by M. Hald.

The fragments were recovered from diamictons; they were not identifiable and they were probably reworked. Thus, the dates give maximum ages. Due to the small sample sizes the dating analyses were performed using the accelerator mass spectrometer in Uppsala run by G. Possnert. The dates have been corrected for isotopic fractionation of *c.* 410 years and reservoir effect of 440 years (Mangerud & Gulliksen 1975).

Site 7120/6 is situated off the mapped boundary for unit 4W. However, unit 4W may well exist as scattered outliers, but the resolution of the seismic lines prevented any reliable conclusions. If we accept these dates, they imply that the younger part of unit 4W is not older than 21 500 BP. Furthermore, the moraine ridges superposed on unit 4W are correlated to the Late Weichselian maximum, i.e. approximately 18 000–16 000 years old (Vorren *et al.* 1988a). Since we infer that units 5E and 6E have been overrun by the glacier during the Late Weichselian maximum advance (see below), this brackets the age of units 5E and 6E to between 21 500 and 18 000 BP.

The outer shelf and shelf break

The extent and thickness of the four major units in the west on the continental shelf are shown on Fig. 5. Due to incomplete seismic records we have not been able to map beyond the shelf edge. The youngest unit, 4W, is generally most extensive. Of the older units, 1W is most extensive in the southeastern area and 2W in the central area, while 3W is confined to the western area (Fig. 6).

Unit 1W. The thickness variation (Fig. 5) of unit 1W is clearly related to the underlying unconformity (URU). At about 72°N, there are two thickness maxima related to infill of a channel in the URU (Figs 3B & 7). Most of unit 1W has a relatively homogeneous acoustic signature, the but in the south there is a series of troughshaped internal reflectors. The troughs are up to 14 km wide, 100 milliseconds TWT deep and they are infilled with acoustically semi-transparent sediments.

Unit 2W. The eastern boundary of unit 2W is irregular. North of 72° the unit fills two depressions in the URU (Figs 5 & 7). Variations in thickness in the south are related to an uneven base comprising several troughs (Fig. 8). There are at least three generations of troughs superimposed on each other. The troughs are normally 3–8 km across and 20–100 milliseconds TWT deep. A preferred orientation towards W and NW was found by Kåsli (1986) and Larsen (1987), respectively. The basal reflectors of

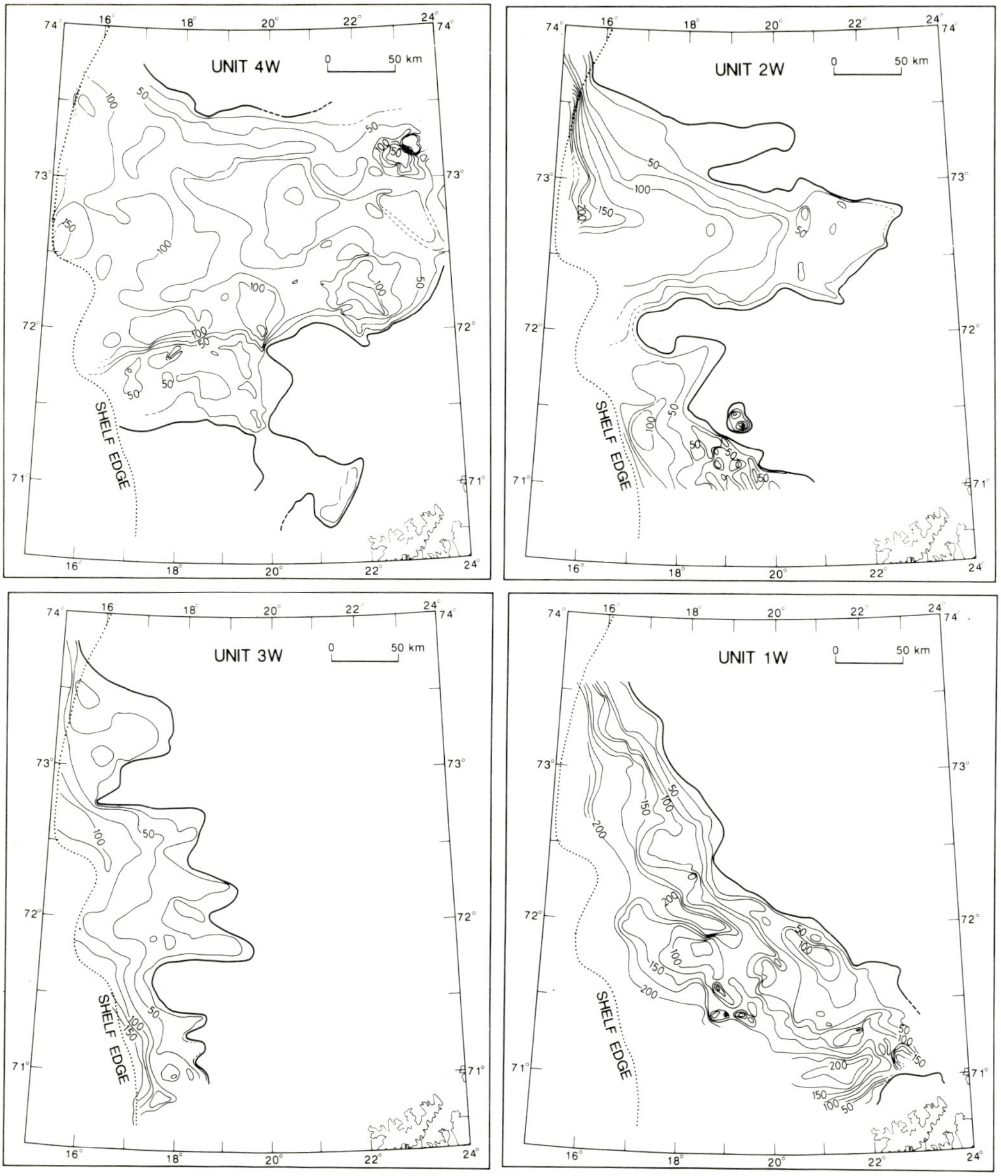

Fig. 5. Isopach maps (in milliseconds TWT) of the four major seismostratigraphic units on the outer continental shelf.

these troughs vary from high-amplitude coherent, with smooth semicircular to discontinuous and somewhat irregular forms.

Unit 3W. The eastern boundary of unit 3W owes its irregular appearance (Fig. 5) to an erosional phase which created the unconformity separating units 3W and 4W. The internal acoustic signature is rather homogeneous, but near the shelf break several internal seaward dipping reflectors occur (Figs 9 & 10).

Unit 4W. The lateral extent of the unit's northeastern boundary (Fig. 5) is uncertain due to scarce data coverage. The boundary in the southeast more or less coincides with the Ingøydjupet Trough margins, and in the southwest there is a gradual wedging out. Outside the

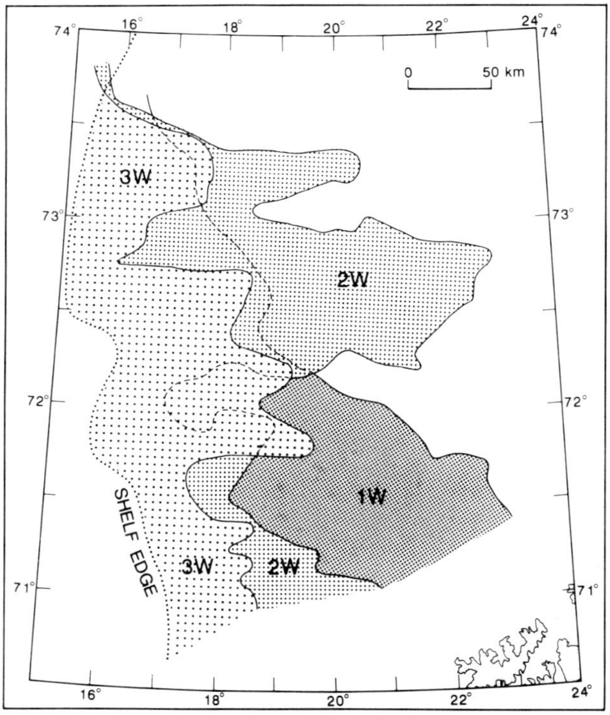

Fig. 6. Subcrop unit 4W, showing the distribution of the three older seismostratigraphic units on the outer shelf.

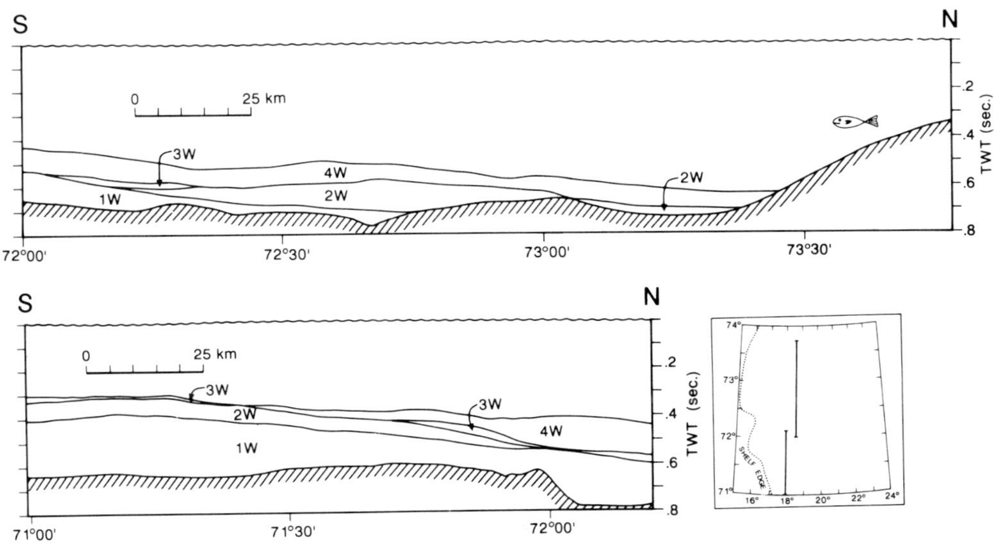

Fig. 7. North–south geoseismic profiles of the upper sedimentary sequence, along 18°30′ and 18°E, respectively (see inset map for location). Note the thickening of unit 1W to the south and of 4W to the north.

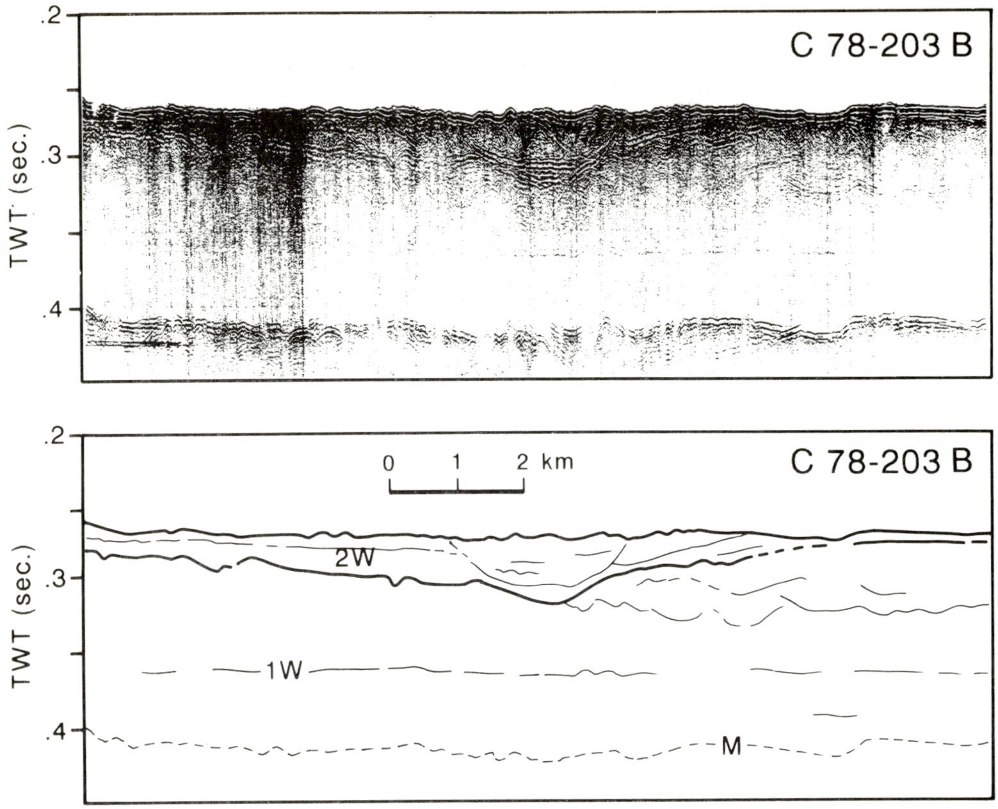

Fig. 8. Sparker profile (IKU C78–203B) showing trough structures in unit 2W (see Fig. 1 for location).

indicated boundaries there may exist outliers or thin parts not detected within the resolution of the sparker profiles (10–20 milliseconds TWT). On the shallow bank in the south the unit is very thin. At 72°N, where the base of the unit dips to greater depths towards the north, it increases rapidly in thickness (Figs 5 & 7).

The seismic signature of the unit is characterized by several discontinuous internal reflectors, except for the western part, where there are some well defined semi-regional subparallel reflectors. The discontinuous reflectors often show an irregular relief. The upper part of the unit is masked by diffraction hyperbolae due to the irregularly iceberg turbated sea floor. Samples recovered from this unit are overconsolidated diamictons (Hald & Vorren 1987). Drill core C85–E (Fig. 1) had two sandy/gravelly samples at the base, the rest of the samples were composed of 32–40% clay, 37–45% silt, 25–15% sand and 5–1% gravel.

The shelf break

The reflection patterns across the shelf break are illustrated by four profiles (Figs 9 & 10); other examples are published by Solheim & Kristoffersen (1984) and Vorren *et al.* (1989).

Most profiles show a complex sigmoid-oblique character. Normally, this implies strata with a history of alternating aggradation and depositional bypass within a high-energy depositional regime (Mitchum *et al.* 1977). As the upper shelf units comprise glacial diamictons and moraine ridges close to the shelf break (Vorren *et al.* 1989), we interpret the truncation ('depositional bypass') of the prograding clinoforms to be due to glacial erosion.

On the shelf break of the Bear Island Trough, units 1W and 2W are thickest, and account for most of the progradation (Figs 5, 9 & 10). Unit 4W is characterized by more aggradation than progradation.

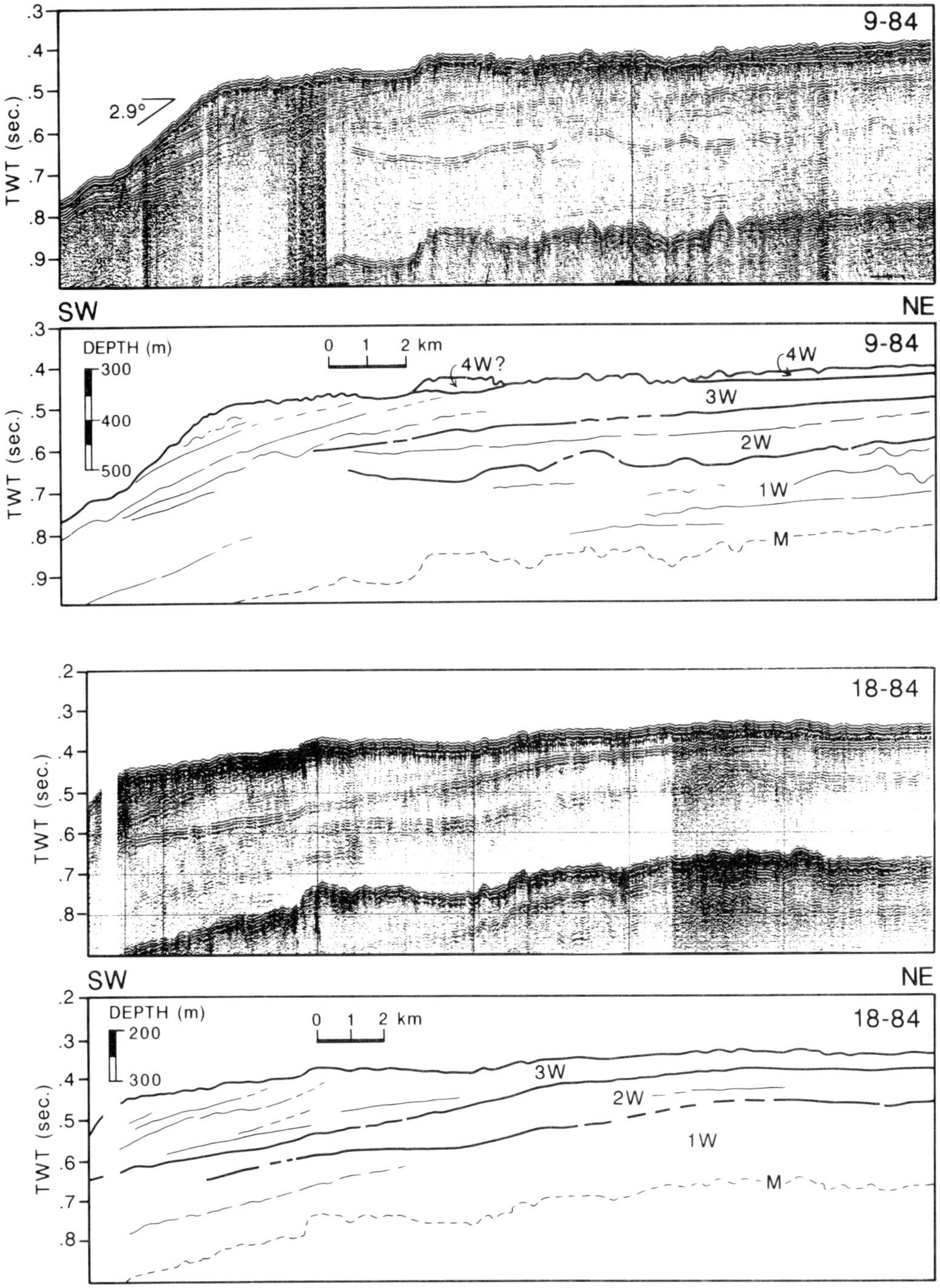

Fig. 9. Sparker profiles across the shelf break south of the Bear Island TMF (see Fig. 1 for location).

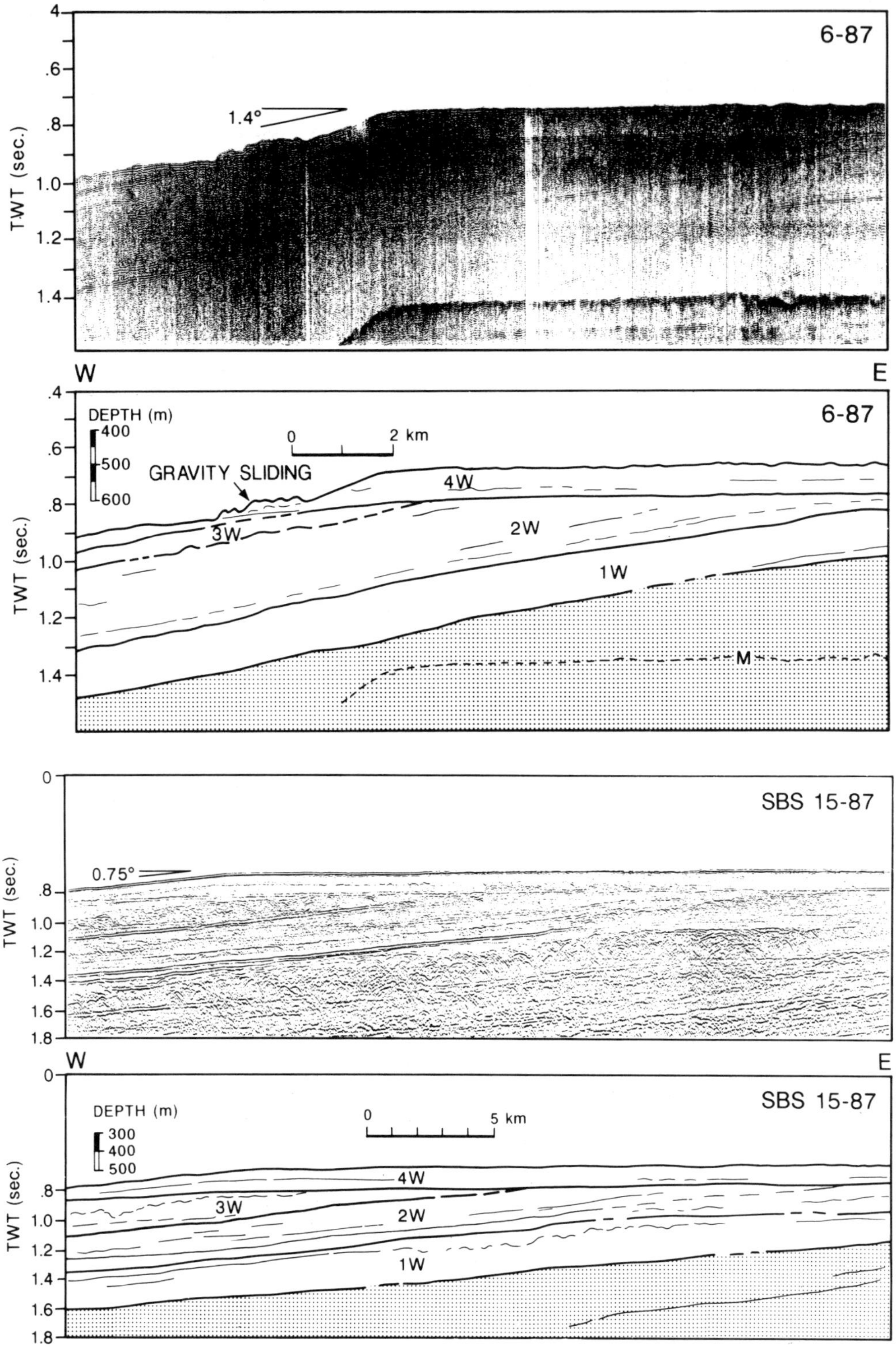

Fig. 10. Seismic profiles across the shelf break of the Bear Island TMF (see Fig. 1 for location). Profile A is an analogue sparker profile while B is a high resolution multichannel profile.

All along the shelf break, there seems to have been sediment gravity sliding. On profile 6–87 (Fig. 10) part of the slide has come to rest just off the shelf break. In the south (profile 9–84, Fig. 9) the prograding clinoforms seem to truncate some of the horizontal reflectors on the shelf; and the clinoforms themselves are truncated on the slope. This indicates quite high mass movement activity on this steeper part of the shelf break.

Origin of the glacigenic sediments on the outer shelf

A high priority problem related to the diamictons on the shelf proper is the question of their primary origin. Considering glaciated shelves in general, several genetic interpretations have been put forward for shelf stratiform diamictons; e.g. basal tills, various types of glacimarine sediments, undermelt diamicton (Gravenor *et al.* 1984), and till tongues (King & Fader 1986, King *et al.* 1987).

The geometry of the outer shelf units (except for the shelf break area) gives little information about their primary origin. This is probably due to severe alteration by erosion. However, the youngest, presumably least modified unit (4W), shows an increase in thickness towards the deepest part of the shelf (Bear Island Trough), suggesting glacimarine deposition in a shelf-embayment of the ice sheet.

The recovered sediment samples from the outer shelf all have the composition of glacigenic diamictons. We are not, at present, able to decide with certainty the primary origin of these sediments, based on the scattered borehole samples themselves. However, most of the samples are dominated totally by glacial foraminiferal faunal assemblages (Hald & Vorren 1987, 1988) indicating a glacimarine origin.

The seismic data show erosional unconformities separating the major units. These are believed to be of glacial origin, representing glacial advances onto the shelf (Vorren *et al.* 1989). Internally disrupted, irregular reflectors and the trough structures also indicate erosional episodes during deposition of the units. Concluding, we suggest that much of the sediment in the outer shelf units was originally of glacimarine origin. Frequent glacier advances have compacted and disturbed these sediments and presumably also deposited basal tills in some areas.

Looking at the individual units, unit 1W and 2W show the most pronounced progradation. Unit 3W is rather thin and confined to the outermost shelf while unit 4W is characterized by aggradation and not so much progradation. This may indicate that during 1W and 2W-time, the glacier margin had a position near the shelf break for longer periods. The limited extent of 3W may be due to severe erosion. During 4W time the limited progradation may be due to a more eastern position of the glacier margin for long periods.

The inner shelf

The mapping of the six major units in the east is not quite completed. In particular there are uncertainties with regard to the extent of units 4E and 6E. The lateral distribution of unit 4E in the northwest is particularly important with respect to correlation to the outer shelf. Tentatively, we suggest that unit 4E is older than unit 4W. Unit 5E is definitely overlapping unit 4W (Fig. 4).

The extent of the individual units varies between 200 and 300 km, and the thickness up to 75 to 200 milliseconds (TWT). Each of the units has a different extent and geometry, but most of them more or less overlap (Fig. 11). A common feature is their near-coastal thickness maxima (within 50–70 km of the coast). This coincides with the large, longitudinal, coast-parallel URU-trough. The three oldest units (1E, 2E and 3E) are in their entirety located in this trough, which seems to have acted both as a sediment trap and as a protection against subsequent erosion. Here, we will discuss units 3E and 5E in more detail.

Unit 3E

Geometry. The base of unit 3E is represented by a relatively plane subsurface except in the south, where it rises sharply towards the coast. Thus, except for the rapid increase in thickness in the south, the isopachs (Fig. 12) represent the morphology of the upper boundary of unit 3E.

East of 27°30′ E there is a marked indentation in the eastern distribution of 3E. This is probably due to pronounced glacial erosion, which occurred after the deposition of unit 4E. The glacier flowed towards NW–NNW, and it probably eroded a substantial amount of the eastern part of unit 3E. Even though there are signs of glacial erosion, the pinch out of the unit in the north and west seems to a large degree to reflect the original extent of unit 3E.

Internal seismic signature. In the area of maximum thickness in the south (Fig. 12), an internal pattern of northward prograding reflectors is observed (Fig. 13). One profile shows a sigmoid-

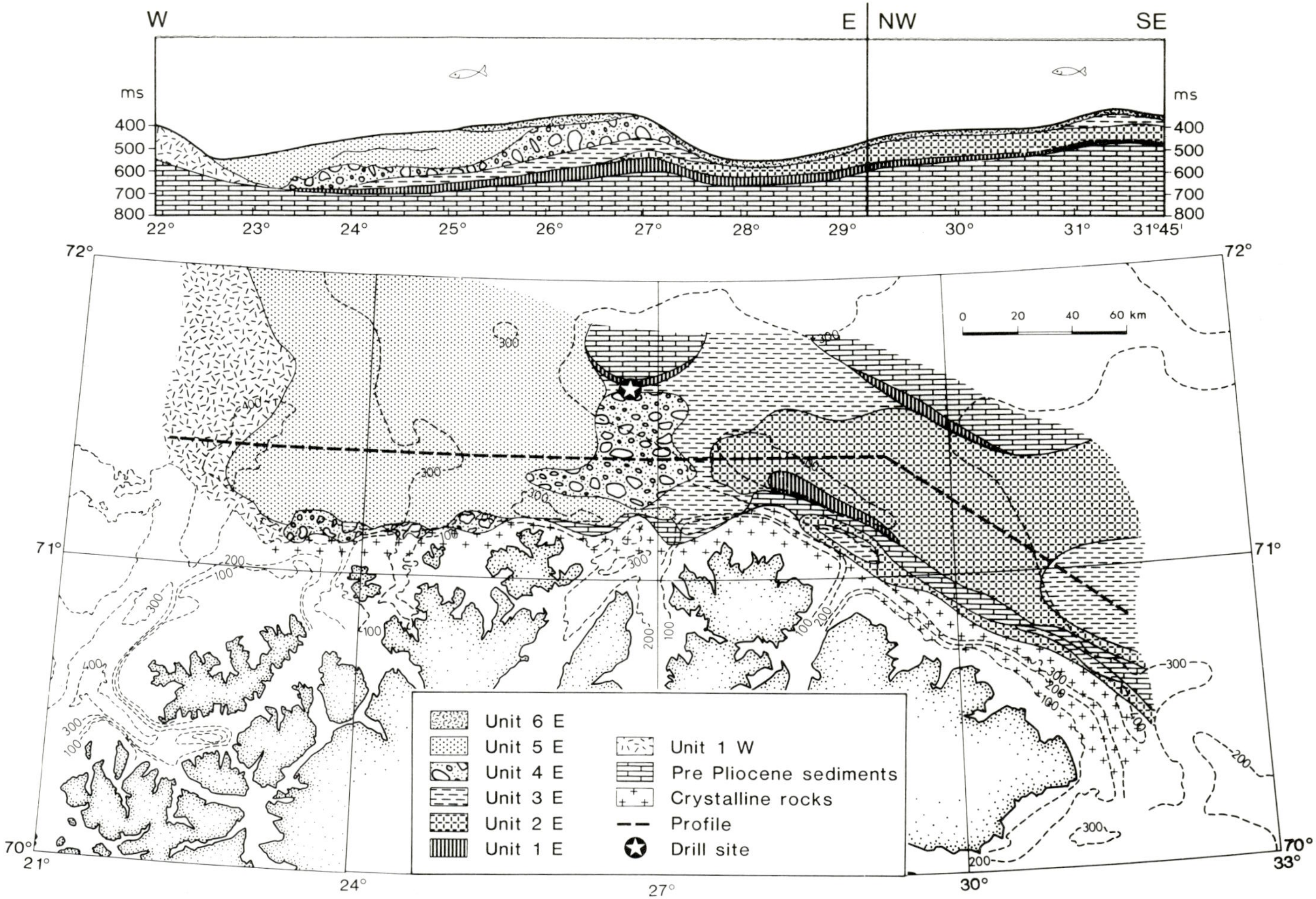

Fig. 11. Map and profile of seismic units on the inner continental shelf. The profile is located along the dashed line. Unit 6E has been omitted from the map (Slightly modified from Vorren *et al.* 1989).

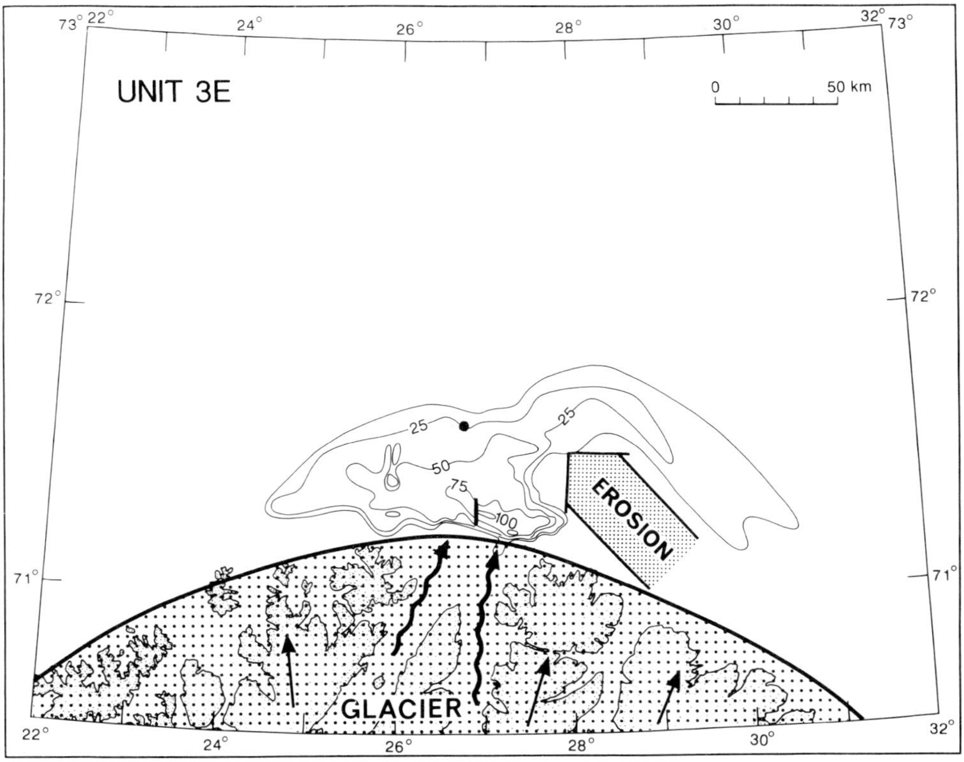

Fig. 12. Isopach map of unit 3E in milliseconds TWT. Approximate position of glacier and main meltwater drainage routes (meandering arrows) during deposition are indicated. Ice flow directions of the subsequent eroding glacier are indicated by large arrow. Dot indicates location of borehole (C85-A), penetrating unit 3E. Bar indicates location of sparker profile shown in Fig. 13.

oblique progradation pattern comprising the whole thickness of the unit (*c*. 100 milliseconds TWT) for a distance of 10 km. Some of the dipping internal reflectors have higher amplitudes than others and show clear evidence of being erosional surfaces.

Sediments. Three samples from a borehole through the northern part of unit 3E (Fig. 12), comprised diamictons of glacial origin (Vorren *et al.* 1989). All of the samples were clearly overconsolidated. Texturally, they comprised about 37% clay, 48% silt, 15% sand and less than 1% gravel. Structurally, one of them was massive with a weak fissility, one showed deformed fine laminae with scattered dropstones, and one was rather massive, containing sand lenses.

Origin of unit 3E. The lateral variation in acoustic character and internal reflection pattern indicates that most of the sediments in unit 3E have been derived from the south. Most of the unit has a semi-transparent acoustic signature indicating relatively fine-grained sediments, as shown by the borehole samples. In the south a pattern of incoherent reflections and much acoustic backscatter indicates coarser sediments deposited in a more proximal position to the glacier.

Two likely mechanisms explaining the deposition of unit 3E will be discussed. (1) Most of the sediments are originally deposited from meltwater plumes or (2) most of the sediments, particularly the proximal part, are deposited as a 'till delta' in the sense of Alley *et al.* (1989). These two alternatives are not wholly mutually exclusive, but alternative (2) implies colder ice and thus less meltwater production.

Re. (1): According to Pfirman & Solheim (1989) subglacial drainage systems at the base of tidewater ice caps may follow bedrock depressions and have a limited number of major stable

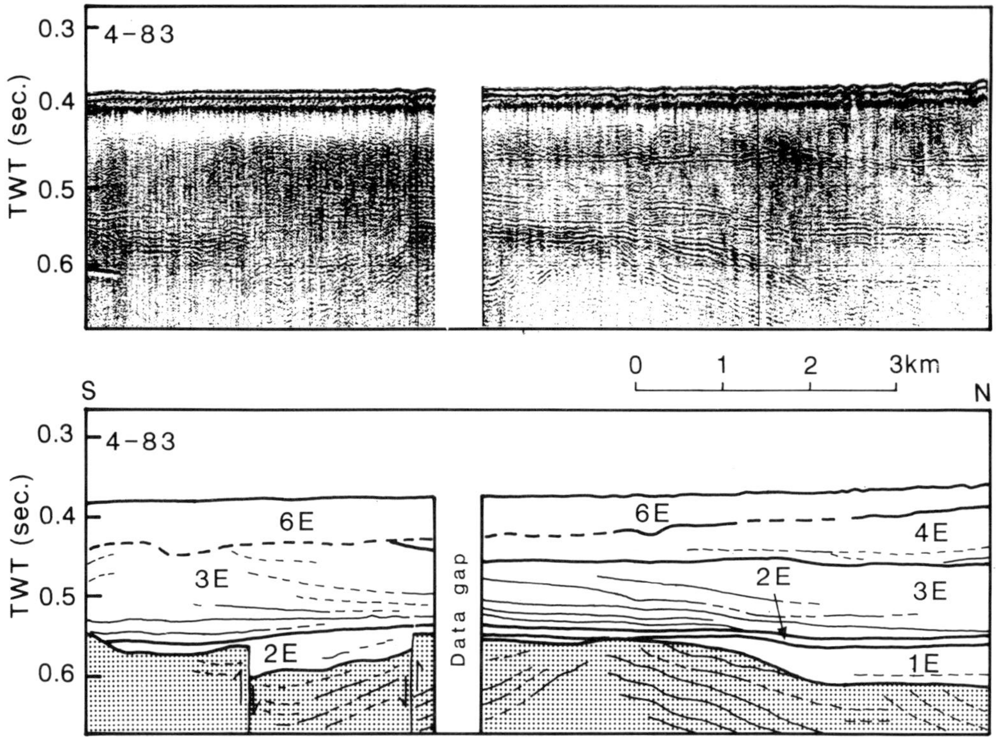

Fig. 13. Sparker profile across part of unit 3E (see Figs 1 & 12 for location). Note internal northward prograding reflectors in unit 3E.

meltwater discharge locations along the ice front, together with numerous minor discharge points. In accordance with this the thickness maximum (apex) of unit 3E is located just at the mouth of the two main fjords (i.e. Laksefjord and Porsangen). The lateral gradation, indicated by the seismics, of coarser grained sediments to the south, may be explained by bedload dumping and sediment gravity flows proximal to the ice front (Powell & Molnia 1989).

Re. (2): 'Till deltas' are believed to comprise basal till topsets with minor sorted sediments, and gravity flow foresets and bottomsets. They are deposited below an ice shelf in front of ice streams moving rapidly on a several-metre-thick water saturated basal till layer (Alley *et al.* 1989).

The internal pattern of prograding reflectors seen in unit 3E (Fig. 13) could possibly be due to the process described and, given the right conditions, the fjords could have acted as drainage routes for high-velocity ice streams. The question is whether there were conditions for ice shelf formation in this area. Ice shelf formation is favoured by low temperatures, confinement in an embayment, pinning points/elevations where the ice shelf becomes grounded, and rapid input from grounded ice. There are no embayments in this area, but as unit 3E is confined to the longitudinal trough just off the coast (Fig. 3B), the northern flank of this may possibly have acted as a grounding line area.

Although we favour mechanism 1, we feel that more data are needed to make a firmer conclusion. The geometry does indicate that during a post-depositional glacial advance from the SE, the eastern part of unit 3E was removed and, therefore, the original geometry was heavily altered in the east.

Unit 5E

Geometry. Unit 5E has a wide lateral extent and a thickness of up to 200 milliseconds TWT (Fig. 14). This unit is denoted by Solheim & Kristoffersen (1984) as the 'Ingøydjupet acoustically transparent lense' and by Vorren *et al.* (1986) as the 'transparent unit'. The 'upper seismic unit' of Sættem & Hamborg (1987) seems to be identical to unit 5E in the central

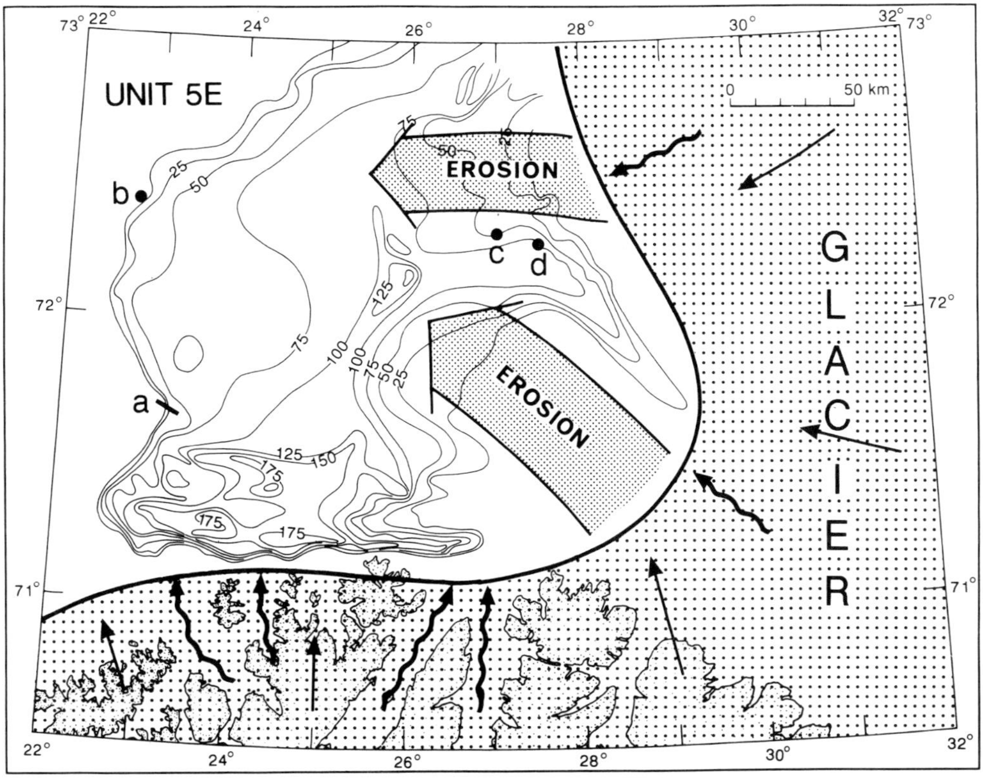

Fig. 14. Isopach map of unit 5E in milliseconds TWT. Approximate position of glacier, ice flow directions (straight arrows) and main meltwater drainage routes (meandering arrows) are indicated. Ice flow directions of subsequently eroding glaciers are indicated by large shaded arrows. a = boomer profile shown in Fig. 15; b, c and d are IKU-boreholes C85-E, 7227/7-U-1 and -U-1B, and 7227/8-U-1 and -2, respectively.

part of the area. Towards the south Sættem & Hamborg seem to correlate to our unit 6E.

The upper boundary of unit 5E is relatively level. The variations in the isopachs are to a large degree a function of the relief of the base of the unit. In the southwest the underlying unit is heavily dissected by glacial and/or subglacial glacifluvial erosion (Vorren *et al.* 1989). In this southwestern area, with maximum thickness, the unit infills the longitudinal coast-parallel trough. Outside this area, unit 5E drapes more or less conformably over the subsurface.

Unit 5E has a very irregular eastern boundary with the same pronounced inward bend as underlying units (e.g. unit 3E). This probably reflects syn- or post-depositional erosion by NW–WNW moving glaciers. A similar inward bend in the isopachs in the NE may indicate a more westerly flow direction in this area.

For long stretches, the western boundary takes the form of an asymmetrical wedge (Fig. 15). The base is more or less horizontal, while the surface dips convexly towards the unit's termination. When discussing the origin of unit 5E it is interesting to note that a similar wedge-geometry as described above is found in other areas, and described as 'till tongues' by King & Fader (1986) and King *et al.* (1987). Furthermore, asymmetrical acoustically transparent wedges found in the northern Barents Sea are interpreted to be formed by rapid glacimarine deposition (Elverhøi & Solheim 1983; Solheim & Kristoffersen 1984). The lower boundary reflector is, in the western part, dominated by a very irregular subsurface (Fig. 15), resembling iceberg ploughed sea-floor relief (Sættem & Hamborg 1987).

Sediments and stratigraphy. Samples from this unit comprise overconsolidated diamictons (Sættem & Hamborg 1987, Hald & Vorren 1988). The samples we have seen (shallow drillings 7227/7-U-1 and-U-1B and 7227/8-U-1 and -2, Fig. 14), have a massive appearance, but with a more or less pronounced fissility. The

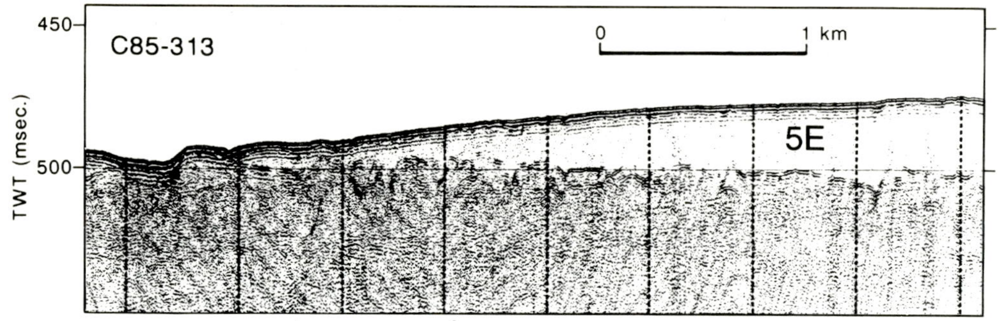

Fig. 15. Deep-tow boomer profile (IKU C85–313) across the western boundary of unit 5E (see Figs 1 & 14 for location). Note the irregular base of unit 5E.

grain-size distribution comprise 35–45% clay, 30–40% silt, 15–25% sand and a few per cent gravel.

The internal seismic signature of unit 5E is transparent/semi-transparent. The acoustic transparency varies laterally; there seems to be a higher level of internal backscatter in the eastern areas than in the west. Some internal reflectors can be followed over shorter distances in this unit (Vorren *et al.* 1986: Figs 4 & 5).

Origin of unit 5E. We have favoured a glacimarine origin for unit 5E (Vorren *et al.* 1989), while Sættem & Hamborg (1987) indicate that their partly equivalent 'upper seismic unit' was mainly deposited as basal till. When looking at the sediment samples recovered from this unit, they certainly have a till-like appearance. However, this is probably due to post-depositional deformation by glacier advances.

We think there are several arguments in favour of a glacimarine origin. First, the transparent acoustic signature probably represents a relatively homogeneous fine grained sediment with few large clasts. Secondly, the lateral variation in acoustic backscatter, although weak, indicates variation in proximity to the glacier. The higher level of backscatter in the east may indicate coarser sediment deposited in a more proximal position. Thirdly, the preservation of what is most likely iceberg plough marks at the units base in the west, would hardly be possible at the base of an advancing tidewater glacier. More likely this preservation is due to settling from suspension. Another question is the relatively abrupt westward termination of the unit. It may be due to oceanic currents and later glacial erosion. Fourthly, the mimicking of the relief of the base of the unit at its top (Vorren *et al.* 1989), indicates a post-depositional compaction of sediments containing more initial pore-water than tills normally have.

In conclusion, we interpret most of unit 5E to have been deposited in an embayment of the merging Fennoscandian and Barents Sea Ice Sheets (Fig. 14). The frontal position of the glacier margins was probably unstable, and short-lived advances across the proglacial glacimarine sediments probably occurred, as witnessed by internal reflectors and sediment deformation. During a subsequent major readvance the whole accumulation was overriden and eroded, particularly in the east.

Discussion

Seismic signature and sedimentary environment

Based on the seismic signatures, we have already (Vorren *et al.* 1989) interpreted several different lithofacies: (semi-)transparent signature representing rapidly deposited glacimarine sediments, mainly from suspension; stratified onlap signature representing slowly deposited distal glacimarine sediments; diverging stratified signature perhaps representing glacier proximal sediments, e.g. submarine outwash fans deposited from meltwater underflows; and chaotic signature representing tills (or gas charged sediments). In this paper we have described internal progradational reflectors, troughs and wedges in the shelf units, and various progradation patterns at the shelf break. The progradational reflectors and troughs require additional discussion.

The progradational pattern found within the shelf units has probably been formed in front of an advancing glacier or as foresets in a till delta under an ice shelf. Most of the reflectors are probably due to changes in the sedimentary regime producing sediments with various impedance characters. But there are also indications that some of the reflectors are partly erosional in origin. These dipping reflectors have higher amplitude than the rest, probably due to glacier-induced compaction, i.e. the glacier/ice shelf must have advanced and retreated over its newly deposited sediments.

The troughs described, particularly frequent in unit 2W, bear some resemblances to valleys incised into Pleistocene and older sediments in the North Sea. Cameron *et al.* (1987) give the following account of mechanisms for the formation of these valleys: 'The valleys have been variously attributed to scour by meltwater ahead of an ice front (Flinn 1967), subglacial meltwater erosion and subsequent tidal scour (Donovan 1973), tidal scour (Thomson & Eden 1977) or subglacial meltwater alone (Flinn 1978)'. King (1980) indicated that exposed and buried channels on the eastern Canadian shelf represent former sub-aerially developed drainage systems possibly modified by subsequent glacigenic processes.

The significant depth, 200–300 m below sea surface, of our troughs is too deep for fluvial erosion, even during the low sea-level of maximum glaciation. The overdeepened troughs indicate a glacigenic origin. The often semicircular or U-shaped form points to glacial erosion. Subglacial meltwater drainage may have initiated the trough erosion (Kåsli 1986; Larsen 1987). Their location in the converging area of the Barents Sea Ice Sheet and the Fennoscandian Ice Sheet (Fig. 16) may support the idea of initial meltwater erosion, since much meltwater must be expected to drain towards such a merging area.

Palaeoenvironment

Based on the interpretation of the youngest seismic units and the determination of these, we suggest the following evolution of the mid/late Weichselian palaeoenvironment in the southern Barents Sea:

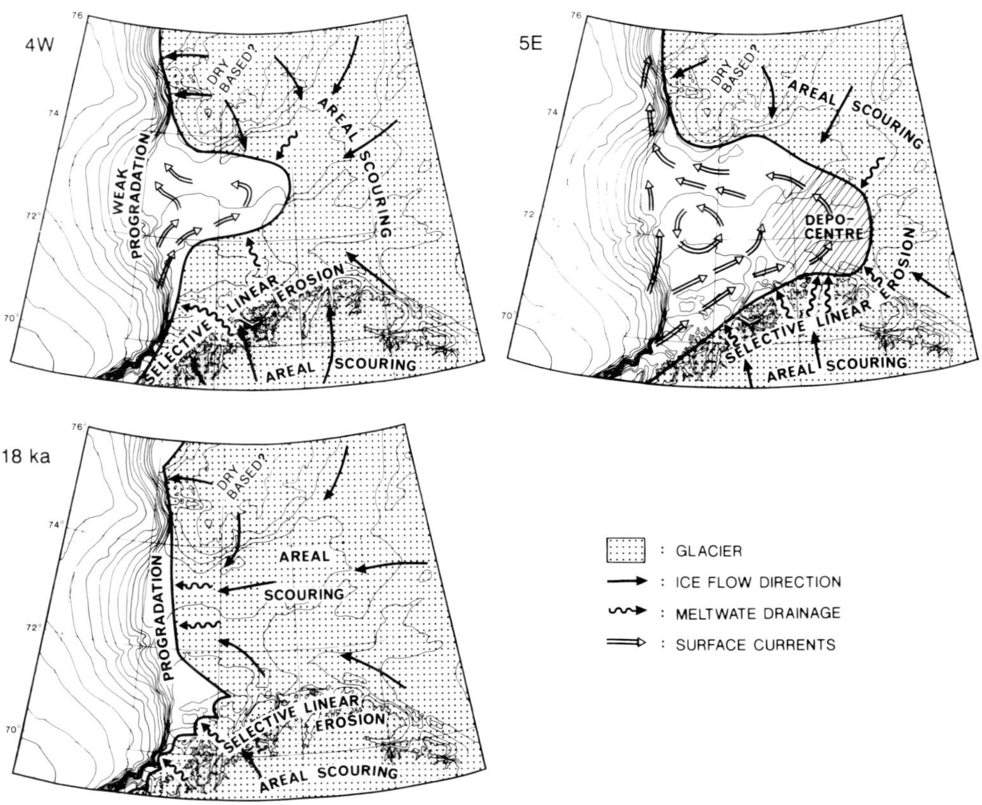

Fig. 16. Schematic development of the Barents Sea during the mid/late-Weichselian. The position of the glacier margin during unit 4W and 5E-time is tentative; large fluctuations in the margin during deposition of these two units is probable. The 18 ka position is after Vorren *et al.* (1986).

Table 1. *Radiocarbon-dates (AMS) on shell fragments from samples from unit 4W in the southwestern Barents Sea.*

Core	Depth (m)	Latitude	Longitude	Age
7120/6-U-1	13.00–13.21	71°37′12″	20°56′30″	> 41 000
7120/6-U-1B	17.74–17.78	71°37′12″	20°56′30″	21 550 ± 550
Dia 84-2	70.25–70.35	72°16′46″	23°09′30″	27 760 ± 735
C85-E	21.75	72°24′19″	22°47′40″	25 320 ± 550

Sample depth is given in metres below sea floor. The samples were recovered from cores by the Continental Shelf Institute (IKU) and dating was performed at the University of Uppsala by G. Possnert.

Unit 4W-time. Base unit 4W is represented by an erosional unconformity. The erosion surface truncates the older units and the erosion seems to have been quite substantial. A minimum age for this erosional period is given by the radiocarbon dates (Table 1), i.e. older than 28 ka, provided the dated shell fragments have not been reworked. However, reworking seems likely, making a younger age possible for the erosional period. During 4W time we infer that glaciers advanced both from the north, east and south (Fig. 16). Sediments from the inner shelf units, i.e. units 4E and older, were eroded. The marked erosional period following the deposition of unit 4E may well correspond to this advance. On the shallower banks in the north, around Bear Island, the ice sheet was probably thin. Thin ice is likely to be dry (cold) based, thus having a reduced eroding capability (Sugden & John 1976). Sedimentation occurred primarily in the outer and deeper part of the Bear Island Trough, probably mostly by settling of suspended sediments supplied by meltwater streams and by icebergs. Iceberg turbation, as witnessed by buried iceberg scouring at the base of unit 5E, and occasional fluctuations of the glacier have disturbed and deformed the unit 4W sediments on the continental shelf.

Unit 5E-time. About 21–22 ka the glacier margin receded to the inner shelf. It is interesting to note that the stratigraphy from Andøya indicates a low to middle Arctic climate prevailing from *c.* 22 ka (T. Alm, pers. comm.) to 18 ka (Vorren *et al.* 1988b) when a high Arctic climate commenced. Thus the period 22–18 ka may have had a more favourable climate than the immediately preceding and following periods. During this period unit 5E was deposited, according to our reconstruction.

The margin of the glacier then formed a huge bay, *c.* 400 km deep and 300 km wide, in the southern Barents Sea (Fig. 16). Meltwater outflows were probably deflected to the right due to the Coriolis effect, setting up a cyclonic (counter clockwise) gyre in the bay. Fossil data from Andøya indicate a partly maritime climate type with an at least seasonally open Norwegian Sea (Vorren *et al.* 1988b).

Meltwater discharge was probably largest in the southern sector. Here, selective linear erosion was producing more sediment than the inferred areal scouring and possibly dry based glaciers in the northern sector (Fig. 16). Pfirman & Solheim (1989) have shown that the distribution of the fine grained components suspended in meltwater discharge from Austfonna in the northern Barents Sea, is governed by coastal circulation. Presently it extends approximately 15 km perpendicular to the ice front and at least 60 km along the front (Pfirman & Solheim 1989). Unit 5E covers an area of at least 200 by 200 km and the question may be raised as to why it has such a great lateral extent. *First*, the position in a bayhead will receive sediments from several surrounding sources. *Second*, the low-density fresh, turbid meltwater may be driven relatively far out into the bay by surface currents set up by katabatic winds blowing out of the glacier. The relatively abrupt western limit may be due to winnowing by bottom currents set up by the central gyre.

The environment outlined should imply corresponding glacimarine accumulation further southwestward along the coast. Such accumulations probably did exist, but they have been removed by later erosion.

Late Weichselian maximum, 18 ka

Submarine moraine ridges suggest an extension of the ice sheet as given in Fig. 16 (Vorren *et al.* 1988a). During this readvance units 4W and 5E on the outer and inner shelf were overridden and eroded. Unit 6E may have been formed during this readvance and subsequent retreat. Probably some till deposition occurred as well. Except for the moraine ridges, these deposits are difficult to distinguish from the older units on the sparker profiles. On the continental slope, accumulation and progradation occurred, adding sediments to unit 4W. The ice margin was probably situated close to the shelf break until 16 ka. Between 16 and 13 ka most of the Barents Sea was deglaciated (Vorren *et al.* 1988a), leaving a relatively thin (< 10 m) sequence of glacimarine sediments in the troughs.

Conclusions

1. The glacigenic sequence (the Barents Synthem) on the Barents Sea continental shelf and shelf break comprises four major seismostratigraphic units on the outer shelf (1W–4W) and six major units on the inner shelf (1E–6E). Most of the units pinch out on the central shelf.
2. The outer shelf units thicken towards the shelf break, except for the youngest one. A complex sigmoid-oblique reflection pattern exists at the shelf break where evidence of frequent mass-movements are seen.
3. The geometry of the outer shelf units is determined partly by the morphology of the basement, but even more so by post-depositional erosion. The oldest, and particularly the second oldest units, are characterized by troughs of uncertain origin. Probably they were formed by subglacial meltwater erosion and later glacially modified.
4. The older inner shelf units are in their entirety located in a coast-parallel trough. The three youngest units also blanket the adjacent areas. The geometry of the younger units gives indications of source areas as well as of ice movement directions of the subsequent eroding ice sheets.
5. Both the outer and inner shelf units have a complex origin. We believe that a substantial part of the sediments were mainly deposited proglacially from turbid meltwater plumes and icebergs. Syn- and post-depositional glacial erosion and till deposition has occurred both on the inner and the outer shelf.
6. The palaeoenvironment and palaeogeography during the mid/late Weichselian is reconstructed. Prior to 21 ka the ice sheet covered most of the Barents Sea, but an embayment existed on the outer shelf. During the time period 21 to 18 ka, the inner area of a bay, 400 by 300 km, received large amounts of glacimarine sediments (unit 5W). During the Weichselian maximum the ice sheet advanced almost to the shelf break.

This work was supported by grants from the Norwegian Research Council for Science and the Humanities. Seismostratigraphic work on the inner shelf is part of E. Lebesbye's dissertation research funded by grants from Norsk Hydro a.s., Saga Petroleum a.s. and Statoil. Many of the seismic profiles were placed at our disposal by the Continental Shelf and Petroleum Technology Research Institute Ltd. (IKU), the Norwegian Petroleum Directorate and the Norwegian Polar Institute. IKU kindly permitted us to publish two of their seismic sections (Figs 8 & 15) and use data from their shallow drillings. The drawings were made by Hilkka Falkseth. A. Elverhøi and L.H. King critically reviewed the manuscript. To all these persons and institutions we offer our sincere thanks.

References

ALLEY, R.B., BLANKENSHIP, D.D., ROONEY, S.T. & BENTLEY, C.R. 1989. Sedimentation beneath ice shelves – the view from ice stream B. *Marine Geology*, **85**, 101–120.

BUGGE, T. & ROKOENGEN, K. 1976. Geologisk kartlegging av de øvre lag på kontinentalsokkelen utenfor Troms. Inst. Kontinentalsokkelunders., Trondheim, 85.

BUGGE, T., LIEN, R.L. & ROKOENGEN, K. 1978. Kartlegging av løsmassene på kontinentalsokkelen utenfor Møre og Trøndelag: seismisk profilering. Inst. Kontinentalsokkelunders., Trondheim, 99.

CAMERON, T.D.J., STOKER, M.S. & LONG, D. 1987. The history of Quaternary sedimentation in the UK sector of the North Sea Basin. *Journal of the Geological Society, London*, **144**, 43–58.

DEKKO, T. 1975. Refleksjonsseismiske undersøkelser i Vestfjorden 1972. Inst. Kontinentalsokkelunders., Trondheim, 77.

DONOVAN, D.T. 1973. The geology and origin of the Silver Pit and other closed basins in the North Sea. *Proceedings of the Yorkshire Geological Society*, **39**, 267–293.

ELVERHØI, A. & SOLHEIM, A. 1983. The problem of the Barents Sea Ice sheet: A sedimentological discussion. *Polar Research*, **1**, 23–42.

ELVERHØI, A., PFIRMAN, S.L., SOLHEIM, A. & LARSSEN, B.B. 1989. Glaciomarine sedimentation in epicontinental seas exemplified by the northern Barents Sea. *Marine Geology*, **85**, 225–250.

FLINN, D. 1967. Ice front in the North Sea. *Nature*, **215**, 1151–1154.

—— 1978. The most recent glaciation of the Orkney-Shetland Channel and adjacent regions. *Scottish Journal of Geology*, **14**, 109–123.

GRAVENOR, C.P., VON BRUNN, V. & DREIMANIS, A. 1984. Nature and classification of waterlain glacigenic sediments, exemplified by Pleistocene, Late Paleozoic and Late Precambrian deposits. *Earth Science Reviews*, **20**, 105–166.

HALD, M. & VORREN, T.O. 1987. Late Cenozoic stratigraphy in shallow drillings in the Barents Sea: Report of investigations on the samples from 1984 and 1985. Report, University of Tromsø, IBG.

___ & ___ 1988. Late Cenozoic stratigraphy in shallow drillings in the southern Barents Sea: Report on the investigation of the boreholes collected by the

IKU 1986. Report, University of Tromsø, IBG.

ISSC (International Subcommission on Stratigraphic Classification) 1987. Unconformity – bounded stratigraphic units. *Geological Society of America Bulletin*, **98**, 232–237.

JANSEN, E., BLEIL, U., HEINRICH, R., KRINGSTAD, L. & SLETTEMARK, B. 1988. Paleoenvironmental changes in the Norwegian Sea and in the northeast Atlantic during the last 2.8 m.y.: Deep Sea Drilling Project/Ocean Drilling Program sites 610, 642, 643 and 644. *Paleoceanography*, **3**, 563–581.

KING, L.H., 1980. Aspects of regional surficial geology related to site investigation requirements – Eastern Canadian shelf. *In*: ARDUS, D.A. (ed.), *Offshore Site Investigation*. Graham & Trotman, London, 37–57.

—— & FADER, G.B.J. 1986. Wisconsinan glaciation of the Atlantic continental shelf of southeast Canada. *Canadian Geological Survey Bulletin*, **33**.

KING, L.H., ROKOENGEN, K. & GUNLEIKSRUD, T. 1987. Quaternary seismo-stratigraphy of the Mid Norwegian Shelf, 65°–67°30′N. A till tongue stratigraphy. Continental Shelf and Petroleum Technology Research Institute A/S, Trondheim, Publication 114.

KÅSLI, K. 1986. *Seismisk stratigrafi og sedimentasjon av de øvre sedimentlag i det sydvestlige Barentshav med hovedvekt på blokk 7120/12*. Thesis, University of Tromsø.

LARSEN, O.A. 1987. *Seismisk stratigrafi og sedimentasjon av de øvre sedimentlag på Tromsøflaket med hovedvekt på blokk 7119/12 og sørlig del av blokk 7120/7*. Thesis, University of Tromsø.

LIEN, R. 1976. Ingeniørgeologisk kartlegging på kontinentalsokkelsen utenfor Lofoten, Vesterålen. Inst. Kontinentalsokkelunders., Trondheim, 78.

MANGERUD, J. & GULLIKSEN, S. 1975. Apparent radiocarbon ages of recent marine shells from Norway, Spitsbergen and Arctic Canada. *Quaternary Research*, **5**, 263–273.

MITCHUM, R.M., VAIL, P.R. & SANGREE, J.B. 1977. Stratigraphic interpretation of seismic reflection patterns in depositional sequences. *American Association of Petroleum Geologists Memoir*, **26**, 117–134.

MOSBY, H. 1968. Surrounding seas. *In*: SØMME, A. (ed.), *A Geography of Norden*, Cappelen, Oslo, 18–26.

PFIRMAN, S. & SOLHEIM, A. 1989. Subglacial meltwater discharge in the tidewater environment: observations from Nordaustlandet, Svalbard archipelago. *Marine Geology*, **86**, 265–281.

POWELL, R.D. & MOLNIA, B.F. 1989. Glacimarine sedimentary processes, facies and morphology of the south-southeast Alaska shelf and fjords. *Marine Geology*, **85**, 359–390.

ROKOENGEN, K. 1980. De øvre lags geologi på kontinentalsokkelen utenfor Møre og Romsdal. Inst. Kontinentalsokkelunders., Trondheim, 105.

—— & RØNNINGSLAND, T.M. 1983. Shallow bedrock geology and Quaternary thickness in the Norwegian sector of the North Sea between 60°30′N and 62°N. *Norsk Geologisk Tidsskrifter*, **63**, 83–102.

SOLHEIM, A. & KRISTOFFERSEN, Y. 1984. Sediment distribution above the upper regional unconformity and the glacial history of western Barents Sea. *Norsk Polarinstitutt Skrifter*, **179**(B).

SPENCER, A.M., HOME, P.C. & BERGLUND, L.T. 1984. Tertiary structural development of western Barents Shelf: Troms to Svalbard. *In*: SPENCER, A.M. (ed.), *Petroleum Geology of the North European Margin*. Graham & Trotman, London, 199–210.

SUGDEN, D.E. & JOHN, B.S. 1976. *Glaciers and Landscape*. Edward Arnold, London.

SÆTTEM, J. & HAMBORG, M. 1987. The geological implications of the upper seismic unit, southeastern Barents Sea. *Polar Research*, **5**(new series), 299–301.

THOMSON, M.E. & EDEN, R.A. 1977. Quaternary deposits of the central North Sea, 3. The Quaternary sequence in the west central North Sea. Report of the Institute of Geological Sciences, 77/12.

VORREN, T.O., HALD, M. & LEBESBYE, E. 1988a. Late Cenozoic environments in the Barents Sea. *Paleoceanography*, **3**, 601–612.

——, —— & THOMSEN, E. 1984. Quaternary sediments and environments on the continental shelf off northern Norway. *Marine Geology*, **57**, 229–257.

—— & KRISTOFFERSEN, Y. 1986. Late Quaternary glaciation in the southwestern Barents Sea. *Boreas*, **15**, 51–59.

——, —— & ANDREASSEN, K. 1986. Geology of the inner shelf west of North Cape, Norway. *Norsk Geologisk Tidsskrift*, **66**, 99–105.

——, LEBESBYE, E., ANDREASSEN, K. & LARSEN, K.-B. 1989. Glacigenic sediments on a passive continental margin as exemplified by the Barents Sea. *Marine Geology*, **85**, 251–272.

——, VORREN, K.-D., ALM, T., GULLIKSEN, S.S. & LØVLIE, R. 1988b. The last deglaciation (20 000 to 11 000 BP) on Andøya, northern Norway. *Boreas*, **17**, 41–77.

Early Holocene morainal bank sedimentology and marine ecology, Skjoldungebrae gorge, North Scoresby Land, East Greenland

D. HUDDART[1] & J.D. PEACOCK[2]

[1] *Department of Outdoor Education and Science, I.M. Marsh Campus, Liverpool Polytechnic, Barkhill Road, Liverpool, L17 6BD, UK*

[2] *Department of Civil Engineering, Heriot-Watt University, Riccarton, Edinburgh, EH14 4AS, UK*

Abstract: Skjoldungebrae gorge sections have allowed a reconstruction of changing sedimentological and palaeoecological environments during the early Holocene in the Kap Petersen area of North Scoresby Land. There is evidence for a former morainal bank and associated glacimarine environments at the grounding line of the early Holocene tidewater Skjoldungebrae. The sequence has been subdivided into: bergstone/suspension, laminite, proximal morainal bank avalanche, distal submarine fan and proximal submarine fan glacimarine facies. There are three molluscan and three foraminiferal assemblages present which indicate likely palaeotemperatures, salinities and water depths. Evidence suggests that sea level fell from 105 m to 77.5 m and as it did so modified the tidewater glacimarine sequence. A readvance of Kong Oscar Fjord ice reworked part of the glacimarine sequence and deposited a basal till over the lower slopes of the Kap Petersen area and in lower Skeldal. Criteria used to distinguish between till and various diamictons are discussed.

Glacimarine sedimentation and landforms are poorly known from Greenland and the eastern Canadian Arctic (Osterman & Andrews 1983). By contrast glacimarine deposition associated with the Antarctic ice sheet is well documented (for example, by Anderson *et al.* (1980) and Drewry & Cooper (1981)). However, models of glacimarine sedimentation developed in Antarctica are not thought to be directly relevant to Arctic conditions as major reworking by meltwater is missing in the Antarctic.

More important to the interpretation of any East Greenland glacimarine sequences are the models and processes described by Powell (1981, 1983, 1984) and Molnia (1983) for modern tidewater glaciers in the Alaskan fjords because they are associated with meltwater reworking, although Alaska is unique because of high rates of tectonic uplift and massive sedimentation rates. Also relevant are the models used for the interpretation of Baffin Island and Svalbard glacimarine successions (Nelson 1981; Mode *et al.* 1983; Osterman & Andrews 1983; Elverhoi *et al.* 1983; Elverhoi 1984).

In North Scoresby Land there is only one tidewater glacier, the Sefströms Gletscher (SG on Fig. 1) which terminates in Alpefjord. However, during the primary mapping of the glacial and marine landforms and sediments in the Kap Petersen area it became clear that the Skjoldungebrae had once been tidewater into Kong Oscar Fjord leaving evidence for early Holocene glacimarine events in the Skjoldungebrae gorge.

The East Greenland Quaternary has been discussed by Bretz (1935), Flint (1948), Funder (1978), Funder & Hjort (1973), Hjort (1973, 1976, 1979, 1981) and Hjort & Bjorck (1983). Funder (1971) describes early Holocene terminal moraines deposited in sea water in Milne Land. These occur as rounded banks up to 50 m high composed of sorted sand and gravel with boulders scattered on their top and are similar to the morainal banks of Powell (1981, 1984). The modern meltwater from the Skjoldungebrae seems to have cut through a similar submarine terminal moraine (Fig. 2). The sediments give detailed information about the depositional environments, processes and ecology of early Holocene glacimarine environments in this part of East Greenland.

To the south of the gorge (Fig. 2) the submarine sediments have been incorporated into historical terminal moraines. To the north these older deglaciation landforms have been modified by a readvance of ice from Kong Oscar Fjord which deposited a till up to a height of at least 62 m in the lower Skjoldungebrae valley and along the lower coastal plain, and by marine erosion as sea level fell from 105 m. The interpretation of this till is controversial and highlights one of the problems in glacimarine sedimentation, that of distinguishing between a till deposited directly from glacier ice and the

From DOWDESWELL, J.A. & SCOURSE, J.D. (eds), 1990, *Glacimarine Environments: Processes and Sediments*. Geological Society Special Publication No 53, pp 289–305.

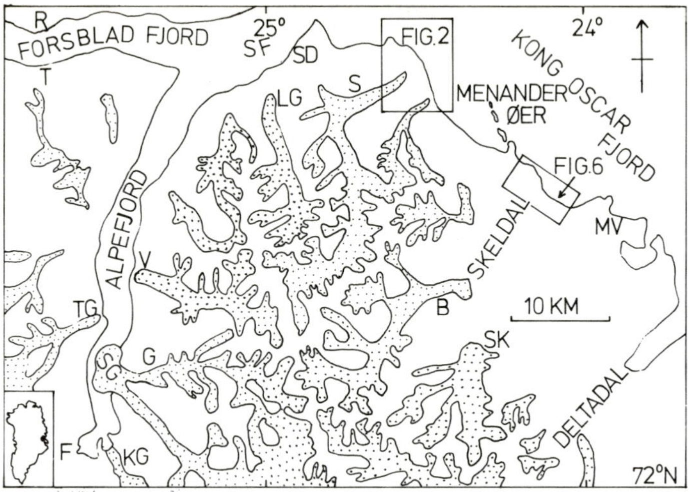

Fig. 1. Location of the Kap Petersen area: R, Randenaes; TG, Trekantgletscher; SG, Sefströmsgletscher; KG, Krabbegletscher; V, Vikingbrae; LG, Linné Gletscher; S, Skjoldungebrae; B, Bersaerkerbrae; SK, Skelbrae; MV, Mesters Vig; T, Taerskeldal; F, Fureso; G, Gulleygletscher; SD, Skipperdal;

various types of diamicton associated with glacimarine environments. Glacimarine diamictons in East Greenland have been described by Nordenskiöld (1907), Washburn (1965), Lasca (1969) and Feyling-Hanssen *et al.* (1983).

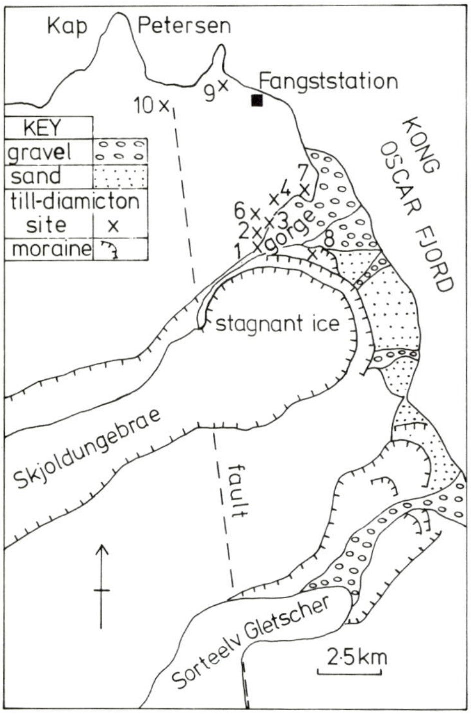

Fig. 2. Location of the Skjoldungebrae and the Kap Petersen area.

Stratigraphy in the Skjoldungebrae gorge

The Skjoldungebrae gorge is located between a fault-guided limestone cliff and the modern coastline (Fig. 2) and has been cut by meltwater from the glacier in historical time. This has exposed a 60 m+ stratigraphic succession which has been modified by marine erosion as sea level progressively lowered. The stratigraphy is given in Figs 3 & 4 and has been divided into eight phases of sedimentation (see Fig. 5). The fauna is shown in Tables 3–6.

Phases of deposition

Phase 1: Suspension/bergstone facies. This sequence is interpreted as a proximal glacimarine suspension and bergstone facies. The depositional environment was dominated by suspension processes producing the parallel-laminated fine sands and silts. At the base of Section 1 the suspension sediment was added to by gravel dropstones from floating ice and dropstone pods and scattered dropstones are common in the lower part of this sequence (Fig. 6). The coarser gravel unit is the distal equivalent of the foreset unit of phase 3 and the medium and coarse sand interbeds are likely to be the result of turbidity current bottomflows running away from the stream input source to the west.

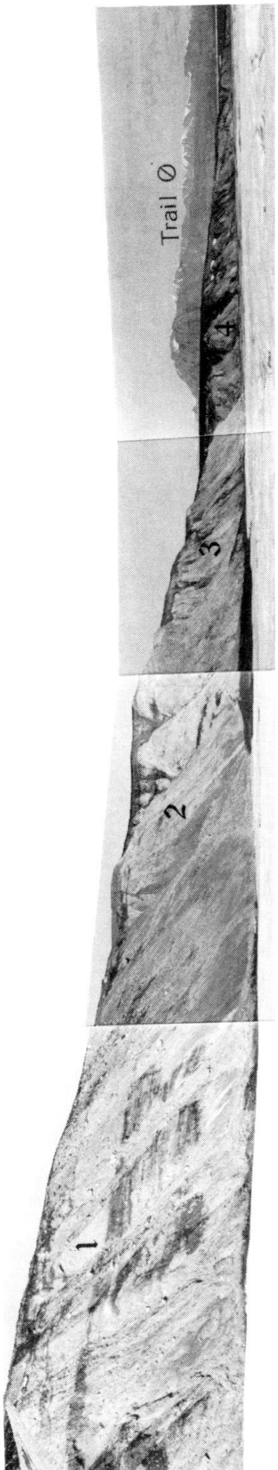

Fig. 3. Panorama of the Skjoldungebrae gorge sections. Trail Ø to the east is divided from the mainland by Kong Oscar Fjord. Location of the stratigraphic logs marked. Approximate distance across the panorama is 3 km. The height of the cliff is at maximum 80 m.

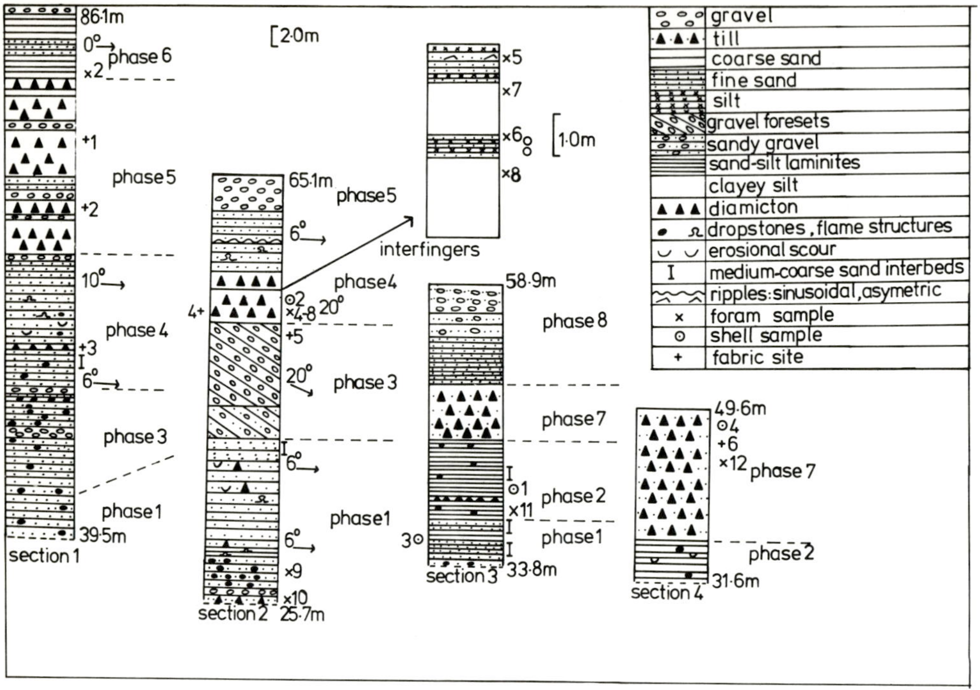

a

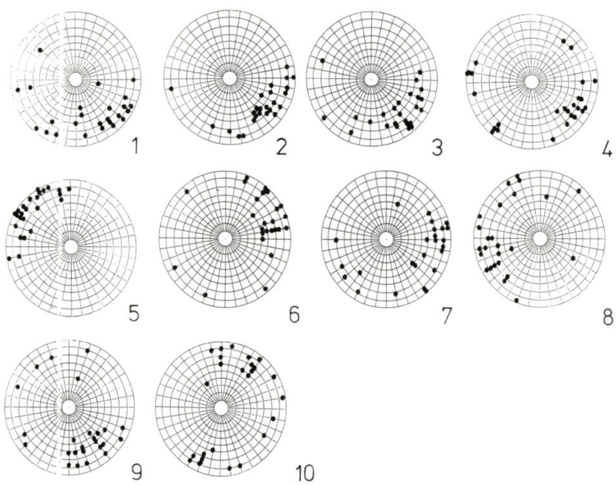

SAMPLE NUMBER	1	2	3	4	5	6	7	8	9	10
MEAN DIP	19.2	19.4	21.6	15.6	11.3	15.4	18.2	17.3	23.3	17.9
HIGHEST DIP	58	40	40	38	24	40	40	50	50	54
			-	+	-	*	*	*	*	*

b

Fig. 4. Stratigraphy in the Skjoldungebrae gorge and diamicton fabric data, Kap Petersen. Heights in metres at the top right of each log.

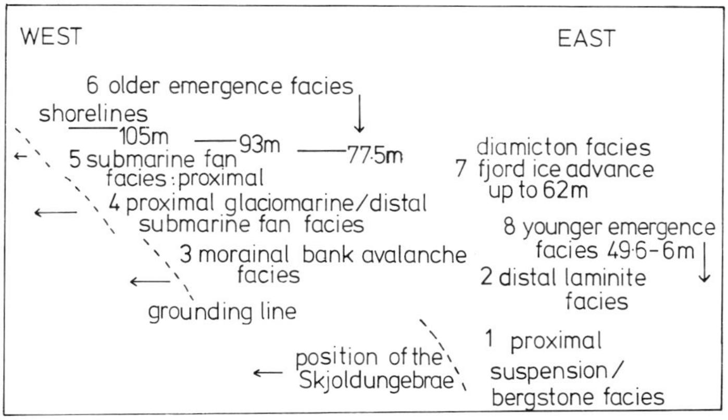

Fig. 5. Inferred phases of sedimentation associated with the Skjoldungebrae gorge area of Kap Petersen. Pecked line represents ice frontal position of the Skjoldungebrae.

a

b

Fig. 6. Suspension/bergstone facies. From Section 2(a) Medium and coarse sand parallel laminated units interbedded with fine sand and silt parallel laminated units. Ripples at the base and gravel dropstones at the top of the photo. Scale 28.0 cm. (b) Parallel laminated fine sand, thin units of interbedded medium and coarse sand. Pebble gravel dropstones and dropstone pods. 6 degree dip to the east. Scale 150 cm.

These could be generated either by slumps on the foreset slope or concentrated suspensions. Water depth was probably considerably less than 30 m.

Phase 2: Laminite facies. The depositional environment was dominated by suspension processes, although the rhythmic sedimentation resembling symmict varves suggests a more distal marine situation than during phase 1 (Figs 7a and 7d). There were sand bottomflows and thin diamicton flows either as flow tills or turbidity currents. The faunal assemblage suggests deeper water than the stratigraphically lower phase 1 or the equivalent stratigraphic level to the west. The Cassidulinae dominate the Elphidiidae.

Phase 3: Morainal bank avalanche facies. This phase indicates avalanching on a morainal bank front into a shallow sea. The top height of the foresets is 51.1 m and the base is 40.5 m indicating a minimum water depth of 11 m.

Phase 4: Proximal glacimarine and distal submarine fan facies. Deposition was in deeper water than in phase 3 and a proximal glacimarine environment was established where the dominant process was suspension. However, there was bottomflow current activity and reworking indicated by the ripples and medium sand laminae with silt clasts. Diamicton units with dips of up to twenty degrees suggest deposition from the ice front which was in close proximity on the

Fig. 7. Laminite and diamicton facies. (a) Section 4 Kap Petersen showing till (maximum thickness 12.0 m) overlying laminites (maximum thickness 6 m). Top of the section at 49.6 m. (b) Vertical jointing and horizontal fissility in the diamicton, Skeldal, location 6 (fabric site D). Scale 5.5 cm. (c) Horizontal fissility in the diamicton, Skeldal, location 6, scale 28.0 cm. (d) Underlying laminites, Skeldal site 6. Scale 5.5 cm.

proximal fan (see phase 5). The foraminiferal assemblage of KP3 seems to indicate a shallower, more brackish water environment, whilst the species from the FKP4–8 assemblage indicate an increased depth of water. However, because the Elphidiidae predominate, water depth was probably always less than 30 m.

Phase 5: Proximal submarine fan facies. Depositional processes indicate sedimentation on a horizontal submarine fan close to ice from the Skjoldungebrae valley. Gravel lithologies in table 1 are dominated by granite from the Staunings Alps and lithologies from the Multicoloured Series and Limestone–Dolomite Series. The sandy gravels represent high flow longitudinal bar transport over the horizontal fan surface. No channels were noted. This was interspersed with suspension sedimentation and dropstone activity.

Phase 6: Older emergence facies. The sediment sequence was modified by shoreline processes at successively lower levels to produce shoreline terraces as outlined in Fig. 5 and Table 2. Many of the boulder gravel erratics resting on the terrace surfaces must have been ice-rafted in at various stages. For example, the highest shoreline in the Kap Petersen area is in the gorge at 105.4 m and is marked by granite boulder er-

Table 1. *Lithologies in the drift of the Kap Petersen area*

sample number	1	2	3	4
Granite	32.7	19.6	27.6	94.0
Grey marble	0.2	6.6	15.4	2.0
Siltstone	0.4	9.5	26.3	4.0
Limestone	—	37.8	15.1	—
Quartzite	—	24.1	14.7	—
'Lamprophyre' Dyke	—	—	0.6	—
Mica schist	—	0.3	0.3	—
Gneiss	66.4	2.7	—	—

All figures are in %.
1. gravel over 1 m in 'b' diameter from diamicton surface near Fangststation; n = 426. 2. pebble gravel from diamicton near Fangststation; n = 296. 3. pebble gravel from section 1 diamicton, Skjoldungebrae gorge; n = 312. 4. gravel over 1 m in diameter from terrace surface at 105 m and 93 m; n = 45.

Table 2. *Deltas and shorelines in the Kap Petersen area*

1. 105.4 m	Older Emergence Facies
2. 93.0 m	Older Emergence Facies
3. 77.5 m	Older Emergence Facies
4. 68.2 m.	
5. 58.9 m shoreline	
6. 52.7 m shoreline	
7. 49.6 m shoreline	
8. 43.1 m	
9. 37.2 m	
10. 31.0 m	
11. 27.7 m	
12. 24.6 m	
13. 23.25 m	
14. 21.7 m	
15. 18.6 m	
16. 15.5–17.05 m	
17. 12.4 m	
18. 6.15 m	

ratics whilst at 93 m below the limestone ridge to the northeast is a similar line of gneiss boulders with 'b' diameters greater than 5 m resting on the shoreline. These older shorelines are probably dated to about 9500–10 000 BP when compared with shorelines of a similar height in the Scoresby Sund area (Funder 1978).

After the erosion of the 77.5 m shoreline, ice readvanced from Kong Oscar Fjord to deposit a till over the lower slopes of the Kap Petersen area and to the south (phase 7).

Phase 7: Diamicton facies. In sections 3 and 4 are grey-brown diamicton sequences extending over the lower slopes of the gorge area up to

Table 3. *Foraminifera, Skjoldungebrae gorge, Section 2*

Sample FKP3: Species present are:	
Elphidium excavatum forma *selseyensis*	31.0
Elphidium excavatum forma *clavata*	26.0
Elphidium incertum	11.8
Nonion orbiculare	11.8
Cassidulina reniforme	8.7
Elphidium umbilicatulum	2.4
Elphidium hallandense	2.4
Islandiella islandica	1.6
Pseudopolymorphina novangliae	1.6
Buccella tenerrima	0.8
Discorbis wrightii	0.8
Elphidium margaritaceum	0.8

All figures in %. Number of species = 12; faunal dominance = 31; faunal diversity = 9; number of foraminifera = 127.

62.0 m. The boundary with the underlying laminated sediment of phase 2 is difficult to trace but there is vertical jointing in the diamicton and little vertical jointing in the laminite facies (Fig. 7). The diamicton matrix is derived from the underlying laminated silt. The unit is relatively pebble free and there are closely spaced fissility planes at an angle of 24 degrees to the east. In places there are small articulated shells but the larger ones are sometimes broken.

The composition of foraminiferal sample FKP12 stratigraphically immediately above the laminite facies in the diamicton is shown in table 4. Only *Quinqueloculina seminulum is* not found in FKP11. In both samples the difference between the faunal diversity of 4 and 5 and the number of species (10 and 8) is insignificant. The foraminifera have been derived by erosion from the laminite facies and incorporated into the diamicton.

Foraminiferal sample FKP15 from a gully in the grey diamicton north of the Fangststation shows a larger number of species (27), although with a faunal diversity of only 8 (table 4). In this assemblage there are Arctic species which are not found in FKP12 or in the laminite facies. The significance of this richer and more Arctic assemblage is debatable. If the diamicton is ice-rafted and the fauna is in situ then there are considerable differences between two samples only 3 km apart. If the diamicton is a till then the differences can be explained by ice incorporation of fjord sediments derived from a more northerly proximal ice position.

This diamicton unit has a marked preferred fabric orientation (Fig. 4b, 5–10). In the pebble gravel fraction the lithologies are dominated by

Table 4. *Foraminifera from the Kap Petersen area*

	FKP9	FKP11	FKP12	FKP14	FKP15
Elphidium excavatum forma *clavata*	49.1	10.4	18.4	27.3	11.3
Elphidium excavatum forma *selseyensis*	24.3	1.0	6.9	14.0	18.4
Cassidulina reniforme	15.0	31.3	27.0	24.6	29.8
Stainforthia loeblichi	—	6.6	4.9	—	*
Nonion orbiculare	2.3	*	—	1.0	1.9
Islandiella islandica	1.5	49.9	43.1	31.3	29.2
Buccella tenerrima	1.3	—	—	—	—
Pseudopolymorphina novangliae	1.3	—	*	—	—
Guttulina lactea	1.1	—	—	—	*
Buccella frigida	1.0	—	—	—	—
Elphidium incertum	*	—	—	—	—
Elphidium hallandense (A)	*	—	—	—	*
Elphidium asklundi (A)	*	—	—	*	*
Lagena clavata	*	—	—	—	—
Elphidium bartletti (A)	*	—	—	—	*
Elphidium albiumbilicatum	*	—	—	—	*
Triloculina trihedra	—	*	*	—	*
Pyrgo williamsoni	—	*	—	—	—
Quinqueloculina seminulum	—	—	*	—	*
Quinqueloculina agglutinata	—	—	—	*	—
Quinqueloculina stalkerri (A)	—	—	—	—	*
Quinqueloculina arctica (A)	—	—	—	—	*
Astrononion gallowayi (A)	—	—	—	*	—
Nonion depressulus	—	—	—	*	—
Nonion labradoricum (A)	—	—	—	*	—
Nonion umbilicatulum	—	—	—	—	*
Nonionella auricula (A)	—	—	—	—	*
Cibicides lobatulus	—	—	—	*	*
Stainforthia schreibersiana	—	—	—	—	2.8
Islandiella helenae	—	—	—	—	*
Cribostomoides crassimargo (A)	—	—	—	—	*
Globigerina spp.	—	*	—	—	*

All figures in %. * refers to species under 1%. (A) refers specifically to Arctic species.

	FKP9	FKP11	FKP12	FKP14	FKP15
Number of species	15	10	8	10	26
Faunal dominance	49	50	43	31	30
Faunal diversity	7	4	5	4	8
Number of foraminifera	1110	1172	1073	501	1094

Sample FKP10 was barren. In sample KP13 there were five foraminifera from three species: *Islandiella islandica, Elphidium asklundi* and *Discorbis wrightii*. In sample FKP16 there were eighteen foraminifera from three species: *Cassidulina reniforma, E. asklundi* and *E. hallandense*.

a local, fine-grained limestone from the Limestone-Dolomite Series (Haller 1971), granite from the Staunings Alps, coloured siltstones from the Multicoloured Series and quartzites from the Quartzite Series or the Multicoloured Series. A minor component is gneiss (Table 1). However, this rock type dominates the lithologies over 1 m in diameter exposed on the surface of the diamicton. According to Haller (1958) the nearest source for this gneiss is on either side of Upper Forsblads Fjord around Randenaes and Taerskeldal or on either side of Alpefjord, between north of Gulleygletscher and Fureso (Fig. 1). Gneiss is missing from the diamicton of phase 4 in section 1 where the erratics are dominated by granite and lithologies from the Upper Eleonore Bay Group. Along with the fabric noted for these sediments this suggests the source for these units was the Upper Skjoldungebrae valley. This suggests that there are two populations from the diamicton: a surface, largely gneiss group ice-floated on to the

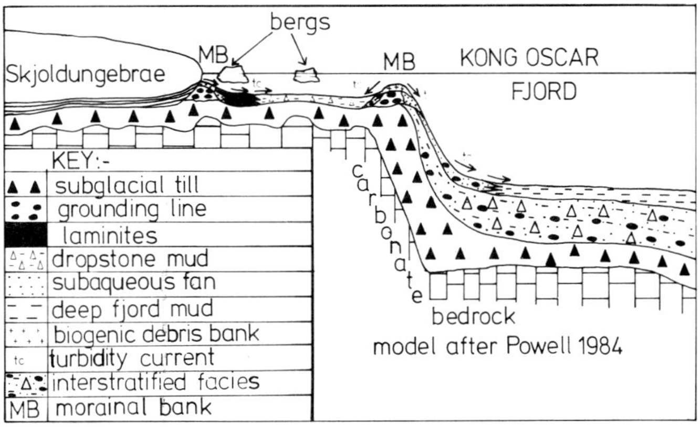

Fig. 8. Model of sedimentation associated with a tidewater Skjoldungebrae, Kong Oscar Fjord.

surface of the diamicton and the diamicton proper derived from the Staunings Alps and the coastal plain sedimentary rocks.

This unit is the equivalent of Washburn (1965) and Lasca's (1969) 'till-like, fjord bottom sediment.' They suggested that in the Mesters Vig region and in Skeldal this was an in situ glacimarine sediment. However, the structure of the deposit and its marked fabric orientation suggest that it is a basal till derived from ice overriding and incorporating the fjord-bottom sediment. The ice moved over the Kap Petersen area from the northeast indicated by the macrofabric diagrams and the lithologies in the till (see later discussion).

Phase 8: Younger emergence facies. South of the Skjoldungebrae gorge the shorelines and sediments deposited in earlier phases have been incorporated into the historical moraines of the Skjoldungebrae. Immediately to the north of Skjoldungebrae gorge is the best preserved shoreline sequence (Table 2) and further to the north around Kap Petersen is located a raised beach sequence cut in diamicton up to a height of 52.7 m. On the headland marking the northern extremity of Kap Petersen Bay there is a cobble gravel beach at 58.9 m and lines of gneiss boulders which are concentrated at a height between 62.0 m and 80.6 m. In Skipperdal there are deltas at 77.5 m and 68.2 m. Above the modern shore is evidence for a similar sequence of events as in the Skjoldungebrae gorge. Above striated red siltstone are 5.3 m of laminated silts with dropstones overlain by 2.5 m of grey, stony, shelly, silt-rich diamicton.

Ecology of the Early Holocene marine environment in Kong Oscar Fjord

Evidence from the Foraminifera

Elphidiidae and Cassidulinidae dominate the assemblages. Elphidiidae are a common constituent of Recent cold, shallow water faunas, characteristic of outer estuarine and baymouth environments with lowered salinities. Cassidulinidae often live in greater abundance in deeper water than the Elphidiidae and the distribution is closely correlated with temperature as they are generally found in shallow cold water and moves to greater depths with increasing surface water temperatures (Balazarini 1983).

The dominance of *Elphidium excavatum* forma *clavata*, indicates a mainly arctic temperature regime, although the mainly Boreal *E. excavatum* forma var. *selseyensis* is often a co-dominant. This dominance indicates that the water was ice-free, at least during the summer (Vilks 1970). *E. excavatum* forma *clavata* composes between 10.4% and 60.5% of the fauna reported here and is usually a characteristic element in Quaternary shallow arctic and sub-arctic waters. It often accounts for between 50% and 100% of the fauna, thrives in under 30 m water depths, with normal marine salinity (30 to 35‰) to approximately 15‰ and where the temperature ranges from 0 to 25°C.

Nonion orbiculare, a common arctic species, is present in the Kap Petersen faunas between 0.6% and 11.8% and is associated with nearshore, intertidal and shallow estuarine environments and salinities under 30‰ (Murray

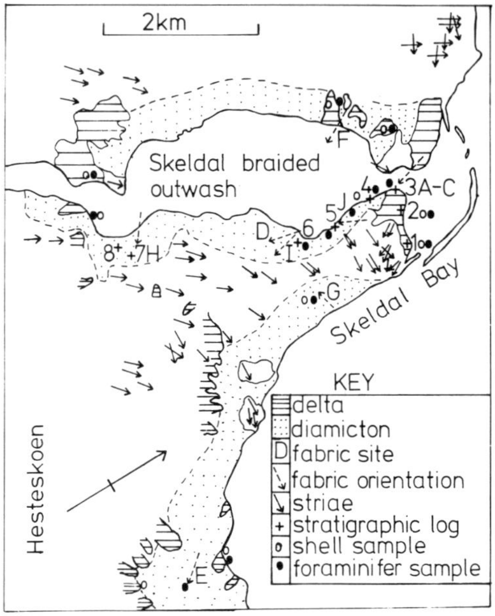

a

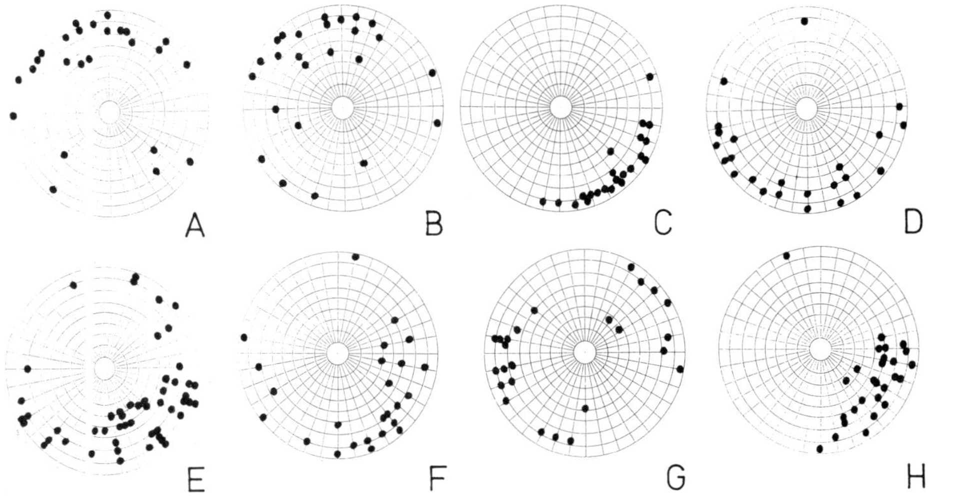

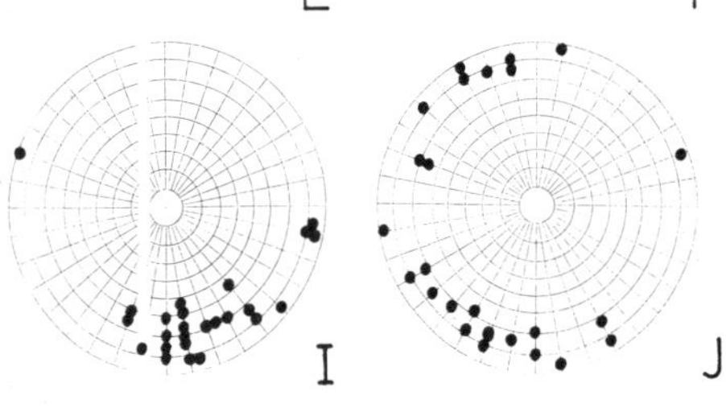

(b) : DIAMICTON FABRIC DATA, SKELDAL

SAMPLE NUMBER	A	B	C	D	E	F	G	H	I	J
HIGHEST DIP	40	48	30	30	52	50	54	54	34	24
TRANSVERSE PEAK	*	*	-	*	*	*	-	-	-	*
CHI SQUARE	34	51.3	77.2	36.9	69.5	48.4	45.6	68.6	112	34
S1 EIGENVALUE	0.599	0.585	0.792	0.531	0.653	0.596	0.664	0.79	0.783	0.627
S3 EIGENVALUE	0.084	0.09	0.011	0.032	0.068	0.075	0.115	0.041	0.021	0.278

b

Fig. 9. Lower Skeldal with (a) log locations, striae and deltas and (b) diamiction fabric data.

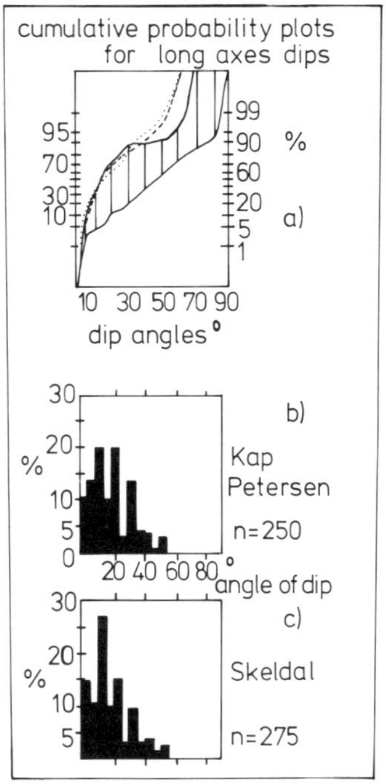

Fig. 10. Cumulative probability distribution plots for long axis dip angles and histograms for long axis dips, Kap Petersen and Skeldal samples and the ice-rafted diamicton data from Domack and Lawson (1985). In (a) the vertical line symbol represents ice-rafted diamictons, the dots: Skeldal and the pecked line: Kap Petersen.

1973). When in abundance with *E. excavatum* forma *clavata* (see KP3, Table 3) it is indicative of cold, shallow waters with salinities ranging from 18 to 24‰ (Cronin 1976).

Both *Buccella frigida* and *Buccella tenerrima* have ecological tolerances similar to those of *E. excavatum* forma *clavata* but they cannot withstand salinities of under 20‰. They are most abundant at depths of 10 m to 50 m. In some Skjoldungebrae samples *Buccella* is common with percentages from 0.8% to 17.1%.

Nonion labradoricum and *Nonionella auricula* are only found in KP 15 and are Arctic cold water species, preferring water depths under 100 m. All of the miliolids are Arctic forms with *Quinqueloculina arctica, Q. seminulum, Q. stalkeri* and *Triloculina trihedra* having ranges into northern temperate waters. They are widely distributed in shallow waters with salinities greater than 32‰ (Murray 1973).

In the Kap Petersen area faunal diversity was never lower than 4 (in the laminite facies). In the till derived from the laminite facies it was 5 and in the shallower water environments of the other facies the faunal diversity was between 7 and 9. Apparently conditions were not as severe as the depositional environments might suggest.

High faunal dominance (Walton 1964) indicates extreme environmental conditions such as low temperatures, low salinity, very shallow or turbid water of low transparency. Again conditions cannot have been extreme as faunal dominance varies between 30 and 61.

There are three foraminiferal assemblages present in the Kap Petersen area (Tables 3–5).

(i) A shallow cold water, relatively near-shore assemblage in samples 3, 4, 6, 8 and 9, dominated by the Elphidiidae but with Cassidulinidae, *Buccella* species and *Pseudopolymorphina novangliae* present. Faunal diversity is 7–9 and the number of species is 12–19. The assemblage coincides with the sedimentary environments associated with the morainal bank construction in shallow water, probably under 15 m in depth, with a salinity between 20 and 30‰.

(ii) A deeper cold water assemblage indicated by samples 11 and 12 from the laminite facies and the till. It is dominated by the Cassidulinidae, has a faunal diversity of between 4 and 5 and the number of species is lower, between 8 and 10. The presence of *Islandiella islandica*, a cold water species common in high latitudes normally indicates 50–70 m depths (Murray 1973) but because Elphidiidae are still present the water is estimated at 30 m.

(iii) A mixed assemblage from the till samples in FKP14 and FKP15 which seems to indicate erosion of both assemblages (i) and (ii) and incorporation in to readvance ice. Cassidulinidae dominate the assemblage although there are significant percentages of Elphidiidae. Faunal diversity is 4 and 8 and the number of species 10 and 26.

Evidence from the low diversity molluscan fauna

There are three assemblages present (Table 6).

(a) The *Hiatella arctica–Mya truncata* assemblage, from pebbly silty, raised beach sands, delta foresets and diamictons (samples 2 and 5–10). It comprises only infaunal species and is similar to Holocene molluscan assemblages found throughout the Arctic (Bennike *et al.* 1986; Feyling Hanssen 1955; Funder 1978). As it is apparently rare today its occurrence has been attributed to selective preservation of the infauna by the enclosing sediment (Funder 1978; Ockelmann 1958). However, as a living *H. arctica–M. truncata* association occurs in glacimarine silt adjacent to tidewater glaciers in Novaya Zemlya (Zenkevitch 1963) and Baffin

Table 5. *Foraminifera from section 2, Skjoldungebrae gorge*

	F4	F6	F8
Elphidium excavatum forma *clavata*	42.5	38.5	60.5
Elphidium excavatum forma *selseyensis*	22.7	30.8	19.8
Buccella tenerrima	13.8	4.4	2.5
Elphidium incertum	4.4	—	*
Cassidulina reniforme	3.9	13.2	4.4
Buccella frigida	3.3	2.2	1.4
Islandiella islandica	2.8	—	4.4
Nonion orbiculare	1.1	3.3	2.5
Elphidium hallandense	1.1	1.6	*
Elphidium asklundi	1.1	—	1.4
Elphidium albiumbilicatum	*	2.7	*
Guttulina lactea	*	*	1.4
Pseudopolymorphina novangliae	*	1.1	2.2
Lagena gracillima	*	—	*
Guttulina glacialis	*	—	—
Silicoigmoilina groenlandica	—	*	—
Gordospira arctica	—	—	*
Eggerella scabra	—	—	*
Discorbis wrightii	—	—	*
Quinqueloculina seminulum	—	—	*
Quinqueloculina lata	—	—	*
Stainforthia loeblichi	—	—	*

	F4	F6	F8
Number of species	14	13	20
Faunal dominance	43	39	61
Faunal diversity	8	7	8
Number of foraminifera	181	182	1306

In sample 5 there were 13 foraminifera present: *E. excavatum* forma *clavata*, *E. excavatum* forma *selseyensis*, *Buccella frigida*, *Pseudopolymorphina novangliae*. In sample 7 there were 44 foraminifera present: *E. excavatum* forma *clavata*, *E. excavatum* forma *selseyensis*, *C. crassa*, *B. tenerrima*, *G. lactea*, *L. gracillima*. All samples in %.

Table 6. *Mollusca from Kap Petersen*

Sample No.	1	2	3	4	5	6	7	8	9	10
Buccinid fragment									+	
Natica affinis					1.7					
Siphonodentalium lobatum	+									
Chlamys islandica							0.4			
Clinocardium ciliatum					3.4					
Hiatella arctica	+	(3)	14.2	49.1	14.5		25.6	73.7	(2)	9.5
Macoma calcarea	0.7		5.0	9.6	7.8		1.3	5.3	(1)	
Mya truncata	+	(14)	10.6	17.4	63.1	(21)	70.9	21.0	(3)	85.7
Mytilus edulis						(1)	1.7		(3)	5.8
Nucula belloti			+							
Nuculana pernula	+									
Portlandia arctica	98.4		70.2	23.9						
Serripes groenlandicus					9.5	(3)				
Tridonta borealis									(6)	
No. of gastropods and bivalves	222	17	109	109	90	25	117	46	(15)	53

Figures in brackets are numbers of valves (counted as whole shells).
Other figures are percentages.
+ Present in low numbers.

Island, it may represent a biocoenosis which was formerly more widespread (Peacock 1989) in an environment which was either of rapid deposition punctuated by erosion or non-deposition or of rapid deposition alone. The occurrence of the epifaunal bivalve *Mytilus edulis* in samples 6, 9 and 10, and of this taxon together with a valve and fragment of *Chlamys islandica* in sample 7 suggests that the strata in which the mollusca lived were deposited during the mid- or low-arctic conditions prevailing in this part of East Greenland between 8400 and 5500 BP.

(b) The second assemblage (sample 1) in phase 2 laminated silt/sand is entirely dominated by paired and single whole valves of *Portlandia arctica*. This almost monospecific assemblage is identical to the *Portlandia arctica* biocoenosis found today in glacimarine conditions over a wide depth range where sedimentation rates are high (Ockelmann 1958) and where water temperatures are less than 4°C. Funder (1978) and Kelly (1985) suggest that the *P. arctica* assemblage is a facies fauna relating to conditions near the ice front and Funder points out that *P. arctica* in the Holocene glacimarine sediments in East Greenland was contemporary with thermophilous faunas. The presence of this assemblage in glacimarine strata implies merely that water temperatures near the ice front were low and not necessarily that conditions generally were high arctic.

(c) The third assemblage (*P. arctica* together with *H. arctica*, *M. truncata* and *M. calcarea*) occurs in sample 3 (laminated silt/sand with diamicton) and sample 4 (diamicton interpreted as till). This fossil fauna, which can be considered as a variant of the *H. arctica*/*M. truncata* assemblage discussed above, is identical to a biocoenosis occurring at a depth of 20–40 m in the glacimarine muds of McBeth Fjord in East Baffin Island (Farrow *et al.* 1984). The latter is therefore taken as a general analogue of the environment in which the fossil fauna flourished. Articulated shells in sample 4 suggest that little deformation of marine strata was required to form the basal till.

Samples 1, 3, 4 and 8 were taken from the laminated silt/sand and the overlying phase 7 diamicton near the base of the sequence. The faunas collectively reflect the cold, near glacier environment, but are not specific to any particular period in East Greenland. Samples 7 and 10, collected from deposits relating to delta tops at 40 to 50 m above sea level and samples 5 and 6 from sediments relating to deltas at 24 to 28 m contain faunas suggesting marine conditions milder than those at present, with deposition during the 8400 to 5500 BP period (Funder 1978; Street 1977). Further age relationships from the Kap Petersen area await C14 dating and will complement the chronology of Håkansson (1974), Funder (1978) and Hjort (1979).

Discussion of the depositional environments suggested by the Skjoldungebrae gorge and Kap Petersen stratigraphy

From the stratigraphy and ecology we are able to interpret the likely depositional environments during sedimentation. The glacimarine sediments were deposited from the tidewater Skjoldungebrae in the ice proximal, relatively shallow water of Kong Oscar Fjord. The sediments can be compared with models of tidewater glacier sedimentation from the Alaskan fjords described by Powell (1981, 1983, 1984) and from eastern Baffin Island by Mode *et al.* (1983).

Models of tidewater glacier sedimentation and proximal glacimarine environments

When a tidewater glacier grounding line is quasi-stable, sediment accumulates to form a morainal bank or submarine bank (Powell 1981). These occur as elongated ridges or isolated mounds, composed of grounding-line melt-out till, dropped compound and residual para-tills, dumped supraglacial debris at the tidewater front, fluvial sediment and sediment gravity flow deposits. This sedimentation occurs in shallow water, under 150 m depth, in Powell's (1981) facies association II.

In the Alaskan model there are several common lithofacies in Powell's (1983) facies II:
(a) subaqueous morainal banks;
(b) ice-contact lateral and central fan deltas;
(c) interstratified silt and sand from underflows and subaqueous sediment gravity flows on morainal bank foreslopes;
(d) bergstone mud which interfingers with (c).

Alternatively, Powell (1983) describes his facies III where the base of the glacier is near tidewater elevation and ice terminates in shallow seawater but calving rates are low. In this facies the dominant sedimentary units are lateral and central deltas, laminites, bergstone mud and gravel/rubble piles. The environment is dominated by large fan deltas that build along a glacier front and/or down fjord. The delta fronts are composed of channel and sheet-sand that interfingers distally with fine-grained sand, silt and clay.

These two divisions of Powell's (1983) model

can be compared with the model developed by Mode *et al.* (1983) in eastern Baffin Island. In their proximal glacial marine zone, within 2–3 km of the ice front there are rapid facies changes, high sedimentation rates and an actively calving ice margin. Where ice is grounded, lodgement and melt-out till are deposited, but along the margin of the glacier moderate to well sorted sands and gravels are deposited as ice-contact fans by meltwater. As the locus of the discharge shifts these fans coalesce into submarine bars. Sediment-laden meltwater is discharged from below, within and from the surface of the glacier. They note four proximal glacimarine zones.

In the light of these models the generalised sedimentological zones for the tidewater front in Kong Oscar Fjord are presented in Fig. 8. The Skjoldungebrae front was apparently fairly stable during the build up of most of the early part of the sequence. The major problem to resolve is the origin of the diamicton on the lower slopes of the Kap Petersen area. Possibly it was part of Mode *et al.*'s (1983) zone c or Powell's (1983) bergstone mud but it thought to be a readvance subglacial till deposited from Kong Oscar Fjord ice after the Skjoldungebrae had downmelted and retreated.

Distinguishing between diamictons and a readvance subglacial till

One of the problems alluded to earlier is the difficulty in distinguishing some glacimarine sediments from subglacial till or sediment flows. In the models above the authors discuss some diamictons which could easily be misinterpreted. How can the bergstone mud of Powell or the 'fjord bottom sediment' of Washburn (1965) and Lasca (1969) be distinguished from a readvance till origin suggested in this paper?

There are four recognisable diamicton types in the Kap Petersen area based on field observations.

(a) Units deposited on the proximal morainal bank. These represent poorly sorted subglacial/englacial sediment which has undergone short-distance fluvial, longitudinal bar transport. Units 1–4 show mean dips between 15–22° and fabric diagrams in Fig. 4 show both transverse peaks and a marked imbrication indicating a consistent flow direction from the Skjoldungebrae valley.

(b) Diamictons of the delta foresets: as fluvial sediment is deposited on the morainal bank it avalanches, with a marked preferred orientation, as in unit 5. There is no transverse peak and a lower mean and lower highest dip value compared with (a) (Fig. 4).

(c) Turbidity current deposits: sediment transported and deposited away from the morainal bank front produces sediment gravity flows as turbidity currents. These give the thin, poorly sorted units in the laminite facies.

(d) A diamicton on the lower slopes of the Skjoldungebrae gorge and in the Kap Petersen area to the north. This has been described further to the south as a 'till-like fjord bottom sediment.' (Washburn 1965; Lasca 1969). In fig. 4 it is represented by samples 6–10 and here in fig. 9 there are marked preferred orientations mainly from the northeast or east, with minor transverse peaks.

Mode *et al.* (1983) state that their subfacies DT with well worn molluscan fragments and a high proportion of gravel could be interpreted as a grounded basal till but that diamicton facies D1–3 with common fossils, stratification and low concentrations of boulders and cobbles would refute a subglacial origin. The Kap Petersen/Skeldal diamicton (d) has worn molluscan fragments, some articulated shells, common macro- and microfossils, no stratification and a low concentration of gravels. It overlaps across the two descriptions by Mode *et al.* (1983). The evidence used to suggest that the Kap Petersen/Skeldal diamicton (d) is a readvance basal till is as follows.

(i) A marked 'a' axis orientation, with a subsidiary transverse peak. Domack (1983) suggested there was a greater central tendency of fabric data in lodgement tills and sediment flows with a greater randomness of long axis orientation in ice-rafted diamictons (Anderson 1983).

(ii) Compactness. Hicock *et al.* (1981) suggested that flow tills are not compacted.

(iii) Marked fissility.

(iv) No vertical long axes (see figures 4 and 9), no stratified matrix, current structures or dropstone structures observed (cf. Christie-Blick 1983; Powell 1984).

(v) The cumulative probability plots and histograms of the long axis dips in Fig. 10 indicate that there are low percentages of dip angles over 45 degrees; that the distribution in both areas is similar; the Kap Petersen and Skeldal till samples are similar to the long axis dip values for the Vashon lodgement till and dissimilar to ice-rafted diamicton (Domack & Lawson 1985; Lavrushin 1968).

(vi) Broken shell debris (although this is common in normal marine deposits), plus articulated shells. Macrofossils can be incorporated

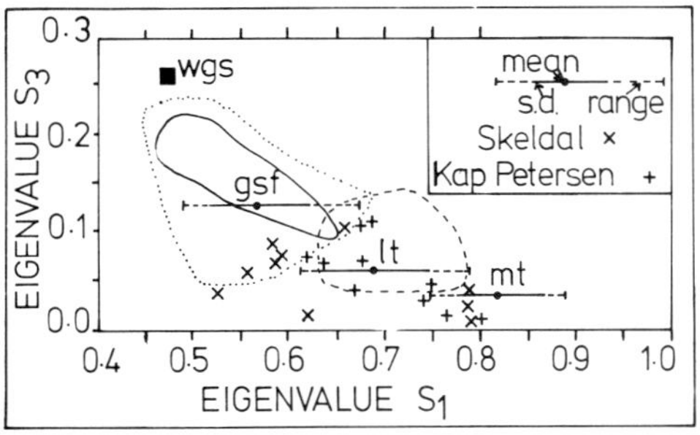

Fig. 11. Plot of S1 against S3 eigenvalues for Kap Petersen and Skeldal diamictons. Other data from modern environments (from Dowdeswell *et al.* 1985, Lawson 1985 and Gibbard (pers. comm.). The variability of fabric strength in modern environments is shown as a standard deviation and ranges about a mean eigenvalue. The envelopes are as follows: dotted line represents glacigenic sediment flows (gsf); continuous line represents fossiliferous diamicton; dashed line represents undeformed lodgement till (lt); mt represents meltout till and wgs represents waterlaid glacigenic sediment.

intact into the basal layers of active glaciers (Boulton 1976) and it is not uncommon for basal tills in coastal areas to contain reworked but little altered shells (Dreimanis 1970; Mangerud 1983; Petersen 1983).

(vii) The conclusions from the distribution of foraminifera in the diamicton and underlying deposits.

(viii) The regional distribution of the diamicton from south of Mesters Vig along the coastal plain, through lower Skeldal and along the coastal plain to Kap Petersen. It has been traced to the north and west as far as Skipperdal. Fabric orientations indicate a regional consistency in this area.

(ix) The eigenvalues (Fig. 11) suggest that the problem diamicton is either a melt-out or lodgement till which has in some locations a secondary flow fabric superimposed (see Domack & Lawson 1985; Dowdeswell *et al.* 1985; Lawson 1985).

Similar problems of interpretation between till and compacted glacimarine sediments have been discussed by Elverhøi (1984). It is apparent that this subtle differentiation between a true till, a glacimarine sediment and an overridden glacimarine sediment is still difficult to make with confidence.

The senior author would like to thank the I.M. Marsh Memorial Fund, Liverpool Polytechnic, the Royal Society for an Overseas Fieldwork grant and the field help from members of the I.M. Marsh East Greenland expedition. This is especially the case for the expedition's organiser Mike Peckham for his field assistance and companionship in the Kap Petersen phase of the expedition. The help and advice of Julian Dowdeswell with regard to eigenvector computer programs is acknowledged. Thanks are also due to D.D. Harkness of the Natural Environment Research Council Radiocarbon Laboratory, Scottish Universities Research and Reactor Centre, East Kilbride, who ^{14}C-dated shell samples. The comments of referees Michael Hambrey and Svend Funder have improved the manuscript considerably.

Note added. ^{14}C dates supplied by the Scottish Universities Research and Reactor Centre, East Kilbride help give a further understanding on the chronology of the sediment sequences and events outlined in the text.

SRR-3590, KP/8 Kap Petersen, north of Fangstation. (72° 24′N 24° 31′W). *Mya truncata* in diamicton interpreted as till at 3.1 m. 8170 ± 65.

SRR-3591, KP/5 Skjoldungebrae Gorge. (72° 23′N 24° 30′W)). *Mya truncata* in fine sand delta foresets at 24.6 m. 6845 ± 75.

SRR-3592, KP/4 Skjoldungebrae Gorge. *Hiatella arctica* and *Mya truncata* in diamicton overlying distal laminites. Diamicton interpreted as a readvance till reworking fjord sediment. 8895 ± 75.

SRR-3593, KP/3 Skjoldungebrae Gorge. *Mya truncata, Hiatella arctica* and *Macoma calcarea* in fjord sediment in a distal environment to the morainal bank sediment assemblage of the tidewater Skjoldungebrae. 8790 ± 75.

SRR-3594, SK/3 Lower Skeldal (72° 18′N 24° 10′W). *Mya truncata* in diamicton. Interpreted as readvance till, overlying laminites. 7670 ± 60.

SRR-3593 from silt laminites in a distal morainal bank sedimentary environment dates a tidewater stage of the Skjoldungebrae into Kong Oscar fjord to approximately 8275 ± 75 (all raw dates need to be corrected by subtracting 515 years as suggested by Håkansson (1983). The dates from the diamicton (SRR-3590, 3592 and 3594) support the view expressed in the text that it is a melt-out or lodgment till deposited from a readvance of ice along the coastal plain. This is because there are considerable age differences (1200 years) in these samples as would be expected from a readvance which reworked fjord bottom sediment of differing age. There is also a considerable altitudinal range between 3.1 m and 49.2 m in these samples. The youngest date in these three till samples gives a minimum age for the readvance of 7155 ± 60 and confirms the readvance as probably equivalent to Funder's (1971) Rødefjord stage in the Scoresby Sund region to the south. SRR-3591 from articulated shells in fine sand delta foresets at 24.6 m gives a date of 6330 + 65 and therefore the ice readvance along Kong Oscar fjord can be placed between approximately 6400–7100 B.P. This suggests a correlation with tills in Vega Sund (Hjort 1973) which are older than 5929 ± 70.

References

ANDERSON, J.B. 1983. Ancient glacial-marine deposits: their spatial and temporal distinction. *In*: MOLNIA, B.F. (ed.) *Glacial-Marine Sedimentation*, Plenum, New York, 3–92.

——, KURTZ, D.D., DOMACK, E.W. & BALSHAW, L.M. 1980. Glacial and glacial marine sediments of the Antarctic continental shelf. *Journal of Geology*, **88**, 399–414.

BALAZARINI, M.A. 1983. Palaeoecology of Late Pleistocene glacial marine sediments in N.W. Washington and W. British Columbia. *In*: MOLNIA, B.F. (ed.) *Glacial-Marine Sedimentation*, Plenum, New York, 571–592.

BENNIKE, O., FUNDER, S. & HJORT, C. 1986. Notes on the Holocene marine fauna of eastern North Greenland. *Bulletin of the Geological Society of Denmark*, **35**, 71–74.

BOULTON, G.S. 1976. A genetic classification of tills and criteria for distinguishing tills of different origin. *In*: STANKOWSKI, W. (ed.) *Till: its genesis and diagenesis*, Adam Mickiewicz University., Poland, Seria Geographia, **12**, 65–80.

BRETZ, H. 1935. Physiographic studies in East Greenland. *In*: BOYD, L.A. (ed.) *The fjord region of East Greenland*. Special Publication American Geographical Society, **18**, 159–267.

CHRISTIE-BLICK, N. 1983. Glacial-marine and subglacial sedimentation, Upper Proterozoic Mineral Fork Formation, Utah. *In*: MOLNIA, B.F. (ed.) *Glacial-Marine Sedimentation*, Plenum, New York, 703–776.

CRONIN, T.M. 1976. Late-Wisconsin marine environments of the Champlain Valley (New York, Quebec). *Quaternary Research*, **7**, 238–253.

DOMACK, E.W. 1983. Facies of Late Pleistocene glacial marine sediments on Whidbey Island, Washington: an isostatic glacial marine sequence. *In*: MOLNIA, B.F. (ed.) *Glacial-Marine Sedimentation*, Plenum, New York, 535–570.

—— & LAWSON, D.E. 1985. Pebble fabric in an ice-rafted diamicton. *Journal of Geology*, **93**, 577–591.

DOWDESWELL, J.A., HAMBREY, M. & WU, R. 1985. A comparison of clast fabric and shape in Late Precambrian and modern glacigenic sediments. *Journal of Sedimentary Petrology*, **55**, 691–704.

DREIMANIS, A. 1970. Are marine fossils in the Quaternary deposits a sufficient evidence for marine deposition? *Baltica*, **4**, 313–322.

DREWRY, D.J. & COOPER, A.P.R. 1981. Processes and models of Antarctic glaciomarine sedimentation. *Annals of Glaciology*, **2**, 177–22.

ELVERHØI, A. 1984. Glacigenic and associated marine sediments in the Weddell Sea, fjords of Spitzbergen and the Barents sea: a review. *Marine Geology*, **57**, 53–88.

——, LØNNE, Ö. & SELAND, R. 1983. Glaciomarine sedimentation in a modern fjord environment, Spitzbergen. *Polar Research*, **1**, 127–149.

FARROW, G., ATKINSON, J. & MOORE, G. 1984. Macrobenthos and bioturbation. *In*: SYVITSKI, J.P.M. (compiler), *Sedimentology of Arctic Fjords Experiment*, **2**, 6–1–6, 35, Canadian data report on hydrography and oceanographic science.

FEYLING-HANSSEN, R.W. 1955. Stratigraphy of the marine Pleistocene of Billefjorden, Vestspitsbergen. *Skrifter Norsk Polarinstitutt*, **107**, 1–186.

——, FUNDER, S. & PETERSEN, K. 1983. The Lodin Elv Formation; a Plio-Pleistocene occurrence in Greenland. *Bulletin of the Geological Society of Denmark*, **31**, 81–106.

FLINT, R.F. 1948. Glacial geology and geomorphology. *In*: BOYD, L.A. (ed.) *The coast of Northeast Greenland*. Special Publication of the American Geographical Society, **30**, 91–210.

FUNDER, S. 1971. Deglaciation of the Scoresby Sund Fjord region, north-east Greenland. *In*: PRICE, R.J. & SUGDEN, D.E. (eds) *Polar Geomorphology*, Institute of British Geographers Special Publication **4**, 33–42.

—— 1978. Holocene stratigraphy and vegetation history in the Scoresby Sund area, East Greenland. *Bulletin of the Grønlands Geologiske Undersøgelse*, **129**.

—— & HJORT, C. 1973. Aspects of the Weichselian chronology in central East Greenland. *Boreas*, **2**, 69–84.

HÅKANSSON, S. 1974. University of Lund radiocarbon dates V11. *Radiocarbon*, **16**, 307–330.

—— 1983. A reservoir age for the coastal waters of Iceland. *Geologiska Föreningens i Stockholm Förhandlingar*, **105**, 64–67.

HALLER, J. 1958. Der 'Zentrale Metamorphe Komplex' von NE-Gronland, teil 11 Die geologische Karte der Staunings Alper und des Forsblads Fjordes. *Meddelelser om Grønland*, **154** (3).

—— 1971. *Geology of the East Greenland Caledonides*, Wiley, London.

HICOCK, S.R., DREIMANIS, A. & BROSTER, B.E. 1981. Submarine flow tills at Victoria, British Columbia. *Canadian Journal of Earth Sciences*, **18**, 71–80.

HJORT, C. 1973. The Vega Transgression–a hypisthermal event in central East Greenland. *Bulletin of the Geological Society of Denmark*, **22**, 25–38.

—— 1976. Remarks on the glacial chronology of the Kong Oscars Fjord and Vega Sund district, central East Greenland, as evidenced by glacial sculpture, striae and similar features. *University of Lund, Department of Quaternary Geology Report* **9**.

—— 1979. Glaciation in northern East Greenland during the Late Weichselian and early Flandrian. *Boreas*, **8**, 281–296.

—— 1981. A glacial chronology of northern East Greenland. *Boreas*, **10**, 259–274.

—— & BJORK, S. 1983. A re-evaluated glacial chronology for northern East Greenland. *Geologiska Föreningens i Stockholm Förhandlingar*, **105**, 235–243.

KELLY, M. 1985. A review of the Quaternary geology of western Greenland. *In*: ANDREWS, J.T. (ed.) *Quaternary Environments: eastern Canadian Arctic, Baffin Bay and western Greenland*, Allen and Unwin, Boston, 461–501.

LASCA, N. 1969. The surficial geology of Skeldal, Mesters Vig, north-east Greenland. *Meddelelser om Grønland*, **176** (3).

LAVRUSHIN, Y.A. 1968. Features of deposition and structure of the glacial marine deposits under conditions of a fjord coast (based on the example of Spitzbergen), (translation), *Litologiya i Poleznye Iskopaemye*, **3**, 63–79.

LAWSON, D.E. 1979. A comparison of the pebble orientations in ice and deposits of the Matanuska Glacier, Alaska. *Journal of Geology*, **87**, 629–645.

MANGERUD, J. 1983. The glacial history of Norway. *In*: EHLERS, J. (ed.) *Glacial Deposits in Northwest Europe*, Balkema, Rotterdam, 3–10.

MODE, W.N., NELSON, A.R. & BINGHAM, J.K. 1983. A facies model of Quaternary glacial-marine cyclic sedimentation along eastern Baffin Island, Canada. *In*: MOLNIA, B.F. (ed.) *Glacial-Marine Sedimentation*, Plenum, New York, 495–534.

MOLNIA, B.F. 1983. Subarctic glacial-marine sedimentation: a model. *In*: MOLNIA, B.F. (ed.) *Glacial-Marine Sedimentation*, Plenum, New York, 95–144.

MURRAY, J.W. 1973. *Distribution and Ecology of Living Benthic Foraminiferids*, Heinemann, London.

NELSON, A.R. 1981. Quaternary glacial and marine stratigraphy of the Qivitu Peninsula, northern Cumberland Peninsula, Baffin Island, Canada. *Bulletin of the Geological Society of America*, **92**, 1143–1261.

NORDENSKIÖLD, O. 1907. On the geology and physical geography of East Greenland. *Meddelelser om Grønland*, **28**, 151–285.

OCKELMANN, W.K. 1958. The zoology of East Greenland: marine lamellibranciata. *Meddelelser om Grønland*, **122** (4), 1–274.

OSTERMAN, L.E. & ANDREWS, J.T. 1983. Changes in glacial-marine sedimentation in Core Hu 77–159, Frobisher Bay, Baffin Island, N.W.T.: a record of proximal, distal and ice-rafting glacial-marine environments. *In*: MOLNIA, B.F. (ed.) *Glacial-Marine Sedimentation*, Plenum, New York, 451–94.

PEACOCK, J.D. 1989. Marine molluscs and late Quaternary environmental studies with particular reference to the Late-glacial period in North West Europe: a review. *Quaternary Science Reviews*, **8**, 179–192.

PETERSEN, K.S. 1983. Redeposited biological material. *In*: EHLERS, J. (ed.) *Glacial Deposits of Northwest Europe*. Balkema, Rotterdam, 203–206.

POWELL, R.D. 1981. A model for sedimentation by tidewater glaciers. *Annals of Glaciology*, **2**, 129–134.

—— 1983. Glacial-marine sedimentation processes and lithofacies of temperate tidewater glaciers, Glacier Bay, Alaska. *In*: MOLNIA, B.F. (ed.) *Glacial-Marine Sedimentation*, Plenum, New York, 185–232.

—— 1984. Glacimarine processes and inductive lithofacies modelling of ice shelf and tidewater glacier sediments based on Quaternary examples. *Marine Geology*, **57**, 1–52.

SMITH, R.K. 1970. Late Glacial foraminifera from South-east Alaska and British Columbia and a world-wide high northern latitude shallow-water faunal province. *Archives des Sciences*, **23**, 675–705.

STREET, F.A. 1977. Deglaciation and marine paleoclimates, Schuchert Dal, Scoresby Sund, East Greenland. *Arctic and Alpine Research*, **9** (4), 421–426.

VILKS, G. 1969. Recent foraminifera in the Canadian Arctic. *Micropaleontology*, **15** (1), 35–60.

WALTON, W.R. 1964. Recent foraminiferal ecology and paleoecology. *In*: IMBRIE, J. & NEWELL, N.D. (eds.) *Approaches to Paleoecology*, Wiley, New York, 151–232.

WASHBURN, A.L. 1965. Geomorphic and vegetational studies in the Mesters Vig District. *Meddelelser om Grønland*, **166** (1).

ZENKEVITCH, L. 1963. *Biology of the seas of the U.S.S.R.*, Allen and Unwin, London.

Proximal and distal glacimarine deposits in southwestern Sweden: contrasts in sedimentation

RODNEY L. STEVENS

Department of Geology, Pomona College, Claremont, California, USA (Present address: Department of Geology, Chalmers University of Technology and University of Göteborg, S-41296 Göteborg, Sweden)

Abstract: The balance between the meltwater supply of suspended sediment from an increasingly distal ice front and the local sources of clastic sediments through turbulent reworking is reflected in the stratigraphy of Weichselian glacimarine deposits near Göteborg. Descriptions from cored and exposed sections and process interpretations are reviewed in order to make generalized comparisons between the proximal to distal trends and the contrasting lithofacies in very distal settings. Since nearly all sediment was initially supplied by the meltwater overflow the distance to ice-front standstills, the seasonal or non-seasonal variability of the discharge and the local drainage were major influences on the textural and geochemical characteristics. This is best illustrated by the rhythmites interpreted to be annual varves. In very distal settings the fine sediments that the meltwater could supply are interrupted in the stratigraphy by silty and sand laminated clays interpreted to be the result of the increased importance of near-shore erosion. Granular sediments could be transported from local sources and concentrated in the stratigraphy during long periods of low meltwater discharge, such as with glacial stadial conditions. During high-discharge phases the fine-grained sedimentation predominated at localities along regional drainage paths for the meltwater and the coarser input from local sources was diluted.

Sedimentation from the meltwaters of temperate glaciers is especially distinctive in its variability. This has been recently documented in modern glacimarine settings (Elverhøi *et al.* 1980, 1983; Syvitski & Murray 1981; Gilbert 1982; Powell 1983; Mackiewicz *et al.* 1984; Cowan *et al.* 1988). Extremely high and very low depositional rates are associated with each other over short geographic distances and time intervals. Similar contrasts occur in texture, such as coarse diamictic gravels, well-sorted sands, lonestones within homogeneous fine-grained clays and rhythmitic stratification interpreted as annual varves, tidal cycles or event sedimentation. The complexity in glacimarine environments is not surprising when one considers the strong, and in many cases decisive, influence that is separately exerted by the climatic and topographic factors within the regional or local environment. Although proximal-distal trends are logically expected to be reflected in the character of glacimarine sediments, and are obvious in many instances, the combination of processes does not always lead to gradual transitions. For instance, the balance between meltwater-supplied sedimentation and deposition from shallow-water reworking is not only dependent upon the distance from the glacier, but can also be significantly influenced by isostatic and eustatic sea-level changes as well as the climatic control on meltwater discharge and local storm turbulence.

Understanding the deviations from the generalized proximal-distal trends in the depositional environment is essential for lithofacies modelling. The interpretation of varve sequences is SW Sweden compliments the ^{14}C-dated stratigraphy with the detail necessary for interpretation of conditions that are typically of short duration in glacimarine environments. For instance, varve thicknesses reflecting annual sediment flux and the internal rhythmite variations in lamina-facies can be combined to help model changes in depositional character during one year and over longer periods. Separate stratigraphical subdivisions representing increasingly distal conditions can be used to compare depositional character if their relationship to ice-front position can be established (cf. Stevens 1986). At the same time as ice-front proximity varies stratigraphically, this approach allows the other geographical factors to be held relatively constant or within known limits.

This paper briefly reviews the descriptions and interpretations of proximal, distal and very distal glacimarine deposits in SW Sweden. The primary objective here is to compare the chang-

From DOWDESWELL, J.A. & SCOURSE, J.D. (eds), 1990, *Glacimarine Environments: Processes and Sediments*. Geological Society Special Publication No 53, pp 307–316.

ing processes and to present a unifying model for their geographic association. The stratigraphical relationships of different lithofacies further allow a discussion of the generalized regional trends reflecting the balance between the influencing factors of deglaciation, climate and sea-level changes.

Geological setting and data character

Deposition of glacimarine sediments during the late Weichselian deglaciation occurred in valleys and lowlands in SW Sweden which were isostatically depressed below sea level (Fig. 1). The topographic relief is usually 50 to 150 m, sufficient to provide a significant influence upon late-stage ice-flow directions and meltwater drainage (Hillefors 1969; Stevens *et al.* 1984). Within fjord-like valleys the glacial retreat toward the NE proceeded step-wise since calving intensity was largely controlled by the variable topography and water depth whereas climate was most important with respect to meltwater discharge and long-term ablation. These localities experienced an early evolution of depositional conditions related to their generally decreasing proximity to ice-front processes, especially the supply of sediment from glacial meltwaters.

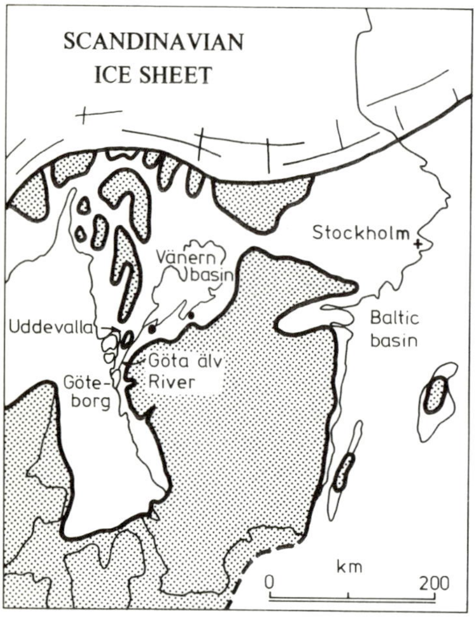

Fig. 1. Areas of glacimarine deposition following retreat of the Scandinavian ice sheet. This is not a paleogeographic map since much of the west-coast terrain had been uplifted and the westward passages along the Göta älv River and at the Uddevalla were considerably more restricted by the time the ice front was positioned as shown. The highest coastline is drawn after Fredén (1982).

Land uplift reflects the predominance of isostatic rebound over the eustatic sea-level rises in this area (cf. Påsse 1983; Svedhage 1985). The emerging topography provided an additional source of sediments through possible reworking of earlier deposits that became increasingly exposed to storm turbulence. At the same time the changing geography concentrated the meltwater drainage along certain pathways while others became more isolated. For those localities still receiving meltwater in very-distal glacimarine settings, changes in outflow volume or direction that were induced by climatic or marine-current influences would have been important for the sediment character, as is considered below.

There is limited exposure today of the poorly consolidated clays in SW Sweden. Excavations and landslide scars erode or deteriorate after a few years and generally offer only a small portion of the glacimarine stratigraphy along the valley slopes. Ten sections of the clay stratigraphy (10–30 m) have been cored using a Swedish piston corer, but neither does this technique allow penetration of the very deepest valley areas where clay thicknesses may exceed 100 m. The lack of direct data concerning the deepest valley in-fills has been compensated to a certain extent by reviewing the abundant geotechnical soundings and non-continuous sampling of deep clay deposits in the Göteborg area. The observations and interpretations in this paper are, nevertheless, primarily based upon the detailed sedimentology of the cored and exposed sections.

Stratigraphy

The Weichselian clay sequences are underlain by either bedrock or relatively thin diamicton and sand. The contacts with the clay deposits are typically abrupt, although the clays can be laminated and interbedded with fine-sand layers in their lowermost portions, lithofacies A (Table 1). Rhythmites defined by the alternation of colour, geochemistry and texture are conspicuous features at many investigated localities (Fig. 2A; detailed documentation in Stevens 1985, 1987). The brown, sub-rhythmite intervals are the finest in texture and have fewer granular (coarse silt and sand) interlaminations compared to the grey portions. Disseminated, fine-grained precipitates of iron-monosulphides within some greyish intervals are marked by

Table 1 *Lithofacies and depositional environments*

Lithofacies	Notes	Interpretation
A. Silty clay with common granular laminae	Silt content and lamina frequency decrease upwards. Rhythmites with textural and colour alternations common.	Proximal glacimarine. Very rapid deposition. Varves.
B. Silty clay	Silt content decreases upwards. Rhythmites less distinct where present.	Distal glacimarine. Rapid deposition.
C. Fine-grained clay	Variable FeS-precipitates may be banded or irregular when present. Grey. Homogenous texture. Marine fauna.	Very distal glacimarine with strong meltwater influence (stadial?). Possible FeS-varves.
D. Clay with common granular laminae	Lamination greatest near valley sides. Shell debris and lonestones.	Very distal glacimarine without water-column stratification.
E. Clays with variable texture and high organic contents	Plant fragments. Some FeS-precipitates. Marine fauna.	Shallow marine. Not considered in the text.

black colour bands that fade with exposure. The carbonate content ranges up to *c*. 10% and is generally higher in the brown intervals compared with the grey intervals. Sequences with over 200 rhythmites have been observed (Caldenius 1951; Stevens 1986). The granular laminae are often graded but the sorting is typically only moderately well developed in most examples. Cross-laminated sand beds within the clays have been observed only in two instances. Lonestones are rare but common enough in some sampled sections to suggest variable stratigraphical distribution. A few, non-articulated mollusc shells have been observed.

Upwards within the lower portions of the clay

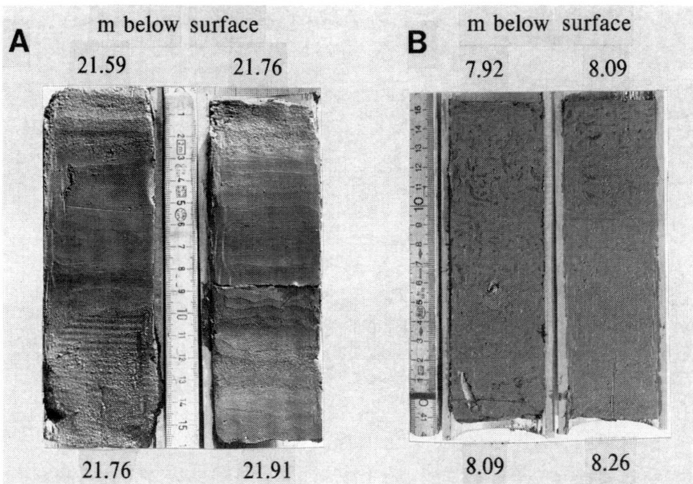

Fig. 2. A. Rhythmite deposits 2 m above the base of a cored clay sequence, 5 km NE of Göteborg (Fig. 1). The sandy intervals at bottom, middle and top (21.91, 21.75 and 21.58 m) represent the coarse-grained deposition during strong discharge ('summer'). These intervals illustrate the moderately poor sorting and transitional contacts that are believed to be most common for suspension deposition. The fine-grained, brownish clay ('winter' deposition) appears darkest. B. Fine clay from distal to very distal deposition in the upper portion of the same sequence. Mollusc shells occur at 8.08 and 8.02 m. Small gas bubbles that give a faint, uneven surface texture after cutting are believed to be hydrogen sulphide, one of the reactants responsible for the precipitation of the finely disseminated iron-monosulphides that have darkened the clay in places. The gas or the concentration of organic material was apparently suitable for precipitation and preservation of the iron-sulphide along the burrow-like tracks.

stratigraphy several trends can be observed: (1) there are fewer and thinner laminae of silt and sand within the clay; (2) grading is less distinct and less common in the upper occurrences of granular laminae, although many lower laminae are also non-graded; (3) rhythmite thicknesses show considerable variation but generally decrease upwards, for instance from 20 cm to less than 3 cm over one 5 m section at Kviberg (Stevens 1987); (4) the details of the rhythmite units internally define "lamina-facies" which show changes in their cyclic associations upward within the rhythmitic stratigraphy (Fig. 3); (5) sand laminae seem to be uncommon above approximately the 20th to 25th rhythmite from the bottom at several localities; (6) intervals of very silty clay are also markedly less represented above this level whereas the silt content in the less silty clays decreases more gradually, especially in the fine-silt fractions; (7) the uppermost rhythmites are often less distinct due to the weaker textural and colour contrasts between the sub-rhythmitic layers; (8) organic contents are low (usually < 1%) but increase slightly upwards, largely corresponding to the changes in grain size; and (9) although palaeontological trends within the lowermost clay sequences are not well known, bioturbation has only been observed above the rhythmitic clays. The detailed lamination in these sediments would have been lost in places had the environment been conducive to an active bottom fauna (cf. Figs 2A and 2B). These stratigraphical changes are continuous over both lithofacies A and B, except for the absence of granular laminae in facies B.

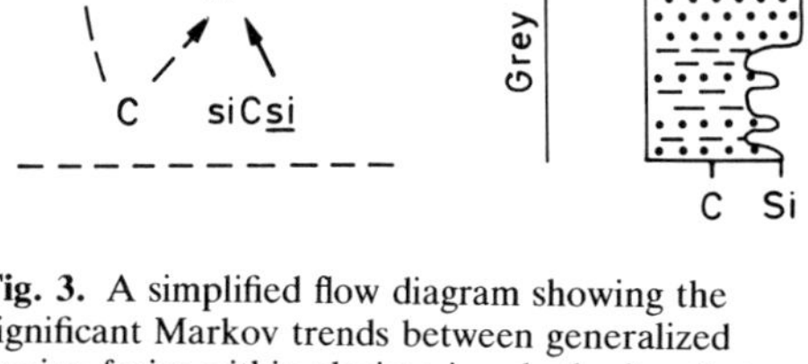

Fig. 3. A simplified flow diagram showing the significant Markov trends between generalized lamina-facies within glacimarine rhythmites that usually range in thickness from 1 to 20 cm. C=clay, Si=silt, si=silty and *si*=silt-laminated. The colour alteration from grey to brown and variations in darkness within each cycle are believed to reflect redox conditions strongly influenced by seasonal variations in water stratification. The arrows reflect the strength of transition based upon the probability differences when compared with a random model (details in Stevens 1987).

The rhythmitic and laminated deposits generally pass upwards into greenish-grey clays with a more homogeneous, very fine-grained texture (often > 50% clay-sized particles) and very high water content, lithofacies C. Both micro- and macrofossil contents are higher as is the organic content (1–2%). Iron-sulphide precipitates are sometimes present so that the greenish grey clay is stained black but becomes lighter in colour with exposure to air (Fig. 2B). Alternating light and dark colour bands, 1–5 cm thick each and comprising sequences of up to 400 band-couplets, have been documented in some early Holocene deposits within the middle of the clay stratigraphy (Mörner 1976; Häger 1981; Cato 1982; Stevens 1987; Klingberg 1989). In sequences where colour banding is not distinct, there is often a tendency for the sulphides to provide an uneven darkening of the fine clays.

Alternating with the fine-clay deposits are one or more intervals of clay with granular laminae, lithofacies D. The clay texture itself is often fine-grained even here although the water content is lower than in the non-laminated portions. The sand and silt laminae are moderately sorted but may contain gravel to cobble sized clasts and considerable shell debris or in situ bivalves. Unsorted, unstratified clayey layers with sand and pebbles have been noted in a few cores. The thickness of the laminated clay intervals is usually less than 0.5 m whereas the fine-clay units can vary up to several metres. The frequency of the laminations as well as their individual thickness and character is quite variable. An apparent tendency was noted in the geotechnical documentation for localities further from valley sides to have less lamination.

Interpretations

Proximal-distal depositional trends

Table 1 and Fig. 4 summarize observed textural trends and interpreted processes within the Weichselian glacimarine environments in SW Sweden. The abrupt transition from glacial and glaciofluvial deposition of the diamicton and sand units to the overlying glacimarine clays is consistent with the quick change to predominately suspension transport and depositional processes once the ice front had passed. The thicker sections near the valley axes, which were not sampled, may have generally been

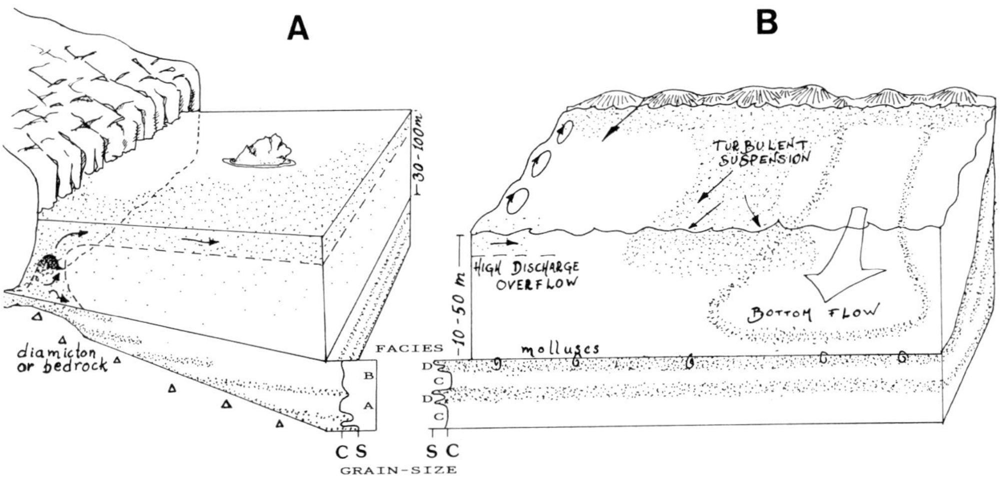

PROXIMAL -- DISTAL GLACIOMARINE (<<100 km)	VERY DISTAL GLACIOMARINE (>100 km)
Granular and fine sediments supplied by meltwater	Only fine grained, suspension transport by meltwater
Lens orientation parallel with valley axis	Lenses extending from valley sides
Proximity to an ice-front standstill important	Proximity to standstill positions not important
Local meltwater drainage decisive	Regional meltwater drainage important
Seasonal variations in sedimentation	Interstadial-stadial variations
Textural rhythmites/varves (1–20 cm)	Silt/sand laminated clays, 10–300 cm (stadial deposits)
FeS-precipitates within some "summer" deposits	FeS-rhythmites/varves developed during interstadials

Fig. 4. A schematic comparison of proximal to distal and very distal glacimarine environments and characteristic lithofacies. Note that the vertical scales are exaggerated and not the same in A and B. The lithofacies in B overlie those in A and are themselves overlain by shallow marine sediments (not shown) in the generalised stratigraphy from this area.

closer to tunnel positions and would presumably have somewhat more gradual transitions. However, the similar trends in proximal glacimarine sediments that are recognizable at most localities suggest that tunnel position was not extremely important since the meltwaters carried a predominantly suspended sediment supply to the valley environments. Although a complicated set of environmental factors was presumably involved in determining the specific direction and extension of the outflow (cf. Gilbert 1982; Powell 1983; Mackiewicz *et al.* 1984; Syvitski 1989), these factors were sufficiently variable to be equalized and allow similar depositional trends in most valley localities. Variation in transport competence of the meltwater overflows is represented by the beds with moderate sorting and variable, normal or reverse grading. Bottom transport, e.g. turbidity currents and sediment-gravity flows, is interpreted to have occurred occasionally and produced the beds with normal grading and good sorting in very proximal positions. Tunnel position and bottom topography would, in these cases, have had an influence analogous to glaciolacustine underflows (De Geer 1940; Smith & Ashley 1985).

At several localities the glacimarine rhythmites are believed to represent annual varves whose textural and geochemical variations are largely related to the seasonal variations induced by meltwater fluctuations (Stevens 1985). A stratified water column, similar to that documented in Glacier Bay (Matthews 1981; Powell & Molnia 1989) is believed to have developed and enhanced both transport competence and sediment sorting by initially limiting the meltwater from mixing with marine water below the pycnocline. Current strength could thereby be maintained longer and flocculation would have been less effective due to flow turbulence, low salinity and the retarded settling across the pycnocline (Syvitski & Murray 1981; Stevens 1985). Stratification hinders downward mixing so that an oxygen-depleted CO_2-zone is interpreted to have periodically underlain the well-oxygenated surface waters and may have accounted for some carbonate dissolution. The carbonate contents are generally highest in the fine-grained 'winter' layers since fewer dissolution opportunities were available without undersaturation in a CO_2-zone and without the permeability for pore-water flux after deposition.

Most important for the clay colour, the stratification and CO_2-zone also helped to slow the oxidation of metabolisable organic matter. Reduction of the iron oxides adsorbed to the clay particles occurred syndepositionally and diagenetically so that the 'summer' sediments obtained a greyish colour (Stevens *et al.* 1987). Reduced iron could have been removed with pore-water flushing, but may also have re-precipitated in sulphidic forms. The black iron-monosulphides observed within some grey layers are metastable and interpreted to have been preserved due to the lack of reactants for their continued transformation to pyrite, which are sulphur, metabolisable organic matter and reactive iron (Berner 1984). Rapid clastic deposition during the spring meltout period would have had a limiting effect upon the first two components so that the black mono-sulphides could at times be preserved (Stevens 1987). The fine-grained, brownish clays reflect the better water-column mixing and oxidation of organics during slow sedimentation and, therefore, the preservation of iron oxides.

The proximal to distal trends within the rhythmitic glacimarine sediments are consistent with the diminishing sediment supply by glacial meltwater. Strong discharge events appear to have been capable of transporting sand to at least 2 km, the estimated distance to the ice front after 20 varve-years with *c.* 100 m/a ice-front retreat and by correlations between varves and ice-marginal deposits (Stevens 1986). The lower silt content beyond this interpreted distance is also indicative of a limit to which discharge currents could effectively transport the denser, granular particles within this particular geographic setting. A somewhat arbitrary but pragmatic boundary based upon the occurrence of granular laminae within the glacimarine sediments of lithofacies A is used here to divide the interpreted proximal and distal sub-environments. The lack of current structures and the generally poor sorting suggests that suspension transport was responsible for most granular laminae as well as the intercalated clay deposits.

Sediment-transport patterns observed in the modern, fjord-like glacimarine environments correspond well with the textural trends observed in SW Sweden (e.g. Queen's Inlet, Alaska, Hoskins & Burrell 1972; Kongsfjord, Spitsbergen, Elverhøi *et al.* 1980, 1983; Baffin Island fjords, Gilbert 1982; Knight Inlet, British Columbia, Farrow *et al.* 1983; Muir Inlet, Alaska, Mackiewicz *et al.* 1984). Horizons blackened with iron-monosulphides are also recorded in sediments from Kongsfjord and Queen's Inlet. In this study, the black mono-sulphides observed within some grey bands may represent incipient 'organic varves'. Had the organic content been higher, the grey layers could have been entirely blackened and the brown banding would have become grey with iron reduction, essentially the same as the 'organic varves' in Kongsfjord and Queen's Inlet cores which were interpreted to be annual concentrations due to algal bloomings, although variations in organic content are not established (Elverhøi *et al.* 1980; Hoskins & Burrell 1972).

Stratigraphically upward, or distally, the colour banding persists further than do the systematic textural variations. It is suggested that within the glacimarine model for this environment, the colour couplets can be used with appropriate discretion for varve interpretations. Distal glacimarine deposition was less dominated by the meltwater influences. The rhythmitic character in some areas records more than 200 varve-years, at which time the ice is interpreted to have been at an approximate distance of 20 km. At localities where the meltwater influence was restricted by topography or wind and marine currents, annual rhythmitic sedimentation was short-lived or indistinct, although frequent granular laminations suggest the initial importance of meltwater fluctuations in the earliest deposits of all cored sections.

Very distal deposition

Above the silty, laminated and sometimes rhythmitic deposits, nearly homogeneous intervals of fine clay (lithofacies C) accumulated when the glacial meltwater still represented the major sediment supply but the seasonal variations had lost their textural and geochemical reflection due to the greater transport distances and the influence of marine processes. On the other hand, changes in the amount of meltwater and, thereby, the supply of fine-grained sediment over longer periods can be expected to have a textural expression if other, more local sources supplied relatively coarser sediments (Fig. 4B). When the ice front was very distal (> 100 km is used here for simplicity), the possibilities for transporting clastic fractions other than clay and fine silt with the westward flowing meltwaters were very limited. For instance, bottom flows would have to cross the 100 km wide Vänern basin (Fig. 1). The variation between narrow passages and wide basins would also have a dispersive effect on overflow currents during major discharge events. Strong water-column stratification was probably required to limit flocculation and to maintain the flow competence necessary to transport significant amounts of even the finest fractions. Upon reaching the west coast the surface currents would be vulnerable to marine processes and mixing which would have aided sediment trapping, analogous to an estuarine turbidity maximum (Cato 1982). When water stratification was only moderately well developed the transport of suspended fines could have been limited by mixing with saline waters in the Vänern basin.

If the iron-monosulphide banding represents varves of 1–5 cm annual accumulation, sedimentation was, in fact, rapid at these localities (Cato 1982; Stevens 1987). Assuming an annual mechanism, sedimentation conditions allowing colour (FeS) banding continued for up to 400 years, essentially the time until the ice front is believed to have retreated from the Vänern basin (Cato 1982). However, recent ^{14}C datings (Helene Bergsten, pers. comm.) suggest that the early-Holocene banded sequences might have been deposited after the ice had retreated from the Vänern basin. In this case, an alternative explanation for these deposits would be that water stratification in the very distal settings remained poor until circulation in the Vänern basin was limited by the emergence of the strait at Uddevalla and those further north where land uplift was greatest (Fig. 1). This development occurred about 9300 BP (Fredén 1982) and would have restricted the influx of marine waters and concentrated the low-salinity outflow to the southern outlet along the Göta-älv-River valley, favouring improved water stratification and transport capability. The closing of the eastern strait (at Närke) connecting the Baltic and the Vänern basins occurred at *c.* 8900 (Olsson & Fredén 1969) so that the approximate period of concentrated, low-salinity outflow could well correspond to the series of 400, interpreted 'FeS-varves'.

The interpretation of the monosulphide bands in these deposits is similar to that for the precipitates in the proximal rhythmites where the environmental combination of rapid deposition, variable organic contents and low-sulphidic surface waters may have prevented the transformation of monosulphides to disulphidic pyrite (Stevens 1987). The pattern of iron-monosulphides in the sediments can also be quite irregular. This may reflect unstable water-column stratification which, in turn, restricted sediment transport and depositional rates. The distribution of reactants within the sediments may also been influenced by pore-water circulation induced by, for instance, bioturbation (Fig. 2B).

During extended periods of low discharge and poor water stratification, such as during glacial stadials, overflow transport of fine sediments was likely to have been ineffective and the local reworking of deposits on the valley slopes became predominantly important. Land uplift successively exposed many slope deposits to waves and marine currents. However, this exposure was locally variable, as the characterization of lithofacies D reflects. In many cases the shorter transport distances and the episodic nature of the processes allowed for coarser deposition than was possible from meltwater suspension. During periods of high meltwater discharge the fine-grained sedimentation would dilute the granular contribution provided by near-shore erosion. In contrast, the laminations and the granular fraction are more evident within certain intervals (facies D) due to their stratigraphical concentration during periods without strong meltwater influx, although the frequency of storms or local, sediment-reworking processes may have been essentially the same. The variable sorting suggests that turbulent suspension, traction currents and sediment gravity flows were all involved in the movement of sediment out from the valley slopes. The cobbles and pebbles associated with the granular laminae would most likely have been sea-ice transported, expectedly a more influential process during cold phases. Although most of the shell debris was probably reworked from near-shore areas, both the more stable, granular sediment and the slower accumulation rates were favourable for mollusc fauna (cf. Farrow *et al.* 1983). Bioturbation does not, however, seem to have been so extensive as to prevent the preservation of many of the laminae. However, the extension of laminae away from the valley sides was limited and at localities further than 1 km from the former shore, it is often difficult to distinguish this laminated facies.

Meltwater drainage over the Vänern basin was diverted southward to the southern outlet of the Baltic basin after *c.* 9000 BP (Fredén 1982, 1988; Dennegård 1984). In addition, the total discharge of the Scandinavian ice sheet had rapidly diminished during this same period in connection with the glacial withdrawal to northern Sweden (Lundqvist 1965). The continued evolution of Holocene environments in SW Sweden was increasingly characterized by the influences of local geography and exposure to marine processes. The isostatic-induced shallowing and eventual land emergence also led to increased isolation and erosional opportunities that could involve the earlier deposited glacimarine deposits. The Holocene, shallow-marine environments reflected in the lithofacies E are not further considered here but are dealt with elsewhere by Påsse (1983), Svedhage (1985) and Stevens (1987).

Comparisons

Trends in the glacimarine lithofacies can be generalised into two environmental settings (Fig. 4). Proximal to distal glacimarine deposits (facies A and B) are taken as one category since they show a transitional relationship within the first tens of kilometres from the ice front. Proximal-glacimarine environments are delimited by the interpreted distance to which granular laminae could be commonly transported. Silty clays are typical for the distal-glacimarine lithofacies, but coarse silt becomes decreasingly represented in this sub-environment. Although there is a gradual change to the very fine clays, which together with the laminated and shelly clay intervals represent a second category of very distal deposition, lithofacies C and D were best developed in SW Sweden when the distance to the ice front was greater than 100 km.

Perhaps the main difference between the two categories is that sediment was initially almost entirely supplied by meltwater whereas in very distal environments the meltwater flow could transport only fine-grained sediments. Furthermore, the balance between meltwater and other clastic sources in the very distal setting was more sensitive to changes in climate and relative sea level. Proximal, granular laminae could have resulted from meltwater deposition following either suspension or bottom transport during major glacial discharges; and these are expected to have an axial orientation within the proximal fjord-like settings. Localities close to ice-front positions during glacial standstills are associated with greater lamination. On the other hand, the granular content of very-distal-glacimarine deposits originated from local erosion and laminae are, therefore, more frequent closer to the valley sides and not influenced by the locality's proximity to former ice-front positions. In contrast to the very rapid deposition of laminated rhythmites in the proximal environment, these laminated intervals were associated with relatively slow accumulation rates.

Meltwater drainage is important in both settings. The subglacial drainage area influences the total amount of discharge that can be supplied to the proximal to distal environments. Seasonal meltwater flux is interpreted to be reflected in the rhythmites of textural and geochemical variations. Brownish and greyish colours are the result of syndepositional and diagenetic iron oxidation or reduction. Within some 'summer' (greyish) deposits the reduction also allowed iron-monosulphides to precipitate.

When the ice was further away the Göta-älv-River valley and the sound at Uddevalla were the two most important pathways for the meltwater from the eastern ice masses due to land emergence (Fredén 1982, 1988; Stevens *et al.* 1984) and the degree of communication with this drainage largely determined the sedimentation character. Seasonal meltwater fluctuations were, in themselves, not a cause of distinct textural variability. Nevertheless, iron-monosulphide rhythmites at some localities are tentatively interpreted to record seasonal changes in the organic matter and sulphur content necessary for their precipitation and preservation. Diminished outflow over longer periods was important in that it allowed the input of sediment from local reworking to be better represented, in the form of granular lamination or silty, sandy clays. The alternation of very fine clay and the intervals with greater granular contents could have been most easily produced by changes from glacial stadial to interstadial conditions, e.g. Younger Dryas to Preboreal. These long-term changes are less evident in proximal to distal environments because of the higher relative sea level and the dominance of meltwater sedimentation even during glacial stadials. The individual rhythmite sequences studied cover limited time spans. However, a regional glacimarine-varve chronostratigraphy seems possible and is necessary for both detailed and extended climatic interpretations.

I extend my thanks to David Penney and Roger Dackombe who have helped improve this manuscript with their suggestions. Financial support has been given by the Swedish Natural Science Research Council and the Swedish National Building Fund.

References

BERNER, R.A. 1984. Sedimentary pyrite formation: an update. *Geochemica et Cosmochimica Acta*, **48**, 605–615.

CALDENIUS, C. 1951. Några geokronologiska profiler i Viskadalen, *Geologiska Föreningens i Stockholm Förhandlingar*, **73**, 423–433.

CATO, I. 1982. Grain-size distribution of the cores – with emphasis on the sedimentary patterns around the Pleistocene/Holocene boundary. *In*: OLAUSSON, E. (ed.) *The Pleistocene/Holocene boundary in south-western Sweden*. Sveriges Geologiska Undersökning C 794, 54–65.

COWAN, E.A., POWELL, R.D. & SMITH, N.D. 1988. Rainstorm-induced event sedimentation at the tidewater front of a temperate glacier. *Geology*, **16**, 409–412.

DENNEGÅRD, B. 1984. Late Weichselian and early Holocene stratigraphy in southwestern Sweden with emphasis on the Lake Vänern area. *Chalmers University of Technology and University of Göteborg (Sweden), Department of Geology Publication*, A 48 (thesis).

DE GEER, G. 1940. Geochronologia Suecica Principles. *Kungliga Svenska vetenskapsakademiens handlingar*, **3:18**, 1–367.

ELVERHØI, A., LIESTØL, Ø. & NAGY, T. 1980. Glacial erosion, sedimentation and microfauna in the inner part of Kongsfjorden, Spitsbergen. *Norsk Polarinstitutt Skrifter*, **172**, 33–58.

——, LØNNE, Ø. & SELAND, R. 1983. Glaciomarine sedimentation in a modern fjord environment, Spitsbergen. *Polar Research*, **1**, 127–149.

FARROW, G.E., SYVITSKI, J.P.M. & TUNNICLIFFE, V. 1983. Suspended particulate loading on the macrobenthos in a highly turbid fjord: Knight Inlet, British Columbia. *Canadian Journal of Fisheries and Aquatic Sciences*, **40**, 273–288.

FREDÉN, C. 1982. An outline of the marine stage of the Väner basin. *In*: OLAUSSON, E. (ed.) *The Pleistocene/Holocene Boundary in south-western Sweden*. Sveriges Geologiska Underökning C 794, 16–26.

—— 1988. Marine life and deglaciation chronology of the Vänern basin southwestern Sweden. *Sveriges Geologiska Undersökning*, Ca 71.

GILBERT, R 1982. Contemporary sedimentary environments on Baffin Island, N.W.T., Canada: glaciomarine processes in fjords of eastern Cumberland Peninsula. *Arctic and Alpine Research*, **14**, 1–12.

HÄGER, K.-O. 1981. Beskrivning av jordlagerföljden i borrkarnorna 18 och 20. *In*: FREDÉN, C., BRUSEWITZ, A.M., CATO, I., FÄLT, L.-M., HÄGER, K.-O., MILLER, U. & SAMUELSSON, L. Geologiska undersökningar *Sveriges Geologiska Undersökning Rapporter och Meddelanden*, **26**, 55–78.

HILLEFORS, Å. 1969. Västsveriges glaciala historia och morfologi, naturgeografiska studier. *Meddelanden från Lunds Universitet (Sweden), Geografiska Institution Avhandling* (thesis).

HOSKINS, C.M. & BURRELL, D.C. 1972. Sediment transport and accumulation in a fjord basin, Glacier Bay, Alaska. *Journal of Geology*, **80**, 539–551.

KLINGBERG, F. 1989. Foraminiferal stratigraphy in late Quaternary marine clay sequencies in southwestern Sweden. *Chalmers University of Technology and University of Göteborg (Sweden), Department of Geology Publication*, A 60 (thesis).

LUNDQVIST, J. 1965. The Quaternary of Sweden. *In*: RANKAMA, K. (ed.) *The Quaternary. Vol. 1*. Interscience Publ., New York, 139–198.

MACKIEWICZ, N.E., POWELL, R.D., CARLSON, P.R. & MOLNIA, B.F. 1984. Interlaminated ice-proximal glacimarine sediments in Muir Inlet, Alaska. *Marine Geology*, **57**, 113–147.

MATTHEWS, J.B. 1981. The seasonal circulation of the Glacier Bay, Alaska fjord system. *Estuarine, Coastal and Shelf Science*, **12**, 679–700.

MÖRNER, N.A. (ed.) 1976. The Pleistocene/Holocene boundary: a proposed boundary-stratotype in Gothenburg, Sweden. *Boreas*, **5**, 193–275.

OLSSON, I.U. & FREDÉN, C. 1969. Radiocarbon measurements from the Degerfors area, Sweden. *Geologiska Föreningens i Stockholm Förhandlingar*, **91**, 201–207.

POWELL, R.D. 1983. Glacial marine sedimentation processes and lithofacies of temperate tidewater glaciers, Glacier Bay, Alaska. *In*: MOLNIA, B.F. (ed.) *Glacial-Marine Sedimentation*. Plenum Press, New York, 185–232.

—— & MOLNIA, B.F. 1989. Glacimarine sedimentary processes, facies and morphology of the south-southeast Alaska shelf and fjords. *Marine Geology*, **85**, 359–390.

PÅSSE, T. 1983. Havsstrandens nivåförändringar i norra Halland under Holocen Tid. *Chalmers University of Technology and University of Göteborg (Sweden), Department of Geology Publication*, A 45 (thesis).

SMITH, N.D. & ASHLEY, G. 1985. Proglacial lacustrine environment, physical processes. *In*: ASHLEY, G.M., SHAW, J. & SMITH, N.D. (eds.) *Glacial Sedimentary Environments*. Society of Economic Paleontologists and Mineralogists Short Course **16**, 135–215.

STEVENS, R.L. 1985. Glaciomarine varves in late-Pleistocene clays near Göteborg, southwestern Sweden. *Boreas*, **14**, 127–132.

—— 1986. Glaciomarine varves and the character of deglaciation, Säveån Valley, southwestern Sweden. *Boreas*, **15**, 289–299.

—— 1987. Glaciomarine fine sediments in southwestern Sweden: late Weichselian – Holocene lithostratigraphy, depositional environments and varve formation. *Chalmers University of Technology and University of Göteborg (Sweden), Department of Geology Publication*, A 54 (thesis).

——, HELLGREN, L.-G. & HÄGER, K.-O. 1984. Depositional environments and general stratigraphy of clay sequences in southwestern Sweden. *In*: CATO, I. (ed.) Rapid mass movements in soils. *Striae*, **19**, 13–18.

——, APRIL, R. & WEDEL, P. 1987. Sediment color and weathered preglacial sources of Quaternary clays in southwestern Sweden, *Geologiska*

Föreningens i Stockholm Förhandlingar, **109**, 241–253.

SVEDHAGE, K. 1985. Shore displacement during late Weichselian and early Holocene in the Risveden area SW Sweden. *Chalmers University of Technology and University of Göteborg (Sweden), Geology Publication*, A 51 (thesis).

SYVITSKI, J.P.M. 1989. On the deposition of sediment within glacier-influenced fjords: oceanographic controls. *Marine Geology*, **85**, 301–329.

—— & MURRAY, J.W. 1981. Particle interaction in fjord suspended sediment. *Marine Geology*, **39**, 215–242.

The Pleistocene–Holocene transition in Southwestern Sweden and the recognition of deglaciation effects in adjacent seas

A.R. LORD

Postgraduate Unit of Micropalaeontology, Department of Geological Sciences, University College London, Gower Street, London WC1E 6BT, UK

Abstract: Detailed investigation of Late Quaternary sequences in southwest Sweden via shallow cored boreholes provides an excellent record of sedimentological patterns and the biological history of the area during regional deglaciation. Climatic amelioration and consequent deglaciation led to the Baltic region being dominated by the Baltic Ice Lake, a major physiographic feature which is estimated to have become undammed at *c.* 10200 years BP, discharging some 10^4 km^3 of freshwater south and west into the Skagerrak and North Sea Basin. Such an event should be easily traced in the marine environment, but away from the immediate onshore area convincing records have yet to be found. The freshwater influx affected the upper part of the water column in the North Sea Basin and should be detectable via an isotopic signature, but good quality cores with planktonic organisms have not yet provided such a record. The effects of Baltic Ice Lake discharge into the North Sea Basin may well be overprinted by contemporary major oceanographic changes in the northeast Atlantic, especially the initiation of the Norwegian Current, and by meltwater input from adjacent land areas.

This paper is based upon the results of a multidisciplinary investigation of the Late Quaternary of southwest Sweden in search of an elusive goal: a stratotype locality for the Pleistocene–Holocene Boundary. The Holocene Commission, at the 1969 INQUA meeting in Paris, defined the Pleistocene–Holocene boundary by means of absolute dating as 10000 ^{14}C years BP and subsequently sought to identify a suitable stratotype section for the boundary. Attention focused on southwest Sweden and a multidisciplinary investigation of a core taken in the Botanic Garden, Göteborg was made (Mörner 1976). The Botanic Garden Core (B873) was not subsequently felt to be entirely suitable as a stratotype section, particularly because a freshwater interval occurred at a critical point in the climatic record as interpreted from the sequence. A new study was made to find a better section, involving specialists from Sweden, Denmark and Great Britain under the chairmanship of Professor Eric Olausson, and a number of sites were investigated with the aid of the Geological Survey of Sweden. In the final report of the working group (Olausson 1982a) eleven sites were investigated including a re-interpretation of the Botanic Garden Core, six of these were studied in detail and three sites were found to be particularly valuable for documenting Late Quaternary events (Figs 1 & 3).

1. Brastad No. 12 Borehole – 58°23′39″ N Lat., 11°31′20″ E Long..
Altitude 45 m above sea level. Penetration 15.1 m.
2. Moltemyr No. 1 Borehole – 58°26′45″ N Lat., 11°32′36″ E Long..
Altitude 55 m above sea level. Penetration 27 m, 16.3 m cored.
3. Solberga No. 7 Borehole – 57°57′05″ N Lat., 11°47′42″ E Long..
Altitude 2 m above sea level. Penetration 27.3 m.

Full details of the cores, localities and regional geological history are given in Olausson (1982a) together with results of investigations of the sediments (grain size, organic carbon, clay minerals, stable isotope record and magnetostratigraphy), their biological content (macrofossils especially molluscs, and microfossils: foraminifera, ostracods, calcareous nannofossils, diatoms, dinoflagellates and pollen) and radiocarbon dating; the present author was concerned with the ostracods.

The palaeogeographical setting of southern Scandinavia and the Skagerrak and Kattegat for the Late Quaternary has been analysed by Stabell & Thiede (1985b, 1986) and Thiede (1987) (Fig. 2).

The Late Quaternary deglaciation of Scand-

From DOWDESWELL, J.A. & SCOURSE, J.D. (eds), 1990, *Glacimarine Environments: Processes and Sediments*. Geological Society Special Publication No 53, pp 317–328.

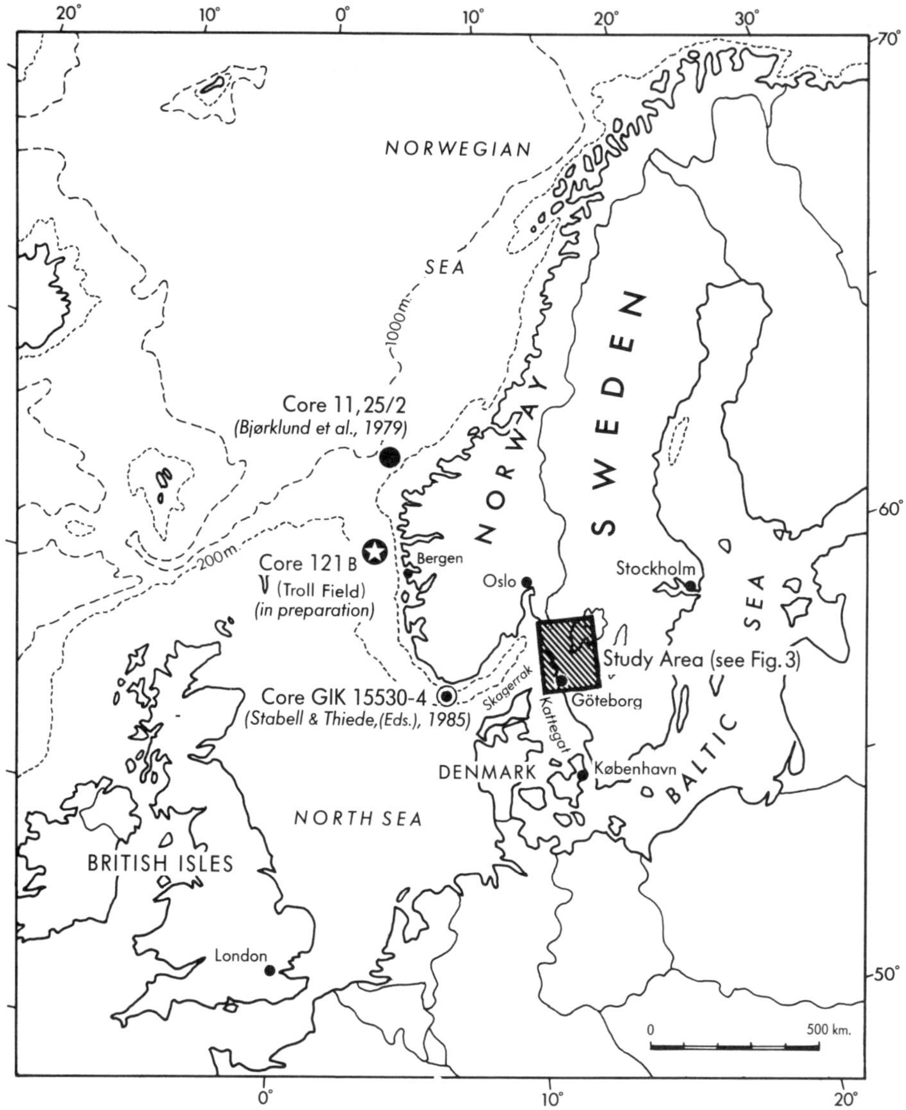

Fig. 1. Location of study area and selected core sites. (Troll Field, *in prep.* refers to J. Nagy *pers. comm.* cited in text.)

inavia was of major importance in that over a relatively short period of time a very large volume of ice decayed, thus providing a substantial quantity of meltwater to adjacent areas especially the Baltic and North Seas. Southwestern Sweden was a critical area in the deglaciation history of the region and the purpose of this paper is to review evidence for the effects of meltwater and freshwater input into adjacent seas.

Late Quaternary history of Southern Sweden

The Weichselian ice sheet completely covered Sweden until the time when climatic amelioration caused wastage of the ice and consequent margin retreat. The ice margin probably reached the modern Swedish west coast at approximately 13 000–12 600 years BP and the subsequent deglaciation history is marked onshore by ice

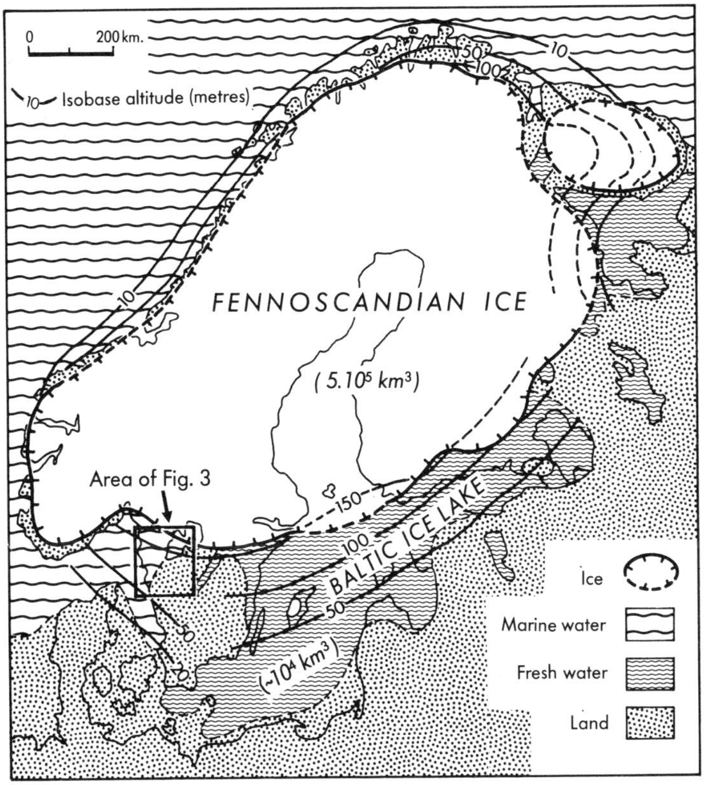

Fig. 2. Palaeogeographical map of Scandinavia at 10 500–10 250 years BP (after Stabell & Thiede 1985b, fig. 3).

retreat features, especially moraines. The moraines can be dated, although opinion differs in detail as to precise ages, and they provide a geographical framework for late-glacial and post-glacial time in the area (Fredén 1982). The named terminal marine features are as follows: Göteborg (12 800–12 600 years BP), Berghem (*c*. 12 300 years BP), Trollhätten (11 900–11 800 years BP), Levene (11 400–11 300 years BP), Skövde (11 000–10 800 years BP) and Billingen (10 400–10 250 years BP) (dates from Fredén 1982, 1988). The Skövde and Billingen moraines are particularly distinctive and form part of the Fennoscandian terminal moraines which can be traced from Norway to Finland. The sea entered the area from the North Sea to the west via the Skagerrak so that by 10 500 years BP a complex geography existed, a coast with many islands and inlets, similar to the modern Swedish west coast north of Göteborg, composed of a hummocky bedrock terrain with glacial sediment infills upon which shallow marine depositional conditions were then imposed.

Marine waters penetrated in a tongue from the north-west to the southeast, along part of what is now the Kattegat, but did not extend far or link through to the Baltic Basin in the east (Stabell & Thiede 1985b, 1986). The nature of the topography of southern Sweden and Denmark was such that meltwater from deglaciation in central Sweden and Finland was ponded up, without a drainage route to the North Sea and Atlantic or to the northeast to the Arctic Ocean. This enormous area of accumulated meltwater, the Baltic Ice Lake, remained in existence until ice margin retreat reached a critical point in relation to regional topography (Fig. 2). The conventional interpretation makes Billingen hill to the northeast of Göteborg the final positive feature damming the Baltic Ice Lake, and by approximately 10 200 years BP the lake was finally able to discharge past the hill into the Vänern Basin and thence to the sea via the line of the modern Göta River and through straits near Uddevalla. The Baltic Ice Lake was possibly lowered by

26–28 m and it is estimated that some 10^4 km^3 of freshwater was discharged into the Skagerrak; a further volume of approximately 5×10^5 km^3 of water followed the same route during the next 1000 years (Olausson 1982b). There are, however, uncertainties about the nature and chronology of the 'Billingen Event' and these are discussed by Fredén (1982, 1988). It is unclear if all or only part of the discharge occurred via Billingen and it is also uncertain whether there was a single event or a series of discharge pulses. Bjørck & Digerfeldt (1984) suggest drainage events during the Allerød and Younger Dryas with final drainage at 10500–10400 years BP, and the isotope data from the present boreholes appear to support this interpretation. The effect of the freshwater entering the Skagerrak can be clearly seen in the cores as a sequence of sediments of non-marine origin; however, erosional features that might be expected to occur as a result of a sudden discharge event are lacking both in the Vänern Basin (Fredén 1982) and to the south west (R.L. Stevens, *pers. comm.*).

Marine waters entered the Baltic Basin after the discharge of the Baltic Ice Lake, the connection from the North Sea being through the Vänern Basin to the Yoldia Sea which occupied much of the area of the former lake. The Scandinavian ice melted rapidly, apparently in about 1000 years. Elevation of topography consequent upon isostatic adjustment resulted in the Vänern Basin becoming dry by *c.* 9000 years BP. A second lake formed, the Ancylus Lake, and it was not until later that marine waters again entered the Baltic Basin with connection to the North Sea via the Kattegat, forming the Litorina Sea as the forerunner of the modern Baltic Sea.

The boreholes (Fig. 3)

Brastad borehole

This site had open marine connections at the end of the Pleistocene, although the small bedrock hills in the vicinity may have formed islands at the time. The core penetrated silts and clays which were sandy in the deepest part. An erosion level was recognized at 2.31 metres below surface (m.b.s.), below which the clay contained scattered exotic clasts (Cato 1982). The biological record (ostracods, foraminifera, diatoms) shows the establishment of cold marine water assemblages following a climatic/environmental extreme, although conditions were always boreoarctic with only subordinate temperate indicators e.g. the proportion of boreal foraminifera does not exceed a maximum of 16% (Knudsen 1982). Both foraminifera and ostracods show signs of having responded to climatic deterioration just below the erosion level at 2.31 metres, but there is a lag factor in the response of the two different types of organism with ostracods more sensitive to changes (in temperature) than foraminifera. The sequence below 2.31 metres represents a relatively temperate phase at some point during the late Weichselian. The Pleistocene–Holocene boundary level falls in an erosional hiatus at this site, which can be interpreted as a reflection of erosion following Baltic Ice Lake discharge through the area.

Moltemyr borehole

This borehole was taken in a small basin surrounded by bedrock hills. Palaeogeographical reconstructions (Fredén 1982) show the site adjacent to an island, but with full marine connections, and other topographic features such as islands or submarine highs may have surrounded the site. The locality was selected on the assumption that the small basin might show short-term fluctuations in the deglaciation pattern of the area. Moltemyr and Brastad are only about 5 km apart and their records should complement each other. Above 2.12 m.b.s. the borehole penetrated fen peat and coarse gyttja, below which were silty clays commonly containing shell fragments. The fossils (ostracods, foraminifera, diatoms, pollen) show a decline in arctic influence up the sequence consequent upon climatic amelioration, and the Pleistocene–Holocene boundary (= Younger Dryas–Preboreal boundary) was recognized at 3.45 m.b.s. In the higher part of the sequence both ostracods and foraminifera show reduced diversity and abundance, although there was no re-establishment of cold indicator species. Thus, if conditions were once again cold in character they did not permit recolonization by cryophillic taxa. The interval above 3.45 m.b.s. has very low ostracod and foraminifera density and diversity and coincides with an increase of clay in the sediments which Cato (1982) tentatively correlates with a phase of meltwater discharge. A distinct decrease in $\delta^{18}O$ between 3.45 and 2.7 metres also indicates meltwater influence. A barren level for ostracods at 3.95 m.b.s., matched by a decline in foraminiferal abundance, may also reflect disruption by meltwater. The broad pattern of ostracod diversity and density, and the appearance of boreal, warmer water species coincides closely with that shown by the foraminifera. The upper part of the

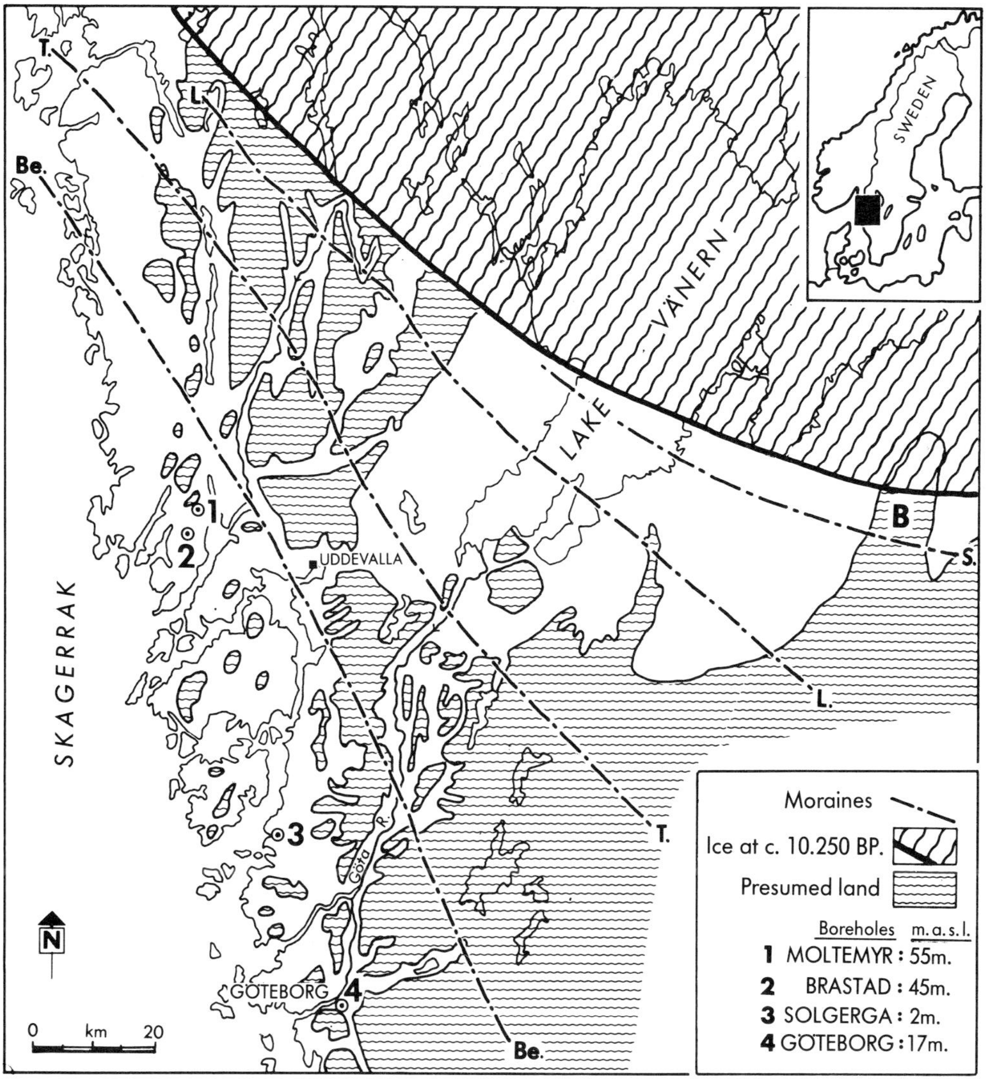

Fig. 3. Location of borehole sites in relation to ice margin features in south western Sweden at *c*. 10 250 years BP (after Fredén 1982, figs. 3.1 & 3.3). B = Billingen.

cored sequence may not represent continuous sedimentation. The foraminifera provide evidence that the interval 3.5 to 2.2 m.b.s. was not simply composed of reworked and redeposited sediment. Boreal species that were relatively rare in the older, colder part of the sequence are common in this interval. However, these later assemblages are less abundant and diverse (Knudsen 1982). The diatom record is also good and indicates increasing isolation from the sea (Miller 1982).

Solberga Borehole

The Solberga Borehole is located in southern Bohuslän about 30 km north of Göteborg, 50 km to the south of the other two boreholes. The site is situated in a wide basin close to the sea and to modern sea level, and yielded a long (> 27 m) clay sequence which reflects a higher sediment accumulation rate than at Moltemyr. Palaeogeographical reconstructions for *c*. 10 250 BP (Fredén 1982) show the site in an open marine setting. The sequence recovered was the fullest obtained, although in other respects data from this and the Brastad and Moltemyr Boreholes complement and supplement each other.

The biological record (ostracods, foraminifera, diatoms, pollen) shows a clear transition from cold water conditions (marine sediments rest on a till at the bottom) to a more temperate situation at 18.00–19.00 m.b.s., where the Pleistocene–Holocene boundary was recognised. Interestingly, the greatest abundance of foraminifera and percentage increase of boreal forms occur slightly before these features are evident in the ostracods, another example of different response rates by different organisms. In this case the foraminifera have reacted to warming before the ostracods, although the richest ostracod sample coincides with the most *diverse* foraminifera assemblage. Diatom and pollen evidences are similar in that they both indicate a critical interval between 19.50 and 19.20 m.b.s.

The interval from *c*. 17.00–5.00 m.b.s. contains poor, low diversity ostracod and foraminifera assemblages and barren intervals. Above 5.00 m diversity and abundance increase in both groups, with assemblages characteristic of modern, shallow marine waters of the Skagerrak and North Sea. The pattern of assemblages documents climatic change, as reflected at the base of water column and by the pattern of sedimentation, from cold water arctic conditions to the establishment of a temperate, essentially modern environment. Fluctuations occur in this overall pattern which may reflect short-term variations in conditions during a warming trend. Despite the establishment of more temperate conditions and the final loss of arctic species by *c*. 18.50 m.b.s., the diversity and abundance decline, and from 17.0 to 5.0 m.b.s. ostracods and foraminifera are relatively less common and diverse. This impoverished interval represents what is in fact an interruption to an established pattern, as essentially modern faunas were present from *c*. 18.25 m.b.s., and represents a significant palaeoenvironmental event.

Particle-size analysis shows a major increase in clay from 18.40 to *c*. 5.0 m.b.s., a dramatic increase in suspended sediment load and rate of sedimentation attributed to the effects of the undamming and discharge of the Baltic Ice Lake. Thus the Solberga sequence provides a good history of regional events, with climatic change indicated by the microfauna and microflora and the effects of the discharge of the Baltic Ice Lake evident. The diatom evidence is particularly interesting (Miller 1982). From 27.0 to 21.5 m.b.s. arctic planktonic diatoms dominate, while from 21.5 to 18.0 m.b.s. arctic forms decline and marine littoral and coastal/temperate planktonic species increase. The interval 18.0 to 17.0 m.b.s. is characterized by a sudden influx of freshwater diatoms, especially the planktonic *Melosira islandica helvetica*, and this flora continues, although in smaller numbers, because of the high sedimentation rate. The uppermost part of the sequence contains littoral brackish and coastal/temperate planktonic diatoms with virtually no freshwater forms. Thus, diatom data complement and support the evidence provided by ostracods and foraminifera.

Sedimentation at the borehole sites

The sediments in all three boreholes are silty clay or clayey silt with some clay and only a small proportion of sand particles. The Solberga Borehole contains the most fine-grained material. The lower part of the sequence (below *c*. 18 metres) has sediment in which on average 50% is less than 2 μm in size, at 17.5 m there is a sand peak and the upper part of the borehole to *c*. 5.00 m.b.s. penetrated finer-grained material with an average of 80% less than 2 μm. This sand peak and overlying homogeneous clay are interpreted by Cato (1982) as a sudden increase in sediment bottom load (sand peak) followed by a longer interval with increased suspended sediment load (homogeneous clay). This event is attributed to the discharge of the Baltic Ice Lake at *c*. 10 200 years BP and the rapid accumulation rate reflects the freshwater-borne clay minerals having flocculated and settled rapidly upon contact with marine saline waters. The clay unit can be traced at a number of sites (including the Göteborg Botanic Garden Core (B873)), and can be related to the drainage channels followed by the lake waters (Cato 1982) as they discharged into the Skagerrak. The erosive effect of the initial event may be reflected by the hiatus in the Brastad Borehole, but in general erosional features indicative of a catastrophic event are lacking. The sediment in the Solberga Borehole between 18.00 and 5.00 m.b.s. must represent not only material from the initial flood but also sediment related to the fluctuating discharge following deglaciation of the region over the next 1000 years. The sporadic occurrences of ostracods and foraminifera probably reflect both reworking and temporary reestablishment of marine conditions before the youngest marine incursion above 5.00 m.b.s. (the Boreal transgression).

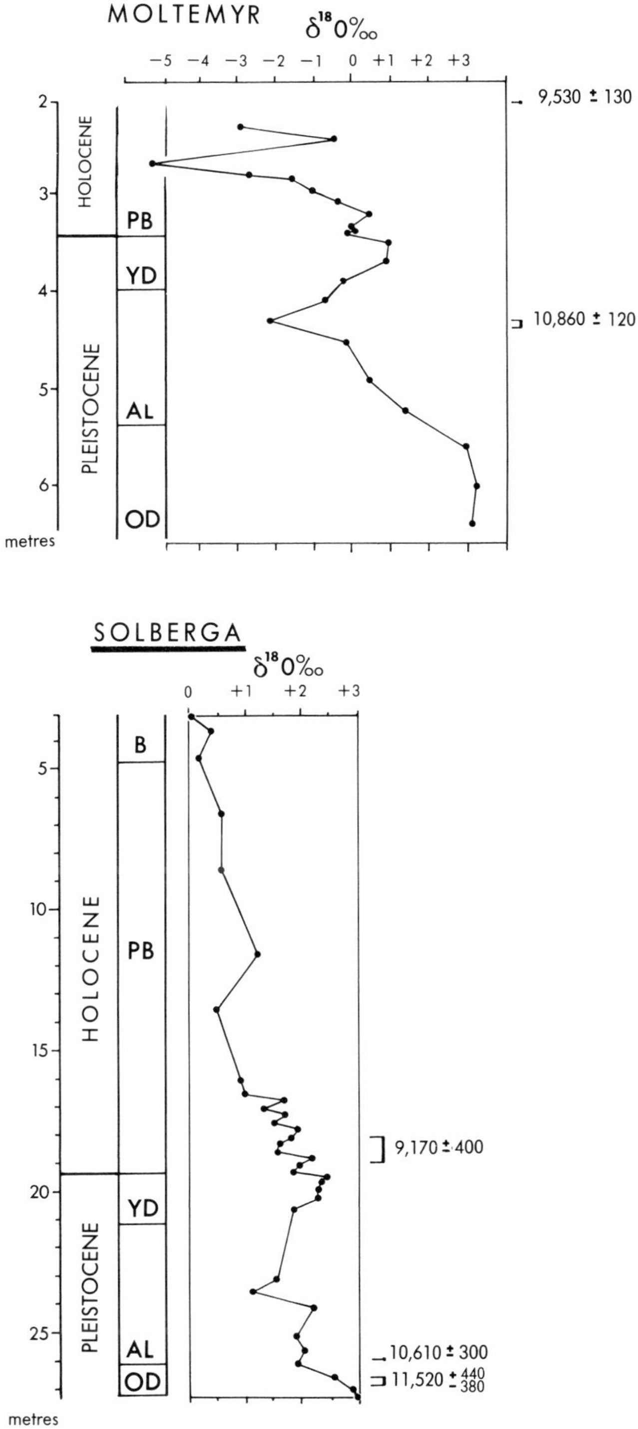

Fig. 4. Chronostratigraphy, $\delta^{18}O$ record and absolute dates for the Moltemyr (above) and Solberga (below) Boreholes, south western Sweden (data from authors in Olausson 1982a).

The sedimentological and biological record is further supported by $\delta^{18}O$ isotopic evidence measured on the benthic foraminifera *Elphidium excavatum*. In the Solberga sequence from *c.* 19.5 to 17.00 m.b.s. the isotope levels fall in a series of fluctuations and are then low to the top of the record (Fig. 4), a pattern attributed to the influx of isotopically light freshwater from the Baltic Ice Lake (Olausson 1982c). The record at Motemyr shows more extreme fluctuations and a generally lower level of $\delta^{18}O$ (Fig. 4). The change in the isotopic composition is recorded in the tests of the bottom-dwelling foraminifera because total mixing of marine and fresh waters occurred in a shallow and topographically restricted situation. This would not occur in deeper marine situations where the freshwater would affect only the upper part of the water column.

Biological evidence from the boreholes and adjacent areas

Information from a number of animal and plant microfossil groups is available. Palynology provided information about terrestrial vegetational history and the correlation of marine and non-marine sequences via pollen spectra (Robertsson 1982). However, dinoflagellates were generally rare (Harland 1982).

Ostracods

Living ostracods inhabit all aquatic environments but in the boreholes only benthic marine species occurred and no non-marine/freshwater forms were recognised. The utility of the group for palaeosalinity and palaeotemperature analysis has been considerably improved in recent years (De Deckker *et al.* 1988), and for Quaternary applications in high northern latitudes new data on living Arctic faunas are especially valuable for understanding the environmental range and tolerance of species occurring in Quaternary cold-water assemblages (Brouwers 1988). Foraminifera are usually more abundant than ostracods in marine environments; however, the fact that ostracods (as crustaceans) grow by moulting means that age/population structures of species/assemblages can be used to assess bottom current transportation.

The assemblages found range from very cold water, Arctic-type associations to ones similar to those living in littoral and inner neritic waters of the modern Skagerrak and North Sea (Lord 1982); the modern Baltic Sea with its restricted geography, salinity gradient, high freshwater inflow and winter ice cover is a special case and the nature of the ostracod faunas has been described by Rosenfeld (1977). In contrast, the ostracods from the Pleistocene–Holocene of core GIK 15530–4 off southern Norway (Fig. 1) analysed by Qvale (1985) were of deeper water origin than the assemblages discussed here and there are few species in common.

Foraminifera

The species found were all benthic in habit and no planktonic forms occurred. The foraminifera were generally much more abundant than ostracods and allowed each borehole sequence to be divided into a number of ecozones (Knudsen 1982), documenting changes in bottom water conditions. The appearance and relative abundance of boreal species is a particularly useful local parameter for gauging climatic change as evidenced in the marine environment (Knudsen 1982), especially when linked to oxygen isotope data from the common benthic foraminifera *Elphidium excavatum* (Cato *et al.* 1982). While these and other lines of evidence document a changing environment it is not always easy to interpret the precise significance of the data. Cato *et al.* (1982) correlate a peak in abundance of boreal foraminifera (41%) with a negative maximum of $-2.25\ \delta^{18}O$ during the late Allerød (in the Moltemyr Borehole) and explain both in terms of bottom water salinity *decrease*. This trend is not indicated by the diatom data. This explanation may be true for the isotope signal but the foraminifera are probably responding to warming encouraging colonization. An immediate decline in abundance of boreal forms above this level probably reflects the influence of lower salinity bottom waters from meltwater and a lag in response may be involved. On the other hand, the isotope data were obtained from one of the foraminifera species concerned, although which one of three ecophenotypes of *E. excavatum* was used is not clear.

More recent work on foraminifera from boreholes in the Göteborg area has provided sequences of assemblages very similar to those recorded at Solberga (Bergsten & Dennegard 1988). The interest of these sites, apart from corroborating the regional pattern already discussed, lies in their proximity to the Botanic Garden Core B873 studied by Mörner *et al.* (1976). The explanation rests in the relative altitudes of the localities, with the Botanic Garden at 17 metres above sea level (m.a.s.l.) and Bergsten and Dennegard's two cores taken in the Göta River valley at 2.7 and 3.6 m.a.s.l.

Benthic foraminifera were common in core GIK 15530–4 from the Skagerrak off southern Norway and in general document a similar story, with a dramatic increase in boreal forms at the Pleistocene–Holocene boundary level, but from a deeper water situation (Nagy & Qvale 1985). A number of species occur that were not seen in Bohuslan, especially *Hyalinea balthica* and *Uvigerina peregrina*. These are both deeper water shelf species.

Foraminifera from the Troll Field (core 121B, Fig. 1) have been analysed by Nagy, providing a record from the late Weichselian and Holocene of the Norwegian Trench. The site is exposed to open ocean influences including the Gulf Stream, but the record of bottom dwelling foraminifera is very similar to that found much further south in the Norwegian Trench in core GIK 15530–4 (Nagy & Qvale 1985).

Records of foraminifera from elsewhere in the North Sea from the Weichselian-Holocene transition are not sufficiently detailed or time-constrained to be useful in this study.

Planktonic foraminifera were absent from the western Swedish boreholes but they occurred in core GIK 15530–4 (Thiede 1985). In this area off southern Norway oceanic water masses bearing planktonic foraminifera were not apparently established until early Preboreal times (*c*. 9500 BP), although radiolaria were present somewhat earlier. It should be noted that the $\delta^{18}O$ record for this core was based on the benthic foraminifera *Elphidium excavatum* and *Cassidulina laevigata* because planktonic forms were relatively rare, especially in the lower part of the core. The results (Erlenkeuser 1985) are directly comparable with those of Olausson (1982c) from western Sweden but from a much deeper bottom water level.

Calcareous nannofossils

Quaternary coccoliths occurred in small numbers, with reworked Cretaceous and Tertiary nannofossils, but were essentially absent in sediments correlated with Baltic Ice Lake discharge (Mikkelsen & Perch-Nielsen 1982). The work of Gard (1986, 1988) on late Quaternary Arctic cores shows what can be achieved with sparse nannofloras, but the western Swedish material was probably too far from open ocean conditions and has been too diluted by high rate sedimentation to be useful.

In core GIK 15530–4 (Mikkelsen 1985) Quaternary coccoliths (*Emiliania huxleyi*) appear near the Pleistocene–Holocene boundary, whereas prior to this the nannofossils were all older, reworked species. Considering the location of this site the calcareous nannofossil record is disappointing.

Diatoms

The diatoms, like ostracods, provide a particularly useful confirmation of the evidence from exclusively marine fossils such as foraminifera with the added dimension of information from non-marine waters. There, the absence of non-marine ostracods may be linked to the temperature and turbidity conditions in the Baltic Ice Lake. The diatom data from the three boreholes (Miller 1982) included both planktonic and benthic species and provided salinity control. The sudden and dominant occurrence of *Melosira islandica helvetica*, a freshwater planktonic diatom, at a number of sites can be directly linked to Baltic Ice Lake discharge, although its presence is not restricted exclusively to the main discharge event. Baltic Ice Lake sediments penetrated by two cores in the western Baltic off Gotland and Öland (Borg 1985) did not contain diatoms (Paabo 1985), but this may reflect post-mortem dissolution of the siliceous frustules.

In core GIK 15530–4 the diatoms also provide a useful record (Stabell 1985) but are exclusively marine. The flora of Stabell's Zone B is comparable with the Younger Dryas (late Weichselian) diatom assemblages of western Sweden (Miller 1982). The effects of Baltic Ice Lake drainage, so clear in western Sweden, were not recognized in the diatom floras off the Norwegian coast (Stabell 1985).

Discussion

The CLIMAP project (McIntyre *et al.* 1976), and subsequent studies, have provided the basis for understanding late Quaternary history in terms of climatic and glacial systems, and oceanic geochemical, nutrient and circulation patterns (Ruddiman 1987), which in turn permit modelling of Quaternary oceans and discussion of their characteristics (Olausson 1985, Ruddiman & McIntyre 1981). The work of the latter authors on the North Atlantic is valuable for the present study.

The reconstructions of Ruddiman & McIntyre (1981) show that following the last glacial maximum (*c*. 18 000 years BP; phase I 20 000–16 000 years BP) the Polar Front withdrew rapidly from the eastern North Atlantic between 13 000 and 11 000 years BP to the vicinity of Iceland, but moved back briefly during a cold phase from 11 000 to 10 000 years BP (= Younger Dryas) before finally retreating northwestwards.

The first significant warming of the southern, central and eastern North Atlantic occurred at *c*. 13 000 years BP, followed by the cooling phase at 11 000 years BP referred to above which involved a return to almost full glacial conditions, and a second warming phase at 10 000 years BP. Jansen & Bjørklund (1985) have documented events in cores from the Norwegian Sea, mostly from deep water, in order to understand changes in surface circulation in this critical area during the past 15 000 years. They recognize the final occurrence of ice-rafted clasts in the south at *c*. 13 000 years BP together with the presence of planktonic foraminifera and a change in benthic foraminifera assemblage composition. At *c*. 12 000 years BP radiolaria, coccoliths and diatoms appear, while at 9800 BP the left-coiling (polar) form of the planktonic foraminifera *Neogloboquadrina pachyderma* gives way to its right-coiling (subpolar) morphotype as the dominant form. In an earlier paper Bjørklund *et al.* (1979) analysed core 11,25/2 from the Norwegian Sea (Fig. 1) which showed the appearance of radiolaria, planktonic foraminifera, diatoms and coccoliths (*E. huxleyi*) in a sandy layer dated by extrapolation as *c*. 11 400 years BP and which was interpreted as reflecting the presence of the Polar Front in the area. Thus the oldest 'post-glacial' levels are chronostratigraphically Weichselian in age and reflect the establishment of essentially modern oceanographic conditions and rich plankton assemblages. Further south off Bergen, in the mouth of the Norwegian Channel, Nagy & Ofstad (1980) have described a series of short cores which collectively provide a late Weichselian-early Holocene record, but which is fragmentary because of local erosion patterns and the nature of the cores. A better series of cores from an adjacent area (Troll Field, especially core 121B) is currently being studied, where the passage of the Polar Front is marked by a diatom abundance peak which in turn correlates closely with a dramatic increase in the boreal element of the benthic foraminifera assemblages (J. Nagy, pers. comm.).

The sites discussed above are from settings with full oceanic connections, but conditions in the more geographically restricted context of the North Sea usually allowed only fragmentary records to be preserved (Sejrup *et al.* 1984). In contrast, the GIK 15530–4 core in the Norwegian Channel-outer Skagerrak area (Stabell & Thiede 1985a) has yielded detailed data for the late Weichselian–Holocene transition. The core documents a change in sediment source, from ice-rafted terrigenous material to the modern pattern with sediment derived from the southern North Sea, but with elements of northern origin carried in by meltwater during the Younger Dryas. The pattern of zoo- and phytoplankton (foraminifera, radiolaria, coccoliths, diatoms, etc.) indicates the arrival of Atlantic surface waters into the Skagerrak and commencement of the Norwegian Coastal Current during and after the Preboreal (*c*. 10 000–9000 BP). The arrival of the Atlantic bottom waters is clearly documented by ostracods and benthic foraminifera, with Arctic assemblages being replaced at approximately the Pleistocene–Holocene boundary level. A number of features observed in the core have been linked by the workers to the effects of meltwater, including the Baltic Ice Lake drainage event, but direct evidence is lacking. Stabell (1985) in her discussion of diatoms from the core suggests that the Baltic Ice Lake meltwater was '. . . probably forced north and westwards along the Norwegian coast and would therefore not affect the conditions at the core site (south side of Norwegian Channel)'. While this conclusion may be correct there is little to substantiate it at present.

Olausson (1985) believes that 80% of Quaternary ice sheets drained into the North Atlantic–Arctic system and that most meltwater discharge into the ocean occurred between 13 000 and 9000 years BP, with a consequent rise of sea level. The Baltic Ice Lake drainage and the subsequent continued deglaciation of Scandinavia were elements of this pattern. There is little evidence for the Baltic Ice Lake water beyond the immediate area of western Sweden, despite the fact that it was discharged into the relatively restricted area of the North Sea prior to formation of the English Channel. The event is picked out by isotopic evidence in western Sweden, where it clearly affected the full water column in a shallow-marine situation so that benthic foraminifera recorded the fresh water isotopic character in their test calcite. In deeper, more open marine waters the freshwater influx would affect the upper water column but evidence is lacking at the core GIK 15530–4 site because planktonic foraminifera were very rare or absent from the critical lower part of the core. Ruddiman & McIntyre (1981) argue that meltwater in the North Atlantic created a strongly stratified water column with reduced productivity, which means that the isotopic record of meltwater influx is poorly represented or unrepresented. In addition, they suggest that planktonic organisms would have concentrated at the base of the meltwater layer at the halocline and that these more saline waters would have given a higher isotopic signature than the meltwaters. It may thus be difficult to detect meltwater via isotopic signatures, at least from marine calcareous plankton. Comparative isotope records from different groups (of $CaCO_3$ or SiO_2 composition) may provide the key, especially if they can be linked to other high-resolution studies of suitable cores, particularly from the North Sea.

At present it appears, from not very good evidence, that the Baltic Ice Lake water was rapidly dissipated and diluted in the North Sea and left little record. However, deglaciation of Britain as well as Scandinavia, coupled with river discharge arising from Alpine deglaciation, must have significantly influenced marine conditions in the restricted area of the North Sea and the longer-term challenge may consist of recognizing the relative impact of these different sources of meltwater and freshwater input in the marine environment.

I am grateful to Professor Eric Olausson (Göteborg) and other colleagues involved with the Pleistocene–Holocene Boundary Working Group, who are in no way responsible for opinions expressed here, and I benefited from discussion with Dr. Hans Petter Sejrup (Bergen) and especially Dr. Jeno Nagy (Oslo). The Royal Society and University College London kindly assisted with travel grants.

References

BERGSTEN, H. & DENNEGARD, B. 1988. Late Weichselian-Holocene foraminiferal stratigraphy and palaeohydrographic changes in the Gothenburg area, southwestern Sweden. *Boreas*, **17**, 229–242.

BJØRCK, S. & DIGERFELDT, G. 1984. Climatic changes at Pleistocene/Holocene boundary in the middle Swedish endmoraine Zone, mainly inferred from stratigraphic indications. *In*: MÖRNER, N.-A. & KARLEN, W. (eds) *Climatic Changes on a Yearly to Millennial Basis*. Reidel Dordrecht, 37–56. Dordrecht, 37–56.

BJØRKLUND, K.R., THIEDE, J. & HOLTEDAHL, H. 1979. Deglaciation of the southeastern Norwegian Sea towards the end of the last glacial age. *Boreas*, **8**, 133–5.

BORG, G.C. 1985. Palaeo-ecology reflected in physical and chemical properties of three baltic cores. *Striae*, **23**, 1–93.

BROUWERS, E.M. 1988. Palaeobathymetry on the continental shelf based on examples using ostracods from the Gulf of Alaska. *In*: DE DECKKER, P., COLIN, J.-P. & PEYPOUQUET, J.-P. (eds) *Ostracoda in the earth sciences*. Elsevier, Amsterdam, 55–76.

CATO, I. 1982. Description of the cores, and grain-size distribution of the cores. *In*: LAUSSON, E. *Sveriges Geologiska Undersökning*, **C794**, 36–65.

——, FREDÉN, C. & OLAUSSON, E. 1982. Summary of investigation. *In*: OLAUSSON, E. *Sveriges Geologiska Undersökning*, **C794**, 253–68.

DE DECKKER, P., COLIN, J.-P. & PEYPOUQUET, J.-P. (eds) 1988. *Ostracoda in the Earth Sciences*. Elsevier, Amsterdam.

ERLENKEUSER, H. 1985. Stable isotopes in benthic foraminifers of Skagerrak core GIK 15530–4: High resolution record of the Younger Dryas & the Holocene. *In*: STABELL, B. & THIEDE, J. *Norsk Geologisk Tidsskrift*, **65**, 49–57.

FREDÉN, C. 1982. An outline of the marine stage of the Väner Basin. *In*: OLAUSSON, E. *Sveriges Geologiska Undersökning*, **C794**, 16–26. Geological notes on the Moltemyr, Brastad and Solberga areas. *In*: OLAUSSON, E. *Sveriges Geologiska Undersökning*, **C794**, 34–5.

—— 1988. Marine life and deglaciation chronology of the Vanern basin, southwestern Sweden. *Sveriges Geologiska Undersökning*, Ca 71, 1–80.

GARD, G. 1986. Calcareous nannofossil biostratigraphy of late Quaternary Arctic sediments. *Boreas*, **15**, 217–29.

—— 1988. Late Quaternary calcareous nannofossil biozonation, chronology and palaeo-oceanography in areas north of the Faeroe-Iceland Ridge. *Quaternary Science Reviews*, **7**, 65–78.

HARLAND, R. 1982. Dinoflagellate cysts. *In*: OLAUSSON, E. *Sveriges Geologiska Undersökning*, **C794**, 211–23.

JANSEN, E. & BJØRKLUND, K.R. 1985. Surface ocean circulation in the Norwegian Sea, 15000 BP to present. *Boreas*, **14**, 243–257.

KNUDSEN, K.L. 1982. Foraminifers. *In*: OLAUSSON, E. *Sveriges Geologiska Undersökning*, **C794**, 148–177.

LORD, A.R. 1982. Ostracods. *In*: OLAUSSON, E. *Sveriges Geologiska Undersökning*, **C794**, 137–147.

MCINTYRE, A., KIPP, N., BÉ, A.W.H., CROWLEY, T., GARDNER, J.V., PRELL, W.L. & RUDDIMAN, W.F. 1976. Glacial North Atlantic 18000 years ago: a CLIMAP reconstruction. *In*: CLINE, R.M. & HAYS, J.D. (eds) *Investigations of Late Quaternary Paleoceanography and Paleoclimatology*. Geological Society of America Memoir, **145**, 43–76.

MIKKELSEN, N. 1985. Late Quaternary evolution of the Skagerrak area as mirrored by calcareous nannoplankton. *In*: STABELL, B. & THIEDE, J. *Norsk Geologisk Tidsskrift*, **65**, 87–90.

—— & PERCH-NIELSEN, K. 1982. Calcareous Nannofossils. *In*: OLAUSSON, E. *Sveriges Geologiska Undersökning*, **C794**, 178–186.

MILLER, U. 1982. Diatoms. *In*: OLAUSSON, E. *Sveriges Geologiska Undersökning*, **C794**, 187–210.

MÖRNER, N.-A. (ed.) 1976. The Pleistocene/Holocene boundary: a proposed boundary-stratotype in Gothenburg, Sweden. *Boreas*, **5**, 193–275.

NAGY, J. & OFSTAD, K. 1980. Quaternary foraminifera and sediments in the Norwegian Channel. *Boreas*, **9**, 39–52.

—— & QVALE, G. 1985. Benthic foraminifers in Upper Quaternary Skagerrak deposits. *In*:

STABELL, B. & THIEDE, J. *Norsk Geologisk Tidsskrift*, **65**, 107–113.

OLAUSSON, E. (ed.) 1982a. The Pleistocene/Holocene boundary in south-west Sweden. *Sveriges Geologiska Undersökning*, **C794**, 1–288.

—— 1982b. General comments on the Pleistocene/Holocene boundary. *In*: OLAUSSON, E. *Sveriges Geologiska Undersökning*, **C794**, 10–15.

—— 1982c. Stable Isotopes. *In*: OLAUSSON, E. *Sveriges Geologiska Undersökning*, **C794**, 82–92.

—— 1985. The glacial oceans. *Palaeogeography, Palaeoclimatology, Palaeoecology*, **50**, 291–301.

PAABO, K. 1985. Diatomological studies of two cores from the western Baltic. *In*: BORG, G.C. *Striae*, **23**, 83–91.

QVALE, G. 1985. Ostracods in Upper Quaternary Skagerrak deposits. *In*: STABELL, B. & THIEDE, J. *Norsk Geologisk Tidsskrift*, **65**, 119–22.

ROBERTSSON, A.-M. 1982. Pollen. *In*: OLAUSSON, E. *Sveriges Geologiska Undersökning*, **C794**, 224–242.

ROSENFELD, A. 1977. Die rezenten Ostracoden-Arten in der Ostsee. *Meyniana*, **29**, 11–49.

RUDDIMAN, W.F. 1987. Atlantic circulation, surface effects on deep flow. *Nature*, **330**, 17–18.

—— & MCINTYRE, A. 1981. The North Atlantic Ocean during the last deglaciation. *Palaeogeography, Palaeoclimatology, Palaeoecology*, **35**, 145–214.

SEJRUP, H.P., AARSETH, I., ELLINGSEN, K.L., REITHER, E., JANSEN, E., LØVLIE, R., BENT, A., BRIGHAM-GRETTE, J., LARSEN, E. & STOKER, M. 1987. Quaternary stratigraphy of the Fladen area, central North Sea: a multidisciplinary study. *Journal of Quaternary Science*, **2**, 35–58.

STABELL, B. 1985. Diatoms in Upper Quaternary Skagerrak sediments. *In*: STABELL, B. & THIEDE, J. *Norsk Geologisk Tidsskrift*, **65**, 91–95.

—— & THIEDE, J. (eds) 1985a. Upper Quaternary marine Skagerrak (NE North Sea) deposits: Stratigraphy & depositional environment. *Norsk Geologisk Tidsskrift*, **65**, 1–149.

—— & —— 1985b. The physiographic evolution of the Skagerrak during the past 15 000 years: Palaeobathymetry and palaeogeography. *In*: STABELL, B. & THIEDE, J. *Norsk Geologisk Tidsskrift*, **65**, 19–22.

—— & —— 1986. Palaeobathymetry and palaeogeography of southern Scandinavia in the late Quaternary. *Meyniana*, **38**, 43–59.

THIEDE, J. 1985. Planktonic foraminifers in Upper Quaternary Skagerrak sediments. *In*: STABELL, B. & THIEDE, J. *Norsk Geologisk Tidsskrift*, **65**, 1–149, 115–118.

—— 1987. The Late Quaternary Skagerrak and its depositional environment. *Boreas*, **16**, 425–432.

Sedimentology and micropalaeontology of glacimarine sediments from the Central and Southwestern Celtic Sea

J.D. SCOURSE[1], W.E.N. AUSTIN[1], R.M. BATEMAN[2], J.A. CATT[2], C.D.R. EVANS[3], J.E. ROBINSON[4] & J.R. YOUNG[5]

[1] *School of Ocean Sciences, University College of North Wales, Menai Bridge, Gwynedd, LL59 5EY, UK*

[2] *Soils Department, AFRC Institute of Arable Crops Research, Rothamsted Experimental Station, Harpenden, Hertfordshire, AL5 2JQ, UK*

[3] *British Geological Survey, Keyworth, Nottingham, NG12 5GG, UK*

[4] *Department of Geological Sciences, University College, Gower Street, London, WC1E 6BT, UK*

[5] *British Museum (Natural History), Cromwell Road, London, SW7 5BD, UK*

Abstract: Thin discontinuous glacigenic sediments occur at or close to the sea bed as far south as 49°N in the Celtic Sea. The northern samples (facies A) are clast-rich, overconsolidated lodgement tills or proximal glacimarine sediments containing sparse reworked microfaunas. The southern samples (facies B) are distal glacimarine plastic silty clays containing abundant cold water microfaunas. Both these facies are correlated with the Late Devensian Scilly Till, enabling a quantitative reconstruction of ice thicknesses, grounding line, sea-level and shoreline elevations in the Celtic Sea at 19 000 years BP. The ice advance terminated in marine waters towards the shelf edge break, and is likely to have constituted a thin lobate surge over deformable marine sediments.

Over the last two decades data has been collected on the Quaternary sediments of the Irish and Celtic Seas by the British Geological Survey (BGS). Several publications arising directly from this work have discussed the environmental and stratigraphical interpretation of the offshore Quaternary sequence, and the correlation of this sequence with the much-studied but controversial onshore stratigraphy (Garrard & Dobson 1974; Delantey & Whittington 1977; Garrard 1977; Pantin & Evans 1984). The occurrence of units interpreted as tills led to a number of suggested ice limits in the region (Fig. 1), but these were mutually conflicting and unclear concerning the limits south of St. George's Channel. Recent data from the Isles of Scilly suggest that the onshore glacigenic sediments there are of Late Devensian age (Scourse 1985, 1989) and not Wolstonian as previously suggested (Mitchell & Orme 1967). This stimulated a detailed reappraisal of thin discontinuous glacigenic sequences found to the south and west of the Isles of Scilly by the BGS during routine sampling surveys between 1974 and 1981 (Fig. 1). This paper presents results of detailed analysis of these sequences which have implications for Late Pleistocene palaeoenvironments in this area. These data come only from the UK sector of the shelf; complementary data from the Irish and French sectors are still needed.

Geographical and geological context

The glacigenic samples were recovered using vibrocorers and were positioned using Decca Main Chain or Pulse 8 navigation systems (Pantin & Evans 1984). They were located between 300 and 500 km southwest of the suggested Devensian limits in the St. George's Channel area (Fig. 1). The fourteen coring sites yielding glacigenic material were situated between the submarine Haig Fras granite outcrop and the shelf edge break to the southwest, which in this region lies at the relatively low level of −185 to −205 m Ordnance Datum (OD; Pantin & Evans 1984). The continental shelf in this area dips gently to the southwest and is almost featureless apart from large linear tidal sand ridges trending SW/NE (Figs 2 & 3; Stride 1963; Bouysse *et al.* 1976; Pantin &

From Dowdeswell, J.A. & Scourse, J.D. (eds), 1990, *Glacimarine Environments: Processes and Sediments*. Geological Society Special Publication No 53, pp 329–347.

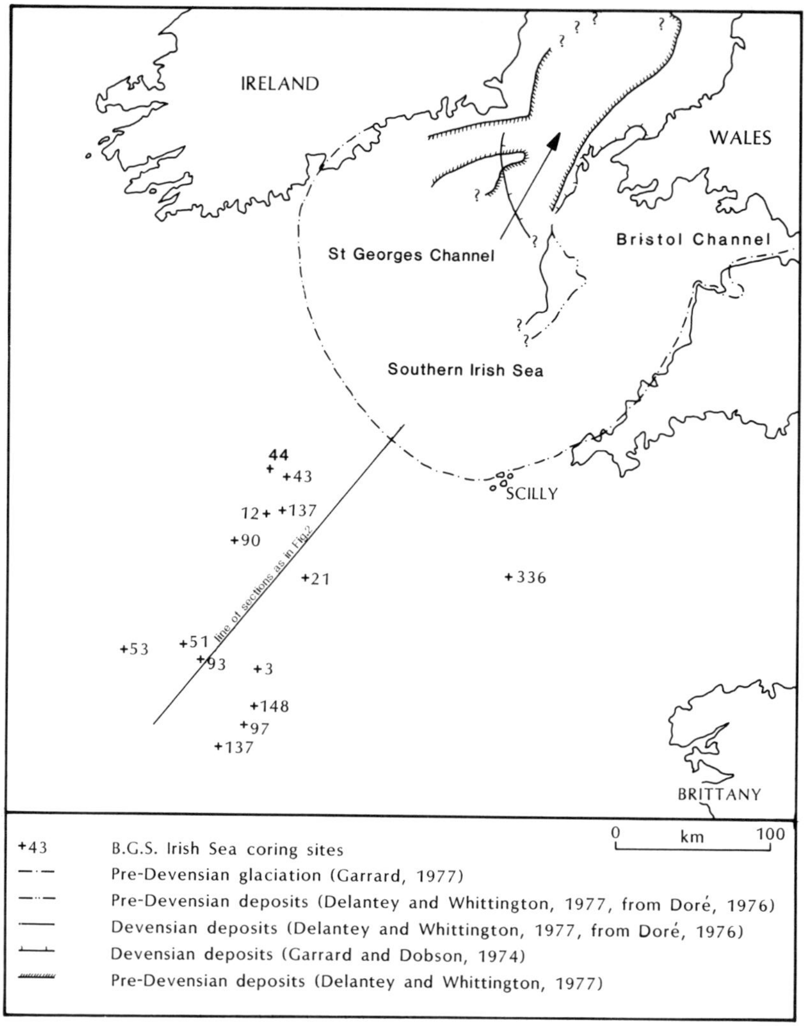

Fig. 1. Central and Southwestern Celtic Sea: location map, BGS vibrocoring sites and postulated glacial limits based on offshore evidence.

Evans 1984; Belderson *et al.* 1986). Pantin & Evans also reported small 'mounds' of possible glacigenic material and scattered boulders on the sea bed as revealed by side-scan sonar, indicating probable ice-rafting. The occurrence of scattered boulders was also reported by Hamilton *et al.* (1980).

Water depth increases from northeast to southwest. The deepest sample, 48/09/137, from beyond the defined 'shelf edge break' (Fig. 2) at 48°N latitude, 9°W longitude, (BGS sample 137), was also the furthest south sample, cored at a depth of −211 m OD; the shallowest, sample 49/09/44, was from a depth of −125 m OD.

There is some evidence for glacimarine sedimentation in deeper water beyond the shelf edge break. Day (1959) reported possible glacimarine sediments at 49°39′N 10°35′W in 1419 m of water on the continental slope, and pebbles have been recovered from the top few metres of late Quaternary age sediment at DSDP site 548 on the Goban Spur in 1251 m of water (Graciansky *et al.* 1985). Ice-rafted material has also been recovered from the open ocean to the west (Ruddiman & McIntyre 1973; 1981), and as far south as the continental slopes off Morocco and Portugal (Kudrass 1973), but the precise age of this material is uncertain.

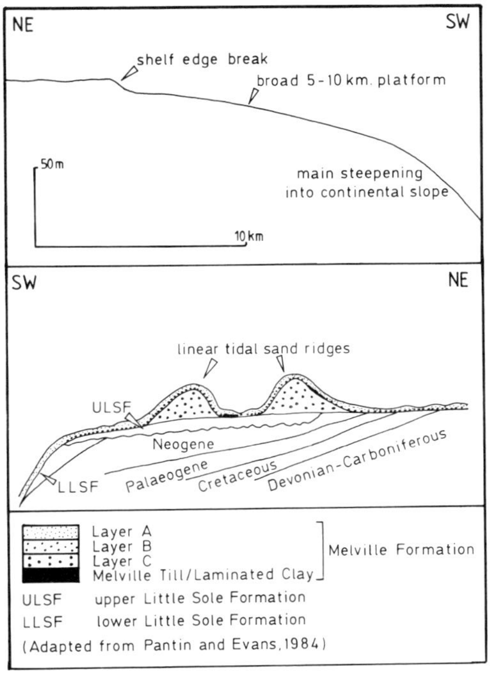

Fig. 2. Central and Southwestern Celtic Sea: continental shelf edge morphology (above) and stratigraphy (below). The Melville Formation consists of four units, layers A, B and C, and the Melville Till/Laminated Clay. Layers A and B overlie the Melville Till/Laminated Clay which is in turn underlain by layer C. The linear tidal sand ridges are largely comprised of layer C. The upper and lower Little Sole Formation are of Pliocene/early Pleistocene age (Evans & Hughes 1984). Vertical scale greatly exaggerated.

Stratigraphy

Pantin & Evans (1984) identified two main Late Cenozoic formations in the Central and Southwestern Celtic Sea, the late Pliocene/early Pleistocene Little Sole Formation and the Late Devensian/early Flandrian Melville Formation (Figs 2 & 3). The Melville Formation consists of the linear tidal sand ridges and intervening sand sheets in addition to the glacigenic material described here. No core penetrated through the glacigenic samples of the Melville Formation into any underlying unit, but it is assumed that the majority of the samples overlie Neogene strata of the Jones (early to mid-Miocene) or Little Sole (Pliocene/early Pleistocene) Formations (Fig. 4; Evans & Hughes 1984).

Most of the glacigenic samples were recovered from bathymetric 'lows' between the linear tidal sand ridges, and one sample, 49/$\overline{09}$/44, was from the flank of a sand ridge (Figs 2 & 3). The recent sediments overlying the glacigenic material have been classified into two layers 'A' and 'B' (Pantin & Evans 1984). Layer A consists of superficial mobile sediments, and layer B is a relatively coarse gravel pavement underlying layer A. An additional layer, C, underlies the glacigenic material and comprises the bulk of material forming the sand ridges. Boulders over 1 m in diameter are widely scattered across the area; these have been retrieved on the anchors of drilling ships (Scourse *et al.* 1989), are identifiable on side-scan sonar and have been observed in submarine photographs (Hamilton *et al.* 1980). They are probably the larger clasts of layer B.

At most sites the vibrocorer was unable to penetrate more than two metres into the sea bed due to the coarse grained character of both layer B and the glacigenic material. At a few sites layers A and B and a few centimetres of the underlying glacigenic sediment were recovered; at 49/$\overline{09}$/44, however, 3.3 m of glacigenic sediment was recovered. This was also the only site which showed any up-core change of lithology; a 15 cm thick lower subunit of overconsolidated clast-rich diamict was overlain by 3.15 m of plastic silty clay.

Both sampling and acoustic devices suggested that the glacigenic sediments do not occur as a sheet, but form isolated patches or mounds, separated by the tidal sand ridges and exposures of non-glacigenic Melville Formation units.

Analyses

The fourteen samples described initially by the BGS as 'glacigenic' were analysed in detail using a variety of techniques as described in Scourse (1985). The results of lithological analyses are presented in Fig. 5, and mineral analyses of the silt fraction in Table 1. Palaeontological results are summarized in Figs 6 (Ostracoda) and 7 (Foraminifera), and Tables 2 (Mollusca) and 3 (Cretaceous calcareous nannoplankton from erratic pebbles of Chalk).

Facies classification and distribution

The various characteristics and analyses of the samples described enable their classification into two facies, 'A' and 'B'. Facies A has been defined lithostratigraphically as the Melville Till and facies B as the Melville Laminated Clay, both members of the Melville Formation (Scourse 1985).

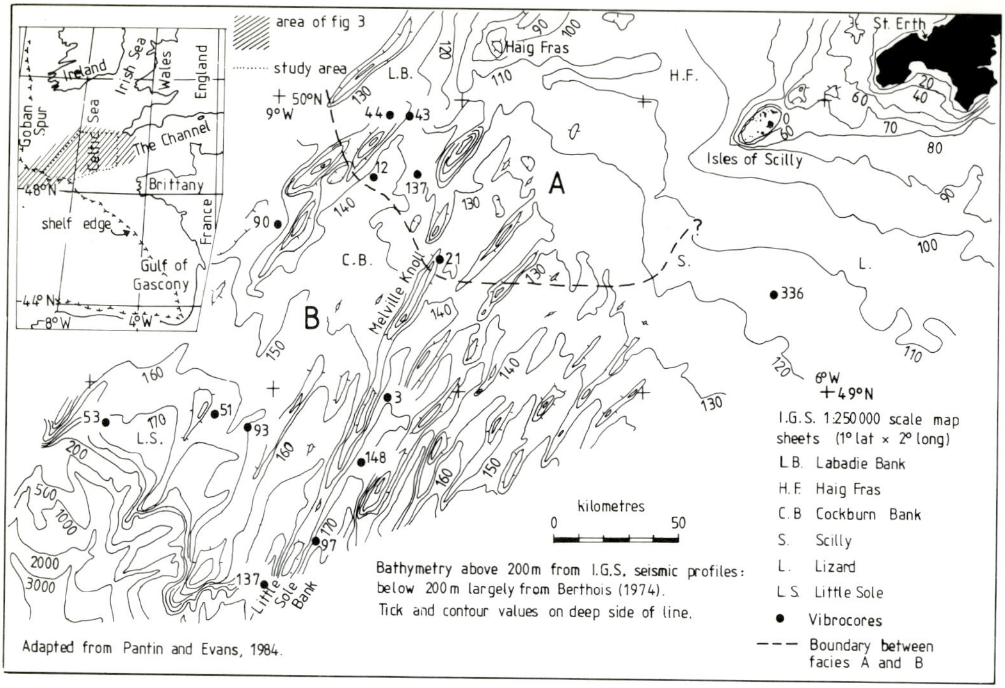

Fig. 3. Central and Southwestern Celtic Sea: distribution of linear tidal sand ridges, boundary between facies A and B, and vibrocoring sites.

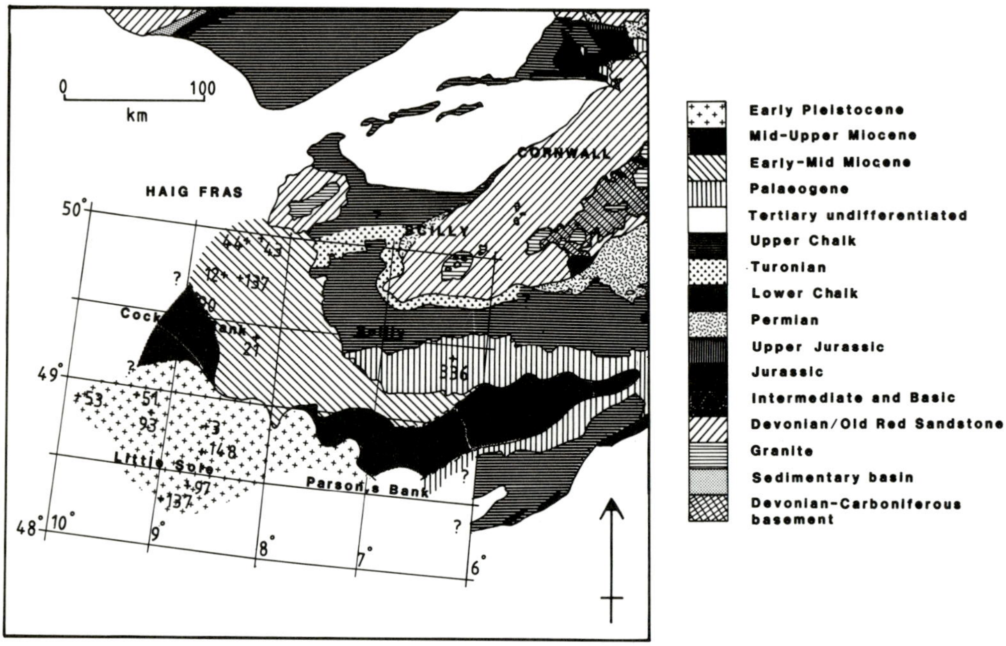

Fig. 4. Central and Southwestern Celtic Sea: solid geology. The Mid–Upper Miocene is the Cockburn Formation, and the Early–Mid Miocene is the Jones Formation (Evans & Hughes 1984). Maps compiled from data in Evans & Hughes (1984) and Martindale & Chesher (1979).

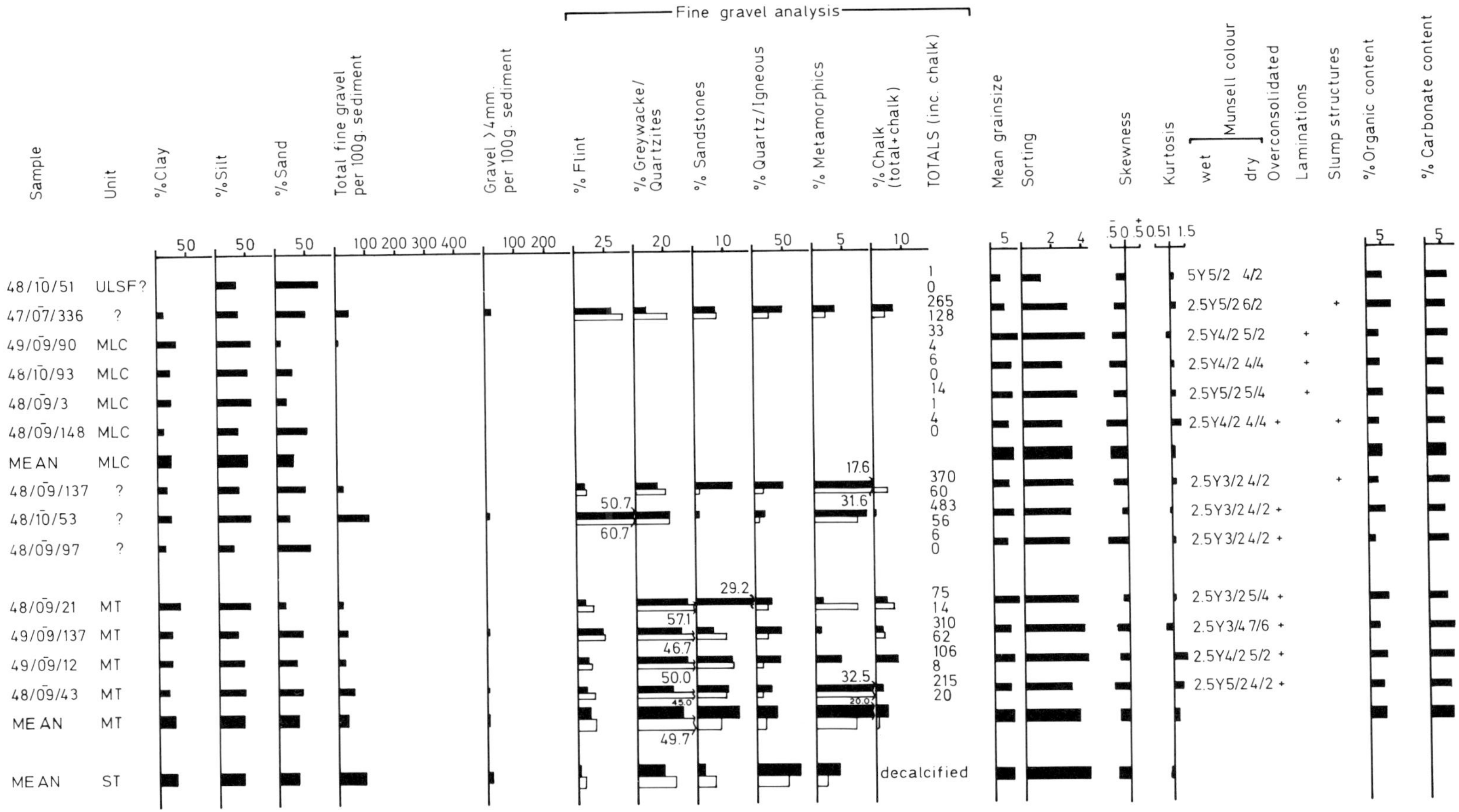

Fig. 5. Granulometric, fine gravel, and other physical data from the Celtic Sea samples, and comparable mean data from the Scilly Till, Isles of Scilly. The Melville Till is facies A and the Melville Laminated Clay is facies B. UISF, Upper Little Sole Formation; MLC, Melville Laminated Clay; MT, Melville Till; ST Scilly Till.

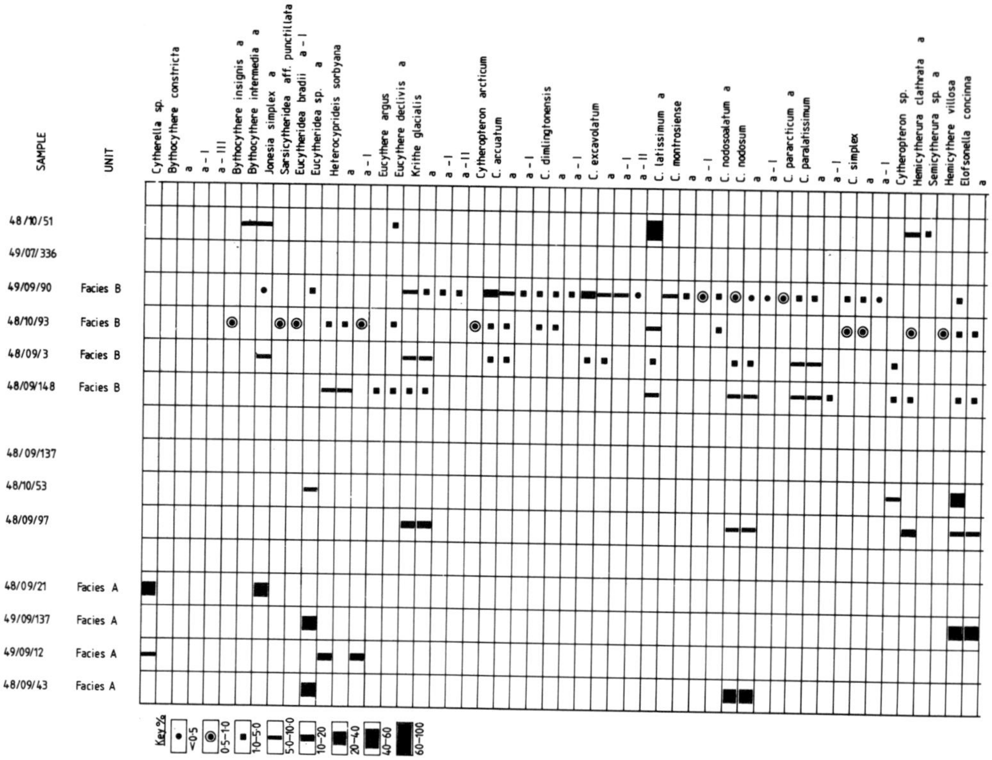
SAMPLE
UNIT
Cytherella sp.
Bythocythere constricta
a
a - I
a - III
Bythocythere insignis a
Bythocythere intermedia a
Jonesia simplex a
Sarsicytheridea aff. punctillata
Eucytheridea bradii a - I
Eucytheridea sp. a
Heterocyprideis sorbyana
a
a - I
Eucythere argus
Eucythere declivis a
Krithe glacialis
a
a - I
a - II
Cytheropteron arcticum
C. arcuatum
a
a - I
C. dimlingtonensis
a
a - I
C. excavolatum
a
a - I
a - II
C. latissimum a
C. montrosiense
a
a - I
C. nodosoalatum a
C. nodosum
a
a - I
C. pararcticum a
C. paralatissimum
a
a - I
C. simplex
a
a - I
Cytheropteron sp.
Hemicytherura clathrata a
Semicytherura sp. a
Hemicythere villosa
Elofsonella concinna
a
48/10/51
49/07/336
49/09/90 Facies B
48/10/93 Facies B
48/09/3 Facies B
48/09/148 Facies B
48/09/137
48/10/53
48/09/97
48/09/21 Facies A
49/09/137 Facies A
49/09/12 Facies A
48/09/43 Facies A
Key %
<0·5
0·5–1·0
1·0–5·0
5·0–10·0
10–20
20–40
40–60
60–100

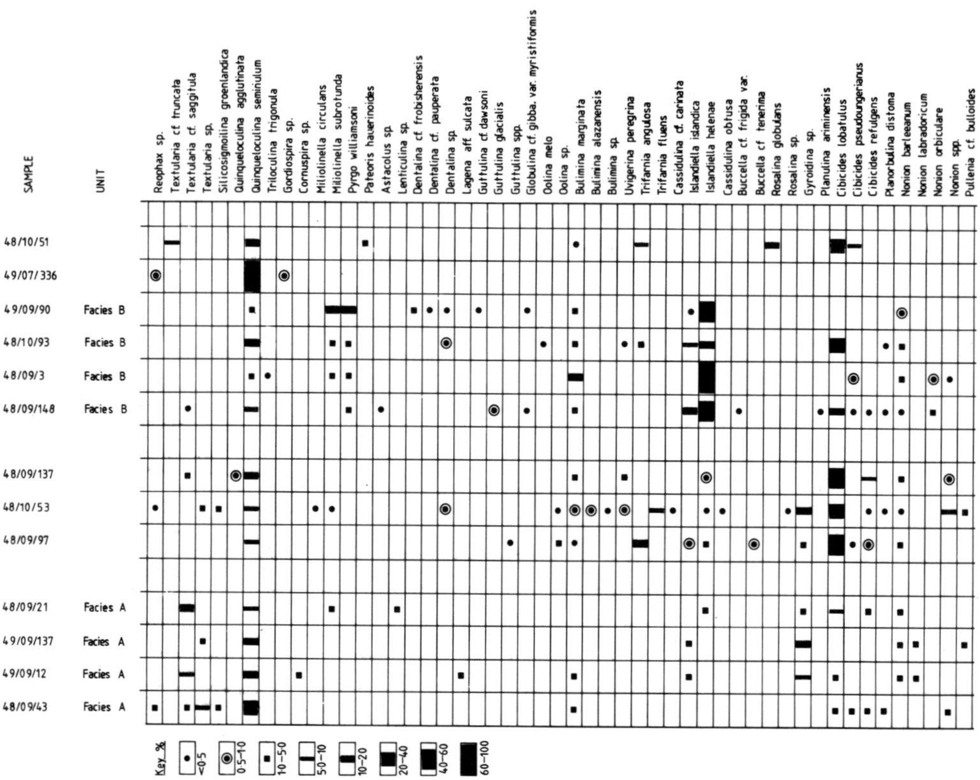
SAMPLE
UNIT
Reophax sp.
Textularia cf. truncata
Textularia cf. saggitula
Textularia sp.
Silicosigmoilina groenlandica
Quinqueloculina agglutinata
Quinqueloculina seminulum
Triloculina trigonula
Gordiospira sp.
Cornuspira sp.
Miliolinella circularis
Miliolinella subrotunda
Pyrgo williamsoni
Pateoris hauerinoides
Astacolus sp.
Lenticulina sp.
Dentalina cf. frobisherensis
Dentalina cf. pauperata
Dentalina sp.
Lagena aff. sulcata
Guttulina cf. dawsoni
Guttulina glacialis
Guttulina spp.
Globulina cf. gibba var. myristiformis
Oolina melo
Oolina sp.
Bulimina marginata
Bulimina alazanensis
Bulimina sp.
Uvigerina peregrina
Trifarina angulosa
Trifarina fluens
Cassidulina cf. carinata
Islandiella islandica
Islandiella helenae
Cassidulina obtusa
Buccella cf. frigida var.
Buccella cf. tenerrima
Rosalina globularis
Rosalina sp.
Gyroidina sp.
Planulina ariminensis
Cibicides lobatulus
Cibicides pseudoungerianus
Cibicides refulgens
Planorbulina distoma
Nonion barleeanum
Nonion labradoricum
Nonion orbiculare
Nonion spp.
Pullenia cf. bulloides
48/10/51
49/07/336
49/09/90 Facies B
48/10/93 Facies B
48/09/3 Facies B
48/09/148 Facies B
48/09/137
48/10/53
48/09/97
48/09/21 Facies A
49/09/137 Facies A
49/09/12 Facies A
48/09/43 Facies A
Key %
<0·5
0·5–1·0
1·0–5·0
5·0–10
10–20
20–40
40–60
60–100

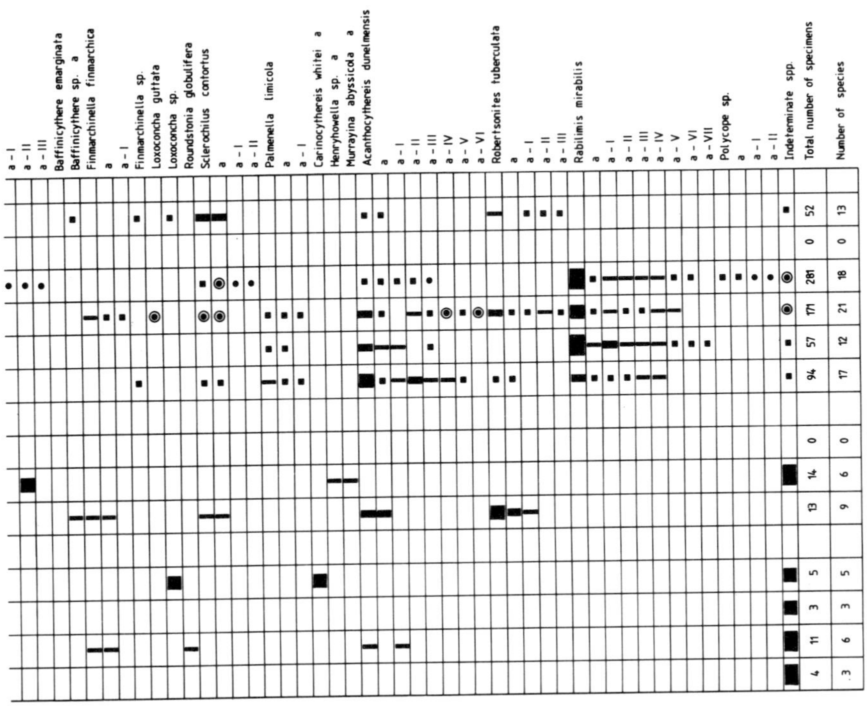

Fig. 6. Ostracoda from the Celtic Sea samples.

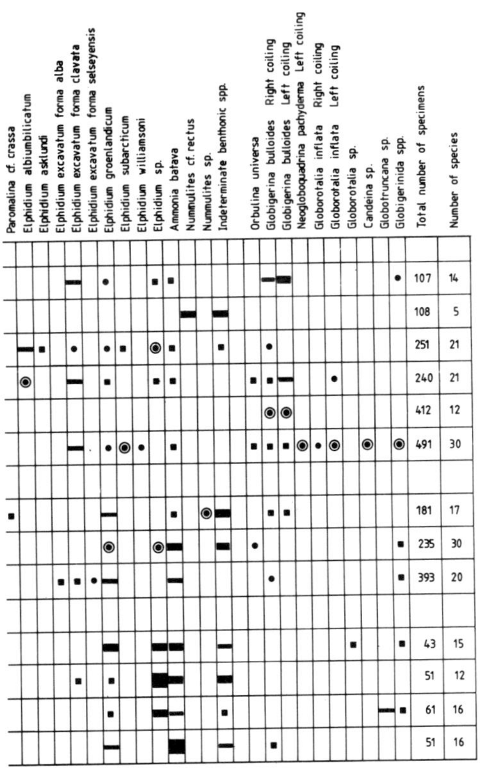

Fig. 7. Foraminifera from the Celtic Sea samples.

Table 1. *Silt mineralogy of the Scilly Till and Celtic Sea samples. Figures in percentages of light and heavy fractions*

	Scilly Till				Celtic Sea Samples							
	St Martin's (SV923171)	St Martin's (SV940158)	St Martin's (SV920158)	Mean	49/$\overline{09}$/12	49/$\overline{09}$/43	49/$\overline{09}$/90	48/$\overline{09}$/3	49/$\overline{09}$/12	48/$\overline{10}$/148	48/$\overline{09}$/137	Mean
(a) *Light fraction* (%)												
Quartz	64.5	63.7	60.9	63.0	75.3	70.6	68.5	71.6	67.7	71.1	66.6	70.2
Felspar	13.7	13.8	12.2	13.2	15.8	17.5	21.8	19.5	20.3	15.7	22.0	18.9
Calcite	0.1	–	0.5	0.2	–	–	–	–	–	–	–	–
Muscovite	9.5	11.5	13.2	11.4	5.0	5.0	7.3	6.1	8.1	9.1	8.3	6.9
Glauconite	7.9	6.2	9.2	7.8	1.2	3.3	0.9	0.1	0.2	1.7	0.4	1.1
Flint	4.3	4.8	4.0	4.4	2.7	3.7	1.6	2.7	3.6	2.4	2.7	2.8
(b) *Heavy fraction* (%)												
Epidote	32.7	20.1	18.2	23.7	14.1	11.8	20.1	18.2	23.5	14.3	10.5	14.7
Zoisite	2.0	1.1	1.0	1.4	3.3	4.0	7.5	7.4	6.6	4.6	6.2	5.7
Zircon	10.3	14.5	10.2	11.7	4.4	10.1	10.6	7.2	10.6	31.6	6.6	11.6
Tourmaline	5.8	3.7	4.8	4.8	2.8	2.9	5.0	6.5	3.5	3.5	3.6	4.0
Chlorite	17.9	42.6	47.0	35.8	17.4	17.9	10.3	24.3	10.7	10.2	11.8	14.6
Biotite	3.1	1.5	2.4	2.3	1.2	0.6	0.2	0.2	0.3	–	1.0	0.7
Green hornblende	6.3	5.9	4.4	5.5	12.5	16.3	25.0	22.7	22.7	11.9	12.1	17.6
Brown hornblende	0.1	0.7	0.2	0.3	1.0	0.2	0.6	0.9	1.0	0.6	0.3	0.7
Tremolite/actinolite	5.1	1.6	2.8	3.2	3.0	5.0	4.6	2.2	4.5	3.1	4.8	3.9
Augite	6.9	–	–	2.3	32.0	20.3	2.8	0.4	2.2	4.9	32.3	13.6
Garnet	2.5	0.6	1.5	1.5	3.3	4.2	6.1	3.8	9.0	6.0	5.5	5.4
Yellow rutile	2.5	5.1	3.3	3.6	1.7	3.0	2.5	1.9	1.9	4.3	1.7	2.4
Brown rutile	0.6	–	–	0.2	1.0	1.4	1.2	0.9	1.7	1.9	1.0	1.3
Anatase	2.2	2.2	1.7	2.0	0.2	0.8	0.6	0.6	0.1	0.3	0.6	0.5
Brookite	0.4	–	0.4	0.3	–	–	–	0.4	–	0.1	0.1	0.1
Brown spinel	–	–	0.4	0.1	–	–	–	–	–	–	–	–
Apatite	0.1	–	–	0.1	–	–	–	–	–	–	–	–
Sphene	1.0	0.4	1.3	0.9	–	–	–	–	–	–	–	–
Olivine	0.1	–	–	0.1	–	–	–	–	–	–	–	–
Staurolite	–	–	0.4	0.1	0.4	0.5	1.4	1.1	1.2	1.6	0.9	1.0
Kyanite	0.4	–	–	0.1	0.7	0.2	1.5	1.3	0.4	1.0	0.3	0.8
Andalusite	–	–	–	–	–	0.2	–	–	–	–	–	0.1
Hypersthene	–	–	–	–	1.0	0.8	–	–	–	0.1	0.7	0.4

Table 2. *Mollusca from the Celtic Sea samples*

Sample	Molluscs present
49/$\overline{09}$/90	Unidentifiable fragments
48/$\overline{10}$/93	*Nuculana pernula* *Nuculana minuta* *Arctica islandica* *Abra prismatica* *Macoma* cf. *calcarea* *Astarte montagui* *Chlamys* cf. *islandica* *Cylichna* sp. *Siphonodentalium* sp.
48/$\overline{09}$/3	*Nuculana* cf. *pernula* *Yoldiella fraterna*
48/$\overline{09}$/148	*Nuculana* cf. *pernula*
49/$\overline{09}$/21	*Hiatella* sp.
49/$\overline{09}$/137	Unidentifiable fragments
40/$\overline{10}$/93	Unidentifiable fragments
49/$\overline{09}$/43	Unidentifiable fragments
48/$\overline{09}$/137	*Spisula* sp. *Nuculana* sp. *Astarte* sp. *Chlamys* sp. *Venus* ovata
48/$\overline{10}$/53	*Astarte sulcata* *Chlamys* ct. *islandica*
48/$\overline{09}$/97	Unidentifiable fragments
48/$\overline{10}$/51	*Colus gracilis* *Nucula nucleus* *Anomia ephippium* *Anomia squamula* *Chlamys distorte* *Modiolus modiolus* *Nassarius incrassatus* *Abra* sp. *Macoma* sp. *Anomia* sp.
47/$\overline{07}$/336	*Emarginula reticulata* *Alvania cancellata* *Gibbula tumida* *Timoclea ovata* *Venus fasciata* *Chlamys opercularis* *Venus ovata* *Glycymeris* sp. *Anomia* sp. *Zizyphinus* sp. *Philbertia* sp.

Facies A

This facies is represented by samples 49/$\overline{09}$/43, 49/$\overline{09}$/12, 49/$\overline{09}$/21 and 49/$\overline{09}$/137 and the lower subunit of 49/$\overline{09}$/44. It is overconsolidated (Lambert & Khowaja 1978), structurally homogeneous, containing abundant fine gravel (between 8 and 54 granules per 100 g sediment), and abundant pebbles $>$ 4 mm. The matrix is very poorly sorted and consistently coarse skewed. The samples are either devoid of microfauna, or contain a few abraded and broken specimens of both temperate and arctic affinities most of which can be attributed to reworking. Two of the samples contain *Hiatella* sp. and three contain barnacle fragments which are probably reworked. The clast lithological assemblage is fairly consistent, with greywackes/quartzites $>$ quartz/igneous $>$ metamorphics $>$ sandstones $>$ flint $>$ chalk.

Facies B

Facies B is represented by samples 48/$\overline{09}$/148, 49/$\overline{09}$/90, 48/$\overline{10}$/93 and 49/$\overline{09}$/3. These samples are not overconsolidated but plastic silty clays, containing very small amounts of fine gravel (between 0 and 6 granules per 100 g sediment), and sometimes display well developed fining-upwards laminae and sand pods. The matrices are moderately to moderately-poorly sorted, and consistently very coarse skewed.

Facies B stands apart from A most distinctively in containing a rich ostracod (Fig. 6) and foraminiferal (Fig. 7) fauna. Samples 48/$\overline{09}$/3, 48/$\overline{09}$/148, 48/$\overline{10}$/93 and 49/$\overline{09}$/90 contain species which today are not found south of the Arctic Circle. The ostracod *Rabilimis mirabilis* (Brady) does not occur south of the fjords of East Greenland (76°N) or the inlets of the White Sea coast (Hazel 1971; Lev 1969; Neale & Howe 1975). *Krithe glacialis* (Brady, Crosskey & Robertson) is similarly distributed, and *Cytheropteron montrosiense* Brady, Crosskey & Robertson is not known alive anywhere today but appears to be Arctic by its association with other organisms elsewhere (Whatley & Masson 1979).

Each of these Arctic species is represented by a wide range of valve sizes constituting a growth series of instars making up almost a complete life history; *R. mirabilis*, for example, is represented by seven moults in sample 48/$\overline{09}$/3 (see Plate 2 in Scourse *et al.* 1989). This indicates deposition under extremely quiet conditions with practically no current activity, otherwise the wide range of valve sizes would have been dispersed. It also suggests that the faunas are

Table 3. *Calcareous nannofossils from erratics of Cretaceous chalk*

Samples		12	21a	21b	137a	137b
Overall abundance		C	A	C	A	F
Preservation		02	01	02	03	03
Species	Range					
Biscutum sp.	EJur –Mst		R	R		
Braarudosphaera sp.	Jur –Trt	R				
Broinsonia enormis	**Cen** –Cmp		F		R	
Chiastozygus literarius	Het –Mst		R		R	
Cribrosphaerella ehrenbergi	Alb –Mst	R	F	F		
Eiffelithus eximius	**LTur** –Cmp		R		R	
Eiffelithus turriseiffelii	LAlb –Mst		C	F	C	
Eprolithus floralis	Alb –**ESnt**	R	F	R	R	F
Gartnerago obliquum	**Cen** –Cmp		R			
Helicolithus trabeculatus	Alb –Mst		F	R		
Lithraphidites carniolensis	Het –Mst	C	C	C	C	
Lucianorhabdus quadrifidus	**Tur** –Cmp		R			
Manivitella pemmatoidea	Het –Mst		R	R		
Microrhabdulus decoratus	**LCen** –Mst		R			
Nannoconus multicadus	**Tur** –Cmp		R	C	F	
Nannoconus spp.	LJur –Mst	F	C	C	C	
Prediscosphaera cretacea	Cen –Mst		F	C		
Quadrum gartneri	**Tur** –**Snt**		R	R		
Retecapsa angustiforata	Het –Mst	R	F	R	F	R
Stradneria crenulata	Het –Mst	R	C	R	F	R
Tranolithus phacelosus	Alb –Cmp	R	C	F	F	
Watznaueria barnesae	Jur –Mst	C	A	C	C	F
Zeugrhabdotus embergeri	LJur –Mst		R	R		
Zygodiscus spp.	Jur –Mst	R	F	R	F	

Abundance symbols (for overall assemblage and individual species): A, abundant, <10 specimens per field of view; C, common, 1–10 specimens per field of view; F, few, 1 specimen per 1–10 fields of view; R, rare, 1 specimen per 10–100 fields of view. *Preservation*: 01–03 mild-severe overgrowth. *Species names*: The nomenclature of Perch-Nielsen (1985) is used, but with somewhat broader species concepts. *Ranges*: Reliable ranges of species within taxonomic concepts adopted. Biostratigraphically useful ranges emboldened; Trt, Tertiary; Mst, Maastrichtian; Cmp, Campanian; Snt, Santonian; Con, Coniacian; Tur, Turonian; Cen, Cenomanian; Alb, Albian; Apt, Aptian; Bar, Barremian; Hau, Hauterivian; Val, Valanginian; Ber, Berriasian; Jur, Jurassic. *Samples*: 12, clast from 49/$\overline{09}$/12; 21a, b two separate clasts from 49/$\overline{09}$/21; 137a, b two separate clasts from 49/$\overline{09}$/137.

not the result of reworking from older deposits. The faunas are undiluted by the addition of any 'temperate' species, but contain additional species of *Cytheropteron, Acanthocythereis, Elofsonella, Heterocyprideis* and *Jonesia* which suggest affinities with the pre-Ipswichian, cold, open sea fauna of the Bridlington Crag of Holderness (Catt & Penny 1966; Neale & Howe 1975), at least in terms of depositional environment, although not necessarily of age.

These faunas are a marked contrast to the species found living in the area today, which are dominated by *Bythocythere turgida, Carinocythereis antiquata, Celtia quadridentata, Loxoconcha multiflora* and *Pterogocythereis jonesi*. None of these were found in facies B.

Cytheropteron arcuatum Brady, Crosskey & Robertson was described from several glacial sites in Scotland (Brady *et al.* 1874) but has never been found living. Other records for the species include the pre-Ipswichian marine Bridlington Crag, the Clyde Beds of the west coast of Scotland including Loch Creran at Shian Ferry (Peacock 1971), and the Pleistocene deposits of the Forties Field in the North Sea (Whatley & Masson 1979).

Cytheropteron excavoalatum Whatley & Masson is a distinctively winged and pitted species described from the Pleistocene of the Forties Field (Whatley & Masson 1979). It is not known as a living species though dead valves have been collected from Cumberland Inlet, Baffin Island (Brady & Norman 1889), and it is undoubtedly an arctic indicator from its associations.

Cytheropteron montrosiense Brady, Crosskey

& Robertson was first described from the Clyde Beds at Campbelltown and the Errol Beds of Tayside (Brady *et al.* 1874). It is also not known as a living species but is clearly an arctic indicator from its associations.

There is much debate about the validity of species of the genus *Krithe* as simple morphological assessments of species are complicated by sexual dimorphism and possibly trimorphism. However, *Krithe glacialis* Brady, Crosskey & Robertson is particularly distinctive, with an abruptly truncated postero-dorsal margin and characteristic duplicature detail. It was originally described from the Errol Clay where it occurs in a turbid-water soft substrate palaeoenvironment close to an ice-front (Paterson *et al.* 1981).

The abundance of the foraminifers *Islandiella helenae* Feyling-Hanssen & Buzas and *I. islandica* (Nørvang), species which constitute the 'High Arctic' assemblage of Feyling-Hanssen (1982), supports the low-water temperatures suggested by the ostracods. Planktonic forams are much more common in facies B than in facies A; the planktonic/benthonic ratio is 0.08 in sample 48/$\overline{09}$/148. Although sample 49/$\overline{09}$/90 contained five valves of the mollusc *Yoldiella (=Portlandia) fraterna* (Verrill & Bush), a boreo-arctic species (Nordsieck 1969), the molluscs are less conclusively cold indicators than the ostracods. In sample 48/$\overline{10}$/93, for instance, the occurrence of *Arctica islandica* in a molluscan assemblage similar to that of the interstadial Clyde Beds of western Scotland (Peacock 1989) indicates relatively warm surface water to allow the planktonic stages to survive. *A. islandica* also prefers clear water rather than the turbid water characteristic of proximal glacimarine conditions, suggesting a distal environment. Moreover, many of the molluscs were either fragmentary or abraded specimens suggesting a degree of reworking (J.D. Peacock and A. Warén, pers. comm. 1984).

Other samples

Some of the samples did not readily fit into facies A or B. These included four samples from near the shelf edge break, 48/$\overline{09}$/137, 48/$\overline{10}$/53, 48/$\overline{09}$/97 and 48/$\overline{10}$/51. Sample 48/$\overline{09}$/97 contained an ostracod and foram fauna which would be consistent with the modern fauna from the area. Sample 48/$\overline{10}$/51 contrasted with all the others in being depleted in clay, green in colour and containing a temperate molluscan assemblage. The micropalaeontological analyses indicate that it is Early Pleistocene in age and can be attributed to the upper Little Sole Formation. As well as being geographically separated from the samples, 49/$\overline{07}$/336 was also palaeontologically dissimilar from facies A and B, containing abraded specimens of *Nummulites* cf. *rectus* Curry, known from the middle/late Eocene boundary of the English Channel, and its occurence here in sands and gravels capping Eocene sediments is interpreted as being due to local reworking.

Palaeoenvironmental and age interpretation

The complexities of the glacimarine sedimentary environment demand caution in the interpretation of suites of interbedded glacial and glacimarine sediments. This caution is even more necessary in situations like this where there is little stratigraphic control. There was insufficient well preserved monospecific mollusc, ostracod or foram material in any of the samples for radiocarbon dating even by the accelerator technique. There is therefore no independent dating evidence to suggest that the separate samples were deposited during the same glacial event, though their geomorphological context would support such an interpretation.

Facies B samples were from depths between −135 and −183 m OD. They are interpreted as distal glacimarine in origin, having been deposited in very quiet conditions beneath icebergs or an ice shelf which provided a fairly continuous rain of sediment to the sea floor. The fining-upwards laminae probably represent discrete sediment input events, and the sand pods frozen sand clasts. Facies A samples were recovered from shallower water, between −127 and −157 m OD, but are less easy to interpret. The sediment would be consistent with either proximal glacimarine conditions or a basal till of lodgement facies. The difference in the ratio of planktonic to benthonic forams between the two facies can be attributed to these differences in depositional environment. It is likely that all the planktonics are derived from open ocean environments suggesting direct contact between the Atlantic and the glacimarine environment of facies B. The planktonic species *Globigerina bulloides* and *Globorotalia inflata* live in temperate water masses and may indicate the proximity of warmer water in the adjacent Atlantic.

There is a transitional zone between the facies A samples and the distal glacimarine sediments of facies B at about 49°30′N between −127 and −145 m OD (Fig. 3). This transition is shown in Fig. 8c–g with facies B samples characterized

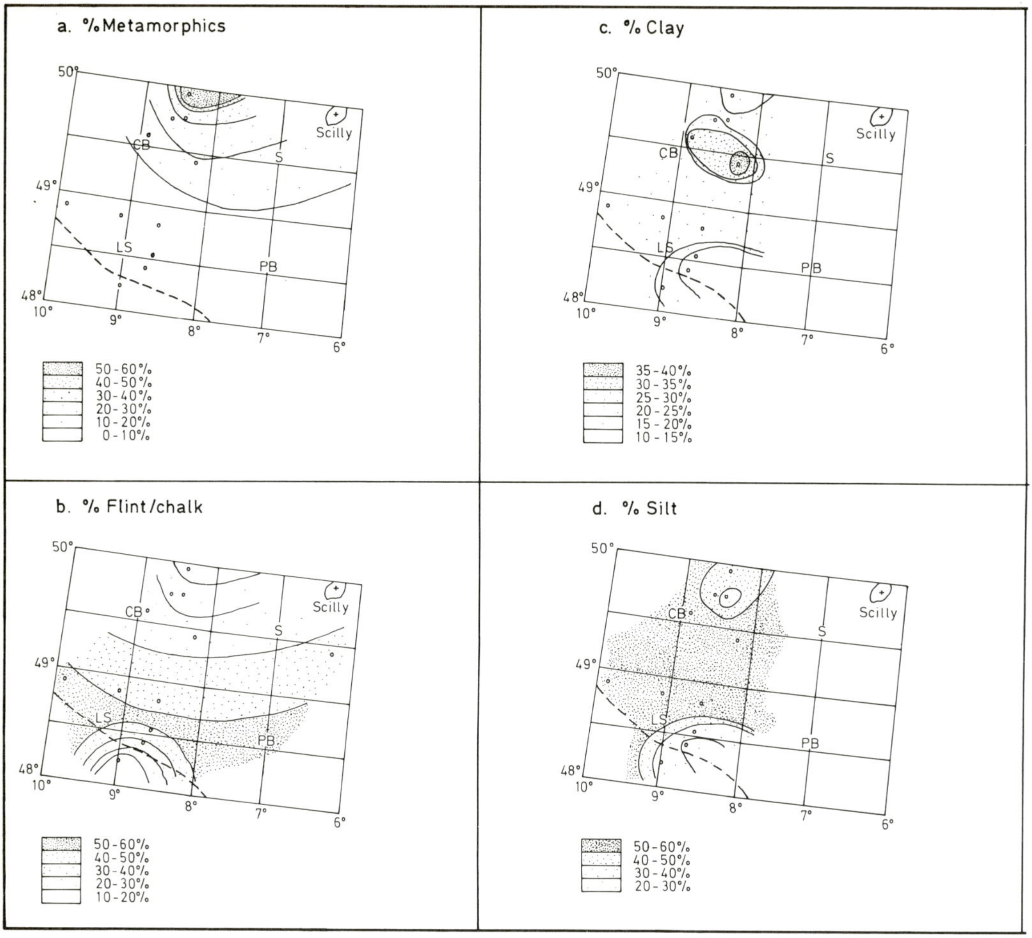

by high clay and silt but low sand values; the transition is also reflected in the sorting coefficient and mean grain size plots. These diagrams also reflect the proximal glacimarine or lodgement till affinities of some of the samples located towards the continental shelf edge, especially 48/$\overline{10}$/53 and 48/$\overline{09}$/137. These samples may represent residual glacimarine sediments containing much reworked material similar to those described from the George V – Adelie continental shelf edge in Antarctica (Domack 1982).

Small chalk clasts were found in samples 49/$\overline{09}$/12, 49/$\overline{09}$/21 and 49/$\overline{09}$/137. In order to constrain the possible origins of these clasts the calcareous nannofossils from them were examined (Table 3). All the samples contain sufficient nannofossils to assign an approximate Cenomanian–early Coniacian age. The maximum, Cenomanian, age is indicated by the general character of the assemblage and the presence of such genera as *Broinsonia* and *Eiffelithus*. The minimum, early Coniacian, age is largely based on the absence of *Micula staurophora*, a robust species which is normally abundant in younger chalks. The better preserved clasts from samples 49/$\overline{09}$/21 and 49/$\overline{09}$/137 contained good nannofloras and can be more closely dated. Within the zonation of Crux (1982) they can both be assigned to the mid Turonian–early Coniacian *Eiffelithus eximius* zone, on the basis of the presence of *E. eximius* and absence of *Micula staurophora*. The assemblages are closely comparable with those reported from the Turonian of Dover by Cooper (in Jarvis *et al.* 1988), notably in the common presence of *Nannoconus* and rarity of *Microrhabdulus decoratus*. A single specimen of the brachiopod *Terebratulina lata* was found associated with the clasts of sample 49/$\overline{09}$/137. This also indicates a Turonian age. In addition, the clasts are all white, which suggests Upper or Middle (i.e. Turonian or younger), rather than

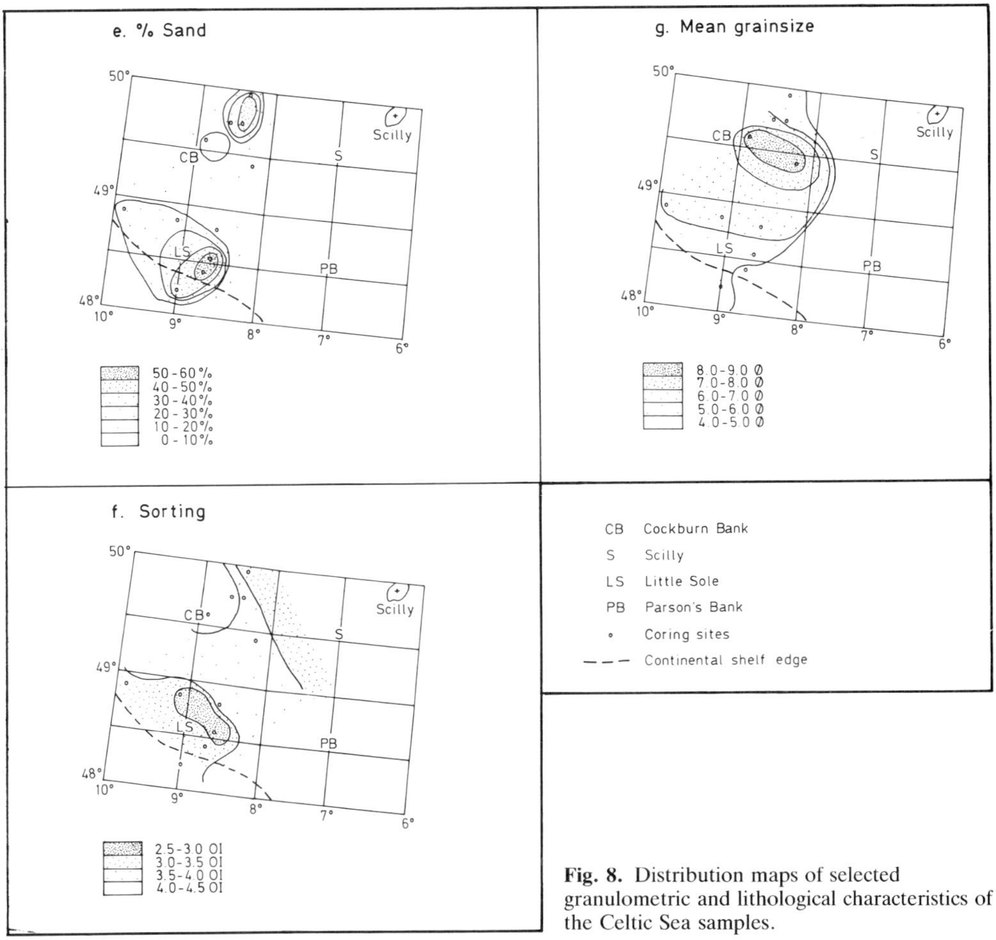

Fig. 8. Distribution maps of selected granulometric and lithological characteristics of the Celtic Sea samples.

Lower Chalk (i.e. Cenomanian) which is typically marly and so off-white or grey. All the evidence suggests that the clasts are of rather uniform age and so probably came from a single source.

The clasts in facies A include many metamorphic types in samples proximal to the Haig Fras granite intrusion and associated metamorphic aureole (Figs 4 and 8a), and Neogene lignite and Miocene glauconitic micrite (A. Morton, pers. comm.) in some samples. These also indicate ice provenance. Taken together they suggest that ice moved from the northeast towards the southwest across the area between Haig Fras and the Isles of Scilly, eroding various lithologies including Turonian and Miocene sediments. There is no evidence of far travelled material (i.e. > 100 km). The sediment cannot, therefore, have been derived from North Atlantic icebergs which were probably responsible for the widespread ice-rafted sediment in the open ocean to the south and west (Kudrass 1973; Ruddiman & McIntyre 1973, 1981).

At around 49°30′N a change in depositional environment occurred, representing either a transition from grounded to floating ice, or simply from proximal to distal glacimarine conditions. This transition occurs between the present-day water depths of −127 and −145 m OD. The upcore change in lithology from sample 49/$\overline{09}$/44 suggests that marine conditions gradually became predominant through time, extending further to the north and perhaps causing former grounded ice to float and calve.

The geomorphological situation of glacigenic samples cored in ‘lows’ between the tidal sand ridges, and in one case on the side of a sand ridge, suggests that glacigenic deposition took place after the main period of formation of the sand ridges. Pantin & Evans (1984) noted that the ice-rafted mounds occur between the sand ridges, and suggested that the sand ridges formed in about 60 m of water by the Huthnance mechanism (Huthnance 1982) during the early stages of the Devensian–Flandrian transgression. Recent numerical modelling of the M_2 tidal streams in the Celtic Sea with sea level lowered by 100 m has suggested tidal currents twice as strong as those at present, certainly sufficient to generate and maintain the ridges now found as fossil forms (Belderson *et al.* 1986). Both these studies therefore imply generation of the sand ridges during the early phase of sea level rise from the extreme minimum reached during the last cold stage (Bloom *et al.* 1974). Sand ridges of this size, up to 60 m high and 50 km long, could not have survived subaerial exposure, so the glacigenic material must have been deposited after the formation of the sand ridges and without any intervening low stand of sea level. This implies a Late Devensian age not only for the sand ridges, but also the overlying glacigenic sediments.

This reconstruction also implies survival of the sand ridges during the succeeding glacial event. The sand ridges stop fairly abruptly along a northwest/southeast line about 100 km to the southwest of the Isles of Scilly in the general vicinity of the facies A/B transition (Fig. 3). This suggests that any sand ridges to the northeast of their present limit were removed by the grounded ice sheet, and that glacimarine sedimentation occurred over, and on the flanks of, the sand ridges to the southwest. Sample 49/$\overline{09}$/44 is crucial in this argument, as it is the only sample recovered from the flank of a sand ridge. It was the shallowest sample to be recovered, at −125 m OD, and was positioned accurately using Decca Pulse 8. The subunit of facies A type material at its base suggests that it may mark the grounded ice limit on the margin of a surviving sand ridge. However, the sand ridges terminate abruptly as far south as 48° off Ushant in the French sector (Bouysse *et al.* 1979), and it seems unlikely, in the absence of sedimentary evidence for glaciation there, that the ice limit extended so far south.

The acoustic character of the sand ridges combines a general absence of continuous reflectors with some point sources and diffuse inclined reflectors, suggesting that they are composed of sands and gravels, though no samples exist to confirm this. A possible explanation for the source of this massive amount of sediment, given the proximity of an ice sheet, must be as subaqueous outwash from the advancing glacimarine terminus. This material may then have been redistributed by shelf processes, such as the Huthnance mechanism or enhanced tidal streams, during the succeeding phase of rising sea level, and the glacigenic sediments deposited during a deglacial phase involving flotation and calving, probably also associated with rising sea level.

Regional correlation and reconstruction

Both the glacigenic samples described here and the Scilly Till (Scourse 1985, 1989) appear to be of Late Devensian age. The age of the Scilly Till is constrained by a series of radiocarbon determinations from organic sediments contained within the underlying Porthloo Breccia (Fig. 9), and by thermoluminescence dates (Wintle 1981) from the Old Man Sandloess which is mineralogically (Catt 1986) and genetically related to the Till (Scourse 1985, 1989). There are also lithological and mineralogical (Table 1) similarities between the Scilly Till and the glacigenic samples described here. Catt (1986) noted a strong similarity between the silt mineralogy of these samples and the Scilly Till, so it is reasonable to assume that the same advance of Irish Sea ice into the Celtic Sea was responsible for both sedimentary bodies.

This correlation enables conclusions to be drawn about sea levels, grounding-line elevation, ice thicknesses and shorelines at around 19 000 years BP in the region. Whereas other authors (e.g. McCabe *et al.* 1986) have found it difficult to evaluate the isostatic and eustatic components associated with glacimarine sequences in areas of multiple glaciation, largely as a result of lags in the visco-elastic response of the crust to ice loading and removal (Andrews 1978), the evidence from this study possesses a number of features which enable simple quantitative reconstructions to be made.

1. The available evidence suggests that this advance was the only glacial event related to an Irish Sea ice stream which affected the Celtic Sea south of 51°N and, therefore, represents the 'feather-edge' of Pleistocene glaciation. There is no evidence that the Southwestern Celtic Sea has been repeatedly glaciated, unlike the Irish Sea Basin further north and the North Sea Basin (Wingfield 1990). This simplifies glacio-isostatic reconstructions and enables the assumption of only one glacial loading event in the region.

2. Ice thickness in the vicinity of the Isles of Scilly cannot have exceeded 100 m otherwise the Islands would have been completely overridden. There is a well documented ice limit across the northern Islands (Mitchell & Orme 1967; Scourse 1985, 1989), the southern part of the archipelago remaining unglaciated. The

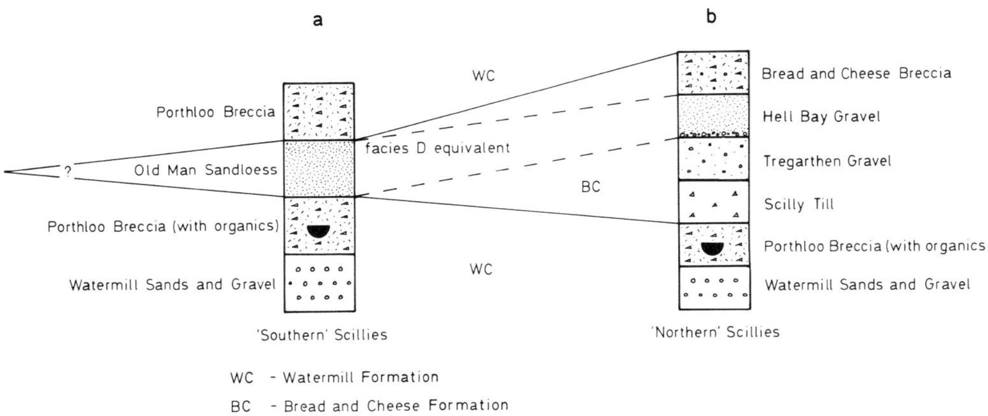

Fig. 9. The Isles of Scilly: lithostratigraphic models for the southern (non-glaciated) and northern (glaciated) Scillies, and their correlation. The Porthloo Breccia has been radiocarbon dated to between $34\,500^{+885}_{-800}$ and $21\,500^{-890}_{-800}$ years BP (Scourse 1989) and the Old Man Sandloess has been thermoluminescence dated to $18\,600^{+3\,700}_{-3\,700}$ years BP (Wintle 1981; Scourse 1989).

highest glaciated parts of the Islands lie at around 38 m OD, and the adjacent sea floor is no deeper than −60 m OD, indicating that the ice was probably somewhat less than 100 m thick.

3. There is evidence for a grounding line between −127 and −145 m OD. The Scilly Till and facies A samples contain material locally eroded from the sea floor indicating that ice was grounded over the northern part of the area, whereas the facies B samples are glacimarine in origin.

By combining this figure for the grounding line elevation with the ice thickness deduced from the Isles of Scilly evidence it is possible to estimate the contemporary sea level using the following equations:

$$a = y - 0.8z + 0.33z \quad (1)$$

hence:

$$a = y - 0.47z \quad (2)$$

$$x = y + 0.33z \quad (3)$$

where x = extra-glacial shoreline elevation post-isostatic rebound, a = grounding-line elevation post-isostatic rebound, z = ice thickness and y = sea-level during glacial event. This assumes submerged ice = $0.8z$; glacio-isostatic depression/rebound = $0.33z$. These last two figures are derived from the relative densities of glacial ice, seawater and average continental lithosphere. Usually only 20% of an ice body floating in seawater is emergent (Paterson 1981), and glacial ice is usually estimated to depress continental crust by one third of its thickness.

These calculations ignore the effects of any migrating forebulge which might give errors of ±10 m on any of the values. These relationships are expressed graphically in Fig. 10. For a = 135 m OD and z = 100 m, the values indicated by the evidence, these calculations therefore predict sea level at around only −88 m OD during the glacial event (y = −88 m OD), and an extra-glacial shoreline at a post-rebound elevation of around −55 m OD (x = −55 m OD).

Several implications follow from these calculations. The advance of ice in this area probably occurred when sea level was somewhat higher than the commonly cited glacio-eustatic minimum of −120±60 m (Shackleton 1987). This suggests that the advance of Irish Sea ice was out of phase with global glacio-eustatic fluctuations. The calculations also explain why there is no 19 000 years BP shoreline analogous to the raised shorelines of Scotland and Northern Ireland above OD on the Isles of Scilly or in Cornwall. There is no evidence in the region for shorelines higher than OD post-dating the Ipswichian interglacial. To generate such a feature, ice thickness would have been in excess of 250 m with sea-level at −88 m OD, in which case the Scillies would have been overridden. If ice thickness was 100 m, sea level would have been at the unrealistically high level of only −32 m OD. The calculations suggest the existence of a submerged shoreline at around −55 m OD, and it is significant that Donovan & Stride (1975) have reported three drowned cliff-lines around the coast of Devon and Cornwall at depths of 44 m, 54 m and 64 m. They attribute these to Late Tertiary events but they could

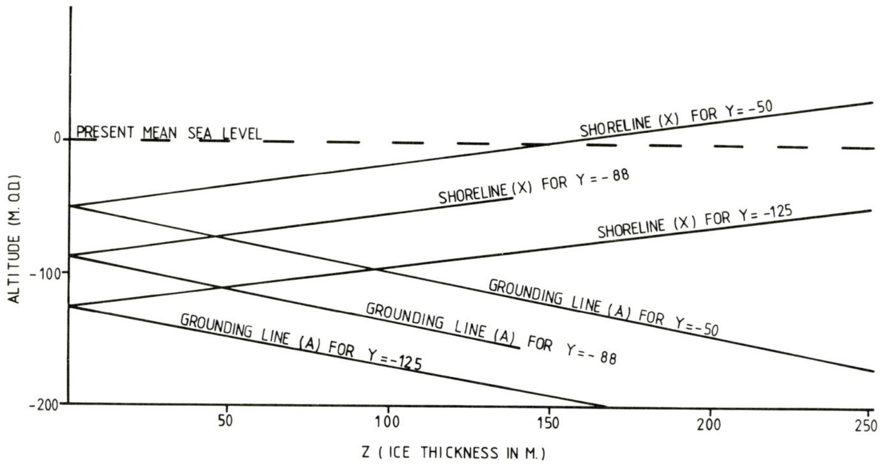

Fig. 10. Graphical representation of shoreline and grounding line elevations in relation to ice thickness at 19 000 years BP in the Celtic Sea. The shoreline and grounding line elevations diverge with increasing ice thickness as a result of increasing postglacial glacio-isostatic rebound.

owe their origin to this phase of the Late Devensian.

These calculations also assist in reconstructing the precise glacimarine environment responsible for the deposition of facies B. Both the biological and lithological evidence from facies B suggests very quiet conditions during deposition; this is surprising given the exceptionally rough conditions, often associated with southwesterly gales, currently experienced on this part of the continental shelf. One possible explanation for this is deposition beneath an ice shelf rather than from icebergs (Vorren *et al.* 1983). Ice shelves have a physically constrained minimum thickness of around 200 m (Paterson 1981), so with a grounding line at −135 m OD, a post-rebound shoreline would occur at around +30 m OD. Ice thicknesses greater than 100 m and a post-rebound shoreline raised above OD are both incompatible with the geological evidence. It is, therefore, difficult to envisage deposition beneath an ice shelf.

The most likely hypothesis for the origin of facies B is, therefore, ice-rafting. The exceptionally quiet conditions indicated by the ostracods suggest that basal melt-out from icebergs was more important than slumping or dumping (Vorren *et al.* 1983). The ice-rafting hypothesis was favoured by Pantin & Evans (1984), who suggested that the mounds of sediment recorded by the side-scan sonar are individual iceberg 'dumps'. Some of the mounds may have such an origin, but the facies B samples suggest quieter conditions and, therefore, an originally more continuous cover. In the absence of an ice shelf it is difficult to provide an explanation for the quiet conditions during glacimarine deposition. During a phase of lowered sea level the wave base would have modified the sea bed frequently, resulting in an even more erosive environment than at present. However, regional climatic conditions may have been fundamentally different during the Late Devensian; a different pattern of depression tracks associated with blocking anticyclones over ice masses (Williams 1975) would have resulted in calmer conditions on this part of the shelf. Alternatively a broad cover of pack ice may have produced tranquil conditions on the sea bed.

The available geological evidence and calculations presented above therefore indicate the existence of a thin ice lobe at the terminus of the Irish Sea ice stream during the Dimlington Stadial of the Late Devensian (Rose 1985). Recent BGS cores and seismic evidence from the Celtic Sea to the north of 50°N supports the hypothesis of an ice lobe advancing from the northeast (Fig. 11), where the occurrence of glacigenic sediment at or near the sea bed strongly suggests a link with the material further south. This inferred lobe originated immediately to the south of St. George's Channel where the 'Celtic Deep succession', a possibly tectonically-controlled basin infill up to 250 m thick, includes a number of glacigenic units (Wingfield 1990).

The existence of an extensive thin ice lobe with low surface gradient suggests that a surge may have been responsible for the glacial event affecting the Isles of Scilly and neighbouring

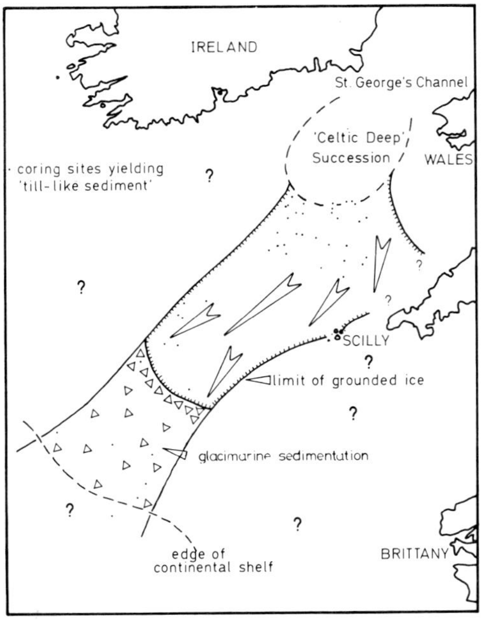

Fig. 11. A reconstruction of the Celtic Sea ice lobe and glacimarine terminus at 19 000 years BP. Dots represent vibrocoring sites yielding glacigenic sediment.

continental shelf. Many of the preconditions now commonly associated with the propagation of glacier surges are fulfilled in this case. In particular the glacier would have advanced over an area covered by marine sediments; if saturated these sediments would have possessed low shear strength and would have been deformable and inherently unstable. Recent studies emphasize the importance of saturated, deformable substrates in surge propagation (Boulton & Jones 1979; Clarke *et al.* 1984), and many hitherto unexplained lobate features at the margins of the Pleistocene ice sheets can now be explained by surges (Frye & Willman 1973; Boulton *et al.* 1977). Three-dimensional modelling of the Weichselian European ice sheet at its maximum, incorporating low basal shear strength over areas of deformable substates (Boulton *et al.* 1985) predicted an 'Irish Sea ice stream' not represented in earlier models which had assumed non-deformable beds. This reconstruction is compatible with the geological evidence presented here.

The authors would like to thank J.D. Peacock and A. Warén for assistance with the identification and interpretation of the mollusc assemblages, and J.R.M. Allen for drawing the diagrams. We are grateful for the helpful comments of the referees, M. Stoker and H.P. Sejrup. James Scourse gratefully acknowledges receipt of an NERC Research Studentship.

References

ANDREWS, J.T. 1978. Sea level history of Arctic coasts during the Upper Quaternary: dating, sedimentary sequences, and history. *Progress in Physical Geography*, **2**, 375–407.

BELDERSON, R.H., PINGREE, R.D. & GRIFFITHS, D.K. 1986. Low sea-level tidal origin of Celtic Sea sand banks — evidence from numerical modelling of M_2 tidal streams. *Marine Geology*, **73**, 99–108.

BERTHOIS, L. 1974. *Carte Bathymetrique du Golfe de Gascogne*. C.N.E.X.O., Paris.

BLOOM, A.L., BROECKER, W.S., CHAPPELL, J.S., MATTHEWS, R.K. & MESOLELLA, K.J. 1974. Quaternary sea-level fluctuations on a tectonic coast: New Th^{230}/U^{234} dates from the Huon Peninsula, New Guinea. *Quaternary Research*, **4**, 185–205.

BOULTON, G.S. & JONES, A.S. 1979. Stability of temperate ice caps and ice sheets resting on beds of deformable sediment. *Journal of Glaciology*, **24**, 29–43.

——, ——, CLAYTON, K.M. & KENNING, M.J. 1977. A British ice sheet model and patterns of glacial erosion and deposition in Britain. *In:* SHOTTON. F.W. (ed.) *British Quaternary Studies*. Oxford University Press. Oxford, 231–246.

——, SMITH, G.D., JONES, A.S. & NEWSOME, J. 1985. Glacial geology and glaciology of the last mid-latitude ice sheets. *Journal of the Geological Society, London*, **142**, 447–474.

BOUYSSE, P., HORN, R., LAPIERRE, F. & LE LANN, F. 1976. Etude des grands bancs de sable du sud-est de la mer Celtique. *Marine Geology*, **20**, 251–275.

BRADY, G.S., CROSSKEY, H.W. & ROBERTSON, D. 1874. A monograph on the Post-Tertiary Entomostraca of Scotland including species from England and Ireland. *Palaeontographical Society*.

—— & NORMAN, A.M. 1889. A monograph of the marine and freshwater Ostracoda of the North Atlantic and of North-Western Europe. *Scientific Transactions of the Royal Dublin Society, Series 2*, **4**, 63–270.

CATT, J.A. 1986. Silt mineralogy of loess and 'till' on the Isles of Scilly. *In*: SCOURSE, J.D. (ed.) *The Isles of Scilly*. Coventry, Quaternary Research Association Field Guide, 134–136.

—— & PENNY, L.F. 1966. The Pleistocene deposits of Holderness, Yorkshire. *Proceedings of the Yorkshire Geological Society*, **35**, 375–420.

CRUX, J.A. 1982. Upper Cretaceous (Cenomanian to Campanian) calcareous nannofossils. *In*: LORD, A.R. (ed.) *A Stratigraphical Atlas of Calcareous Nannofossils*. Ellis Horwood, Chichester 81–135.

CLARKE, G.K.C., COLLINS, S.G. & THOMPSON, D.E. 1984. Flow, thermal structure and subglacial conditions of a surge type glacier. *Canadian Journal of Earth Sciences*, **21**, 232–240.

DAY, A.A. 1959. The continental margin between Brittany and Ireland. *Deep-Sea Research*, **5**, 249–265.

DELANTEY, L.J. & WHITTINGTON, R.J. 1977. A reassessment of the "Neogene" deposits of the South Irish Sea and Nymphe Bank. *Marine Geology*, **24**, 23–30.

DOMACK, E.W. 1982. Sedimentology of glacial and glacial marine deposits on the George V – Adelie continental shelf, East Antarctica. *Boreas*, **11**, 79–97.

DONOVAN, D.T. & STRIDE, A.H. 1975. Three drowned coastlines of probable Late Tertiary age around Devon and Cornwall. *Marine Geology*, **19**, M35–M40.

EVANS, C.D.R. & HUGHES, M.J. 1984. The Neogene succession of the South Western approaches, Great Britain. *Journal of the Geological Society, London*, **141**, 315–326.

FEYLING-HANSSEN, R. 1982. Molluscs and other megafossils. *In*: OLAUSSON, E. (ed.) The Pleistocene-Holocene boundary in South West Sweden. *Sveriges geologiska undersökning*, **794**, 120–136.

FRYE, J.C. & WILLMAN, H.B. 1973. Wisconsinan climatic history interpreted from Lake Michigan lake deposits and soils. *In*: BLACK, R.F., GOLDTHWAIT, R.P. & WILLMAN, H.B. (eds) *The Wisconsinan Stage*. Memoir of the Geological Society of America, **136**, 71–106.

GARRARD, R.A. 1977. The sediments of the South Irish Sea and Nymphe Bank area of the Celtic Sea. *In*: KIDSON, C. & TOOLEY, M.J. (eds) *The Quaternary History of the Irish Sea*. Liverpool, Seel House Press, 69–92.

—— & DOBSON, M.R. 1974. The nature and maximum extent of glacial sediments off the West Coast of Wales. *Marine Geology*, **16**, 31–44.

GRACIANSKY, P.C. de, POAG, C.W. *et al*. 1985. *Initial Reports of the Deep Sea Drilling Project*. **80**. Washington: US Government Printing Office.

HAZEL, J.E. 1971. Ostracode zoogeography in Southern Nova Scotian and Northern Virginian faunal provinces. *United States Geological Survey, Professional Paper*, **529–E**, 1–21.

HAMILTON, D., SOMERVILLE, J.H. & STANFORD, P.H. 1980. Bottom currents and shelf sediments, southwest of Britain. *Sedimentary Geology*, **26**, 115–138.

HUTHNANCE, J.M. 1982. On one mechanism forming linear tidal sand banks. *Estuarine, Coastal and Shelf Science*, **14**, 79–99.

JARVIS, I., CARSON, G.A., COOPER, M.K.E., HART, M.B., LEARY, P., TOCHER, B.A., HORNE, D. & ROSENFELD, A. 1988. Microfossil assemblages and the Cenomanian – Turonian (late Cretaceous) oceanic anoxic event. *Cretaceous Research*, **9**, 3–104.

KUDRASS, H-R. 1973. Sedimentation am Kontinentalhang vor Portugal und Marokko im Spätpleistozän und Holozan. *'Meteor' Forsch. – Ergebnisse*, **C13**.

LAMBERT, J.T. & KHOWAJA, Z.M. 1978. Geotechnical analysis of till-like material from the S.W. Approaches. *Institute of Geological Sciences Engineering Geology Unit Internal Report No. 78/23*.

LEV, O.M. 1969. Analiz i Sopostavlenie Komplexov Ostakod iz Pozdnekainozoiskikh Otlozhenii Nizovev Rek Eniseya, Obi, Pechory, Mezeni i Severnoi Dviny. *Uchenye Zapiski Nauchno-Issledovatel'skogo Instituta Geologii Arktiki* **28**, 25–34.

MARTINDALE, W. & CHESHER, J.A. 1979. *1:2 500 000 Map of the Sub-Pleistocene Geology of the British Isles and the Adjacent Continental Shelf*. Institute of Geological Sciences, NERC.

MCCABE, A.M., HAYNES, J.R. & MACMILLAN, N.F. 1986. Late-Pleistocene tidewater glaciers and glaciomarine sequences from north County Mayo, Republic of Ireland. *Journal of Quaternary Science*, **1**, 73–84.

MITCHELL, G.F. & ORME, A.R. 1967. The Pleistocene deposits of the Isles of Scilly. *Quarterly Journal of the Geological Society, London*, **123**, 59–92.

NEALE, J. & HOWE, H.V. 1975. The marine Ostracoda of Russian Harbour, Nova Zemlya and other high latitude faunas. *In*: SWAIN, F.M. (ed.) *Biology and Palaeobiology of Ostracoda*. Bulletins of American Palaeontology, **65**.

NORDSIECK, F. 1969. *Die europaischen Meeresmuscheln (Bivalvia) vom Eismeer bis Kapverden, Mittelmeer und Schwarzes Meer*. Gustav Fischer, Stuttgart.

PANTIN, H.M. & EVANS, C.D.R. 1984. The Quaternary history of the Central and Southwestern Celtic Sea. *Marine Geology*, **57**, 259–293.

PATERSON, W.S.B. 1981. *The Physics of Glaciers*. 2nd edition. Pergamon, Oxford.

PATERSON, I.B., ARMSTRONG, M. & BROWN, M.A.E. 1981. Quaternary estuarine deposits in the Tay-Earn area, Scotland. *Institute of Geological Sciences Report*, **81/7**.

PEACOCK, J.D. 1971. Terminal features of the Creran Glacier of Loch Lomond Readvance age in western Benderloch, Argyll, and their significance in the late-glacial history of the Loch Linnhe area. *Scottish Journal of Geology*, **7**, 349–356.

—— 1989. Marine molluscs and late Quaternary environmental studies with particular reference to the Late-glacial period in Northwest Europe: a review. *Quaternary Science Reviews*, **9**, 179–192.

PERCH-NIELSEN, K. 1985. Mesozoic calcareous nannofossils. *In*: BOLLI, H.M., SAUNDERS, J.B. & PERCH-NIELSEN, K. (eds) *Plankton Stratigraphy*. Cambridge University Press, 329–426.

ROSE, J. 1985. The Dimlington Stadial/Chronozone: a proposal for naming the main glacial episode of the Late Devensian in Britain. *Boreas*, **14**, 225–230.

RUDDIMAN, W.F. & MCINTYRE, A. 1973. Time-transgressive deglacial retreat of polar waters from the North Atlantic. *Quaternary Research*, **3**, 117–130.

—— & MCINTYRE, A. 1981. The North Atlantic Ocean during the last deglaciation. *Palaeogeography, Palaeoclimatology, Palaeoecology*, **3**, 145–214.

SCOURSE, J.D. 1985. *Late Pleistocene Stratigraphy of the Isles of Scilly and Adjoining Regions*. University of Cambridge, Ph.D. thesis.

—— 1990. The Isles of Scilly. *In*: EHLERS, J., GIBBARD, P.L. & ROSE, J. (eds) *Glacial Deposits in Great Britain and Ireland*. Rotterdam, A.A. Balkema.

——, ROBINSON, J.E. & EVANS, C.D.R. 1990. Central and Southwestern Celtic Sea. *In*: EHLERS, J., GIBBARD, P.L. & ROSE, J. (eds) *Glacial Deposits in Great Britain and Ireland*. Rotterdam, A.A. Balkema.

SHACKLETON, N.J. 1987. Oxygen isotopes, ice volume and sea level. *Quaternary Science Reviews*, **6**, 183–190.

STRIDE, A.H. 1963. North-east trending ridges of the Celtic Sea. *Proceedings of the Ussher Society*, **1**, 62–63.

VORREN, T.O., HOLD, M., EDVARDSEN, M. & LIND-HANSEN, O-W. 1983. Glacigenic sediments and sedimentary environments on continental shelves: General principles with a case study from the Norwegian shelf. *In*: EHLERS, J. (ed.) *Glacial Deposits in North West Europe*. Rotterdam, A.A. Balkema, 61–76.

WHATLEY, R.C. & MASSON, D.G. 1979. The ostracod genus *Cytheropteron* from the Quarternary (sic) and Recent of Great Britain. *Revista Española de Micropaleontología*, **11**, 223–277.

WILLIAMS, R.B.G. 1975. The British climate during the Last Glaciation; an interpretation based on periglacial phenomena. *In*: WRIGHT, A.E. & MOSELEY, F. (eds) *Ice Ages: Ancient and Modern*, Liverpool, Seel House Press, 95–117.

WINGFIELD, R.T.R. 1990. The origin of major incisions within the Pleistocene deposits of the North Sea. *Marine Geology*, **91**, 31–52.

WINTLE, A.G. 1981. Thermoluminescence dating of Late Devensian loesses in southern England. *Nature*, **289**, 479–480.

Glacially-influenced sedimentation on the Hebridean slope, northwestern United Kingdom continental margin

M.S. STOKER

British Geological Survey, Murchison House, West Mains Road, Edinburgh, EH9 3LA, UK

Abstract: Pleistocene sediments on the northern Hebrides Slope form a localized, mounded slope-front deposit up to 200 m thick of limited downslope and alongslope extent. In seismic profiles, this slope-front deposit is characterized by a prograding, oblique-parallel reflection configuration. The bulk of the succession is characterized by three distinct depositional packages, each comprising an amalgamated sequence of mass flow deposits consisting primarily of debris flow diamicts with turbidite sands and muds. They represent re-sedimented glacimarine deposits. The mass flow packages are bounded by discrete, slope-wide, moderate-to high-amplitude reflectors. These reflections may be due to different degrees of consolidation of the sediments and/or the presence of thin, interbedded units of glacimarine dropstone mud. They indicate periods of reduced sediment supply to the slope. The slope facies interdigitate with a submarine end-moraine at the shelf-edge, and deposition probably occurred during a single glacial cycle. The restricted occurrence of the debris flows is thus attributed to a local source reflecting the shelf-edge position of an ice-sheet in this area at the time of sedimentation. Fluctuations in the sedimentation rate, independent of major climatic variation, suggest that the ice-sheet may have experienced alternate periods of basal melting and freezing. Ice-marginal processes have contributed significantly to the construction of the slope apron on the northern Hebrides Slope. During this period of sedimentation, the shelf-edge has prograded seaward, to the NW, by up to 5 km.

On a stable continental margin, glacio-eustatic falls in sea level may result in grounded ice extending out to the shelf-edge and depositing coarse debris directly onto the slope (Eyles & Miall 1984; Powell 1984). In this setting, the slope apron is likely to grow by lateral accretion accompanied by significant shelf-edge progradation producing a clinoform stratigraphy (Eyles *et al.* 1985; Piper 1988). Slope sedimentation rates may be high due to the proximity of the ice-margin (Piper *et al.* 1987; Piper 1988), and this may have a direct bearing on the geometry of the slope apron. The Scotian Slope, off eastern Canada, provides an example of an ice-proximal slope where ice-marginal processes were directly responsible for the present slope profile, which is characterized by erosional gullies, sediment mounds and slumps (Piper & Sparkes 1987; Piper 1988). This paper describes Pleistocene ice-marginal sediments from the Hebrides Slope, off NW Scotland, in an attempt to determine the nature of the sediments and processes of deposition, and their influence on the present slope morphology.

The Hebrides Shelf and Slope lie on the eastern flank of the northern Rockall Trough (Fig. 1). A major bulge on the margin NW of St. Kilda, owes its origin to a buried Tertiary volcanic centre on the outer shelf (Evans *et al.* 1989). A smaller unrelated bulge occurs to the NE of this feature, between 59° and 59° 20′N (Fig. 1). This area has experienced ice-marginal sedimentation (Stoker 1988c, 1990) and forms the basis of the present study. West of the island of Sula Sgeir, the Hebrides Shelf is relatively smooth and flat-lying although some mounds and ridges are present on the outer shelf. The shelf break lies between 140 and 200 m, although this increases to 400 m on the Wyville-Thompson Ridge (Fig. 2). The slope displays convex/concave profiles with slope angles ranging from 1°–4° (Stoker 1990). On the upper slope, down to 450 m, the seabed has been intensely scoured by icebergs (Belderson *et al.* 1973; Stoker 1988a). Below this depth, the seabed is relatively smooth and featureless, although towards the base of the slope and into the northern Rockall Trough it is locally mounded and hummocky, due to a combination of contouritic and mass flow processes (Richards *et al.* 1987; Stoker 1988b, 1990).

This paper is based primarily on a study of high-resolution seismic profiles, i.e. 2 × 40 cu. in. airgun, 1 Kj sparker and deep-tow boomer, together with core data. These data are integrated into the regional stratigraphic and depo-

From DOWDESWELL, J.A. & SCOURSE, J.D. (eds), 1990, *Glacimarine Environments: Processes and Sediments*. Geological Society Special Publication No 53, pp 349–362.

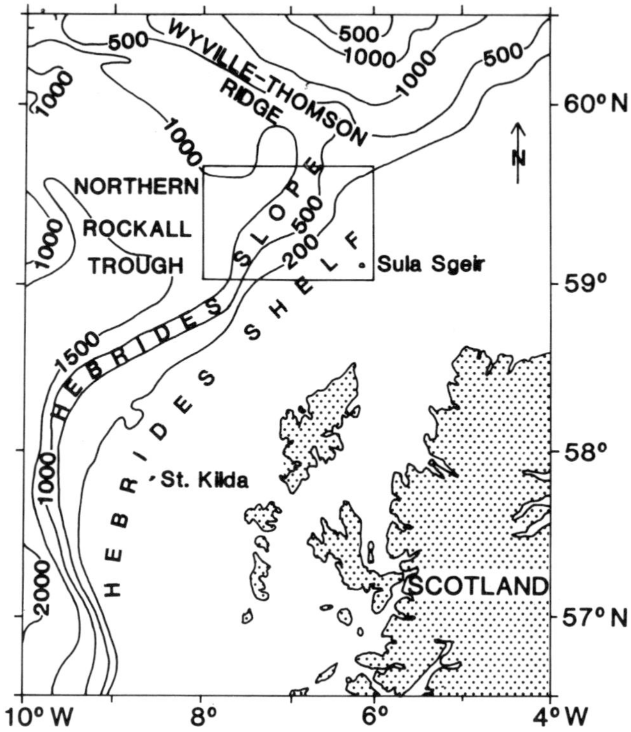

Fig. 1. Bathymetric setting of the Hebridean margin after Roberts *et al.* (1977), (contours in metres). Box shows the location of the study area which is expanded in Figs 2 and 3.

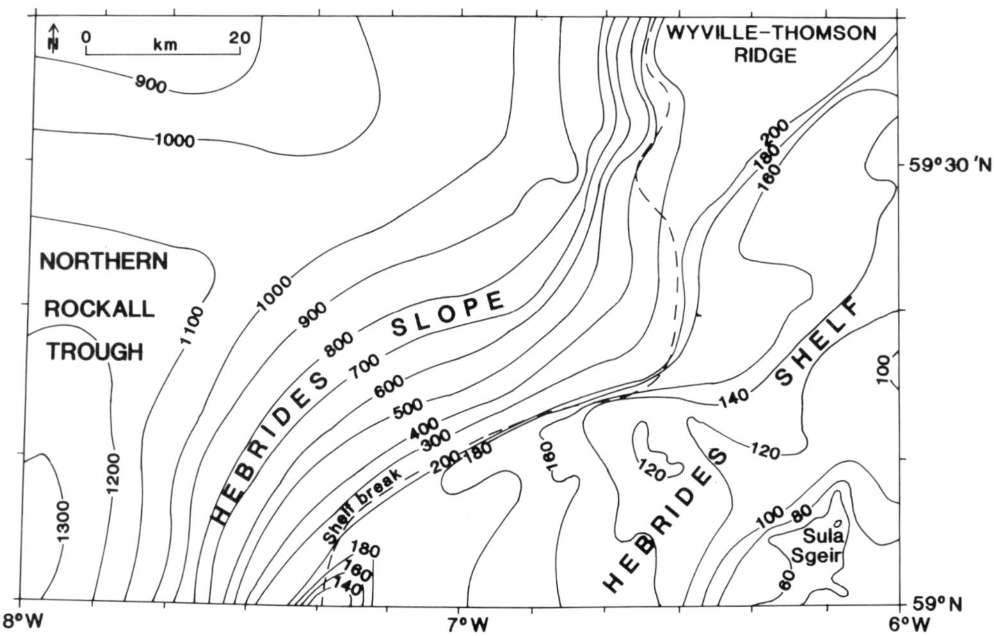

Fig. 2. Detailed bathymetry of the northern Hebridean Margin based on well navigated BGS Atlas Deso 20 echo-sounder profiles. Contours in metres assuming $V = 1480\ m\ s^{-1}$.

sitional setting which is summarized in 1:250 000 scale mapping of the northern Hebridean margin (Stoker 1990).

Regional setting

The sedimentary sequence described forms part of the MacLeod sequence, a Plio-Pleistocene sediment pile built out on the outer shelf and slope by lateral accretion of prograding units (Figs 3 & 4) (Stoker 1988b, 1989). On the outer shelf and upper slope, this sediment pile rests unconformably on tilted and deformed late Palaeocene to Oligocene sediments. Towards the base of the slope, around the margin of the northern Rockall Trough, an extensive deep-water Miocene succession onlaps the Palaeogene sediments; the Plio-Pleistocene is conformable on the Miocene with limited onlap onto the Palaeogene (Fig. 4).

The MacLeod sequence is up to 600 m in thickness in the mid-slope region, but thins into the northern Rockall Trough where it is generally less than 200 m thick. It has been subdivided into two main depositional units (Fig. 4): (1) a Plio-Pleistocene, pre-glacial, slope-front/basinal marine succession, associated with deep-water sediment waves and contourite mounding in the northern Rockall Trough (Richards *et al.* 1987; Stoker 1988b, 1990); and, (2) a Pleistocene slope-front mound of glacigenic origin (Stoker 1990).

This study concentrates on the glacigenic unit, which locally exceeds 200 m in thickness in the mid-slope area, but pinches out downslope and alongslope (Figs 3 & 4). Its SW limit lies just

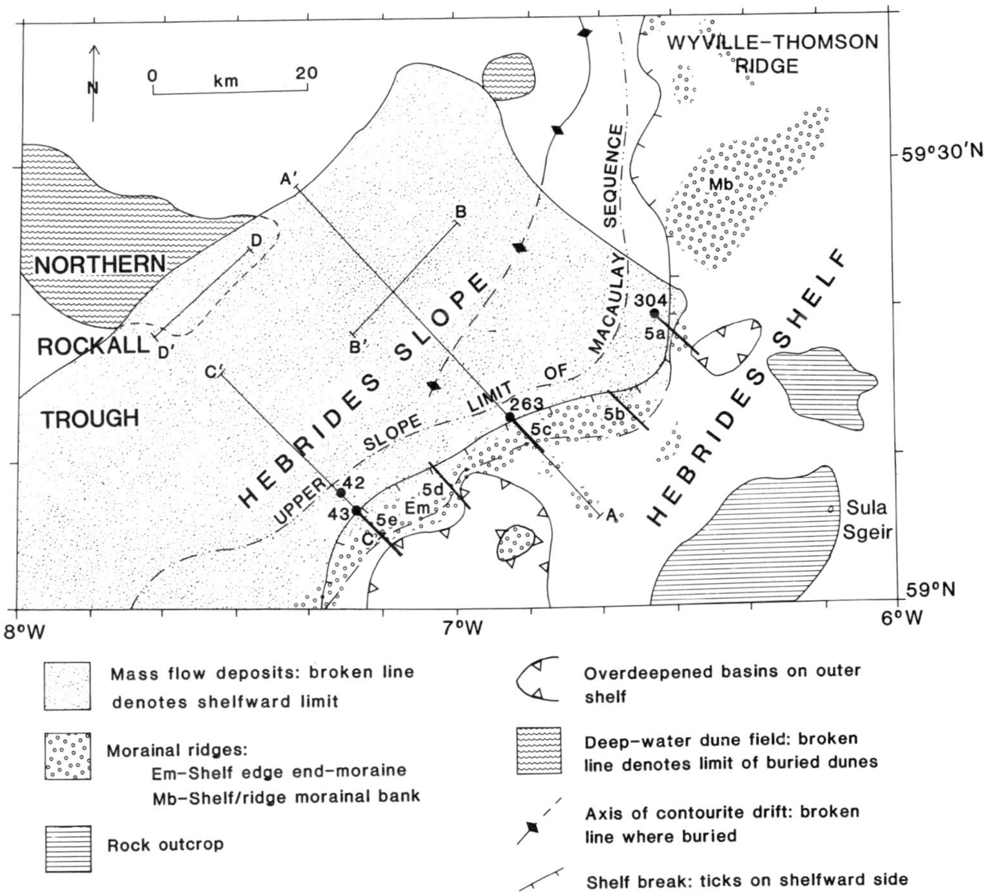

Fig. 3. Map showing principal morphological features of the study area, and location of seismic reflection profiles illustrated in other figures. Profile A–A′ in Fig. 4; B–B′ in Fig. 6; C–C′ in Fig. 7; D–D′ in Fig. 8. Profiles 5a–e in Fig. 5. Locations of cores described in text indicated by numbered solid circles.

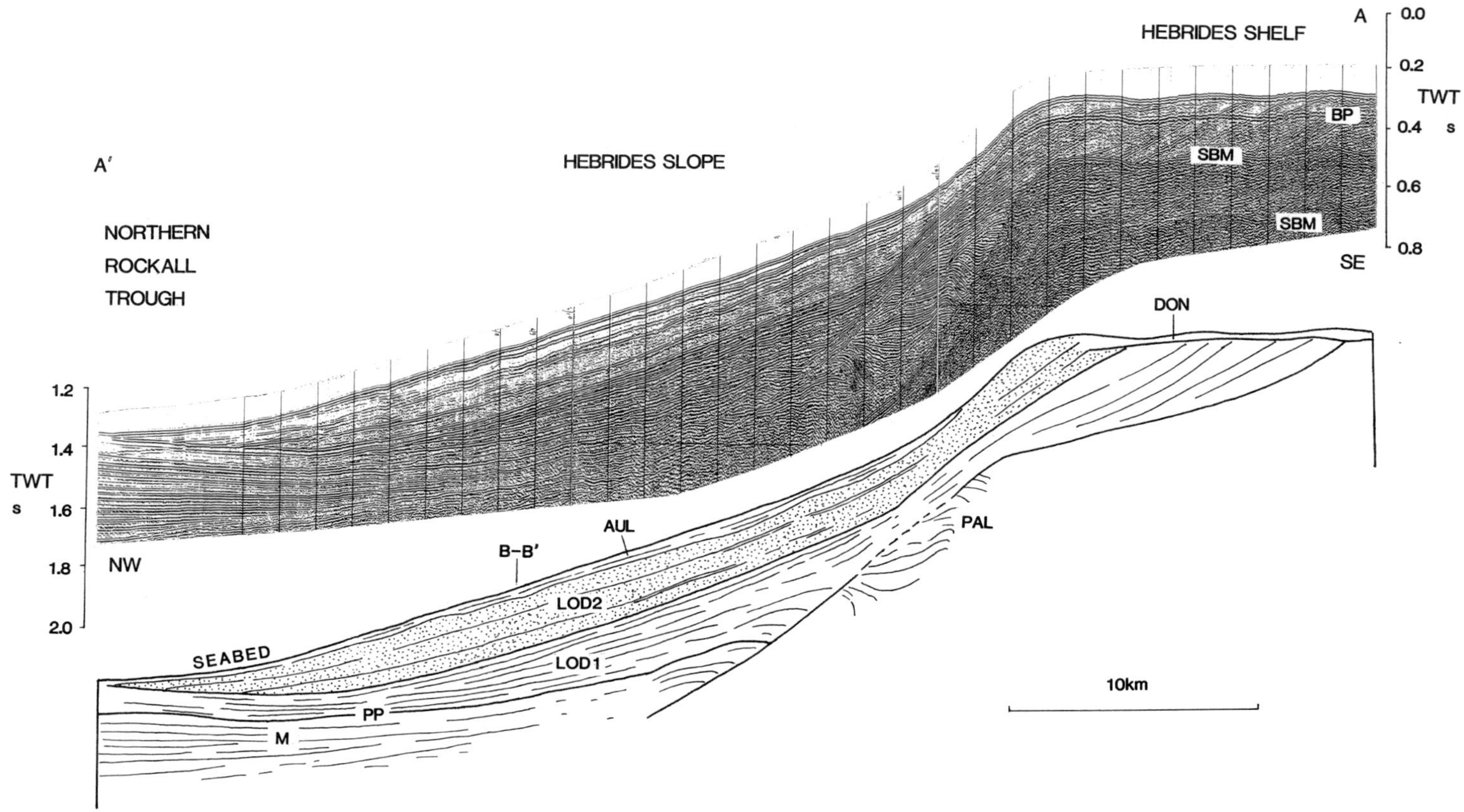

Fig. 4. Single-channel airgun line 83/04–6 and interpreted line drawing illustrating the Cainozoic setting of the northern Hebridean Margin. PAL: Palaeogene; M: Miocene; PP: base of Plio-Pleistocene; LOD1: MacLeod sequence, pre-glacial unit; LOD2: Macleod sequence, glacigenic unit (stippled); AUL: MacAulay sequence; DON: MacDonald sequence; BP: Bubble pulse; SBM: Seabed multiple. Intersection with section B–B′ (Fig. 6) indicated. Profile location shown as A–A′ in Fig. 3.

beyond the limit of the study area (D. Evans pers. comm.). Over most of the slope, its base appears conformable on the pre-glacial sediments although erosion has occurred locally on the mid-slope, and downlap occurs towards the base of the slope. On the upper slope, the top of the unit is at seabed and has been intensely scoured by icebergs. Below 300 m water depth, however, the unit is overlain by the MacAulay sequence, a thin glacimarine/marine veneer of Pleistocene to Holocene age (Stoker *et al.* 1989). At the base of the slope, the glacigenic unit pinches out and the MacAulay sequence rests on the pre-glacial sediments (unit 1) of the MacLeod sequence.

On the outer shelf, a diamict veneer, the MacDonald sequence, covers much of the area and has been moulded into a number of morainal ridges, the most distinctive of which forms a submarine end-moraine at the shelf-edge (Figs 3 & 4) (Stoker 1988c). This feature is 20–30 m thick, up to 4.5 km wide and has been traced laterally for 50 km. At the junction of the Hebrides Shelf and the Wyville–Thomson Ridge, this feature passes laterally into a wedge-shaped morainal bank consisting of ice-proximal glacimarine sediments (Fig. 3) (Stoker 1988c). Associated with the diamict are several glacially scoured and over-deepened basins (Fig. 3) (Holtedahl 1952; Sutherland 1987) which have subsequently been infilled by glacimarine argillites. Thus, the physiography of the outer shelf is comparable with other northern hemisphere, high-latitude glaciated shelves, e.g. Scotian Shelf (King *et al.* 1972; King & Fader 1986) and Norwegian Shelf (Andersen 1979; King *et al.* 1987), and is considered a relict glaciated terrain.

At the shelf-edge, the relationship between the glacigenic slope succession and the end-moraine varies from profile to profile (Fig. 5). In some areas the glacigenic unit is partly truncated by the moraine, elsewhere the moraine is truncated or appears interbedded with the slope deposits. Taken collectively, these profiles suggest an interdigitating contact at the shelf-edge. Moreover, the sediment build-out has a transgressive aspect with evidence for both erosional and depositional processes. These data, although not conclusive, are interpreted in terms of sedimentation associated with an advancing ice-sheet during a single glacial cycle (cf. King *et al.* 1987, Fig. 23).

This relationship suggests that at some time during the Pleistocene an ice-sheet extended out to the shelf-edge resulting in ice-marginal sedimentation on the slope. The timing of this glaciation is uncertain. Regional stratigraphic evidence suggests that it is probably not older than Middle Pleistocene (Cameron *et al.* 1987), although it may predate the last Scottish ice-sheet (late Devensian/late Weichselian) which probably did not extend beyond the Outer Hebrides (Sutherland & Walker 1984).

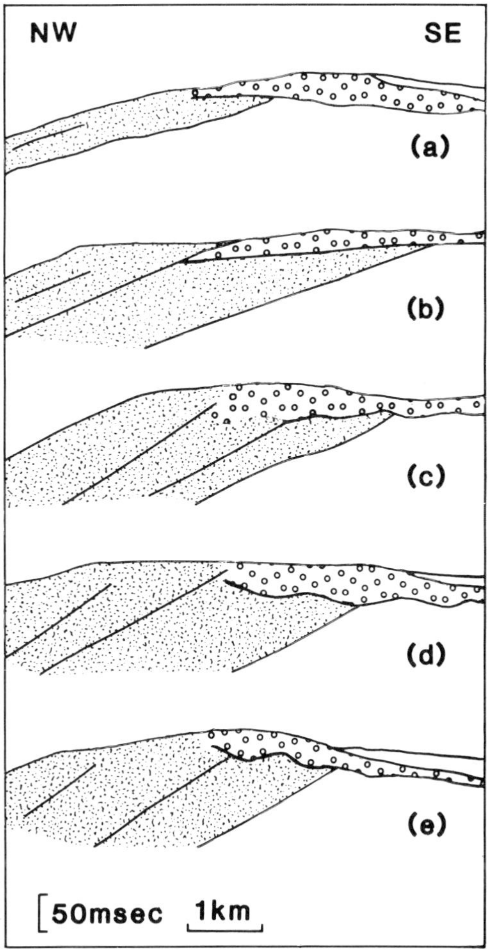

Fig. 5. Schematic interpretations of sparker profiles at the shelf break illustrating the contact relationships between the shelf-edge end-moraine (MacDonald sequence, open circles) and the glacigenic slope unit (MacLeod sequence, stippled). Location of profiles shown in Fig. 3.

Glacigenic slope succession

Seismic characteristics

On airgun profiles, the glacigenic succession is characterised by a prograding, oblique-parallel reflection configuration (Fig. 4). The bulk of the succession is acoustically structureless with few coherent internal reflections and occurs as three distinct packages, 60–70 m thick (Figs 4 & 6). However, on sparker and deep-tow boomer profiles from the mid-to-lower slope, each structureless package can be seen to consist of an amalgamated sequence of lensoid bodies (Figs 6 & 7). Individual lenses are laterally discontinuous and range from 3–20 m thick and up to 12 km in downslope length (Fig. 7) and 7.5 km alongslope. They appear to have an overall lobate form. Reflectors separating lenses are of low amplitude and discontinuous. The tops and bases of the lenses vary from planar to irregular and hummocky but it is uncertain to what extent these reflect erosional and/or depositional contacts. In the mid-slope region, reflectors in the underlying pre-glacial unit may have been partly eroded and truncated prior to deposition of the glacigenic unit (Fig. 7). At the base of the slope, these lenses overlie, and may have partly eroded a Miocene to Pleistocene deep-water dune field as some of the dune crests appear truncated (Fig. 8). Hyperbolic reflections occur locally within the lenses and along their bounding surfaces. On the upper slope, internal details are less clear on the boomer profiles (Fig. 7) although the sparker and airgun profiles suggest that these packages extend to the top of the slope. The lack of resolution on the upper slope is probably due to overcompaction and increased density of the sediment as a result of iceberg scouring, thus inhibiting acoustic penetration and forming a more reflective seabed.

The structureless packages are separated by discrete high amplitude reflectors which are comformable with substrate irregularities and have a planar to slightly irregular form (Figs 4, 6 & 7). They are laterally continuous over the slope, extending from the shelf-edge to the base of the slope where they downlap onto the underlying pre-glacial deposits.

Core control

Lithological control is restricted to the upper slope where the uppermost structureless package crops out at seabed (Figs 3 & 7). Four sediment cores, 42, 43, 263 and 304, have an intercorrelatable upper unit of Holocene muddy gravelly sands, ranging from 0.01–1.15 m thick, which overlies diamict and interbedded muds and sands of Pleistocene age (Fig. 9). The Holocene sediments form part of a widespread veneer which covers most of the outer shelf and slope (Stoker 1988a, 1990) and are not considered further here.

The diamict occurs in all of the cores and is at least 4.75 m thick in core 304 (Fig. 9). It is predominantly firm to stiff, olive grey to dark grey brown, very poorly sorted, massive and matrix-supported with clasts showing no discernible orientation (Fig. 10). The clast content is variable and includes mylonite, psammite, quartzite and pelite of NW Highland aspect, Torridonian and/or Devonian sandstones and grits and Tertiary basalt, up to 8 cm in length. About 40% of the clasts are angular to subangular and faceted and the bulk of them are fresh in appearance. Some clast-clast contacts occur in core 43, and thin sandy laminae impart a crude stratification on core 42 which locally appears deformed. Black sulphidic knots and streaks occur sporadically throughout the diamict. Grain size analysis of core 43 indicates roughly equal proportions of sand and mud with an average composition of 12% gravel, 42% sand and 46% mud of which 15% is clay fraction. The diamict appears to be predominantly non-bioturbated although faint mottling is observed on X-radiographs from the top of core 43 (Fig. 10a). Dinoflagellate cyst analysis of core 43 revealed very poor assemblages of round brown *Protoperidinium spp.* cysts and *Spiniferites spp.*, with the bulk of the core being barren of cysts (R Harland pers. comm.). Although shell fragments are scattered throughout the diamict, rare whole shells of *Chlamys islandica* (Müller) were found in core 263 (D K Graham pers. comm.). The flora and fauna are indicative of a cold to arctic depositional environment.

Interbedded muds and sands occur in core 263, above the diamict, and form a unit about 2.8 m thick (Fig. 9). The contact with the diamict is gradational. The sands are olive grey to dark grey and vary from 30–70 cm thick. Internally they appear massive although the tops and bases of each bed are gradational; the base of a sand bed generally shows inverse-grading over an interval of 1–3 cm, while the top displays normal grading over a thicker interval up to 10 cm. Outsized gravel clasts occur sporadically within and at the tops of beds. The clast content is as for the diamict and many of these clasts are encrusted by worm tubes and bryozoa. The dark grey to olive grey muds, range from 10–40 cm thick and are bioturbated with scattered matrix-supported clasts. Shells and shell fragments occur throughout these deposits, and occasionally form thin shell bands in the muds.

Interpretation

The lensoid units, with a hummocky to smooth profile and lack of internal structure, display

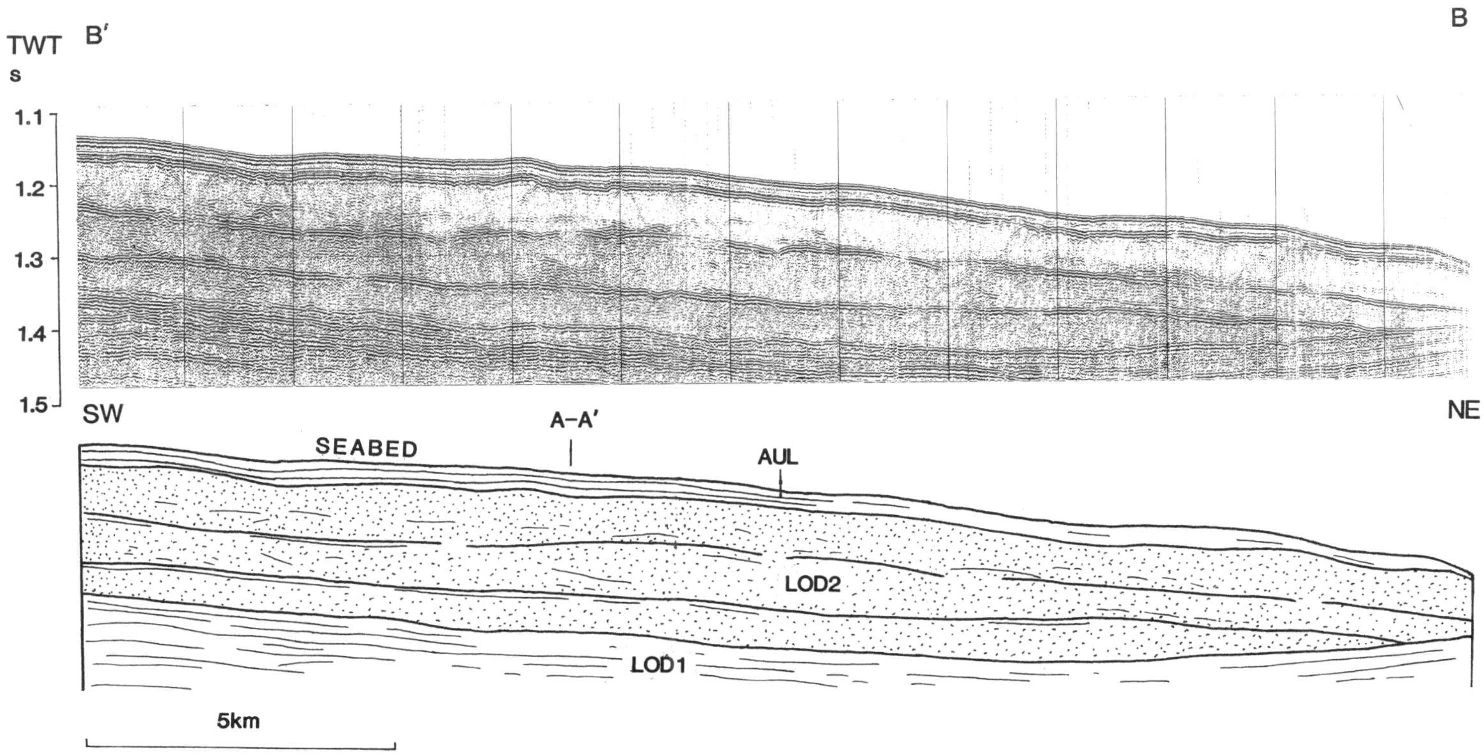

Fig. 6. Sparker profile 84/06–24 and interpreted line drawing showing seismic characteristics and alongslope pinch-out of glacigenic unit (stippled). Abbreviations as in Fig. 4. Intersection with section A–A′ (Fig. 4) indicated. Profile location shown as B–B′ in Fig. 3.

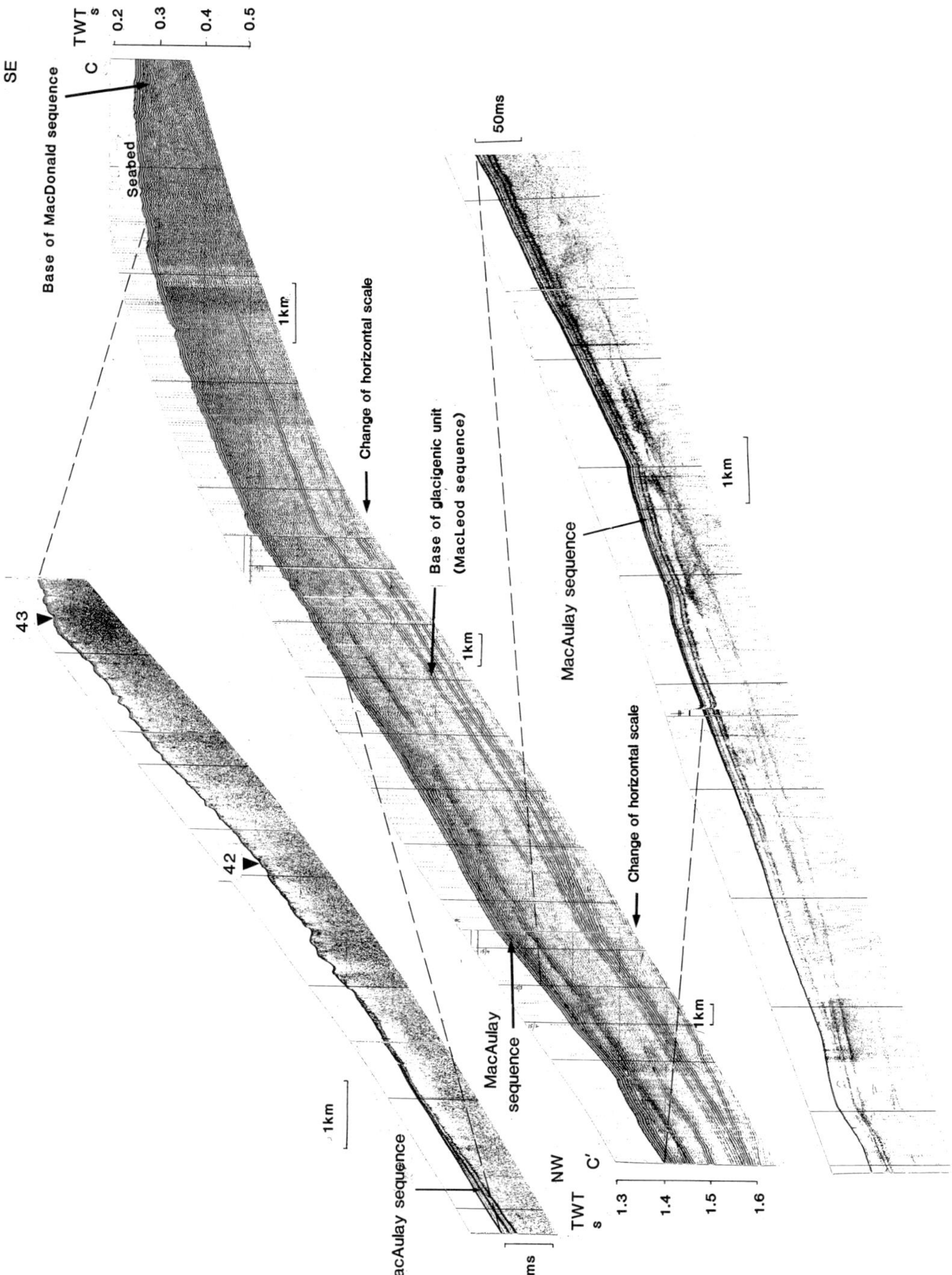

Fig. 7. Sparker profile 83/04–15 (located as C–C′ in Fig. 3) and shorter sections of deep-tow boomer profile 85/07–8 run along the same line illustrating the high-resolution seismic characteristics of the slope deposits, in particular the amalgamated succession of lensoid bodies. Locations of cores 42 and 43 also indicated. Note also the iceberg-scoured upper slope and possible erosive base of the glacigenic unit in the mid-slope region. The downslope scale change on the sparker section is a result of changes in the firing rate of the sparker.

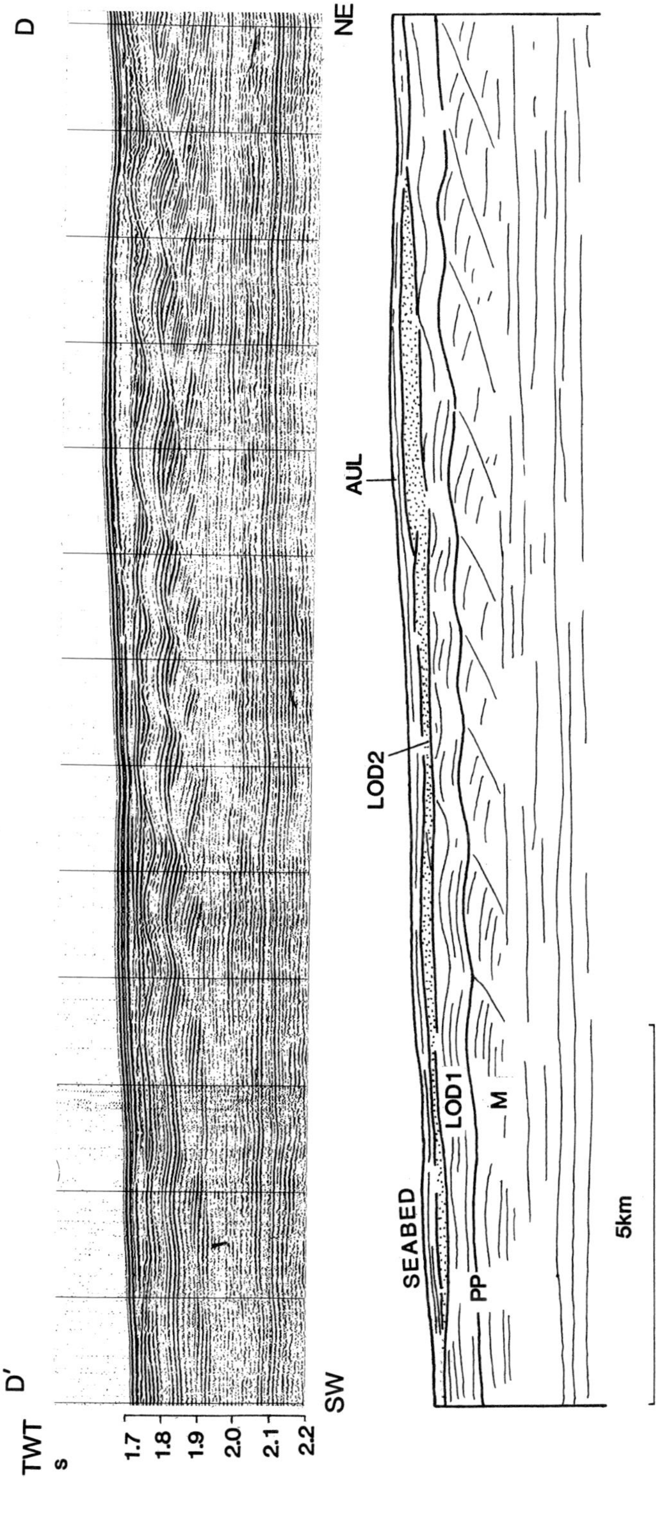

Fig. 8. Airgun profile 85/05–4 and interpreted line drawing showing distal-edge of glacigenic unit (stippled) overlying and possibly locally eroding part of a deep-water dune field. Abbreviations as in Fig. 4. Profile location shown as D–D′ in Fig. 3.

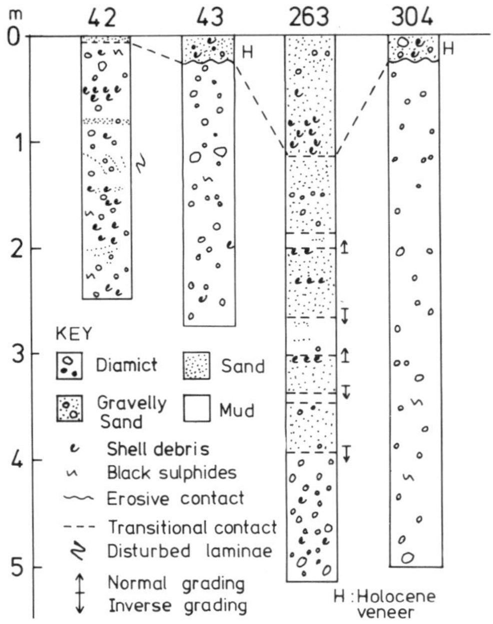

Fig. 9. Graphic logs of the shallow cores. For locations see Fig. 3.

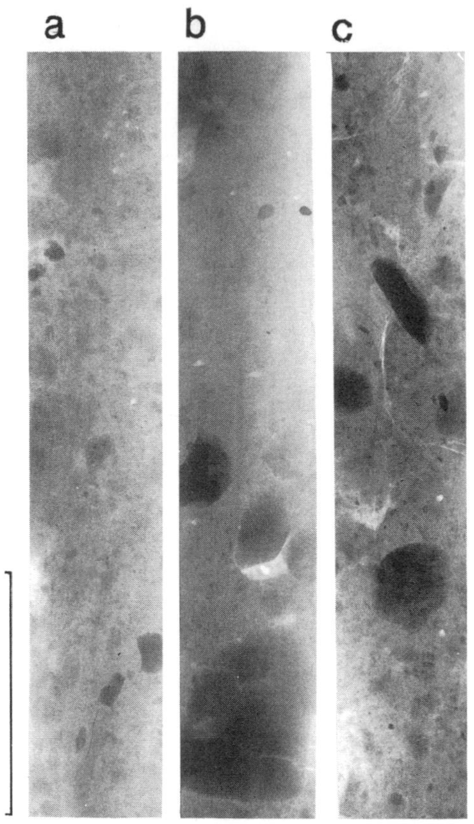

Fig. 10. X-radiographs of diamict in core 43 showing abundant matrix supported clasts and a featureless to slightly mottled background texture. (a) 0.17–0.33 m; (b) 1.53–1.69 m; (c) 2.44–2.60 m. Scale bar is 5 cm.

the characteristic acoustic response of debris flow deposits (Nardin *et al.* 1979; Damuth 1980; Embley 1980). The structureless internal character has been attributed to the scattering of incident sound energy by chaotic bedding and flow structures, while impedance contrasts at the top and bottom of the flows return distinct echoes or reflections (Simm & Kidd 1984). In unconsolidated sediments even a short displacement will be enough to disrupt the continuity of internal reflectors (Embley 1980).

Flow dimensions are consistent with examples described from other settings. On the Mississippi delta, depositional lobes vary in thickness from a few metres to 30 m and have a downslope length up to 10 km. Where lobes coalesce, the resulting overlapping or amalgamated sequence may total a thickness of 50–60 m (Prior & Coleman 1984). On the continental margins off NW Africa (Embley 1976; Jacobi 1976; Simm & Kidd 1984; Kidd *et al.* 1987), eastern N America (Embley 1980), Israel (Almagor & Wiseman 1982), S Korea (Chough 1984) and California (Thornton 1984) debris flows extend down the slope into deep water rise and abyssal plain settings, and range from 1–60 m thick and from several kilometres to several hundred kilometres in length.

The sediment cores support this interpretation. The massive, matrix-supported character of the diamict, together with its random fabric, uniform texture, polymodal clast content, deformed laminae and near absence of fauna and flora are typical of a cohesive debris flow (Hampton 1972; Nardin *et al.* 1979; Flood *et al.* 1979; Embley 1980; Lowe 1982; Hill *et al.* 1982; Pickering *et al.* 1986). Although some lodgement tills and most ice-rafted deposits are lithologically similar (Eyles *et al.* 1983), deformation structures and the association with graded sands (see below) together with the seismic data support a mass flow origin.

The inversely to normally graded sands are comparable to the S2 and S3 divisions of Lowe (1982) associated with deposition from a sandy high-density turbidity current. In such deposits, a thin inversely graded basal zone, S2, often passes up into a masive to normally graded deposit, S3, as the turbidity current passes through traction carpet and suspension sedimentation stages, respectively. Large floating or outsized clasts are a characteristic feature of such deposits (Postma *et al.* 1988). The interbedded muds suggest an intimate association with the sands and may represent suspension sediments deposited by residual low-density turbidity currents (Te division of Lowe 1982), although deposition from sediment plumes cannot be discounted.

The transitional contact between the diamict and the graded sand in core 263 suggests that the two lithologies formed in close association. Hampton (1972) observed that flow separation in the snout of a debris flow winnows out fine sediment which then moves away as a turbidity flow. This mechanism has been used to explain downslope changes in sediment gravity flow deposits in recent cores from the continental slope of the Weddell Sea, Antarctica (Wright & Anderson 1982; Wright *et al.* 1983), and in early Proterozoic glacigenic slope deposits in Canada (Miall 1985).

The lensoid bodies are therefore interpreted as high-energy mass flow deposits consisting primarily of debris flow diamicts with turbidite sands and muds. The lithological and palaeontological data suggest that these sediments represent re-sedimented glacimarine deposits.

The moderate to high amplitude reflectors, separating the mass flow packages, form distinct slope-wide acoustic interfaces. The strength of the reflections suggests a marked contrast in the acoustic impedance across each boundary. Changes in acoustic impedance may be a function of lithological variation and/or changes in the degree of consolidation of the sediments in the succession. If the reflections are a result of lithological variation, their character suggests that they represent thin (*c.* 10 m) sedimentary units, whose lateral continuity and conformable nature imply continuous strata deposited in a relatively widespread and uniform environment, independent of bottom relief (Mitchum *et al.* 1977; Sangree & Widmier 1977). This pattern is generally indicative of low energy, hemipelagic sediments deposited primarily from suspension which, in the present context, could be interpreted as glacimarine, argillite sheet-drapes. The overlying MacAulay sequence, which consists primarily of dropstone muds (Stoker *et al.* 1989), may be a thicker accumulation of such deposits.

Different degrees of consolidation may result from post-depositional changes in the sediment. In the present setting, mass flow deposits are likely to have an excess pore pressure upon deposition. The expulsion of excess pore water through seabed dewatering would result in the reduction of void ratios and lead to an increase in density and shear strength of the deposits. Consequently, the acoustic response may be a result of contrasts in the overall geotechnical character of the depositional packages. Such contrasts between packages are likely to develop if sedimentation was intermittent rather than continuous.

In the absence of direct core information, the reason for the strong reflections remains unknown. Nevertheless, both potential causes, individually or in combination, suggest periods of reduced sedimentation, possibly even non-deposition, on the slope. Sporadic breaks in reflection continuity and/or amplitude changes may be a consequence of erosion by the overlying mass flow deposits.

Discussion

Within the study area, the debris flow succession has well defined downslope and alongslope limits (Fig. 3). The overall extent of the succession is indicated by the distinct bathymetric bulge on the slope (Fig. 1). Thus, their restricted occurrence is attributed to a local debris source, that source being an ice-sheet which reached the shelf-edge in this area. This supports the evidence, presented earlier, that sedimentation occurred during a single glacial cycle. If this part of the Hebridean margin had been subjected to multiple glaciation it seems improbable that the limits of the debris flow packages would coincide so closely during successive glacial cycles.

The slope succession displays a distinct cyclicity with three major episodes of mass flow deposition separated by two relatively quiescent intervals when sedimentation was greatly reduced. In the absence of evidence for multiple glaciation, the cyclicity is attributed to a variable sedimentation rate associated with a single ice-sheet, independent of major climatic variation. Thus, the amount of sediment available to the system was probably dependent upon the state of the grounded ice, i.e. cold- or warm-based, which can vary over daily, seasonal or longer time periods and which controls the productivity of basal meltwater (Eyles *et al.* 1985; Powell 1984). Deposition rates in marine environments are much greater around thawing ice margins than around cold ones (Eyles *et al.* 1985), although it appears that the largest volumes of sediment are associated with ice-sheets which have alternate areas of melting and freezing (Boulton 1974; Clapperton 1975; Mills 1978; Eyles 1979; Shaw 1979). If changes in basal ice state persist throughout the duration of a glaciation, deposition is likely to be episodic with periods of high sedimentation alternating with periods of reduced sediment supply.

In a melting-base regime, it is envisaged that large amounts of ice-proximal, bedload and suspended glacimarine sediment were delivered directly to the outer shelf and upper slope and were subsequently redeposited over the length of the slope. Sediment deposited on the outer shelf may have been redistributed on the slope through ice-push from grounding line fluctuations and surges (Powell 1984). Repeated fluctuations in ice-front location, throughout glaciation, may be responsible for the interdigitating nature of the outer shelf/slope sequences (Fig. 5). Similar relations associated with grounding line movements have been described from the Scotian and mid-Norwegian shelves (King & Fader 1986; King *et al.* 1987; Piper 1988). At the shelf-edge and on the upper slope, the sediment pile may have been unstable due to high sedimentation rates resulting in slope oversteepening and downslope re-sedimentation of the glacimarine sediments by mass flow processes. If the sedimentation rate was high enough, underconsolidation of the sediments may have been a further contributory factor to the triggering of downslope movement (Prior *et al.* 1982). Gravity flowage may also have been triggered by marine geostrophic currents and shock-wave induced liquefaction through iceberg calving and/or sporadic storm wave activity (Powell 1984). Moreover, seismic activity cannot be discounted (Bugge *et al.* 1987; Jansen *et al.* 1987). The variable distribution of debris flows on the slope reflects differing lengths of transport of the material. Whilst some flows stabilised on the mid-to-upper slope, others travelled the length of the slope before accumulating at its base.

During periods of basal freezing, the sediment supply to the outer shelf and slope either ceased or was greatly reduced, as was the potential for sediment redistribution by mechanisms such as ice-push. Such periods may have been of sufficient duration to allow significant post-depositional modifications to occur to the degree of consolidation of the existing slope deposits. Alternatively or additionally, sporadic sediment plumes and transport of material by sea-ice and icebergs (cf. Clark & Hanson 1983) were probably the main contributory sources for any sediment that was deposited on the slope. The restricted sediment supply may have resulted in the deposition of thin, slope-wide units of fine suspension sediments probably with a subsidiary input of coarse ice-rafted detritus. It should be emphasized that this style of sedimentation most likely occurred throughout the depositional history of the glacigenic slope succession, but that its presence is generally masked by the substantial accumulation of debris flows. Any such deposits may originally have been more extensively developed on the outer shelf, but were reworked by subsequent ice-front movements.

Following this glaciation, the outer shelf and slope have been essentially starved of sediment. As noted earlier, the last Scottish ice-sheet is not thought to have extended beyond the Outer Hebrides (Sutherland & Walker 1984), thus any subsequent sedimentation on the slope at this time was probably ice-distal in character. This is confirmed by core analysis of the MacAulay sequence which overlies the debris flows (Stoker *et al.* 1989).

Ice-marginal processes have clearly played a significant role in the construction of the slope apron. Debris flow accumulation on the northern Hebrides Slope has resulted in the shelf-edge locally prograding 5 km seaward. The episodic nature of deposition suggests that progradation was unlikely to have been a uniform process. It is more probable that specific phases of shelf-edge progradation were fairly rapid in association with high-energy mass flow sedimentation.

I would like to thank Dr. R. Harland and Mr. D.K. Graham for providing the palaeontological data, Dr. D. Evans and Mr. D. Long for their critical appraisal of an early draft of the manuscript, and Miss E.J. Gillespie for help in draughting the figures. This paper is published with the permission of the Director, British Geological Survey, NERC.

References

Almagor, G. & Wiseman, G. 1982. Submarine slumping and mass movements on the continental slope of Israel. *In*: Saxov, S. & Nieuwenhuis, J.K. (eds) *Marine slides and other mass movements*. Plenum, New York, 95–128.

Andersen, B.G. 1979. The deglaciation of Norway 15 000–10 000 B.P. *Boreas*, **8**, 79–87.

Belderson, R.H., Kenyon, N.H. & Wilson, J.B. 1973. Iceberg plough marks in the northeast Atlantic. *Palaeogeography, Palaeoclimatology, Palaeoecology*, **13**, 215–224.

Boulton, G.S. 1974. Processes and patterns of glacial erosion. *In*: Coates, D.R. (ed.) *Glacial Geomorphology*, State University of New York, Binghampton, N.Y., 41–87.

Bugge, T., Befring, S., Belderson, R.H., Eidvin, T., Jansen, E., Kenyon, N.H., Holtedahl, H. & Sejrup, H.P. 1987. A giant three-stage submarine slide off Norway. *Geo-Marine Letters*, **7**, 191–198.

Cameron, T.D.J., Stoker, M.S. & Long, D. 1987. The history of Quaternary sedimentation in the

UK sector of the North Sea Basin. *Journal of the Geological Society, London*, **144**, 43–58.

CHOUGH, S.K. 1984. Fine-grained turbidites and mass-flow deposits in the Ulleung (Tsushima) Back-arc Basin, East Sea (Sea of Japan). *In*: STOW, D.A.V. & PIPER, D.J.W. (eds.) *Fine grained sediments: Deep-water processes and facies*, Geological Society Special Publication **15**, 185–196.

CLAPPERTON, C.M. 1975. The debris content of surging glaciers in Svalbard and Iceland. *Journal of Glaciology*, **14**, 395–406.

CLARK, D.L. & HANSON, A. 1983. Central Arctic Ocean sediment texture: a key to ice transport mechanisms. *In*: MOLNIA, B.F. (ed.) *Glacial-Marine Sedimentation*, Plenum, New York, 301–330.

DAMUTH, J.E. 1980. Use of high frequency (3.5 khz–12 khz) echograms in the study of near-bottom sedimentation processes in the deep sea: A review. *Marine Geology*, **38**, 51–75.

EMBLEY, R.W. 1976. New evidence for the occurrence of debris flow deposits in the deep sea. *Geology*, **4**, 371–374.

—— 1980. The role of mass transport in the distribution and character of deep ocean sediments with special reference to the North Atlantic. *Marine Geology*, **38**, 23–50.

EVANS, D., ABRAHAM, D.A. & HITCHEN, K. 1989. The Geikie Intrusive Centre, west of Lewis: Its structure and influence on Tertiary geology. *Scottish Journal of Geology*, **25**, 339–352.

EYLES, C.H., EYLES, N. & MIALL, A.D. 1985. Models of glaciomarine sedimentation and their application to the interpretation of ancient glacial sequences. *Palaeogeography, Palaeoclimatology, Palaeoecology*, **51**, 15–84.

EYLES, N. 1979. Facies of supraglacial sedimentation on Icelandic and Alpine temperate glaciers. *Canadian Journal of Earth Sciences*, **16**, 1341–1361.

—— & MIALL, A.D. 1984. Glacial facies models. *In*: WALKER, R.G. (ed.) *Facies Models*, Geoscience Canada Reprint Series, Geologists Association of Canada, Toronto, 15–38.

—— EYLES, C.H. & MIALL, A.D. 1983. Lithofacies types and vertical profile models: an alternative approach to the description and environmental interpretation of glacial diamict and diamictite sequences. *Sedimentology*, **30**, 393–410.

FLOOD, R.D., HOLLISTER, C.D. & LONSDALE, P. 1979. Disruption of the Feni sediment drift by debris flows from Rockall Bank. *Marine Geology*, **32**, 311–334.

HAMPTON, M.A. 1972. The role of subaqueous debris flow in generating turbidity currents. *Journal of Sedimentary Petrology*, **42**, 775–793.

HILL, P.R., AKSU, A.E. & PIPER, D.J.W. 1982. The deposition of thin bedded subaqueous debris flow deposits. *In*: SAXOV, S. & NIEUWENHUIS, J.K. (eds) *Marine slides and other mass movements*, Plenum Press, New York, 273–287.

HOLTEDAHL, O. 1952. A comparison of a Scottish and a Norwegian shelf area. *Transactions of the Geological Society of Edinburgh*, **15**, 214–220.

JACOBI, R.D. 1976. Sediment slides on the northwestern continental margin of Africa. *Marine Geology*, **22**, 157–173.

JANSEN, E., BEFRING, S., BUGGE, T., EIDVIN, T., HOLTEDAHL, H. & SEJRUP, H.P. 1987. Large submarine slides on the Norwegian continental margin: sediments, transport and timing. *Marine Geology*, **78**, 77–107.

KIDD, R.B., HUNTER, P.M. & SIMM, R.W. 1987. Turbidity-current and debris-flow pathways to the Cape Verde Basin: status of long-range side-scan sonar (GLORIA) surveys. *In*: WEAVER, P.P.E. & THOMSON, J. (eds) *Geology and Geochemistry of Abyssal Plains*, Geological Society Special Publication **31**, 33–48.

KING, L.H. & FADER, G.B.J. 1986. Wisconsinan glaciation of the Atlantic Continental Shelf of Southeast Canada. *Geological Survey of Canada Bulletin*, **363**.

——, MACLEAN, B. & DRAPEAU, G. 1972. The Scotian Shelf submarine end-moraine complex. *International Geological Congress XXIV*, Canada, 237–249.

——, ROKOENGEN, K. & GUNLEIKSRUD, T. 1987. Quaternary seismostratigraphy of the Mid Norwegian Shelf, 65°–67°30′N – a till tongue stratigraphy. *Institutt for Kontinentalsokkelundersøkelser og petroleumsteknologi A/S Report no. 114*.

LOWE, D.R. 1982. Sediment gravity flows: ii. Depositional models with special reference to the deposits of high-density turbidity currents. *Journal of Sedimentary Petrology*, **52**, 279–297.

MIALL, A.D. 1985. Sedimentation on an early Proterozoic continental margin under glacial influence: the Gowganda Formation (Huronian), Elliot Lake area, Ontario, Canada. *Sedimentology*, **32**, 763–788.

MILLS, H.H. 1978. Some characteristics of glacial sediments on Mount Rainier, Washington. *Journal of Sedimentary Petrology*, **48**, 1345–56.

MITCHUM, JR., R.M. VAIL, P.R. & SANGREE, J.B. 1977. Seismic stratigraphy and global changes of sea level. Part 6: Stratigraphic interpretation of seismic reflection patterns in depositional sequences. *In*: PAYTON, C.E. (ed.) *Seismic Stratigraphy- applications to hydrocarbon exploration*. American Association of Petroleum Geologists Memoir, **26**, 117–133.

NARDIN, T.R., HEIN, F.J., GORSLINE, D.S. & EDWARDS, B.D. 1979. *A review of mass movement processes, sediment and acoustic characteristics, and contrasts in slope and base-of-slope systems versus canyon-fan-basin floor systems*. Society of Economic Palaeontologists and Mineralogists Special Publication **27**, 61–73.

PICKERING, K.T., STOW, D.A.V., WATSON, M.P. & HISCOTT, R.N. 1986. Deep-water facies, processes and models: A review and classification scheme for modern and ancient sediments. *Earth Science Reviews*, **23**, 75–174.

PIPER, D.J.W. 1988. Glaciomarine sedimentation on the Continental Slope off Eastern Canada.

Geoscience Canada, **15**, 23–28.
—— & SPARKES, R. 1987. Proglacial sediment instability features on the Scotian Slope at 63°W. *Marine Geology*, **76**, 15–31.
——, NORMARK, W.R. & SPARKES, R. 1987. Late Cenozoic stratigraphy of the Central Scotian Slope, Eastern Canada. *Bulletin of Canadian Petroleum Geology*, **35**, 1–11.
POSTMA, G., NEMEC, W. & KLEINSPEHN, K.L. 1988. Large floating clasts in turbidites: a mechanism for their emplacement. *Sedimentary Geology*, **58**, 47–61.
POWELL, R.D. 1984. Glacimarine processes and inductive lithofacies modelling of ice shelf and tidewater glacier sediments based on Quaternary examples. *Marine Geology*, **57**, 1–52.
PRIOR, D.B., BORNHOLD, B.D., COLEMAN, J.M. & BRYANT, W.R. 1982. Morphology of a submarine slide, Kitimat Arm, British Columbia. *Geology*, **10**, 588–592.
—— & COLEMAN, J.M. 1984. Submarine slope instability. *In*: BRUNSDEN, D. & PRIOR, D.B. (eds) *Slope Instability*, Wiley, 419–455.
RICHARDS, P.C., RITCHIE, J.D. & THOMSON, A.R. 1987. Evolution of deep-water climbing dunes in the Rockall Trough-Implications for overflow currents across the Wyville-Thomson Ridge in the (?) late Miocene. *Marine Geology*, **76**, 177–183.
ROBERTS, D.G., HUNTER, P.M. & LAUGHTON, A.S. 1977. Bathymetry of the Northeast Atlantic, sheet 2: Continental margin around the British Isles. *Institute of Oceanographic Sciences*.
SANGREE, J.B. & WIDMIER, J.M. 1977. Seismic stratigraphy and global changes of sea level. Part 9: Seismic interpretation of clastic depositional facies. *In*: PAYTON, C.E. (ed.) *Seismic stratigraphy- applications to hydrocarbon exploration*, American Association of Petroleum Geologists Memoir **26**, 165–184.
SHAW, J. 1979. Genesis of the Sveg tills and Rogen moraines of Central Sweden: a model of basal melt-out. *Boreas*, **8**, 409–426.
SIMM, R.W. & KIDD, R.B. 1984. Submarine debris flow deposits detected by long-range side-scan sonar 1000 kilometres from source. *Geo-Marine Letters*, **3**, 13–16.
STOKER, M.S. 1988a. *Sula Sgeir (59°N, 8°W): Sea Bed Sediments*. British Geological Survey 1:250 000 Offshore Map Series.
——, 1988b. *Sula Sgeir (59°N, 8°W): Solid Geology*. British Geological Survey, 1:250 000 Offshore Map Series.
——, 1988c. Pleistocene ice-proximal glaciomarine sediments in boreholes from the Hebrides Shelf and Wyville-Thomson Ridge, NW UK Continental Shelf. *Scottish Journal of Geology*, **24**, 249–262.
——, 1990. *Sula Sgeir (59°N, 8°W): Quaternary Geology*. British Geological Survey, 1:250 000 Offshore Map Series.
——, HARLAND, R., MORTON, A.C. & GRAHAM, D.K. 1989. Late Quaternary stratigraphy of the northern Rockall Trough and Faeroe-Shetland Channel, northeast Atlantic Ocean. *Journal of Quaternary Science*, **4**, 211–222.
SUTHERLAND, D.G. 1987. Submerged rock platforms on the continental shelf west of Sula Sgeir. *Scottish Journal of Geology*, **23**, 251–260.
—— & WALKER, M.J.C. 1984. A late Devensian ice-free area and possible interglacial site on the Isle of Lewis, Scotland. *Nature*, **309**, 701–703.
THORNTON, S.E. 1984. Basin model for hemipelagic sedimentation in a tectonically active continental margin: Santa Barbara Basin, California Continental Borderland. *In*: STOW, D.A.V. & PIPER, D.J.W. (eds) *Fine-grained sediments: deep-water processes and facies*, Geological Society Special Publication **15**, 377–394.
WRIGHT, R. & ANDERSON, J.B. 1982. The importance of sediment gravity flow to sediment transport and sorting in a glacial-marine environment: Weddell Sea, Antarctica. *Geological Society of America Bulletin*, **93**, 951–963.
——, ANDERSON, J.B. & FISCO, P.P. 1983. Distribution and association of sediment gravity deposits and glacial/glacial-marine sediments around the Continental Margin of Antarctica. *In*: MOLNIA, B.F. (ed.) *Glacial-Marine Sedimentation*, Plenum, New York 265–300.

Sedimentation patterns and facies geometries on a temperate glacially-influenced continental shelf: the Yakataga Formation, Middleton Island, Alaska

CAROLYN H. EYLES[1] & MARTIN B. LAGOE[2]

[1] *Department of Geography, McMaster University, Hamilton, Ontario, Canada L8S 4K1*

[2] *Department of Geological Sciences, The University of Texas at Austin, Austin, Texas 78713, USA*

Abstract: The 5 km thick Yakataga Formation of the Gulf of Alaska provides the longest and most accessible record of late Cenozoic glaciation in the world. The formation consists of interbedded marine and glacimarine sediments of late Miocene to Holocene age and is well exposed on offshore islands and in the coastal mountain ranges as a result of the convergent margin setting of the Gulf. About 1.3 km of early Pleistocene Yakataga Formation sediments are exposed on extensive intertidal platforms around Middleton Island which is sited close to the modern continental shelf edge. Variably graded gravels and sands at the base of the Middleton Island succession were deposited by sediment gravity flows and represent the fill of a large submarine channel. The remainder of the succession is dominated by thick (up to 100 m) planar tabular units of fossiliferous diamict containing a series of striated boulder pavements and shell-rich beds (coquinas). Diamict facies formed predominantly by the settling out of fines from suspended sediment plumes together with ice-rafting of coarser debris; foraminiferal biofacies data indicate palaeobathymetries of between 20 and 250 m. Boulder pavements are lag surfaces which developed in relatively shallow water and were subsequently faceted and striated by a grounding ice sheet extending to the continental shelf edge; coquinas record episodes of current winnowing in shallow water (20–50 m) but also contain ice-rafted debris. Extensive mud units formed in relatively deep water during periods of reduced ice volume and are similar to 'mud blankets' accumulating on the modern shelf. The Yakataga Formation on Middleton Island contains a rich record of changing palaeobathymetry and ice volumes in the Gulf of Alaska.

Glacimarine sedimentation began along the Gulf of Alaska continental margin 5–6 million years ago in response to tectonic uplift of the Chugach and St. Elias Mountains along the Alaskan coast and late Cenozoic climatic cooling (Armentrout 1983) and continues to the present day (Plafker & Addicott 1976; Molnia 1983). This interval is recorded by the Yakataga Formation, a 5 km thick succession of marine and glacimarine clastic rocks of late Miocene to Holocene age. These rocks are exposed on offshore islands and on mountain slopes around the northeastern margins of the Gulf of Alaska basin (Fig. 1), as a result of strong uplift in a complex convergent plate margin setting (Bruns & Schwab 1983; Plafker 1987). The Yakataga Formation is argued to be the longest and most complete sedimentary record of late Cenozoic glaciation in the world (Plafker & Addicott 1976) and spans the development of the high latitude northern hemisphere ice sheets around 2.5–3 million years ago (Kennett 1982). The Yakataga Formation thus has the potential to reveal a long and detailed record of climate, tectonic activity and sea level change in the Gulf of Alaska during the late Cenozoic and may be used as a high resolution reference for glaciation and climate history in the North Pacific Ocean.

This paper describes the results of combined studies of sedimentary facies and microfaunal biofacies contained within a thick and extensive succession of Yakataga Formation sediments exposed on Middleton Island (Fig. 1). The succession consists predominantly of stacked tabular glacimarine diamict facies which contain coquinas (shell-rich beds, N. Eyles & Lagoe 1989) and boulder pavements (C.H. Eyles 1988); diamicts are intimately associated with marine muds and sandy gravelly submarine channel-fill deposits (C.H. Eyles 1987). Analysis of microfaunas within the diamict facies indicates sedimentation in water depths that varied between 20 m and 250 m (Lagoe *et al.* 1989).

Shallow marine shelves fringing glaciated continents are major depositories for glacially-

From Dowdeswell, J.A. & Scourse, J.D. (eds), 1990, *Glacimarine Environments: Processes and Sediments*. Geological Society Special Publication No 53, pp 363–386.

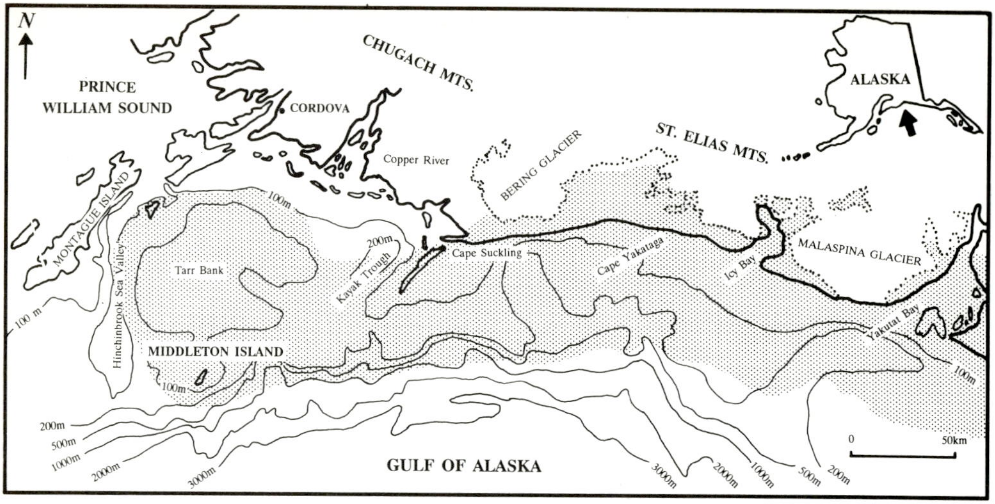

Fig. 1. (a) Location map, Gulf of Alaska region showing outcrop and subsea distribution of the Yakataga Formation (stippled area). Bathymetric contours are given in metres.

produced sediments (Bell & Laine 1985). Sediments accumulating in these environments are likely to be selectively preserved to form the dominant glacigenic sediment type in the rock record (C.H. Eyles *et al.* 1985). Although there is a reasonable understanding of processes operating in modern glacimarine environments (Anderson *et al.* 1979, 1980, 1984; Powell 1981, 1983; Mackiewicz *et al.* 1984), little is known of the nature of sedimentary sequences accumulating in such environments. The Yakataga Formation on Middleton Island provides a unique opportunity to document lithofacies types, associations and geometries, and associated biofacies assemblages, on a temperate glacially-influenced continental shelf. This is particularly significant given the continuity of glacimarine processes to the present day in the Gulf.

Age and regional setting

The Yakataga Formation consists of a series of marine and glacimarine clastic sediments of late Miocene to Holocene age and underlies a large area of the Gulf of Alaska continental shelf and adjacent mainland (Miller 1953; Plafker & Addicott 1976; Armentrout 1983). The formation crops out on the mainland in a broad belt along the southern margin of the Chugach and St. Elias mountains, extending from the Suckling Hills in the west to Icy Point in the east (Fig. 1). It was named by Taliaferro (1932) after its recognition in the coastal area of the Yakataga district (Plafker & Addicott 1976). Yakataga Formation sediments are also exposed on a number of offshore islands in the Gulf of Alaska (Miller 1953; Plafker & Addicott 1976; C.H. Eyles 1987, 1988).

The regional tectonic setting of the Gulf of Alaska is one of a complex convergent plate margin in which the northward moving Pacific plate is undergoing subduction below the North American plate along the Aleutian trench in the western part of the Gulf. In the eastern sector, complex transpressive movements are occurring between the two plates along major right-lateral transform faults (e.g. Denali, Fairweather, Totschunda faults; Rogers 1977; Bruns 1983; Bruns & Schwab 1983; Plafker 1987). The Yakataga Formation forms part of the thick Tertiary basin fill of the eastern Gulf of Alaska (Plafker 1967; Plafker & Addicott 1976; Bruns 1985) and was deposited in a complex linear shelf basin which has undergone extensive deformation due to convergence tectonics. Compressional stresses operate along a northwest–southeast trend (Bruns & Schwab 1983; Stephens *et al.* 1984) producing broad folds and thrusts below the continental shelf and causing pulsatory uplift of the mountain belt fringing the Gulf coast (Armentrout 1983).

Mainland exposures of Yakataga Formation sediments are assigned a middle Miocene age on the basis of molluscan megafossils (Allison 1978) but analysis of planktonic foraminifera and radiometric ages from glauconites in the underlying Poul Creek Formation rocks indicate a late Miocene age for the earliest Yakataga Formation sediments (Armentrout *et al.* 1978). More accessible sections through younger Yakataga Formation rocks are found offshore on Middleton Island (Figs 1 & 2) and have been dated by palaeomagnetism at between 1.8 and 0.7 million years (Plafker & Addicott 1976; Allison 1978). Middleton Island sits on the shelf break structural high (von Huene 1972; Plafker 1987; Fig. 1) and is undergoing strong pulsatory uplift and tilting, recorded by a series of terraces on the island, the oldest of which is older than 4000 years old (Miller 1953; Plafker 1965, 1984; Plafker & Rubin 1967; Plafker & Addicott 1976). The most recent uplift, of 3.5 m, occurred during the 1964 Alaskan earthquake (Plafker 1965, 1984; Prescott & Lisowski 1977) and resulted in the exposure of a total thickness of over 1.25 km of Yakataga Formation sediments across extensive wave cut platforms (Fig. 2). The sediments are partially lithified and strike northeast–southwest with dips of approximately 28° to the northwest; most of the north-westward tilting of the Middleton Island succession is thought to have occurred during the Holocene (Plafker 1987).

Lithofacies descriptions

The Yakataga Formation sediments on Middleton Island were first described by Miller (1953); Plafker & Addicott (1976) also gave brief descriptions of these sediments in a regional summary of the Yakataga Formation. This study presents a geological map of Middleton Island based on 1982 air photographs and field mapping (Fig. 2) as well as sedimentological logs summarizing the lithofacies sequences occurring along the northern, western and southern shores of the island, which were uplifted and exposed in 1964 (Fig. 3).

The Middleton Island succession is about 1.25 km thick (Fig. 3) and is dominated by thick (up to 100 m) units of massive, stratified, and deformed diamict facies containing boulder pavements and shell bands (coquinas). The diamict units are interbedded with muddy sands, silts and gravels (Figs 2 & 3). A 200 m thick sequence of channelized gravels and sands forms the base of the exposed succession.

Diamict facies

Diamict facies are poorly lithified and consist of scattered angular to rounded clasts within a sandy mud matrix (Figs 4–6). The concentration of clasts is extremely variable and ranges from less than 10 clasts m^{-2} to 280 clasts m^{-2}, with clast size ranging from granules (> 2 mm) to boulders more than 4 m in diameter. Striated and/or faceted clasts are present (Fig. 4c). Barnacles and calcareous worm tubes are common on the upper surfaces of clasts (Fig. 4d); molluscs and microfossils, together with wood fragments, are locally abundant (Fig. 4e). Clast lithologies include sandstones, siltstones, granites and mafic extrusive and intrusive igneous rocks all derived from source terrains in the Chugach and St. Elias mountains (Miller 1953; Plafker & Addicott 1976). Diamict facies can be subdivided into massive and stratified types.

Massive diamicts (Facies Dmm; see key, Fig. 3). Massive diamicts are ungraded and unstratified and typically show a random distribution of clasts within the matrix (Fig. 4a, b), although angular clasts of the same lithology may be grouped together as clast 'clusters' (Fig. 4f). Clast long axes are randomly oriented and no distinct clast fabric is apparent. Textural analyses indicate that massive diamict facies have a poorly sorted sandy mud matrix (Fig. 5). Basal contacts with underlying muds or diamicts are typically conformable, and often show well developed loading structures; upper contacts are either transitional into massive muddy sands, muds or channelized laminated muds with dropstones, or are sharp and overlain by other diamicts (Fig. 3). Massive diamict facies form extensive stratiform beds up to 100 m thick and are particularly common along the northern shore of the island (Fig. 2).

Stratified diamicts (Facies Dms). Stratification is defined in diamict facies either by variability in matrix texture, concentration of clasts into lines or bands, or the presence of abundant sand stringers and gritty lenses (Fig. 6a). Clast fabric is generally random but some units show preferential alignment of clast long axes parallel to the trend of stratification. Shell fragments and microfossils are locally abundant and can be concentrated in clast-rich bands (Fig. 6b). Stratified facies are coarser grained and more poorly sorted than massive diamicts (Fig. 5) and often infill broad shallow channels cut into underlying massive facies (Fig. 6b). Basal contacts may either be flat-lying and conformable or channelized and slightly erosive; the latter are marked by concentrations of clasts or shell fragments. Channelized units of stratified diamict are most common at the southern end of the island (Fig. 2b).

Massive and stratified diamict facies commonly contain deformed masses of sand and gravel which form a variety of pillow, dyke and wedge-like injection structures (Clark 1984; Fig. 7). Large pillows of fine-grained muddy sand are common in several massive diamict beds in the northern part of the island (Fig. 2 & 7a),

(a)

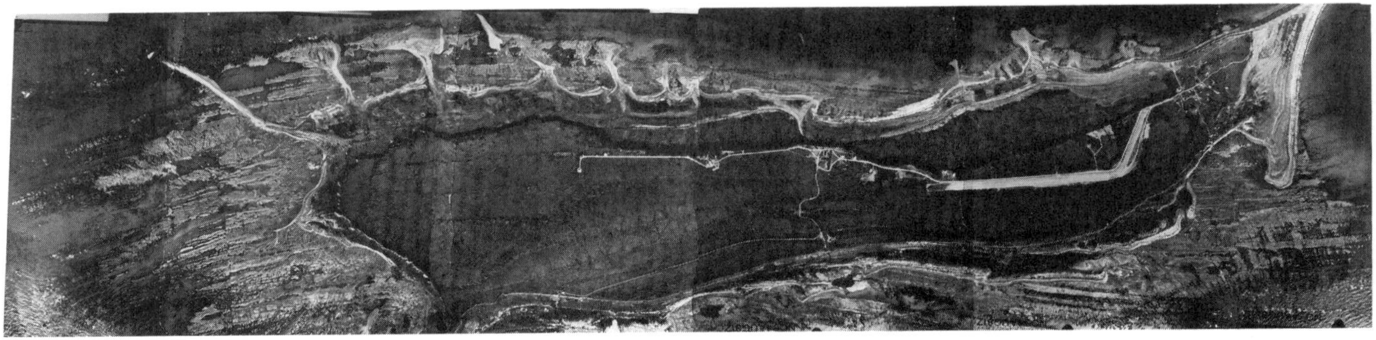

(b)

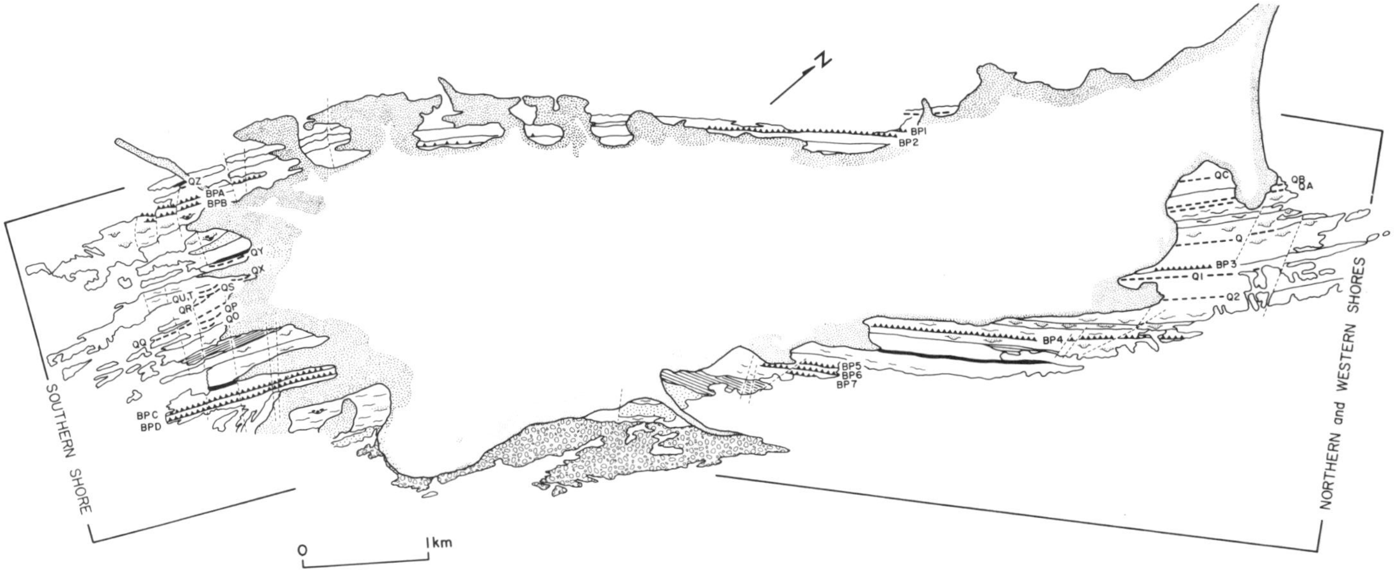

Fig. 2. (a) Composite aerial photograph of Middleton Island taken by the National Oceanographic Administration, August 1981. The island consists of a series of uplifted marine terraces, the oldest of which has been dated at c. 4000 years BP. (Plafker & Rubin 1967). (b) Geological map of Middleton Island based on field logging and air photo interpretation. Yakataga Formation sediments exposed on intertidal platforms surrounding the island strike NE–SW with dips of between 22–37° to the northwest. Location of measured sections used to draw sedimentological logs (Fig. 3) shown. Symbols as in Fig. 3 except for coquinas shown by heavy broken lines, faults shown by fine broken lines.

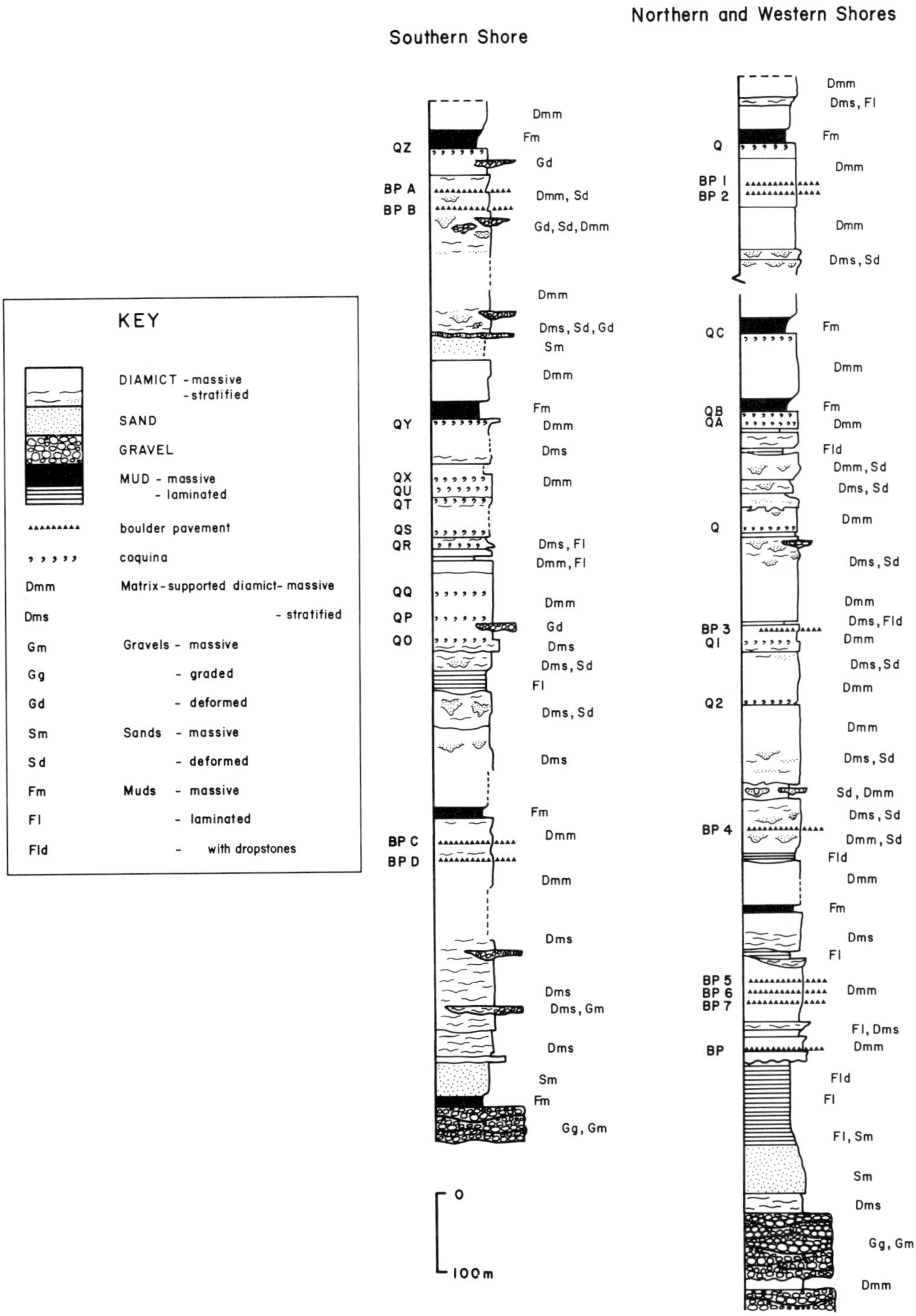

Fig. 3. Sedimentological logs through the Yakataga Formation on Middleton Island drawn from traverses of the northern, western (composite log) and southern shores of the island.

Fig. 4. Massive diamict facies. (a) massive, clast-poor diamicts exposed on the northern shore. Stratigraphic top to upper left. Ice axe is approximately 1 m long. (b) clast-rich, massive diamict, northern shore. Note absence of well developed clast fabric. (c) striated, bullet-shaped clast in massive diamict. (d) calcareous worm tubes on the upper surface of a clast in a stony diamict unit. (e) unbroken mollusc shell (*Neptunea?*) in massive diamict. (f) 'cluster' of angular clasts of the same lithology (sandstone) within massive diamict. The clasts are not shaped or faceted and probably originated as ice-rafted englacial debris dumped from a melting iceberg.

whereas deformed gravel beds are more common in stratified diamicts exposed along the southern shores (Figs 2 & 3). Diamict beds containing deformed sand and gravel masses form relatively thick, stratiform units (up to 30 m thick; Fig. 3), which can be traced for several kilometres along strike (Fig. 2).

Boulder pavements and coquinas. Boulder pavements and shell bands (coquinas) both form extensive planar surfaces within diamict facies (Fig. 2). Boulder pavements, with flattened and striated upper surfaces, can be traced over several kilometres along strike and show a consistent striation direction (Fig. 8; C.H. Eyles 1988). Few clasts on these pavements have the characteristics bullet shape or long axis alignment typical of debris transported and deposited from the basal traction zone of grounded glacier ice as reported by Boulton (1978). Instead, boulder shape varies from sub-rounded to sub-

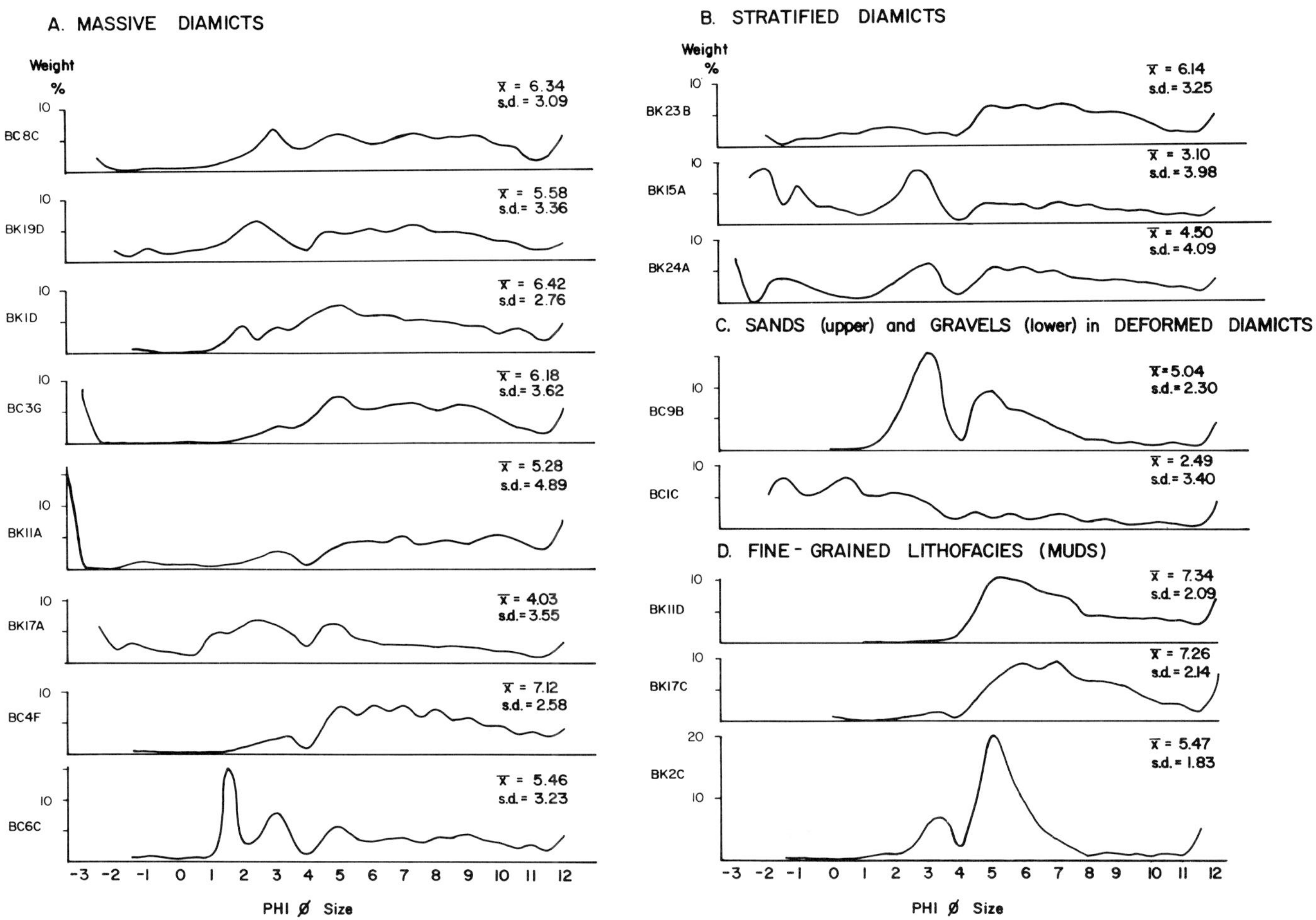

Fig. 5. Textural analyses of diamict, sand and mud lithofacies on Middleton Island. × = mean grain size; s.d. = standard deviation. Samples were analysed at half phi intervals using automated 'sedigraph' (Duncan & Lahaie 1979) and seiving techniques.

(a) (b)

Fig. 6. Stratified diamict facies. (a) bands of clast-rich and clast-poor diamict defining stratification. Matrix texture is coarser in the clast-rich bands. (b) stratified diamict filling channel cut into underlying massive diamict facies. The clast-rich bands of the stratified unit contain abundant shell fragments. Stratigraphic top to upper right.

(a) (b)

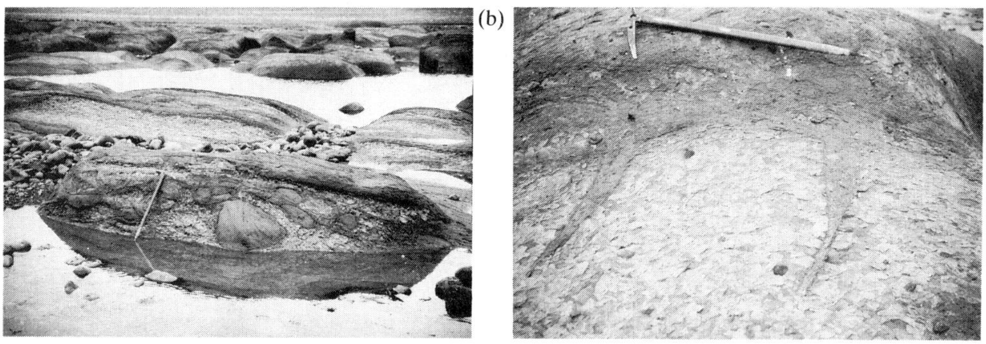

Fig. 7. Deformation structures within diamict facies. (a) pillows of fine-grained muddy sand in clast-poor diamict. Ice axe is approximately 1 m long. (b) wedge-shaped injections of muddy sand into underlying massive diamict.

angular and long axis orientation is random (Fig. 8). Boulder pavements occur within both massive and stratified diamicts that contain concentrations of whole, unabraded microfossils and lack any structures suggestive of deformation by horizontal shear stresses. Concentrations of shell fragments and burrow structures can be seen along several pavements (Fig. 8c). Nine pavements have been identified on Middleton Island, some of which may be tentatively traced from the northern to the southern outcrops (Fig. 2). Detailed descriptions of these boulder pavements are given by C.H. Eyles (1988).

Shell rich bands (coquinas) also form extensive surfaces within diamict facies and consist of whole and broken bivalves, gastropods, brachiopods, worm tubes and barnacles in a matrix of either calcareous sand or mud (Fig. 9; N. Eyles & Lagoe 1989). Coquinas cropping out along the northern shore of the island consist predominantly of pecten shells (*Chlamys* sp.) in a coarse, sandy matrix and form beds up to 1 m thick (Fig. 9a). These can be traced as planar surfaces for up to 3 km along strike (Fig. 2). Shell-rich bands of the southern shores are typically channelized and discontinuous and form shelly lags in a muddy matrix (Fig. 9b).

Gravel facies (Facies Gm, Gms, Gg, Gi)

Gravel facies on Middleton Island are concentrated at the base of the succession on the eastern shore in a 200 m-thick sequence of broad, stacked channel fills (Fig. 3). The gravels show variable grading characteristics and eight distinct facies types can be identified, including massive, inversely and normally graded facies (Fig. 10; C.H. Eyles 1987). Clasts are commonly striated and faceted; occasional outsize clasts

(a)
(b)
(c)

Fig. 8. Boulder pavements. (a) B.P. 2, western shore. This pavement consists of large clasts up to 3 m in diameter and can be traced for several kilometres along strike. Figure (arrowed) for scale. Stratigraphic top to right. (b) Faceted and striated upper surface of boulders in B.P. 4, northern shore. (c) Gravel filled U-shaped burrows (immediately above hammer) along B.P. 2, western shore. Stratigraphic top to upper right.

occur. The gravels fill channels up to 500 m wide and 70 m deep and are associated with graded sands. Lenticular units of massive diamict consisting of scattered clasts in a muddy sand matrix, are associated with the channelized gravels. Channel fills consisting of poorly sorted and crudely graded angular gravels in a muddy sand matrix, are also exposed locally in cliffs on the southwestern shore (Fig. 10c) and crop out patchily within diamicts in the nearby intertidal zone (Fig. 10d). These isolated gravel channel fills are restricted to exposures along the southern shore and do not occur in the northern part of the island (Fig. 2b).

Sand facies (Sm, Sd)

Discrete sand beds are a minor component of the Middleton Island succession and account for less than 5% of the sediments logged (Fig. 3). The most extensive and thickest sand unit (75 m thick) occurs immediately above channelized gravels at the base of the sequence and consists of massive to crudely stratified silty sand (Figs 3 & 11a). This unit contains abundant burrows containing simple retrusive spreiten structures (e.g. Eckdale *et al.* 1984; Fig. 11b), occasional small, randomly oriented cylindrical burrows, and thin (< 3 cm), irregular beds of gritty, grey limestone composed predominantly of echinoid spines and bryozoan material (Fig. 11c).

Sand facies also occur as normally and inversely graded facies associated with the coarse, gravelly channel fills at the base of the succession (Fig. 3; C.H. Eyles 1987).

Fine-grained facies (Fl, Fld, Fm)

Fine-grained facies occur either as laminated units of silt, fine sand and clay or as massive muds (Fig. 12). Laminated or thin-bedded facies consist of stacked graded beds, ranging in thickness from less than 1 cm to around 20 cm, which contain isolated dropstones (Fld; Fig. 12a). These facies fill broad shallow channels on the upper surface of diamict units (Fig. 12b) and are commonly associated with channelized stratified diamicts on the southern shores of the island (Fig. 2b). A thick unit (75 m) of laminated fine sand and mud forms the upper part of the fining upwards sequence at the base of the Middleton Island succession on the eastern shore (Fig. 3). The laminae have a gradational basal contact with crudely bedded sands and fine and thin upwards, with a progressive increase upsection in the number of dropstones and diamict *pellets*. They are sharply overlain by clast-rich massive diamicts (Fig. 3).

Massive muds occur as stone-free interbeds (< 25 m thick) within diamict facies and often overlie coquina beds (Fig. 12c). These facies are laterally continuous and several massive mud beds can be traced from one side of the island to the other, particularly in the upper part of the succession (Fig. 3). The 23 m thick

Fig. 9. Coquinas (shell-rich beds). (a) View along Q1, northern shore, showing planar outcrop geometry. Stratigraphic top to left. (b) Channelized coquina consisting of abundant shell fragments in a mud matrix. QR, southern shore.

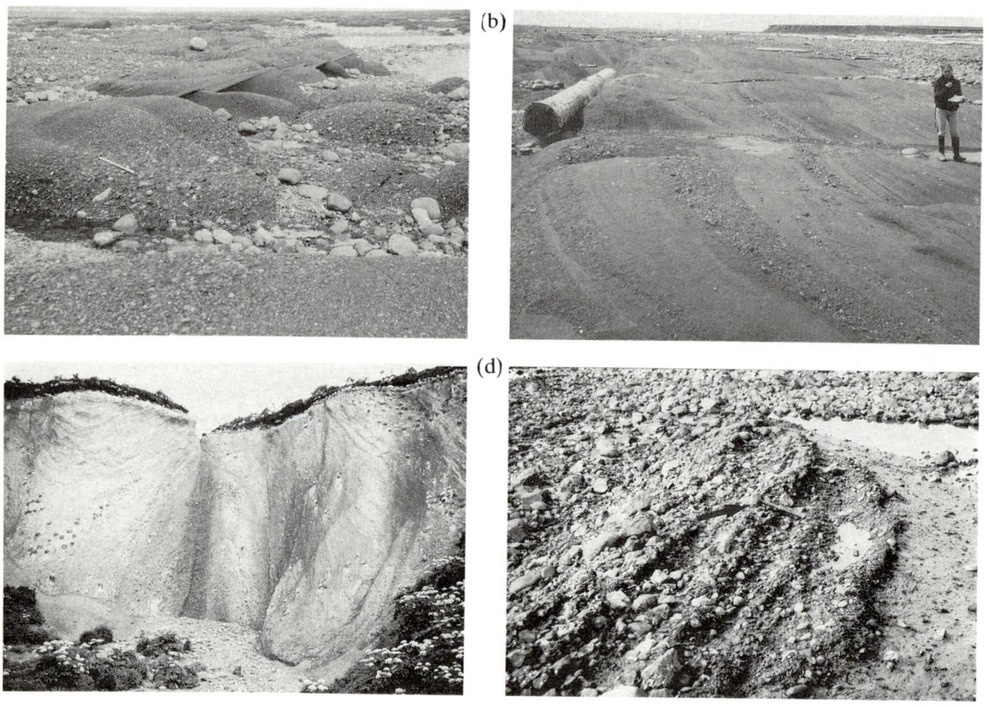

Fig. 10. Gravel facies. Graded gravel channel fills. (a) normally and inversely graded gravels filling broad channels up to 500 m wide at the base of the Middleton Island succession. Stratigraphic top to upper left. (b) normally graded gravels and sands forming a broad channel fill. Stratigraphic top to right. (c) steeply dipping gravels filling a channel exposed in cliff face on the southwestern shore. Stratigraphic top towards viewer. Cliff is approximately 20 m high. (d) poorly sorted gravels filling a narrow channel (< 30 m wide) cut into diamict exposed in the intertidal zone of the southern shore. Hammer (arrowed) for scale.

Fig. 11. Sand facies (stratigraphic top to upper right in all photographs). (a) crudely bedded muddy sand overlying gravel channel fills at the base of the Middleton Island succession. The sands contain abundant burrows and shell fragments. (b) single retrusive spreiten burrows in muddy sands. The burrows are vertical to near vertical and may be echinoid burrows. (c) massive to crudely bedded muddy sands above the channel fills on the southern shore. The horizontal lines (arrowed) are thin, irregular limestone beds which consist of echinoid spines and bryozoan fragments.

unit of sandy mud exposed at the top of the succession on the western and southern shores contains abundant shell fragments, sparse ice-rafted clasts and a series of horizontal pebble bands which form the upper surface of underlying burrowed zones (Fig. 12d).

Foraminiferal biofacies

Foraminifera are common throughout most of the Yakataga Formation. Faunal analysis can provide crucial paleoenvironmental information for the interpretation of depositional environments represented by Yakataga lithofacies (*e.g.* Lagoe 1983; Lagoe *et al.* 1989). Most species of foraminifera found in the Yakataga Formation on Middleton Island are still living in the modern Gulf of Alaska. The paleoenvironmental analysis of the Yakataga biofacies is therefore based on an understanding of modern foraminiferal distributions.

Lagoe *et al.* (1989) provide a quantitative analysis of calcareous benthic foraminiferal death assemblages in the modern Gulf of Alaska, based on the census data of Echols & Armentrout (1980). Modern Gulf of Alaska faunas also contain large numbers of agglutinated species and analyses of total benthic faunas (calcareous and agglutinated) are utilized by Zellers & Lagoe (1990) and Zellers (1989) to address paleoenvironmental problems in the Yakataga Formation in the onshore Robinson Mountains. Middleton Island foraminiferal faunas contain very rare or no agglutinated species, hence the modern biofacies model of Lagoe *et al.* (1989) is utilized here. Biofacies patterns based on total faunas (calcareous and agglutinated) are easily related to the patterns seen in the calcareous species alone (Zellers & Lagoe 1990).

Five major and three minor biofacies are recognized from quantitative analysis (cluster and factor analysis) of calcareous benthic foraminiferal distributions in the modern Gulf of Alaska (Fig. 13; Lagoe *et al.* 1989). The *Elphidium excavatum clavatum* biofacies represents inner neritic water depths (0–80 metres) and sandy substrates. The *Cassidulina* spp. biofacies represents outer neritic water depths (80–170 metres) and muddy substrates just seaward of the 'mudline' (see Armentrout 1980). The *Epistominella pacifica* biofacies characterizes upper bathyal water depths (170–500 metres) and muddy substrates. The *Nonionellina labradorica* biofacies is found sporadically in upper bathyal water depths on sandy substrates in Yakutat Seavalley. The final biofacies is dominated by *Cassidulina californica* and is found in sediment-starved areas (relict sediments, bedrock highs, etc.). The biofacies is not bathymetrically restricted but is most

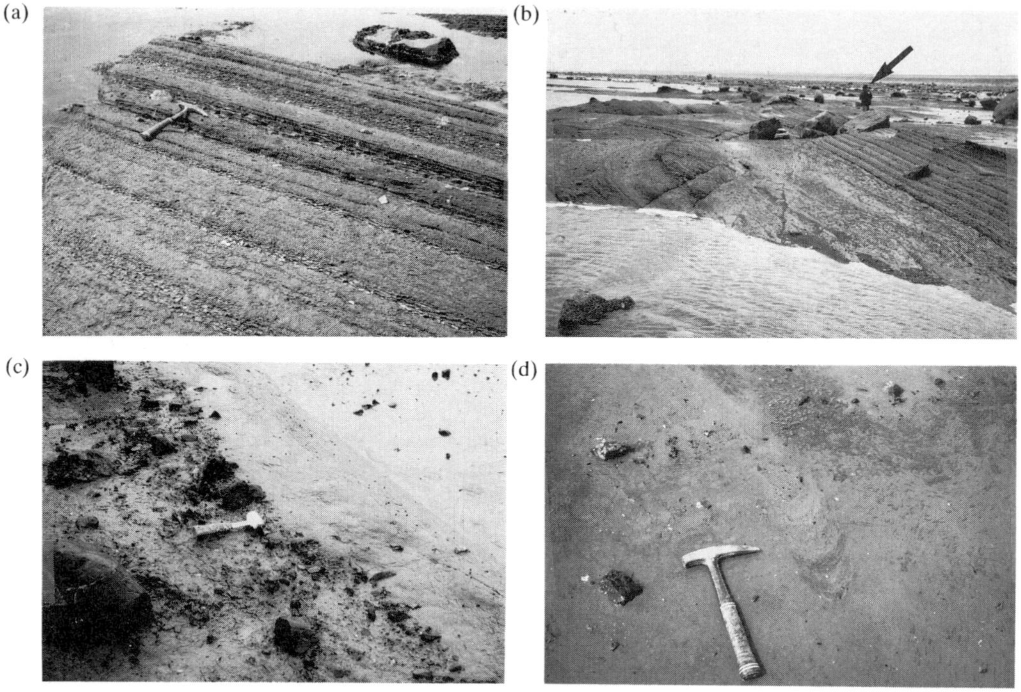

Fig. 12. Fine-grained facies. (a) stacked graded beds of fine sand to mud containing scattered dropstones. These beds fill a broad channel (> 100 m wide) on the upper surface of a diamict unit. (b) thinly-bedded fine-grained facies with dropstones filling a channel on the surface of a stratified diamict unit. Stratigraphic top to right. Figure (arrowed) for scale. (c) contact between shelly lag on upper surface of stratified diamict unit and massive to crudely bedded muds above. Stratigraphic top to upper right. (d) sandy muds on the western shore of the island containing large burrow structures which originate from shelly and clast-rich lag surfaces. Stratigraphic top to upper left.

sensitive to sedimentation rates, preferring areas of little or no net sediment accumulation (Lagoe *et al.* 1989; Bergen & O'Neil, 1979; Quinterno *et al.* 1980). Less commonly encountered biofacies (factors 6, 7 and 8 on Fig. 13) are dominated by *Quinqueloculina* spp., *Trichohyalus ornatissima* and *Elphidium oregonese* respectively and represent innermost neritic assemblages reflecting the environmental variability in the nearshore region. The relationships summarized above demonstrate that the distribution of benthic foraminiferal biofacies in the modern Gulf of Alaska exhibit important correlations with water depth, substrate, and sedimentation rate. It should be emphasized that the current modern biofacies model is most representative of interglacial conditions with no significant tidewater ice margins located on the open continental shelf.

Analyses to date of benthic foraminiferal faunas from Middleton Island samples indicate the presence of four major biofacies (Lagoe *et al.* 1989), all of which show close resemblance to modern biofacies (Table 1). The *Elphidium excavatum clavatum* biofacies represents inner neritic water depths; the *Cassidulina norcrossi* s.l. biofacies is very similar to the modern *Cassidulina* sp. biofacies and represents outer neritic water depths; the *Uvigerina juncea* biofacies is most similar to the modern *Epistominella pacifica* biofacies and represents outermost neritic to uppermost bathyal waterdepths; and the *Cassidulina californica* biofacies is similar to its modern namesake, representing sediment-starved areas.

The distribution of these biofacies provides a palaeoenvironmental framework for the Yakataga Formation on Middleton Island. Foraminifera are generally most abundant in muds, but are found in all lithofacies, including diamicts. The general distribution of major biofacies and inferred water depths in the sections studied are shown on Fig. 14. Maximum water depth fluctuations indicated by the biofacies

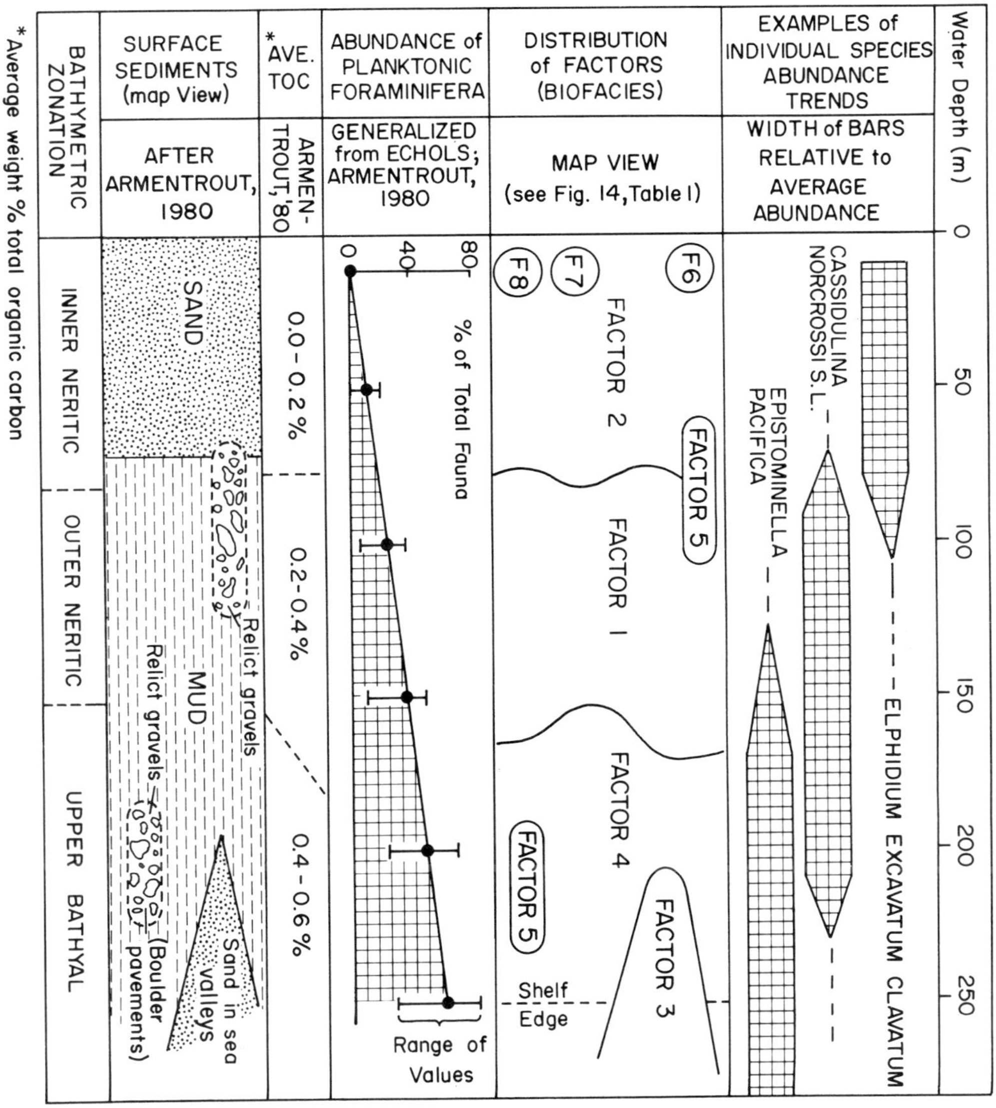

Fig. 13. Summary diagram of major modern foraminiferal biofacies trends (calcareous species only) on the Yakataga continental margin, Gulf of Alaska. Figure is based on quantitative analysis by Lagoe *et al.* (1989) of data in Echols & Armentrout (1980). Factor 2 = *Elphidium excavatum clavatum* biofacies; Factor 1 = *Cassidulina* spp. biofacies; Factor 4 = *Epistominella pacifica* biofacies; Factor 3 = *Nonionellina labradorica* biofacies; and Factor 5 = *Cassidulina californica* biofacies. Factors 6,7,8 are sporadically-distributed, innermost neritic biofacies dominated by *Quinqueloculina* spp., *Trichohyalus ornatissima* and *Elphidium oregonense* respectively. (Figure from Lagoe *et al.* 1989.)

patterns are on the order of 100–200 metres. In addition to these paleobathymetric trends, the occurrence of the *Cassidulina californica* biofacies in coquina beds and some boulder pavements indicates that periods of at least local sediment-starvation punctuated the depositional history of the Yakataga Formation on Middleton Island.

Interpretation of facies

Diamict facies

Miller (1953) proposed the term *Yakatagite* to describe the diamict facies of the Yakataga Formation, suggesting they formed in a shallow marine environment by the combined processes

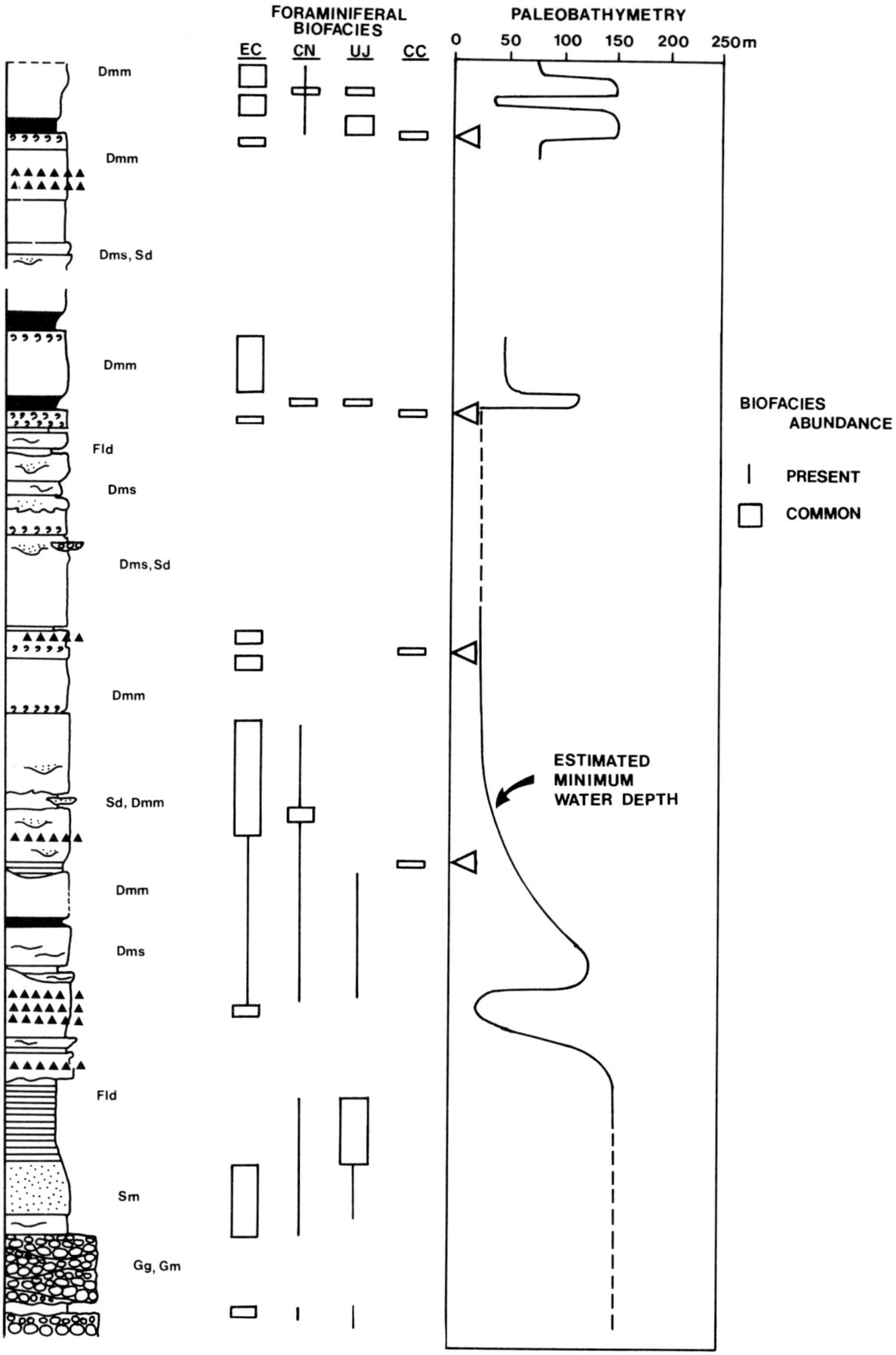
FORAMINIFERAL BIOFACIES
EC
CN
UJ
CC
PALEOBATHYMETRY
0
50
100
150
200
250m
Dmm
Dmm
Dms, Sd
Dmm
Fld
Dms
Dms, Sd
Dmm
Sd, Dmm
Dmm
Dms
Fld
Sm
Gg, Gm
BIOFACIES ABUNDANCE
PRESENT
COMMON
ESTIMATED MINIMUM WATER DEPTH

Table 1. *Biofacies types identified from Middleton Island*

1. *Elphidium excavatum clavatum* Biofacies (Abr. EC): characteristic taxa are *Elphidium excavatum clavatum, Elphidiella hannai, Elphidium bartletti, Buccella* sp., *Quinqueloculina* spp. Inferred palaeoenvironment: Inner neritic (*c.* 20–80 m); sandy substrates; photic zone.
2. *Cassidulina norcrossi* s.l. Biofacies (Abr. CN): characteristic taxa are *Cassidulina norcrossi* s.l., *Cassidulina limbata, Cassidulina minuta, Cassidulina islandica*, subordinate *Cassidulina californica*. Inferred palaeoenvironment: Outer neritic (*c.* 80–150 m); muddy substrates; subphotic zone.
3. *Cassidulina californica* Biofacies (Abr. CC): characteristic species are dominant *Cassidulina californica, Cibicides* cf. *conoideus*, subordinate *Cassidulina norcrossi* s.l. *Cassidulina limbata.* This biofacies often grades into the *Cassidulina norcrossi s.l.* biofacies. Inferred palaeoenvironment: Neritic (not strongly restricted bathymetrically); firm to hard substrates in areas with little or no net sediment accumulation.
4. *Uvigerina juncea* Biofacies (Abr. UJ): characteristic species are *Uvigerina juncea, Epistominella pacifica, Pullenia salisburyi*, subordinate *Epistominella naraensis*. Inferred palaeoenvironment: Outermost neritic/uppermost bathyal (*c.* 150–250 m); muddy substrates; subphotic zone.

Environmental interpretations based on modern benthic foraminifera which represent interglacial conditions (Lagoe *et al.* 1989).

of suspension deposition, ice rafting and highly variable traction current activity. Analysis of molluscan assemblages from diamicts in the Middleton Island succession suggests deposition in water depths ranging between 15 m and 136 m (Allison 1978). Foraminiferal data suggest water depths varied from 20 to 250 m (Table 1). Diamicts in the lower part of the Middleton Island succession appear to have formed in deeper water than those of the middle part (Fig. 14).

The sandy mud matrix of the diamict lithofacies on Middleton Island (Fig. 5) is similar to glacially-derived mud accumulating in quiet nearshore environments of the modern Gulf of Alaska by settling of suspended fines from suspension (Plafker & Addicott 1976; Molnia & Carlson 1978; Plafker 1981). However, deposition of ice-rafted debris in the modern gulf is restricted to fjord environments with tidewater ice margins and very little coarse-grained material is rafted out onto the open shelf (Molnia & Carlson 1978; Powell 1981, 1983; Molnia 1981, 1983; Powell & Molnia 1989). The thick diamict sequences on Middleton Island were deposited during episodes of expanded ice cover around the Gulf of Alaska when ice margins reached tidewater and supplied abundant ice-rafted material to the open shelf environment (Fig. 15; Plafker & Addicott 1976). High rates of mud accumulation (averaging 4.5 m/1000 years; Molnia *et al.* 1980) are characteristic of the modern Gulf as a result of the very large quantities of fine-grained sediment brought in by glacially-fed river systems draining the coastal mountain ranges (*e.g.* Copper River; Galloway 1977). Even higher sedimentation rates may have occurred during times of expanded ice cover.

Massive diamicts (Facies Dmm). Massive diamict facies on Middleton Island probably formed as a result of rapid mud deposition and ice rafting on a shallow marine shelf under quiet water conditions in which little to no sorting of the accumulating sediment occurred (Fig. 15b; Miller 1953; Plafker & Addicott 1976). Such conditions exist in water depths below the lower limit of wave influence or in shallower water depths below an ice shelf or a dense cover of icebergs or seasonal ice which dampen wave activity (Miller 1953). Modern winter storm waves in the Gulf of Alaska can disturb seafloor sediments at shelf edge depths of up to 200 m (Carlson *et al.* 1977). This suggests that floating ice may have had a significant dampening effect on wave energies during Yakataga Formation time. Conformable basal contacts indicate that massive diamict facies accumulated passively by settling from suspension.

Stratified diamicts (Facies Dms). Stratified diamicts accumulate where some size sorting of the constituent materials occurs either as a result of traction current winnowing, selective depo-

Fig. 14. Stratigraphic summary of measured sections through Yakataga Formation on the northern and western shores of Middleton Island (for lithofacies key see Fig. 3). Foraminiferal biofacies abbreviations explained in Table 1. Paleobathymetric curves are minimum estimates, see Table 1 for ranges of possible water depths. Triangles on left of paleobathymetric column indicate periods of sediment-starvation corroborated by foraminiferal biofacies analysis. Ongoing work will likely document additional horizons associated with coquinas and some boulder pavements.

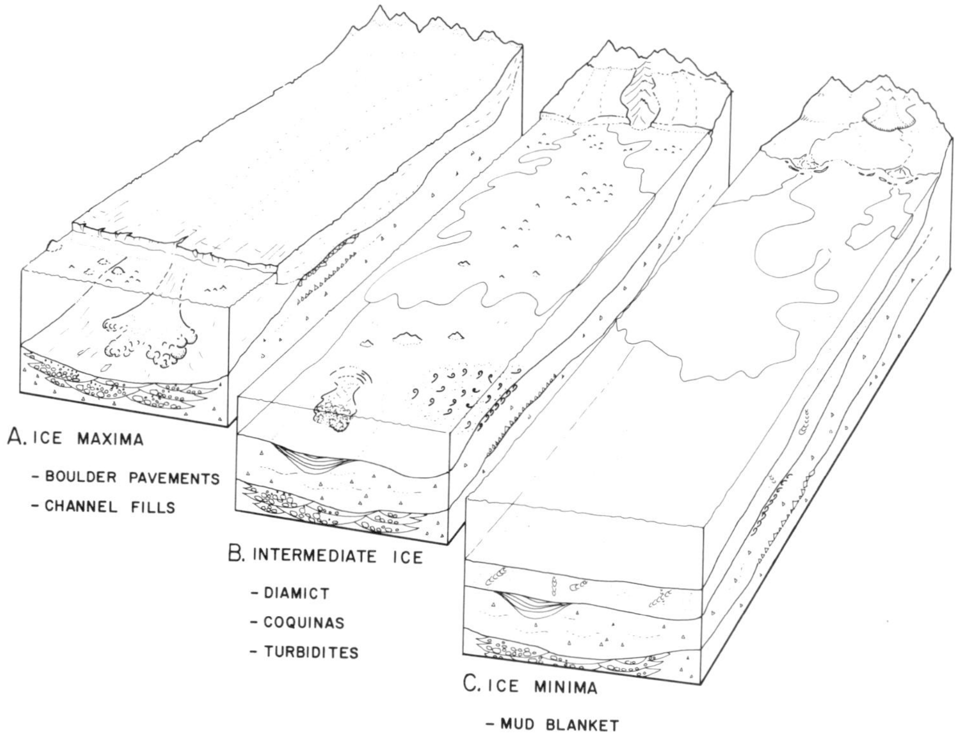

Fig. 15. Summary of depositional processes responsible for formation of the Middleton Island succession. (a) maximum ice cover on the shelf; winnowed boulder lags are striated to form boulder pavements on banks and shoals; density underflows and turbidites derived from subglacial meltwaters fill channels on the shelf edge/ upper slope with thick sequences of variably graded gravels and sands. (b) intermediate ice cover: setting of fine-grained sediment from meltwater plumes and coarser debris from floating icebergs produces extensive tabular diamict units. Localized resedimentation events produce turbidite basin-fills; sediment starvation on banks allows coquinas to develop. (c) minimal ice cover: supply of IRD is removed and extensive mud blankets form (e.g. modern Gulf of Alaska shelf).

sition of certain grain sizes (Anderson *et al.* 1980), or by downslope movement as sediment gravity flows (Fig. 15b; Kurtz & Anderson 1979; Anderson *et al.* 1979; Wright & Anderson 1982). Slow moving currents with velocities as low as 1 cm s^{-1} (Hollister & Heezen 1972) can cause separation of coarser and finer size grades during settling of suspended sediment and produce textural banding in the diamict matrix (Barrett 1975; Elverhoi & Roaldset 1983; Orheim & Elverhoi 1981; Anderson *et al.* 1984). Higher velocity currents can winnow both the substrate and accumulating sediment to form either shelly or pebbly lags (e.g. Fig. 6b) or relatively coarse-grained stratified diamicts (e.g. *residual* facies; Anderson *et al.* 1980). Sand and gravel may accumulate as discontinuous beds or lenses. Bottom currents capable of eroding and transporting fine-grained sediment can be generated by wave and tidal processes and are particularly effective during storms. The modern longshore Coastal Current and shelf edge Alaska Current have velocities averaging 20 cm s^{-1} and 30 cm s^{-1} respectively (Reed & Schumacher 1987).

Stratified diamicts that show juxtaposed beds with variable matrix textures, flow structures and banding, and preferential alignment of clast long axes parallel to flow direction, are the result of downslope resedimentation of material, probably as cohesive debris flows (Lowe 1979; Visser 1983). Sediment gravity flows are common on the modern shelf and are triggered by a variety of processes including oversteepening of depositional slopes, earthquake shocks, isostatic readjustments, storm wave shock or iceberg turbation (Reimnitz 1972; Shepard & Emery 1973; Ness & Kulm 1973; Carlson & Molnia 1977; Anderson *et al.* 1979; Schwab & Lee 1983, 1988; Postma 1984; Clukey *et al.* 1985).

Large deformed masses of sand and gravel

contained within some diamict facies most likely originated by the soft sediment loading of relatively coarse-grained material into a poorly consolidated diamict substrate (e.g. Anketell *et al.* 1970; Mills 1983; N. Eyles & Clark 1985). The origin of these sediments and associated structures is not fully understood as yet. Sands and gravels may have been derived from shallow water areas and rapidly deposited by sediment gravity flows or traction currents generated by storms or periods of particularly high discharge of glacial meltwater. In addition, the Gulf of Alaska is one of the most seismically active areas in North America and experiences frequent earthquake shocks which are known to generate reverse density loading and downslope failure of poorly consolidated sediments on the Alaskan continental shelf (Waller 1968; Page 1975; Sangrey & Molnia 1979; Bruns & Schwab 1983; Plafker 1984; Schwab & Lee 1988). At least 95 earthquakes greater than magnitude 6 on the Richter Scale have occurred on the Kodiak Shelf in the Gulf of Alaska since 1902 (Hampton 1983). Deformation structures and associated 'chaotic' masses of gravel and sand may alternatively be the product of ice berg grounding and turbation (e.g. Vorren *et al.* 1983; Barnes & Lien 1988). These diamicts and included sand and gravel masses accumulated in relatively shallow water (Fig. 14) and are therefore very likely to have been influenced both by floating ice and storms.

Boulder pavements and coquinas. The formation of striated boulder pavements is generally attributed to lodgement processes at the base of moving glaciers (Dreimanis & Reavely 1953; Dreimanis 1976; Boulton 1978). Boulder pavements on Middleton Island, however, sit in a sequence of undisturbed glacimarine sediments and individual clasts lack the distinctive shaping and long axis orientation of boulders deposited directly from basal ice (Boulton 1978). An alternate mechanism for their formation is suggested by C.H. Eyles (1987) and involves the development of submarine boulder lag surfaces by storm wave and current winnowing of underlying diamict lithofacies (e.g. Molnia & Carlson 1980; Carlson 1989) and the subsequent abrasion of the lag by a grounding marine ice sheet (Fig. 15a). The presence of thin coquinas and burrows along several of the pavements gives strong support to their origin as lag surfaces. Consistently oriented striae on the boulder surface indicate abrasion by glacier ice moving toward the south rather than by berg or seasonal ice which are known to produce more variable striae (Plafker & Addicott 1976; McKay & McKay 1977; Hansom 1983). Microfaunal data from the diamict immediately below and above the pavements indicate that boulder pavements formed in water depths of between 20 and 150 m and rapid deepening immediately followed the formation of some boulder pavements (e.g. B.P. 5; Fig. 14). It is most likely that ice that abraded the pavements was part of the fringing ice sheet around the Gulf that expanded at several different times to the continental shelf edge. The absence of glacitectonic structures and sediments characteristic of subglacial sedimentation below a grounded ice sheet (e.g. N. Eyles *et al.* 1982) strongly suggests that the ice sheet was partially floating at its outer margins and only locally touched down on bank highs. The expanding ice sheet was probably in a surging state, moved very rapidly to the shelf edge and was broken up by the subsequent rise in sea level.

Microfaunal data indicate that the coquinas on Middleton Island formed in relatively shallow water depths (20–50 m) and in some cases were rapidly covered by deeper water muds (Coquinas A, B and C; Fig. 14). The *Cassidulina californica* biofacies is characteristically associated with coquinas (Fig. 14); this biofacies is also found on shallow shoal and bank areas of the modern Gulf of Alaska and is typical of areas with firm to hard substrates and little or no net sediment accumulation (Lagoe *et al.* 1989; Bergen & O'Neil 1979; Quinterno *et al.* 1980; Echols & Armentrout 1980). This suggests a model of coquina formation involving reduced water depths and increased energy regimes which prevented mud deposition and caused the formation of lag surfaces (Bloos 1982; Kaye 1984; N. Eyles & Lagoe 1989). These sediment-starved lag surfaces would then be available for colonization by shelly faunas. Similar shelly lags form on bank areas winnowed by marine currents in modern Alaskan fjords (Powell & Molnia 1989). The presence of abundant ice-rafted clasts within the coquinas indicates that water depths were sufficient to allow the free passage of floating ice over the depositional site; in regions of temperate glaciers, such as the Gulf of Alaska, relatively small bergs are calved from tidewater ice margins and can move freely in shallow water depths in excess of approximately 15 m (B.F. Molnia, pers. comm., 1985). Channelized muddy coquinas, common on the southern shore of the island, probably represent shell-rich lags formed as channelized currents scoured and winnowed shell-bearing diamict lithofacies (N. Eyles & Lagoe 1989). Such currents may have originated as sediment gravity flows or density underflows hugging lows on the basin floor.

Sand and Gravel Facies

The 200 m thick suite of massive, normally and inversely graded gravels and sands forming the base of the Middleton Island succession (Fig. 3) was deposited by channelized sediment gravity flows ranging from debris flows to turbidites, gravity flows similar to those described by Walker (1975, 1978), Lowe (1979, 1982) and Nardin *et al.* (1979) (Fig. 15a; C.H. Eyles 1987). Mapping shows that these coarse-grained facies crop out as a series of stacked channel fills within a multiple channel system (C.H. Eyles 1987). A glacial sediment source is indicated by the common occurrence of striated and faceted clasts within the gravels; background ice-rafting is suggested by scattered *outsize* clasts and interbedded lenticular units of massive diamict. The massive diamict units formed as abandoned channels were filled passively with pelagic mud and ice-rafted debris and contain foraminiferal assemblages characteristic of outer shelf and upper slope environments (*Uvigerina juncea* Biofacies; Fig. 14). The lowermost coarse-grained facies on Middleton Island thus appear to represent the infill of a submarine channel system cut into the outer continental shelf edge and upper slope (C.H. Eyles 1987). This channel system was probably supplied with large quantities of glacial and fluvioglacial sediment by grounded and partially floating ice sheets that extended out on to the Gulf of Alaska continental shelf from the coastal mountain ranges (Fig. 15a).

The coarse-grained channel fills form the base of a thick fining upwards sequence (Fig. 3), being overlain by a thin unit of stratified diamict and a thick sequence of massive to crudely bedded muddy sands. Similar poorly sorted sand facies have been attributed to rapid deposition of suspended sediment from decelerating turbidity currents (Middleton 1970; Hiscott & Middleton 1979; Walker 1984). The sands probably fill a broad channel overlying the channelized gravels and formed as cold, dense, sediment-laden underflows or turbidity currents swept down the channel. Modern meltwater streams, draining the coastal glaciers around the Gulf of Alaska, feed large sea valleys that cross the continental shelf (Fig. 1) in which thick wedges of similar silty sand and clayey silt are accumulating (Carlson *et al.* 1977; Molnia & Carlson 1980; Carlson *et al.* 1982). The absence of ice-rafted material in both the modern sands and the muddy sands and muds on Middleton Island implies that ice margins did not reach tidewater during deposition of the latter. A reduction in regional ice volumes would explain the overall fining upwards trend at the base of the Middleton Island succession (Fig. 3). Benthic foraminifera contained within sand facies overlying the basal gravels indicate outer shelf conditions and water depths of between 150 m and 250 m (Fig. 14) while planktic foraminifera (*Neoglobigerina pachyderma*) show a right-coiling pattern indicative of cool temperate water masses. In contrast, modern planktic foraminifera from the Gulf of Alaska show left-coiling patterns characteristic of sub-arctic water masses (Bradshaw 1959). These data support the notion that the fining upwards trend at the base of the Middleton Island succession resulted from a reduction in regional ice volumes and reduced sediment supply associated with eustatic sea level rise.

Fine-grained facies

The thick package of laminated silts forming the upper part of the fining upwards sequence at the base of the Middleton Island succession (Fig. 3), appears to record the final stages of infilling of a broad submarine channel by turbidites. The fining upwards trend and absence of ice-rafted material in the sands and lower silts suggests, as stated above, the progressive withdrawal of a glacial sediment source. Broad channel fills of graded silts and clays that occur on the upper surface of diamict units (Fig. 3) are also interpreted as turbidite deposits.

Massive muds record quiet-water sedimentation of suspended sediment during times of reduced ice cover and are analogous to the thick postglacial muds that blanket much of the modern Alaskan continental shelf (Molnia & Carlson 1980; Molnia 1983; Carlson 1989). Several massive mud units can be correlated from one end of the island to the other (Figs 2 & 3). Foraminiferal biofacies indicate deposition in relatively deep water (150–250 m; Fig. 14); Planktic foraminifera (*Neoglobigerina pachyderma*) found in one massive mud unit close to the top of the succession (Fig. 14), show right-coiling patterns characteristic of relatively warm (cool temperate) water masses. These patterns are typical of 'interglacial' episodes when sea level was eustatically raised.

Horizontal shell-rich pebble bands and burrowed horizons, separating individual massive sandy mud units, probably represent lag horizons produced by periodic erosion of the substrate or variable sedimentation rates and interrupted mud accumulation (Wilson 1982; Eckdale *et al.* 1984).

Lateral Facies Relationships and Geometries

The fining upwards sequence at the base of the exposed section on the northern shore of Middleton Island is not exactly matched by exposures on the southern shore which pass from graded gravels into massive sands and a thick sequence of stratified diamicts (Fig. 3). The lateral facies relationships between these two sequences are difficult to establish as exposures are limited and the upper contact of the gravel channel fills is extensively faulted; much of this faulting is probably due to synsedimentary growth faulting and slumping caused by rapid sedimentation of the channel fills. Such complex lateral facies relationships are typical of submarine channel sequences (Massari 1978; Hein & Walker 1982; Clifton 1984).

Correlation of the measured sections from the remainder of the northern/western and southern shores of the island (Fig. 3) is based on a 'best fit' matching of massive mud beds, boulder pavements and coquinas (Figs 2 & 3). However, few of these features continue the whole length of the island; only boulder pavements A/B and 1/2 and C/D and 4 can be correlated from one end of the island to the other (Figs 2 & 3). Planar bounding surfaces are most common in the sequences exposed on the northern and western shores of the island where tabular diamict beds predominate and show little lateral facies variability over more than 5 km along strike (Fig. 2b). The only slight irregularities in this simple layered stratigraphy are the broad shallow channels on the upper surface of the diamict units which are filled with laminated muds (Figs 2b & 12b).

Facies geometries are most difficult to determine on the southern shores due to extensive faulting (Fig. 2b). However, boulder pavements A/B and C/D define a thick, tabular diamict package containing a series of channelized stratified diamict units and coquinas (Figs 2 & 3). In general it is difficult to correlate any bed along the length of the island as the lateral transition between the tabular outcrop forms of the north and the more channelized and faulted forms of the south is obscured below the vegetated part of the island (Fig. 2b). The same architectural form, consisting of tabular packages of diamict with internal complexity produced by locally channelized and lenticular facies, is also shown by Yakataga Formation sediments exposed on coastal mountain ranges around the Gulf (Fig. 16; Armentrout 1983).

Fig. 16. Yakataga Formation sediments exposed in the Richardson Mountains of the Gulf of Alaska coastal ranges. The sediments consist of interbedded marine mudstones, sandstones and glacimarine diamictites (Armentrout 1983). The sections shown here are approximately 1200 m high.

Lateral variability of facies types and geometries within an overall 'layer-cake' stratigraphy can be explained by differences in local depositional conditions. On Middleton Island, tabular diamicts of the northern shore accumulated on a relatively flat substrate periodically exposed to areal scour by traction currents. In contrast, the more channelized facies of the southern exposures formed in an environment where sediment sorting by channelized currents predominated; this environment may record the presence of a topographic low on the basin floor along which density underflows and turbidity currents were concentrated. When boulder pavements 5, 6 and 7 (Figs 2 & 3) were forming at the northern end of the island, it is possible that channelized turbid meltwaters were scouring the substrate on the southern shores to produce the well stratified diamicts and discontinuous gravel pockets (Figs 2 & 3). Likewise, planar coquinas on the northern shores of the island (Q2 through QC; Figs 2 & 3) may be equivalent to channelized coquinas on the southern shores (Q0 through QY; Figs 2 & 3).

Discussion

Figure 15 summarizes the depositional processes responsible for the formation of the Middleton Island succession. Development of features such as boulder pavements and gravel channel fills occurred when ice volumes were at their maximum and a partially grounded/partially floating ice sheet extended to the continental shelf edge (Fig. 15a). This situation may have existed only for relatively short periods of time when ice streams from the coastal mountains coalesced in the Gulf and surged rapidly across the shelf. During periods of 'intermediate' ice volume, when many coastal glaciers reached tidewater but did not extend far onto the shelf, extensive units of diamict accumulated (Fig. 15b). Localized sediment starvation on banks and shoals, caused by relative sea level lowering and relatively high energy regimes, allowed colonization of lags by shelly faunas and the development of coquinas. Further reduction of ice volumes removed the ice-rafted component of sedimentation and resulted in the accumulation of extensive mud blankets (Fig. 15c). This situation is well represented by the modern 'interglacial' Gulf of Alaska continental shelf (Molnia & Carlson 1980; Molnia 1983).

The Middleton Island succession contains a particularly rich faunal record which allows detailed reconstruction of changing water depths through time. Water depths ranged from 250 m during deposition of the massive mud beds and lenticular diamicts within the gravel channel fills to less than 30 m during formation of some coquinas and diamicts (Fig. 14). Such changes in water depth may have been caused by glacio-isostatic or tectonic movements of the basin floor or by eustatic sea level changes related to changing global ice volumes; all of these factors are known to have affected relative water depths across the Gulf of Alaska in post-glacial times but their individual effects cannot as yet be firmly distinguished in the Middleton Island succession. Identification of right- and left-coiling planktic foraminifera may help discriminate major eustatic controls on sea level elsewhere in the Yakataga Formation, particularly in mainland outcrops which record deeper water depositional conditions (Lagoe 1983).

Local tectonic movements of the basin were probably important in controlling water depths and sedimentation style on Middleton Island as the island sits on a fault-bounded horst block which forms part of the tectonically-active outer shelf high (von Heune 1972). Coquinas and boulder pavements probably record changes in relative water depth over this paleobathymetrically sensitive 'high' but are unlikely to be represented in laterally-equivalent deeper water sediments. Such features are therefore probably unsuitable for use as interregional marker horizons.

The characteristics of Yakataga Formation sedimentation on Middleton Island are controlled to a large extent by the convergent plate margin setting of the basin. Changes in plate movement and uplift of the coastal mountain ranges in late Miocene times are thought to have been a major factor in the initiation of glaciation in southern Alaska (Hopkins 1972; Plafker & Addicott 1976); continued, punctuated uplift of the mountains has probably influenced glacial ice volumes in the region throughout the late Cenozoic. Tectonic activity has also caused repeated changes in relative sea level and sediment supply across the continental shelf (e.g. Plafker 1965), events recorded by lithological changes and erosion surfaces in the resulting sedimentary successions. Active mountain building, resulting from convergent margin tectonics around the Gulf of Alaska, has been instrumental in releasing large volumes of sediment to be transported by meltwaters issuing from wet-based (temperate) glaciers, out onto the continental shelf. As a result, even the modern Gulf of Alaska region, with reduced ice volumes, is characterized by very high sedimentation rates (Molnia *et al.* 1980).

The data presented in this paper are the results of the first stage of a detailed regional sedimentological and micropalaeontological study of Yakataga Formation sediments. They serve to illustrate the enormous value of combining two fields of study to maximize the amount of palaeoenvironmental information that can be obtained from any succession. Future plans are to extend this work to mainland outcrops and incorporate stable isotope and palaeomagnetic analyses in the hope of resolving some of the climatic, tectonic and bathymetric changes that have occurred in the Gulf of Alaska during the late Cenozoic. These data will serve as a yardstick with which to compare similar data from deep sea cores to be collected in the northern Pacific in the 1990s, and from mid-latitude Pacific regions. By applying this interdisciplinary approach to the Yakataga Formation, which is the largest, most continuous and most accessible record of glacially-influenced marine sedimentation on earth, it is hoped that a better understanding of global change through the late Cenozoic will be obtained.

This work was funded by a Natural Science and Engineering Research Council of Canada University Research Fellowship and Operating Grant to Eyles and a National Science Foundation grant (Grant #EAR 8720823) to Lagoe. We are grateful to the Alaskan Chugach Native Corporation for permission to work on Middleton Island and to the Federal Aviation Administration for logistic support. The manuscript was reviewed by Nick Eyles, Paul Carlson and Martin Sharp.

References

ALLISON, R.C. 1978. Late Oligocene through Pliocene molluscan faunas in the Gulf of Alaska region. *The Veliger*, **21**, 171–188.

ANDERSON, J.B., BRAKE, C.F. & MYERS, N.C. 1984. Sedimentation on the Ross Sea continental shelf, Antarctica. *Marine Geology*, **57**, 295–334.

——, KURTZ, D.D., DOMACK, E.W. & BALSHAW, K.M. 1980. Glacial and glacial marine sediments of the Antarctic continental shelf. *Journal of* Geology, **88**, 399–414.

——, —— & WEAVER, F.M. 1979. *Sedimentation on the Antarctic Continental Slope*. Society of Economic Paleontologists and Mineralogists Special Publication **27**, 265–283.

ANKETELL, J.M., CEGLA, J. & DZULYNSKI, S. 1970. On the deformational structures in systems with reversed density gradients. *Annales Societatis Geologorum Poloniae*, **40**, 3–29.

ARMENTROUT, J.M. 1980. Surface sediments and associated faunas of upper slope, shelf and bay environments, Yakataga-Yakutat area, northern Gulf of Alaska. *In*: FIELD, M.E. *et al.* (eds). *Quaternary Deposits and Environments of the Pacific Coast*. Pacific Section, Society of Economic Paleontologists and Mineralogists, 241–255.

—— 1983. Glacial lithofacies of the Neogene Yakataga Formation, Robinson Mountains, Southern Alaska Coast Range, Alaska, *In*: MOLNIA, B.F. (ed.), *Glacial-Marine Sedimentation*. Plenum Press, New York, 629–666.

——, ECHOLS, R.J. & NASH, K.W. 1978. *Late Neogene climatic cycles of the Yakataga Formation, Gulf of Alaska*. Stanford University Publications, Geological Sciences **14**, 3–4.

BARNES, P.W. & LIEN, R. 1988. Icebergs rework shelf sediments to 500 m off Antarctica. *Geology*, **16**, 1130–1133.

BARRETT, P.J. 1975. Textural characteristics of Cenozoic preglacial and glacial sediments at site 270, Ross Sea, Antarctica. *In*: *Initial Reports, Deep Sea Drilling Project, 28*. United States Government Printing Office, Washington, D.C., 757–765.

BELL, M. & LAINE, E.P. 1985. Erosion of the Laurentide region of North America by glacial and glaciofluvial processes. *Quaternary Research*, **23**, 154–174.

BERGEN, F.W. & O'NEIL, P. 1979. Distribution of Holocene foraminifera in the Gulf of Alaska. *Journal of Paleontology*, **53**, 1267–1292.

BLOOS, G. 1982. Shell beds in the Lower Lias of South Germany–Facies and Origin. *In*: EINSELE, G. and SEILACHER, A. (eds) *Cyclic and Event Stratification*. Springer-Verlag, Berlin, 223–239.

BOULTON, G.S. 1978. Boulder shapes and grain size distribution of debris as indicators of transport paths through a glacier till genesis. *Sedimentology*, **25**, 773–799.

BRADSHAW, J.S. 1959. Ecology of living planktonic foraminifera in the north and equatorial Pacific. *Contributions, Cushman Foundation for Foraminiferal Research* **10**, 25–64.

BRUNS, T.R. 1983. Model for the origin of the Yakutat block, an accreting terrane in the northern Gulf of Alaska. *Geology*, **11**, 718–721.

—— 1985. *Tectonics of the Yakutat block, an allochthonous terrane in the northern Gulf of Alaska*. U.S. Geol. Survey, O.F. Rpt. 85–13, 112p.

—— & SCHWAB, W.C. 1983. *Structure and seismic stratigraphy of the Yakataga segment of the continental margin, northern Gulf of Alaska*. United States Geological Survey Miscellaneous Field Studies Map, MF-1424, 4 sheets, scale 1:250 000.

CARLSON, P.R. 1989. Seismic reflection characteristics of glacial and glacimarine sediment in the Gulf of Alaska and adjacent fjords. *Marine Geology*, **85**, 391–416.

——, BRUNS, T.R., MOLNIA, B.R. & SCHWAB, W.C. 1982. Submarine valleys in the northeastern Gulf of Alaska: characteristics and probable origin. *Marine Geology*, **47**, 217–242.

—— & MOLNIA, B.F. 1977. Submarine faults and slides on the continental shelf, northern Gulf of Alaska. *Marine Geotechnology*, **2**, 275–290.

——, ——, KITTLESON, S.C. & HAMPSON, J.C. JR. 1977. *Distribution of bottom sediments on the continental shelf, northern Gulf of Alaska*. United States Geological Survey Miscellaneous Field Studies Map MF-876.

CLARK, B.M. 1984. *Periglacial structures in Late Proterozoic diamictites of the Port Askaig Formation, Scotland and their relation to soft sediment structures in the Late Cenozoic Yakataga Formation, Alaska*. Thesis, Department of Geology, University of Toronto, Toronto, Ontario Canada.

CLIFTON, H.E. 1984. Sedimentation units in stratified resedimented conglomerate, Paleocene submarine canyon fill, Point Lobos, California. *In*: KOSTER, E.H. & STEEL, R.J. (eds). *Sedimentology of Gravels and Conglomerates*. Canadian Society of Petroleum Geologists, Memoir **10**, 429–441.

CLUKEY, E.C., KULHAWY, F.H., LIU, P.L-F. & TATE, G.B. 1985. The impact of wave loads and pore water pressure generation on initiation of sediment transport. *Geomarine Letters* **5**, 177–183.

DREIMANIS, A. 1976. Tills: their origins and properties. *In*: LEGGET, R.F. (ed.), *Glacial Till*, Royal Society of Canada Special Publication **12**, 11–49.

—— & REAVELY, G.H. 1953. Differentiation of the Lower and Upper till along the north shore of Lake Erie. *Journal of Sedimentary Petrology*, **23**, 238–259.

DUNCAN, G.A. & LAHAIE, G.G. 1979. Size analysis procedures used in the sedimentology laboratory: NWRI, CCIW, Hydraulics Division Manual.

ECHOLS, R.J. & ARMENTROUT, J.M. 1980. Holocene foraminiferal distribution patterns on the shelf and slope, Yakataga-Yakutat area, northern Gulf of Alaska, *In*: FIELD, M.E. *et al.* (eds) *Quaternary Depositional Environments of the Pacific Coast*. Pacific Section, Society of Economic

Paleontologists and Mineralogists, 281–303.

Eckdale, A.A., Bromley, R.G. & Pemberton, S.G. 1984. *Ichnology: Trace fossils in sedimentology and stratigraphy*. Society of Economic Paleontologists and Mineralogists, Short Course 15, Tulsa, Oklahoma.

Elverhoi, A. & Roaldset, E. 1983. Glaciomarine sediments and suspended particulate matter, Weddell Sea Shelf, Antarctica. *Polar Research* **1**, 1–21.

Eyles, C.H. 1987. Glacially-influenced submarine channel sedimentation in the Yakataga Forma*tion, Middleton Island, Alaska. Journal of Sedimentary Petrology*, **57**, 1004–1017.

—— 1988. A model for striated boulder pavement formation on glaciated shallow marine shelves, an example from the Yakataga Formation, Alaska. *Journal of Sedimentary Petrology*, **58**, 62–71.

——, Eyles, N. & Miall, A.D. 1985. Models of glaciomarine sedimentation and their application to the interpretation of ancient glacial sequences. *Palaeogeography, Palaeoclimatology, Palaeoecology*, **51**, 15–84.

Eyles, N. & Clark, B.M. 1985. Gravity-induced soft-sediment deformations in glaciomarine sequences of the Upper Proterozoic Port Askaig Formation, Scotland. *Sedimentology*, **32**, 789–814.

—— & Lagoe, M.B. 1989. Sedimentology of shell-rich deposits (coquinas) in the glaciomarine upper Cenozoic Yakataga Formation, Middleton Island, Alaska. *Bulletin of the Geological Society of America*, **101**, 129–142.

——, Sladen, J.A. & Gilroy, S. 1982. A depositional model for stratigraphic complexes and facies superimposition in lodgement tills. *Boreas*, **11**, 317–333.

Galloway, W.E. 1977. Sediments and stratigraphic framework of the Copper River fandelta, Alaska. *Journal of Sedimentary Petrology*, **46**, 726–737.

Hampton, M.A. 1983. Geology of Kodiak Shelf, Alaska: environmental consideration for resource development. *Continental Shelf Research*, **1**, 253–281.

Hanson, J.D. 1983. Ice-formed intertidal bolder pavements in the sub-Antarctic. *Journal of Sedimentary Petrology* **53**, 135–145.

Hein, F.J. & Walker, R.G. 1982. The Cambro-Ordovician Cap Enrage Formation, Quebec, Canada: conglomeritic deposits of a braided submarine channel with terraces. *Sedimentology*, **29**, 309–330.

Hiscott, R.N. & Middleton, G.V. 1979. Depositional mechanics of thick-bedded sandstones at the base of a submarine slope, Tourelle Formation (Lower Ordovician). Quebec, Canada. *In*: Doyle, L.J. & Pilkey, O.H. Jr. (eds) *Geology of Continental Slopes*. Society of Economic Paleontologists and Mineralogists Special Publication **27**, 307–326.

Hollister, C.D. & Heezen, B.C. 1972. Geologic effects of ocean bottom currents. *In*: Gordon, A.L. (ed.) *Studies in Physical Oceanography*. Gordon & Breach, New York, 37–66.

Hopkins, D.M. 1972. The paleogeography and climatic history of Beringia during late Cenozoic time. *Internard*, **12**, 121–150.

Kaye, B.G. 1984. *Shell bands in glaciomarine sediments of the Yakataga Formation on Middleton Island, Gulf of Alaska: Characteristics, origin and stratigraphic significance*. Thesis: Department of Geology, University of Toronto, Toronto, Ontario, Canada.

Kennett, J.P. 1982. *Marine Geology*. Prentice-Hall, Englewood Cliffs, New Jersey.

Kurtz, D.D. & Anderson, J.B. 1979. Recognition and sedimentologic description of recent debris flow deposits from the Ross Sea and Weddell Sea; Antarctica. *Journal of Sedimentary Petrology*, **49**, 1159–1170.

Lagoe, M.B. 1983. Oligocene through Pliocene foraminifera from the Yakataga Reef section, Gulf of Alaska Tertiary Province, Alaska. *Micropaleontology*, **29**, 202–222.

——, Eyles, C.H. & Eyles, N. 1989. Paleoenvironmental significance of foraminiferal biofacies in the glaciomarine Yakataga Formation, Middleton Island, Gulf of Alaska. *Journal of Foraminiferal Research*, **19**, 194–209.

Lowe, D.R. 1979. Sediment gravity flows; their classification and some problems of application to natural flows. *In*: Doyle, L.J. & Pilkey, O.H. Jr. (eds) *Geology of Continental Slopes*. Society of Economic Paleontologists and Mineralogists Special Publication **27**, 75–82.

—— 1982. Sediment-gravity flows. II. Depositional models with special reference to the deposits of high-density turbidity currents. *Journal of Sedimentary Petrology*, **52**, 279–297.

McKay, J.R. & McKay, D.R. 1977. The stability of ice-push features, Mackenzie River, Canada. *Canadian Journal of Earth Sciences*, **14**, 2213–2225.

Mackiewicz, N.E., Powell, R.D. Carlson, P.R. & Molnia, B.F. 1984. Interlaminated ice-proximal glaciomarine sediments in Muir Inlet, Alaska. *Marine Geology*, **57**, 113–147.

Massari, S.F. 1978. High-constructive coarse-textured delta systems, Tortonian, Southern Alps. Evidence of lateral deposits in delta slope channels. *Societa Geologica Italiana Memorie*, **18**, 93–124.

Middleton, G.V. 1970. Experimental studies related to problems of flysch sedimentation. *In*: Lajoie, J. (ed.) *Flysch Sedimentology in North America*. Geological Association of Canada Special Paper **7**, 253–272.

Miller, D.J. 1953. Late Cenozoic marine glacial sediments and marine terraces of Middleton Island, Alaska. *Journal of Geology*, **61**, 17–40.

Mills, P.C. 1983. Genesis and diagnostic value of soft-sediment deformation structures–a review. *Sedimentary Geology*, **35**, 83–104.

Molnia, B.F. 1981. Sedimentation in coastal embayments, Northeastern Gulf of Alaska. *In*: *Proceedings of the 11th Annual Offshore Technology Conference, Houston*, Paper OTC 3435,

665–676.

—— 1983. Subarctic glacial-marine sedimentation: a model. *In*: Molnia, B.F. (ed.) *Glacial-Marine Sedimentation*. Plenum, New York, 95–144.

—— & Carlson, P.R. 1978. Surface sedimentary units ıof the northern Gulf of Alaska continental shelf. Bulletin of American Association of Petroleum Geologists **62**, 633–643.

—— & —— 1980. Quaternary sedimentary facies on the continental shelf of the northeast Gulf of Alaska. *In*: Colburn, I., Field, M., Ingle, J. & Douglas, R. (eds) *Quaternary depositional environments of the Pacific Coast*. Society of Economic Paleontologists and Mineralogists Symposium Volume 157–168.

Levy, W.P. & Carlson, P.R. 1980. *Map showing sedimentation rates in the northeastern Gulf of Alaska*. United States Geological Survey Miscellaneous Field Studies Map MF-1170, 1 sheet.

Nardin, T.R., Hein, F.J., Gorsline, D.S. & Edwards, B.D. 1979. A review of mass movement processes, sediment and acoustic characteristics and contrast in slope and base of slope systems versus canyon-fan-basin floor systems. *In*: Doyle, L.J. & Pilkey, O.H., Jr. (eds) *Geology of Continental Slopes*. Society of Economic Paleontologists and Mineralogists Special Publication **27**, 61–73.

Ness, G.E. & Kulm, L.D. 1973. Origin and development of Surveyor deep-sea channel. *Bulletin of the Geological Society of America*, **84**, 3339–3354.

Orheim, O. & Elverhoi, A. 1981. Model for submarine glacial deposition. *Annals of Glaciology*, **2**, 123–128.

Page, R.A. 1975. Evaluation of seismicity and earthquake shaking of offshore sites. *Offshore Technology Conference Dallas, Texas*, **3**, 1979–1990.

Plafker, G. 1965. Tectonic deformation associated with the 1964 Alaska earthquake. *Science*, **148**, 1675–1678.

—— 1967. *Geologic map of the Gulf of Alaska Tertiary province, Alaska*. United States Geological Survey Miscellaneous Geological Investigations Map I-484, Scale 1:500 000.

—— 1981. Late Cenozoic glaciomarine deposits of the Yakataga Formation, Alaska. *In*: Hambrey, M.J. & Harland, W.B. (eds) *Earth's Pre-Pleistocene Glacial Record*. Cambridge University Press, Cambridge, 694–699.

—— 1984. Tectonics of the great 27 March 1964 Alaska earthquake. Geological Society of America Regional Meeting, Anchorage, Alaska, Abstracts with Programs 16.

—— 1987. Regional geology and petroleum potential of the northern Gulf of Alaska. *In*: Sholle, D.W., Grantz, A. & Vedder, J.G. (eds) *Geology and resource potential of the continental margin of the western North American and adjacent ocean basins – Beaufort Sea to Baja California*. Circum-Pacific Council for Energy and Mineral Resources, Earth Science Series, **6**, 229–268.

—— & Addicott, W.O. 1976. Glaciomarine deposits of Miocene through Holocene age in the Yakataga Formation along the Gulf of Alaska margin, Alaska. *In*: Miller, T.P. (ed.) *Recent and ancient sedimentary environments in Alaska*, Symposium Proceedings, Alaska Geological Society, Q1–Q19.

—— & Rubin, M. 1967. Vertical tectonic displacement in South Central Alaska during and prior to the Great 1964 Earthquake. *Journal of Geosciences*, Osaka City University, **10**, 53–66.

Postma, G. 1984. Mass-flow conglomerates in a submarine canyon. Abrioja fan-delta, Pliocene, southeast Spain. *In*: Koster, E.H. & Steel, R.J. (eds) *Sedimentology of gravels and conglomerates*. Canadian Society of Petroleum Geologists Memoir **10**, 237–258.

Powell, R.D. 1981. A model for sedimentation by tidewater glaciers. *Annals of Glaciology*, **2**, 129–134.

—— 1983. Glacial-marine sedimentation processes and lithofacies of temperate tidewater glaciers. *In*: Molnia, B.F. (ed.) *Glacial-Marine Sedimentation*. Plenum, New York, 185–232.

—— & Molnia, B.F. 1989. Glacimarine sedimentary processes, facies and morphology of the south-southeast Alaska shelf and fjords. *Marine Geology*, **85**, 359–390.

Prescott, W.H. & Lisowski, M. 1977. Deformation at Middleton Island, Alaska, during the decade after the Alaska Earthquake of 1964. *Bulletin of the Seismological Society of America*, **67**, 579–586.

Quinterno, P., Carlson, P.R. & Molnia, B.F. 1980. Benthic foraminifers from eastern Gulf of Alaska. *In*: Field, M.E. *et al.* (eds) *Quaternary Depositional Environments of the Pacific Coast*. Pacific Section, Society of Economic Paleontologists and Mineralogists, 13–21.

Reed, R.K. & Schumacher, J.D. 1987. Physical Oceanography. *In*: Hood, D.W. & Zimmerman, S.T. (eds) *The Gulf of Alaska*. National Oceanic and Atmospheric Administration (United States) 57–91.

Reimnitz, E. 1972. Effects in the Copper River Delta. *In*: *The Great Alaska Earthquake of 1964: Oceanography and Coastal Engineering*. National Academy of Sciences, Washington D.C., 290–302.

Rogers, J.F. 1977. Implications of plate tectonics for offshore Gulf of Alaska petroleum exploration. Offshore Technology Conference, Houston, 1977, 11–16.

Sangrey, D.A. & Molnia, B.F. 1979. Geotechnical Engineering Analysis of Unconsolidated Sediments from Alaska Coastal Waters. Proceedings of the 11th Annual *Offshore Technology Conference, Houston*, Paper OTC 3436, 677–682.

Schwab, W.C. & Lee, H.J. 1983. Geotechnical analyses of submarine landslides in glacial-marine sediment, northeast Gulf of Alaska. *In*: Molnia, B.F (ed.) *Glacial-Marine Sedimentation*. Plenum, New York, 145–184.

—— & —— 1988. Causes of two slope failure types in continental shelf sediments, NE Gulf of Alaska.

Journal of Sedimentary Petrology **58**, 1–11.

SHEPARD, F.P. & EMERY, K.O. 1973. Congo submarine canyon and fan valley. *American Association of Petroleum Geologists Bulletin*, **57**, 1679–1691.

STEPHENS, C.D., FOGELMAN, K.A., LAHR, J.C. & PAGE, R.A. 1984. Wrangell Benioff Zone, southern Alaska. *Geology*, **12**, 373–376.

TALIAFERRO, N.L. 1932. Geology of the Yakataga, Katalla and Nichalak Districts, Alaska. *Geological Society of America Bulletn*, **43**, 749–782.

VISSER, J.N.J. 1983. Submarine debris flow deposits from the Upper Carboniferous Dwyka Tillite Formation in the Kalahari Basin, South Africa. *Sedimentology*, **30**, 511–524.

VON HUENE, R. 1972. Structure of the Continental margin and tectonism at the Eastern Aleutian Trench. *Bulletin of the Geological Society of America*, **83**, 3613–3626.

VORREN, T.O., HALD, M., EDVARDSEN, M. & LIND-HANSEN, O. 1983. Glacigenic sediments and sedimentary environments on continental shelves: General principles with a case study from the Norwegian shelf. *In*: EHLERS, J. (ed.) *Glacial Deposits of Northwest Europe*. Balkema, Rotterdam, 61–73.

WALKER, R.G. 1975. Generalized facies models for resedimented conglomerates of turbidite association. *Bulletin of the Geological Society of America*, **86**, 737–748.

—— 1978. Deep-water sandstone facies and ancient submarine fans: models for exploration for stratigraphic traps. *Bulletin of the American Association of Petroleum Geologists*, **62**, 932–966.

WALKER, R.G. 1984. Turbidites and associated coarse clastic deposits. *In*: WALKER, R.G. (ed.) *Facies Models*. Geoscience Canada Reprint Series **1**, 171–188.

WALLER, R.M. 1968. Water-sediment ejections in the Great Alaska Earthquake of 1964. Publication 1603. National Academy of Sciences, Washington, D.C., 97–116.

WILSON, J.B. 1982. Shelly faunas associated with temperate offshore tidal deposits. *In*: STRIDE, A.H. (ed.) *Offshore Tidal Sands: Processes and Deposits*. Chapman and Hall, London, 126–171.

WRIGHT, R. & ANDERSON, J.B. 1982. The importance of sediment gravity flow to sediment transport and sorting in a glacial marine environment: Eastern Weddell Sea, Antarctica. *Bulletin of the Geological Society of America*, **93**, 951–963.

ZELLERS, S. 1989. Foraminiferal biofacies analysis of the Yakataga Formation, Icy Bay, Alaska. Geological Society of America, Abstracts with Programs, **21**, 162.

—— 1990. Foraminiferal biofacies analysis of the Yakataga Formation, Icy Bay, Alaska: Insights into Pliocene glaciomarine paleoenvironments of the Gulf of Alaska. *Palaeos*.

—— & LAGOE, M.B. 1990. Quantitative multivariate analysis of Holocene foraminifera and bivalves from the Gulf of Alaska: Insights into paleoenvironmental interpretation of the Late Cenozoic Yakataga Formation. *Journal of Foraminiferal Research*.

Marginal-marine glacial sedimentation in the late Precambrian succession of East Greenland

A.C.M. MONCRIEFF[1] & M.J. HAMBREY[2]

Department of Earth Sciences, University of Cambridge, Downing Street, Cambridge CB2 3EQ, UK

[1] Current address: British Antarctic Survey, Natural Environment Research Council, High Cross, Madingley Road, Cambridge, CB3 OET, UK

[2] Current address: Scott Polar Research Institute, University of Cambridge, Lensfield Road, Cambridge CB2 1ER, UK

Abstract: The late Precambrian succession of the fjord region of East Greenland is characterized by two glacigenic formations. The lower one, the Ulvesø Formation, rests on different members of the Limestone–Dolomite 'series' (Eleonore Bay Group), and comprises diamictite, conglomerate, sandstone, and rhythmite, all with a high dolomite content. Few beds can be traced for more than a few hundred metres, and erosional breaks are sometimes present. Clasts are predominantly intrabasinal. A turbiditic dolomitic sandstone–mudstone succession, the Arena Formation, lies above. The upper glacigenic horizon, the Storeelv Formation, contains a similar range of facies as the Ulvesø Formation, but is less dolomitic and contains exotic gneiss, granite and volcanic clasts, as well as intrabasinal carbonates.

The Ulvesø and Storeelv formations contain facies characteristic of a low-level terrestrial to marine glacial environment, including: basal tillite, waterlain tillite, debris-flow deposits, ice proximal to distal glacimarine sediments, glaciolacustrine and glacimarine rhythmites with ice-rafted material, permafrost features, aeolian sands and evaporites.

The bulk of the glacigenic sediment appears to have been associated with an ice sheet of continental proportions, which in terms of its thermal regime was probably below the pressure melting point but sliding on its bed. Our interpretation of large parts of the succession as waterlain till implies that floating, debris-bearing ice was an important part of the system, probably in the form of outlet glaciers or ice streams, coalescing as an ice shelf in a tectonically controlled marine basin beyond an irregular coastline. Considerable current activity, both beneath the ice shelf, and sedimentation controlled by iceberg drifting in distinct lanes existed. The succession further illustrates the response of glacimarine sedimentation to ice-retreat onto land, with marked changes from facies with obvious glacial affinities to those devoid of them.

Late Precambrian diamictites have now been documented on all continents including Antarctica (Stump *et al.* 1988). After much debate in the 1970s when a glacial origin was not highly favoured (e.g. Schermerhorn 1974), many have been shown to be glacigenic (see reviews in Hambrey & Harland 1981).

The preservation of glacigenic sediments in the geological record is often selective, and those that have had the best chance of survival occur in marine sequences (Nystuen 1985). However, compared with the terrestrial environment our knowledge of processes in the less accessible glacimarine environment is relatively sketchy, and is based mainly on shallow coring on the sea floor beyond present ice margins (see Dowdeswell 1987 for a recent review of relevant literature). From such data it is often difficult to define the geometry of facies, a problem only partly offset by shallow seismic and side-scan sonar observations. In contrast, good exposures of ancient glacimarine sequences provide insights into facies geometry unobtainable in contemporary environments. The late Precambrian sequence of East Greenland is interpreted here as just such an example.

East Greenland is one of several areas in the Arctic–North Atlantic region where thick complexes of marine to low-level terrestrial glacigenic sediments are well exposed. Indeed, East Greenland has the most extensive and finest outcrops in the entire region. Here, diamictites crop out from 71° 30'–74°N, a distance

From Dowdeswell, J.A. & Scourse, J.D. (eds), 1990, *Glacimarine Environments: Processes and Sediments*. Geological Society Special Publication No 53, pp 387–410.

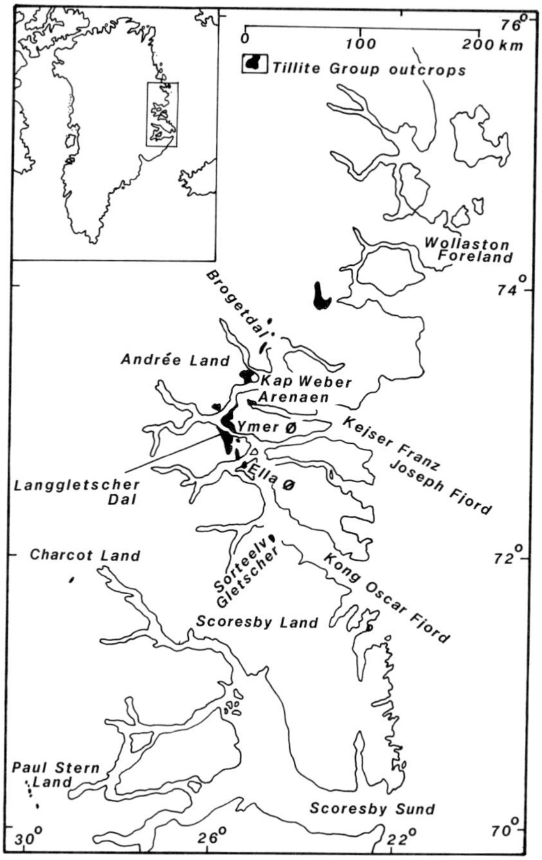

Fig. 1. Distribution of Tillite Group and equivalent rocks in East Greenland.

of some 500 km (Fig. 1). The rocks occur in two main stratigraphic settings. In the fjord region they occur within the main part of the Caledonian fold belt, and belong to a >20 km thick sedimentary pile of Late Riphean to Ordovician age. The diamictites form the bulk of two formations in the Tillite Group (the Ulvesø and Storeelv), each up to about 200 m thick (Table 1) (see Henriksen & Higgins 1976, Higgins 1981, Hambrey & Spencer 1987, for details of stratigraphy throughout the fjord region and summaries of earlier work). The rocks in this area have been affected by high crustal level thrusting and low-amplitude folding (Manby & Hambrey 1989).

To the SW of the main outcrops, in Charcot Land and Gåseland (Fig. 1), small outcrops of diamictite are in direct contact with Early Proterozoic crystalline basement (Table 1), the contact in some places representing a palaeoland surface (Moncrieff 1989a). These outcrops traditionally have been said to belong to the foreland, west of the Caledonian fold belt (e.g. Henriksen & Higgins 1976), but recent work (Manby & Hambrey 1989) has shown that these rocks too were probably allochthonous.

The contrasts between the fjord region and the so-called foreland have previously been discussed from a palaeogeographical perspective (Moncrieff 1989a). The land-mass bearing the late Precambrian ice lay in the 'foreland' areas. From here, ice extended northeastwards into the fjord region where there existed a subsiding basin of complex topography and bathymetry.

A wide range of glacially-influenced facies were deposited in a marginal-marine to terrestrial setting, including distal to proximal glacimarine sediments, waterlain tills, basal tills, and sediments of glaciolacustrine, fluvial, aeolian, evaporitic and periglacial origin (Hambrey & Spencer 1987, Moncrieff 1989a, Moncrieff & Hambrey 1988). In this paper

Table 1. *Stratigraphic setting of Late Precambrian diamictites of East Greenland*

SW 'Foreland'	NE Fold belt	
	Cambro-Ordovician	
<<<<<< thrust <<<<<<		
–		Spiral Creek Fm.
–	Tillite	Canyon Fm.
Støfanget Fm.		Storeelv Fm. (tillite-bearing)
– – – – –	Group	Arena Fm.
Early Proterozoic gneissic basement		Ulvesø Fm. (tillite-bearing)
	Upper Eleonore Bay Gp.	Limestone-Dolomite 'series' Multicoloured 'series' Quartzite 'series'
	Lower Eleonore Bay Group	

(See Hambrey & Spencer 1987 for full details of stratigraphy).

we describe these facies and their associations within the context of developing a depositional model for the fjord region.

Having examined the evidence obtained from contemporary glacimarine environments (Antarctica, Alaska and Svalbard) we conclude that the East Greenland sequence does not have a suitable modern anologue. The nearest equivalents seem to be the composite ice shelf setting around James Ross Island, Antarctic Peninsula, or the subpolar tidewater glaciers of Svalbard.

Nomenclature

A terminological difficulty concerning the designation of glaciers as temperate, polar or subpolar has arisen, since the late Precambrian glaciers appear, from palaeomagnetic evidence, to have formed in low palaeolatitudes, but with 'subpolar' characteristics. Thus for ice masses that were generally below the pressure-melting point, but which were sliding on their beds in association with meltwater, we shall use the term 'cold, wet-based glaciers'.

We have noted the absence of a suitable descriptive textural classification for poorly sorted sediments, including diamictites. Such a classification was developed for describing the East Greenland rocks (Moncrieff 1989b), and is followed in this paper, although no distinction is made in the abbreviations between sandy, intermediate and muddy diamictites (Table 2).

There are numerous genetic classifications of glacigenic sediments, but the INQUA Commission on the Genesis and Lithology of Glacial Quaternary Deposits has done much to clarify our thinking over the past two decades. However, we are still a long way from a comprehensive, universally agreed classification, as the Commission's final report makes clear (Dreimanis 1988). An earlier INQUA classification (1979), which met with considerable support was assembled by A. Dreimanis and applied to pre-Pleistocene deposits by Hambrey & Harland (1981). In modified form (Dreimanis & Lundquist 1984) the INQUA classification represents a consensus among many glacial geologists, but there are some (especially those working on glacial processes) for whom this scheme is unacceptable (Lawson 1979, 1981). The INQUA definition of till is 'a sediment that has been transported and is subsequently deposited by or from glacier ice with little or no reworking by water'. 'Glaciomarine (glacial-marine) sediment includes a mixture of glacial detritus and marine sediment deposited more or less contemporaneously. The glacier component may be released directly from glaciers and ice shelves or delivered to the marine depositional site from those sources by gravity, moving fluids or iceberg rafting...' (Borns & Matsch 1988). Beyond this, there is no consensus. Our own classification has evolved following work in sediments of various ages, and deposited in different types of marine setting (Fairchild & Hambrey 1984, Barrett *et al.* 1990, Hambrey *et al.* 1990).

In the marine environment we define three main genetic varieties of glacigenic sediment based on relative position to the glacier.

Waterlain till is deposited by continuous rain-out of basal glacial debris just seaward of the grounding line, with little or no reworking by bottom currents. The marine component is neg-

Table 2. *Lithological definitions used in this paper (after Moncrieff 1989b)*

Increasing gravel content →
%GRAVEL (> 2 mm) in whole rock, as determined in outcrop

<table>
<tr><th>Increasing mud content →
sand to mud ratio of matrix</th><th>trace (<0.01%)</th><th>< 1</th><th>1–5</th><th>5–80</th><th>80–95</th><th>95–100</th></tr>
<tr><td>MUD
< 0.06 mm</td><td>mudstone
(M)</td><td>mudstone with
dispersed clasts
(Mc)</td><td rowspan="2">clast-poor
muddy diamictite
(M[D])</td><td rowspan="2">clast-rich
muddy diamictite
(MD)</td><td>muddy
conglomerate
(MC)</td><td rowspan="6">conglomerate

(C)</td></tr>
<tr><td>0.11</td><td rowspan="2">sandy mudstone

(SM)</td><td rowspan="2">sandy mudstone
with
dispersed clasts
(SMc)</td><td rowspan="4">conglomerate

(C)</td></tr>
<tr><td>0.54</td><td rowspan="2">clast-poor
intermediate
diamictite
(I[D])</td><td rowspan="2">clast-rich
intermediate
diamictite
(ID)</td></tr>
<tr><td>1</td><td rowspan="2">muddy
sandstone
(MS)</td><td rowspan="2">muddy sandstone
with
dispersed clasts
(MSc)</td></tr>
<tr><td>1.86</td><td>clast-poor
sandy diamictite

(S[D])</td><td>clast-rich
sandy diamictite

(SD)</td></tr>
<tr><td>9

SAND
2–0.06 mm</td><td>sandstone
(S)</td><td>sandstone with
dispersed clasts
(Sc)</td><td colspan="2">gravelly sandstone (GS)</td><td>sandy
conglomerate
(SC)</td></tr>
</table>

75%

ligible, so it would be inappropriate to classify the deposit as glacimarine sediment.

Proximal glacimarine sediment is transitional with waterlain till and is deposited beneath a glacier tongue or ice shelf, or close to a calving frontal cliff. It is formed by rain-out of basal debris but has strong indications of current reworking, accompanied by considerable background marine sedimentation. This latter may include sediment transported in suspension and biogenic material.

Distal glacimarine sediment is formed beyond the glacier front and is dominated by marine sedimentary processes, with only a small ice-rafted gravel component. In Phanerozoic sediments bioturbation may be important.

This classification does not take into account reworking of debris, while the classification of Anderson *et al.* (1980, 1983) does. They suggested the terms compound and residual, which are adopted here. Compound refers to a deposit which reflects the composition of sediment released by ice and because of its high mud content, indicates low current activity. A residual deposit is one in which current activity prevents the finer part of the sediment in the water column from being deposited. Residual deposits may form in areas of strong current activity beneath an ice shelf, or in shoal areas beyond the ice front. To fully describe the glacimarine environment a consideration of both the proximity of ice and reworking processes is needed. We propose the terms ice-proximal, ice-transitional and ice-distal compound or residual glacimarine sediment. The terms proximal and distal need not have direct implications for the proximity of land; they refer instead to the concentration of floating, debris-bearing ice.

Approaches to defining glacigenic facies

In recent years there has been much discussion on the use of facies in interpreting glacigenic sediments (e.g. Eyles *et al.* 1983, Dreimanis 1984, Karrow 1984). In particular, the use of facies codes to define an horizon or unit has become popular. In this study, we distinguish descriptive and interpretative facies, where the descriptive facies are denoted by a simple two-part abbreviation, intended only as shorthand. The descriptive facies is a recurring unit with a range of distinctive features, detailed enough to be used for simple interpretation. This flexible approach has been followed by various authors (Miall 1977, 1978, Eyles *et al.* 1983; Christie-Blick 1983) each using different criteria to define their facies. We do not feel that the construction of a set of standard facies codes (Eyles *et al.* 1983) is a desirable goal because minor, but significant variations are concealed. Starting afresh with each sequence encourages a reappraisal of the features involved. The descriptive facies used here (Table 3) take the form of an uppercase letter denoting the lithology (as defined in Table 2), together with a lowercase letter indicating a distinguishing feature. If one descriptive facies is followed by another in brackets, then the latter is interbedded with, but subordinate to the former. These descriptive facies are intended as a simplified guide, not a complete description of the rock.

Interpretative facies take account of all the features which cannot reasonably be contained in abbreviations, for example fabric data, nature of contacts, petrology and lateral facies changes. In this paper the relevant observational data, including the descriptive facies, are presented in lithological logs (Figs 2–4). These data form the basis for determining the mode of deposition of individual units and for defining the interpretative facies which follow.

Table 3. *Key to descriptive facies and their abbreviations in the Tillite Group*

Lithology	
C	Conglomerate
D	Clast-rich diamictite
[D]	Clast-poor diamictite
S	Sandstone
GS	Gravelly sandstone
MS	Muddy sandstone
SM	Sandy mudstone
M	Mudstone
Structure	
m	Massive (no compositional variation, no depositional foliation).
b	Bedded (compositional variation or depositional foliation up to the point where they become an interbedded facies).
l	Laminated (bedded on a scale of millimetres)
x	cross-bedded or laminated (size and style of cross-sets unspecified)
r	Rhythmically laminated (containing a regularly repeating pattern of lamination).
c	With dispersed clasts (containing between 1 and 0.01% gravel grade sediment).
d	Discontinuous (lithology extends less than 5 m laterally).
t	Tabular (lithology has roughly parallel upper and lower surfaces and extends (further than 5 m laterally).

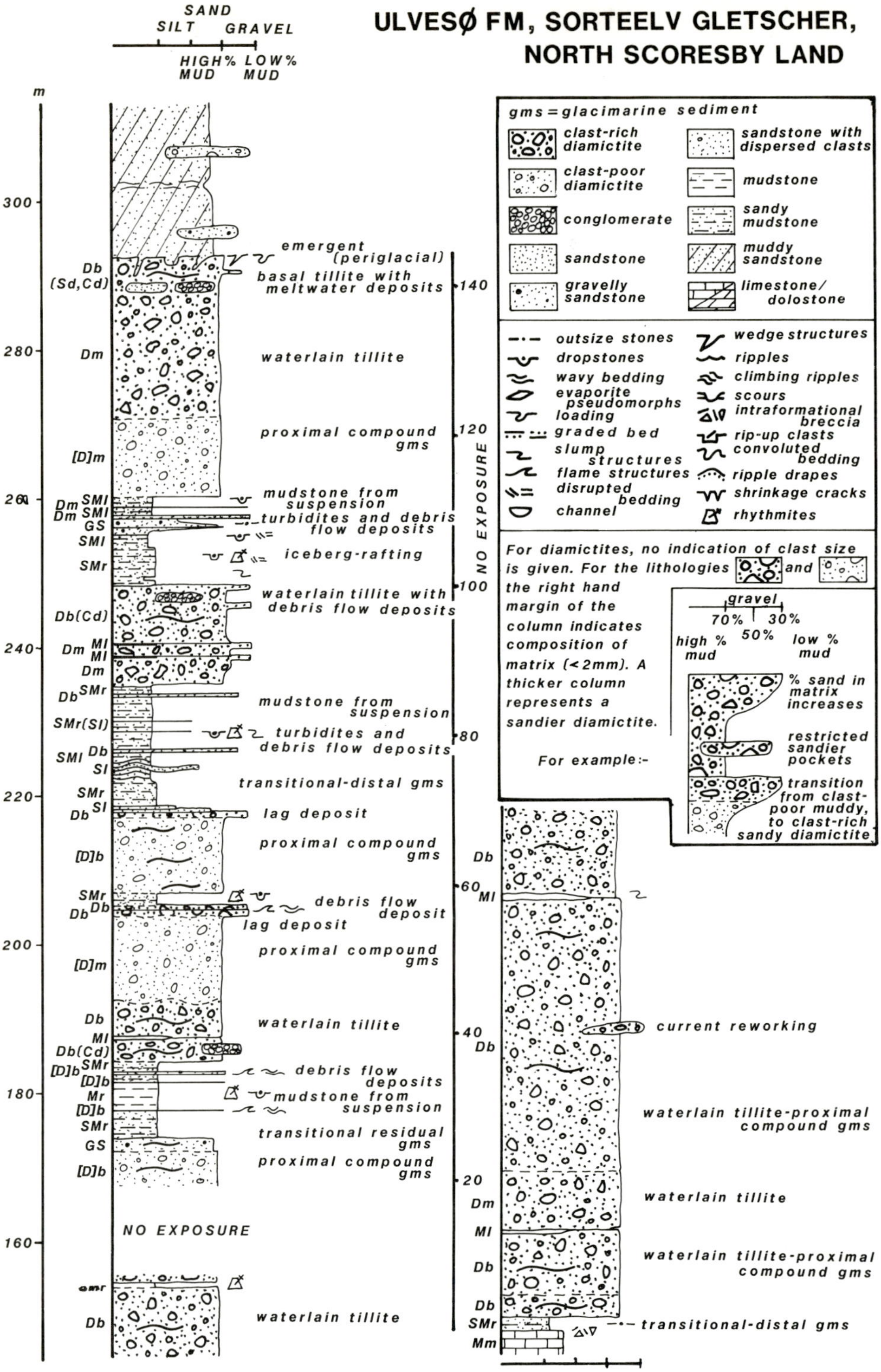

Fig. 2. Lithological log of the Ulvesø Formation at Sorteelv Gletscher.

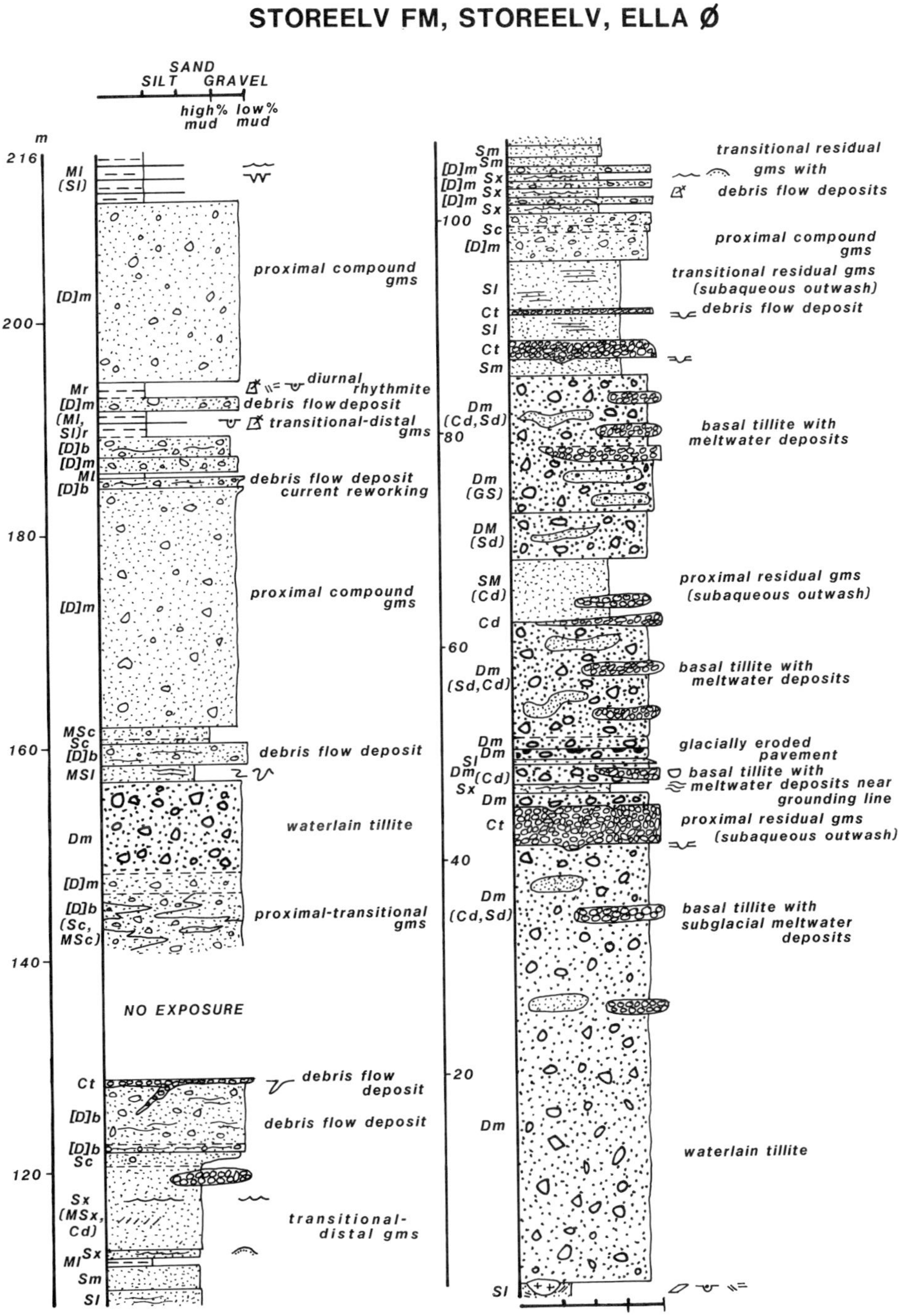

Fig. 3. Lithological log of the Storeelv Formation at Storeelv. Key as in Fig. 2.

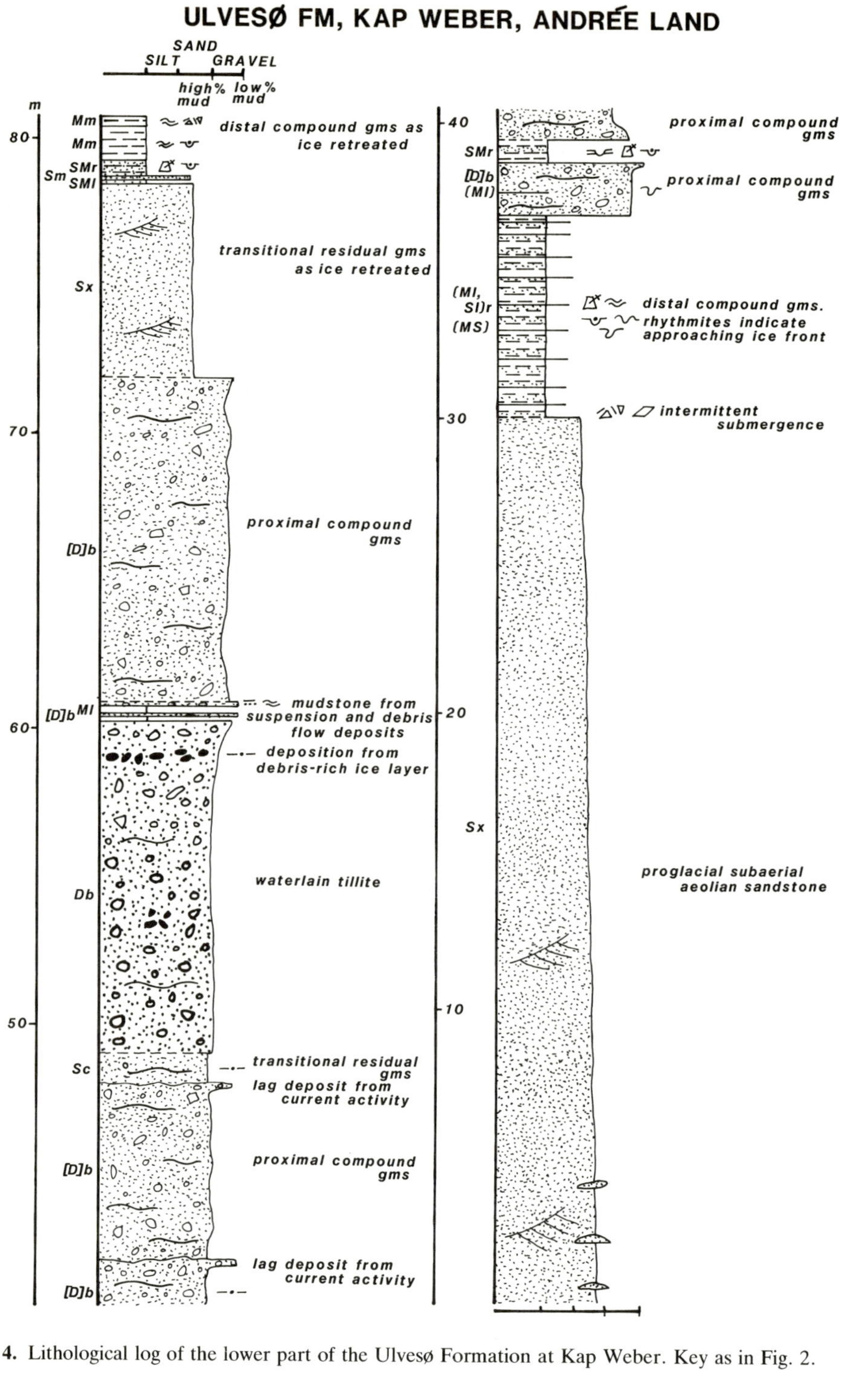

Fig. 4. Lithological log of the lower part of the Ulvesø Formation at Kap Weber. Key as in Fig. 2.

Interpretative facies in the glacigenic units of the Tillite Group

In this section, we describe the characteristics of the interpretative facies together with the basis for their interpretation.

Basal tillite facies

This facies comprises clast-rich muddy diamictite (D) beds up to several tens of metres thick. These diamictites are massive or display a variety of bedding structures, ranging from poorly developed bedding-parallel depositional foliation, to limited compositional variation in the form of thin sandier horizons. A distinguishing feature of the basal tillite facies is the development of large numbers of discontinuous conglomerate (Cd) and discontinuous sandstone (Sd) horizons within the diamictites. Such lenses and pods are common in the Storeelv Formation at Arenaen (Fig. 5a) and on Ella Ø, where they are roughly stratiform. Well preserved examples sometimes have a roughly elliptical cross section (Fig. 5b). In other cases they bear no relation to bedding planes. Structures within them are rare and when they do occur are severely distorted by soft-sediment deformation (Fig. 5c, d). We interpret these Cd and Sd units as subglacial drainage deposits, since they are too discontinuous and poorly developed for proglacial outwash. The disturbed structures are interpreted as meltout-induced collapse and loading features, possibly with a component of glaciotectonic deformation.

Two features of these D units indicate a subglacial origin more clearly: (1) Pebble fabrics in tills are often strongly aligned. Weathering and lithification both tend to reduce the strength of clast fabric patterns and none of the fabric data from the Tillite Group approach the strength of those seen in modern lodgement tills (Dowdeswell *et al.* 1985). However, some fabric patterns fall within the fields designated as lodgement tills by Dowdeswell and Sharp (1986) and are interpreted as such. (2) One such lodgement fabric, in the Storeelv Formation at Storeelv (Fig. 3), is associated with a striated boulder pavement. Two striated horizons occur, one at the base and one 52 m above the base of the formation (Moncrieff & Hambrey 1988). On the basis of lateral and vertical facies associations, notably the lack of any signs of emergence, these striated surfaces may have formed from a glacier grounded below sea level.

Waterlain tillite facies

The upper part of the Ulvesø Formation at Sorteelv Gletscher is characterised by Dm or Db (massive to weakly bedded varieties) units up to about 10 m in thickness and containing a variety of bedding structures, including wispy discontinuous bands of sandstone and mudstone. Compositional banding is more diffuse and elongate than that found in the basal tillite facies and the bedding is often disrupted by larger stones and compositionally distinct lumps of diamictite. Although the lithologies are not suited to the preservation of perfect drop structures we interpret these clasts as dropstones and till pellets respectively (Fig. 5e, f). Diffuse bedding is interpreted as the product of short-term changes in the mud: sand ratio of the sediment supply. Such changes may have resulted from the removal of mud by currents, or by a temporary halting of sediment supply (such as a patch of debris-free ice), during which only mud in suspension was deposited. This mechanism also explains the mud-free tops which occur extensively on many waterlain tillite units.

The distinct Cd and Sd units, so common in the basal tillite facies, do not occur, but the D units are occasionally punctuated by laterally more extensive conglomerate and sandstone units such as those 68 m above the base of the Storeelv Formation at Storeelv (Fig. 3) also Hambrey & Spencer 1987, Fig 19). The greater lateral extent of conglomerate and sandstone units, together with signs of widespread current activity suggest deposition in a proglacial subaqueous setting. Boulder horizons also occur, but vary from those in the basal tillite facies in that they lack striated tops and preferred azimuth. Instead they show a typical waterlain clast fabric, with random azimuth and a low angle of long axis to bedding.

These waterlain tillites formed immediately seaward of the grounding line of a floating, debris-laden ice shelf or glacier tongue. Further away from the grounding line, input of debris was reduced, reworking became significant and the name ‘waterlain tillite’ is no longer appropriate. At Kap Weber, the Storeelv Formation contains a boulder horizon which occurs just above a contact interpreted as a transition from basal to waterlain tillite. This horizon may be the result of melting out of the basal, debris-rich ice, in which boulders are likely to have been concentrated, as it floats clear of the sea bed. The waterlain tillite facies is often associated with basal tillite facies, as shown by the Storeelv Formation in Storeelv (Fig. 3), and

a

b

c

d

e

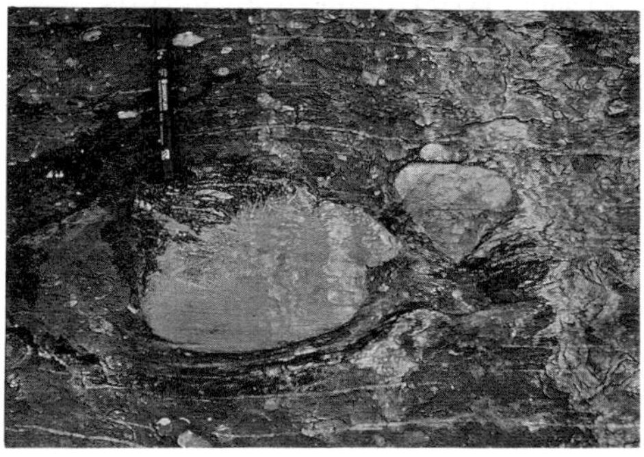

Fig. 5. (a) Photomontage of the Storeelv Formation exposed in the back wall of Arenaen. The dark lithology is maroon diamictite which contains numerous sorted sand bodies interpreted as subglacial meltwater deposits. Some are stratiform, but not continuous enough to be proglacial outwash deposits. The major channel sandstone at the top left is 15 m thick, probably of subglacial or immediately proglacial origin. A few metres above this is well laminated mudstone of the Canyon Formation. (b) Well preserved intratillite sandstone body, Storeelv Formation, Arenaen. Note the strongly curved base and top, where the margin of the sandstone rises to just beneath the hammer head. (c) Intratillite sandstone body with distorted bedding, probably due to the movement of saturated sediment; Storeelv Formation, Bastionbugt, Ella Ø. Scale is given by the pen. (d) Meltwater-washed horizon in basal till facies; Storeelv Formation, Kap Weber. Scale is in cm. (e) Proximal compound glacimarine deposit in which larger clasts disrupt previously formed bedding structures in a manner suggestive of dropstones; Ulvesø Formation, Langgletscher Dal. Pen for scale. (f) Lensoidal diamictite pod in well-laminated mudstone sequence. This is interpreted as a till 'pellet', a cohesive lump of basal debris dropped by an iceberg; Ulvesø Formation, Sorteelv Gletscher. Scale is in cm. (g) Rhythmically laminated mudstone with ice-rafted dolostone boulder 0.8 m long; Ulvesø Formation, Sorteelv Gletscher. (h) Proximal residual glacimarine deposit in which current activity during settling has removed mud and led to the concentration of sand and gravel in the residual sediment. Storeelv Formation, Storeelv. Scale is in cm. (i) Cross-laminated interval in gravelly sandstone resulting from current activity during a period of low ice-rafting intensity; Ulvesø Formation, Kap Weber. The laminated interval is 5 cm thick.

with a variety of glacimarine facies as shown by the Ulvesø Formation at Sorteelv Gletscher (Fig. 2).

Debris-flow facies

Units interpreted as debris-flow deposits are common in the Ulvesø and Storeelv formations and normally consist of muddy, sandy, clast-rich or clast-poor diamictite ([D]). Sometimes they are clearly identified as a distinct unit a few metres thick, with an erosional base, rip-up clasts and flow-banding in an otherwise homogeneous matrix.

In other cases, identification depends on indirect means. Kurtz & Anderson (1979) and Wright & Anderson (1982) stressed the importance of debris-flow deposits on the continental slope of Antarctica and stated that there was little lithological difference between sediments interpreted as tills and those as debris-flow deposits, and that identification depended on faunal characteristics. The Tillite Group lacks fossils other than rare acritarchs, but unlike the

Antarctic diamictites referred to above, the Precambrian diamictites were strongly influenced by meltwater. Mud, sand and gravel lenses produced by meltwater or current activity in basal and waterlain tillites were destroyed during debris flowage, resulting in an homogeneous deposit. Nevertheless, there are numerous cases in which the interpretation remains open to question.

Glacimarine facies

Interaction of marine and glacial processes in glacimarine environments. Moving seawards from a grounding line the contribution of sediment from basal ice is progressively reduced and marine processes become dominant. Sediments become increasingly stratified, with fewer large clasts, and pass somewhat arbitrarily from waterlain till into our classes of proximal then distal glacimarine sediment.

Such sediments would include clast-poor diamictiton, gravels, sands, muds and all combinations of them as given in Table 2. Many glacimarine sediments carry no direct signs of deposition from ice but a glacial influence can often be inferred from association with more obviously glacigenic sediments. Although in this paper we use the term 'glacimarine', deposition in a large lake is also possible.

Normal marine sedimentation is influenced by sorting of sediment by currents, both within the water column and along the seabed. Under glacimarine conditions these processes are supplemented by ice-rafting. Marine and ice-rafting mechanisms often act independently of one another. Thus, a fine grained, well-laminated mudstone deposited at some considerable distance from the sediment source, may contain numerous boulders released from a single passing iceberg (Fig. 5g). In general, however, the two processes are related so that with increasing distance from the grounding line, the percentage of gravel decreases in parallel with the sand: mud ratio (Barrett 1975, Piper & Brisco 1975, Kurtz & Anderson 1979, Anderson *et al.* 1980, 1983). Reversals in both these trends may be brought about by far-travelled debris-laden icebergs, or by increasing current strength offshore.

Glacimarine facies in the Tillite Group

Ice-proximal glacimarine facies. Both the Ulvesø and Storeelv formations contain numerous clast-poor diamictite ([D]) units with only 1–5% gravel. These units contain frequent bedding structures in the form of isolated mud pockets and sand laminae, both of which are frequently punctured by clasts interpreted as dropstones. On the assumption that the percentage of gravel decreases away from the grounding line, while bedding becomes better developed, these sediments are inferred to have formed seaward of the zone of waterlain tillite, with which they are strongly associated.

There can be no way of knowing whether [D] units in the Tillite Group were deposited beneath an ice shelf or beneath heavy concentrations of icebergs. As will be discussed later, the former possibility is preferred.

[D] units vary widely in the grain size of the matrix (Figs 3–4 and lower part of key in Fig. 2). We suggest that this variation from muddy [D] to sandy [D] enables compound and residual proximal glacimarine sediments to be distinguished. Residual ice-proximal glacimarine sediments develop as currents winnow out the mud fraction during settling, leading to the development of sandy [D], sandy D, and GS beds as more intense currents concentrate clasts in the residual sediment (Fig. 5h). Cross-bedding and limited wavy to parallel bedding, with frequent, localised mud pockets and laminae, further indicate the importance of winnowing and scouring (Fig. 5i). Debris-flow deposits are also common in the ice-proximal glacimarine facies.

Transitions between the different types of marine to glacial sediments, both vertically and laterally are common in the Tillite Group. A particularly striking example is in the Ulvesø Formation at Kap Weber. The lowermost [D] unit, an ice-proximal glacimarine sediment, is replaced 30 m laterally by an ice-transitional residual glacimarine sediment in the form of a GS unit with an SC top. Over another 20 m the lithology reverts to [D].

Figure 6a summarizes the relationship between basal till, waterlain till and glacimarine sediment, as deposited beneath the Tillite Group ice shelf. It emphasizes the gradational and arbitrary nature of their contacts. Transition between these types is often rapid. Comparison with Fig. 6b shows that extensive areas of waterlain till cannot form beyond a grounded ice front.

Ice-proximal to ice-distal transitional glacimarine facies. Seven lithologies in the Tillite Group which have a lower gravel content than [D] are classified as transitional glacimarine facies. These are: sandstone with dispersed clasts (Sc), muddy sandstone with dispersed clasts (MSc), sandy mudstone (SM), sandy mudstone with dispersed clasts (SMc) and sandstone (S). Minor

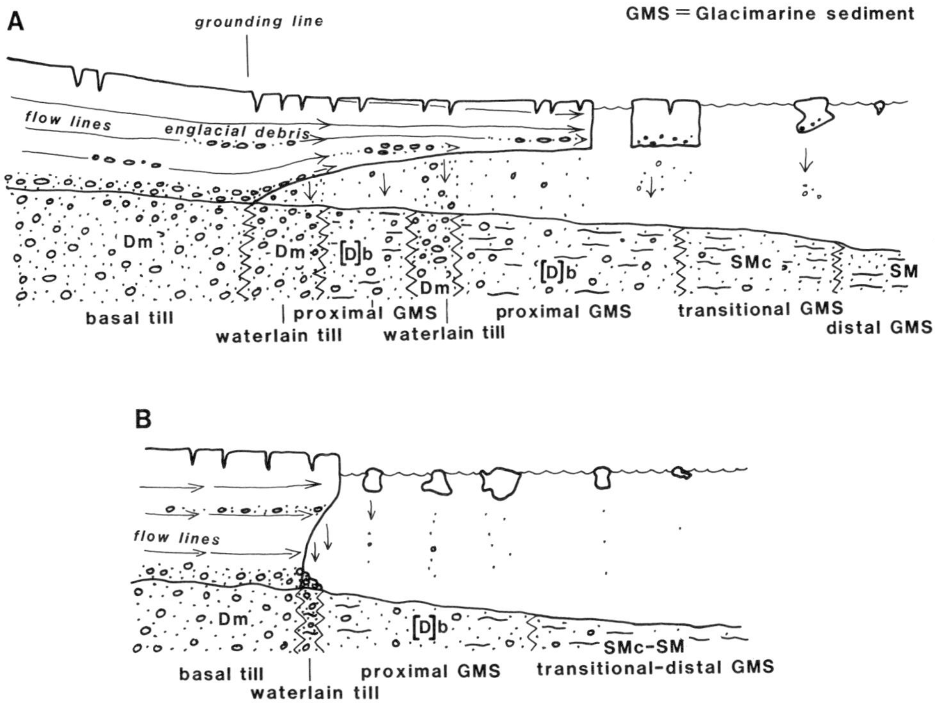

Fig. 6. Schematic cross section of (A) a floating ice shelf and (B) a grounded ice front showing the relationship between basal till, waterlain till and glacimarine sediment in the type of setting envisaged for the Tillite Group. Comparison of the two figures shows how an area of floating, debris-bearing ice is required for the development of extensive waterlain tills.

amounts of gravelly sandstone (GS) and [D] are also included. Many of these lithologies occur in the upper half of the Ulvesø Formation at Sorteelv Gletscher (Fig. 2). Their interpretation as glacimarine is justified by the occurrence of numerous dropstones.

Lithologies GS and Sc (Fig. 7a) are classified as residual glacimarine sediment and represent deposition of sand and gravel together, in contrast to sandy conglomerates (SC) in which the sand is likely to be a later infiltrate (Dyer 1970, Moncrieff 1989b). Consequently, GS and Sc are thought to reflect a certain amount of ice-rafting. Where a debris-flow origin for a unit is unlikely from the nature of grading, geometry or the occurrence of fine lamination, coarse sandstones with abundant gravel are interpreted as ice-proximal residual glacimarine sediment, formed under the influence of strong current conditions which removed the mud component. As gravel content decreases, so the influence of ice-rafting is inferred to decrease. Much of the sand component is likely to be the result of bedload transport, but the gravel component indicates ice-rafting intensity since alternative transport mechanisms are unlikely. Coarse-grained sandstones containing rare clasts are consequently interpreted as ice-distal residual deposits, for example, beach, tidal or shallow marine environments away from major iceberg lanes. As sand grain-size decreases and mud becomes increasingly dominant, so the distance from the sediment source or pathway is inferred to increase.

The higher mud content of lithologies MSc, MS, SMc, SM and Mc imply less current activity, suggesting that gravel is once again proportional to ice-rafting intensity. However, coarse sand is also likely to be a result of ice-rafting. Therefore, we suggest that in well laminated glacimarine strata of these lithologies, gravel and coarse sand are proportional to ice-proximity. Variations in clay, silt and fine sand content distinguish compound and residual types.

Ice-distal glacimarine facies. In the Tillite Group, the lithologies SMc, Mc, SM and M are interpreted as belonging to a distal glacimarine environment. Massive lithologies are rare, possibly due to the lack of burrowing organisms. Most are finely laminated as a result either of rhythmically varying sedimentation or of depo-

Fig. 7. (a) Slumped sandstone with dispersed clasts. This transitional residual glacimarine deposit represents limited ice-rafting onto a sea bed dominated by sand deposited from currents and suspension. Ulvesø Formation, Kap Weber. Scale is in cm. (b) Thin section of ice-proximal rhythmites representing daily input variations; Storeelv Formation, Storeelv, Ella Ø. Field of view is 2 mm across. (c) Ice-distal rhythmite with ice-rafted component; Ulvesø Formation, Kap Weber. Scale is in cm. (d) Thin section of two graded couplets from unit shown in Fig. 7c. Grading is not continuous, but frequently reverses as input of sediment varied with meltwater fluctuations. Field of view is 10 mm across. (e) Thin section of ice-distal rhythmite stratigraphically lower than that shown in figures 7c, d. This represents a more distal position. Field of view is 4 mm across.

sition from fine grained turbidity currents. However, ice-distal facies form only a small part of the Storeelv and Ulvesø formations. The Arena Formation, between the two glacigenic formations, contains little sign of glacial activity. This formation, is largely turbiditic and probably represents a period of complete ice withdrawal.

Rhythmite facies

Rhythmically laminated deposits are common in the Ulvesø and Storeelv formations and span both ice-distal (most common), transitional and ice-proximal settings. Rhythmites are graded bed which form as a result of *rhythmically* varying sediment input. The criteria used to distinguish them from the deposits of slump-triggered (non-rhythmic) turbidites are summarised as follows.

Structure. Beds deposited by turbidity currents contain, or are associated with, current-induced structures, especially cross-lamination, rippled surfaces, rip-up clasts and scours. Beds deposited by suspension settling contain only planar laminae. However, current indicators are of no use in distinguishing between turbidites and rhythmites deposited by continuous sediment underflows such as those common in glacial lakes (Gustavson 1975; Ashley 1975).

Grading. Fine-grained turbidites are normally graded throughout (Fig. 8a), whereas beds formed by seasonal discharge (either underflow or suspension settling from turbid overflows) are not. The latter commonly have a lower, coarse lamina which is poorly sorted and represents a whole range of grain sizes settling through the water from the continuous input at the surface. Seasonal variations in discharge also result in finer scale grading within a couplet. For instance, a mud-rich lamina may interrupt the coarse lower horizon, representing a temporary decrease in sediment input (Fig. 8a).

Couplet grouping. Gravity-driven turbidity currents occur irregularly and a sequence of beds produced by individual flows are separated by varying thicknesses of mud. True rhythmites, formed by seasonal or daily discharge variations are more regularly spaced, although some variation arises from annual weather fluctuations (Fig. 8c and d).

Ice-proximal rhythmites. Ice-proximal rhythmites occur 197 m above the base of the Storeelv Formation at Storeelv (Fig. 3). Sandy laminae about 0.3 mm thick and containing grains up to 0.15 mm in diameter with 10% clay are interbedded with muddy laminae 0.2 mm thick containing fine sand grains up to 0.03 mm in

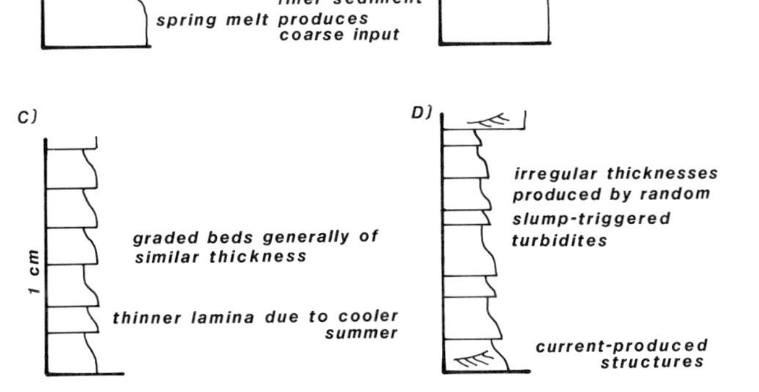

Fig. 8. Features developed in graded couplets; (a, b) Comparison of features developed in a couplet produced by seasonal discharge (a), and a fining upwards unit deposited by a single low density turbidity current (b). (c, d) Comparison of couplet grouping and grading in a sequence controlled by seasonal discharge where each graded couplet represents the passage of a year (c), and a sequence of beds deposited by random, slump-triggered turbidites (d).

diameter (Fig. 7b). Sorting is moderate and laminae are sometimes discontinuous over distances greater than 10 cm. These couplets are grouped into packets up to 10 cm in thickness and are interrupted by 1–5 cm thick bands of diamictite containing dropstones and till pellets. Their grain size indicates a position close to the ice, while their thickness suggests a short period of deposition. These thin rhythmites are therefore interpreted as having formed in response to daily fluctuations in glacier melt-water discharge. Sand and mud were deposited during peak discharge, followed by mud alone during low discharge periods. The laminae are not always continuous because of the turbulent nature of the site in which they were deposited. Such discharge-related rhythmites are similar to the cyclopels of Mackiewicz *et al.* (1984). The thicker diamictite bands show no evidence of debris flowage and are thought to be of ice-rafted origin.

Ice-distal rhythmites. Ice-distal rhythmites interpreted as suspension deposits are best developed 30–40 m above the base of the Ulvesø Formation at Kap Weber (Fig. 4), where several metres of graded laminae occur. These laminae are perfectly planar and continuous over at least 10 m. In the upper part of the 30–40 m sequence, graded units are typically 5–10 mm thick. Maximum grain size is 0.08 mm and ice-rafted fragments occur throughout (Fig. 7c). On the basis of thickness and grain size, the rhythmites are likely to represent annual cycles. However, grading is not continuous from bottom to top. Each lamina contains sub-laminae which are also graded; these are thought to represent sub-seasonal discharge variations (Fig. 7d). There are no silt-free laminae and the tops consist predominantly of claystone with numerous grains of fine silt. Neither are there any clay-free laminae. Clay occurs in small amounts right at the base of each lamina, then becomes more abundant upwards.

These rhythmites are different in the lower part of the 30–40 m unit. Laminae are between 1 and 3 mm thick and begin with an unsorted, ungraded section containing grains up to 0.08 mm in diameter. At the top of this sandy/silty section is a zone comprising up to a third of the total thickness in which clay gradually becomes dominant over silt, followed by the top third which consists of claystone with virtually no silt (Fig. 7e). Outsized ice-rafted grains are present throughout, while thin, graded, muddy sandstone horizons up to 1 cm thick and containing grains up to 1 mm in diameter also occur; these sandstones are interpreted as debris-flow deposits. The rhythmites are interpreted as suspension-settling deposits, formed further from the ice front than those stratigraphically higher. They represent a site of deposition so far from the ice front that for most of the year silt and fine sand were not being deposited. Clay settled throughout the year, with coarser material arriving during the relatively short peak discharge period of summer. The sequence at Kap Weber is therefore interpreted as recording the approach of an ice front.

Aeolian facies

At the base of the Ulvesø Formation, beneath the rhythmites at Kap Weber and in Brogetdal are some 30 m of a light orange-weathering sandstone which we interpret as being aeolian in origin. Sedimentary structures are well-exposed at the base but become concealed upwards due to jointing. Grain size decreases slightly up the unit from a modal 0.3–0.4 mm at the base to a modal 0.2–0.3 mm at the top. Sorting is moderate throughout. Bedding occurs on a centimetre scale, defined by slight grading and by concentrations of carbonate grains. Tabular cross-bedding occurs on a scale of several metres, but the exact geometry is unclear due to the fractured exposure. At the base of the unit, 1 m thick cross-bedding is truncated at the top by gently dipping, parallel-laminated sandstone. Foresets dip at an angle of about 30° and many are picked out by centimetre-scale lenses of well rounded sand grains between 0.5 and 1 mm in diameter. These lenses thicken towards the top of the foreset and have very high porosities. They also occur within horizontally laminated units, as 1–2 cm thick beds, continuous for up to 15 cm, or more commonly as lenses with convex tops and flat bases. Such lenses occur along specific horizons and up to six occur at any one level (Fig. 9a, b). Grains are mostly of quartz, with about 5% carbonate and 1–2% interstitial clay.

The lenses of coarse, well rounded sand are thought to be avalanche deposits or the remnants of coarse-crested granule ripples. The petrology, sorting and roundness in the bulk of the unit are not typically aeolian, suggesting a short period of transportation. Consequently, this unit is interpreted as an immature aeolian deposit.

Permafrost facies

Permafrost features, such as ice-contractional wedge fillings were described from the Tillite Group by Hambrey & Spencer (1987). Another,

(a) (b)

Fig. 9. (a) Aeolian sandstone with coarse-grained sandstone lenses interpreted as granule ripples (arrowed), Ulvesø Formation, Kap Weber. Scale is in cm. (b) Coarse, well-sorted sandstone laminae within aeolian sandstone, Ulvesø Formation, Kap Weber. Scale is in cm.

particularly well developed set of such features is described here.

The youngest diamictite of the Ulvesø Formation at Sorteelv Gletscher (Fig. 2) has a 10 m thick unit at the top, containing bedded gravel and sandstone lenses. Piercing the top of this bed along the sharp contact with the overlying muddy sandstone are sandstone and gravelly sandstone bodies of various shapes. There are three types as seen in cross section, semicircular, sinuous elongate and sharp elongate (Fig. 10).

This horizon shows contrasting features. The semicircular units are load structures of a dense sandstone into a water-saturated underlying bed, in this case a diamictite. The sinuous elongate wedges also appear to have had a soft host sediment. However, the sharp elongate wedges must represent the infilling of a brittle type of crack.

This unit is tentatively interpreted as a permafrost horizon which has thawed in places. The semicircular loads formed in a thawed mobile layer, while the longer wedges formed as infillings to frost contraction wedges. The massive nature of these infillings indicates that the cracks were fully open when the sand was deposited during a single event. The two elongate forms share so many features such as shape, style of infilling and size, that the sinuous forms are interpreted as partially thawed versions of the sharp wedges. The exposed diamictite was cracked by frost contraction and the resulting wedges subsequently infilled by the overlying sandstone. Partial thawing then occurred, resulting in the distortion of some, but not all, wedges. Where thawing was complete, sandstone load structures developed.

Evaporitic facies

Pseudomorphs after evaporitic minerals occur at two horizons, 30 m above the base of the Ulvesø Formation at Kap Weber and at the top of the Arena Formation on Ella Ø (Moncrieff and Hambrey 1988).

Evaporites in the Ulvesø Formation occur above the Kap Weber aeolian sandstone, in a poorly developed rhythmite (Fig. 4). Scattered quartz grains up to 0.5 mm in diameter form discontinuous laminae about 4 mm apart in a background of massive, very fine-grained ferroan dolomite. This background dolomite is extensively replaced by two main types of non-ferroan dolomite pseudomorphs after gypsum.

(i) Elongate laths ranging in size from 0.1×0.05 mm to 1.5×0.1 mm. These have blunt, pointed, or ragged (swallow-tail) ends and often have delicate feathery needles growing off them (Fig. 11a).

(ii) Amorphous rounded crystals, pseudomorphing alabastrine gypsum (< 0.05 mm in diameter) (Fig. 11b).

Ferroan dolomite also occurs as a replacement in much larger, euhedral or anhedral pseudomorphs up to 3 mm in diameter. They typically have ferroan dolomite rims with non-ferroan centres (Fig. 11b). All have druzy infillings and some have sharp square cross sections, suggesting a cubic mineral, possibly halite. However, most are anhedral or six-sided and a

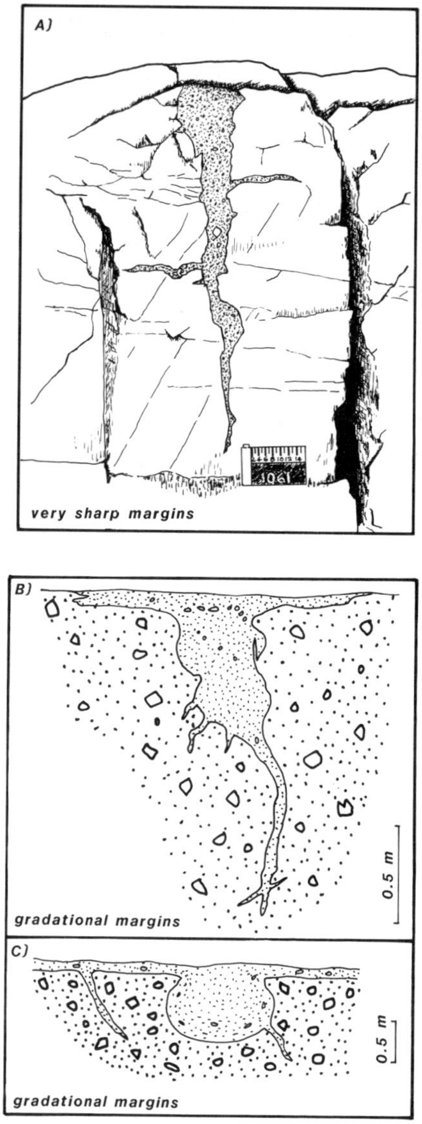

Fig. 10. Features developed near the top of the Ulvesø Formation at Sorteelv Gletscher, in a horizon interpreted as periglacial in origin. (a) Sketch from photograph of an elongate, sharp-sided sandstone wedge, penetrating a diamictite unit. The wedge is interpreted as of frost contraction origin; Ulvesø Formation, Sorteelv Gletscher. Scale is in cm. (b) Sinuous gravelly sandstone wedge penetrating diamictite. This wedge shares so many features with (a) that a similar origin is inferred, with post-formational shifting as the permafrost thawed. (c) Load structure of gravelly sandstone into diamictite, showing that in some areas, thawing of the permafrost layer was extensive.

gypsum precursor seems more likely. The great abundance of gypsum pseudomorphs, which make up virtually the entire rock at some horizons, suggests that these may have formed gypsum crusts (Watson 1985).

The close association with aeolian deposits suggests that these evaporites were deposited in shallow periglacial lakes or marine embayments. They may have formed as a result of crystals forming in surface waters, then falling to the bottom. Some of the horizons of pseudomorphs are graded, which supports this interpretation.

Evaporite pseudomorphs at the top of the Arena Formation were described in detail by Moncrieff & Hambrey (1988) and take the form of roughly spherical calcite pods up to 0.02 m in diameter (Fig. 11c) which form pseudomorphs after synsedimentary gypsum growths.

Depositional model for the late Precambrian tillites of East Greenland

Moncrieff (1989a) discussed the nature of the Tillite Group ice, concluding that it was an ice sheet of continental proportions, that developed on a recently emergent but subdued land mass. The thermal nature of the ice was cold, wet-based (as defined earlier), as demonstrated by the occurrence of meltwater activity, rhythmites and periglacial features. In this paper, we are more concerned with developing a depositional model for the Tillite Group.

Much of the glacigenic sequence is interpreted as waterlain, so that floating, debris-bearing ice, probably in the form of a continuous ice shelf, or series of ice shelves is inferred to have prevailed during much of the time represented by the Ulvesø and Storeelv formations. Waterlain tillite was deposited in areas of high sedimentation rate with little current reworking. It is likely that basal melting close to the grounding line was responsible. The combination of extensive deposition and a widespread floating ice shelf (Moncrieff 1989a) in association with abundant meltwater is one that has few modern analogues. Sub-polar and temperate glaciers such as those in Svalbard and Alaska respectively, are debris-bearing, but do not float as a continuous sheet when they reach the sea (Fig. 12). Indeed, only northern Ellesmere Island in the Arctic has floating ice shelves, and these are backed by a mountainous hinterland. Many subpolar glaciers in Greenland do float, but in deep fjords with high surrounding relief.

Antarctica is the only region with extensive ice shelves today, but these carry little debris as far as their seaward limits. Although many ice

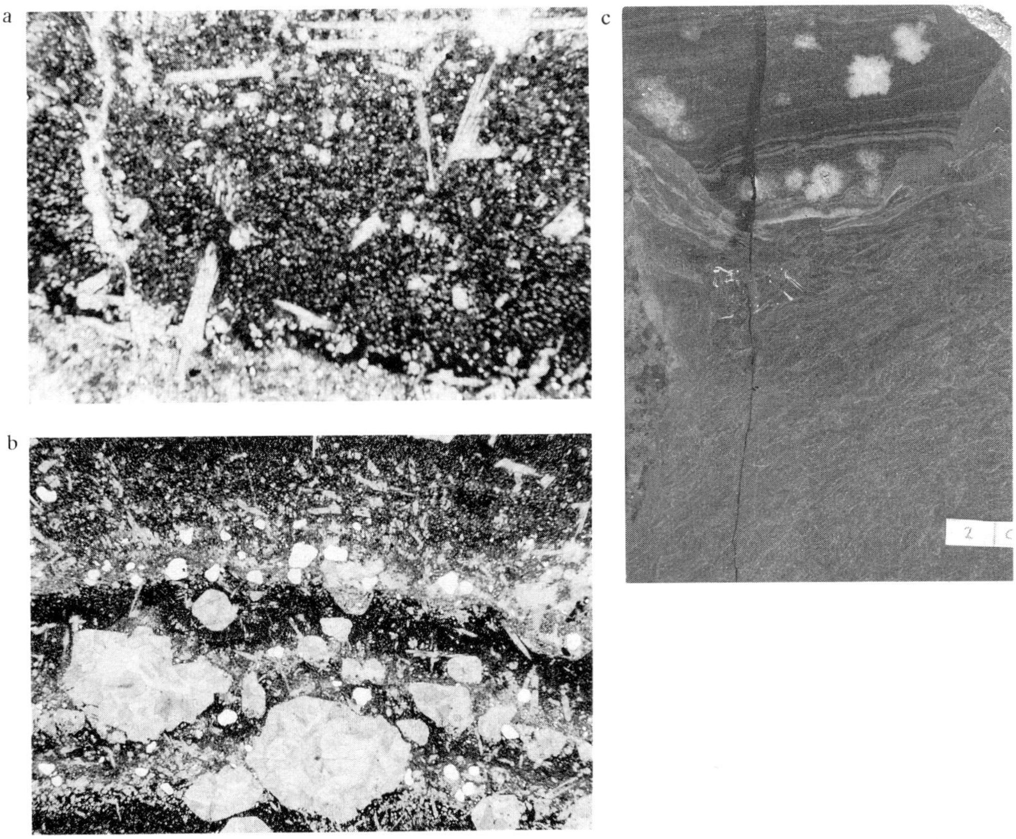

Fig. 11. Evaporitic facies. (a) Thin section of pseudomorphs after gypsum. Large elongate and feathery laths are surrounded by numerous finer-grained pseudomorphs after alabastrine gypsum. Ulvesø Formation, Kap Weber. Field of view is 4 mm across. (b) Thin section of large euhedral or anhedral pseudomorphs after gypsum; Ulvesø Formation, Kap Weber. Field of view is 10 mm across. (c) Arena/Storeelv Formation contact. Centimetre scale pods of calcite, pseudomorphing gypsum, can be seen displacing laminae, indicating early growth. Ella Ø.

streams feeding ice shelves such as the Ross, Ronne-Filchner and Amery may be sliding on their bed, much of the Antarctic ice sheet is not, and rates of basal debris entrainment cannot compare with those in Alaska or Svalbard. Basal layers of ice shelves such as the Ross, Ronne-Filchner and Amery ice shelves tend to be clean (Gow *et al.* 1968), with any sediment contained in the ice being rapidly deposited in a narrow zone as the basal layer melts close to the grounding line (Morgan 1972, Zotikov *et al.* 1980) (Fig. 13). A complicating factor is that basal freeze-on of saline ice may occur beneath some ice shelves and delay the release of debris as far as the calving zone (e.g. the Ronne-Filchner. Lange and MacAyeal 1986, Thyssen 1988). However, most of the mass of ice shelves comes from direct snowfall, so that flow lines are consistently downwards. Antarctic ice shelves are therefore relatively clean and incapable of depositing the volume of sediment seen in the Tillite Group. It is outlet glaciers and ice streams rather than ice shelves which contain the bulk of the basal debris in Antarctica today (Drewry & Cooper 1981, Drewry 1986), notably areas like James Ross and Vega Islands in the northern part of the Antarctic Peninsula, where there is significant meltwater activity (Anderson *et al.* 1983). Interlocking, floating glacier tongues which form when these outlet glaciers flow into the sea are probably the closest analogy for the Tillite Group ice, although on a smaller scale.

In the Tillite Group, current activity was significant beneath the floating ice. This is shown by the rapid lateral transitions between com-

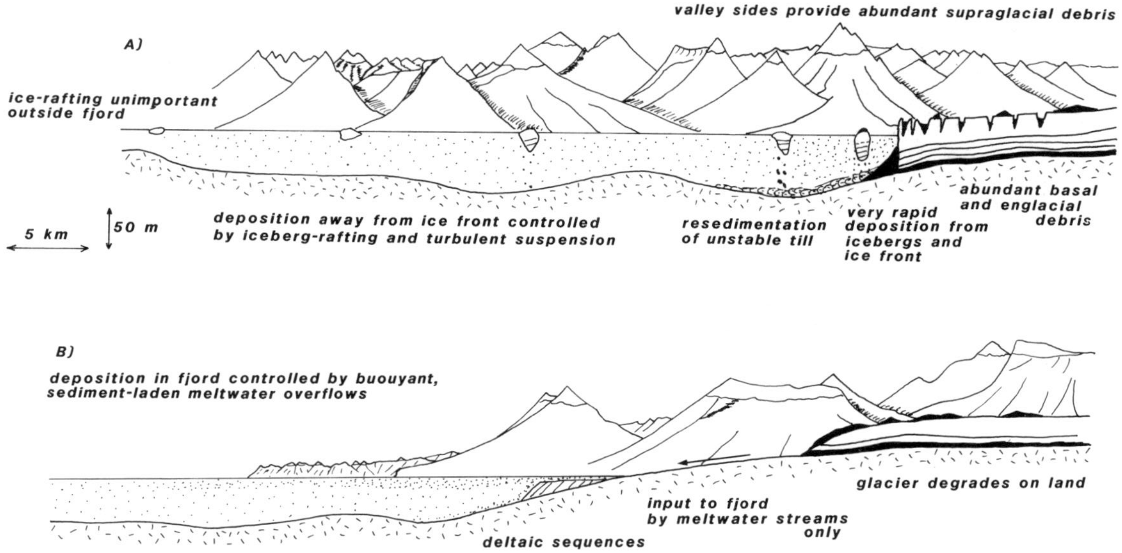

Fig. 12. General sedimentological models of glacimarine environments. (a) Cross section through an ideal tidewater glacier in Alaska, grounded on the sea floor, and characterized by vertical or overhanging frontal cliff. Sedimentation is by direct melt-out at the front and by iceberg-rafting and suspended sediment further out. (b) Cross section through an idealised turbid outwash fjord. The glacier terminates on land and does not enter the fjord. Ice-rafting cannot occur and the only sediment input is by meltwater over an outwash plain and delta complex.

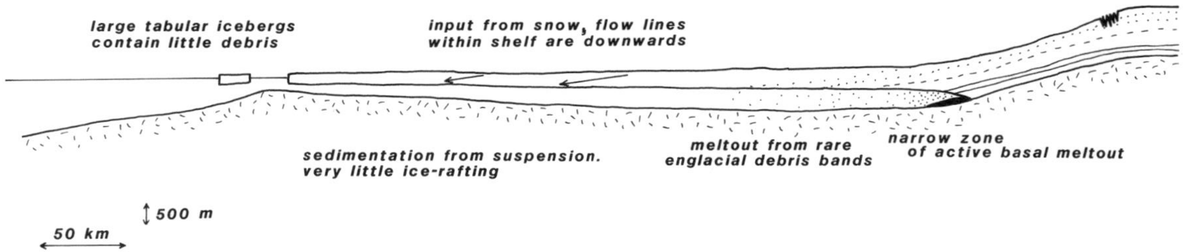

Fig. 13. Cross section through an idealized Antarctic ice shelf. The debris content of the ice is low and most sediment is deposited within a few kilometres of the grounding line.

pound and residual glacimarine sediment. Rapid variation in dropstone densities (notably in the Ulvesø Formation at Sorteelv Gletscher) suggest that distinct iceberg 'drift lanes' existed, indicating current activity in the open sea. The East Greenland sequences also demonstrate the effect that ice-retreat onto land has on glacimarine sedimentation. The 108–125 m interval in the Storeelv Formation at Storeelv (Fig. 3) and the upper part of the Storeelv Formation at Sorteelv Gletscher (Hambrey & Moncrieff 1985) show rapid transitions from lithologies with obvious glacial affinities, to apparently non-glacial facies. At Storeelv, the transition involves basal tillite. This pattern can be explained by the fluctuation of vertical grounded ice cliffs around sea level. At any given site, direct glacial deposition into the sea takes place up to the moment when the ice front retreats above sea level. After this, no ice-rafting can occur and the only means of transporting sediment from the ablating ice front to the sea is by carrying it in meltwater streams. Thus normal fluvial, deltaic and shallow marine sedimentary processes appear to be acting, and the close proximity of an ice front is disguised. A generalized model for the sedimentary environments occurring in the glacimarine parts of the Ulvesø and Storeelv formations is shown in Fig. 14. Only one part of a glacierized coastline is represented. An idealized vertical section representing the advance of a Tillite Group type of ice margin as derived from Figs. 2–4, is given in Fig. 15.

Conclusions

The Late Precambrian succession of East Greenland contains the best preserved and most extensive glacigenic strata of this age in the Arctic-North Atlantic region, cropping out over a distance of 500 km north–south. These strata occur in the upper part of a Riphean-Ordovician sequence more than 20 km thick, in the Caledonian fold belt of the fjord region. Other outcrops of glacigenic strata occur sporadically to the SW, where they rest on basement that was formerly considered to belong to the foreland, but has now been shown to be affected by Caledonian deformation.

Glacigenic sediments form the bulk of two of the five formations in the Tillite Group (the Ulvesø and Storeelv formations). Although each major stratigraphic unit can be traced throughout much of the fjord region, facies variations within each glacigenic formation are pronounced and reflect deposition in a marginal glacimarine to low-level terrestrial environment. The southwestern outcrops represent

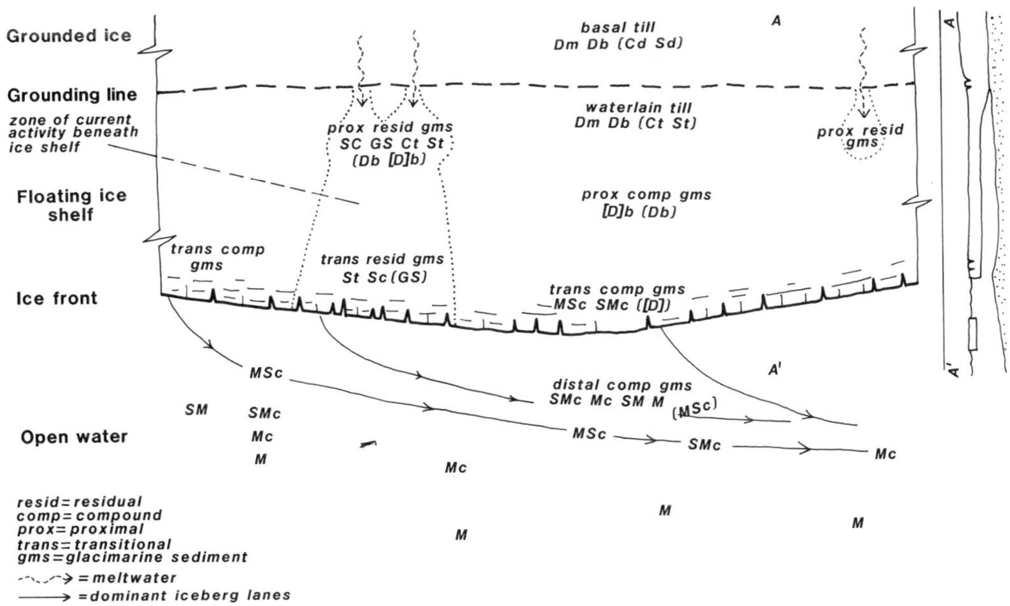

Fig. 14. Proposed model for deposition of the glacial facies in the Tillite Group with a plan view of an ice shelf showing grounding line and ice front. Likely positions of various descriptive and interpretative facies are marked, assuming sub-ice shelf currents and a major iceberg drift lane, outside of which ice-rafting is less important.

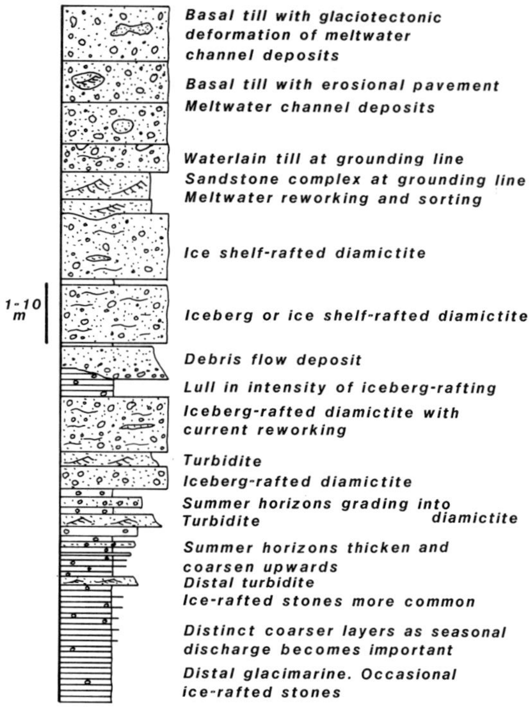

Fig. 15. Hypothetical sedimentary sequence deposited by an advancing ice front of the Tillite Group type, in water depths of not less than 20 m.

sedimentation closer to the source glaciers, in an area of more pronounced relief.

Glacigenic sediments deposited in the marine environment include tillite deposited from ice grounded below sea level, waterlain tillite, ice-proximal to ice-distal glacimarine sediment, debris-flow deposits, and rhythmites with ice-rafted material. Terrestrial glacigenic and related facies include basal (lodgement) tillite, fluvioglacial sands and gravels, proglacial aeolian sands, periglacial wedge fillings and evaporites.

This facies association, and the wide extent here and in other parts of the Arctic–North Atlantic region, suggests that the ice cover was of continental proportions. The ice was probably cold (below the pressure-melting point), but was sliding on its bed and producing limited quantities of meltwater before entering the sea. The abundance of waterlain tillite suggests that there was widespread deposition from floating ice close to the grounding line, and that the tidewater glaciers coalesced to form an ice shelf along an irregular coastline. Considerable current activity with reworking of the glacimarine sediment, resedimentation as debris-flows and iceberg drifting in distinct lanes were all important processes.

By comparison with previous studies of glacimarine environments in Alaska, Svalbard and Antarctica, it is clear that few directly analogous situations to that represented by the Tillite Group have been investigated to date. We cite the ice shelves at the NE tip of the Antarctic Peninsula, and those on the north and NE coasts of Ellesmere Island as being the nearest modern equivalents.

Understanding glacimarine environments generally is hindered by the difficulty of gaining access to presently forming sediments. Ancient, well exposed successions, such as that in East Greenland, can aid interpretation of modern sediments because facies relationships and geometry are much clearer.

The bulk of this work was undertaken in the Department of Earth Sciences, University of Cambridge, whom we thank for use of facilities. The project was funded by the U.K. Natural Environment Research Council (Grant no. GR3/5438 to W.B. Harland and Studentship to ACMM). We thank N. Henriksen and A.K. Higgins of the Greenland Geological Survey for help in planning the work; Nordisk Mineselskab, Copenhagen for logistical support, and the Commission for Scientific Research in Greenland for permission to work in the Northeast Greenland National Park during 1984 and 1985 when the field work was undertaken.

References

ANDERSON, J.B., KURTZ, D.D., DOMACK, E.W. & BALSHAW, K.M. 1980. Glacial and glacial marine sediments of the Antarctic continental shelf. *Journal of Geology*, **88**, 399–414.

——, BRAKE, C.F., DOMACK, E.W., MYERS, N.C. & WRIGHT, R. 1983. Development of a polar glacial-marine sedimentation model from Antarctic Quaternary deposits and glaciological information. *In*: MOLNIA, B.F. (ed.) *Glacial-marine sedimentation*. Plenum, New York, 322–364.

ASHLEY, G.M. 1975. Rhythmic sedimentation in Glacial Lake Hitchcock, Massachusetts – Connecticut. *In*: JOPLING, A.V. & MCDONALD, B.C. (eds), *Glaciofluvial and glaciolacustrine sedimentation*. Society of Economic and Petroleum Geologists, Special Publication **23**, 304–320.

BARRETT, P.J. 1975. Characteristics of pebbles from Cenozoic marine glacial sediments in the Ross Sea (DSDP sites 270–274) and the south Indian Ocean (site 268). *In*: HAYES, D.E., FRAKES, L.A., *et al.* (eds) *Initial reports of the Deep Sea Drilling Project 28*. U.S. Government Printing Office, Washington D.C., 769–784.

——, HAMBREY, M.J. & ROBINSON, P.H. 1990. Cenozoic glacial and tectonic history from CIROS-1, McMurdo Sound. *In*: THOMSON, M.R.A., CRAME, J.A. & THOMSON, J.W. (eds) *Geological evolution of Antarctica*, Cambridge University Press, Cambridge.

BORNS, H.W. & MATSCH, C.L. 1988. A provisional genetic classification of glaciomarine environments, processes and sediments. *In*: GOLDTHWAIT, R.P. & MATSCH, C.L. (eds) *Genetic Classification of Glacigenic Deposits*, Balkema, Rotterdam, 261–266.

CHRISTIE-BLICK, N. 1983. Glacial-marine and subglacial sedimentation. Upper Proterozoic Mineral Fork Formation, Utah. *In*: MOLNIA, B.F. (ed.), *Glacial-marine sedimentation*. Plenum Press, New York and London, 703–776.

DOWDESWELL, J.A. 1987. Processes of glacimarine sedimentation. *Progress in Physical Geography*, **11**, 52–90.

—— & SHARP, M.J. 1986. Characterisation of pebble fabrics in modern terrestrial glacigenic sediments. *Sedimentology,* **33**, 699–710.

——, HAMBREY, M.J. & WU, R.T. 1985. A comparison of clast fabric and shape in late Precambrian and modern glacigenic sediments. *Journal of Sedimentary Petrology*, **55**, 691–704.

DREIMANIS, A. 1984. Discussion of 'Lithofacies types and vertical profile models; an alternative approach to the description and environmental interpretation of glacial diamict and diamictite sequences.' *Sedimentology*, **31**, 883–884.

——, 1988. Tills: Their genetic terminology and classification. *In*: GOLDTHWAIT, R.P. & MATSCH, C.L. (eds) *Genetic Classification of Glacigenic Deposits*, Balkema, Rotterdam, 17–83.

—— & LUNDQUIST, J. 1984. What should be called till? *In*: KONIGSSON, L.K. (ed.). Ten years of Nordic till research. *Striae*, **20**, 5–10.

DREWRY, D.J. 1986. *Glacial geologic processes*. Edward Arnold, London.

—— & COOPER, A.P.R. 1981. Processes and models of Antarctic glaciomarine sedimentation. *Annals of Glaciology*, **2**, 117–122.

DYER, K.R. 1970. Grain size parameters for sandy gravels. *Journal of Sedimentary Petrology,* **40**, 616–620.

EYLES, N., EYLES, C.H. & MIALL, A.D. 1983. Lithofacies types and vertical profile models. An alternative approach to the description and environmental interpretation of glacial diamict and diamictite sequences. *Sedimentology*, **30**, 393–410.

FAIRCHILD, I.J. & HAMBREY, M.J. 1984. The Vendian succession of northeastern Spitsbergen: petrogenesis of a dolomite-tillite association. *Precambrian Research*, **26**, 111–67.

GOW, A.J., UEDA, H.T. & GARFIELD, D.E. 1968. Antarctic Ice Sheet: preliminary results of first core hole to bedrock. *Science*, **161**, 1011–1013.

GUSTAVSON, T.C. 1975. Sedimentation and physical limnology in proglacial lake Malaspina, southeastern Alaska. *In*: JOPLING, A.V. & MCDONALD, B.C. (eds) *Glaciofluvial and glaciolacustrine sedimentation*. Society of Economic and Petroleum Geologists, Special Publication, **23**, 249–263.

HAMBREY, M.J. & HARLAND, W.B. 1981. (eds). *Earth's pre-Pleistocene glacial record*, Cambridge University Press, Cambridge.

—— & MONCRIEFF, A.C.M. 1985. Vendian stratigraphy and sedimentology of the East Greenland Caledonides. *Rapport Grønlands Geologiske Undersøgelse*, **321**, 88–94.

—— & SPENCER, A.M. 1987. Late Precambrian glaciation of central East Greenland. *Meddelelser om Grønland*, Geoscience **19**, Copenhagen.

——, BARRETT, P.J. & ROBINSON, P.H. 1990. Stratigraphy and sedimentology. *In*: BARRETT, P.J. (ed.) *Cenozoic glacial history from CIROS-1 drill-hole, McMurdo Sound, Antarctica*. DSIR Miscellaneous Bulletin, Wellington.

HENRIKSEN, N. & HIGGINS, A.K. 1976. East Greenland Caledonian fold belt. *In*: ESCHER, A. & WATT, W.S. (eds) *Geology of Greenland*, Greenland Geological Survey, Copenhagen.

HIGGINS, A.K. 1981. The Late Precambrian Tillite Group of the Kong Oscars Fjord and Kejser Franz Josefs Fjord region of East Greenland. *In*: HAMBREY, M.J. & HARLAND, W.B. (eds) *Earth's pre-Pleistocene glacial record*. Cambridge University Press, Cambridge, 778–781.

KARROW, P.F. 1984. Discussion of 'Lithofacies types and vertical profile models; an alternative approach to the description and environmental interpretation of glacial diamict and diamictite sequences'. *Sedimentology*, **31**, 883–884.

KURTZ, D.D. & ANDERSON, J.B. 1979. Recognition and sedimentologic description of recent debris flow deposits from the Ross and Weddell seas, Antarctica. *Journal of Sedimentary Petrology*, **49**, 1159–1170.

LANGE, M. & MACAYEAL, D.R. 1986. Numerical models of the Filchner-Ronne ice shelf: an assessment of reinterpreted ice thickness distributions. *Journal of Geophysical Research*, **91**, 10457–10462.

LAWSON, D.E. 1979. Sedimentological analysis of the western terminus region of the Matanuska Glacier, Alaska. *U.S. Cold Regions Research and Engineering Laboratory*, Report 79–9.

—— 1981. Sedimentological characteristics and classification of depositional processes and deposits in the glacial environment. *U.S. Cold Regions Research and Engineering Laboratory*, Report 81–27.

MACKIEWICZ, N.E., POWELL, R.D., CARLSON, P.R. & MOLNIA, B.F. 1984. Interlaminated ice-proximal glacimarine sediments in Muir Inlet, Alaska. *Marine Geology*, **57**, 113–147.

MANBY, G.M. & HAMBREY, M.J. 1989. The structural setting of the Late Proterozoic tillites of East Greenland. *In*: GAYER, R.A. (ed.) *The Caledonide geology of Scandinavia*, Graham & Trotman, London, 299–312.

MIALL, A.D. 1977. A review of the braided river depositional environment. *Earth Science Reviews*, **13**, 1–62.

—— 1978. Lithofacies types and vertical profile models in braided rivers: a summary. *In*: MIALL, A.D. (ed.), *Fluvial Sedimentology*, Memoir Canadian Society Petroleum Geologists, **5**, 597–604.

MONCRIEFF, A.C.M. 1989a. The Tillite Group and related rocks of East Greenland: implications for Late Proterozoic palaeogeography. *In*: GAYER, R.A. (ed.) *The Caledonide geology of Scandinavia*, Graham & Trotman, London, 285–297.

—— 1989b. Classification of poorly sorted sediments. *Sedimentary Geology*, **65**, 191–194.

—— & HAMBREY, M.J. 1988. Late Precambrian glacially related grooved and striated surfaces in the Tillite Group of central East Greenland. *Palaeogeography, Palaeoclimatology & Palaeoecology*, **65**, 183–200.

MORGAN, V.I. 1972. Oxygen isotope evidence for bottom freezing on the Amery Ice Shelf. *Nature*, **238**, 393–4.

NYSTUEN, J.P. 1985. Facies and preservation of glaciogenic sequences from the Varanger Ice Age in Scandinavia and other parts of the North Atlantic region. *Palaeogeography, Palaeoclimatology & Palaeoecology*, **51**, 209–230.

PIPER, D.J.W. & BRISCO, C.D. 1975. Deep-water continental-margin sedimentation, DSDP leg 28, Antarctica. *In*: HAYES, D.A., FRAKES, L.A. *et al.* (eds.). Initial reports of the Deep Sea Drilling Project 28. U.S. Government Printing Office, Washington D.C, 727–68.

SCHERMERHORN, L.J.G. 1974. Late Precambrian mixtites: glacial and/or non-glacial. *American Journal of Science*, **27**, 673–824.

STUMP, E., MILLER, J.M.G., KORSCH, R.J. & EDGERTON, D.G. 1988. Diamictite from Nimrod Glacier area, Antarctica: possible Proterozoic glaciation on the seventh continent. *Geology*, **16**, 225–228.

THYSSEN, F. 1988. Special aspects of the central part of the Filchner-Ronne ice shelf, Antarctica. *Annals of Glaciology*, **11**, 173–179.

WATSON, A. 1985. Structure, chemistry and origins of gypsum crusts in southern Tunisia and the central Namib Desert. *Sedimentology*, **32**, 855–875.

WRIGHT, R. & ANDERSON, J.B. 1982. The importance of sediment gravity flow to sediment transport and sorting in a glacial marine environment: eastern Weddell Sea, Antarctica. *Geological Society of America Bulletin*, **93**, 951–63.

ZOTIKOV, I.A., ZAGORODNOV, V.S. & RAIKOVSKY, J.V. 1980. Core drilling through the Ross Ice Shelf (Antarctica) confirmed basal freezing. *Science*, **207**, 1463–1465.

Subject Index

Geographical Index

Special Publications of The Geological Society

52 Phosphorite Research and Development
51 The Cadomian Orogeny
50 Classic Petroleum Provinces
49 The Geology and Tectonics of the Oman Region
48 Geological Applications of Wireline Logs
47 Origins and Evolution of the Antarctic Biota
46 Phanerozoic Ironstones
45 Alpine Tectonics
44 Inversion Tectonics
43 Evolution of Metamorphic Belts
42 Magmatism in the Ocean Basins
41 Deltas: Sites and Traps for Fossil Fuels
40 Lacustrine Petroleum Source Rocks
39 Early Tertiary Volcanism and the Opening of the NE Atlantic
38 The Caledonian-Appalachian Orogen
37 Gondwana and Tethys*
36 Diagenesis of Sedimentary Sequences
35 Desert Sediments: Ancient and Modern
34 Fluid Flow in Sedimentary Basins and Aquifers
33 Geochemistry and Mineralization of Proterozoic Volcanic Suites
32 Coal and Coal-bearing Strata: Recent Advances
31 Geology and Geochemistry of Abyssal Plains
30 Alkaline Igneous Rocks
29 Deformation of Sediments and Sedimentary Rocks
28 Continental Extensional Tectonics
27 Evolution of the Lewisian and Comparable Precambrian High Grade Terrains
26 Marine Petroleum Source Rocks
25 Sedimentation in the African Rifts
24 The Nature of the Lower Continental Crust
23 Habitat of Palaeozoic Gas in North West Europe (Scottish Academic Press)
22 The English Zechstein and Related Topics
21 North Atlantic Palaeoceanography
20 Palaeoecology and Biostratigraphy of Graptolites
19 Collision Tectonics
18 Sedimentology: Recent Developments and Applied Aspects
14 Variscan Tectonics of the North Atlantic Region
12 Petroleum Geochemistry and Exploration of Europe
11 Residual Deposits: Surface Related Weathering Processes and Materials
10 Trench-Forearc Geology
8 The Caledonides of the British Isles — reviewed
6 Geological Background to Fossil Man
2 The Fossil Record
1 The Phanerozoic Time-Scale

Special Reports of The Geological Society

19 Magnetostratigraphy
18 Geophysical Logs in British Stratigraphy
17 Acritarchs in British Stratigraphy
16 Trilobites in British Stratigraphy
15 A Correlation of Jurassic Rocks in the British Isles, Pt 2, Middle and Upper Jurassic
14 A Correlation of Jurassic Rocks in the British Isles, Pt 1, Lower Jurassic
13 A Correlation of Triassic Rocks in the British Isles
12 A Correlation of Tertiary Rocks in the British Isles
10 A Correlation of Silesian Rocks in the British Isles
9 A Correlation of Cretaceous Rocks in the British Isles
8 A Correlation of Devonian Rocks in the British Isles
7 A Correlation of Dinantian Rocks in the British Isles
6 A Correlation of Precambrian Rocks in the British Isles
2 A Correlation of Cambrian Rocks in the British Isles
1 A Correlation of Silurian Rocks in the British Isles

Memoirs of The Geological Society

12 Palaeozoic Palaeogeography and Biogeography
11 The Ophiolite of Northern Oman
10 The Chronology of the Geological Record
9 The Nature and Timing of Orogenic Activity in the Caledonian Rocks of the British Isles
8 A Palaeogeological Map of the Lower Palaeozoic Floor below the cover of Upper Devonian, Carboniferous and Later Formations
6 Late Pre-Cambrian Glaciation in Scotland
5 Shallow-water Sedimentation, as illustrated in the Upper Devonian Baggy Beds
4 The Geology of Portuguese Timor

Geological Society Engineering Geology Special Publications

6 Field Testing in Engineering Geology
5 Engineering Geology of Underground Movements
4 Planning and Engineering Geology
3 Groundwater in Engineering Geology
2 Site Investigation Practice: Assessing BS 5930

Titles not listed are out of print
* *Available from Oxford University Press*